Atmospheric Circulation Dynamics and General Circulation Models

Springer
Berlin
Heidelberg
New York
Hong Kong
London
Milan
Paris
Tokyo

Masaki Satoh

Atmospheric Circulation Dynamics and General Circulation Models

Springer

Published in association with
Praxis Publishing
Chichester, UK

PRAXIS

Professor Masaki Satoh
Department of Mechanical Engineering
Saitama Institute of Technology
Saitama
Japan

SPRINGER–PRAXIS BOOKS IN ENVIRONMENTAL SCIENCES
SUBJECT *ADVISORY EDITOR*: John Mason B.Sc., M.Sc., Ph.D.

ISBN 3-540-42638-8 Springer-Verlag Berlin Heidelberg New York

Springer-Verlag is a part of Springer Science+Business Media (springeronline.com)
Bibliographic information published by Die Deutsche Bibliothek
Die Deutsche Bibliothek lists this publication in the Deutsche Nationalbibliografie; detailed bibliographic data are available from the Internet at http://dnb.ddb.de

A catalogue record for this book is available from the Library of Congress

Printed in Germany

Cover design: Jim Wilkie
Project Management: Originator Publishing Services, Gt Yarmouth, Norfolk, UK

Printed on acid-free paper

Contents

Preface

This book describes the fundamentals of atmospheric dynamics, the theories behind atmospheric circulation structures, and modeling of the general circulation and its applications. It consists of three parts: I, II, and III. Part I summarizes the principle ideas of atmospheric dynamics. Part II describes the theories of atmospheric structures from various perspectives. Part III describes basic concepts for making a general circulation model of the atmosphere and its applications to the study of atmospheric structures.

Part I deals with the fundamentals of atmospheric dynamics, such as equation sets, approximations of equation sets, basic balances of the atmosphere, waves, and instability. These topics are already fully discussed in standard textbooks on dynamic meteorology. Thus, summaries of these important topics are presented in this part complete with the necessary principles and basics to derive each of the equations. It also describes elemental topics on physical processes, such as the moisture, radiation, and turbulent processes that are required for understanding atmospheric structures.

The main part of this book is Part II which describes the various perspectives of atmospheric structures. It discusses how the general circulation of the atmosphere is maintained and how the circulation depends on external parameters. Atmospheric structures are examined by means of global-averaged properties, horizontally one-dimensional structure, vertically one-dimensional structure, meridionally two-dimensional structure, and horizontally two-dimensional or spherical structure. Their roles in moist circulations are also examined. In this book the author tries to focus on the more basic subjects that can be confirmed by readers from first principles or well-defined assumptions, rather than describing many observational facts or various applications. Of course, more advanced theories are developed to describe the properties of the general circulation of the atmosphere. The author wishes to present one of the frameworks for understanding the general circulation of the atmosphere. The subjects in Parts I and II cannot be clearly divided; some of the chapters in Part II are closely related to those in Part I (e.g., Chapters 7 and 18, Chapters 9 and 15, and Chapters 10 and 14).

Part III describes a method for construction of a general circulation model. This part focuses on the spectral model that is the most standard dynamical framework of the general circulation model of the atmosphere. Equation sets, numerical discretization in the spherical domain and in the vertical direction, and time integration schemes are described. Some applications of general circulation models are also presented. This part has a slightly different character from that of Parts I and II, since it describes numerical techniques and discretization. The author believes that the understanding of atmospheric dynamics and atmospheric structures is a prerequisite for the development of a general circulation model and that the use of general circulation models is a prerequisite for the study of the general circulation of the atmosphere. As Chapter 20 highlights, there are ongoing efforts to come up with new types of atmospheric general circulation models and to bring about improvements in existing models. If one aims to develop an atmospheric general circulation model, one needs to verify the performance of the model by using the well-known characteristics of atmospheric dynamics. Results obtained using atmospheric general circulation models should be validated with those reached theoretically. In addition to this, through numerical experiments with general circulation models, one may gain further insights into the understanding of the atmospheric general circulation structures examined in Part II. In the author's view, after reading this book students who want to study atmospheric dynamics should have the ability to construct a general circulation model by means of their own programming, though various atmospheric general circulation models are available. It is hoped that they will be able to study the properties of the atmosphere using a numerical model whose details and limitations are thoroughly familiar to themselves. The three-part structure of this book is intended to bring this about.

General circulation models are nowadays used in various fields, such as weather forecasting, climate prediction, and environmental estimations. Recently, despite the urgent social needs relating to global warming, understanding of the general circulation of the atmosphere is still not fully established. This is because consideration needs to given to the more complicated climate system: the atmosphere, ocean, land, ice, chemical composition, and biosphere. Since incorporation of many complicated parts are required for applications to the real climate or environmental systems, it is very difficult for beginners to understand the roles of the dynamics of the atmosphere, even when they are the central part of the models. On understanding the contents of this book, the reader should have a good idea of the dynamical reasons for the response of general circulation to external disturbances like CO_2 increase and ozone depletion.

This book is designed for the reader to reach the entrance level for a graduate course in Atmospheric Science and Environmental Science. It is particularly recommended for researchers of mathematical physics, such as fluid dynamics, chaos, fractals, and nonlinear physics. One of the approaches to dynamical meteorology involves laboratory experiments (e.g., using a rotating annulus). In such laboratory experiments, the general characteristics of rotating stratified fluids are explored using a wider range of parameters than those used for the real atmosphere. Understanding the general characteristics of fluid motions provides us with comprehensive insight into the behavior of the real atmosphere. Today, general circulation models can be used just like the rotating annulus in laboratories. Of course, numerical errors or the use of experimental laws that cannot be derived from the first principles of physical laws (e.g., eddy viscosity) are inevitable, so that the results of general circulation models cannot be regarded as objects of mathematical physics in a pure sense. However, if one views atmospheric motions as one of the realizations of rotating stratified fluids, there exist many unsolved problems that demand mathematical and physical treatment. If those mathematical physicists who are not familiar with dynamic meteorology are interested in general circulation models because of reading this book and if they can easily use them in their work stations, there will be much valuable feedback to our field of atmospheric science. This is my hope in writing this book.

Most of this book is based on seminar notes that were made when the author joined a series of seminars held at the Meteorological Research Laboratory, Department of Geophysics, University of Tokyo. Parts I, II, and III, respectively, correspond to the master seminar (M-seminar), the atmospheric structure seminar, and the atmospheric general circulation model seminar. The M-seminar was a unique seminar for the education of graduate students and covered important topics of meteorology; that is, waves, instability, moist dynamics, radiation, turbulence, and statistical analysis were studied in each term with six to eight weeks devoted to each topic. Among them, the topics of waves, instability, and moist dynamics constitute the main contribution to Part I. Radiation and turbulence are briefly touched on at the end of this part (in Chapters 10 and 11, respectively).

Part II is devoted to the study of atmospheric structures. Besides the notes from the atmospheric structure seminar, this part is also based on the author's research into atmospheric general circulation. In this part, physical mechanisms rather than observational facts are mainly described for the purpose of understanding atmospheric general circulation. The author adopted this strategy because useful textbooks on observations of the general circulation, such as Peixoto and Oort (see Chapter 13), are widely available. In addition, digitally tabulated meteorological data are nowadays easily accessible through the World Wide Web such that one can access observational data (Kalnay et al.; see Appendix A3). In this sense, this book is not intended to be self-contained; readers are advised to check observational evidence by means of available textbooks or digital data.

Part III describes the theoretical basis of the atmospheric general circulation model (AGCM). The description is based on the dynamical core of the spectral model that was developed at the University of Tokyo by the late Dr. Numaguti during the 1990s. The author studied with Dr. Numaguti at the University of Tokyo and made descriptive notes of the dynamics of AGCM. Part III mainly follows the author's personal seminar notes. This AGCM was further developed at the Center for Climate System Research (CCSR), University of Tokyo, and the National Institute of Environmental Studies (NIES) and is called the CCSR/NIES AGCM. The manual of CCSR/NIES AGCM is also referred to in Part III. The CCSR/NIES

AGCM was then introduced to the Earth Simulator Center and was improved in terms of computer performance for it to be run on the Earth Simulator; the model is called AFES (AGCM for the Earth Simulator). The dynamical core of the original AGCM has been further developed by the GFD-Dennou Club and is available online at http://www.gfd-dennou.org. The work of the GFD-Dennou Club is very useful, and the products of GFD seminars archived at this website can be found in Part I.

The author was taken aback when he was asked to write a book by the publisher, especially as many good textbooks on dynamic meteorology have already been written by such established researchers as Holton, Pedlosky, Gill, Andrew et al., Lindzen, Salmon, and Salby. As far as atmospheric modeling is concerned, this subject warrants a book on its own; in fact, there are already useful textbooks such as Haltiner and Williams, Durran, and Krishnamurti et al. (see Chapter 20). It is not the author's intention to write a book that is in competition with theirs. However, the author has learned a great deal from the above systematic seminars and considers that such material will be useful to those students interested in becoming researchers on the atmosphere. The seminar notes were compiled from reviews of many standard papers, textbooks, and some original work from our group. Through the seminars, differences between the nomenclatures have been standardized and errors in the original works amended.

The author is grateful to the following for reading the draft manuscripts: Yoshi-Yuki Hayashi, Masaki Ishiwatari, Keita Iga, Masatsugu Odaka, Keiichi Ishioka, Wataru Ohfuchi, Yoshihisa Matsuda, Masahiro Takagi, Yuuji Kitamura, Yosuke Kosaka, and group members on turbulence at the University of Tokyo and collaborators of the GFD-Dennou Club. The author also thanks Hirofumi Tomita and Koji Goto for preparing some figures in Chapter 24, including the one on the front cover that was made with the Earch Simulator. The numerical calculations of the author's research cited in this book were done using the HITACH SR8000 at the University of Tokyo under the cooperative research efforts of the Center for Climate System Research, University of Tokyo, the NEC SX5 at the National Institute of Environmental Studies, the parallel computers at the High-Tech Research Center for the Saitama Institute of Technology, and the NEC SX5 and the Compaq cluster system at the Frontier Research System for Global Change.

Figures

Tables

Part I

Principle Ideas

In this Part I, the basic mathematical and physical ideas are summarized. These materials are required for studying the atmospheric general circulation in the following two parts. The first half of this part describes the basic properties of dry air, and the latter half is devoted to those of moist air and physical processes.

In Chapter 1 the thermodynamic properties and the basic equations for dry air are described. Since these equations are the same as those used for the fluid dynamics, they are too generous to describe specific atmospheric motions. In general, we introduce various kinds of approximations and assumptions under the basic balances realized in the atmosphere. Thus, we consider the balances of atmospheric motions and their stabilities in Chapter 2, and various forms of approximated equation sets are introduced based on the basic balances in Chapter 3. These equation sets are used in the following chapters.

The concepts of waves and instabilities are most essential for description of atmospheric motions. The basic theories and properties of waves seen in the atmosphere and are summarized in Chapter 4, and those of instabilities are summarized in Chapter 5. As applications of waves, forced motions (i.e., the atmospheric responses to forcing) are described in Chapter 6. Chapter 7 relates the disturbance fields and mean fields and offers new equation sets suitable for description of disturbance fields and its effect on the mean fields. The equation sets introduced in this chapter play important roles in meridional circulations of the atmosphere.

In Chapters 8 and 9, the thermodynamics properties and the basic equations of moist air are summarized. These correspond to Chapter 1 for dry air. Studies of moist circulations, which are given in Chapter 15 in Part II, are based on the concepts of these two chapters. We briefly summarize the radiation process in Chapter 10, and the turbulent process in Chapter 11. These are required to construct physical models used for atmospheric general circulation models, and are basic conceptual notions to study structures of atmospheric general circulations.

1

Basic equations

Many characteristics of atmospheric general circulation, particularly the mid-latitude circulation, are described by the equations for dry air, where air that contains no water vapor is referred to as the *dry air*. This chapter summarizes the basic physical properties of dry air. Most of the contents in this chapter are general characteristics of fluid dynamics, and are not necessarily specific to atmospheric motions. However, the basic equations described in this chapter are frequently invoked for consideration of atmospheric circulations, and are used in the following chapters.

We focus on expressions and manipulations of basic equations for practical use, rather than explaining the elemental principles of thermodynamics, fluid mechanics, or vector analysis. Readers can find more appropriate textbooks for these elements. It is hoped that practical learning of the equations in this chapter will lead to applications of these equations to many atmospheric fields.

We start from the thermodynamic relations of dry air in general forms. Next, the governing equations of dry air are presented. The equations of mass, momentum and energy conservations are formulated in the conservative form. Other useful relations, such as the equations of angular momentum and vorticity, are also derived. As a supplement to this chapter, the general formulas used in the transformation of the coordinate system and basic equations in various coordinate systems are summarized in the appendix of this book: Appendix A1.

1.1 Dry air

1.1.1 Equation of state and thermodynamic variables

In general, air can be regarded as an ideal gas for practical use in meteorology. Deviation from the ideal gas is almost negligible not only in the earth atmosphere but also in the atmospheres of other planets. On the assumptions of the ideal gas, however, theoretically different quantities sometimes degenerate into a similar quantity, which may cause confusion. Thus, we first derive the basic relations of dry air without using the assumptions of an ideal gas in order to clarify what quantities are related with each other.

Throughout this book, the atmosphere is regarded as a fluid in which temperature and other thermodynamic quantities are defined at any point. This means that the fluid is in *local thermodynamic equilibrium.* If a fluid is composed of one component and is in thermodynamic equilibrium, the thermodynamic state of the fluid is determined by the values of temperature T [K] and pressure p [Pa]. The volume of the fluid in a given mass is determined by the equation of state which relates the volume to temperature and pressure. Let v [$\mathrm{m^3\ kg^{-1}}$] denote the volume per unit mass of a fluid (i.e., the *specific volume*). The equation of state is formally written as

$$v = v(p, T). \tag{1.1.1}$$

The density ρ [$\mathrm{kg\ m^{-3}}$] is an inverse of the specific volume $\rho = 1/v$. Any thermodynamic variables can be expressed by p and T using the equation of state $v(p,T)$ and the specific heat at constant pressure C_p. In this section, we first derive expressions of the thermodynamic variables in case of a general form of the equation of state. Note that, even when the fluid consists of multiple components, it can be regarded as one component system if each component is well mixed and has a fixed composition in the fluid. A more general formulation of the multiple component system will be described in Chapter 8.

Let u and s denote internal energy and entropy per unit mass, respectively. Following the first and second laws of thermodynamics, the changes in u and s in a quasi-static process are related as

$$du = Tds - pdv. \tag{1.1.2}$$

Enthalpy h, Helmholtz's free energy f, and Gibbs' free energy g per unit mass are respectively defined by

$$h = u + pv, \tag{1.1.3}$$

$$f = u - Ts, \tag{1.1.4}$$

$$g = h - Ts = u + pv - Ts. \tag{1.1.5}$$

From these definitions and (1.1.2), the changes in these energies are expressed as

$$dh = Tds + vdp, \tag{1.1.6}$$

$$df = -sdT - pdv, \tag{1.1.7}$$

$$dg = -sdT + vdp. \tag{1.1.8}$$

Thus we have

$$\left(\frac{\partial u}{\partial s}\right)_v = T, \qquad \left(\frac{\partial u}{\partial v}\right)_s = -p, \tag{1.1.9}$$

$$\left(\frac{\partial h}{\partial s}\right)_p = T, \qquad \left(\frac{\partial h}{\partial p}\right)_s = v, \tag{1.1.10}$$

$$\left(\frac{\partial f}{\partial T}\right)_v = -s, \qquad \left(\frac{\partial f}{\partial v}\right)_T = -p, \tag{1.1.11}$$

$$\left(\frac{\partial g}{\partial T}\right)_p = -s, \qquad \left(\frac{\partial g}{\partial p}\right)_T = v, \tag{1.1.12}$$

from which we obtain the *Maxwell relations*:

$$-\left(\frac{\partial^2 u}{\partial s \partial v}\right)^{-1} = \left(\frac{\partial s}{\partial p}\right)_v = -\left(\frac{\partial v}{\partial T}\right)_s, \tag{1.1.13}$$

$$\left(\frac{\partial^2 h}{\partial s \partial p}\right)^{-1} = \left(\frac{\partial s}{\partial v}\right)_p = \left(\frac{\partial p}{\partial T}\right)_s, \tag{1.1.14}$$

$$-\frac{\partial^2 f}{\partial T \partial v} = \left(\frac{\partial s}{\partial v}\right)_T = \left(\frac{\partial p}{\partial T}\right)_v, \tag{1.1.15}$$

$$-\frac{\partial^2 g}{\partial T \partial p} = \left(\frac{\partial s}{\partial p}\right)_T = -\left(\frac{\partial v}{\partial T}\right)_p. \tag{1.1.16}$$

The specific heat at constant volume C_v [J kg^{-1} K^{-1}] and the specific heat at constant pressure C_p [J kg^{-1} K^{-1}] per unit mass are defined by

$$\left(\frac{\partial s}{\partial T}\right)_v = \frac{C_v}{T}, \tag{1.1.17}$$

$$\left(\frac{\partial s}{\partial T}\right)_p = \frac{C_p}{T}. \tag{1.1.18}$$

Using (1.1.2) and (1.1.6) and the above definitions, specific heats are related as

$$\left(\frac{\partial u}{\partial T}\right)_v = C_v, \tag{1.1.19}$$

$$\left(\frac{\partial h}{\partial T}\right)_p = C_p. \tag{1.1.20}$$

Generally, these specific heats are thermodynamic variables that depend on temperature and pressure. The value of C_p at any thermodynamic state can be determined if its dependency on T at any pressure p_0 is known. Using (1.1.18) and (1.1.16), we have the relation:

$$\begin{aligned} C_p(p,T) &= C_p(p_0,T) + \int_{p_0}^{p} \left(\frac{\partial C_p}{\partial p}\right)_T dp' \\ &= C_p(p_0,T) - T \int_{p_0}^{p} \left(\frac{\partial^2 v}{\partial T^2}\right)_{p'} dp', \end{aligned} \tag{1.1.21}$$

which expresses the dependency of C_p on p and T with the equation of state (1.1.1).

Expressions of the other thermodynamic variables can be determined if the equation of state and the specific heat at constant pressure are known. First, the change in entropy is given by

$$ds = \left(\frac{\partial s}{\partial T}\right)_p dT + \left(\frac{\partial s}{\partial p}\right)_T dp = \frac{C_p}{T} dT - \left(\frac{\partial v}{\partial T}\right)_p dp, \tag{1.1.22}$$

where (1.1.16) and (1.1.18) are used. In a similar way, the changes in internal energy and enthalpy are expressed respectively as

$$du = \left[C_p - p\left(\frac{\partial v}{\partial T}\right)_p\right] dT - \left[-T\left(\frac{\partial v}{\partial T}\right)_p + p\left(\frac{\partial v}{\partial p}\right)_T\right] dp, \quad (1.1.23)$$

$$dh = C_p dT + \left[v - T\left(\frac{\partial v}{\partial T}\right)_p\right] dp = C_p dT - T^2\left[\frac{\partial}{\partial T}\left(\frac{v}{T}\right)\right]_p dp, \quad (1.1.24)$$

where (1.1.2), (1.1.6), and (1.1.22) are used. Letting s_0 and h_0 denote entropy and enthalpy at a specified temperature T_0 and pressure p_0, respectively, we have the following expressions:

$$s(p,T) = s_0 + \int_{T_0}^{T} \frac{C_p(p_0,T')}{T'} dT' - \int_{p_0}^{p} \left(\frac{\partial v(p',T)}{\partial T}\right)_p dp', \quad (1.1.25)$$

$$h(p,T) = h_0 + \int_{T_0}^{T} C_p(p_0,T') dT' - T^2 \int_{p_0}^{p} \left[\frac{\partial}{\partial T}\left(\frac{v(p',T)}{T}\right)\right]_p dp', \quad (1.1.26)$$

$$u(p,T) = h(p,T) - p\,v(p,T). \quad (1.1.27)$$

Using (1.1.23) and (1.1.19), the specific heat at constant volume C_v can be expressed with the equation of state and C_p; that is,

$$C_p - C_v = p\left(\frac{\partial v}{\partial T}\right)_p. \quad (1.1.28)$$

We conclude this section by introducing other useful quantities and relations. The *expansion coefficient* is defined by

$$\alpha \equiv \frac{1}{v}\left(\frac{\partial v}{\partial T}\right)_p = -\frac{1}{\rho}\left(\frac{\partial \rho}{\partial T}\right)_p. \quad (1.1.29)$$

Using α, the change in entropy (1.1.22) is rewritten as

$$ds = \frac{C_p}{T} dT - \frac{\alpha}{\rho} dp. \quad (1.1.30)$$

Using (1.1.14), the change in entropy is also rewritten in terms of the changes in pressure and density as

$$\begin{aligned} ds &= \left(\frac{\partial s}{\partial p}\right)_v dp + \left(\frac{\partial s}{\partial v}\right)_p dv = \frac{1}{\rho^2}\left(\frac{\partial p}{\partial T}\right)_s \left[\left(\frac{\partial \rho}{\partial p}\right)_s dp - d\rho\right] \\ &= \frac{C_p}{\rho \alpha T}\left(\frac{1}{c_s^2} dp - d\rho\right), \end{aligned} \quad (1.1.31)$$

where we define

$$c_s^2 \equiv \left(\frac{\partial p}{\partial \rho}\right)_s; \quad (1.1.32)$$

c_s is the *speed of sound*, as can be found in Section 4.2. It is convenient to define

$$\gamma_d \equiv \left(\frac{\partial T}{\partial p}\right)_s = -\frac{\left(\frac{\partial s}{\partial p}\right)_T}{\left(\frac{\partial s}{\partial T}\right)_p} = \frac{\left(\frac{\partial v}{\partial T}\right)_p}{\frac{C_p}{T}} = \frac{\alpha T}{\rho C_p}, \qquad (1.1.33)$$

where γ_d is interpreted as the *dry adiabatic lapse rate* with respect to pressure (Section 2.1).

1.1.2 Thermodynamic variables of the ideal gas

Now that general expressions of the thermodynamic variables are derived, we then introduce the equation of state for an ideal gas. Based on the Boyle-Charles law, the equation of state for an ideal gas is written as

$$v^I = \frac{R^* T}{p}, \qquad (1.1.34)$$

where v^I [m^3 mol^{-1}] denotes a volume of the ideal gas per one mol and $R^* = 8.3144$ J mol^{-1} K^{-1} is the universal gas constant. Letting m [kg mol^{-1}] denote the molecular weight of the ideal gas, the specific volume, that is the volume per unit mass, is written as

$$v = \frac{v^I}{m} = \frac{R^* T}{mp}. \qquad (1.1.35)$$

Eq. (1.1.34) expresses the equation of state of any kinds of the ideal gas. It is also applicable to a mixture of ideal gases. According to the assumption of the ideal gas, a volume of the mixture is given by the sum of volume of the individual gases when each gas is in the state of having the same temperature and pressure. Letting n_k [mol kg^{-1}] denote the mol number per unit mass of the k-th component, we can express the specific volume of the mixed gas as

$$v = \sum_k n_k v^I = \sum_k n_k \frac{R^* T}{p}. \qquad (1.1.36)$$

Note that we have the relation

$$\sum_k n_k m_k = \sum_k q_k = 1, \qquad (1.1.37)$$

where m_k is the molecular weight of the k-th component and

$$q_k = n_k m_k \qquad (1.1.38)$$

is the *mass concentration* of the k-th component. The atmosphere of the earth is generally described by the equation of state for an ideal gas as long as the release of the latent heat is regarded as the external heating. The most variable components of the earth atmosphere is water vapor. All the other major components of the atmosphere have almost uniform concentrations. Thus, it is convenient to define *dry air*, which is a mixture of the gases in the atmosphere except water vapor. The mean molecular weight of dry air m_d [kg mol^{-1}] is given by

$$m_d \equiv \frac{1}{\sum_k n_k}, \tag{1.1.39}$$

where the subscript k denotes each component of dry air. The equation of state for dry air is written as

$$v = \frac{R_d T}{p}, \tag{1.1.40}$$

or

$$p = \rho R_d T, \tag{1.1.41}$$

where

$$R_d \equiv \frac{R^*}{m_d} = \sum_k n_k R^* \tag{1.1.42}$$

is the gas constant for dry air: $R_d = 287.04$ J kg^{-1} K^{-1}.

It is easy to see that the internal energy and the enthalpy of ideal gas depend only on temperature. Actually, substituting the equation of state (1.1.40) into (1.1.23) and (1.1.24), respectively, we express their dependencies on pressure as

$$\left(\frac{\partial u}{\partial p}\right)_T = -T\left(\frac{\partial v}{\partial T}\right)_p - p\left(\frac{\partial v}{\partial p}\right)_T = -T\frac{R_d}{p} + p\frac{R_d T}{p^2} = 0, \tag{1.1.43}$$

$$\left(\frac{\partial h}{\partial p}\right)_T = -T^2\left[\frac{\partial}{\partial T}\left(\frac{v}{T}\right)\right]_p = 0. \tag{1.1.44}$$

In practice, we can assume that the specific heat is constant irrespective of temperature in the atmosphere. In this case, (1.1.28) becomes

$$C_p - C_v = R_d, \tag{1.1.45}$$

that is, C_v is also constant. In this case, integrating (1.1.19) and (1.1.20) over temperature gives

$$u = C_v T, \qquad h = C_p T, \tag{1.1.46}$$

where we have assumed that the internal energy at $T = 0$ K is zero. We often use the ratio of the two specific heats γ and the ratio of the gas constant to the specific heat at constant pressure κ:

$$\gamma \equiv \frac{C_p}{C_v}, \qquad \kappa \equiv \frac{R_d}{C_p} = 1 - \gamma^{-1}. \tag{1.1.47}$$

From (1.1.29) and (1.1.40), the expansion coefficient of the ideal gas is simply expressed as

$$\alpha = \frac{1}{T}. \tag{1.1.48}$$

Substituting (1.1.40) into (1.1.22), we have the change in entropy as

$$ds = C_p \frac{dT}{T} - R_d \frac{dp}{p}. \tag{1.1.49}$$

Hence, the entropy can be expressed as

$$s = C_p \ln T - R_d \ln p + s_0, \tag{1.1.50}$$

where s_0 is an arbitrary constant. In atmospheric dynamics, the *potential temperature* is frequently used in place of entropy. The potential temperature θ is related to entropy as

$$s \equiv C_p \ln \theta. \tag{1.1.51}$$

If we set $s_0 = R_d \ln p_0$ in (1.1.50), we obtain

$$\theta = T \left(\frac{p_0}{p} \right)^{\kappa}, \tag{1.1.52}$$

where p_0 is a reference pressure at which potential temperature becomes equal to temperature. Normally, $p_0 = 1000$ hPa is used.

The adiabatic process often appears in atmospheric motions. In the adiabatic process, the change in thermodynamic variables occurs at constant entropy. Using (1.1.13)–(1.1.16), we can derive the change in specific volume with respect to temperature and pressure at constant entropy, respectively, as

$$\left(\frac{\partial v}{\partial T} \right)_s = -\frac{\left(\frac{\partial s}{\partial T} \right)_v}{\left(\frac{\partial s}{\partial v} \right)_T} = -\frac{\frac{C_v}{T}}{\left(\frac{\partial p}{\partial T} \right)_v} = -\frac{C_v v}{R_d T} = -\frac{C_v}{p}, \tag{1.1.53}$$

$$\left(\frac{\partial v}{\partial p} \right)_s = -\frac{\left(\frac{\partial s}{\partial p} \right)_v}{\left(\frac{\partial s}{\partial v} \right)_p} = \frac{\left(\frac{\partial v}{\partial T} \right)_s}{\left(\frac{\partial p}{\partial T} \right)_s} = -\frac{C_v v}{C_p p}, \tag{1.1.54}$$

from which the changes in density are given by

$$\left(\frac{\partial \rho}{\partial T} \right)_s = \frac{1}{\gamma - 1} \frac{\rho}{T}, \tag{1.1.55}$$

$$c_s^2 \equiv \left(\frac{\partial p}{\partial \rho} \right)_s = \gamma \frac{p}{\rho}, \tag{1.1.56}$$

where c_s is the speed of sound defined by (1.1.32). The dry adiabatic lapse rate γ_d, (1.1.33), is expressed as

$$\gamma_d \equiv \left(\frac{\partial T}{\partial p} \right)_s = \frac{1}{\rho C_p}. \tag{1.1.57}$$

Using (1.1.31) with (1.1.48), the change in entropy is written as

$$ds = \frac{C_p}{\rho} \left(\frac{1}{c_s^2} dp - d\rho \right), \tag{1.1.58}$$

or the change in the potential temperature is

$$\frac{\rho}{\theta}d\theta = \frac{1}{c_s^2}dp - d\rho. \tag{1.1.59}$$

Using (1.1.56), in the adiabatic process $ds = d\theta = 0$, pressure and density are related as

$$\frac{p}{\rho^\gamma} = \text{const.} \tag{1.1.60}$$

1.2 Conservation laws and basic equations

Now we turn to the formulation of the governing equations of a fluid. Fluid motions are described by the conservation laws of mass, momentum, and energy. We first describe the conservative forms of each equation and then rewrite them in various forms.

1.2.1 Conservation law and conservation of mass

Let us consider a domain which has a volume V and is surrounded by a surface S. This domain is fixed with respect to the space. Let A denote an arbitrary physical quantity per unit volume. The conservation law states that the change in A is given by the sum of fluxes which go through the surface S and the source within the domain V. Hence, the balance equation for A is written as

$$\frac{d}{dt}\int A dV = -\int F_n dS + \int \sigma[A] dV, \tag{1.2.1}$$

where F_n is an outward normal component of a flux density $\boldsymbol{F}$ of A on the surface S, and $\sigma[A]$ is a source of A per unit volume and unit time (Fig. 1.1).

Since the domain V can be arbitrarily chosen, it can be fixed in the space independent of time. Using Gauss's law

$$\int F_n dS = \int \nabla \cdot \boldsymbol{F} dV, \tag{1.2.2}$$

the differential expression of (1.2.1) can be given by

$$\frac{\partial}{\partial t}A + \nabla \cdot \boldsymbol{F} = \sigma[A]. \tag{1.2.3}$$

Let us divide the flux density $\boldsymbol{F}$ into two parts:

$$\boldsymbol{F} = A\boldsymbol{v} + \boldsymbol{F}', \tag{1.2.4}$$

where the first term on the right-hand side is the advective part and the second term is the rest of the flux, such as the contribution of diffusion. Substitution of this into (1.2.3) yields

$$\frac{\partial}{\partial t}A + \nabla \cdot (A\boldsymbol{v} + \boldsymbol{F}') = \sigma[A]. \tag{1.2.5}$$

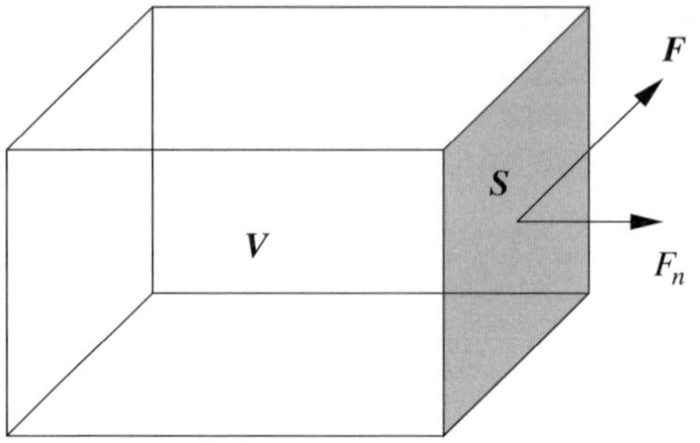

FIGURE 1.1: A control volume V with the surface S. The outward normal component of the flux $\boldsymbol{F}$ at one side of V is F_n.

Letting a quantity per unit mass a be $A = \rho a$, we rewrite (1.2.5) as

$$\frac{\partial}{\partial t}(\rho a) + \nabla \cdot (\rho a \boldsymbol{v} + \boldsymbol{F}') \quad = \quad \sigma[\rho a]. \tag{1.2.6}$$

The *conservation of mass* is given by setting $A = \rho$ and $a = 1$. In this case,

$$\boldsymbol{F} \quad = \quad \rho \boldsymbol{v}, \qquad \sigma[\rho] \quad = \quad 0,$$

where the latter means no source nor sink of mass (i.e, the conservation of mass). Substituting these equations into (1.2.3), we obtain the conservation of mass, or the *continuity equation* as

$$\frac{\partial \rho}{\partial t} + \nabla \cdot (\rho \boldsymbol{v}) \quad = \quad 0, \tag{1.2.7}$$

or

$$\frac{d\rho}{dt} + \rho \nabla \cdot \boldsymbol{v} \quad = \quad 0, \tag{1.2.8}$$

where

$$\frac{d}{dt} \quad = \quad \frac{\partial}{\partial t} + \boldsymbol{v} \cdot \nabla \tag{1.2.9}$$

is the material derivative.

Using the conservation of mass, the time dependence of an arbitrary quantity a is written as

$$\rho \frac{da}{dt} \quad = \quad \frac{\partial}{\partial t}(\rho a) + \nabla \cdot (\rho a \boldsymbol{v}). \tag{1.2.10}$$

Using this, the balance equation (1.2.6) is rewritten as

$$\rho \frac{da}{dt} + \nabla \cdot \boldsymbol{F}' \quad = \quad \sigma[\rho a]. \tag{1.2.11}$$

If there is no source nor sink of a, $\sigma[\rho a] = 0$, and there is no flux other than the advective part, $\boldsymbol{F}' = 0$, (1.2.11) becomes

$$\frac{da}{dt} \quad = \quad 0.$$

This means that a conserves along fluid motion.

1.2.2 Conservation of momentum

Let the velocity of the fluid be $\boldsymbol{v}$ and its components in the Cartesian coordinates be v_i ($i = 1, 2, 3$). The balance equation for the momentum is given by setting $A = \rho v_i$ and appropriate expressions of the flux and source terms. A general form of the flux of the momentum is given by a tensor

$$\Pi_{ij} = \rho v_i v_j - \sigma_{ij}, \tag{1.2.12}$$

which is called the *momentum flux density tensor*: $\rho v_i v_j$ is the advective part, and σ_{ij} is the stress tensor. The source term of the momentum is given as

$$\sigma[\rho v_i] = -\rho g_i, \tag{1.2.13}$$

where g_i is the i-th component of an external force $\boldsymbol{g}$. As the external force, we only consider a potential force that is derived from a potential Φ:

$$\boldsymbol{g} = \nabla\Phi. \tag{1.2.14}$$

Thus, the balance equation for the momentum is expressed as

$$\frac{\partial}{\partial t}(\rho v_i) + \frac{\partial}{\partial x_j}\Pi_{ij} = -\rho g_i, \tag{1.2.15}$$

which is called the *conservation of momentum* or the *equations of motion*. If the external force g_i vanishes, a domain integral of the momentum ρv_i is conserved.

The stress tensor σ_{ij} is expressed as

$$\sigma_{ij} = -p\delta_{ij} + \sigma'_{ij}, \tag{1.2.16}$$

where $-p\delta_{ij}$ is the pressure tensor and σ'_{ij} is the viscous stress tensor.† From the requirement of the symmetry, it can be shown that the viscous stress tensor must take the form

$$\sigma'_{ij} = \eta\left(\frac{\partial v_i}{\partial x_j} + \frac{\partial v_j}{\partial x_i} - \frac{2}{3}\delta_{ij}\nabla\cdot\boldsymbol{v}\right) + \zeta\delta_{ij}\nabla\cdot\boldsymbol{v}, \tag{1.2.17}$$

where η and ζ are the *coefficients of viscosity*. The coefficients of viscosity are thermodynamic variables and as such are functions of pressure and temperature in general. As can be shown by (1.2.52), because of the requirement that entropy must increase the coefficients of viscosity are non-negative.‡ Substituting (1.2.16) and

---

†δ_{ij} is the unit tensor or Kronecker's delta, defined by

$$\delta_{ij} = \begin{cases} 1, & i = j, \\ 0, & i \neq j. \end{cases}$$

We also use abbreviated expressions; the differential along the i-th direction $\frac{\partial A}{\partial x_i}$ is written as $A_{,i}$, and Einstein's summation convention is used where the summation from 1 to 3 for the suffix is implied if a suffix is repeated in a term. Hence, for instance, $v_{i,i} = \sum_{i=1}^{3}\frac{\partial v_i}{\partial x_i} = \nabla\cdot\boldsymbol{v}$.

‡The symmetry $\sigma'_{ij} = \sigma'_{ji}$ is derived from the conservation of angular momentum. See Section 1.3.1.

(1.2.17) into Π_{ij} and using (1.2.14), we can rewrite the conservation of momentum (1.2.15) as

$$\frac{\partial}{\partial t}(\rho v_i) + \frac{\partial}{\partial x_j}\left(\rho v_i v_j + p\delta_{ij} - \sigma'_{ij}\right) = -\rho\frac{\partial \Phi}{\partial x_i}, \tag{1.2.18}$$

or

$$\frac{d\boldsymbol{v}}{dt} = -\frac{1}{\rho}\nabla p - \nabla\Phi + \boldsymbol{f}, \tag{1.2.19}$$

where $\boldsymbol{f}$ is the frictional force, whose components are defined by

$$f_i = \frac{1}{\rho}\frac{\partial}{\partial x_j}\sigma'_{ij}. \tag{1.2.20}$$

In the case that the coefficients of viscosity are constant, substitution of (1.2.17) yields

$$\boldsymbol{f} = \frac{\eta}{\rho}\nabla^2\boldsymbol{v} + \frac{1}{\rho}\left(\zeta + \frac{1}{3}\eta\right)\nabla(\nabla\cdot\boldsymbol{v}). \tag{1.2.21}$$

Thus, in the special case when the fluid is non-divergent $\nabla\cdot\boldsymbol{v} = 0$, (1.2.19) reduces to

$$\frac{d\boldsymbol{v}}{dt} = -\frac{1}{\rho}\nabla p - \nabla\Phi + \nu\nabla^2\boldsymbol{v}, \tag{1.2.22}$$

where

$$\nu \equiv \frac{\eta}{\rho} \tag{1.2.23}$$

is the *kinematic viscosity*. This form of the equations of motion is called the *Navier-Stokes equation.*

We then derive the momentum equation in a rotating frame. We define the angular velocity of the rotating frame by $\boldsymbol{\Omega}$, which is constant irrespective of time. We also designate a quantity in the inertial frame by a subscript a and that in the rotating frame by a subscript r. A time derivative of a vector $\boldsymbol{A}$ is transformed as

$$\left(\frac{d\boldsymbol{A}}{dt}\right)_a = \left(\frac{d\boldsymbol{A}}{dt}\right)_r + \boldsymbol{\Omega}\times\boldsymbol{A}. \tag{1.2.24}$$

If a position vector $\boldsymbol{x}$ is substituted into $\boldsymbol{A}$, we have

$$\left(\frac{d\boldsymbol{x}}{dt}\right)_a = \left(\frac{d\boldsymbol{x}}{dt}\right)_r + \boldsymbol{\Omega}\times\boldsymbol{x}, \tag{1.2.25}$$

that is,

$$\boldsymbol{v}_a = \boldsymbol{v}_r + \boldsymbol{\Omega}\times\boldsymbol{x}, \tag{1.2.26}$$

where $\boldsymbol{v}_a$ is the velocity in the inertial frame and $\boldsymbol{v}_r$ is the velocity in the rotating frame. In the same way, the transformation of the time derivative of $\boldsymbol{v}_a$ (i.e., the acceleration vector) is

$$\begin{aligned} \left(\frac{d\boldsymbol{v}_a}{dt}\right)_a &= \left(\frac{d\boldsymbol{v}_r}{dt}\right)_r + 2\boldsymbol{\Omega}\times\boldsymbol{v}_r + \boldsymbol{\Omega}\times(\boldsymbol{\Omega}\times\boldsymbol{x}) \\ &= \left(\frac{d\boldsymbol{v}_r}{dt}\right)_r + 2\boldsymbol{\Omega}\times\boldsymbol{v}_r - \nabla\left(\frac{1}{2}\Omega^2 X^2\right), \end{aligned} \tag{1.2.27}$$

where

$$X = \boldsymbol{x} - \frac{(\boldsymbol{x}\cdot\boldsymbol{\Omega})\boldsymbol{\Omega}}{|\boldsymbol{\Omega}^2|} \tag{1.2.28}$$

is the distance to the axis of rotation. The second and third terms on the right-hand side of (1.2.27) are the inertial forces; $-2\boldsymbol{\Omega}\times\boldsymbol{v}_r$ is the *Coriolis force* and $-\boldsymbol{\Omega}\times(\boldsymbol{\Omega}\times\boldsymbol{x})$ is the *centrifugal force*, which is the gradient of the *centrifugal potential energy* $-\frac{1}{2}\Omega^2X^2$. Substituting (1.2.27) into (1.2.19), we have the momentum equation in the rotating frame as

$$\frac{d\boldsymbol{v}_r}{dt} + 2\boldsymbol{\Omega}\times\boldsymbol{v}_r = -\frac{1}{\rho}\nabla p - \nabla\Phi_r + \boldsymbol{f}, \tag{1.2.29}$$

where

$$\Phi_r = \Phi - \frac{1}{2}\Omega^2 X^2 \tag{1.2.30}$$

is the *geopotential* or the *gravitational potential energy*, defined as the sum of the potential energy for the attractive force and the centrifugal potential energy. The equation of motion in the rotating frame (1.2.29) is formally the same as the equation of motion in the inertial frame (1.2.19) except for the Coriolis force if $\boldsymbol{v_a}$ is replaced by $\boldsymbol{v_r}$ and Φ is replaced by Φ_r. Hereafter, the subscript r for the quantities of the rotating frame is arbitrarily omitted.

1.2.3 Conservation of energy

The conservation of energy is expressed as the balance equation for the total energy. For application to the atmosphere, the total energy per unit volume E can be defined as the sum of kinetic energy $\frac{1}{2}\rho\boldsymbol{v}^2$, potential energy $\rho\Phi$ (or $\rho\Phi_r$ in the case of the rotating frame), and internal energy ρu, where u is the specific internal energy per unit mass:

$$E = \frac{1}{2}\rho\boldsymbol{v}^2 + \rho\Phi + \rho u. \tag{1.2.31}$$

The conservation of total energy is described by

$$\sigma[E] = 0, \tag{1.2.32}$$

that is, there is no source of total energy. Thus, the balance equation for the total energy is expressed as

$$\frac{\partial}{\partial t}E + \nabla\cdot\boldsymbol{F}^E = 0, \tag{1.2.33}$$

where $\boldsymbol{F}^E$ designates the flux density vector of total energy. The expression of $\boldsymbol{F}^E$ can be determined by the balance equations for each component of the total energy E.

First, the balance equation for the kinetic energy is given by an inner product of $\boldsymbol{v}$ and the equation of momentum (1.2.19), or (1.2.29) in the case of the rotating frame:

$$\rho \frac{d}{dt}\frac{\boldsymbol{v}^2}{2} = -\boldsymbol{v}\cdot\nabla p - \rho\boldsymbol{v}\cdot\nabla\Phi + v_i \frac{\partial}{\partial x_j}\sigma'_{ij}. \tag{1.2.34}$$

This can be rewritten in the flux form as

$$\frac{\partial}{\partial t}\left(\frac{1}{2}\rho\boldsymbol{v}^2\right) + \frac{\partial}{\partial x_j}\left\{\left(\frac{1}{2}\rho\boldsymbol{v}^2 + p\right)v_j - \sigma'_{ij}v_i\right\} = p\nabla\cdot\boldsymbol{v} - \varepsilon - \rho\boldsymbol{v}\cdot\nabla\Phi, \tag{1.2.35}$$

where

$$\varepsilon \equiv \sigma'_{ij}\frac{\partial}{\partial x_j}v_i, \tag{1.2.36}$$

which can be rewritten by using (1.2.17), as

$$\varepsilon = \frac{1}{2}\eta\left(\frac{\partial v_i}{\partial x_j} + \frac{\partial v_j}{\partial x_i} - \frac{2}{3}\delta_{ij}\nabla\cdot\boldsymbol{v}\right)^2 + \zeta(\nabla\cdot\boldsymbol{v})^2. \tag{1.2.37}$$

ε is the dissipation of the kinetic energy and is called the *dissipation rate.* Since the coefficients η and ζ are non-negative, ε is also non-negative. The right-hand side of (1.2.35) is the source term of the kinetic energy $\sigma[\frac{1}{2}\rho\boldsymbol{v}^2]$.

Second, the balance equation for the potential energy is rather trivial. Since Φ is independent of time $\frac{\partial\Phi}{\partial t} = 0$, we have

$$\rho\frac{d\Phi}{dt} = \rho\boldsymbol{v}\cdot\nabla\Phi, \tag{1.2.38}$$

the flux form of which is given by

$$\frac{\partial(\rho\Phi)}{\partial t} + \nabla\cdot(\rho\boldsymbol{v}\Phi) = \rho\boldsymbol{v}\cdot\nabla\Phi. \tag{1.2.39}$$

The right-hand side is the source term of the potential energy $\sigma[\rho\Phi]$. This term is the work done by gravity.

Finally, the balance equation for the internal energy can be given as follows. The source term of the internal energy $\sigma[\rho u]$ must be specified from the requirement of the conservation of total energy (1.2.32):

$$\sigma[E] = \sigma\left[\frac{1}{2}\rho\boldsymbol{v}^2\right] + \sigma[\rho\Phi] + \sigma[\rho u] = 0. \tag{1.2.40}$$

Thus, using (1.2.35) and (1.2.39), we have

$$\sigma[\rho u] = -p\nabla\cdot\boldsymbol{v} + \varepsilon. \tag{1.2.41}$$

The balance equation for the internal energy is, therefore, written as

$$\frac{\partial(\rho u)}{\partial t} + \nabla\cdot(\rho u \boldsymbol{v} + \boldsymbol{F}^{ene}) \quad = \quad -p\nabla\cdot\boldsymbol{v} + \varepsilon, \tag{1.2.42}$$

where $\boldsymbol{F}^{ene}$ is the flux density vector of internal energy other than the advection term of internal energy. $\boldsymbol{F}^{ene}$ may be called the *heat flux* for simplicity. Using the continuity equation (1.2.7), (1.2.42) is rewritten in the advective form as

$$\rho\left(\frac{du}{dt} + p\frac{dv_s}{dt}\right) \quad = \quad \varepsilon - \nabla\cdot\boldsymbol{F}^{ene}, \tag{1.2.43}$$

where $v_s = 1/\rho$ is the specific volume.

For dry air, in general, the heat flux $\boldsymbol{F}^{ene}$ is the sum of energy fluxes due to radiation and conduction of heat:

$$\boldsymbol{F}^{ene} \quad = \quad \boldsymbol{F}^{rad} + \boldsymbol{F}^{therm}, \tag{1.2.44}$$

where $\boldsymbol{F}^{rad}$ is the radiative flux and $\boldsymbol{F}^{therm}$ is the flux due to conduction of heat or the thermal diffusion flux. According to Fourier's law, the thermal diffusion flux is proportional to the gradient of temperature,

$$\boldsymbol{F}^{therm} \quad = \quad -\kappa_T\nabla T, \tag{1.2.45}$$

where κ_T is the thermal conductivity. The *thermal diffusivity* or the *thermometric conductivity*

$$k \quad \equiv \quad \frac{\kappa_T}{\rho C_p}, \tag{1.2.46}$$

is also used instead of κ_T. In general, κ_T is positive because of the second law of thermodynamics (the principle of increase of entropy), as will be shown by (1.2.52). The thermal diffusion flux in a dry atmosphere is referred to as the *sensible heat flux*.

The explicit form of the balance equation for the total energy is given by the sum of (1.2.35), (1.2.38), and (1.2.42) as

$$\begin{aligned}\frac{\partial}{\partial t}&\left\{\rho\left(\frac{\boldsymbol{v}^2}{2} + \Phi + u\right)\right\} \\ &+\nabla\cdot\left\{\rho\boldsymbol{v}\left(\frac{\boldsymbol{v}^2}{2} + \Phi + u\right) + p\boldsymbol{v} - v_j\sigma'_{ij} + \boldsymbol{F}^{ene}\right\} \quad = \quad 0.\end{aligned} \tag{1.2.47}$$

The advective form of the total energy is also given as

$$\rho\frac{d}{dt}\left(\frac{\boldsymbol{v}^2}{2} + \Phi + u\right) + \nabla\cdot\left(p\boldsymbol{v} - v_j\sigma'_{ij} + \boldsymbol{F}^{ene}\right) \quad = \quad 0. \tag{1.2.48}$$

From the comparison between (1.2.47) and (1.2.33), we have the expression of the total energy flux $\boldsymbol{F}^E$ as

$$\boldsymbol{F}^E \quad = \quad \rho\boldsymbol{v}\left(\frac{\boldsymbol{v}^2}{2} + \Phi + u\right) + p\boldsymbol{v} - v_j\sigma'_{ij} + \boldsymbol{F}^{ene} \tag{1.2.49}$$

$$\quad = \quad \rho\boldsymbol{v}\left(\frac{\boldsymbol{v}^2}{2} + \sigma\right) - v_j\sigma'_{ij} + \boldsymbol{F}^{ene}, \tag{1.2.50}$$

where σ is called the *static energy* or the *dry static energy*, defined by

$$\sigma \equiv u + \frac{p}{\rho} + \Phi = h + \Phi. \tag{1.2.51}$$

h is specific enthalpy per unit mass.

1.2.4 Entropy balance

We describe here the balance of entropy. If local thermodynamic equilibrium is satisfied, we have from the thermodynamic relation (1.1.2)

$$T\frac{ds}{dt} = \frac{du}{dt} + p\frac{dv_s}{dt}, \tag{1.2.52}$$

where s is the specific entropy per unit mass. Using (1.2.43), we obtain

$$\rho T\frac{ds}{dt} = \varepsilon - \nabla \cdot \boldsymbol{F}^{ene}. \tag{1.2.53}$$

This is the equation for the production of entropy. The right-hand side of this equation expresses diabatic change. Thus, if we define diabatic heating as

$$Q \equiv \frac{1}{\rho C_p}(\varepsilon - \nabla \cdot \boldsymbol{F}^{ene}), \tag{1.2.54}$$

the change in entropy is expressed as

$$\frac{ds}{dt} = \frac{C_p Q}{T}. \tag{1.2.55}$$

In the case of the ideal gas, (1.2.55) is rewritten by using the potential temperature (1.1.52) as

$$\frac{d\theta}{dt} = \frac{\theta}{T}Q, \tag{1.2.56}$$

which is the equation of potential temperature.

Let us consider the second law of thermodynamics or the principle of increase of entropy. The equation of entropy (1.2.53) can be rewritten as

$$\begin{aligned} \rho\frac{ds}{dt} &= \frac{\varepsilon}{T} - \frac{\nabla \cdot \boldsymbol{F}^{ene}}{T} \\ &= \frac{\varepsilon}{T} + \frac{1}{T^2}\boldsymbol{F}^{ene} \cdot \nabla T - \nabla \cdot \frac{\boldsymbol{F}^{ene}}{T}. \end{aligned} \tag{1.2.57}$$

The first and second terms on the right-hand side represent the production of entropy, and the third term is the convergence of entropy flux density. Then, the change in entropy is partitioned as

$$\frac{ds}{dt} = \frac{d_i s}{dt} + \frac{d_e s}{dt}, \tag{1.2.58}$$

where

$$\frac{d_i s}{dt} = \frac{\varepsilon}{\rho T} + \frac{1}{\rho T^2}\boldsymbol{F}^{ene} \cdot \nabla T, \tag{1.2.59}$$

$$\frac{d_e s}{dt} = -\frac{1}{\rho}\nabla \cdot \frac{\boldsymbol{F}^{ene}}{T}. \tag{1.2.60}$$

$d_i s/dt$ is the production of entropy due to internal entropy, and $d_e s/dt$ is the change due to the convergence of entropy flux. The second law of thermodynamics requires the inequality

$$\frac{d_i s}{dt} \geq 0. \tag{1.2.61}$$

The expressions of ε and $\boldsymbol{F}^{ene}$ in (1.2.59) must satisfy this requirement.

The dissipation rate ε is given by (1.2.36). The expression of ε is determined so as not to be negative. From the symmetry of the indices, the viscous stress tensor is expressed as (1.2.17). As a result, ε has the form (1.2.37). From the requirement $\varepsilon \geq 0$, the coefficients of viscosity must be positive: $\eta \geq 0$ and $\zeta \geq 0$.

As for the second term on the right-hand side of (1.2.59), the heat flux $\boldsymbol{F}^{ene}$ has contributions of the radiative flux and the thermal diffusion flux as given by (1.2.44). The term involving the thermal diffusion flux can be rewritten as

$$\frac{1}{T^2}\boldsymbol{F}^{therm} \cdot \nabla T = \frac{\kappa_T}{T^2}|\nabla T|^2, \tag{1.2.62}$$

where (1.2.45) is used. Since this must be non-negative, it is concluded that the thermal conductivity κ_T must not be negative. On the other hand, the term involving the radiative flux $\boldsymbol{F}^{rad}$ does not satisfy the inequality (1.2.61), in general. This comes from the fact that local thermodynamic equilibrium is not generally satisfied for the photon gas in the atmosphere.

1.2.5 Enthalpy balance and Bernoulli's theorem

From the transformation between enthalpy and kinetic energy, we obtain Bernoulli's theorem. Using (1.1.3), the change in enthalpy h is expressed as

$$\frac{dh}{dt} = \frac{du}{dt} + \frac{d(pv_s)}{dt} = \frac{du}{dt} + p\frac{dv_s}{dt} + v_s\frac{dp}{dt}. \tag{1.2.63}$$

Thus, substituting (1.2.43) into this, we obtain the equation of enthalpy

$$\rho\frac{dh}{dt} = \frac{dp}{dt} + \varepsilon - \nabla \cdot \boldsymbol{F}^{ene}. \tag{1.2.64}$$

Summing up this with the equation of kinetic energy (1.2.34), we obtain

$$\rho\frac{d}{dt}\left(\frac{\boldsymbol{v}^2}{2} + h\right) = \frac{\partial p}{\partial t} - \rho\boldsymbol{v} \cdot \nabla\Phi + \frac{\partial}{\partial x_j}(\sigma'_{ij}v_i) - \nabla \cdot \boldsymbol{F}^{ene}. \tag{1.2.65}$$

Furthermore, summing up this equation and the equation of potential energy (1.2.38), we obtain a generalized form of *Bernoulli's equation*

$$\rho\frac{d}{dt}\left(\frac{\boldsymbol{v}^2}{2} + \sigma\right) = \frac{\partial}{\partial x_j}(\sigma'_{ij}v_i) - \nabla \cdot \boldsymbol{F}^{ene} + \frac{\partial p}{\partial t}, \tag{1.2.66}$$

where σ is the static energy defined by (1.2.51). We can also obtain this expression by adding $\frac{\partial p}{\partial t}$ to both sides of (1.2.47). If the pressure change is negligible ($\frac{\partial p}{\partial t} = 0$), and there is no dissipation ($\sigma'_{ij} = 0$) and no heat flux ($\nabla \cdot \boldsymbol{F}^{ene} = 0$), (1.2.66) becomes

$$\frac{d}{dt}\left(\frac{\boldsymbol{v}^2}{2} + \sigma\right) = 0. \tag{1.2.67}$$

That is, the sum of the kinetic energy and the static energy is conserved along fluid motion. In the case of a steady flow, the sum of the kinetic energy and the static energy is constant along streamlines:

$$\frac{\boldsymbol{v}^2}{2} + \sigma = \frac{\boldsymbol{v}^2}{2} + u + \frac{p}{\rho} + \Phi = \text{const.} \tag{1.2.68}$$

In a special case under the condition of no gravity and uniform temperature, the sum of the kinetic energy and p/ρ is constant:

$$\frac{\boldsymbol{v}^2}{2} + \frac{p}{\rho} = \text{const.}, \tag{1.2.69}$$

which is a familiar form of *Bernoulli's theorem.* If $\boldsymbol{v}^2/2 \ll \sigma$ is satisfied, (1.2.67) is approximated by

$$\frac{d\sigma}{dt} = 0. \tag{1.2.70}$$

It can be said that the static energy is approximately conserved if the kinetic energy is comparatively smaller than the static energy. In such a system, the static energy σ behaves as a conservative quantity like entropy or potential temperature.

1.3 Angular momentum, vorticity, and divergence

Various equations can be derived from the equation of motion. In the first place, we obtain the conservation of angular momentum. Second, the circulation theorem is derived through the balance of momentum of forces along the boundary of a finite domain. In the limit of an infinitesimal domain, the circulation theorem reduces to the vorticity equations. Third, the equation of potential vorticity is derived from the vorticity equations and the equation of a scalar quantity, such as potential temperature. Under an appropriate condition, the potential vorticity is a materially conserved quantity. At the end of this section, the divergence equation is introduced as a pair of vorticity equations.

1.3.1 Conservation of angular momentum

The angular momentum per unit mass is defined by

$$\boldsymbol{l} \equiv \boldsymbol{x} \times \boldsymbol{v}, \tag{1.3.1}$$

where $\boldsymbol{x}$ is a position vector. The origin of the position vector is arbitrarily fixed. In the case of the earth, it is convenient to define the origin at the center of the earth. The i-th component of the angular momentum is denoted by $l_i \equiv \varepsilon_{ijk} x_j v_k$.[†] Each component of the equation of motion (1.2.18) is written as

$$\rho \frac{dv_i}{dt} = \partial_j \sigma_{ij} - \rho g_i, \tag{1.3.2}$$

where $\sigma_{ij} = -p\delta_{ij} + \sigma'_{ij}$, $g_i = \partial_i \Phi$, and $\partial_i = \partial/\partial x_i$. From the outer product of the position vector $\boldsymbol{x}$ and the above equation, we obtain the equation of angular momentum

$$\rho \frac{dl_i}{dt} = \partial_l (\varepsilon_{ijk} x_j \sigma_{kl}) - \rho \varepsilon_{ijk} x_j g_k \tag{1.3.3}$$

or

$$\rho \frac{dl_i}{dt} = -\varepsilon_{ijk} x_j \partial_k p + \partial_l (\varepsilon_{ijk} x_j \sigma'_{kl}) - \rho \varepsilon_{ijk} x_j g_k. \tag{1.3.4}$$

In the above derivation, we have assumed that the stress tensor is symmetric: $\sigma_{ij} = \sigma_{ji}$.[‡] Eq. (1.3.3) can be rewritten in the flux form, that is the conservation of angular momentum

$$\frac{\partial}{\partial t}(\rho l_i) + \partial_l (\rho l_i v_l - \varepsilon_{ijk} x_j \sigma_{kl}) = -\rho \varepsilon_{ijk} x_j g_k. \tag{1.3.5}$$

Remember that the angular momentum is a vector quantity. If the forcing term does not exist on the right-hand side and the stress flux at the boundary is identically zero, the domain integral of the angular momentum is conserved. In the case of the earth, the gravity force has only a radial component in the approximate sense and the contribution of the right-hand side vanishes. Thus, the three components of angular momentum are conserved if the stress flux at the boundary does not exist.

For application to the earth atmosphere, in practice, only the component parallel to the rotation axis is important as the angular momentum. We define the axial component of angular momentum as l_z, where z is the coordinate along the rotation axis in the direction toward the north pole. Using the spherical coordinates (λ, φ, r), where λ is longitude, φ is latitude, and r is the distance from the center of the earth,

---

[†] ε_{ijk} is the antisymmetric tensor, defined by

$$\varepsilon_{ijk} = \begin{cases} 1, & (i,j,k) = (1,2,3) \text{ or } (2,3,1) \text{ or } (3,1,2), \\ -1, & (i,j,k) = (3,2,1) \text{ or } (2,1,3) \text{ or } (1,3,2), \\ 0, & \text{the other combinations of } (i,j,k). \end{cases}$$

A vector product is expressed as $(\boldsymbol{A} \times \boldsymbol{B})_i = \varepsilon_{ijk} A_j B_k$. We also have the following useful relation:

$$\varepsilon_{ijk}\varepsilon_{ilm} = \delta_{jl}\delta_{km} - \delta_{jm}\delta_{kl},$$

from which we have generally $[\boldsymbol{A} \times (\boldsymbol{B} \times \boldsymbol{C})]_i = A_j B_i C_j - A_j B_j C_i$.

[‡] This comes from the requirement that the contribution of spin angular momentum of fluid particles is negligible and that entropy must increase in the viscosity fluid (de Groot and Mazur, 1984). Expression of the viscous stress tensor (1.2.17) is a result of this requirement.

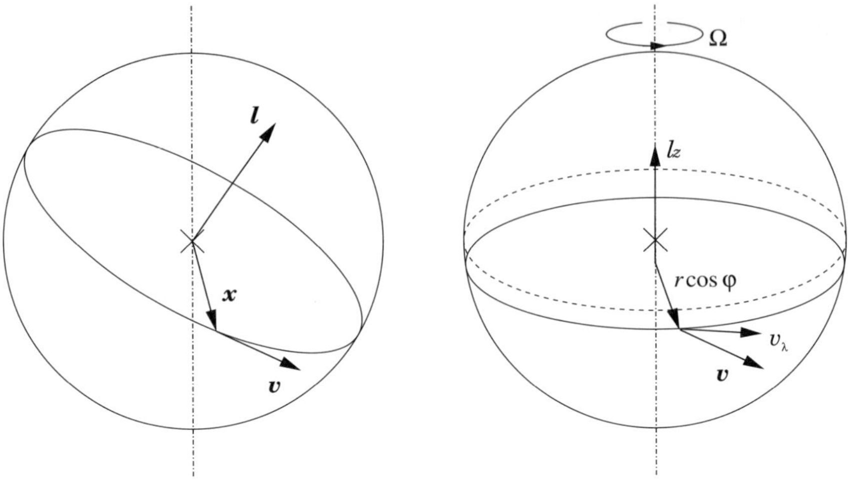

FIGURE 1.2: Explanation of angular momentum. Left: $\boldsymbol{v}$ is the velocity vector, $\boldsymbol{x}$ is the position vector, and $\boldsymbol{l}$ is the angular momentum vector. Right: l_z is the axial component of angular momentum, v_λ is the longitudinal component of velocity, r is radius, φ is latitude, and Ω is the angular velocity.

we have

$$l_z = xv_y - yv_x = r\cos\varphi(v_\lambda + \Omega r\cos\varphi). \tag{1.3.6}$$

v_λ is the longitudinal component of the velocity. Fig. 1.2 shows the relation between the angular momentum vector $\boldsymbol{l}$ and its axial component l_z. The corresponding expression of the conservation of angular momentum in these coordinates is given in Section A1.5.

1.3.2 The circulation theorem

Integration of the equation of motion along a closed curve gives the circulation theorem. The circulation can be related to vorticity. We use both the velocity in the inertial frame $\boldsymbol{v}_a$ and that in the rotating frame $\boldsymbol{v}_r$ to explain the circulation theorem. We refer to a quantity in the inertial frame with a subscript a and that in the rotating frame with r. We take an arbitrarily closed domain A on a material surface and define a closed circuit C of the domain A. The *circulation* in the inertial frame is defined by

$$\Gamma_a \equiv \oint_C \boldsymbol{v}_a \cdot d\boldsymbol{x}. \tag{1.3.7}$$

According to Stokes' theorem, the circulation can be rewritten as

$$\Gamma_a = \int_A \boldsymbol{\omega}_a \cdot \boldsymbol{n}\, dA, \tag{1.3.8}$$

where $\boldsymbol{n}$ is a vector normal to the surface A and vorticity

$$\boldsymbol{\omega}_a \equiv \nabla \times \boldsymbol{v}_a \tag{1.3.9}$$

is introduced. In particular, vorticity in the inertial frame is called *absolute vorticity.* Since A and C are a material surface and a material curve, respectively, both the shapes of A and C change with time as the fluid moves.

The time derivative of the circulation is calculated as

$$\begin{aligned} \frac{d\Gamma_a}{dt} &= \frac{d}{dt}\int \boldsymbol{v}_a \cdot d\boldsymbol{x} = \int \frac{d}{dt}\boldsymbol{v}_a \cdot d\boldsymbol{x} + \int \boldsymbol{v}_a \cdot \frac{d}{dt}d\boldsymbol{x} \\ &= \int \frac{d}{dt}\boldsymbol{v}_a \cdot d\boldsymbol{x} + \int \boldsymbol{v}_a \cdot d\boldsymbol{v}_a = \int \frac{d}{dt}\boldsymbol{v}_a \cdot d\boldsymbol{x}. \end{aligned} \tag{1.3.10}$$

Substituting the equation of motion (1.2.19) into $\frac{d}{dt}\boldsymbol{v}_a$, we have the equation of circulation

$$\frac{d\Gamma_a}{dt} = -\oint_C \frac{\nabla p}{\rho} \cdot d\boldsymbol{x} + \oint_C \boldsymbol{f} \cdot d\boldsymbol{x}, \tag{1.3.11}$$

which is referred to as the *circulation theorem.*

In a similar way, we define circulation and vorticity in the rotating frame by

$$\Gamma_r \equiv \oint_C \boldsymbol{v}_r \cdot d\boldsymbol{x} = \int_A \boldsymbol{\omega}_r \cdot \boldsymbol{n}\, dA, \tag{1.3.12}$$

$$\boldsymbol{\omega}_r \equiv \nabla \times \boldsymbol{v}_r. \tag{1.3.13}$$

Vorticity in the rotating frame ω_r is called *relative vorticity.* Relative vorticity is related to absolute vorticity as

$$\begin{aligned} \boldsymbol{\omega}_a &= \nabla \times \boldsymbol{v}_a = \nabla \times (\boldsymbol{v}_r + \Omega \times \boldsymbol{x}) \\ &= \nabla \times \boldsymbol{v}_r + \nabla \times (\Omega \times \boldsymbol{x}) = \boldsymbol{\omega}_r + 2\Omega. \end{aligned} \tag{1.3.14}$$

Then, the relation between circulation in the inertial frame Γ_a and circulation in the rotating frame Γ_r is given as

$$\Gamma_a = \Gamma_r + \int_A 2\boldsymbol{\Omega} \cdot \boldsymbol{n}\, dA. \tag{1.3.15}$$

The first term on the right-hand side of the equation of circulation (1.3.11) is called the *baroclinic term.* It is rewritten as

$$\begin{aligned} -\oint_C \frac{\nabla p}{\rho} \cdot d\boldsymbol{x} &= -\int_A \nabla \times \frac{\nabla p}{\rho} \cdot \boldsymbol{n}\, dA \\ &= \int_A \frac{\nabla \rho \times \nabla p}{\rho^2} \cdot \boldsymbol{n}\, dA. \end{aligned} \tag{1.3.16}$$

If the fluid is *barotropic* (i.e., the density of fluid ρ is a function of pressure p)

$$\rho = \rho(p), \tag{1.3.17}$$

then $\nabla\rho \times \nabla p = 0$ is satisfied and the baroclinic term does not contribute to the change in circulation. On the other hand, if ρ does not solely depend on p, the fluid is called *baroclinic*; a fluid that is not barotropic is baroclinic. In this case,

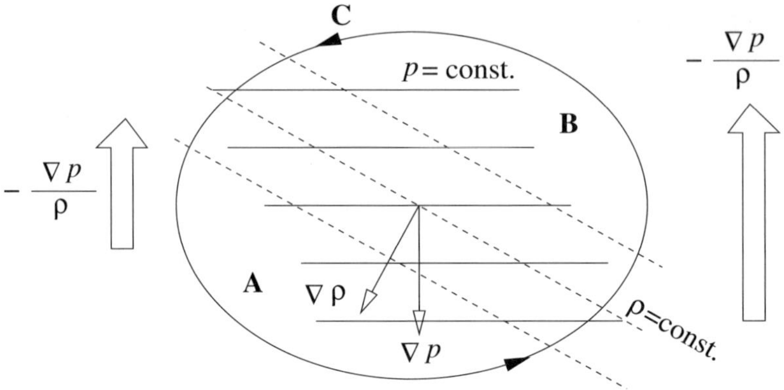

FIGURE 1.3: The baroclinic term and the direction of the pressure gradient.

since $\nabla\rho \times \nabla p \neq 0$, the baroclinic term is not zero in general. Nevertheless, we can always eliminate the baroclinic term by taking the surface A parallel to $\nabla\rho \times \nabla p$. For example, if a thermodynamic variable $\Psi = \Psi(\rho, p)$ is chosen, since

$$\nabla\Psi(\rho, p) = \frac{\partial\Psi}{\partial\rho}\nabla\rho + \frac{\partial\Psi}{\partial p}\nabla p, \tag{1.3.18}$$

then $\nabla\Psi$ is normal to $\nabla\rho\times\nabla p$. Therefore, if A is defined on a surface of a constant Ψ, the contribution of the baroclinic term becomes zero. (See also the section of potential vorticity: Section 1.3.4.)

Let us consider the roles of the terms of the circulation theorem using the circulation along C shown in Fig. 1.3. The baroclinic term is defined by the line integral of the pressure gradient along C. Since density ρ is larger in the region A than in region B, the magnitude of the pressure gradient $|\nabla p/\rho|$ is larger in B than A. Thus, the baroclinic term increases circulation toward the anti-clockwise direction. The second term on the right-hand side of (1.3.11) is the dissipation term due to friction. This term does not necessarily weaken circulation in general.

1.3.3 Vorticity equations

The circulation theorem is given by integration of the equation of motion along a closed curve. In this subsection, the vorticity equations are derived as a derivative form of the circulation theorem. We start from the equation of motion in a tensor form, (1.3.2); that is

$$\rho\left(\frac{\partial v_i}{\partial t} + v_j\partial_j v_i\right) = \partial_j\sigma_{ij} - \rho\partial_i\Phi. \tag{1.3.19}$$

We use the following identity to rewrite the advection term

$$\varepsilon_{ijk}\omega_j v_k = \varepsilon_{ijk}(\varepsilon_{jlm}\partial_l v_m)v_k = v_j\partial_j v_i - \partial_i\frac{\boldsymbol{v}^2}{2}, \tag{1.3.20}$$

where the vorticity is written as $\omega_i = \varepsilon_{ijk}\partial_j v_k$; that is

$$\boldsymbol{v} \cdot \nabla \boldsymbol{v} = \boldsymbol{\omega} \times \boldsymbol{v} + \nabla \frac{\boldsymbol{v}^2}{2}. \tag{1.3.21}$$

The right-hand side is sometimes called the vector invariant form of the advection term. Using the identity (1.3.20), the equation of motion (1.3.19) is rewritten as

$$\frac{\partial v_i}{\partial t} = -\varepsilon_{ijk}\omega_j v_k + \frac{1}{\rho}\partial_j \sigma_{ij} - \partial_i \left(\Phi + \frac{\boldsymbol{v}^2}{2} \right), \tag{1.3.22}$$

or

$$\frac{\partial \boldsymbol{v}}{\partial t} = -\boldsymbol{\omega} \times \boldsymbol{v} - \frac{\nabla p}{\rho} - \nabla \left(\Phi + \frac{\boldsymbol{v}^2}{2} \right) + \boldsymbol{f}, \tag{1.3.23}$$

where $\boldsymbol{f}$ is the frictional force defined by (1.2.20).

To derive the vorticity equations, we calculate $\varepsilon_{ijk}\partial_j(1.3.22)_k$. We immediately obtain

$$\frac{\partial \omega_i}{\partial t} + \partial_j \left(\omega_i v_j - \omega_j v_i - \varepsilon_{ijk} \frac{1}{\rho} \partial_l \sigma_{kl} \right) = 0. \tag{1.3.24}$$

This is the flux form of *vorticity equations.* Using the identity

$$\partial_j(\omega_i v_j - \omega_j v_i) = v_j \partial_j \omega_i + \omega_i \partial_j v_j - \omega_j \partial_j v_i,$$

we obtain the advective form of vorticity equations:

$$\frac{d\omega_i}{dt} = \omega_j \partial_j v_i - \omega_i \partial_j v_j + \varepsilon_{ijk}\partial_j \left(\frac{1}{\rho} \partial_l \sigma_{kl} \right), \tag{1.3.25}$$

or

$$\frac{d\boldsymbol{\omega}}{dt} = \boldsymbol{\omega} \cdot \nabla \boldsymbol{v} - \boldsymbol{\omega}(\nabla \cdot \boldsymbol{v}) - \nabla \frac{1}{\rho} \times \nabla p + \nabla \times \boldsymbol{f}. \tag{1.3.26}$$

Eq. (1.3.26) represents vorticity equations in an inertial frame. Vorticity in (1.3.26) is interpreted as absolute vorticity $\boldsymbol{\omega}_a$. The vorticity equations in the rotating frame are directly derived from (1.3.26). Using (1.2.24) and substituting (1.2.26) and (1.3.14) into (1.3.26), we have

$$\begin{aligned} \frac{d\boldsymbol{\omega}_r}{dt} = &\ (\boldsymbol{\omega}_r + 2\boldsymbol{\Omega}) \cdot \nabla \boldsymbol{v}_r - (\boldsymbol{\omega}_r + 2\boldsymbol{\Omega})(\nabla \cdot \boldsymbol{v}_r) \\ &- \nabla \frac{1}{\rho} \times \nabla p + \nabla \times \boldsymbol{f}. \end{aligned} \tag{1.3.27}$$

By comparison with the circulation theorem (1.3.11), the vorticity equations (1.3.26) formally have additional terms:

$$\boldsymbol{\omega} \cdot \nabla \boldsymbol{v} - \boldsymbol{\omega}(\nabla \cdot \boldsymbol{v}). \tag{1.3.28}$$

Let us consider the meanings of these terms by decomposing their components into the Cartesian coordinates. We take the z-direction as the direction parallel to

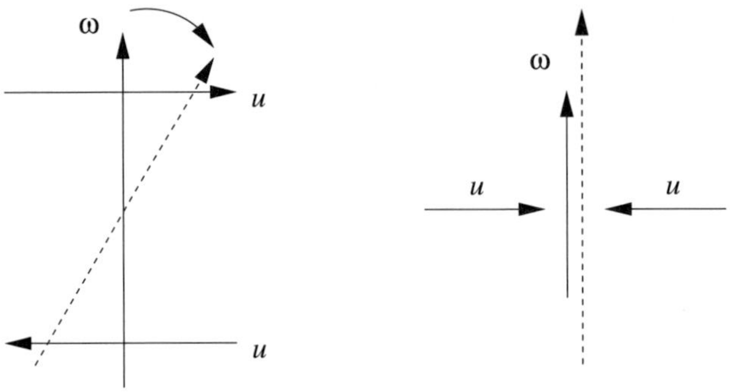

FIGURE 1.4: The schematic figures of the tilting term (left) and the stretching term (right). ω is vorticity and u is velocity. Vorticity ω changes as the arrow with the dotted line by shear (left) or convergence (right).

the vector $\boldsymbol{\omega}$. Letting $\boldsymbol{i}$, $\boldsymbol{j}$, and $\boldsymbol{k}$ be unit vectors of the x-, y-, and z-directions, respectively, we rewrite (1.3.28) as

$$\begin{aligned} &\omega\frac{\partial}{\partial z}(u\boldsymbol{i}+v\boldsymbol{j}+w\boldsymbol{k})-\omega\boldsymbol{k}(\nabla\cdot\boldsymbol{v}) \\ &= \boldsymbol{i}\omega\frac{\partial u}{\partial z}+\boldsymbol{j}\omega\frac{\partial v}{\partial z}-\boldsymbol{k}\omega\left(\frac{\partial u}{\partial x}+\frac{\partial v}{\partial y}\right). \end{aligned} \tag{1.3.29}$$

The first two terms on the right-hand side generate vorticity components perpendicular to $\boldsymbol{\omega}$. These are called the *tilting terms.* For example, if the velocity u has a shear in the z-direction, the vorticity vector is tilted in the x-direction and the x-component of vorticity is generated. The third term on the right of (1.3.29) is related to the change in the component of the $\boldsymbol{\omega}$'s own direction. If there is convergence in a plane perpendicular to the vector $\boldsymbol{\omega}$, the vorticity increases (Fig. 1.4). This term is referred to as the *stretching term.*

1.3.4 Potential vorticity

In the circulation theorem (1.3.11), the baroclinic term can be eliminated by appropriately choosing the integral path. Similarly, a particular direction of the vorticity has a special change in vorticity equations. To show it in a general form, we use an arbitrary scalar field $\Psi(\boldsymbol{x},t)$. Let us consider a vorticity component perpendicular to a surface of $\Psi=$ const. The inner product of flux form vorticity equations (1.3.24) and $\partial_i\Psi$ gives

$$\partial_i\Psi\left[\frac{\partial\omega_i}{\partial t}+\partial_j\left(\omega_i v_j-\omega_j v_i-\varepsilon_{ijk}\frac{1}{\rho}\partial_l\sigma_{kl}\right)\right] = 0. \tag{1.3.30}$$

The left-hand side can be rewritten as

$$\begin{aligned} &\frac{\partial}{\partial t}(\omega_i\partial_i\Psi)-\omega_i\frac{\partial}{\partial t}(\partial_i\Psi)+\partial_j(\omega_i\partial_i\Psi\cdot v_j) \\ &\quad -\omega_i v_j\partial_j\partial_i\Psi-\partial_i\Psi\cdot\omega_j\partial_j v_i-\partial_i\left[\Psi\varepsilon_{ijk}\partial_j\left(\frac{1}{\rho}\partial_l\sigma_{kl}\right)\right] \end{aligned}$$

$$
\begin{aligned}
&= \frac{\partial}{\partial t}(\omega_i \partial_i \Psi) + \partial_j(\omega_i \partial_i \Psi \cdot v_j) - \partial_i \left(\omega_i \frac{d\Psi}{dt} \right) \\
&\quad - \partial_i \left[\Psi \varepsilon_{ijk} \partial_j \left(\frac{1}{\rho} \partial_l \sigma_{kl} \right) \right]. \qquad (1.3.31)
\end{aligned}
$$

Thus, we obtain the following conservation equation

$$
\frac{\partial}{\partial t}(\rho \Pi) + \partial_i \left[\rho \Pi v_i - \frac{d\Psi}{dt}\omega_i - \Psi \varepsilon_{ijk} \partial_j \left(\frac{1}{\rho} \partial_l \sigma_{kl} \right) \right] = 0, \qquad (1.3.32)
$$

where

$$
\Pi \equiv \frac{\boldsymbol{\omega} \cdot \nabla \Psi}{\rho} \qquad (1.3.33)
$$

is called *potential vorticity.* Note that the dimension of potential vorticity depends on the choice of scalar Ψ. The flux term of (1.3.32) involving the stress tensor σ_{kl} has different forms such as

$$
\begin{aligned}
\partial_i \left[-\Psi \varepsilon_{ijk} \partial_j \left(\frac{1}{\rho} \partial_l \sigma_{kl} \right) \right] &= \nabla \cdot \left[\Psi \left(\nabla \frac{1}{\rho} \times \nabla p - \nabla \times \boldsymbol{f} \right) \right], \qquad (1.3.34) \\
(\partial_i \Psi) \varepsilon_{ijk} \partial_j \left(-\frac{1}{\rho} \partial_l \sigma_{kl} \right) &= \nabla \Psi \cdot \left(\nabla \frac{1}{\rho} \times \nabla p - \nabla \times \boldsymbol{f} \right), \qquad (1.3.35) \\
\partial_i \left(\varepsilon_{ijk} \partial_j \Psi \cdot \frac{1}{\rho} \partial_l \sigma_{kl} \right) &= \nabla \cdot \left[\nabla \Psi \times \left(-\frac{1}{\rho} \nabla p + \boldsymbol{f} \right) \right]. \qquad (1.3.36)
\end{aligned}
$$

Thus, the advective form of the equation of potential vorticity is written as

$$
\rho \frac{d\Pi}{dt} = \partial_i \left[\frac{d\Psi}{dt} \omega_i + \Psi \varepsilon_{ijk} \partial_j \left(\frac{1}{\rho} \partial_l \sigma_{kl} \right) \right], \qquad (1.3.37)
$$

or

$$
\rho \frac{d\Pi}{dt} = \boldsymbol{\omega} \cdot \nabla \frac{d\Psi}{dt} - \nabla \Psi \cdot \left(\nabla \frac{1}{\rho} \times \nabla p - \nabla \times \boldsymbol{f} \right). \qquad (1.3.38)
$$

If one chooses a thermodynamic variable as scalar Ψ such that

$$
\Psi = \Psi(\rho, p), \qquad (1.3.39)
$$

then $\nabla \Psi$ becomes normal to $\nabla(1/\rho) \times \nabla p$ because of (1.3.18). Furthermore, if we assume

$$
\begin{aligned}
\frac{d\Psi}{dt} &= 0, \qquad (1.3.40) \\
\boldsymbol{f} &= 0, \qquad (1.3.41)
\end{aligned}
$$

the right-hand side of (1.3.38) vanishes. In this case, the potential vorticity is materially conserved

$$
\frac{d\Pi}{dt} = 0. \qquad (1.3.42)
$$

If potential temperature (or entropy) is chosen as Ψ, (1.3.39) and (1.3.40) are satisfied under the adiabatic condition. In the case of $\Psi = \theta$ (potential temperature)

$$P \equiv \frac{\boldsymbol{\omega} \cdot \nabla\theta}{\rho} \tag{1.3.43}$$

is called *Ertel's potential vorticity.* The expressions of (1.3.32) and (1.3.38) become

$$\frac{\partial}{\partial t}(\rho P) + \nabla \cdot \left(\rho P \boldsymbol{v} - \dot{\theta}\boldsymbol{\omega} - \theta \cdot \nabla \times \boldsymbol{f}\right) = 0, \tag{1.3.44}$$

$$\frac{dP}{dt} = \frac{1}{\rho}\boldsymbol{\omega} \cdot \nabla\dot{\theta} + \frac{1}{\rho}\nabla\theta \cdot \nabla \times \boldsymbol{f}. \tag{1.3.45}$$

From (1.2.56), the change in the potential temperature is given by

$$\dot{\theta} = \frac{\theta}{T}Q, \tag{1.3.46}$$

where Q is the diabatic term. We note that Ertel's potential vorticity in the rotating frame is expressed as

$$P = \frac{(\boldsymbol{\omega}_r + 2\boldsymbol{\Omega}) \cdot \nabla\theta}{\rho}, \tag{1.3.47}$$

where relative vorticity is used in (1.3.43). Potential vorticity plays an important role as a Lagrangian conservative quantity in both meteorology and oceanography.

The equation of potential vorticity can be derived in a different way that is also instructive. For instance, Pedlosky (1987) derives the conservation of potential vorticity (1.3.42) from the circulation theorem (1.3.11) as a special case. The mathematical basis of potential vorticity is given by Hamiltonian fluid dynamics (Shepherd 1990; Salmon 1998). It can also be shown that the conservation of potential vorticity (1.3.38) is derived from the advective form of vorticity equations (1.3.26). Combining (1.3.26) with the continuity equation (1.2.7), we have

$$\frac{d}{dt}\frac{\boldsymbol{\omega}}{\rho} = \frac{\boldsymbol{\omega}}{\rho} \cdot \nabla\boldsymbol{v} + \frac{\nabla\rho \times \nabla p}{\rho^3} + \frac{1}{\rho}\nabla \times \boldsymbol{f}. \tag{1.3.48}$$

The inner product of this with the gradient of a scalar Ψ yields

$$\frac{d}{dt}\left(\frac{\boldsymbol{\omega}}{\rho} \cdot \nabla\Psi\right) = \nabla\Psi \cdot \frac{d}{dt}\left(\frac{\boldsymbol{\omega}}{\rho}\right) + \frac{\boldsymbol{\omega}}{\rho}\frac{d}{dt}\nabla\Psi. \tag{1.3.49}$$

Since

$$\nabla\Psi \cdot \frac{d}{dt}\frac{\boldsymbol{\omega}}{\rho} = \nabla\Psi \cdot \left(\frac{\boldsymbol{\omega}}{\rho} \cdot \nabla\boldsymbol{v}\right) + \nabla\Psi \cdot \frac{\nabla\rho \times \nabla p}{\rho^3} + \frac{\nabla\Psi}{\rho} \cdot \nabla \times \boldsymbol{f}, \tag{1.3.50}$$

$$\frac{\boldsymbol{\omega}}{\rho}\frac{d}{dt}\nabla\Psi = \frac{\boldsymbol{\omega}}{\rho}\nabla\frac{d}{dt}\Psi - \left(\frac{\boldsymbol{\omega}}{\rho} \cdot \nabla\boldsymbol{v}\right) \cdot \nabla\Psi. \tag{1.3.51}$$

Then, we have the equation of potential vorticity

$$\frac{d\Pi}{dt} = \frac{\boldsymbol{\omega}}{\rho} \cdot \nabla\frac{d\Psi}{dt} + \nabla\Psi \cdot \frac{\nabla\rho \times \nabla p}{\rho^3} + \frac{\nabla\Psi}{\rho} \cdot \nabla \times \boldsymbol{f}, \tag{1.3.52}$$

which is equivalent to (1.3.38).

1.3.5 Divergence equation

In Section 1.3.3 vorticity equations are derived by applying the rotation operator to the equation of motion. As a counterpart to vorticity equations, the divergence equation can be derived by applying the divergence operator to the equation of motion. The divergence of velocity is defined as

$$D \equiv \partial_i v_i = \nabla \cdot \boldsymbol{v}. \tag{1.3.53}$$

Applying the divergence operator to the equation of motion (1.3.22) yields

$$\frac{\partial D}{\partial t} = -\partial_i(\varepsilon_{ijk}\omega_j v_k) + \partial_i\left(\frac{1}{\rho}\partial_j\sigma_{ij}\right) - \partial_i^2\left(\Phi + \frac{\boldsymbol{v}^2}{2}\right). \tag{1.3.54}$$

The first term on the right-hand side is rewritten as

$$\begin{aligned}\partial_i(\varepsilon_{ijk}\omega_j v_k) &= -\omega_j^2 + \varepsilon_{ijk}(\partial_i\omega_j)v_k \\ &= -\omega_j^2 - (\partial_i\partial_i v_k)v_k + (\partial_k\partial_i v_i)v_k \\ &= -|\boldsymbol{\omega}|^2 - \boldsymbol{v}\cdot\nabla^2\boldsymbol{v} + \boldsymbol{v}\cdot\nabla D.\end{aligned} \tag{1.3.55}$$

Hence, we have the divergence equation as

$$\frac{dD}{dt} = |\boldsymbol{\omega}|^2 + \boldsymbol{v}\cdot\nabla^2\boldsymbol{v} - \nabla^2\left(\frac{\boldsymbol{v}^2}{2} + \Phi\right) - \nabla\cdot\left(\frac{\nabla p}{\rho}\right) + \nabla\cdot\boldsymbol{f}. \tag{1.3.56}$$

The divergence equation in the rotating frame is also derived from the equation of motion in the rotating frame (1.2.29). Since divergence in the rotating frame is equal to divergence in the inertial frame D, we can show in the same way as the derivation of (1.3.56) that

$$\nabla\cdot\frac{d\boldsymbol{v}_r}{dt} = \frac{dD}{dt} - |\boldsymbol{\omega}_r|^2 - \boldsymbol{v}_r\cdot\nabla^2\boldsymbol{v}_r + \nabla^2\frac{v_r^2}{2}. \tag{1.3.57}$$

We also have

$$\nabla\cdot(2\boldsymbol{\Omega}\times\boldsymbol{v}_r) = -2\boldsymbol{\Omega}\cdot\boldsymbol{\omega}_r. \tag{1.3.58}$$

Thus, the divergence equation in the rotating frame is written as

$$\begin{aligned}\frac{dD}{dt} &= (\boldsymbol{\omega}_r + 2\boldsymbol{\Omega})\cdot\boldsymbol{\omega}_r + \boldsymbol{v}_r\cdot\nabla^2\boldsymbol{v}_r - \nabla^2\left(\frac{\boldsymbol{v}_r^2}{2} + \Phi_r\right) \\ &\quad -\nabla\cdot\left(\frac{\nabla p}{\rho}\right) + \nabla\cdot\boldsymbol{f}.\end{aligned} \tag{1.3.59}$$

The divergence equation (1.3.59) is a prognostic equation for the divergence of the three-dimensional velocity field. In practice, this form of the equation is rarely used. Instead, an equation for the horizontal divergence of the two-dimensional velocity is usually used for the governing equations of atmospheric general circulation models (see Chapter 20). Nevertheless, we should point out the relevance of the three-dimensional divergence equation to the sound wave.

Let us consider perturbations from the state at rest ($\boldsymbol{v} = 0$) and linearize (1.3.56). We neglect the gravitational potential Φ and assume that basic states of the density and pressure are constant $\rho = \rho_0$ and $p = p_0$. By dividing density and pressure into constant basic state portions and perturbation portions, $\rho' = \rho - \rho_0$ and $p' = p - p_0$, respectively, we have the linearized perturbation equation for divergence as

$$\frac{\partial D}{\partial t} = -\frac{1}{\rho_0}\nabla^2 p'. \tag{1.3.60}$$

Linearizing the continuity equation (1.2.7), the divergence is related as

$$D = -\frac{1}{\rho_0}\frac{\partial \rho}{\partial t}. \tag{1.3.61}$$

Substituting this into (1.3.60), we obtain

$$\frac{\partial^2 \rho'}{\partial t^2} = \nabla^2 p'. \tag{1.3.62}$$

If the flow is adiabatic, the perturbation of pressure is written as

$$p' = \left(\frac{\partial p}{\partial \rho}\right)_s \rho' = c_s^2 \rho', \tag{1.3.63}$$

where c_s is the speed of sound defined by (1.1.32). Thus, by eliminating ρ' and assuming c_s^2 to be constant, we finally have

$$\frac{\partial^2 p'}{\partial t^2} = c_s^2 \nabla^2 p'. \tag{1.3.64}$$

This is a wave equation describing the propagation of sound waves. We will explain the characteristics of the sound wave in Section 4.2.

Generally, flow fields $\boldsymbol{v}$ are expressed by using *velocity potential* χ and *streamfunction vector* $\boldsymbol{\psi}$ as

$$\boldsymbol{v} = \nabla\chi - \nabla \times \boldsymbol{\psi}. \tag{1.3.65}$$

χ and $\boldsymbol{\psi}$ are also called scalar potential and vector potential, respectively. In this case, the vorticity vector and the divergence are respectively expressed as

$$\boldsymbol{\omega} = \nabla \times \boldsymbol{v} = -\nabla \times \nabla \times \boldsymbol{\psi} = -\nabla(\nabla \cdot \boldsymbol{\psi}) + \nabla^2 \boldsymbol{\psi}, \tag{1.3.66}$$

$$D = \nabla \cdot \boldsymbol{v} = \nabla^2 \chi. \tag{1.3.67}$$

Velocity potential and streamfunction vector are not uniquely determined from the velocity field. The velocity field is not affected by the transformation $\chi' = \chi + \nabla \times \boldsymbol{a}$ and $\boldsymbol{\psi}' = \boldsymbol{\psi} + \nabla b$ where $\boldsymbol{a}$ is an arbitrary vector and b is an arbitrary scalar. Thus, the streamfunction vector can always be chosen such that $\nabla \cdot \boldsymbol{\psi} = 0$, and then $\boldsymbol{\omega} = \nabla^2 \boldsymbol{\psi}$. In the special case for an irrotational field with $\boldsymbol{\omega} = 0$, the velocity is simply expressed as $\boldsymbol{v} = \nabla\chi$, and thus described by the divergence equation for D.

In practice, the combination of the vorticity equation and the divergence equation is used to describe two-dimensional horizontal flow fields. In Section 3.4, the vorticity equation and the divergence equation for the shallow water model are introduced to describe two-dimensional flow fields. In Chapter 17, the spherical motions on earth are studied using the equations of the shallow water model. The equation set of general circulation models is summarized in Chapter 20. Since (for spectrum models) the streamfunction and the velocity potential are easily calculable from vorticity and divergence, respectively, the vorticity and divergence equations are generally used as the prognostic equations of this type of general circulation models.

References and suggested reading

Most of the topics in this chapter are generally described as the fundamentals of fluid dynamics. Among the many good textbooks on fluid dynamics are Landau and Lifshitz (1987) and Batchelor (1967) for the foundations of the fluid dynamics. For the empirical laws of the non-equilibrium system, such as the expression of the stress tensor and entropy production rate, see de Groot and Mazur (1984). There is another approach that uses Hamiltonian mechanics for the foundation of fluid dynamics. Hamiltonian fluid mechanics is reviewed by Shepherd (1990) and chapter 7 of Salmon (1998).

Batchelor, G. K., 1967: *Fluid Dynamics.* Cambridge University Press, Cambridge, UK, 615 pp.

de Groot, S. R. and Mazur, P., 1984: *Non-equilibrium thermodynamics.* Dover, New York, 510 pp.

Landau, L. and Lifshitz, E. M., 1987: *Fluid Mechanics*, 2nd ed. Butterworth-Heinemann, Oxford, UK, 539 pp.

Pedlosky, J., 1987: *Geophysical Fluid Dynamics*, 2nd ed. Springer-Verlag, New York, 710 pp.

Salmon, R., 1998: *Lectures on Geophysical Fluid Dynamics.* Oxford University Press, New York, 378 pp.

Shepherd, T. G., 1990: Symmetric conservation laws, and Hamiltonian structure in geophysical fluid dynamics. *Advances in Geophys.*, **32**, 287–338.

2

Basic balances and stability

Large-scale motions of the atmosphere can be viewed in a balanced state in an approximate sense. The primary balances of the atmosphere are the hydrostatic balance and the geostrophic balance. In this chapter the basic properties of these balances are summarized and their stabilities examined.

First, the hydrostatic balance and its various expressions are described in Section 2.1. In a strict sense, the hydrostatic balance is a dynamic balance between the pressure gradient force and the gravitational force without any motion. In this section, approximation of the geopotential is introduced. If viewed in a rotating frame, the effect of a component of a rigid body rotation is counted as a potential force due to the centrifugal potential and can be included as part of the geopotential. Then, in Section 2.2, the geostrophic balance and the thermal wind balance are described. The geostrophic balance is defined as a static balance with wind fields in the rotating frame. The thermal wind balance is the corresponding balance of vorticity. In the following two sections, the stability of these balanced states is examined. The stability of the hydrostatic balance is studied using the parcel method in Section 2.3. The stability of the thermal wind balance is argued in Section 2.4 on the assumption that the perturbation has a two-dimensional axisymmetric flow. In the last section (Section 2.5), balances and stabilities in a more simplified system (i.e., a constant rotating plane (f-plane)), are summarized.

These balances are fundamental in all respects both for atmospheric structure and atmospheric modeling. The statistically averaged states or the meridional structure of the atmosphere are thought to be hydrostatically and geostrophically balanced (Chapters 14, 16, and 18). To derive the various approximate equations in the next chapter, balanced states are regarded as the zero-th order approximation or the reference fields. In Chapters 4 and 5, perturbations to balanced states are examined: waves and instability. Although stability of the atmosphere is considered both in the present chapter and in Chapter 5, the present chapter is devoted to studying the conditions of stability for balanced states, while in Chapter 5 the structures of unstable waves are particularly considered.

2.1 Hydrostatic balance

We start from the equation of motion in the rotating frame (1.2.29) to consider the basic balance of the atmosphere:†

$$\frac{d\boldsymbol{v}}{dt} + 2\boldsymbol{\Omega} \times \boldsymbol{v} = -\frac{1}{\rho}\nabla p - \nabla\Phi + \boldsymbol{f}. \qquad (2.1.1)$$

If the atmosphere is at rest with $\boldsymbol{v} = 0$, the first and second terms on the right-hand side are balanced:

$$0 = -\frac{1}{\rho}\nabla p - \nabla\Phi. \qquad (2.1.2)$$

That is, the gravity force is balanced by the pressure gradient force. This balance is called the *hydrostatic balance*. Let us define the z coordinate parallel to $\nabla\Phi$ with $\frac{\partial\Phi}{\partial z} > 0$ and let the unit vector in the z direction be denoted by $\boldsymbol{k}$. A constant z surface coincides with the surface $\Phi =$ const. Using the acceleration due to gravity g defined by

$$g \equiv \frac{\partial\Phi}{\partial z}, \qquad (2.1.3)$$

the hydrostatic balance reads

$$0 = -\frac{1}{\rho}\frac{\partial p}{\partial z} - g. \qquad (2.1.4)$$

In this state, the pressure p is uniform on the surface $z =$ const. The hydrostatic balance is approximately satisfied for large-scale fields, even when the atmosphere has non-zero velocity. The condition when the hydrostatic balance is satisfied will be considered in Section 3.2, where quasi-geostrophic approximation is introduced.

In general, the z coordinate thus defined is not a straight line, since the distribution of Φ is not spherically symmetric about the center of the earth. However, the deviation of Φ from its spherically symmetric state is usually neglected. In this case, the z-coordinate coincides with the radial direction. The potential Φ is the sum of the potential energy for the attractive force Φ_a and the centrifugal potential energy given by (1.2.30): $\Phi = \Phi_a + \Phi_c$. For the earth, these potential energies are expressed as

$$\Phi_a = GM\left(\frac{1}{R} - \frac{1}{r}\right), \qquad (2.1.5)$$

$$\Phi_c = -\frac{1}{2}\Omega^2 X^2, \qquad (2.1.6)$$

where G is the gravitational constant, M is the mass of the earth, R is the mean radius of the earth, and $X = r\cos\varphi$ is the distance from the rotation axis: $G = 6.673 \times 10^{-11}$ m^3 kg^{-1} s^{-2}, $M = 5.9736 \times 10^{24}$ kg, $R = 6371$ km. Fig. 2.1 shows

†The subscript r for the values in the rotating frame is omitted in this chapter.

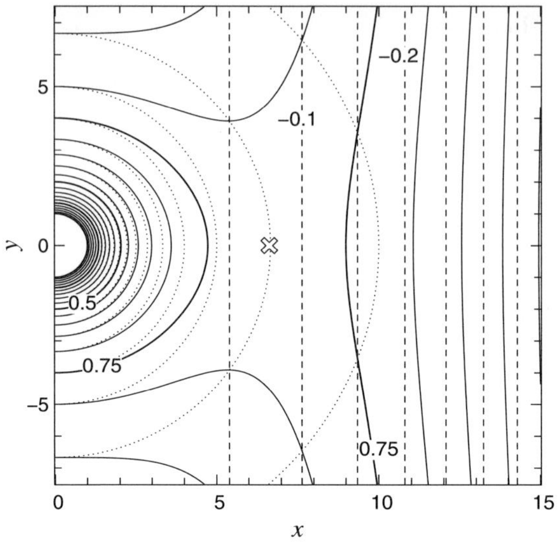

FIGURE 2.1: Distributions of gravitational potential energy (solid), centrifugal potential energy (dashed), and potential energy for the attractive force (dotted). The ordinate and the abscissa are in unit of the earth's radius R. The equator corresponds to $y = 0$. The contour interval is $0.1GM/R$, and the contours inside the earth are omitted. The cross is the point where the acceleration due to gravity is zero in the rotating frame with the rotation rate Ω.

the distributions of the potential energies around the earth. There is a special point on the plane intersecting the equator at the radius

$$r = R_c \equiv \left(\frac{GM}{\Omega^2}\right)^{1/3} \approx 4.224 \times 10^7 \text{ m}, \tag{2.1.7}$$

where no gravity force works if an object rotates at the same rate as the earth's rotation Ω. At the equator on the earth's surface, in contrast, the ratio between the acceleration due to the centrifugal force and that due to the attractive force is given by

$$\frac{\Omega^2 R}{GM/R^2} = \left(\frac{R}{R_c}\right)^3 \approx 0.00343 \tag{2.1.8}$$

(i.e., the contribution of the centrifugal force is small). From this fact, it is generally assumed for the study of atmospheric dynamics that gravitational potential energy is spherically symmetric.

The sea surface on the earth agrees with an iso-surface of potential energy Φ, which is called the *geoid*. From (2.1.8), we can assume that the geoid of the earth is an exact sphere with a constant radius R. The distance from the center of the earth r is written as

$$r = R + z, \tag{2.1.9}$$

where z is the height from the ocean surface (i.e., the geoid given by a constant surface of Φ). In addition, the atmosphere can be assumed to be thin; that is,

$$\frac{z}{R} \ll 1. \tag{2.1.10}$$

In this case, the acceleration due to gravity is assumed to be constant; from (2.1.5), the approximated value is given as

$$g = \frac{GM}{R^2} \approx 9.82 \text{ m s}^{-2}, \tag{2.1.11}$$

and the gravitational potential is given by

$$\Phi = gz. \tag{2.1.12}$$

Hereafter, we use this approximation for the geopotential.

The *scale height* of pressure is defined as

$$H_p \equiv \left(-\frac{1}{p}\frac{\partial p}{\partial z}\right)^{-1} = \frac{R_d T}{g}, \tag{2.1.13}$$

where the hydrostatic balance and the equation of state $p = \rho R_d T$ (1.1.41) are used. If the profile of temperature $T(z)$ is given, the distribution of pressure is expressed as

$$p(z) = p_s \exp\left(-\int_0^z \frac{dz}{H(z)}\right) = p_s \exp\left(-\int_0^z \frac{g}{R_d T(z)}\,dz\right), \tag{2.1.14}$$

where p_s is the surface pressure. If the temperature T is uniform, the pressure is given as $p = p_s e^{-z/H}$.

Now, let us derive different expressions of the hydrostatic balance. Using the thermodynamic relation (1.1.6)

$$dh = Tds + \frac{1}{\rho}dp, \tag{2.1.15}$$

the hydrostatic balance (2.1.2) can be expressed as

$$0 = T\nabla s - \nabla\sigma, \tag{2.1.16}$$

where $\sigma = h + \Phi$ is the static energy introduced by (1.2.51).[†] Eq. (2.1.16) implies that static energy is uniform if the atmosphere is isentropic and in hydrostatic balance. Using (2.1.12) and enthalpy $h = C_p T$ for an ideal gas with constant specific heat, static energy is written as

$$\sigma = C_p T + gz. \tag{2.1.17}$$

[†]In the literature, σ is referred to as the *Montgomery function*. The pressure gradient term in the entropy or potential temperature coordinate is expressed by the gradient of the Montgomery function. The symbol Ψ or M is used to denote the Montgomery function in general. See Section 3.3.5.

From (2.1.16), we have

$$T\frac{\partial s}{\partial z} = \frac{\partial \sigma}{\partial z} = C_p\frac{\partial T}{\partial z} + g. \tag{2.1.18}$$

From this, if the atmosphere is particularly isentropic, the gradient of temperature is given by

$$\frac{\partial T}{\partial z} = -\frac{g}{C_p} \equiv -\Gamma_d, \tag{2.1.19}$$

where $\Gamma_d = g/C_p$ is called the *dry adiabatic lapse rate.*

The hydrostatic balance is also rewritten by using the potential temperature (1.1.52) for an ideal gas with constant specific heat. We use the *Exner function* π, defined by

$$\pi \equiv C_p\left(\frac{p}{p_0}\right)^\kappa = C_p\frac{T}{\theta}, \tag{2.1.20}$$

where $\kappa = R_d/C_p$. Since $\theta\pi = h$, we have

$$\frac{1}{\rho}dp = \theta d\pi = dh - \pi d\theta. \tag{2.1.21}$$

Hence, the hydrostatic balance (2.1.2) can be rewritten as

$$0 = -\theta\nabla\pi - \nabla\Phi = \pi\nabla\theta - \nabla\sigma. \tag{2.1.22}$$

From this, the vertical gradient of the potential temperature is related as

$$\frac{\partial\theta}{\partial z} = \frac{1}{\pi}\frac{\partial\sigma}{\partial z} = \frac{\theta}{T}\left(\frac{\partial T}{\partial z} + \frac{g}{C_p}\right). \tag{2.1.23}$$

2.2 Geostrophic balance and thermal wind

Let us assume that the effect of rotation is large in the equation of motion (2.1.1) and that the Coriolis term on the second term on the left-hand side is balanced by the first and second terms on the right-hand side:

$$2\boldsymbol{\Omega} \times \boldsymbol{v} = -\frac{1}{\rho}\nabla p - \nabla\Phi, \tag{2.2.1}$$

which is called the *geostrophic balance.* Applying the rotation operator to this gives

$$0 = 2\boldsymbol{\Omega}\cdot\nabla\boldsymbol{v} - 2\boldsymbol{\Omega}(\nabla\cdot\boldsymbol{v}) - \nabla\frac{1}{\rho}\times\nabla p. \tag{2.2.2}$$

In the case that the flow is non-divergent $\nabla\cdot\boldsymbol{v} = 0$, (2.2.2) reduces to

$$0 = 2\boldsymbol{\Omega}\cdot\nabla\boldsymbol{v} - \nabla\frac{1}{\rho}\times\nabla p, \tag{2.2.3}$$

which is called the *thermal wind balance.*

Eq. (2.2.2) can also be given from the balance of the vorticity equation in the rotating frame. In (1.3.27)

$$\begin{aligned} \frac{d\boldsymbol{\omega}_r}{dt} &= (\boldsymbol{\omega}_r + \underline{2\boldsymbol{\Omega}}) \cdot \nabla \boldsymbol{v} - (\boldsymbol{\omega}_r + \underline{2\boldsymbol{\Omega}})(\nabla \cdot \boldsymbol{v}) \\ &\quad \underline{-\nabla \frac{1}{\rho} \times \nabla p} + \nabla \times \boldsymbol{f}, \end{aligned} \tag{2.2.4}$$

the balance between the underlined three terms corresponds to (2.2.2). These terms are the tilting term, the stretching term, and the baroclinic term, respectively. In particular, (2.2.3) means the balance between the tilting term and the baroclinic term, which is shown in Fig. 2.2. From (2.1.15) and (2.1.21), the baroclinic term is rewritten as

$$\begin{aligned} \boldsymbol{B} &= -\nabla \times \frac{\nabla p}{\rho} &&= -\nabla \frac{1}{\rho} \times \nabla p \\ &= \nabla \times (T \nabla s) &&= \nabla T \times \nabla s \\ &= -\nabla \times (\theta \nabla \pi) &&= -\nabla \theta \times \nabla \pi. \end{aligned} \tag{2.2.5}$$

A state in which the baroclinic term $\boldsymbol{B}$ vanishes is called *barotropic* (i.e., iso-surfaces of density are parallel to iso-surfaces of pressure such that density is a function of pressure: $\rho = \rho(p)$). In this case, (2.2.2) becomes

$$0 = 2\boldsymbol{\Omega} \cdot \nabla \boldsymbol{v} - 2\boldsymbol{\Omega}(\nabla \cdot \boldsymbol{v}). \tag{2.2.6}$$

Using a Cartesian coordinate by setting the z axis in the direction of the vector $\boldsymbol{\Omega}$, each component of (2.2.6) is written as

$$\left(\frac{\partial u}{\partial z}, \frac{\partial v}{\partial z}, \frac{\partial u}{\partial x} + \frac{\partial v}{\partial y} \right) = (0, 0, 0).$$

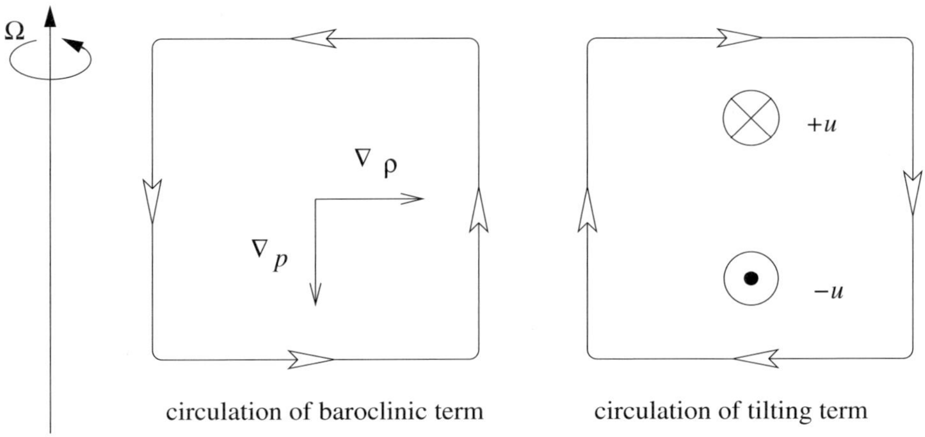

FIGURE 2.2: The thermal wind balance. The circulation of the baroclinic term is balanced by the circulation of the tilting term.

In particular, in the case of non-divergent $\nabla \cdot \boldsymbol{v} = 0$, we obtain

$$\frac{\partial \boldsymbol{v}}{\partial z} = \left(\frac{\partial u}{\partial z}, \frac{\partial v}{\partial z}, \frac{\partial w}{\partial z} \right) = 0 \tag{2.2.7}$$

(i.e., the velocity does not change in the direction of the rotation vector). This is an expression of the *Taylor-Proudman theorem.*

Here, we use *scale analysis* to consider conditions under which the underlined terms in the vorticity equation (2.2.4) predominate the other terms. Let U, L, and τ be typical scales of velocity, length, and time, respectively. We define the *Rossby number* ε, the *time Rossby number* ε_T, and the *Ekman number* E by

$$\varepsilon \equiv \frac{U}{2\Omega L}, \tag{2.2.8}$$

$$\varepsilon_T \equiv \frac{1}{2\Omega \tau}, \tag{2.2.9}$$

$$E \equiv \frac{\nu}{2\Omega L^2}, \tag{2.2.10}$$

where ν is the kinematic viscosity and Ω is the angular velocity of the rotation. Ratios between magnitudes of various terms in (2.2.4) can be expressed by using the above non-dimensional numbers. Using these numbers, the underlined terms in (2.2.4) are larger than the other terms if the following conditions are satisfied:

1. The relative vorticity $\boldsymbol{\omega}_r$ is smaller than $2\boldsymbol{\Omega}$

$$\frac{|\boldsymbol{\omega}_r|}{|2\boldsymbol{\Omega}|} \approx \frac{U}{2\Omega L} = \varepsilon \ll 1.$$

2. The time derivative on the left-hand side is smaller than the tilting term on the right-hand side

$$\frac{\left| \left(\frac{\partial}{\partial t} + \boldsymbol{v} \cdot \nabla \right) \boldsymbol{\omega}_r \right|}{|2\boldsymbol{\Omega} \cdot \nabla \boldsymbol{v}|} \approx \frac{\max\left(U/\tau L, U^2/L^2 \right)}{2\Omega U/L} = \max\left(\frac{1}{2\Omega\tau}, \frac{U}{2\Omega L} \right) = \max(\varepsilon_T, \varepsilon) \ll 1.$$

3. The friction term is negligible when compared with the tilting term. From (1.2.21)

$$\frac{|\nabla \times \boldsymbol{f}|}{|2\boldsymbol{\Omega} \cdot \nabla \boldsymbol{v}|} \approx \frac{\nu U/L^3}{2\Omega U/L} = \frac{\nu}{2\Omega L^2} = E \ll 1.$$

Now, for application to the real atmosphere, we express the geostrophic balance (2.2.1) in spherical coordinates (λ, φ, r), where λ is longitude, φ is latitude, and

r is the distance from the center of the earth. The velocity vector is denoted by $\boldsymbol{v} = (u, v, w)$ and the angular velocity vector by

$$\boldsymbol{\Omega} = (0, \Omega\cos\varphi, \Omega\sin\varphi), \tag{2.2.11}$$

and assume that Φ is spherically symmetric: $\Phi = \Phi(r)$. Then, (2.2.1) is written as

$$-2\Omega\sin\varphi\, v + 2\Omega\cos\varphi\, w = -\frac{1}{\rho r\cos\varphi}\frac{\partial p}{\partial\lambda}, \tag{2.2.12}$$

$$2\Omega\sin\varphi\, u = -\frac{1}{\rho r}\frac{\partial p}{\partial\varphi}, \tag{2.2.13}$$

$$-2\Omega\cos\varphi\, u = -\frac{1}{\rho}\frac{\partial p}{\partial r} + g. \tag{2.2.14}$$

From (A1.5.2) in the appendix, the non-divergent condition in the spherical coordinates is expressed as

$$\nabla\cdot\boldsymbol{v} = \frac{1}{r\cos\varphi}\frac{\partial u}{\partial\lambda} + \frac{1}{r\cos\varphi}\frac{\partial}{\partial\varphi}(\cos\varphi\, v) + \frac{1}{r^2}\frac{\partial}{\partial r}(r^2 w) = 0. \tag{2.2.15}$$

Applying the rotation operator to (2.2.12)–(2.2.14) and using the non-divergent condition, we have an expression for thermal wind balance (2.2.3) using spherical coordinates:†

$$-\frac{2\Omega\cos\varphi}{r}\frac{\partial u}{\partial\varphi} - 2\Omega\sin\varphi\frac{\partial u}{\partial r} = -\frac{1}{r}\left(\frac{\partial\rho^{-1}}{\partial\varphi}\frac{\partial p}{\partial r} - \frac{\partial\rho^{-1}}{\partial r}\frac{\partial p}{\partial\varphi}\right), \tag{2.2.16}$$

$$-\frac{2\Omega\cos\varphi}{r}\left(\frac{\partial v}{\partial\varphi} + w\right) - 2\Omega\sin\varphi\frac{\partial v}{\partial r} = -\frac{1}{r\cos\varphi}\left(\frac{\partial\rho^{-1}}{\partial r}\frac{\partial p}{\partial\lambda} - \frac{\partial\rho^{-1}}{\partial\lambda}\frac{\partial p}{\partial r}\right), \tag{2.2.17}$$

$$-\frac{2\Omega\cos\varphi}{r}\left(\frac{\partial w}{\partial\varphi} - v\right) - 2\Omega\sin\varphi\frac{\partial w}{\partial r} = -\frac{1}{r^2\cos\varphi}\left(\frac{\partial\rho^{-1}}{\partial\lambda}\frac{\partial p}{\partial\varphi} - \frac{\partial\rho^{-1}}{\partial\varphi}\frac{\partial p}{\partial\lambda}\right). \tag{2.2.18}$$

Let us introduce a few more assumptions to help us describe the real atmosphere. Since the depth of the atmosphere is very thin, we replace the variable r by z where $r = R + z$ and R is the mean radius of the earth and assume $z \ll R$. The non-divergent condition is approximated to

$$\nabla\cdot\boldsymbol{v} = \frac{1}{R\cos\varphi}\frac{\partial u}{\partial\lambda} + \frac{1}{R\cos\varphi}\frac{\partial}{\partial\varphi}(\cos\varphi\, v) + \frac{\partial w}{\partial z} = 0. \tag{2.2.19}$$

---

†We should compare (2.2.16)–(2.2.18) with the right-hand side of Eqs. (A1.5.11). Since there are metric terms associated with a change in the bases, we should note $\frac{d\boldsymbol{\omega}}{dt} \neq \left(\frac{d\omega_\lambda}{dt}, \frac{d\omega_\varphi}{dt}, \frac{d\omega_r}{dt}\right)$.

Let the horizontal and vertical scales of motion be L and H, respectively, the scale of the horizontal velocity u and v be U, and the scale of the vertical velocity w be W. From (2.2.19), we have

$$\frac{W}{U} \approx \frac{H}{L}. \tag{2.2.20}$$

Hence, if the *aspect ratio* δ is small enough such that

$$\delta \equiv \frac{H}{L} \ll 1, \tag{2.2.21}$$

the term involving w in (2.2.12) can be neglected. In this case, (2.2.12) and (2.2.13) become

$$-fv = -\frac{1}{\rho R\cos\varphi}\frac{\partial p}{\partial\lambda}, \qquad fu = -\frac{1}{\rho R}\frac{\partial p}{\partial\varphi}, \tag{2.2.22}$$

where

$$f \equiv 2\Omega\sin\varphi \tag{2.2.23}$$

is the *Coriolis parameter.* The velocity components u and v in the geostrophic balance are called the *geostrophic winds* or the *geostrophic flows.* Furthermore, if $\Omega U/g \ll 1$ is satisfied, the balance in the vertical component (2.2.14) is reduced to the hydrostatic balance (2.1.4). In this case, the right-hand side of (2.2.16), for instance, can be rewritten as

$$\begin{aligned}\frac{\partial\rho^{-1}}{\partial\varphi}\frac{\partial p}{\partial z} - \frac{\partial\rho^{-1}}{\partial z}\frac{\partial p}{\partial\varphi} &= \left(\frac{\partial\rho^{-1}}{\partial\varphi}\right)_p\frac{\partial p}{\partial z}\\ &= g\left(\frac{\partial\ln\rho}{\partial\varphi}\right)_p = -g\left(\frac{\partial\ln T}{\partial\varphi}\right)_p,\end{aligned} \tag{2.2.24}$$

where we used the equation of state $p = \rho R_d T$ and the general formula:

$$\begin{aligned}\frac{\partial A}{\partial y}\frac{\partial B}{\partial z} - \frac{\partial A}{\partial z}\frac{\partial B}{\partial y} &= \frac{\partial(A,B)}{\partial(y,z)} = \left(\frac{\partial A}{\partial y}\right)_B\frac{\partial B}{\partial z} = -\left(\frac{\partial A}{\partial z}\right)_B\frac{\partial B}{\partial y}\\ &= -\left(\frac{\partial B}{\partial y}\right)_A\frac{\partial A}{\partial z} = \left(\frac{\partial B}{\partial z}\right)_A\frac{\partial A}{\partial y}.\end{aligned} \tag{2.2.25}$$

Thus, we can rewrite the thermal wind balance (2.2.16) and (2.2.17) as

$$-f\frac{\partial u}{\partial z} = \frac{g}{R}\left(\frac{\partial\ln T}{\partial\varphi}\right)_p, \tag{2.2.26}$$

$$-f\frac{\partial v}{\partial z} = -\frac{g}{R\cos\varphi}\left(\frac{\partial\ln T}{\partial\lambda}\right)_p. \tag{2.2.27}$$

These equations imply that if temperature has a gradient on a constant pressure surface (isobaric surface), the horizontal velocity component normal to the temperature gradient has a shear in the vertical direction. This is the original idea behind the name of the thermal wind.

2.3 Stability of hydrostatic balance

In the scaling argument, the atmospheric vertical structure is approximately in hydrostatic balance and the horizontal velocity field is approximately in geostrophic balance with the pressure field. These balanced states become unstable when a certain condition is satisfied. This and the next sections are devoted to the consideration of stability of the basic balances of the atmosphere. The following three methods are mainly used to consider the stability of a balanced state: the *parcel method*, the *linear stability analysis*, and the *calculus of variation*. Here we use the parcel method to derive a criterion of the stability of hydrostatic balance. In the next section, stability of geostrophic balance will be considered in a more general method of the calculus of variation. Linear stability analysis will be used to examine the structure of unstable waves in Chapter 4.

Let us first consider a basic state in hydrostatic balance and its stability. The relation between the density ρ and pressure p of a fluid in hydrostatic balance is expressed by (2.1.4). Criterion of the stability of a fluid in balanced state can be derived by considering forces acting on a fluid parcel. Suppose that a fluid parcel undergoes a small displacement in one direction without disturbing the environment. The basic state is regarded as being unstable if the parcel's displacement is accelerated. On the other hand, the basic state is thought to be stable if the forces acting on the parcel tend to return it to the original position. The resulting forces on the parcel come from the deviation in hydrostatic balance. The pressure and density of the parcel are denoted by $\tilde{p}$ and $\tilde{\rho}$, respectively, and we define deviations from those of the environment as $p' = \tilde{p} - p$ and $\rho' = \tilde{\rho} - \rho$, respectively. Multiplying the equation of motion (2.1.1) by $\tilde{\rho}$ and subtracting the hydrostatic balance (2.1.4), we obtain

$$\tilde{\rho}\frac{d\boldsymbol{v}}{dt} = -2\tilde{\rho}\boldsymbol{\Omega} \times \boldsymbol{v} - \nabla p' - \rho'\nabla\Phi + \tilde{\rho}\boldsymbol{f}. \tag{2.3.1}$$

In general, therefore, the sum of the Coriolis force, the pressure gradient force, the buoyancy force, and the frictional force determines the direction of displacement of the fluid. We neglect the effects of rotation and friction for simplicity for the following stability analysis with the parcel method.

In hydrostatic balance, both pressure p and density ρ depend only on height z if Φ is a function of z. Let a fluid parcel at height z be slowly moved with a small interval ζ in the vertical direction without exchange of heat with the environment. During the movement, the pressure of the parcel is kept the same as that of the environment: $p' = 0$. After the movement, only the buoyancy force $-\rho' g$ is acting on the parcel. If this buoyancy force acts toward the original position, the parcel will return to the original level. In this case, the environment is said to be *stably stratified*. Thus, the stability condition is expressed as

$$\rho'\zeta > 0. \tag{2.3.2}$$

For instance, in the case that a fluid parcel undergoes an upward displacement and the density of the parcel becomes heavier than that of the environment, the buoyancy acting on the parcel is downwards. If we express density as a function of pressure p and entropy s such that $\rho = \rho(p, s)$, we can expand the deviation of the density as

$$\begin{aligned} \rho' &= \tilde{\rho}(z+\zeta) - \rho(z+\zeta) \\ &= \rho(\tilde{p}(z+\zeta), \tilde{s}(z+\zeta)) - \rho(p(z+\zeta), s(z+\zeta)) \\ &= \rho(p(z+\zeta), s(z)) - \rho(p(z+\zeta), s(z+\zeta)) \\ &= -\left(\frac{\partial \rho}{\partial s}\right)_p \frac{\partial s}{\partial z}\zeta, \end{aligned} \tag{2.3.3}$$

where the displacement ζ is assumed to be small and $\tilde{s}(z+\zeta) = s(z)$ since the parcel motion is adiabatic. Then, we can rewrite the condition (2.3.2) as

$$\left(\frac{\partial \rho}{\partial s}\right)_p \frac{\partial s}{\partial z}\zeta^2 < 0. \tag{2.3.4}$$

We note

$$\left(\frac{\partial \rho}{\partial s}\right)_p = \left(\frac{\partial T}{\partial s}\right)_p \left(\frac{\partial \rho}{\partial T}\right)_p = -\frac{T}{C_p}\alpha\rho, \tag{2.3.5}$$

where the definitions of the specific heat at constant pressure (1.1.18) and the expansion coefficient (1.1.29) are used. In general, the expansion coefficient is positive: $\alpha > 0$. From (2.3.4) and (2.3.5), therefore, the stability condition reads

$$\frac{\partial s}{\partial z} > 0 \tag{2.3.6}$$

(i.e., the basic state is stable if the entropy of the environment increases with height z). This condition can be written using the potential temperature (1.1.51) as

$$\frac{\partial \theta}{\partial z} > 0. \tag{2.3.7}$$

We should note that the above condition cannot be directly applicable if the composition of a fluid is inhomogeneous. Suppose that the density of a fluid is expressed as $\rho = \rho(p, s, q)$, where q is the mass concentration of an inhomogeneous component of the fluid. q is defined as the ratio of the mass of this component to the total mass of a fluid parcel. If the concentration in the environment varies with height z, the dependency of density on concentration modifies the stability condition. If a fluid parcel is not mixed with the environment and its composition does not change during its displacement, the stability condition is written as

$$\begin{aligned} \rho'\zeta &= [\rho(p(z+\zeta), s(z), q(z)) - \rho(p(z+\zeta), s(z+\zeta), q(z+\zeta))]\,\zeta \\ &\approx -\left[\left(\frac{\partial \rho}{\partial s}\right)_{p,q} \frac{\partial s}{\partial z} + \left(\frac{\partial \rho}{\partial q}\right)_{p,s} \frac{\partial q}{\partial z}\right]\zeta > 0. \end{aligned} \tag{2.3.8}$$

This means that, even when the entropy of the environment increases with height, the balanced state could be unstable if the concentration q increases with height on the condition $\left(\frac{\partial \rho}{\partial q}\right)_{p,s} > 0$.

Next, we study motions of a fluid parcel based on the assumptions used for the parcel method. If just the buoyancy force acts on the fluid parcel, the vertical component of the equation of motion is expressed as

$$\tilde{\rho}\frac{d^2\zeta}{dt^2} = -\rho' g, \tag{2.3.9}$$

where $w = \frac{d\zeta}{dt}$ is used. Expanding ρ' by ζ as (2.3.3) and using (2.3.5), we obtain

$$\frac{d^2\zeta}{dt^2} = -N^2\zeta, \tag{2.3.10}$$

where

$$N^2 \equiv \frac{g\alpha T}{C_p}\frac{\partial s}{\partial z}. \tag{2.3.11}$$

N is called the *buoyancy frequency*, or the *Brunt-Väisälä frequency*. The stability condition (2.3.6) is expressed as $N^2 > 0$. In this case, the buoyancy force acts on the parcel as a restoring force, so that the parcel oscillates with frequency N. If $N^2 < 0$, on the other hand, the buoyancy force amplifies the displacement of the parcel (i.e., the stratification is unstable). If $N^2 = 0$, the parcel undergoes no further displacement. This state is called *neutral.*

The Brunt-Väisälä frequency can be written in various forms. Using the thermodynamic formula, we have

$$\begin{aligned} N^2 &= \frac{g\alpha T}{C_p}\left[\left(\frac{\partial s}{\partial T}\right)_p \frac{\partial T}{\partial z} + \left(\frac{\partial s}{\partial p}\right)_T \frac{\partial p}{\partial z}\right] \\ &= g\alpha\left(\frac{\partial T}{\partial z} + \frac{g\alpha T}{C_p}\right). \end{aligned} \tag{2.3.12}$$

In the case of the ideal gas, since $\alpha = 1/T$, we have

$$N^2 = \frac{g}{C_p}\frac{\partial s}{\partial z} = \frac{g}{\theta}\frac{\partial \theta}{\partial z} = \frac{g}{T}\left(\frac{\partial T}{\partial z} + \frac{g}{C_p}\right). \tag{2.3.13}$$

Using (1.1.58), it is also written as

$$N^2 = -\frac{g}{\rho}\left(\frac{\partial \rho}{\partial z} - \frac{1}{c_s^2}\frac{\partial p}{\partial z}\right) = \frac{g}{H_\rho} - \frac{g^2}{c_s^2}, \tag{2.3.14}$$

where

$$H_\rho \equiv \left(-\frac{1}{\rho}\frac{\partial \rho}{\partial z}\right)^{-1} \tag{2.3.15}$$

is the *scale height* of density and c_s is the speed of sound defined by (1.1.32).

2.4 Stability of axisymmetric flows†

In this section, we consider the stability of the thermal wind field. In particular, we assume that the flow is zonally axisymmetric about a rotation axis and that the structure of the perturbation is also axisymmetric. The cylindrical coordinates (r, φ, z) are used, where r is the distance from the rotation axis, φ is the azimuthal coordinate, and z is the axial coordinate (Fig. 2.3). The flow is assumed to be frictionless and adiabatic. Using (A1.4.7) in the appendix, the equations for the axisymmetric flows with respect to the inertial frame are written as

$$\frac{dv_r}{dt} - \frac{v_\varphi^2}{r} = -\frac{1}{\rho}\frac{\partial p}{\partial r} - \frac{\partial \Phi}{\partial r}, \tag{2.4.1}$$

$$\frac{dv_\varphi}{dt} + \frac{v_r v_\varphi}{r} = -\frac{1}{\rho r}\frac{\partial p}{\partial \varphi} - \frac{1}{r}\frac{\partial \Phi}{\partial \varphi}, \tag{2.4.2}$$

$$\frac{dv_z}{dt} = -\frac{1}{\rho}\frac{\partial p}{\partial z} - \frac{\partial \Phi}{\partial z}, \tag{2.4.3}$$

$$\frac{d\rho}{dt} + \rho\left[\frac{1}{r}\frac{\partial}{\partial r}(rv_r) + \frac{1}{r}\frac{\partial}{\partial \varphi}v_\varphi + \frac{\partial}{\partial z}v_z\right] = 0, \tag{2.4.4}$$

$$\frac{ds}{dt} = 0, \tag{2.4.5}$$

where v_r, v_φ, and v_z are the velocity components in the r-, φ-, and z-directions, respectively, and s is entropy per unit mass. Φ is gravitational potential in the inertial frame and does not contain centrifugal potential. By omitting the φ-dependence, the time derivative is written as

$$\frac{d}{dt} = \frac{\partial}{\partial t} + v_r\frac{\partial}{\partial r} + v_z\frac{\partial}{\partial z}. \tag{2.4.6}$$

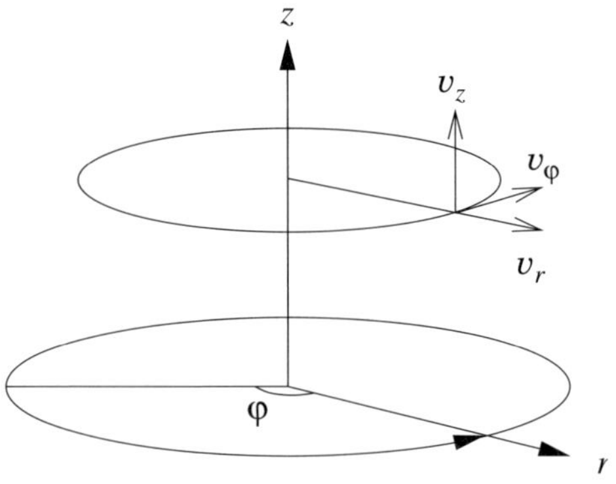

FIGURE 2.3: The cylindrical coordinates and an axisymmetric flow.

†The calculus of variation in this section follows Charney (1973) .

It is convenient to define the angular momentum

$$l \equiv v_\varphi r, \tag{2.4.7}$$

and a two-dimensional vector in the meridional section $(r,\ z)$ as

$$\boldsymbol{v}_m = (v_r, v_z). \tag{2.4.8}$$

Let us first define the total energy of the system. Using (2.4.1)–(2.4.5) and the thermodynamic relation (1.2.52), we write conservation of total energy (1.2.48) as

$$\rho\frac{d}{dt}\left(\frac{1}{2}\boldsymbol{v}^2 + u + \Phi\right) = -\nabla\cdot p\boldsymbol{v}, \tag{2.4.9}$$

where u is internal energy. Kinetic energy is expressed using the angular momentum (2.4.7) as

$$\boldsymbol{v}^2 \equiv v_r^2 + v_\varphi^2 + v_z^2 = \frac{1}{2}\boldsymbol{v}_m^2 + \frac{l^2}{2r^2} = \frac{1}{2}\boldsymbol{v}_m^2 + l^2\chi, \tag{2.4.10}$$

where

$$\chi \equiv \frac{1}{2r^2}. \tag{2.4.11}$$

We then integrate the conservation of total energy (2.4.9) in the whole domain. Using Gauss's law and assuming that the normal components of the velocity at the boundaries of the domain are zero, we have

$$\frac{d}{dt}(K_m + K_\varphi + U + G) = 0, \tag{2.4.12}$$

where

$$K_m \equiv \int \frac{1}{2}\rho(v_r^2 + v_z^2)\, dV = \int \frac{1}{2}\rho\boldsymbol{v}_m^2\, dV, \tag{2.4.13}$$

$$K_\varphi \equiv \int \frac{1}{2}\rho v_\varphi^2\, dV = \int \rho\frac{l^2}{2r^2}\, dV = \int \rho l^2\chi\, dV, \tag{2.4.14}$$

$$U \equiv \int \rho u\, dV, \tag{2.4.15}$$

$$G \equiv \int \rho\Phi\, dV. \tag{2.4.16}$$

Thus, we have

$$E = K_m + K_\varphi + U + G = K_m + J = \text{const.}, \tag{2.4.17}$$

where

$$J \equiv K_\varphi + U + G = \int \rho\left(l^2\chi + u + \Phi\right)\, dV \tag{2.4.18}$$

is further introduced.

In the following discussion, we consider perturbations axisymmetric about the z-coordinate. (2.4.1)–(2.4.4) are rewritten as

$$\frac{dl}{dt} = 0, \tag{2.4.19}$$

$$\frac{d\boldsymbol{v}_m}{dt} + l^2\nabla\chi = -\frac{1}{\rho}\nabla p - \nabla\Phi, \tag{2.4.20}$$

$$\frac{d\rho}{dt} + \rho\left[\frac{1}{r}\frac{\partial}{\partial r}(rv_r) + \frac{\partial}{\partial z}v_z\right] = 0, \tag{2.4.21}$$

where

$$\frac{d\boldsymbol{v}_m}{dt} = \left(\frac{dv_r}{dt}, \frac{dv_z}{dt}\right), \tag{2.4.22}$$

$$\nabla = \left(\frac{\partial}{\partial r}, \frac{\partial}{\partial z}\right). \tag{2.4.23}$$

Since (2.4.22) does not include metric terms, it is different from the time derivative of a three-dimensional vector given by (A1.4.5).

Let us assume that the basic state has a steady zonal flow with no meridional flow: $\boldsymbol{v}_m = 0$. From (2.4.20), the basic state satisfies

$$l^2\nabla\chi = -\frac{1}{\rho}\nabla p - \nabla\Phi, \tag{2.4.24}$$

which corresponds to geostrophic balance. In this case, however, the Coriolis force does not appear since it is viewed from the inertial frame. This states the balance between centrifugal force, the pressure gradient force, and gravity, and is called the *cyclostrophic balance*. Applying the rotation operator to this, we have the thermal wind balance as

$$\nabla l^2 \times \nabla\chi = \nabla\frac{1}{\rho} \times (-\nabla p). \tag{2.4.25}$$

(2.4.24) and (2.4.25) are explicitly written as

$$-\frac{l^2}{r^3} = -\frac{1}{\rho}\frac{\partial p}{\partial r} - \frac{\partial\Phi}{\partial r}, \tag{2.4.26}$$

$$0 = -\frac{1}{\rho}\frac{\partial p}{\partial z} - \frac{\partial\Phi}{\partial z}, \tag{2.4.27}$$

$$\frac{1}{r^3}\frac{\partial l^2}{\partial z} = -\frac{\partial\rho^{-1}}{\partial r}\frac{\partial p}{\partial z} + \frac{\partial\rho^{-1}}{\partial z}\frac{\partial p}{\partial r}. \tag{2.4.28}$$

The quantities of this basic field will be referred to using a subscript 0.

We add an axisymmetric perturbation to the basic state to study the stability of the basic field. The conservation of energy is used to examine whether the perturbation grows or not. We assume that the perturbation is expressed by a small displacement represented by a vector $\boldsymbol{\delta r} = (\delta r, \delta z)$.† Corresponding to the

†Since the perturbation is axisymmetric, it is more like a fluid ring than a fluid parcel.

initial displacement, the density ρ, the pressure p, and the angular momentum l have perturbations, as have all kinds of energies K_φ, K_m, U, and G. Here, the first-order change of an individual quantity is denoted by δ, and the second-order change by δ^2. For example, the energy (2.4.18) is expanded as

$$\begin{aligned} J - J_0 &= \frac{dJ}{dt}dt + \frac{1}{2}\frac{d^2J}{dt^2}dt^2 + O(dt^3) \\ &= \delta J + \frac{1}{2}\delta^2 J + O(dt^3). \end{aligned} \tag{2.4.29}$$

From the conservation of energy (2.4.17), we have

$$E - J = K_m \geq 0. \tag{2.4.30}$$

The total energy E is constant regardless of the evolution of the perturbation. Since $K_m = 0$ at the equilibrium state, J takes the maximum value at the equilibrium state. Hence, the first variation must satisfy

$$\delta J = 0. \tag{2.4.31}$$

If the inequality

$$\delta^2 J < 0 \tag{2.4.32}$$

is satisfied, the perturbation is amplified. This is the unstable condition of the basic state.

Let the difference between values at two different points be denoted by d, and a material change of the perturbation by δ. Using the displacement vector $\delta\boldsymbol{r}$, we have

$$\delta\chi = d\chi = \nabla\chi \cdot \delta\boldsymbol{r}, \tag{2.4.33}$$
$$\delta\Phi = d\Phi = \nabla\Phi \cdot \delta\boldsymbol{r}. \tag{2.4.34}$$

On the other hand, from (2.4.19) and (2.4.5), we have

$$\delta l = 0, \tag{2.4.35}$$
$$\delta s = 0. \tag{2.4.36}$$

From the continuity equation (2.4.21), we have

$$\delta\rho = -\rho\nabla\cdot\delta\boldsymbol{r} = -\rho D, \tag{2.4.37}$$

where

$$D \equiv \nabla\cdot\delta\boldsymbol{r}. \tag{2.4.38}$$

Thus, we obtain

$$\delta\frac{1}{\rho} = -\frac{\delta\rho}{\rho^2} = \frac{D}{\rho}, \tag{2.4.39}$$
$$\delta u = \frac{p}{\rho^2}\delta\rho + T\delta s = -\frac{p}{\rho}D, \tag{2.4.40}$$
$$\delta p = \left(\frac{\partial p}{\partial\rho}\right)_s \delta\rho + \left(\frac{\partial p}{\partial s}\right)_\rho \delta s = c_s^2\delta\rho = -c_s^2\rho D. \tag{2.4.41}$$

Using these relations and $\delta(\rho dV) = 0$, the first variation in J, (2.4.18), is expressed as

$$\begin{aligned}
\delta J &= \int \left[\delta(l^2\chi) + \delta u + \delta\Phi\right] \rho dV \\
&= \int \left(l^2\nabla\chi \cdot \delta\boldsymbol{r} - \frac{p}{\rho}\nabla \cdot \delta\boldsymbol{r} + \nabla\Phi \cdot \delta\boldsymbol{r}\right) \rho dV \\
&= \int \left(l^2\nabla\chi + \frac{1}{\rho}\nabla p + \nabla\Phi\right) \cdot \delta\boldsymbol{r}\rho dV, \qquad (2.4.42)
\end{aligned}$$

where $\delta\boldsymbol{r} = 0$ is assumed at the boundaries when integrating by parts. From the constraint $\delta J = 0$, (2.4.31), at the equilibrium state, the relation (2.4.24) is recovered (i.e., the geostrophic balance).

The second variation in J can be calculated as

$$\begin{aligned}
\delta^2 J &= \delta(\delta J) \\
&= \int \left[l^2\delta(\nabla\chi \cdot \delta\boldsymbol{r}) + \delta\left(\frac{1}{\rho}\nabla p \cdot \delta\boldsymbol{r}\right) + \delta(\nabla\Phi \cdot \delta\boldsymbol{r})\right] \rho dV \\
&= \int \left(l^2\nabla\chi + \frac{1}{\rho}\nabla p + \nabla\Phi\right) \cdot \delta^2\boldsymbol{r}\rho dV \\
&\quad + \int \left[l^2\nabla(\delta\chi) + \delta\left(\frac{1}{\rho}\nabla p\right) + \nabla(\delta\Phi)\right] \cdot \delta\boldsymbol{r}\rho dV \\
&= \int \left[l^2\nabla(\nabla\chi \cdot \delta\boldsymbol{r}) + \frac{D}{\rho}\nabla p \right. \\
&\quad \left. + \frac{1}{\rho}\nabla(-c_s^2\rho D) + \nabla(\nabla\Phi \cdot \delta\boldsymbol{r})\right] \cdot \delta\boldsymbol{r}\rho dV \\
&= \int \nabla \cdot \left\{\left[\rho l^2(\nabla\chi \cdot \delta\boldsymbol{r}) + \rho(\nabla\Phi \cdot \delta\boldsymbol{r}) - \rho c_s^2 D\right] \delta\boldsymbol{r}\right\} dV \\
&\quad - \int \left(l^2\nabla\chi + \nabla\Phi + \frac{1}{\rho}\nabla p\right) \cdot \delta\boldsymbol{r}DdV \\
&\quad - \int \left\{\left[\nabla(\rho l^2) \cdot \delta\boldsymbol{r}\right] (\nabla\chi \cdot \delta\boldsymbol{r}) + (\nabla\rho \cdot \delta\boldsymbol{r})(\nabla\Phi \cdot \delta\boldsymbol{r}) \right. \\
&\quad \left. - \rho c_s^2 D^2 + 2D(\nabla p \cdot \delta\boldsymbol{r})\right\} dV \\
&= \int \left\{- \left[\nabla(\rho l^2) \cdot \delta\boldsymbol{r}\right] (\nabla\chi \cdot \delta\boldsymbol{r}) - (\nabla\rho \cdot \delta\boldsymbol{r})(\nabla\Phi \cdot \delta\boldsymbol{r}) - \frac{1}{c_s^2\rho}(\nabla p \cdot \delta\boldsymbol{r})^2 \right. \\
&\quad \left. + \frac{1}{c_s^2\rho}(\delta p - dp)^2\right\} dV, \qquad (2.4.43)
\end{aligned}$$

where $dp = \nabla p \cdot \delta\boldsymbol{r}$ and the relation

$$\frac{1}{c_s^2\rho}(\delta p - dp)^2 = \rho c_s^2 D^2 + \frac{1}{c_s^2\rho}(\nabla p \cdot \delta\boldsymbol{r})^2 - 2D\nabla p \cdot \delta\boldsymbol{r}, \qquad (2.4.44)$$

is used. This term vanishes because of the assumption of the parcel method where the pressure of the fluid parcel is always the same as that of the environment. Thus, using (2.4.24), the terms within the last integral of (2.4.43) are rewritten as

$$
\begin{aligned}
& -\left[\nabla(\rho l^2)\cdot\delta\boldsymbol{r}\right](\nabla\chi\cdot\delta\boldsymbol{r}) - (\nabla\rho\cdot\delta\boldsymbol{r})(\nabla\Phi\cdot\delta\boldsymbol{r}) - \frac{1}{c_s^2\rho}(\nabla p\cdot\delta\boldsymbol{r})^2 \\
&= -\rho(\nabla l^2\cdot\delta\boldsymbol{r})(\nabla\chi\cdot\delta\boldsymbol{r}) - (\nabla\rho\cdot\delta\boldsymbol{r})\left[(l^2\nabla\chi+\nabla\Phi)\cdot\delta\boldsymbol{r})\right] \\
&\quad -\frac{1}{c_s^2\rho}(\nabla p\cdot\delta\boldsymbol{r})^2 \\
&= -\rho(\nabla l^2\cdot\delta\boldsymbol{r})(\nabla\chi\cdot\delta\boldsymbol{r}) + \frac{1}{\rho}(\nabla p\cdot\delta\boldsymbol{r})\left[(\nabla\rho - \frac{1}{c_s^2}\nabla p)\cdot\delta\boldsymbol{r})\right] \\
&= -\rho(\nabla l^2\cdot\delta\boldsymbol{r})(\nabla\chi\cdot\delta\boldsymbol{r}) - \frac{\alpha T}{C_p}(\nabla s\cdot\delta\boldsymbol{r})(\nabla p\cdot\delta\boldsymbol{r}),
\end{aligned}
\tag{2.4.45}
$$

where we have used

$$
\nabla\rho - \frac{1}{c_2^2}\nabla p = -\frac{\rho\alpha T}{C_p}\nabla s, \tag{2.4.46}
$$

which is given from (1.1.31). In the case of the ideal gas, since

$$
\frac{\alpha T}{C_p}\nabla s = \nabla\ln\theta, \tag{2.4.47}
$$

the second variation in J is expressed as

$$
\delta^2 J = \int\left[(\nabla l^2\cdot\delta\boldsymbol{r})(-\nabla\chi\cdot\delta\boldsymbol{r}) + (\nabla\ln\theta\cdot\delta\boldsymbol{r})\left(-\frac{1}{\rho}\nabla p\cdot\delta\boldsymbol{r}\right)\right]\rho dV. \tag{2.4.48}
$$

If $\delta^2 J < 0$, (2.4.32), is satisfied, the perturbation is amplified. Thus, the necessary and sufficient condition for the instability of the basic state is that the contents of the square brackets in (2.4.48) be negative for at least one direction of $\delta\boldsymbol{r}$. On the other hand, the basic state is stable if the contents of the square brackets in (2.4.48) are positive for any $\delta\boldsymbol{r}$ at any position.

The terms within the square brackets of (2.4.48) are written as

$$
\delta\boldsymbol{r}M\delta\boldsymbol{r} = (\nabla l^2\cdot\delta\boldsymbol{r})(-\nabla\chi\cdot\delta\boldsymbol{r}) + (\nabla\ln\theta\cdot\delta\boldsymbol{r})\left(-\frac{1}{\rho}\nabla p\cdot\delta\boldsymbol{r}\right) \tag{2.4.49}
$$

$$
= M_{11}\delta r^2 + M_{22}\delta z^2 + (M_{12}+M_{21})\delta r\delta z, \tag{2.4.50}
$$

where M is a two-by-two matrix, defined by

$$
M = \begin{pmatrix} M_{11} & M_{12} \\ M_{21} & M_{22}\end{pmatrix} = \begin{pmatrix} \frac{1}{r^3}\frac{\partial l^2}{\partial r} - \frac{1}{\rho}\frac{\partial\ln\theta}{\partial r}\frac{\partial p}{\partial r} & -\frac{1}{\rho}\frac{\partial\ln\theta}{\partial r}\frac{\partial p}{\partial z} \\ \frac{1}{r^3}\frac{\partial l^2}{\partial z} - \frac{1}{\rho}\frac{\partial\ln\theta}{\partial z}\frac{\partial p}{\partial r} & -\frac{1}{\rho}\frac{\partial\ln\theta}{\partial z}\frac{\partial p}{\partial z}\end{pmatrix}. \tag{2.4.51}
$$

The basic state is stable if

$$
\delta\boldsymbol{r}M\delta\boldsymbol{r} > 0 \tag{2.4.52}
$$

is satisfied for any $\delta \boldsymbol{r}$. This condition is satisfied if the following two inequalities hold:

$$\det M = M_{11}M_{22} - M_{12}M_{21} > 0, \tag{2.4.53}$$
$$\mathrm{Trace} M = M_{11} + M_{22} > 0. \tag{2.4.54}$$

Using (2.4.51), we have

$$\begin{aligned} \det M &= (\nabla l^2 \times \nabla \ln \theta) \cdot \left[(-\nabla \chi) \times \left(\frac{1}{\rho}\nabla p\right)\right] \\ &= -\frac{1}{r^3}\frac{1}{\rho}\frac{\partial p}{\partial z}\left(\frac{\partial l^2}{\partial r}\frac{\partial \ln \theta}{\partial z} - \frac{\partial l^2}{\partial z}\frac{\partial \ln \theta}{\partial r}\right) \\ &= \frac{2l}{r^3\theta}\frac{\partial \Phi}{\partial z}\left(\frac{\partial l}{\partial r}\frac{\partial \theta}{\partial z} - \frac{\partial l}{\partial z}\frac{\partial \theta}{\partial r}\right), \end{aligned} \tag{2.4.55}$$

$$\begin{aligned} \mathrm{Trace} M &= \nabla l^2 \cdot (-\nabla \chi) + \nabla \ln \theta \cdot \left(-\frac{1}{\rho}\nabla p\right) \\ &= \frac{2l}{r^3}\frac{\partial l}{\partial r} - \frac{1}{\rho\theta}\frac{\partial \theta}{\partial r}\frac{\partial p}{\partial r} - \frac{1}{\rho\theta}\frac{\partial \theta}{\partial z}\frac{\partial p}{\partial z}, \end{aligned} \tag{2.4.56}$$

where (2.4.27) is used. The inequality (2.4.53) is related to the sign of potential vorticity. Using the potential temperature θ, the potential vorticity of the axisymmetric flow can be expressed as

$$P \equiv \frac{1}{\rho}\boldsymbol{\omega} \cdot \nabla \theta = \frac{1}{\rho}\left(\omega_r \frac{\partial \theta}{\partial r} + \omega_z \frac{\partial \theta}{\partial z}\right). \tag{2.4.57}$$

Since the two components of the vorticity are given by

$$\omega_r = \frac{1}{r}\frac{\partial v_z}{\partial \varphi} - \frac{\partial v_\varphi}{\partial z} = -\frac{1}{r}\frac{\partial l}{\partial z}, \tag{2.4.58}$$
$$\omega_z = \frac{1}{r}\frac{\partial (v_\varphi r)}{\partial \varphi} - \frac{1}{r}\frac{\partial v_z}{\partial \varphi} = \frac{1}{r}\frac{\partial l}{\partial r}, \tag{2.4.59}$$

potential vorticity is expressed as

$$P = \frac{1}{\rho r}\left(\frac{\partial l}{\partial r}\frac{\partial \theta}{\partial z} - \frac{\partial l}{\partial z}\frac{\partial \theta}{\partial r}\right). \tag{2.4.60}$$

Therefore, the inequality (2.4.53) is equivalent to

$$Pl\frac{\partial \Phi}{\partial z} > 0. \tag{2.4.61}$$

If the atmospheric circulation is completely zonally symmetric about the rotation axis and has no meridional motion, the stability condition is simply $P > 0$ in the northern hemisphere and $P < 0$ in the southern hemisphere. This is because, if we take the z-axis in the direction of the rotation vector, we have $l\frac{\partial \Phi}{\partial z} > 0$ in the northern hemisphere and $l\frac{\partial\Phi}{\partial z} > 0$ in the southern hemisphere. The potential vorticity (2.4.60) is rewritten by using (2.2.25) as

$$P = \frac{1}{\rho r}\left(\frac{\partial l}{\partial r}\right)_\theta \frac{\partial\theta}{\partial z}. \tag{2.4.62}$$

Hence, the condition (2.4.61) becomes

$$Pl\frac{\partial\Phi}{\partial z} = \frac{l}{\rho r}\left(\frac{\partial l}{\partial r}\right)_\theta \frac{\partial\theta}{\partial z}\frac{\partial\Phi}{\partial z} > 0. \tag{2.4.63}$$

If both θ and Φ increase with z (as in the northern hemisphere), (2.4.63) is reduced to

$$l\left(\frac{\partial l}{\partial r}\right)_\theta = \frac{1}{2}\left(\frac{\partial l^2}{\partial r}\right)_\theta > 0. \tag{2.4.64}$$

That is, the basic state is stable if the angular momentum decreases as the distance from the rotation axis becomes larger on an isentropic surface.

Since the special case $P = 0$ is not included in the above condition, we need to consider it separately. We particularly examine the case where the angular momentum of the basic state is uniform $l = \text{const}$. A static state with no motion is an example of this case. The geostrophic balance (2.4.24) simply becomes the hydrostatic balance

$$0 = -\frac{1}{\rho}\nabla p - \nabla\Phi_r, \tag{2.4.65}$$

where

$$\Phi_r = \Phi - l^2\chi = \Phi - \frac{l^2}{2r^2} = \Phi - \frac{v_\varphi^2}{2} \tag{2.4.66}$$

is the geopotential. In this case, p, ρ, and θ are constant on the surfaces of constant Φ_r. If we define ζ as a length measured along the direction of $\nabla\Phi_r$ and define $g = \frac{\partial\Phi}{\partial\zeta} > 0$ such that

$$\nabla\Phi_r = \frac{\partial\Phi_r}{\partial\zeta}\nabla\zeta = g\nabla\zeta, \tag{2.4.67}$$

we can write (2.4.49) as

$$\delta\boldsymbol{r}M\delta\boldsymbol{r} = (\nabla\ln\theta\cdot\delta\boldsymbol{r})(\nabla\Phi_r\cdot\delta\boldsymbol{r}) = \frac{g}{\theta}\frac{\partial\theta}{\partial\zeta}\delta\zeta^2. \tag{2.4.68}$$

Therefore, the stability condition of the basic state (2.4.52) becomes

$$\frac{\partial\theta}{\partial\zeta} > 0. \tag{2.4.69}$$

This condition is the same as (2.3.6) in Section 2.3.

In the case $P = 0$, on the other hand, if the potential temperature of the basic state is uniform $\theta = \text{const.}$, (2.4.49) becomes

$$\delta \boldsymbol{r} M \delta \boldsymbol{r} = (\nabla l^2 \cdot \delta \boldsymbol{r})(-\nabla \chi \cdot \delta \boldsymbol{r}) = \frac{2l}{r^3}\frac{\partial l}{\partial r}\delta r^2. \tag{2.4.70}$$

Thus, the basic state is stable if

$$l\frac{\partial l}{\partial r} = \frac{1}{2}\frac{\partial l^2}{\partial r} > 0. \tag{2.4.71}$$

This stability condition is for the angular momentum on the surface perpendicular to the rotating axis. This implies that the flow is stable if the angular momentum increases with r. Contrary to this, perturbations to the zonal flow amplify if the angular momentum decreases with r; this is called the *inertial instability.* Let u denote a relative azimuthal velocity in the rotating frame with the angular velocity Ω. The angular momentum is written as $l = ur + \Omega r^2$. Then inequality (2.4.71) becomes

$$(u + \Omega r)\left[\frac{1}{r}\frac{\partial}{\partial r}(ur) + 2\Omega\right] > 0. \tag{2.4.72}$$

Using (2.4.59), this is rewritten as a condition for the vorticity

$$l\omega_z > 0 \tag{2.4.73}$$

(i.e., the flow is stable if the vorticity has the same sign as the angular momentum).

2.5 Balances and stability of parallel flows on the f-plane

A rotating system whose vertical component of the rotation vector is regarded as constant is called an *f-plane.* The precise definition of the f-plane depends on how the orientation of the vertical direction of the system varies. In the case of the atmosphere on earth, the f-plane may be defined as a plane tangential to a geopotential surface. In such a case, the f-plane approximation is applicable only in a local region where the curvature of the earth is negligible. (See the β-plane in Section 3.2.2.) In this subsection, we examine the stability of parallel flows on the f-plane using the stability condition of axisymmetric flows.

In the last part of the previous section, stability was considered on a plane perpendicular to the rotation axis. The f-plane is different from such a plane, since the f-plane is introduced as a plane tangential to a geopotential surface. If the change in geopotential is negligible over the range considered in the system, however, the stability condition of the axisymmetric flows can be used to study that of a parallel flow on the f-plane.

Let us consider a region where the distance from the rotating axis is large enough. We consider a system in a rotating frame whose angular velocity is Ω and set the Coriolis parameter $f = 2\Omega$. We define R as a typical distance from the rotating axis, and define a local coordinate y by $y = R - r$, where r is the distance from the rotating axis. The absolute azimuthal velocity is written as

$$v_\varphi = u + \Omega r, \tag{2.5.1}$$

where u is the relative velocity. The geopotential is written as

$$\Phi_r = \Phi - \frac{\Omega^2 r^2}{2}, \tag{2.5.2}$$

where the second term on the right-hand side is the centrifugal potential energy. In this rotating frame, (2.4.26) becomes

$$-fu - \frac{u^2}{r} = -\frac{1}{\rho}\frac{\partial p}{\partial r} - \frac{\partial \Phi_r}{\partial r}. \tag{2.5.3}$$

We need to assume that the radial component of the gradient of Φ_r and the metric term u^2/r are negligible to obtain the geostrophic balance on the f plane. To satisfy these requirements, we need $4g/f^2 \gg R \gg u/f$ where $g = \frac{\partial \Phi}{\partial z} > 0$. One might think that this condition is rarely satisfied; the first inequality is not important, however, if one considers a plane tangential to a surface of constant Φ_r instead of a plane normal to the rotation axis. In any case, it can be possible to relate a parallel flow on the f-plane to an axisymmetric flow.

The geostrophic balance and the hydrostatic balance on the f-plane are written as

$$fu = -\frac{1}{\rho}\frac{\partial p}{\partial y}, \qquad 0 = -\frac{1}{\rho}\frac{\partial p}{\partial z} - g. \tag{2.5.4}$$

Using (2.2.24), the thermal wind balance is written as

$$\begin{aligned} f\frac{\partial u}{\partial z} &= \frac{\partial \rho^{-1}}{\partial y}\frac{\partial p}{\partial z} - \frac{\partial \rho^{-1}}{\partial z}\frac{\partial p}{\partial y} = g\left(\frac{\partial \ln \rho}{\partial y}\right)_p = -g\left(\frac{\partial \ln T}{\partial y}\right)_p \\ &= -\frac{g}{\theta}\left(\frac{\partial \theta}{\partial y}\right)_p. \end{aligned} \tag{2.5.5}$$

In a rotating frame with angular velocity Ω, the angular momentum is originally given by

$$l = r(u + \Omega r). \tag{2.5.6}$$

Under the condition $y \ll R$, angular momentum can be approximated to

$$\begin{aligned} l &= (R-y)[\Omega(R-y) + u] \\ &= R\left\{\left[1 + \left(\frac{y}{R}\right)^2\right]\Omega + \left(1 - \frac{y}{R}\right)u - 2y\Omega\right\} \\ &\approx R(R\Omega + u - fy). \end{aligned} \tag{2.5.7}$$

Thus, a quantity

$$L = u - fy \tag{2.5.8}$$

can be used as an angular momentum in the f-plane. This is validated if $L \ll R\Omega$ is satisfied. The dimension of L is different from that of angular momentum by length R. The corresponding potential vorticity is

$$P = \frac{1}{\rho}\left(\frac{\partial L}{\partial z}\frac{\partial \theta}{\partial y} - \frac{\partial L}{\partial y}\frac{\partial \theta}{\partial z}\right) = \frac{1}{\rho}\left[\frac{\partial L}{\partial z}\left(\frac{\partial \theta}{\partial y}\right)_p - \left(\frac{\partial L}{\partial y}\right)_p\frac{\partial \theta}{\partial z}\right], \tag{2.5.9}$$

which is similar to (2.4.60). Using (2.3.13), (2.5.5), and (2.5.8), we have the following expression

$$
\begin{aligned}
P &= \frac{\theta}{\rho g}\left\{\left[f-\left(\frac{\partial u}{\partial y}\right)_p\right]N^2 - f\left(\frac{\partial u}{\partial z}\right)^2\right\} \\
&= \frac{\theta}{\rho g}N^2\left[f(1-Ri^{-1})-\left(\frac{\partial u}{\partial y}\right)_p\right],
\end{aligned}
\tag{2.5.10}
$$

where Ri is the *Richardson number* defined by

$$Ri = \frac{N^2}{\left(\frac{\partial u}{\partial z}\right)^2}. \tag{2.5.11}$$

The stability condition (2.4.63) is

$$lP > 0. \tag{2.5.12}$$

Since l has the same sign as Ω or f, we have from (2.5.10)

$$N^2\left[f^2(1-Ri^{-1})-f\left(\frac{\partial u}{\partial y}\right)_p\right] > 0. \tag{2.5.13}$$

In the case that $N^2 > 0$ and $\left(\frac{\partial u}{\partial y}\right)_p$ is negligible, the stability condition is reduced to the condition of the Richardson number:

$$Ri > 1. \tag{2.5.14}$$

References and suggested reading

The basic balances of the atmosphere are described in detail in standard textbooks on dynamic meteorology; we refer to Chapter 4 of Lindzen (1990), for instance. The stability analysis of the axisymmetric flow in Section 2.4 almost follows Charney (1973). Stability conditions of various types of flows are summarized in Chandrasekhar (1961) and Drazin and Reid (1981) and will be further described in Chapter 5.

Chandrasekhar, S., 1961: *Hydrodynamic and Hydromagnetic Stability.* Oxford University Press, New York, 654 pp.

Charney, J. G., 1973: Planetary fluid dynamics. In: P. Morel (ed.), *Dynamical Meteorology.* Reidel, Dordrecht, Netherlands, pp. 97–351.

Drazin, P. G. and Reid, W. H., 1981: *Hydrodynamic Stability.* Cambridge University Press, Cambridge, UK, 527 pp.

Lindzen, R. S., 1990: *Dynamics in Atmospheric Physics.* Cambridge University Press, Cambridge, UK, 310 pp.

3

Approximations of equations

In this chapter, we describe various approximations of the governing equations of the atmosphere. Although the governing equations are shown in a general form in Chapter 1, they are not appropriate for describing particular motions on spatial or time scales; suitable approximations to the spatial or time scales of motions need to be considered. Approximate forms of the governing equations are constructed based on the fundamental balances of the atmosphere considered in Chapter 2 (i.e., hydrostatic balance and geostrophic balance).

First, in Sections 3.1 and 3.2, according to the time scales of motions, two types of approximations are explained: the Boussinesq approximation and the quasi-geostrophic approximation. Next, primitive equations are introduced as the governing equations for the global motions on the earth in Section 3.3. The equation set considered here uses the approximation based on the difference between spatial scales of horizontal and vertical motions. Primitive equations in various vertical coordinates are also summarized. At the end of this chapter, shallow water equations are described in Section 3.4.

The approximations to the equations are related to the waves that will be categorized in Chapter 4. The Boussinesq approximation is applicable to the motions in which time scales of gravity waves are relevant. Boussinesq equations describe relatively slower motions related to gravity waves and do not contain sound waves. The quasi-geostrophic approximation is appropriate to much slower motions in which only the time variations of geostrophic winds are relevant. We consider under what conditions these approximations are applicable. Primitive equations comprise the equation set employed in numerical models without simplification or approximation. In general, however, based on the assumption of hydrostatic balance, by primitive equations we particularly mean *hydrostatic* primitive equations. By transforming the vertical coordinate using the hydrostatic balance, we have various expressions for primitive equations.

The approximate equations introduced in this chapter will be fully used in the theoretical consideration of atmospheric dynamics in the following chapters.

Waves (Chapter 4) and instabilities (Chapter 5) in the atmosphere are based on the approximate equations suitable for respective motions. Atmospheric general circulation models in Part III will be formulated based on primitive equations. The shallow water equations model is a first step toward the construction of atmospheric general circulation models.

3.1 Boussinesq approximation

The *Boussinesq approximation* is used when one is interested in convective motions or gravity waves that originate from buoyancy forces and that are slower than sound waves. The Boussinesq approximation is based on the incompressible approximation, since sound waves are derived from the compressibility of fluids. In the Boussinesq approximation, the difference in density is considered in order to take account of the effect of buoyancy or gravity waves. The properties of gravity waves and sound waves will be discussed in Chapter 4.

It is relatively troublesome to clarify the situations to which the Boussinesq approximation is applicable. Ogura and Phillips (1962) tried to derive the Boussinesq approximation of the ideal gas using scale analysis. In this section, following the concept of Ogura and Phillips, we introduce the various steps of the Boussinesq approximation used in meteorology, though we do not intend to describe details of the derivations.

We consider a deviation field from a basic state at rest. Variables in the basic state are denoted by a subscript s, and those of the deviation field are denoted by a prime:

$$A = A_s + A'.$$

By setting the basic state velocity $\boldsymbol{v}_s = 0$ in the equation of motion, we obtain the hydrostatic balance in the basic state as (2.1.2):

$$0 = -\frac{1}{\rho_s}\nabla p_s - \nabla\Phi. \tag{3.1.1}$$

Using this equation, the sum of the pressure gradient force and gravity is rewritten as

$$\begin{aligned} -\frac{1}{\rho}\nabla p - \nabla\Phi &= -\frac{1}{\rho}(\nabla p_s + \nabla p') - \nabla\Phi \\ &= -\frac{1}{\rho}\nabla p' - \frac{\rho'}{\rho}\nabla\Phi, \end{aligned} \tag{3.1.2}$$

where the second term on the right-hand side represents the buoyancy. Using (2.1.22), the hydrostatic balance and its departure, (3.1.1) and (3.1.2), are rewritten in terms of potential temperature θ and the Exner function π as

$$0 = -\theta_s\nabla\pi_s - \nabla\Phi, \tag{3.1.3}$$

$$-\theta\nabla\pi - \nabla\Phi = -\theta\nabla\pi' + \frac{\theta'}{\theta}\nabla\Phi. \tag{3.1.4}$$

In the Boussinesq approximation, density or potential temperature in the equation of motion is treated as constant except for the buoyancy term; the deviation of density or potential temperature, ρ' or θ', appears only in the buoyancy term and ρ or θ in the other terms are replaced by constants ρ_0 or θ_0. In the continuity equation, however, we have different degrees of approximation depending on the treatment of density. By omitting the time derivative of density in the continuity equation, we obtain the equation of the non-divergence condition:

$$\nabla \cdot (\rho \boldsymbol{v}) = 0. \tag{3.1.5}$$

We have two categories of the Boussinesq approximation for this equation; one is given by replacing density ρ by the basic state density $\rho_s(z)$ and the other given by replacing ρ by a constant ρ_0.

According to Ogura and Phillips (1962), the equation set of the Boussinesq approximation can be derived under the following conditions: (i) the variation of potential temperature is small, (ii) the speed of sound is faster than a typical scale of velocity of fluids and a typical scale of phase speed, and (iii) the vertical scale of motions is smaller than the scale height of the atmosphere. The first category of the Boussinesq approximation, which is often used for the description of atmospheric convection, is given under the conditions (i) and (ii):

$$\frac{d\boldsymbol{v}}{dt} = -\theta_0 \nabla \pi' + \frac{\theta'}{\theta_0} \nabla \Phi + \boldsymbol{f}, \tag{3.1.6}$$

$$\nabla \cdot (\rho_s \boldsymbol{v}) = 0, \tag{3.1.7}$$

$$\frac{d\theta}{dt} = \frac{\theta}{T} Q. \tag{3.1.8}$$

The above equations are called *anelastic equations*, since the sound waves are not included in this system. If condition (iii) is also imposed in addition to (i) and (ii), the equation set of the Boussinesq approximation in the usual context is given as

$$\frac{d\boldsymbol{v}}{dt} = -\frac{1}{\rho_0} \nabla p' - \frac{\rho'}{\rho_0} \nabla \Phi + \boldsymbol{f}, \tag{3.1.9}$$

$$\nabla \cdot \boldsymbol{v} = 0, \tag{3.1.10}$$

$$\frac{dT}{dt} = Q, \tag{3.1.11}$$

which we call *Boussinesq equations*. Since the depth of the atmosphere is shallow, temperature T is used in the thermodynamic equation (3.1.11) instead of potential temperature.

In the case that the speed of sound c_s approaches infinity, the thermodynamic relation (1.1.31) or (1.1.59) becomes

$$d\rho \approx -\frac{\rho_0 \alpha T}{C_p} ds = -\frac{\rho_0 \alpha T}{\theta} d\theta, \tag{3.1.12}$$

where density is assumed to be constant ρ_0 and the expansion coefficient of the ideal gas is $\alpha = T^{-1}$ from (1.1.29). Under condition (iii), since the effect of pressure is negligible in the potential temperature, we have an approximation

$$d\rho \approx -\frac{\rho_0 \alpha T}{\theta} d\theta \approx -\alpha \rho_0 dT. \tag{3.1.13}$$

Under this condition, (3.1.11) becomes equivalent to (3.1.8). In this case, the density variation is approximated as

$$\rho = \rho_0[1-\alpha(T-T_0)], \tag{3.1.14}$$

where T_0 is a reference value of temperature. Using the deviation of temperature $T' = T - T_0$, the Boussinesq equations (3.1.9)–(3.1.11) are rewritten as

$$\frac{d\boldsymbol{v}}{dt} = -\frac{1}{\rho_0}\nabla p' + \alpha T' \nabla \Phi + \boldsymbol{f}, \tag{3.1.15}$$

$$\nabla \cdot \boldsymbol{v} = 0, \tag{3.1.16}$$

$$\frac{dT'}{dt} = Q. \tag{3.1.17}$$

In the case that the coefficient of viscosity is constant, the frictional force (1.2.21) is written as

$$\boldsymbol{f} = \nu \nabla^2 \boldsymbol{v}, \tag{3.1.18}$$

where ν is kinematic viscosity. If thermal diffusion alone contributes to the heat flux, (1.2.45) and (1.2.54) give

$$Q = \frac{1}{\rho C_p}(\varepsilon - \nabla \cdot \boldsymbol{q}) = \frac{1}{\rho C_p}(\varepsilon - \nabla \cdot \kappa_T \nabla T) = \frac{\varepsilon}{\rho C_p} - k\nabla^2 T, \tag{3.1.19}$$

where thermal conductivity κ_T is assumed to be constant. k is the thermal diffusivity defined by (1.2.46). The dissipation rate due to the friction ε is, from (1.2.37),

$$\varepsilon = \frac{1}{2}\eta \left(\frac{\partial v_i}{\partial x_j} + \frac{\partial v_j}{\partial x_i} \right)^2 . \tag{3.1.20}$$

ε is sometimes omitted in Boussinesq equations.

Hence, as the most familiar form, the Boussinesq equations in Cartesian coordinates are given by

$$\frac{du}{dt} = -\frac{1}{\rho_0}\frac{\partial p'}{\partial x} + \nu \nabla^2 u, \tag{3.1.21}$$

$$\frac{dv}{dt} = -\frac{1}{\rho_0}\frac{\partial p'}{\partial y} + \nu \nabla^2 v, \tag{3.1.22}$$

$$\frac{dw}{dt} = -\frac{1}{\rho_0}\frac{\partial p'}{\partial z} + \alpha g T' + \nu \nabla^2 w, \tag{3.1.23}$$

$$\frac{\partial u}{\partial x} + \frac{\partial v}{\partial y} + \frac{\partial w}{\partial z} = 0, \tag{3.1.24}$$

$$\frac{dT'}{dt} = \kappa \nabla^2 T'. \tag{3.1.25}$$

We have assumed that the geopotential is written as $\Phi = gz$.

3.2 Quasi-geostrophic approximation

3.2.1 Scaling

The evolution of geostrophic winds is described by a set of equations called *quasi-geostrophic equations*, which are given by the quasi-geostrophic approximation. We introduce the quasi-geostrophic approximation using the spherical coordinates (λ, φ, z) where λ is longitude, φ is latitude, and z is height with $z = r - R$ and R being the radius of the earth. Let us define a basic state which is at rest in hydrostatic balance. The potential temperature of the basic state is denoted by θ_s and the Brunt-Väisälä frequency by

$$N^2 \equiv \frac{g}{\theta_s}\frac{d\theta_s}{dz}. \tag{3.2.1}$$

We use scale analysis to derive conditions when the quasi-geostrophic approximation is valid. We designate the horizontal scale of motion by L, the vertical scale by H, the time scale by τ, the horizontal and vertical velocity scales by U and W, the typical value of potential temperature by Θ, and that of the Brunt-Väisälä frequency by N. In order to derive the quasi-geostrophic approximation, we need assumptions for the Boussinesq approximation (3.1.6)–(3.1.8). In addition, we require the following assumptions:

$$\delta \equiv \frac{H}{L} \ll 1, \tag{3.2.2}$$

$$\max(\varepsilon_T, \varepsilon) \equiv \max\left(\frac{1}{2\Omega\tau}, \frac{U}{2\Omega L}\right) \ll 1, \tag{3.2.3}$$

$$S \equiv \left(\frac{NH}{2\Omega L}\right)^2 \gg O(\delta\varepsilon), \tag{3.2.4}$$

$$F \equiv \frac{(2\Omega L)^2}{gH} \leq O(1), \tag{3.2.5}$$

where we call δ the *aspect ratio*, ε the *Rossby number*, ε_T the time Rossby number, S the stability parameter, and F the rotation *Froude number*. We further restrict consideration in motions in the mid-latitudes: $\sin\varphi \approx 1$ and $\cos\varphi \approx 1$.

In order to obtain the relation between the magnitudes of variables in the geostrophic field, we assume that deviation from the basic hydrostatic balance is in the generalized geostrophic balance, (2.2.12)–(2.2.14). If the Boussinesq approximation is applicable, using the expression for hydrostatic balance (3.1.4), these are rewritten as

$$-2\Omega\sin\varphi\, v + 2\Omega\cos\varphi\, w = -\frac{\theta_0}{R\cos\varphi}\frac{\partial\pi'}{\partial\lambda}, \tag{3.2.6}$$

$$2\Omega\sin\varphi\, u = -\frac{\theta_0}{R}\frac{\partial\pi'}{\partial\varphi}, \tag{3.2.7}$$

$$-2\Omega\cos\varphi\, u = -\theta_0\frac{\partial\pi'}{\partial z} + \frac{\theta'}{\theta_0}g, \tag{3.2.8}$$

where θ_0 is a reference potential temperature and the prime $'$ is omitted for velocity fields, since the velocity of the basic state is zero: $u = u'$, $v = v'$, and $w = w'$. From (3.2.7), the typical scale of the deviation of the Exner function π' is estimated as

$$\pi' \approx \frac{2\Omega U L}{\Theta}. \tag{3.2.9}$$

Substituting this into the balance in (3.2.8), we have the ratio between each term of (3.2.8) as

$$\begin{aligned} -2\Omega\cos\varphi\, u &= -\theta_0 \frac{\partial \pi'}{\partial z} + \frac{\theta'}{\theta_0} g, \\ 2\Omega U \quad &: \quad \frac{2\Omega U}{\delta} \quad : \quad \frac{\Delta\Theta}{\Theta} g, \end{aligned}$$

where the typical value of θ' is estimated as $\Delta\Theta$. Under the requirement $\delta \ll 1$, (3.2.2), the Coriolis term (the left-hand side) is much smaller than the pressure gradient force (the first term on the right-hand side). Thus, we also have the hydrostatic balance for the deviation field:

$$0 = -\theta_0 \frac{\partial \pi'}{\partial z} + \frac{\theta'}{\theta_0} g. \tag{3.2.10}$$

From this, the magnitude of the deviation of potential temperature is given by

$$\theta' \approx \Delta\Theta = \frac{2\Omega U \Theta}{\delta g} = \varepsilon F \Theta. \tag{3.2.11}$$

If we assume adiabatic motion in the equation of energy:

$$\frac{d\theta'}{dt} + w \frac{\partial \theta_s}{\partial z} = 0. \tag{3.2.12}$$

This gives the scale of vertical velocity as

$$w \approx W = \frac{1}{\tau} \frac{\Delta\Theta}{\frac{\partial \theta_s}{\partial z}} = \frac{U}{L} \frac{\varepsilon F \Theta}{\frac{\partial \theta_s}{\partial z}} = \frac{\delta\varepsilon}{S} U, \tag{3.2.13}$$

where the time scale is estimated as $\tau \approx L/U$. Since $\delta\varepsilon/S \ll 1$ from (3.2.4), vertical velocity W is smaller than that of horizontal velocity U. Therefore, (3.2.6) is approximated as

$$-2\Omega \sin\varphi\, v = -\frac{\theta_0}{R\cos\varphi} \frac{\partial \pi'}{\partial \lambda}. \tag{3.2.14}$$

That is, the deviation field satisfies the geostrophic balance of horizontal winds given by (3.2.7) and (3.2.14).

3.2.2 Synoptic-scale quasi-geostrophic equations

Quasi-geostrophic equations have different forms depending on the horizontal scales of motions. First, we consider the case when the horizontal scale of motions is much smaller than the radius of the earth:

$$\frac{L}{R} = O(\varepsilon) \ll 1. \tag{3.2.15}$$

The horizontal length of this scale is in the range of 100–1000km for the case of the atmospheric motions of the earth; it is called the *synoptic scale.* We expand each variable in a series of the Rossby number ε. According to the assumptions on the scaling (3.2.2)–(3.2.5), we will see that leading order terms are reduced to the geostrophic balance and evolution equations of the deviation field are given from the following terms. We expand

$$\begin{aligned} u &= u^{(0)} + \varepsilon u^{(1)} + \cdots, & v &= v^{(0)} + \varepsilon v^{(1)} + \cdots, \\ w &= w^{(0)} + \varepsilon w^{(1)} + \cdots, & \pi' &= \pi^{(0)} + \varepsilon \pi^{(1)} + \cdots, \\ \theta' &= \theta^{(0)} + \varepsilon \theta^{(1)} + \cdots, \end{aligned}$$

and also expand latitude φ about a reference latitude $\varphi^{(0)}$ with the definitions

$$x = \lambda R \cos\varphi^{(0)}, \qquad y = R(\varphi - \varphi^{(0)}).$$

Using y, the Coriolis parameter is expanded as

$$\begin{aligned} f \equiv 2\Omega \sin\varphi &= 2\Omega \sin\varphi^{(0)} + 2\Omega \frac{y}{R} \cos\varphi^{(0)} + \cdots \\ &= f^{(0)} + \beta y + \cdots, \end{aligned} \tag{3.2.16}$$

where

$$f^{(0)} = 2\Omega \sin\varphi^{(0)}, \qquad \beta = \frac{\partial f}{\partial y} = \frac{2\Omega}{R} \cos\varphi^{(0)}.$$

If just the leading term of the Coriolis parameter is used, it is called the f-plane approximation in which the Coriolis parameter is constant. If we take second-order terms, the Coriolis parameter linearly depends on y. Such an approximation is called the *β-plane approximation.*

From leading order $O(\varepsilon^0)$ terms, the equations of motion are written as geostrophic and hydrostatic balances, (3.2.7), (3.2.14), and (3.2.10):

$$-f^{(0)} v^{(0)} = -\theta_0 \frac{\partial \pi^{(0)}}{\partial x}, \tag{3.2.17}$$

$$f^{(0)} u^{(0)} = -\theta_0 \frac{\partial \pi^{(0)}}{\partial y}, \tag{3.2.18}$$

$$0 = -\theta_0 \frac{\partial \pi^{(0)}}{\partial z} + \frac{\theta^{(0)}}{\theta_0} g. \tag{3.2.19}$$

It can be found from (3.2.17) and (3.2.18) that the horizontal wind components are non-divergent

$$\frac{\partial u^{(0)}}{\partial x} + \frac{\partial v^{(0)}}{\partial y} = 0. \tag{3.2.20}$$

From this with the continuity equation (3.1.7), we obtain

$$\frac{1}{\rho_s} \frac{\partial}{\partial z} (\rho_s w^{(0)}) = 0. \tag{3.2.21}$$

Then, if we impose an appropriate boundary condition, we obtain

$$w^{(0)} = 0. \tag{3.2.22}$$

Eq. (3.2.21) implies $W \ll \delta U$. In order to satisfy (3.2.13), therefore, $S \gg \varepsilon$ is required; that is

$$S \geq O(1). \tag{3.2.23}$$

The equations of first-order $O(\varepsilon^1)$ terms in the Rossby number expansion are written as follows

$$\frac{\partial u^{(0)}}{\partial t} + u^{(0)}\frac{\partial u^{(0)}}{\partial x} + v^{(0)}\frac{\partial u^{(0)}}{\partial y} - f^{(0)}v^{(1)} - \beta y v^{(0)} = -\theta_0 \frac{\partial \pi^{(1)}}{\partial x}, \tag{3.2.24}$$

$$\frac{\partial v^{(0)}}{\partial t} + u^{(0)}\frac{\partial v^{(0)}}{\partial x} + v^{(0)}\frac{\partial v^{(0)}}{\partial y} + f^{(0)}u^{(1)} + \beta y u^{(0)} = -\theta_0 \frac{\partial \pi^{(1)}}{\partial y}, \tag{3.2.25}$$

$$0 = -\theta_0 \frac{\partial \pi^{(1)}}{\partial z} + \frac{\theta^{(1)}}{\theta_0} g, \tag{3.2.26}$$

$$\frac{\partial u^{(1)}}{\partial x} + \frac{\partial v^{(1)}}{\partial y} + \frac{1}{\rho_s}\frac{\partial}{\partial z}(\rho_s w^{(1)}) = 0, \tag{3.2.27}$$

$$\frac{\partial \theta^{(0)}}{\partial t} + u^{(0)}\frac{\partial \theta^{(0)}}{\partial x} + v^{(0)}\frac{\partial \theta^{(0)}}{\partial y} + w^{(1)}\frac{\partial \theta_s}{\partial y} = 0. \tag{3.2.28}$$

The vorticity equation is derived from (3.2.24) and (3.2.25) as

$$\begin{aligned} &\frac{\partial \zeta^{(0)}}{\partial t} + u^{(0)}\frac{\partial \zeta^{(0)}}{\partial x} + v^{(0)}\frac{\partial \zeta^{(0)}}{\partial y} + \beta v^{(0)} \\ &\quad = -f^{(0)}\left(\frac{\partial u^{(1)}}{\partial x} + \frac{\partial v^{(1)}}{\partial y}\right), \end{aligned} \tag{3.2.29}$$

where

$$\zeta^{(0)} = \frac{\partial v^{(0)}}{\partial x} - \frac{\partial u^{(0)}}{\partial x} \tag{3.2.30}$$

is the vorticity. Combining the vorticity equation (3.2.29) with the continuity equation (3.2.27), we have

$$\frac{d^{(0)}}{dt}(\zeta^{(0)} + \beta y) = f^{(0)}\frac{1}{\rho_s}\frac{\partial}{\partial z}(\rho_s w^{(1)}), \tag{3.2.31}$$

where

$$\frac{d^{(0)}}{dt} = \frac{\partial}{\partial t} + u^{(0)}\frac{\partial}{\partial x} + v^{(0)}\frac{\partial}{\partial y}.$$

From the thermodynamic equation (3.2.28) and (3.2.17)–(3.2.19), we have

$$\begin{aligned}\frac{1}{\rho_s}\frac{\partial}{\partial z}(\rho_s w^{(1)}) &= -\frac{1}{\rho_s}\frac{\partial}{\partial z}\left[\frac{1}{\rho_s}\left(\frac{\rho_s}{\frac{\partial\theta_s}{\partial z}}\theta^{(0)}\right)\right] \\ &= -\frac{d^{(0)}}{dt}\left[\frac{1}{\rho_s}\frac{\partial}{\partial z}\left(\frac{\rho_s}{\frac{\partial\theta_s}{\partial z}}\theta^{(0)}\right)\right] + \frac{1}{\frac{\partial\theta_s}{\partial z}}\left(\frac{\partial u^{(0)}}{\partial z}\frac{\partial\theta^{(0)}}{\partial x} + \frac{\partial v^{(0)}}{\partial z}\frac{\partial\theta^{(0)}}{\partial y}\right) \\ &= -\frac{d^{(0)}}{dt}\left[\frac{1}{\rho_s}\frac{\partial}{\partial z}\left(\frac{\rho_s}{\frac{\partial\theta_s}{\partial z}}\theta^{(0)}\right)\right].\end{aligned}$$

Using this equation and (3.2.31), we obtain the *quasi-geostrophic potential vorticity equation*:

$$\frac{d^{(0)}}{dt}\Pi_g = 0, \tag{3.2.32}$$

where Π_g is the *quasi-geostrophic potential vorticity* defined by

$$\Pi_g \equiv \zeta^{(0)} + f^{(0)}\frac{1}{\rho_s}\frac{\partial}{\partial z}\left(\frac{\rho_s}{\frac{\partial\theta_s}{\partial z}}\theta^{(0)}\right) + f^{(0)} + \beta y \tag{3.2.33}$$

$$= \frac{\theta_0}{f^{(0)}}\left[\left(\frac{\partial^2}{\partial x^2} + \frac{\partial^2}{\partial y^2}\right)\pi^{(0)} + \frac{1}{\rho_s}\frac{\partial}{\partial z}\left(\rho_s\frac{f^{(0)2}}{N^2}\frac{\partial}{\partial z}\pi^{(0)}\right)\right] + f^{(0)} + \beta y. \tag{3.2.34}$$

Since the geostrophic winds $u^{(0)}$ and $v^{(0)}$ are non-divergent (3.2.20), the streamfunction ψ can be introduced as

$$u^{(0)} = -\frac{\partial\psi}{\partial x}, \qquad v^{(0)} = \frac{\partial\psi}{\partial y}. \tag{3.2.35}$$

From comparison between (3.2.17), (3.2.18) and (3.2.35), we have

$$\pi^{(0)} = \frac{f^{(0)}}{\theta_0}\psi, \qquad \theta^{(0)} = -\frac{\theta_0}{g}f^{(0)}\frac{\partial\psi}{\partial z}. \tag{3.2.36}$$

Eq. (3.2.32) can be written by using ψ as

$$\begin{aligned}\frac{d^{(0)}}{dt}\Pi_g &= \left(\frac{\partial}{\partial t} - \frac{\partial\psi}{\partial y}\frac{\partial}{\partial x} + \frac{\partial\psi}{\partial x}\frac{\partial}{\partial y}\right)\left[\left(\frac{\partial^2}{\partial x^2} + \frac{\partial^2}{\partial y^2}\right)\psi\right. \\ &\left. + \frac{1}{\rho_s}\frac{\partial}{\partial z}\left(\rho_s\frac{f^{(0)2}}{N^2}\frac{\partial}{\partial z}\psi\right) + f^{(0)} + \beta y\right] = 0,\end{aligned} \tag{3.2.37}$$

where from (3.2.34)

$$\Pi_g = \left(\frac{\partial^2}{\partial x^2} + \frac{\partial^2}{\partial y^2}\right)\psi + \frac{1}{\rho_s}\frac{\partial}{\partial z}\left(\rho_s\frac{f^{(0)2}}{N^2}\frac{\partial}{\partial z}\psi\right) + f^{(0)} + \beta y, \tag{3.2.38}$$

$$\frac{d^{(0)}}{dt} = \frac{\partial}{\partial t} - \frac{\partial\psi}{\partial y}\frac{\partial}{\partial x} + \frac{\partial\psi}{\partial x}\frac{\partial}{\partial y}. \tag{3.2.39}$$

The thermodynamic equation (3.2.28) is also expressed by ψ using (3.2.36) as

$$\frac{d^{(0)}}{dt}\frac{\partial \psi}{\partial z} + N^2 w^{(1)} = 0. \tag{3.2.40}$$

3.2.3 Planetary-scale quasi-geostrophic equations

We have different types of quasi-geostrophic equations, if the horizontal scale is comparable with the radius of the earth:

$$\frac{L}{R} = O(\varepsilon^0). \tag{3.2.41}$$

These types of equations are used for the description of large-scale motions, such as oceanic circulations. (In the oceanic case, however, the thermodynamic equation takes a different form from that shown below.) In this case, quasi-geostrophic equations are given by

$$-2\Omega \sin\varphi\, v^{(0)} = -\frac{\theta_0}{R\cos\varphi}\frac{\partial \pi^{(0)}}{\partial \lambda}, \tag{3.2.42}$$

$$2\Omega \sin\varphi\, u^{(0)} = -\frac{\theta_0}{R}\frac{\partial \pi^{(0)}}{\partial \varphi}, \tag{3.2.43}$$

$$0 = -\theta_0 \frac{\partial \pi^{(0)}}{\partial z} + \frac{\theta^{(0)}}{\theta_0} g, \tag{3.2.44}$$

$$\frac{1}{R\cos\varphi}\frac{\partial u^{(0)}}{\partial \lambda} + \frac{1}{R\cos\varphi}\frac{\partial}{\partial y}(\cos\varphi v^{(0)}) + \frac{1}{\rho_s}\frac{\partial}{\partial z}(\rho_s w^{(0)}) = 0, \tag{3.2.45}$$

$$\frac{\partial \theta^{(0)}}{\partial t} + \frac{u^{(0)}}{R\cos\varphi}\frac{\partial \theta^{(0)}}{\partial \lambda} + \frac{v^{(0)}}{R}\frac{\partial \theta^{(0)}}{\partial \varphi} + w^{(0)}\frac{\partial \theta_s}{\partial y} = 0. \tag{3.2.46}$$

Substituting the geostrophic flows $u^{(0)}$ and $v^{(0)}$ from (3.2.42) and (3.2.43) into the continuity equation (3.2.45), we have

$$\begin{aligned} 0 &= \frac{1}{\rho_s}\frac{\partial}{\partial z}(\rho_s w^{(0)}) - \frac{\theta_0}{2\Omega R \sin^2\varphi}\frac{\partial \pi^{(0)}}{\partial \lambda} \\ &= \frac{1}{\rho_s}\frac{\partial}{\partial z}(\rho_s w^{(0)}) - \frac{\cos\varphi}{R\sin\varphi} v^{(0)}. \end{aligned} \tag{3.2.47}$$

It can be seen that this equation corresponds to the vorticity equation of this approximation. From this, we get the scaling of vertical motion: $W = \delta U$. Hence, from (3.2.13), we obtain

$$S \geq O(\varepsilon). \tag{3.2.48}$$

The corresponding potential vorticity equation becomes

$$\frac{d^{(0)}}{dt}\Pi_g = 0, \tag{3.2.49}$$

where

$$\Pi_g = 2\Omega \sin\varphi \frac{1}{\rho_s}\frac{\partial}{\partial z}\left(\frac{\rho_s}{\frac{\partial \theta_s}{\partial z}}\theta^{(0)}\right), \tag{3.2.50}$$

$$\frac{d^{(0)}}{dt} = \frac{\partial}{\partial t} + \frac{u^{(0)}}{R\cos\varphi}\frac{\partial}{\partial \lambda} + \frac{v^{(0)}}{R}\frac{\partial}{\partial \varphi} \tag{3.2.51}$$

(i.e., potential vorticity is expressed only by thermal stratification). The contribution of relative vorticity is smaller than that of thermal stratification.

3.3 Primitive equations

3.3.1 The equations in spherical coordinates

Primitive equations are the basic equations for the atmosphere or the ocean and are employed in numerical models without any approximations. In practice, however, the set of equations based on the shallow atmosphere approximation and the hydrostatic balance are called primitive equations. To derive the normally used form of primitive equations, we start with the equations in latitude-longitude spherical coordinates in a rotating frame. The equations in spherical coordinates are described in Appendix A1.5. Here, we introduce the effect of rotation into these equations. Letting (u, v, w) be the zonal, north-south, and vertical winds, respectively, we write the equations of motion, mass, and entropy in spherical coordinates in a rotating frame as

$$\begin{aligned}\frac{du}{dt} &= \frac{uv}{r}\tan\varphi - \frac{uw}{r} - 2\Omega\cos\varphi \cdot w + 2\Omega \sin\varphi \cdot v \\ &\quad - \frac{1}{\rho r\cos\varphi}\frac{\partial p}{\partial \lambda} - \frac{1}{r\cos\varphi}\frac{\partial \Phi}{\partial \lambda} + f_\lambda,\end{aligned} \tag{3.3.1}$$

$$\frac{dv}{dt} = -\frac{u^2}{r}\tan\varphi - \frac{vw}{r} - 2\Omega\sin\varphi \cdot u - \frac{1}{\rho r}\frac{\partial p}{\partial \varphi} - \frac{1}{r}\frac{\partial \Phi}{\partial \varphi} + f_\varphi, \tag{3.3.2}$$

$$\frac{dw}{dt} = -\frac{u^2+v^2}{r} + 2\Omega\cos\varphi \cdot u - \frac{1}{\rho}\frac{\partial p}{\partial r} - \frac{\partial \Phi}{\partial r} - g + f_r, \tag{3.3.3}$$

$$\frac{d\rho}{dt} = -\rho\left[\frac{1}{r\cos\varphi}\frac{\partial u}{\partial \lambda} + \frac{1}{r\cos\varphi}\frac{\partial}{\partial \varphi}(v\cos\varphi) + \frac{1}{r^2}\frac{\partial}{\partial r}(wr^2)\right], \tag{3.3.4}$$

$$\frac{ds}{dt} = \frac{C_p Q}{T}, \tag{3.3.5}$$

where the material derivative is given by

$$\frac{d}{dt} = \frac{\partial}{\partial t} + \frac{u}{r\cos\varphi}\frac{\partial}{\partial \lambda} + \frac{v}{r}\frac{\partial}{\partial \varphi} + w\frac{\partial}{\partial r} \tag{3.3.6}$$

and $(f_\lambda, f_\varphi, f_z)$ is the frictional force. In the thermodynamic equation (3.3.5), s is entropy and Q is the diabatic term. Using potential temperature, it is expressed as

$$\frac{d\theta}{dt} = \frac{\theta}{T}Q. \tag{3.3.7}$$

This expression remains unchanged in the following argument.

We approximate the above equations because the atmosphere is very thin in depth in comparison with its horizontal dimensions. As described in Section 2.1, under the condition $r = R + z$ and $z/R \ll 1$, the geopotential can be given by $\Phi = gz$ as in (2.1.12). Accordingly, the radius r in equations (3.3.1) to (3.3.5) is replaced by the constant R. In this system, we need to change the definition of angular momentum as

$$l = uR\cos\varphi + \Omega R^2 \cos^2\varphi. \qquad (3.3.8)$$

The first line of the equation of u, (3.3.1), corresponds to the time derivative of angular momentum. To be consistent with (3.3.8), therefore, the two terms proportional to w in (3.3.1) should be omitted. In a similar way, the term proportional to w in the equation of v, (3.3.2), is also omitted. By conserving kinetic energy $\frac{1}{2}(u^2+v^2+w^2)$, we need to drop the corresponding terms in the equations of w: the first and second terms on the right-hand side of (3.3.3). Furthermore, in the continuity equation, w/R is negligible in comparison to $\frac{\partial w}{\partial z}$. Thus, Eqs. (3.3.1)–(3.3.4) are approximated as

$$\frac{du}{dt} = \frac{uv}{R}\tan\varphi + 2\Omega\sin\varphi\cdot v - \frac{1}{\rho R\cos\varphi}\frac{\partial p}{\partial\lambda} + f_\lambda, \qquad (3.3.9)$$

$$\frac{dv}{dt} = -\frac{u^2}{R}\tan\varphi - 2\Omega\sin\varphi\cdot u - \frac{1}{\rho R}\frac{\partial p}{\partial\varphi} + f_\varphi, \qquad (3.3.10)$$

$$\frac{dw}{dt} = -\frac{1}{\rho}\frac{\partial p}{\partial z} - g + f_r, \qquad (3.3.11)$$

$$\frac{d\rho}{dt} = -\rho\left[\frac{1}{R\cos\varphi}\frac{\partial u}{\partial\lambda} + \frac{1}{R\cos\varphi}\frac{\partial}{\partial\varphi}(v\cos\varphi) + \frac{\partial w}{\partial z}\right], \qquad (3.3.12)$$

where

$$\frac{d}{dt} = \frac{\partial}{\partial t} + \frac{u}{R\cos\varphi}\frac{\partial}{\partial\lambda} + \frac{v}{R}\frac{\partial}{\partial\varphi} + w\frac{\partial}{\partial z}. \qquad (3.3.13)$$

At this point, the approximation used to derive this equation set is called the *traditional approximation.*†

We further assume hydrostatic balance (2.1.4) in the equation of vertical motion (3.3.11). Generally, hydrostatic balance is thought to be applicable when the horizontal scale of motion L is larger than the vertical scale H; that is

$$\delta \equiv \frac{H}{L} \ll 1, \qquad (3.3.14)$$

where δ is called the *aspect ratio.*‡ Assuming hydrostatic balance with the tradi-

†We have introduced the traditional approximation using the assumption $\frac{z}{R} \ll 1$ and the consistency of the definition of angular momentum. Note that neglect of terms proportional to $2\Omega\cos\varphi$ is not justified by the assumption $\frac{z}{R} \ll 1$. See Veronis (1968).

‡Precisely, in order to use the hydrostatic balance, one needs to examine conditions under which the other terms in (3.3.11) are negligible. One of the conditions is that the ratio of the angular velocity Ω to the Brunt-Väisälä frequency N is small. See Phillips (1968).

tional approximation, we obtain

$$\frac{du}{dt} = \frac{uv}{R}\tan\varphi + 2\Omega\sin\varphi\cdot v - \frac{1}{\rho R\cos\varphi}\frac{\partial p}{\partial\lambda} + f_\lambda, \quad (3.3.15)$$

$$\frac{dv}{dt} = -\frac{u^2}{R}\tan\varphi - 2\Omega\sin\varphi\cdot u - \frac{1}{\rho R}\frac{\partial p}{\partial\varphi} + f_\varphi, \quad (3.3.16)$$

$$0 = -\frac{1}{\rho}\frac{\partial p}{\partial z} - g, \quad (3.3.17)$$

$$\frac{d\rho}{dt} = -\rho\left[\frac{1}{R\cos\varphi}\frac{\partial u}{\partial\lambda} + \frac{1}{R\cos\varphi}\frac{\partial}{\partial\varphi}(v\cos\varphi) + \frac{\partial w}{\partial z}\right]. \quad (3.3.18)$$

These equations together with (3.3.5) or (3.3.7) are called *primitive equations*, or *hydrostatic primitive equations*. Since vertical velocity w is not a predictable variable in primitive equations, it should be calculated using a diagnostic formula.

3.3.2 Transformation of the vertical coordinate

When the atmosphere is in hydrostatic balance, the governing equations can be written in a simplified form by transforming the vertical coordinate from the height z to an appropriate function of z. In the following subsections, we first derive governing equations of the atmosphere in a generalized vertical coordinate ζ and, then, as examples of the vertical coordinate, we list equations in the pressure p coordinates, the σ coordinates (where $\sigma = p/p_s$ is pressure divided by the surface pressure p_s), and potential temperature θ coordinates.

We consider the primitive equations in the Cartesian coordinates, which are given from (3.3.15)–(3.3.18) and (3.3.5), by dropping the metric terms and setting $dx = R\cos\varphi d\lambda$, $dy = Rd\varphi$, and $f = 2\Omega\sin\varphi$ as

$$\frac{\partial\rho}{\partial t} + \nabla_z\cdot(\rho\boldsymbol{v}_H) + \frac{\partial}{\partial z}(\rho w) = 0, \quad (3.3.19)$$

$$\frac{d\boldsymbol{v}_H}{dt} + f\boldsymbol{k}\times\boldsymbol{v}_H = -\frac{1}{\rho}\nabla_z p - \nabla_z\Phi + \boldsymbol{f}_H, \quad (3.3.20)$$

$$0 = -\frac{1}{\rho}\frac{\partial p}{\partial z} - \frac{\partial\Phi}{\partial z}, \quad (3.3.21)$$

$$\frac{ds}{dt} = \frac{C_p Q}{T}, \quad (3.3.22)$$

where ∇_z is the gradient on a constant z surface, $\boldsymbol{k}$ is a unit vector in the vertical direction, and f is the Coriolis parameter. In (3.3.20), since Φ depends only on z, we generally have $\nabla_z\Phi = 0$. We keep this term, however, since expressions of the pressure gradient force in different coordinates can be understood through this term.

As a vertical coordinate, we use a variable ζ that monotonically changes in the vertical direction. ζ can be related to z by the function

$$\zeta = \zeta(x, y, z, t), \quad (3.3.23)$$

where ζ is a monotonic function with respect to z. It is convenient to define density in ζ coordinates. Let dm be mass in a column that has a height interval dz with unit area. We may write

$$dm \quad = \quad \rho dz \quad = \quad \rho_\zeta d\zeta, \tag{3.3.24}$$

where ρ_ζ is regarded as density in ζ coordinates. Using the hydrostatic balance (3.3.21), density is expressed as

$$\rho_\zeta \quad = \quad \rho\frac{\partial z}{\partial \zeta} \quad = \quad -\frac{1}{g}\frac{\partial p}{\partial \zeta}, \tag{3.3.25}$$

where $g = \frac{\partial \Phi}{\partial z}$.

Relations between derivatives on the z surface and those on the ζ surface are expressed as follows. Any variable A can be expressed in two coordinates: $A(x, y, z, t)$ or $A(x, y, \zeta, t)$, These two functions are related as

$$A(x,y,z,t) \quad = \quad A(x,y,\zeta(x,y,z,t),t). \tag{3.3.26}$$

The partial derivatives of A with respect to x are related as

$$\left(\frac{\partial A}{\partial x}\right)_z \quad = \quad \left(\frac{\partial A}{\partial x}\right)_\zeta + \frac{\partial A}{\partial \zeta}\left(\frac{\partial \zeta}{\partial x}\right)_z, \tag{3.3.27}$$

$$\left(\frac{\partial A}{\partial x}\right)_\zeta \quad = \quad \left(\frac{\partial A}{\partial x}\right)_z + \frac{\partial A}{\partial z}\left(\frac{\partial z}{\partial x}\right)_\zeta. \tag{3.3.28}$$

Similar relations hold for the derivatives with respect to y and t. Using

$$\left(\frac{\partial z}{\partial x}\right)_\zeta \quad = \quad -\frac{\partial z}{\partial \zeta}\left(\frac{\partial \zeta}{\partial x}\right)_z \tag{3.3.29}$$

and (3.3.27), we have

$$\begin{aligned}\frac{\partial z}{\partial \zeta}\left(\frac{\partial A}{\partial x}\right)_z \quad &= \quad \frac{\partial z}{\partial \zeta}\left(\frac{\partial A}{\partial x}\right)_\zeta - \left(\frac{\partial z}{\partial x}\right)_\zeta \frac{\partial A}{\partial \zeta} \\ &= \quad \frac{\partial}{\partial x}\left(\frac{\partial z}{\partial \zeta}A\right)\bigg|_\zeta - \frac{\partial}{\partial \zeta}\left[\left(\frac{\partial z}{\partial x}\right)_\zeta A\right],\end{aligned} \tag{3.3.30}$$

which are rewritten using (3.3.25) and (3.3.29) as

$$\frac{\rho_\zeta}{\rho}\left(\frac{\partial A}{\partial x}\right)_z \quad = \quad \frac{\partial}{\partial x}\left(\frac{\rho_\zeta}{\rho}A\right)\bigg|_\zeta + \frac{\partial}{\partial \zeta}\left[\left(\frac{\partial \zeta}{\partial x}\right)_z \frac{\rho_\zeta}{\rho}A\right]. \tag{3.3.31}$$

If we write $A = \rho a$ where a is a specific value of A per unit volume, we have the transformation of the flux-form equation of a:

$$\begin{aligned}&\frac{\rho_\zeta}{\rho}\left[\frac{\partial}{\partial t}(\rho a) + \nabla_z \cdot (\rho \boldsymbol{v}_H a) + \frac{\partial}{\partial z}(\rho w a)\right] \\ &\quad = \quad \frac{\partial}{\partial t}(\rho_\zeta a) + \nabla_\zeta \cdot (\rho_\zeta \boldsymbol{v}_H a) + \frac{\partial}{\partial \zeta}(\rho_\zeta \dot{\zeta} a),\end{aligned} \tag{3.3.32}$$

where ∇_ζ is the gradient on a constant ζ surface and the time derivative of the vertical coordinate $\dot{\zeta}$ is related to w as

$$w = \frac{dz}{dt} = \left(\frac{\partial z}{\partial t}\right)_\zeta + \boldsymbol{v}_H \cdot \nabla_\zeta z + \dot{\zeta}\frac{\partial z}{\partial \zeta}, \tag{3.3.33}$$

$$\dot{\zeta} = \frac{d\zeta}{dt} = \left(\frac{\partial \zeta}{\partial t}\right)_z + \boldsymbol{v}_H \cdot \nabla_z \zeta + w\frac{\partial \zeta}{\partial z}. \tag{3.3.34}$$

$\dot{\zeta}$ is regarded as the vertical velocity in ζ coordinates. Using this relation, the Lagrangian derivative of a is written as

$$\begin{aligned} \frac{da}{dt} &= \left(\frac{\partial a}{\partial t}\right)_z + \boldsymbol{v}_H \cdot \nabla_z a + w\frac{\partial a}{\partial z} \\ &= \left(\frac{\partial a}{\partial t}\right)_\zeta + \boldsymbol{v}_H \cdot \nabla_\zeta a + \dot{\zeta}\frac{\partial a}{\partial \zeta}. \end{aligned} \tag{3.3.35}$$

Now we can write down the governing equations in ζ coordinates. We transform equations (3.3.19)–(3.3.22) using the above relations. First, the equation of entropy (3.3.22) is unchanged except for the Lagrangian derivative given by (3.3.35) with $a = s$. Using (3.3.32), the continuity equation (3.3.19) is written in ζ coordinates as

$$\frac{\partial \rho_\zeta}{\partial t} + \nabla_\zeta \cdot (\rho_\zeta \boldsymbol{v}_H) + \frac{\partial}{\partial \zeta}(\rho_\zeta \dot{\zeta}) = 0, \tag{3.3.36}$$

or

$$\frac{d\rho_\zeta}{dt} + \rho_\zeta\left(\nabla_\zeta \cdot \boldsymbol{v}_H + \frac{\partial \dot{\zeta}}{\partial \zeta}\right) = 0. \tag{3.3.37}$$

Next, the equation of motion (3.3.20) becomes in ζ coordinates

$$\frac{d\boldsymbol{v}_H}{dt} + f\boldsymbol{k} \times \boldsymbol{v}_H = -\frac{1}{\rho}\nabla_\zeta p - \nabla_\zeta \Phi + \boldsymbol{f}_H, \tag{3.3.38}$$

where the pressure gradient force is transformed using (3.3.27) with $A = p$ and the hydrostatic balance (3.3.21). Multiplying this equation by ρ_ζ, we obtain the conservation of momentum in flux form:

$$\begin{aligned} &\frac{\partial}{\partial t}(\rho_\zeta \boldsymbol{v}_H) + \nabla_\zeta \cdot (\rho_\zeta \boldsymbol{v}_H \boldsymbol{v}_H) + \frac{\partial}{\partial \zeta}(\rho_\zeta \dot{\zeta} \boldsymbol{v}_H) + \rho_\zeta f\boldsymbol{k} \times \boldsymbol{v}_H \\ &= -\nabla_\zeta\left(\frac{\rho_\zeta}{\rho}p\right) + \frac{\partial}{\partial \zeta}(p\nabla_\zeta z) + \rho_\zeta \boldsymbol{f}_H, \end{aligned} \tag{3.3.39}$$

where the pressure gradient force on the right-hand side is transformed using (3.3.30) or (3.3.31). This form is useful if one takes an average along the x-direction on a particular ζ surface, for instance. If the ζ surface does not intersect with the ground, the average along the x-direction of the first term on the right-hand side vanishes so that only the second term remains as the pressure gradient force. This contribution exists unless the ζ surface is horizontally flat $\nabla_\zeta z \neq 0$; this term is called the *form drag*.

For later use, we give various expressions of the pressure gradient force. Using (2.1.22), we can rewrite

$$-\frac{1}{\rho}\nabla_z p - \nabla_z \Phi \quad = \quad -\frac{1}{\rho}\nabla_\zeta p - \nabla_\zeta \Phi \quad = \quad \pi \nabla_\zeta \theta - \nabla_\zeta \Psi, \tag{3.3.40}$$

where Ψ is the Montgomery function or the static energy, π is the Exner function.† In particular, substituting $\zeta = z$, $\zeta = p$, $\zeta = \sigma$, and $\zeta = \theta$, we obtain expressions of the pressure gradient force in the respective coordinates:

$$-\frac{1}{\rho}\nabla_z p \quad = \quad \nabla_p \Phi \quad = \quad R_d T \nabla_\sigma \ln p_s + \nabla_\sigma \Phi \quad = \quad \nabla_\theta \Psi. \tag{3.3.41}$$

3.3.3 Pressure coordinates

Using pressure as a vertical coordinate, $\zeta = p$, we write down the equations in *pressure coordinates* or *isobaric coordinates*. From (3.3.25), the density in this coordinate is $\rho_p = -1/g = \text{const}$. Thus, using (3.3.38), (3.3.21), and (3.3.36), we obtain

$$\frac{d\boldsymbol{v}_H}{dt} + f\boldsymbol{k} \times \boldsymbol{v}_H \quad = \quad -\nabla_p \Phi + \boldsymbol{f}_H, \tag{3.3.42}$$

$$0 \quad = \quad -\frac{\partial \Phi}{\partial p} - \alpha, \tag{3.3.43}$$

$$\nabla_p \cdot \boldsymbol{v}_H + \frac{\partial \omega}{\partial p} \quad = \quad 0, \tag{3.3.44}$$

where $\alpha = 1/\rho$ is the specific volume.‡ We define pressure velocity as

$$\omega \quad \equiv \quad \frac{dp}{dt}. \tag{3.3.45}$$

From (3.3.33), the relation between ω and w is given by

$$w \quad = \quad \frac{dz}{dt} \quad = \quad \left(\frac{\partial z}{\partial t}\right)_p + \boldsymbol{v}_H \cdot \nabla_p z + \omega \frac{\partial z}{\partial p}. \tag{3.3.46}$$

Using the thermodynamic relation (1.1.6), we have

$$T\frac{ds}{dt} \quad = \quad \frac{dh}{dt} - \omega\alpha, \tag{3.3.47}$$

then we have from (3.3.22),

$$\frac{dh}{dt} \quad = \quad \omega\alpha + C_p Q. \tag{3.3.48}$$

†In (2.1.22), we have used the symbol σ for static energy. However, since we use the definition $\sigma = p/p_s$ here, a different symbol Ψ is used to represent the Montgomery function. In some literature, the Montgomery function is denoted by the symbol M.

‡In Chapter 2, the specific volume is denoted by the symbol v_s, while α is used to denote the expansion coefficient.

For the ideal gas with a constant specific heat, since $h = C_p T$, we have an expression for the equation of temperature:

$$\frac{dT}{dt} = \frac{\kappa T \omega}{p} + Q. \tag{3.3.49}$$

From the above equations, the transformations of energy are expressed as follows. By summing up the inner product of $\boldsymbol{v}_H$ with (3.3.42) and (3.3.43) multiplied by ω, we have the equation of kinetic energy

$$\begin{aligned}\frac{d}{dt}\frac{\boldsymbol{v}_H^2}{2} &= -\boldsymbol{v}_H \cdot \nabla_p \Phi - \omega \frac{\partial \Phi}{\partial p} - \omega\alpha + \boldsymbol{v}_H \cdot \boldsymbol{f}_H \\ &= -\nabla_p \cdot (\boldsymbol{v}_H \Phi) - \frac{\partial}{\partial p}(\omega \Phi) - \omega\alpha + \boldsymbol{v}_H \cdot \boldsymbol{f}_H. \end{aligned} \tag{3.3.50}$$

Adding this equation to (3.3.48), we obtain the conservation of total energy in pressure coordinates as

$$\frac{d}{dt}\left(\frac{\boldsymbol{v}_H^2}{2} + h\right) = -\nabla_p \cdot (\boldsymbol{v}_H \Phi) - \frac{\partial}{\partial p}(\omega \Phi) + \boldsymbol{v}_H \cdot \boldsymbol{f}_H + C_p Q. \tag{3.3.51}$$

This form of the conservation of energy is different from the conservation of total energy given by (1.2.47). The total energy per unit mass is defined as $v^2/2 + \Phi + u$ in general, whereas the total energy in pressure coordinates shown above is defined as $\boldsymbol{v}_H^2/2 + h$. First, there is no contribution from vertical velocity to total energy in pressure coordinates, since hydrostatic balance is assumed in primitive equations. Although the two energies are still different other than their vertical velocity, it will be shown in Section 12.1.3 that a vertical integral of the total energy in pressure coordinates is equivalent to that of the total energy in z coordinates.

Pressure coordinates have the following characteristics. Among their advantages: the continuity equation is a simple form as given by (3.3.44); it is a pseudo non-divergent-form equation; it is easy to relate quantities in pressure coordinates with those obtained from observations, since the altitudes of observation points of radiosondes are reported by their pressure values; in general, the inclinations of constant pressure surfaces are small, so that the physical position of constant pressure surfaces are close to constant z surfaces; and the mass between two constant pressure surfaces remains the same (pressure coordinates can be viewed as mass coordinates). Among their disadvantages, constant pressure surfaces (isobaric surfaces) may intersect with the ground at the lower boundary of the atmosphere.

3.3.4 Sigma coordinates

To overcome the disadvantage of pressure coordinates (i.e., isobaric surfaces may intersect with the ground), one can choose terrain-following coordinates in which the ground surface agrees with a constant coordinate surface. Using the surface pressure p_s, one can achieve this by choosing the vertical coordinate as

$$\sigma = \frac{p}{p_s}. \tag{3.3.52}$$

In this case, $\sigma = 1$ at the ground level and $\sigma = 0$ at the top of the atmosphere. These are called σ coordinates (or *sigma coordinates*). The definition of σ can be more generalized; any function $\sigma = \sigma(p)$ with $\sigma = 1$ at the lower boundary can be used as a vertical coordinate. Here, we concentrate on the simplest case (3.3.52). From (3.3.25), the density in σ coordinates is defined by

$$\rho_\sigma = \rho \frac{\partial z}{\partial \sigma} = -\frac{p_s}{g}. \tag{3.3.53}$$

Using (3.3.37), (3.3.38), (3.3.41), and (3.3.53), primitive equations in σ coordinates are written as

$$\frac{d\boldsymbol{v}_H}{dt} + f\boldsymbol{k} \times \boldsymbol{v}_{\boldsymbol{H}} = -\nabla_\sigma \Phi - R_d T \nabla_\sigma \ln p_s + \boldsymbol{f}_H, \tag{3.3.54}$$

$$0 = \frac{\partial \Phi}{\partial \sigma} + \frac{R_d T}{\sigma}, \tag{3.3.55}$$

$$\frac{d \ln p_s}{dt} = -\nabla_\sigma \cdot \boldsymbol{v}_H - \frac{\partial \dot\sigma}{\partial \sigma}. \tag{3.3.56}$$

A diagnostic equation for vertical velocity $\dot\sigma$ (or *sigma velocity*) is given from (3.3.56); integration from $\sigma = 1$ to σ of (3.3.56) gives

$$(\sigma - 1)\frac{\partial \ln p_s}{\partial t} = -\int_1^\sigma \boldsymbol{v}_{\boldsymbol{H}} \cdot \nabla_\sigma \ln p_s \, d\sigma - \int_1^\sigma \nabla_\sigma \cdot \boldsymbol{v}_H \, d\sigma - \dot\sigma, \tag{3.3.57}$$

while integration from $\sigma = 1$ to $\sigma = 0$ gives

$$\frac{\partial \ln p_s}{\partial t} = \int_1^0 \ln p_s \, d\sigma + \int_1^0 \nabla_\sigma \cdot \boldsymbol{v}_H \, d\sigma. \tag{3.3.58}$$

By eliminating the time derivative of $\ln p_s$ from these two equations, we obtain the expression of the vertical velocity as

$$\begin{aligned} \dot\sigma &= \left[(1-\sigma) \int_1^0 \nabla_\sigma \cdot \boldsymbol{v}_H \, d\sigma - \int_1^\sigma \nabla_\sigma \cdot \boldsymbol{v}_H \, d\sigma \right] \\ &\quad + \left[(1-\sigma) \int_1^0 \boldsymbol{v}_H \, d\sigma - \int_1^\sigma \boldsymbol{v}_H \, d\sigma \right] \cdot \nabla_\sigma \ln p_s. \end{aligned} \tag{3.3.59}$$

We also have

$$\dot\sigma = \frac{d}{dt}\left(\frac{p}{p_s}\right) = \frac{\omega}{p} - \sigma \frac{d \ln p_s}{dt}. \tag{3.3.60}$$

Using this and (3.3.56), therefore, the equation of temperature (3.3.49) is written as

$$\frac{dT}{dt} = \kappa T \left(\frac{\dot\sigma}{\sigma} - \frac{\partial \dot\sigma}{\partial \sigma} - \nabla_\sigma \cdot \boldsymbol{v}_{\boldsymbol{H}} \right) + Q. \tag{3.3.61}$$

While an advantage of σ coordinates is that the ground surface always corresponds to the coordinate surface $\sigma = 1$ so that constant σ surfaces never intersect with the ground, σ surfaces may be distorted in the area of steep topography irrespective of atmospheric structure.

3.3.5 Isentropic coordinates

If the atmosphere is statically stable, the potential temperature monotonically increases with height as shown in (2.3.7). In this case, the potential temperature θ can be used as a vertical coordinate. For large-scale motions where the hydrostatic approximation is valid such that primitive equations are applicable, vertical structure is thought to be statically stable in general. Thus, *potential temperature coordinates* (θ coordinates), or *isentropic coordinates*, can be used to describe the motions. The density in θ coordinates is defined by[†]

$$\rho_\theta \equiv -\frac{1}{g}\frac{\partial p}{\partial \theta}. \tag{3.3.62}$$

From (3.3.37), (3.3.38), (3.3.41), (3.3.40), and (3.3.22), the equations in θ coordinates are written as

$$\frac{d\boldsymbol{v}_H}{dt} + f\boldsymbol{k}\times\boldsymbol{v}_H = -\nabla_\theta\Psi + \boldsymbol{f}_H, \tag{3.3.63}$$

$$\frac{\partial\Psi}{\partial\theta} = \pi, \tag{3.3.64}$$

$$\frac{\partial\rho_\theta}{\partial t} + \nabla_\theta\cdot(\rho_\theta\boldsymbol{v}_H) + \frac{\partial}{\partial\theta}(\rho_\theta\dot{\theta}) = 0, \tag{3.3.65}$$

$$\dot{\theta} = \mathcal{Q}, \tag{3.3.66}$$

where π is the Exner function (2.1.20), and the diabatic term in (3.3.66) is expressed as

$$\mathcal{Q} \equiv \frac{\theta}{T}Q = \frac{\pi}{C_p}Q. \tag{3.3.67}$$

The flux-form equation of motion is given by multiplying ρ_θ by (3.3.63) and using (3.3.39)

$$\frac{\partial}{\partial t}(\rho_\theta\boldsymbol{v}_H) + \nabla_\theta\cdot(\rho_\theta\boldsymbol{v}_H\boldsymbol{v}_H) + \frac{\partial}{\partial\theta}(\rho_\theta\dot{\theta}\boldsymbol{v}_H) + \rho_\theta f\boldsymbol{k}\times\boldsymbol{v}_H$$
$$= -\nabla_\theta\left(\frac{\rho_\zeta}{\rho}p\right) + \frac{\partial}{\partial\theta}(p\nabla_\theta z) + \rho_\theta\boldsymbol{f}_H. \tag{3.3.68}$$

If use is made of an alternative flux form of the pressure gradient force (3.5.2) in the appendix of this chapter (Section 3.5) with (3.3.64), the momentum equation is rewritten as

$$\frac{\partial}{\partial t}(\rho_\theta\boldsymbol{v}_H) + \nabla_\theta\cdot(\rho_\theta\boldsymbol{v}_H\boldsymbol{v}_H) + \frac{\partial}{\partial\theta}(\rho_\theta\dot{\theta}\boldsymbol{v}_H) + \rho_\theta f\boldsymbol{k}\times\boldsymbol{v}_H$$
$$= -\frac{1}{g}\nabla_\theta\left(\frac{\kappa}{\kappa+1}p\pi\right) + \frac{1}{g}\frac{\partial}{\partial\theta}(p\nabla_\theta\Psi) + \rho_\theta\boldsymbol{f}_H. \tag{3.3.69}$$

[†]In the literature, the symbol σ is frequently used to represent the density in isentropic coordinates instead of ρ_θ.

The potential vorticity equation in isentropic coordinates has various useful forms. Using (3.3.62) under hydrostatic balance, potential vorticity (1.3.33) is rewritten as

$$\begin{aligned} P &= \frac{1}{\rho}\boldsymbol{\omega}_a \cdot \nabla\theta = \frac{1}{\rho}\left(\boldsymbol{\omega}_{aH} \cdot \nabla_H\theta + \omega_{az}\frac{\partial\theta}{\partial z}\right) \\ &= \frac{1}{\rho}\left(-\boldsymbol{\omega}_{aH} \cdot \nabla_\theta z + \omega_{az}\right)\frac{\partial\theta}{\partial z} = -g\omega_{a\theta}\frac{\partial\theta}{\partial p} = \frac{\omega_{a\theta}}{\rho_\theta}, \end{aligned} \tag{3.3.70}$$

where $\boldsymbol{\omega}_a$ is absolute vorticity vector in the hydrostatic balance, and $\omega_{a\theta}$ is absolute vorticity evaluated on isentropic surfaces. ω_{az} and $\boldsymbol{\omega}_{aH}$ are the vertical component and the horizontal two-dimensional vector of absolute vorticity, respectively; that is

$$\boldsymbol{\omega}_a = \left(-\frac{\partial v}{\partial z}, \frac{\partial u}{\partial z}, \frac{\partial v}{\partial x} - \frac{\partial u}{\partial y} + f\right) = (\boldsymbol{\omega}_{aH}, \omega_{az}). \tag{3.3.71}$$

Denoting the relative vorticity evaluated on isentropic surfaces by ω_θ and using (3.3.28), we have

$$\begin{aligned} \omega_{a\theta} &= \omega_\theta + f = \left(\frac{\partial v}{\partial x}\right)_\theta - \left(\frac{\partial u}{\partial y}\right)_\theta + f \\ &= \frac{\partial v}{\partial x} - \frac{\partial u}{\partial y} + f + \frac{\partial v}{\partial z}\left(\frac{\partial z}{\partial x}\right)_\theta - \frac{\partial u}{\partial z}\left(\frac{\partial z}{\partial y}\right)_\theta \\ &= \omega_{az} - \boldsymbol{\omega}_{aH} \cdot \nabla_\theta z. \end{aligned} \tag{3.3.72}$$

The equation of potential vorticity in isentropic coordinates can be derived in the following manner. Starting with the vector-invariant form of the advection term

$$\boldsymbol{v}_H \cdot \nabla_\theta \boldsymbol{v}_H = \omega_\theta \boldsymbol{k} \times \boldsymbol{v}_H + \nabla_\theta \frac{v_H^2}{2}, \tag{3.3.73}$$

which is similar to that given by (1.3.21), we can rewrite the equation of motion (3.3.63) as

$$\frac{\partial \boldsymbol{v}_H}{\partial t} + \nabla_\theta\left(\frac{\boldsymbol{v}_H^2}{2} + \Psi\right) + \omega_{a\theta}\boldsymbol{k} \times \boldsymbol{v}_H + \dot{\theta}\frac{\partial \boldsymbol{v}_H}{\partial \theta} = \boldsymbol{f}_H. \tag{3.3.74}$$

By applying rotation operator $\boldsymbol{k} \cdot \nabla_\theta \times$ to this equation, we obtain the flux-form vorticity equation

$$\frac{\partial \omega_{a\theta}}{\partial t} + \nabla_\theta \cdot (\omega_{a\theta}\boldsymbol{v}_H + \boldsymbol{J}_{\dot{\theta}} + \boldsymbol{J}_F) = 0, \tag{3.3.75}$$

where

$$\boldsymbol{J}_{\dot{\theta}} = \left(\dot{\theta}\frac{\partial v}{\partial \theta},\ -\dot{\theta}\frac{\partial u}{\partial \theta},\ 0\right), \tag{3.3.76}$$

$$\boldsymbol{J}_f = (-f_y,\ f_x,\ 0). \tag{3.3.77}$$

Using the expression of P (3.3.70), Eq. (3.3.75) immediately yields the flux-form potential vorticity equation as

$$\frac{\partial}{\partial t}(\rho_\theta P) + \nabla_\theta \cdot (\rho_\theta P \boldsymbol{v}_H + \boldsymbol{J}_{\dot{\theta}} + \boldsymbol{J}_f) \quad = \quad 0. \tag{3.3.78}$$

It can be seen from (3.3.76) and (3.3.77) that vertical component of the flux in (3.3.78) does not exist. That is, the component of the potential vorticity flux in the cross isentropic direction is identically zero, if the atmosphere is in hydrostatic balance. This is a general characteristic of potential vorticity in isentropic coordinates, and it always holds even if frictional or diabatic terms exist.

Eq. (3.3.75) gives the advective-form vorticity equation, as

$$\frac{d_\theta}{dt}\omega_{a\theta} + \omega_{a\theta}\nabla_\theta \cdot \boldsymbol{v}_H \quad = \quad -\nabla_\theta \cdot (\boldsymbol{J}_{\dot{\theta}} + \boldsymbol{J}_f), \tag{3.3.79}$$

where

$$\frac{d_\theta}{dt} \quad \equiv \quad \frac{\partial}{\partial t} + \boldsymbol{v}_H \cdot \nabla_\theta. \tag{3.3.80}$$

The continuity equation (3.3.65) is also rewritten as

$$\frac{d_\theta}{dt}\rho_\theta + \rho_\theta \nabla_\theta \cdot \boldsymbol{v}_H \quad = \quad -\frac{\partial}{\partial \theta}(\rho_\theta \dot{\theta}). \tag{3.3.81}$$

Eliminating $\nabla_\theta \cdot \boldsymbol{v}_H$ from (3.3.79) and (3.3.81), and using the expression of P (3.3.70), we have a change in potential vorticity on the isentropic surface:

$$\frac{d_\theta}{dt}P \quad = \quad -\frac{1}{\rho_\theta}\nabla_\theta \cdot (\boldsymbol{J}_{\dot{\theta}} + \boldsymbol{J}_f) + \frac{P}{\rho_\theta}\frac{\partial}{\partial \theta}(\rho_\theta \dot{\theta}). \tag{3.3.82}$$

Eq. (3.3.82) is further rewritten in three-dimensional advective form, as

$$\frac{dP}{dt} \quad = \quad P\frac{\partial \dot{\theta}}{\partial \theta} + \frac{1}{\rho_\theta}\left(\frac{\partial \dot{\theta}}{\partial y}\frac{\partial u}{\partial \theta} - \frac{\partial \dot{\theta}}{\partial x}\frac{\partial v}{\partial \theta} + \boldsymbol{k} \cdot \nabla \times \boldsymbol{f}_H\right). \tag{3.3.83}$$

Isentropic coordinates have a distinct advantage. That is, isentropic surfaces represent nearly material surfaces; an air parcel stays on an isentropic surface under the adiabatic condition $\mathcal{Q} = 0$. In this case, since vertical velocity is $\dot{\theta} = 0$ from (3.3.66), the air parcel does not exit from the original isentropic surface. This means that the motions of the air parcel are two dimensional on the isentropic surface. One of the disadvantages of isentropic coordinates is that the isentropic surfaces generally intersect with the ground. In mid-latitudes, in particular, θ surfaces have a large inclination to z surfaces. As a result, the divergence $\nabla_\theta \cdot \boldsymbol{v}_H$ or the rotation $\boldsymbol{k} \cdot \nabla_\theta \times \boldsymbol{v}_H$ on the θ surface may be very different from those on the z surface. Another disadvantage is that ambiguity of the definition of isentropic surfaces arises when stratification gets close to neutral. In addition, calculations on isentropic surfaces are more or less complicated compared with other coordinates.

3.4 Shallow water equations

In this final section of this chapter, we describe shallow water equations. Shallow water equations can be viewed as mathematical models of horizontal motions of the atmosphere in which the air has almost vertically uniform motions. Originally, shallow water equations were used to describe motions in a shallow fluid layer, in particular for the elevation of the fluid surface. We introduce shallow water equations in this original context

Let us consider a fluid on a horizontally uniform bottom. We define the height of the fluid surface as η and introduce $\Phi = g\eta$. Shallow water equations are given by

$$\frac{D}{Dt}\boldsymbol{v}_H + f\boldsymbol{k} \times \boldsymbol{v}_H = -g\nabla_H \eta + \boldsymbol{f}_H, \tag{3.4.1}$$

$$\frac{\partial \eta}{\partial t} + \nabla_H \cdot (\eta \boldsymbol{v}_H) = 0, \tag{3.4.2}$$

where $\boldsymbol{v}_H = (u, v)$ is the two-dimensional velocity vector, ∇_H is the horizontal gradient operator, $\boldsymbol{f}_H$ is the two-dimensional forcing, and

$$\frac{D}{Dt} = \frac{\partial}{\partial t} + \boldsymbol{v}_H \cdot \nabla_H. \tag{3.4.3}$$

Multiplying (3.4.1) by $\boldsymbol{v}_H$ and (3.4.2) by g, and summing up the two, we obtain the energy equation of shallow water equations:

$$\frac{D}{Dt}\left(\Phi + \frac{v_H^2}{2}\right) + \nabla_H \cdot (\Phi \boldsymbol{v}_H) = \boldsymbol{v}_H \cdot \boldsymbol{f}_H. \tag{3.4.4}$$

Vorticity and divergence equations are frequently used for shallow water equations. The relative vorticity and divergence of shallow water are defined, respectively, as

$$\omega = \boldsymbol{k} \cdot \nabla_H \times \boldsymbol{v}_H = \frac{\partial v}{\partial x} - \frac{\partial u}{\partial y}, \tag{3.4.5}$$

$$\delta = \nabla_H \cdot \boldsymbol{v}_H = \frac{\partial u}{\partial x} + \frac{\partial v}{\partial y}. \tag{3.4.6}$$

Using the identity

$$\boldsymbol{v}_H \cdot \nabla_H \boldsymbol{v}_H = \omega \boldsymbol{k} \times \boldsymbol{v}_H + \nabla_H \frac{v_H^2}{2}, \tag{3.4.7}$$

we can rewrite (3.4.1) as

$$\frac{\partial \boldsymbol{v}_H}{\partial t} = -(\omega + f)\boldsymbol{k} \times \boldsymbol{v}_H - \nabla_H \left(\Phi + \frac{v_H^2}{2}\right) + \boldsymbol{f}_H. \tag{3.4.8}$$

Applying rotation and divergence operators to this equation gives the vorticity and divergence equations, respectively:

$$\frac{\partial \omega}{\partial t} = -\nabla_H \cdot (\omega + f)\boldsymbol{v}_H + \boldsymbol{k} \cdot \nabla_H \times \boldsymbol{f}_H, \tag{3.4.9}$$

$$\frac{\partial \delta}{\partial t} = \boldsymbol{k} \cdot \nabla_H \times (\omega + f)\boldsymbol{v}_H - \nabla_H^2 \left(\Phi + \frac{v_H^2}{2}\right) + \nabla_H \cdot \boldsymbol{f}_H. \tag{3.4.10}$$

Using the absolute vorticity $\omega_a = \omega + f$, (3.4.9) and (3.4.2) are rewritten as

$$\frac{D\omega_a}{Dt} = -\omega_a\delta + \boldsymbol{k}\cdot\nabla_H \times \boldsymbol{f}_H, \tag{3.4.11}$$

$$\frac{D\eta}{Dt} = -\eta\delta. \tag{3.4.12}$$

The above two equations are combined to make the potential vorticity equation:

$$\frac{D\Pi}{Dt} = \frac{\boldsymbol{k}\cdot\nabla_H \times \boldsymbol{f}_H}{\eta}, \tag{3.4.13}$$

where

$$\Pi = \frac{\omega_a}{\eta} = \frac{\omega + f}{\eta}. \tag{3.4.14}$$

is potential vorticity. This has a different dimension from that of (1.3.43).

In special cases, atmospheric motions can be described by shallow water equations. First, shallow water equations are analogous to primitive equations in isentropic coordinates, Second, a two-layer model of atmospheric motions is described by shallow water equations. Third, when atmospheric motions in hydrostatic balance are linearized, the equations for horizontal motion are equivalent to linearized shallow water equations. The third property will be examined in Section 4.7.1.

Here in this section, we only show how the equations in isentropic coordinates are related to shallow water equations. Under adiabatic conditions, the equations in isentropic coordinates (3.3.63)–(3.3.65) are written as

$$\frac{d_\theta}{dt}\boldsymbol{v}_H + f\boldsymbol{k}\times\boldsymbol{v}_H = -\nabla_\theta\Psi + \boldsymbol{f}_H, \tag{3.4.15}$$

$$\frac{\partial\rho_\theta}{\partial t} + \nabla_\theta\cdot(\rho_\theta\boldsymbol{v}_H) = 0, \tag{3.4.16}$$

where we have used the fact that d/dt is equal to d_θ/dt under the adiabatic condition $\dot{\theta} = 0$ since

$$\frac{d}{dt} = \frac{\partial}{\partial t} + \boldsymbol{v}_H\cdot\nabla_\theta + \dot{\theta}\frac{\partial}{\partial\theta} = \frac{d_\theta}{dt} + \dot{\theta}\frac{\partial}{\partial\theta}. \tag{3.4.17}$$

The two equations (3.4.15) and (3.4.16) are equivalent to the shallow water equations, (3.4.1) and (3.4.2), if we can assume that Ψ is proportional to ρ_θ such that Ψ is identical with Φ. In general, however, such a simple proportional relation between Ψ and ρ_θ does not exist. We nevertheless can relate them in the following manner (Held and Phillip, 1990). Let us consider a layer between two isentropic surfaces, and assume that velocities are vertically uniform in this layer. Let the differences of potential temperature, pressure, and π at the two isentropic surfaces be denoted by $\Delta\theta$, Δp, and $\Delta\pi$, respectively. From (3.3.62) and (3.3.64), we have

$$\nabla_\theta\Psi \approx \Delta\theta\nabla_\theta\pi, \tag{3.4.18}$$

$$\rho_\theta \approx -\frac{1}{g}\Delta p \approx -\frac{1}{g}\frac{dp}{d\pi}\Delta\pi, \tag{3.4.19}$$

where $\nabla_\theta \pi$ and $\frac{dp}{d\pi}$ take representative values in this layer. If the change in π on the upper isentropic surface, say, is negligible compared with that on the lower isentropic surfaces, and the change in $\frac{dp}{d\pi}$ is small within the layer, (3.4.15) and (3.4.16) become (3.4.1) and (3.4.2) with the relation

$$\Phi = \Delta\theta\Delta\pi. \tag{3.4.20}$$

If there is a vertical velocity $\dot{\theta}$, the contribution of this term is added to (3.4.16). Correspondingly, a source or sink term Q is added to the right-hand side of (3.4.2).

3.5 Appendix: Derivation of a generalized pressure gradient

We derive a generalized form of the pressure gradient in (3.3.69) in this appendix. In the case that g is constant, the flux form of the pressure gradient force in (3.3.39) is written using (3.3.25) as

$$-\nabla_\varsigma\left(\frac{\rho_\varsigma}{\rho}p\right)+\frac{\partial}{\partial\varsigma}(p\nabla_\varsigma z) = -\frac{1}{g}\nabla_\varsigma\left(\frac{\partial\Phi}{\partial\varsigma}p\right)+\frac{1}{g}\frac{\partial}{\partial\varsigma}(p\nabla_\varsigma\Phi). \tag{3.5.1}$$

In order to generalize this expression, we define a function $H = Z(\varsigma)P(p)$, which satisfies in general

$$\begin{aligned}\frac{\partial}{\partial\varsigma}(p\nabla_\varsigma H) &= \frac{\partial}{\partial\varsigma}(pZ\nabla_\varsigma P) = \frac{\partial}{\partial\varsigma}(pZP'\nabla_\varsigma p)\\ &= \frac{\partial}{\partial\varsigma}\left(Z\nabla_\varsigma\int^p p_1P'(p_1)dp_1\right) = \nabla_\varsigma\frac{\partial}{\partial\varsigma}\left(Z\int^p p_1P'(p_1)dp_1\right).\end{aligned}$$

On the other hand, we have

$$\begin{aligned}\nabla_\varsigma\left(\frac{\partial H}{\partial\varsigma}p\right) &= \nabla_\varsigma\left(Z'Pp+ZP'\frac{\partial p}{\partial\varsigma}p\right)\\ &= \nabla_\varsigma\left(Z'Pp+Z\frac{\partial}{\partial\varsigma}\int^p p_1P'(p_1)dp_1\right)\\ &= \nabla_\varsigma\left[Z'\left(pP-\int^p p_1P'(p_1)dp_1\right)+\frac{\partial}{\partial\varsigma}\left(Z\int^p p_1P'(p_1)dp_1\right)\right].\end{aligned}$$

Thus, (3.5.1) can be written as

$$\begin{aligned}&-\frac{1}{g}\nabla_\varsigma\left(\frac{\partial\Phi}{\partial\varsigma}p\right)+\frac{1}{g}\frac{\partial}{\partial\varsigma}(p\nabla_\varsigma\Phi)\\ &= -\frac{1}{g}\nabla_\varsigma\left[\frac{\partial}{\partial\varsigma}(\Phi+H)p-Z'\left(pP-\int^p p_1P'(p_1)dp_1\right)\right]\\ &\quad+\frac{1}{g}\frac{\partial}{\partial\varsigma}\left[p\nabla_\varsigma(\Phi+H)\right].\end{aligned}$$

Noting that enthalpy is written as $h = \theta\pi$, we can set $Z = \theta$, $P = \pi(p)$, and $H = h$ with $\Phi + H = \Psi$. In this case, we have

$$\begin{aligned} Z'\left(pP - \int^p p_1 P'(p_1)dp_1\right) &= p\pi - \int^p p_1\pi'(p_1)dp_1 \\ &= p\pi - \frac{\kappa}{\kappa+1}p\pi \;=\; \frac{1}{\kappa+1}p\pi, \end{aligned}$$

where $\kappa = R_d/C_p$. We therefore have an alternative flux form of the pressure gradient force:

$$\begin{aligned} &-\frac{1}{g}\nabla_\zeta\left(\frac{\partial\Phi}{\partial\zeta}p\right) + \frac{1}{g}\frac{\partial}{\partial\zeta}\left(p\nabla_\zeta\Phi\right) \\ &= -\frac{1}{g}\nabla_\zeta\left[\left(\frac{\partial\Psi}{\partial\zeta} - \frac{\pi}{\kappa+1}\right)p\right] + \frac{1}{g}\frac{\partial}{\partial\zeta}\left(p\nabla_\zeta\Psi\right). \end{aligned} \tag{3.5.2}$$

References and suggested reading

The derivation of Boussinesq equations in addition to their use in atmospheric applications is discussed in Ogura and Phillips (1962). We closely follow Pedlosky (1987) in our derivation of quasi-geostrophic equations. The various expressions in isentropic coordinates are summarized, for instance, in Andrews et al. (1987) and Haynes and McIntyre (1990). Interesting arguments on the treatment of boundary in isentropic coordinates are found in Schneider et al. (2002). The applicability of shallow water equations can be seen in Held and Phillips (1990), in which the relation between primitive equations in isentropic coordinates and shallow water equations are discussed.

Andrews, D. G., Holton, J. R., and Leovy, C. B., 1987: *Middle Atmosphere Dynamics*. Academic Press, San Diego, 489 pp.

Haynes, P. H., McIntyre, M. E., 1990: On the conservation and impermeability theorems for potential vorticity. *J. Atmos. Sci.*, **47**, 2021–2031.

Held, I. M. and Phillips, P. J., 1990: A barotropic model of the interaction between the Hadley cell and a Rossby wave. *J. Atmos. Sci.*, **47**, 856–869.

Ogura, Y. and Phillips, A., 1962: Scale analysis of deep and shallow convection in the atmosphere. *J. Atmos. Sci.*, **19**, 173–179.

Pedlosky, J., 1987: *Geophysical Fluid Dynamics*, 2nd ed., Springer-Verlag, New York, 710 pp.

Phillips, N. A., 1968: Reply. *J. Atmos. Sci.*, **25**, 1155–1157.

Schneider, T., Held, I. M., and Garner, S. T., 2002: Boundary effects in potential vorticity dynamics. *J. Atmos. Sci.*, **60**, 1024–1040.

Veronis, G., 1968: Large-amplitude Bénard convection in a rotating fluid. *J. Fluid Mech.*, **31**, 113–139.

4

Waves

Constituting the basis of atmospheric dynamics, the fundamental properties of waves are described in this chapter. Waves are themselves important atmospheric motions and play roles in transporting energy, momentum, and tracers. Waves also have remote effects through their propagations. In this chapter, to begin with, we briefly review wave theory. Then, the governing equations of the atmosphere are linearized under various conditions including sound waves, gravity waves, inertial waves, Rossby waves, spherical waves, and equatorial waves. The structures and propagations of these waves are explained using the linear system. We mainly consider cases when the basic properties are spatially uniform. Wave propagations in an inhomogeneous flow will not be described in this chapter, although they are important and have interesting behaviors.

The concept of wave propagation is also very important for the construction of numerical models. Mathematically, waves can be defined as neutral eigenmodes of linearized governing equations. In contrast, instability (discussed in Chapter 5) can be viewed as unstable eigenmodes. The roles of wave transport are considered in Chapter 7. If diabatic or mechanical forcing is applied to the atmosphere, waves are excited at the forced region. This kind of circulation is called a forced motion, and is discussed in Chapter 6. Forced motions statistically establish the equilibrium states of the atmosphere through the propagation of excited waves by being balanced by the dissipation process in the atmosphere. If the forcing is steady, a statistically steady circulation is realized as a result. Some aspects of atmospheric general circulation can be viewed as this type of forced motion.

4.1 Wave theory

In general, partial differential equations can be solved by using a systematic method called wave theory. With this method, an approximate solution of partial differential equations is given in the form of sinusoidal wavetrain. We consider a general method of solving linear partial differential equations in this section.

We consider a single linear partial differential equation or a set of linear partial differential equations. Letting L be a linear operator, we write a linear partial differential equation as

$$L[\phi(\boldsymbol{x},t)] \quad = \quad 0, \tag{4.1.1}$$

where $\phi(\boldsymbol{x},t)$ is a solution to the equation. We assume a wavy form of the solution

$$\phi(\boldsymbol{x},t) \quad = \quad A(\boldsymbol{x},t)e^{i\theta(\boldsymbol{x},t)}, \tag{4.1.2}$$

where θ is called the *phase* and A is the *amplitude*. In order that ϕ is wavy, the amplitude A must be a slowly changing function of $\boldsymbol{x}$ and t compared with θ. Under this condition, ϕ almost keeps the same value when the phase θ is increased by 2π; this implies that ϕ is approximately periodic.

Wave number and *frequency* are defined by the derivatives of the phase with respect to space and time, respectively

$$\boldsymbol{k} \;\equiv\; \nabla\theta, \qquad \omega \;\equiv\; -\frac{\partial\theta}{\partial t}. \tag{4.1.3}$$

$\boldsymbol{k}$ and ω are normally slowly changing functions with respect to $\boldsymbol{x}$ and t. When they are not, the wave number and the frequency defined above change rapidly such that they lose their wave characteristics. From (4.1.3), we have

$$\frac{\partial\boldsymbol{k}}{\partial t} + \nabla\omega \quad = \quad 0, \tag{4.1.4}$$

which represents the *conservation of wave numbers*. The wave number per unit length is given by $k_i/2\pi$ for each of the directions x_i ($i = 1$, 2, and 3) and that per unit time is given by $\omega/2\pi$, respectively. Their inverses are the *wavelength* and the *period*:

$$\lambda_i \;\equiv\; \frac{2\pi}{k_i}, \qquad \text{for} \quad i = 1,2,3; \qquad \tau \;\equiv\; \frac{2\pi}{\omega}. \tag{4.1.5}$$

The surface with constant phase $\theta =$ const. is called the phase surface. The phase surface is normal to the direction of $\boldsymbol{k}$, and its propagation speed in the direction of $\boldsymbol{k}$ is

$$c_p \quad = \quad \frac{\omega}{|\boldsymbol{k}|}, \tag{4.1.6}$$

which is called *phase speed*. Phase speed with its normal direction to the phase surface, $c_p\boldsymbol{k}/|\boldsymbol{k}|$, is called the *phase velocity*. The propagation speed of the phase surface in any direction can also be defined; the phase speeds in the x_i direction are given by

$$c_{pi} \quad = \quad \frac{\omega}{k_i}, \qquad \text{for} \quad i = 1,2,3. \tag{4.1.7}$$

It should be noted that $c_{p1}^2 + c_{p2}^2 + c_{p3}^2 \neq c_p^2$, but $c_{p1}^{-2} + c_{p2}^{-2} + c_{p3}^{-2} = c_p^{-2}$, in general. The left panel of Fig. 4.1 shows the relation between the phase speed and phase lines in two-dimensional space.

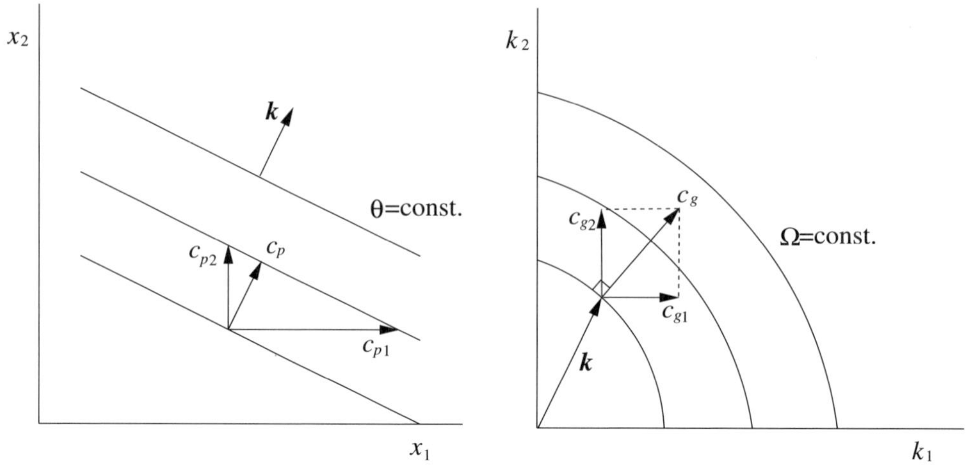

FIGURE 4.1: (Left) Schematic relation between the phase speed c_p and phase lines $\theta = $ const. The inclined solid lines are constant phase lines and $\boldsymbol{k}$ is the wave number vector and is normal to a phase line. c_{p1} and c_{p2} are phase speeds in the directions of x_1 and x_2, respectively, and satisfy $1/c_{p1}^2 + 1/c_{p2}^2 = 1/c_p^2$. (Right) Schematic relation between the group velocity $\boldsymbol{c_g}$ and the phase relation Ω in the wave number space. The solid curves are constant lines of frequency $\Omega = $ const. and are normal to the group velocity $\boldsymbol{c_g}$. $\boldsymbol{k}$ is the wave number vector. c_{g1} and c_{g2} are components of the group velocity in the directions of x_1 and x_2, respectively, and satisfy $c_{g1}^2 + c_{g2}^2 = \boldsymbol{c_g}^2$.

Substituting (4.1.2) into (4.1.1), and neglecting small terms such as derivatives of A, $\boldsymbol{k}$, and ω, we obtain a relation between $\boldsymbol{k}$ and ω in the form

$$\omega = \Omega(\boldsymbol{k}; \boldsymbol{x}, t), \tag{4.1.8}$$

which is called the *dispersion relation.* We can see by substituting (4.1.3) into (4.1.8) that the dispersion relation is a partial differential equation of phase θ. Once the dispersion relation is given, the equations for changes in wave number and frequency can be constructed. Substituting (4.1.8) into (4.1.4), we obtain

$$\frac{\partial \boldsymbol{k}}{\partial t} + \boldsymbol{c}_g \cdot \nabla \boldsymbol{k} = -\nabla\Omega(\boldsymbol{k}; \boldsymbol{x}, t), \tag{4.1.9}$$

where

$$\boldsymbol{c}_g \equiv \frac{\partial}{\partial \boldsymbol{k}}\Omega(\boldsymbol{k}; \boldsymbol{x}, t), \tag{4.1.10}$$

which is called the *group velocity.* In the same way, since

$$\frac{\partial \omega}{\partial t} = \frac{\partial \boldsymbol{k}}{\partial t} \cdot \frac{\partial \Omega}{\partial \boldsymbol{k}} + \frac{\partial \Omega}{\partial t} = -\boldsymbol{c}_g \cdot \nabla\omega + \frac{\partial \Omega}{\partial t}, \tag{4.1.11}$$

we have

$$\frac{\partial \omega}{\partial t} + \boldsymbol{c}_g \cdot \nabla\omega = \frac{\partial}{\partial t}\Omega(\boldsymbol{k}; \boldsymbol{x}, t). \tag{4.1.12}$$

If the signs of the wave number and the frequency are reversed, $\omega \to -\omega$ and $\boldsymbol{k} \to -\boldsymbol{k}$, the dispersion relation remains the same. Then, we can restrict the range of frequency as $\omega \geq 0$ without loss of generality. The right panel of Fig. 4.1 shows the schematic relation between the group velocity and the dispersion relation in the wave number space. The group velocity is normal to the constant line of frequency $\Omega =$ const. in this space.

The dispersion relation is more systematically derived using the WKBJ (Wentzel-Kramer-Brillouin-Jeffreys) method, by which we also obtain the equation for amplitude A. The WKBJ method can be used under the condition that the change in A is slower than that in phase θ. Introducing a small parameter $\varepsilon \ll 1$, we write

$$\boldsymbol{X} = \varepsilon \boldsymbol{x}, \qquad T = \varepsilon t, \tag{4.1.13}$$

and

$$\theta(\boldsymbol{x}, t) = \frac{\Theta(\boldsymbol{X}, T)}{\varepsilon}. \tag{4.1.14}$$

The wave number and frequency, (4.1.3), are written as

$$\boldsymbol{k} = \frac{\partial \Theta}{\partial \boldsymbol{X}}, \qquad \omega = -\frac{\partial \Theta}{\partial T}. \tag{4.1.15}$$

The phase θ is a rapidly changing function of $\boldsymbol{x}$ and t, while the amplitude, the wave number, and the frequency are slowly changing functions of $\boldsymbol{X}$ and T. Thus, Θ defined above becomes a slowly changing function of $\boldsymbol{X}$ and T.

We express the linear partial differential equation (4.1.1) in the form,

$$L\left(\frac{\partial}{\partial t}, \frac{\partial}{\partial x_1}, \frac{\partial}{\partial x_2}, \frac{\partial}{\partial x_3}; \boldsymbol{x}, t\right) \phi(\boldsymbol{x}, t) = 0, \tag{4.1.16}$$

where L is a polynomial of $\frac{\partial}{\partial t}$, $\frac{\partial}{\partial x_1}$, $\frac{\partial}{\partial x_2}$, and $\frac{\partial}{\partial x_3}$, the coefficients of which may depend on $\boldsymbol{x}$ and t. We expand ϕ and L by a series of ε as

$$\phi(\boldsymbol{x}, t) = \sum_{n=0}^{\infty} \varepsilon^n A_n(\boldsymbol{X}, T)\, e^{i\frac{\Theta(\boldsymbol{X}, T)}{\varepsilon}}, \tag{4.1.17}$$

$$L = \sum_{n=0}^{\infty} \varepsilon^n L_n. \tag{4.1.18}$$

The operator L_n is a polynomial whose coefficients are the $O(\varepsilon^n)$ terms of L. Substituting (4.1.17) and (4.1.18) into (4.1.16), and using (4.1.13), the phase Θ and the amplitude A_0 can be expressed as functions of $\boldsymbol{X}$ and T.

The phase Θ is determined by the dispersion relation. Substituting (4.1.15) into the $O(\varepsilon^0)$ terms of (4.1.16) and dividing by A_0, we obtain the dispersion relation:

$$L_0(-i\omega, ik_1, ik_2, ik_3) = 0. \tag{4.1.19}$$

If (4.1.15) is substituted into this equation, this can be viewed as a partial differential equation for Θ.

The equation for the amplitude A_0 can be given by the next order equation of the series, if the operator L is well behaved (i.e., self-adjoint). Writing out the

$O(\varepsilon^1)$ terms of (4.1.16) and using (4.1.19), we obtain after a lengthy calculation:

$$\frac{\partial}{\partial T}\mathcal{A} + \nabla_X \cdot (\boldsymbol{c_g}\mathcal{A}) = 0, \tag{4.1.20}$$

where $\nabla_X = (\frac{\partial}{\partial X_1}, \frac{\partial}{\partial X_2}, \frac{\partial}{\partial X_3})$ and

$$\mathcal{A} = \frac{\partial L_0(-i\omega, ik_1, ik_2, ik_3)}{\partial \omega} A_0 A_0^* \tag{4.1.21}$$

is called the *wave action.* This equation means that the wave action is transported with the group velocity $\boldsymbol{c_g}$. The change in the amplitude A is described by the equation of the wave action.

4.2 Sound waves

In the following sections, we examine wave properties based mainly on the dispersion equations. We consider a frictionless and adiabatic system and use the governing equations for stratified dry air in the rotating frame:

$$\frac{d\rho}{dt} + \rho\nabla \cdot \boldsymbol{v} = 0, \tag{4.2.1}$$

$$\frac{d\boldsymbol{v}}{dt} + 2\boldsymbol{\Omega} \times \boldsymbol{v} = -\frac{1}{\rho}\nabla p - \nabla\Phi, \tag{4.2.2}$$

$$\frac{ds}{dt} = 0. \tag{4.2.3}$$

These are the equation of density ρ, velocity $\boldsymbol{v}$, and entropy s, respectively. We will successively investigate various types of waves by introducing approximations to these equations.

First, in this section, we consider the sound waves in the system without gravity or rotation. By setting $\boldsymbol{\Omega} = 0$ and $\Phi = 0$ in (4.2.1)–(4.2.3), we have

$$\frac{d\rho}{dt} + \rho\nabla \cdot \boldsymbol{v} = 0, \tag{4.2.4}$$

$$\frac{d\boldsymbol{v}}{dt} = -\frac{1}{\rho}\nabla p, \tag{4.2.5}$$

$$\frac{ds}{dt} = 0. \tag{4.2.6}$$

We consider perturbations from a basic state that is at rest with uniform pressure p and entropy s. Density ρ and temperature T are also uniform in the basic state. Let overbar $(\bar{\ })$ denote a quantity of the basic state and prime $(')$ denote that of perturbation, such that

$$p = \overline{p} + p', \quad \rho = \overline{\rho} + \rho', \quad \boldsymbol{v} = \boldsymbol{v}'.$$

Substituting these decompositions into (4.2.4)–(4.2.6), and omitting the nonlinear terms, we have the linearized equations for the perturbation field:

$$\frac{\partial \rho'}{\partial t} + \overline{\rho} \nabla \cdot \boldsymbol{v}' = 0, \tag{4.2.7}$$

$$\frac{\partial \boldsymbol{v}'}{\partial t} = -\frac{1}{\overline{\rho}} \nabla p', \tag{4.2.8}$$

$$\frac{\partial s'}{\partial t} = 0. \tag{4.2.9}$$

Using (1.1.58), (4.2.9) becomes

$$\frac{\partial p'}{\partial t} = c_s^2 \frac{\partial \rho'}{\partial t}, \tag{4.2.10}$$

where c_s is the speed of sound defined by (1.1.56); for dry air it is expressed as

$$c_s^2 \equiv \left(\frac{\partial p}{\partial \rho} \right)_s = \gamma \frac{\overline{p}}{\overline{\rho}} = \gamma R_d \overline{T}. \tag{4.2.11}$$

Eqs. (4.2.7), (4.2.8), and (4.2.10) provide five prognostic equations for five variables (ρ', p', u, v, w). If five independent initial values are given, therefore, a solution can be completely determined. This implies that the dispersion relation has five solutions for frequency. To obtain the dispersion relation, let us decompose the variables into Fourier components. Substituting

$$(\rho', p', \boldsymbol{v}') = (\hat{\rho}, \hat{p}, \hat{\boldsymbol{v}})\, e^{i(\boldsymbol{k}\cdot\boldsymbol{x} - \omega t)} \tag{4.2.12}$$

into the above equations where $\boldsymbol{k} = (k, l, m)$, we obtain the coefficient matrix for $(\hat{\rho}, \hat{p}, \overline{\rho}\hat{\boldsymbol{v}})$ as

$$\begin{pmatrix} -i\omega & 0 & ik & il & im \\ ic_s^2\omega & -i\omega & 0 & 0 & 0 \\ 0 & ik & -i\omega & 0 & 0 \\ 0 & il & 0 & -i\omega & 0 \\ 0 & im & 0 & 0 & -i\omega \end{pmatrix} \begin{pmatrix} \hat{\rho} \\ \hat{p} \\ \overline{\rho}\hat{u} \\ \overline{\rho}\hat{v} \\ \overline{\rho}\hat{w} \end{pmatrix} = \begin{pmatrix} 0 \\ 0 \\ 0 \\ 0 \\ 0 \end{pmatrix}. \tag{4.2.13}$$

We obtain the dispersion relation by setting the determinant of this matrix to zero:

$$\omega^3(\omega^2 - c_s^2 \boldsymbol{k}^2) = 0. \tag{4.2.14}$$

This equation has two kinds of solutions:

$$\omega^2 - c_s^2 \boldsymbol{k}^2 = 0, \tag{4.2.15}$$

$$\omega^3 = 0. \tag{4.2.16}$$

The former corresponds to the sound wave mode, or sound waves, whereas the latter corresponds to the *vortical mode*. Eqs. (4.2.7), (4.2.10), and (4.2.8) can be combined to a single equation for pressure:

$$\frac{\partial^2 p'}{\partial t^2} - c_s^2 \nabla^2 p' = 0. \tag{4.2.17}$$

The dispersion relation of sound wave mode (4.2.15) is readily seen from this equation. Eq. (4.2.17) means that a disturbance propagates with the speed of sound c_s. On the other hand, rotation of (4.2.8) yields

$$\frac{\partial}{\partial t}(\nabla \times \boldsymbol{v}') \quad = \quad 0. \tag{4.2.18}$$

This gives either $\nabla \times \boldsymbol{v}' = 0$ or $\omega = 0$. If $\nabla \times \boldsymbol{v}' = 0$, we obtain $\omega \neq 0$, which corresponds to sound wave mode. In the case $\omega = 0$, on the other hand, vorticity is constant: $\nabla \times \boldsymbol{v}' = \text{const}$. This is why the solution for $\omega = 0$ is called the vortical mode.

For the positive frequency $\omega \geq 0$, the dispersion relation (4.2.15) reduces to

$$\omega \quad = \quad c_s|\boldsymbol{k}|. \tag{4.2.19}$$

The phase speed in the direction of $\boldsymbol{k}$ becomes

$$c \quad = \quad \frac{\omega}{|\boldsymbol{k}|} \quad = \quad c_s, \tag{4.2.20}$$

and the group velocity is

$$\boldsymbol{c}_g \quad = \quad \frac{\partial \omega}{\partial \boldsymbol{k}} \quad = \quad c_s \frac{\boldsymbol{k}}{|\boldsymbol{k}|}. \tag{4.2.21}$$

The magnitude of group velocity is the same as the phase speed c_s and is independent of wave number. This means that the sound waves are not dispersive.

The structure of sound waves is described by the phase relations. The phase relations are determined by substituting (4.2.12) into (4.2.10) and (4.2.8),

$$\hat{p} \quad = \quad c_s^2 \hat{\rho}, \tag{4.2.22}$$

$$\hat{\boldsymbol{v}} \quad = \quad \frac{\boldsymbol{k}}{\omega \overline{\rho}} \hat{p} \quad = \quad \frac{\boldsymbol{k} c_s^2}{\omega \overline{\rho}} \hat{\rho}. \tag{4.2.23}$$

The direction of fluid velocity is parallel to that of group velocity. Thus, we can see that sound waves have no vorticity. Fig. 4.2 shows the structure of a sound wave propagating toward the right-hand side.

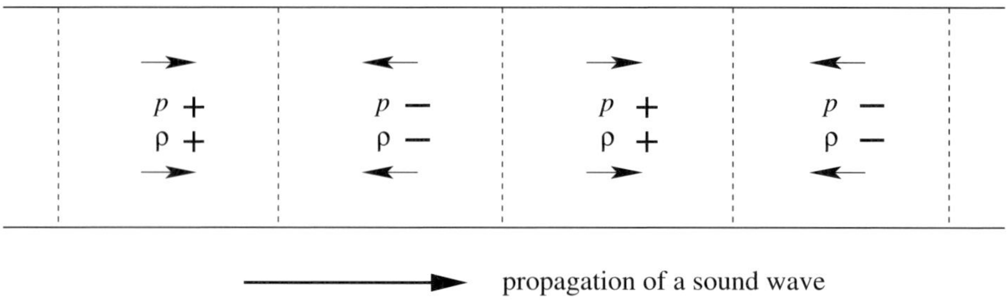

FIGURE 4.2: Structure of a sound wave in a one-dimensional tube. Relations between pressure p, density ρ, and velocity (arrows) for the sound wave propagating toward the right-hand side. Nodal points are indicated by dashed lines.

4.3 Gravity waves

4.3.1 Dispersion relation

We next examine waves in a fluid with gravity and no rotation. In this case, the governing equations (4.2.1)–(4.2.3) become

$$\frac{d\rho}{dt} + \rho\nabla\cdot\boldsymbol{v} = 0, \tag{4.3.1}$$

$$\frac{d\boldsymbol{v}}{dt} = -\frac{1}{\rho}\nabla p - g\hat{\boldsymbol{z}}, \tag{4.3.2}$$

$$\frac{ds}{dt} = 0, \tag{4.3.3}$$

where $\hat{\boldsymbol{z}}$ is the unit vector in the vertical direction and g is the acceleration due to gravity. We consider a basic state that has no motion and satisfies the hydrostatic balance:

$$0 = -\frac{1}{\overline{\rho}}\frac{\partial\overline{p}}{\partial z} - g. \tag{4.3.4}$$

The linearized equations for the perturbations from the basic state are given by

$$\frac{\partial\rho'}{\partial t} + w'\frac{\partial\overline{\rho}}{\partial z} + \overline{\rho}\nabla\cdot\boldsymbol{v}' = 0, \tag{4.3.5}$$

$$\frac{\partial\boldsymbol{v}'}{\partial t} = -\frac{1}{\overline{\rho}}\nabla p' - \frac{\rho'}{\overline{\rho}}g\hat{\boldsymbol{z}}, \tag{4.3.6}$$

$$\frac{\partial s'}{\partial t} + w'\frac{\partial\overline{s}}{\partial z} = 0. \tag{4.3.7}$$

The equation of entropy (4.3.7) is rewritten by using (1.1.58) as

$$\frac{1}{c_s^2}\frac{\partial p'}{\partial t} - \frac{\partial\rho'}{\partial t} + \frac{N^2}{g}\overline{\rho}w' = 0, \tag{4.3.8}$$

where N is the Brunt-Väisälä frequency defined by (2.3.11):

$$N^2 = -\frac{g}{\overline{\rho}}\left(\frac{\partial\rho}{\partial s}\right)_p\frac{\partial\overline{s}}{\partial z} = -\frac{g}{\overline{\rho}}\left(\frac{\partial\overline{\rho}}{\partial z} - \frac{1}{c_s^2}\frac{\partial\overline{p}}{\partial z}\right). \tag{4.3.9}$$

Here, to simplify the coefficients of the perturbation equations, we transform the variables of the perturbation field as

$$\tilde{\rho} = \frac{1}{\sqrt{\overline{\rho}}}\rho', \qquad \tilde{p} = \frac{1}{\sqrt{\overline{\rho}}}p', \qquad \tilde{\boldsymbol{v}} = \sqrt{\overline{\rho}}\boldsymbol{v}'. \tag{4.3.10}$$

Substituting these relations into (4.3.5), (4.3.6), and (4.3.8), we have

$$\frac{\partial\tilde{\rho}}{\partial t} + \nabla\cdot\tilde{\boldsymbol{v}} - \frac{1}{2H_\rho}\tilde{w} = 0, \tag{4.3.11}$$

$$\frac{\partial\tilde{\boldsymbol{v}}}{\partial t} = -\nabla\tilde{p} + \frac{1}{2H_\rho}\tilde{p}\hat{\boldsymbol{z}} - \tilde{\rho}g\hat{\boldsymbol{z}}, \tag{4.3.12}$$

$$\frac{\partial\tilde{p}}{\partial t} - c_s^2\frac{\partial\tilde{\rho}}{\partial t} + \frac{c_s^2N^2}{g}\tilde{w} = 0, \tag{4.3.13}$$

where H_ρ is the scale height of density, defined by (2.3.15). From (4.3.4) and (4.3.9), H_ρ is related to the Brunt-Väisälä frequency as shown by (2.3.14).

If we assume that all the coefficients of (4.3.11)–(4.3.13) are constant, we obtain the dispersion relation. In the case of the ideal gas, this assumption is equivalent to the isothermal basic state. Letting the temperature of the basic state be $T = T_0$, we have

$$c_s^2 = \gamma R_d T_0 = \gamma g H, \tag{4.3.14}$$

$$H_\rho = \frac{R_d T_0}{g} = H, \tag{4.3.15}$$

$$N^2 = \frac{g}{H_\rho} - \frac{g^2}{c_s^2} = \frac{\kappa g}{H}, \tag{4.3.16}$$

where $\gamma = c_p/c_v$, $\kappa = R_d/c_p$, and

$$H \equiv \left(-\frac{1}{\overline{p}} \frac{\partial \overline{p}}{\partial z} \right)^{-1} = \frac{\overline{p}}{\overline{\rho} g} = \frac{R_d T_0}{g} \tag{4.3.17}$$

is the scale height of pressure.

Substituting the Fourier components

$$(\tilde{\rho}, \tilde{p}, \tilde{\boldsymbol{v}}) = (\hat{\rho}, \hat{p}, \hat{\boldsymbol{v}})\, e^{i(\boldsymbol{k}\cdot\boldsymbol{x} - \omega t)} \tag{4.3.18}$$

into (4.3.11)–(4.3.13), we obtain the coefficient matrix for $(\hat{\rho}, \hat{p}, \hat{\boldsymbol{v}})$:

$$\begin{pmatrix} -i\omega & 0 & ik & il & im - \frac{1}{2H_\rho} \\ ic_s^2\omega & -i\omega & 0 & 0 & \frac{c_s^2 N^2}{g} \\ 0 & ik & -i\omega & 0 & 0 \\ 0 & il & 0 & -i\omega & 0 \\ g & im - \frac{1}{2H_\rho} & 0 & 0 & -i\omega \end{pmatrix} \begin{pmatrix} \hat{\rho} \\ \hat{p} \\ \hat{u} \\ \hat{v} \\ \hat{w} \end{pmatrix} = \begin{pmatrix} 0 \\ 0 \\ 0 \\ 0 \\ 0 \end{pmatrix}. \tag{4.3.19}$$

By setting the determinant of this matrix to zero and using (4.3.16), we obtain

$$\omega \left[\omega^4 - c_s^2 \left(\boldsymbol{k}^2 + \frac{1}{4H_\rho^2} \right) \omega^2 + c_s^2 N^2 k_H^2 \right] = 0, \tag{4.3.20}$$

where $k_H = (k^2 + l^2)^{1/2}$. This relation has five solutions for ω: $\omega = 0$ and four square roots of

$$\omega^2 = \frac{c_s^2}{2} \left(\boldsymbol{k}^2 + \frac{1}{4H_\rho^2} \right) \pm \sqrt{\frac{c_s^2}{4} \left(\boldsymbol{k}^2 + \frac{1}{4H_\rho^2} \right)^2 - 4N^2 k_H^2} \tag{4.3.21}$$

$$= \frac{c_s^2}{2} \left(\boldsymbol{k}^2 + \frac{1}{4H_\rho^2} \right) (1 \pm A^{\frac{1}{2}}), \tag{4.3.22}$$

where

$$A \equiv 1 - \frac{4N^2 k_H^2}{c_s^2 \left(\boldsymbol{k}^2 + \frac{1}{4H_\rho^2} \right)^2}. \tag{4.3.23}$$

This equation expresses the dispersion relations of two types of waves. In the case of no gravity, $g \to 0$ (i.e., $N^2 \to 0$ and $H_\rho \to \infty$), one of the dispersion relations approaches $\omega^2 = c_s^2 \boldsymbol{k}^2$, which agrees with the dispersion relation of sound waves, or sound wave mode, (4.2.15). The other approaches $\omega^2 = 0$ as $g \to 0$. It is called gravity wave mode and expresses the dispersion relation of *gravity waves.* The remaining solution $\omega = 0$ is vortical mode. In general, the horizontal vorticity of vortical mode stays constant irrespective of time.

The vertical wave number m can either be a real or imaginary number. The vertical structure of waves depends on the sign of m^2. The solutions with $m^2 > 0$ are called *internal waves*, whereas those with $m^2 < 0$ are called *external waves.* The vertical structure of internal waves is sinusoidal, whereas that of external waves is exponential, and the amplitude increases or decreases with height. This means that appropriate upper or lower boundary conditions are required for the external waves to be the solutions. Specifically, the gravity waves for $m^2 > 0$ and $m^2 < 0$ are called internal gravity waves and external gravity waves, respectively. Internal sound waves and external sound waves are also defined for sound wave mode with $m^2 > 0$ and $m^2 < 0$, respectively.

If $\boldsymbol{k}^2 \gg 4N^2/c_s^2$, A given by (4.3.23) approaches one. In this limit, the square of the frequency of internal sound waves ω_s approaches

$$\omega_s^2 = c_s^2 \left(\boldsymbol{k}^2 + \frac{1}{4H_\rho^2} \right), \tag{4.3.24}$$

while the square of the frequency of internal gravity waves ω_g approaches

$$\omega_g^2 = \frac{N^2 k_H^2}{\boldsymbol{k}^2 + \frac{1}{4H_\rho^2}}. \tag{4.3.25}$$

It can be shown from (4.3.22) that the frequency of internal sound waves is always higher than that of internal gravity waves:

$$\omega_s^2 \geq \frac{c_s^2}{4H_\rho^2} > N^2 \geq \omega_g^2, \tag{4.3.26}$$

since we have $c_s^2/4H_\rho^2 N^2 = \gamma/4\kappa > 1$ using (4.3.14)–(4.3.16).

The inside of the square in (4.3.21) is rewritten as

$$D = \frac{c_s^4}{4} \left[\left(k_H^2 + m^2 + \nu^2 - \frac{N^2}{c_s^2} \right)^2 + 4N^2(m^2 + \nu^2) \right], \tag{4.3.27}$$

where

$$\nu \equiv -\frac{1}{2H_\rho} + \frac{g}{c_s^2} = \frac{1}{2}\left(\frac{g}{c_s^2} - \frac{N^2}{g} \right) = \frac{2-\gamma}{2\gamma} \frac{g}{R_d T_0}, \tag{4.3.28}$$

which is positive in general. In the special case with

$$m^2 = -\nu^2 = -\frac{1}{4H_\rho^2} + \frac{N^2}{c_s^2}, \tag{4.3.29}$$

the frequencies are given by

$$\omega^2 = c_s k_H^2, \tag{4.3.30}$$

$$\omega^2 = N^2. \tag{4.3.31}$$

The former is equivalent to the dispersion relation of sound wave mode with no gravity. The waves corresponding to this mode are called *Lamb waves.* Since $m^2 = -\nu^2 < 0$, Lamb waves are the external waves. The second solution (4.3.31) expresses oscillation by buoyancy. The frequency of Lamb waves (4.3.30) is equal to buoyancy frequency (4.3.31) at $k_H^2 = N^2/c_s^2$. In this case, since $D = 0$, sound wave mode agrees with gravity wave mode. Lamb waves have the characteristics of gravity waves if $k_H^2 < N^2/c_s^2$, while they have the characteristics of sound waves if $k_H^2 > N^2/c_s^2$.

Let us nondimensionalize the dispersion relation (4.3.20) by setting $k_H^* = c_s k_H/N$, $m^* = c_s m/N$, $\omega^* = \omega/N$, and $\delta = c_s/2H_\rho N$. The nondimensional dispersion relation is written as

$$\omega^{*4} - (k_H^{*2} + m^{*2} + \delta^2)\omega^{*2} + k_H^{*2} = 0. \tag{4.3.32}$$

Fig. 4.3 shows the dispersion relation for $\delta = \sqrt{\gamma/4\kappa} = 1.107$. (a) is the relation between k_H^* and ω^* for various values of m^* and (b) is the relation between m^* and ω^* for various values of k_H^*. In (a), the sloping dashed-dotted line represents Lamb waves with $m^{*2} = 1 - \delta^2$. The frequency of gravity wave mode is always $\omega^* < 1$ and $\omega^* \to 1$ as $k_H^* \to \infty$. (c) and (d) show the relations between k_H^* and m^* for sound wave mode and the gravity wave mode, respectively.

4.3.2 Gravity waves in the hydrostatic Boussinesq approximation

To investigate the characteristics of gravity waves, we consider the systems that do not contain sound waves. The following two kinds of approximations are introduced. First is the system in the hydrostatic balance in the vertical direction. In this case, the vertical component of (4.3.6) is replaced by

$$0 = -\frac{\partial p'}{\partial z} - \rho' g. \tag{4.3.33}$$

The dispersion relation is given from the horizontal components of (4.3.6), (4.3.5), and (4.3.8) in the same way as in the previous subsection. Since the time derivative does not exist in (4.3.33), the dispersion relation becomes a cubic polynomial of ω. It does not contain sound wave mode except for the Lamb waves as a compressible mode. The solutions for the frequency of the dispersion relation have gravity wave mode and vortical mode with $\omega = 0$. The dispersion relation of gravity waves is given by

$$\omega^2 = \frac{N^2 k_H^2}{m^2 + \frac{1}{4H_\rho^2}}. \tag{4.3.34}$$

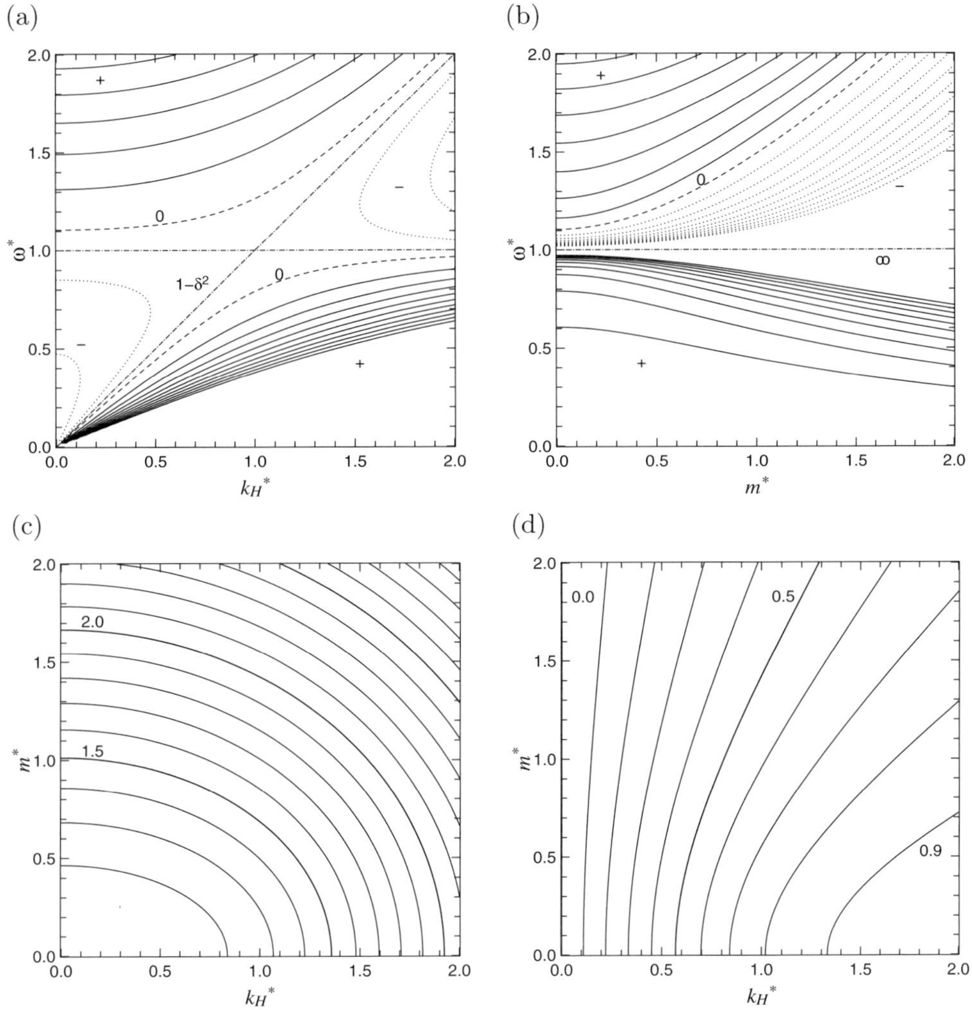

FIGURE 4.3: The dispersion relations of sound waves and gravity waves for the isothermal atmosphere for $\delta = \sqrt{\gamma/4\kappa} = 1.107$. (a) shows contours of $m^{*2} = n/2$ (n is an integer) as a function of k_H^* and ω^*. The dashed-dotted line represents Lamb waves. (b) shows the contours of $k_H^{*2} = n/2$ as a function of m^* and ω^*. (c) represents the contours of $\omega^* = n/10$ as a function of k_H^* and m^* for sound waves. (d) is the same as (c) but for gravity waves. In (a) and (b), positive values are shown by solid curves, negative values by dotted curves, and zero by dashed curves. In (c) and (d), only positive values of ω^* are shown.

Second, we consider the Boussinesq system which is anelastic and does not contain sound waves. With the Boussinesq approximation, change in density is neglected except for the buoyancy term. The equations for perturbation fields are written as

$$\nabla \cdot \boldsymbol{v}' = 0, \tag{4.3.35}$$

$$\frac{\partial \boldsymbol{v}'}{\partial t} = -\frac{1}{\overline{\rho}}\nabla p' - \frac{\rho'}{\overline{\rho}} g\hat{\boldsymbol{z}}, \tag{4.3.36}$$

$$\frac{\partial \rho'}{\partial t} - \frac{N^2}{g}\overline{\rho} w' = 0, \tag{4.3.37}$$

where $\overline{\rho}$ is constant. Eq. (4.3.37) is derived from the equation of temperature (3.1.17) and the relation between density and temperature (3.1.14) and $N^2 = g\alpha\frac{\partial T'}{\partial z}$. The dispersion relation is again a cubic polynomial of ω. In particular, the dispersion relation of gravity waves is given by

$$\omega^2 = \frac{N^2 k_H^2}{\boldsymbol{k}^2}. \tag{4.3.38}$$

This is formally derived from (4.3.20) with the limit $c_s^2 \to \infty$ and $H_\rho \to \infty$.

4.3.3 Group velocity and phase speed

The group velocity of gravity waves is given by differentiating the dispersion relation (4.3.20) with respect to k, l, and m. Using (4.3.22), we obtain the group velocity as

$$c_{gx} = \frac{\partial \omega}{\partial k} = \mp \frac{k}{\omega}\frac{N^2 - \omega^2}{\boldsymbol{k}^2 + \frac{1}{4H_\rho^2}} A^{-\frac{1}{2}} = \mp c_{px}\frac{(N^2-\omega^2)k^2}{\omega^2\left(\boldsymbol{k}^2 + \frac{1}{4H_\rho^2}\right)} A^{-\frac{1}{2}}, \tag{4.3.39}$$

$$c_{gz} = \frac{\partial \omega}{\partial m} = \pm \frac{\omega m}{\boldsymbol{k}^2 + \frac{1}{4H_\rho^2}} A^{-\frac{1}{2}} = \pm c_{pz}\frac{m^2}{\boldsymbol{k}^2 + \frac{1}{4H_\rho^2}} A^{-\frac{1}{2}}. \tag{4.3.40}$$

c_{gy} is also given by replacing k with l in c_{gx}. The phase speeds in the x- and z-directions are respectively given by

$$c_{px} = \frac{\omega}{k} = \frac{c_s}{k}\sqrt{\left(\boldsymbol{k}^2 + \frac{1}{4H_\rho^2}\right)\frac{1 \pm A^{\frac{1}{2}}}{2}}, \tag{4.3.41}$$

$$c_{pz} = \frac{\omega}{m} = \frac{c_s}{m}\sqrt{\left(\boldsymbol{k}^2 + \frac{1}{4H_\rho^2}\right)\frac{1 \pm A^{\frac{1}{2}}}{2}}. \tag{4.3.42}$$

For sound wave mode, we generally have $c_{gx}c_{px} > 0$ and $c_{gz}c_{pz} > 0$, since $\omega^2 > c_s^2/4H_\rho^2 > N^2$. For gravity wave mode, on the other hand, we have $c_{gx}c_{px} > 0$ and $c_{gz}c_{pz} < 0$, since $\omega^2 < N^2$. This means that the vertical direction of the group velocity is opposite to that of phase velocity.

Using the Boussinesq approximation, the dispersion relation of gravity waves becomes $\omega = Nk_H/|\boldsymbol{k}|$ from (4.3.38). The group velocity is given by

$$c_{gx} = \frac{k}{\omega}\frac{N^2-\omega^2}{\boldsymbol{k}^2} = \frac{Nkm^2}{k_H|\boldsymbol{k}|^3}, \quad c_{gy} = \frac{l}{\omega}\frac{N^2-\omega^2}{\boldsymbol{k}^2} = \frac{Nlm^2}{k_H|\boldsymbol{k}|^3},$$
$$c_{gz} = -\frac{m\omega}{\boldsymbol{k}^2} = -\frac{Nk_H m}{|\boldsymbol{k}|^3}. \tag{4.3.43}$$

These correspond to the limit $c_s^2 \to \infty$ and $H_\rho \to \infty$ in (4.3.39) and (4.3.40). In this case, we have

$$\boldsymbol{c_g} \cdot \boldsymbol{k} \quad = \quad c_{gx}k + c_{gy}l + c_{gz}m \quad = \quad 0 \tag{4.3.44}$$

(i.e., the direction of the group velocity of gravity waves is perpendicular to the direction of phase velocity). Phase speeds are given by

$$c_{px} = \frac{\omega}{k} = \frac{Nk_H}{k|\boldsymbol{k}|}, \qquad c_{py} = \frac{\omega}{l} = \frac{Nk_H}{l|\boldsymbol{k}|}, \qquad c_{pz} = \frac{\omega}{m} = \frac{Nk_H}{m|\boldsymbol{k}|}. \tag{4.3.45}$$

In the case when phase velocity is parallel to the xz plane and vertical wave number is large (i.e., $m^2 \gg k_H^2$ and $l = 0$), group velocity and phase speeds are simply given by

$$c_{gx} = c_{px} = \frac{N}{m}, \qquad c_{gz} = -c_{pz} = -\frac{Nk}{m^2}. \tag{4.3.46}$$

These expressions are useful for the schematic illustration of the propagation of gravity waves (see Fig. 4.4).

4.3.4 The structure of gravity waves

To obtain the structure of gravity waves, we consider the phase relations of the variables by expressing $\hat{u}$, $\hat{v}$, $\hat{w}$, and $\hat{\rho}$ in terms of $\hat{p}$. From the horizontal components of the equation of motion (4.3.12), we have

$$\frac{\partial \tilde{u}}{\partial t} = -\frac{\partial \tilde{p}}{\partial x}, \qquad \frac{\partial \tilde{v}}{\partial t} = -\frac{\partial \tilde{p}}{\partial y}. \tag{4.3.47}$$

Differentiating (4.3.11) and (4.3.13) with respect to time, and using (4.3.11)–(4.3.13), we have the following two relations between $\tilde{w}$ and $\tilde{p}$:

$$\frac{\partial}{\partial t}\left(\frac{\partial}{\partial z} - \nu\right)\tilde{w} = \left(\frac{\partial^2}{\partial x^2} + \frac{\partial^2}{\partial y^2} - \frac{1}{c_s^2}\frac{\partial^2}{\partial t^2}\right)\tilde{p}, \tag{4.3.48}$$

$$\left(\frac{\partial^2}{\partial t^2} + N^2\right)\tilde{w} = -\frac{\partial}{\partial t}\left(\frac{\partial}{\partial z} + \nu\right)\tilde{p}. \tag{4.3.49}$$

Eliminating $\tilde{w}$ from (4.3.48) and (4.3.49), we obtain the equation for $\tilde{p}$ as

$$\left[\left(\frac{\partial^2}{\partial t^2} + N^2\right)\left(\frac{\partial^2}{\partial x^2} + \frac{\partial^2}{\partial y^2} - \frac{1}{c_s^2}\frac{\partial^2}{\partial t^2}\right) + \frac{\partial^2}{\partial t^2}\left(\frac{\partial^2}{\partial z^2} - \nu^2\right)\right]\tilde{p} = 0. \tag{4.3.50}$$

We can see that the dispersion relation (4.3.20) is given by substituting the sinusoidal form (4.3.18) into this equation. Next, we have a relation between $\tilde{\rho}$ and $\tilde{p}$ using (4.3.12) and (4.3.11) and eliminating $\tilde{u}$, $\tilde{v}$, and $\tilde{w}$:

$$\left[\frac{\partial^2}{\partial t^2} + \left(\frac{\partial}{\partial z} - \frac{1}{2H_\rho}\right)g\right]\tilde{\rho} = \left[\frac{\partial^2}{\partial x^2} + \frac{\partial^2}{\partial y^2} + \left(\frac{\partial}{\partial z} - \frac{1}{2H_\rho}\right)^2\right]\tilde{p}. \tag{4.3.51}$$

We also have another relation between $\tilde{\rho}$ and $\tilde{p}$ using (4.3.13) and (4.3.12) and eliminating $\tilde{w}$:

$$\left(\frac{\partial^2}{\partial t^2}+N^2\right)\tilde{\rho} = \left[\frac{1}{c_s^2}\frac{\partial^2}{\partial t^2}-\frac{N^2}{g}\left(\frac{\partial}{\partial z}-\frac{1}{2H_\rho}\right)\right]\tilde{p}. \qquad (4.3.52)$$

Substituting (4.3.18) into (4.3.47), (4.3.48), (4.3.49), (4.3.51), and (4.3.52), we have phase relations:

$$\omega\hat{u} = k\hat{p}, \qquad \omega\hat{v} = l\hat{p}, \qquad (4.3.53)$$

$$\omega(m+i\nu)\hat{w} = \frac{1}{c_s^2}(\omega^2-c_s^2k_H^2)\hat{p}, \qquad (4.3.54)$$

$$(\omega^2-N^2)\hat{w} = \omega(m-i\nu)\hat{p}, \qquad (4.3.55)$$

$$\left[\omega^2-i\left(m+\frac{i}{2H_\rho}\right)g\right]\hat{\rho} = \left[k_H^2+\left(m+\frac{i}{2H_\rho}\right)^2\right]\hat{p}, \qquad (4.3.56)$$

$$(\omega^2-N^2)\hat{\rho} = \frac{1}{c_s^2}\left[\omega^2+i\left(m+\frac{i}{2H_\rho}\right)\frac{c_s^2N^2}{g}\right]\hat{p}. \qquad (4.3.57)$$

If $\omega \neq 0$ and $\omega^2 \neq N^2$, $\hat{u}$, $\hat{v}$, $\hat{w}$, and $\hat{\rho}$ are expressed in terms of $\hat{p}$ as

$$\hat{u} = \frac{k}{\omega}\hat{p}, \qquad \hat{v} = \frac{l}{\omega}\hat{p}, \qquad (4.3.58)$$

$$\hat{w} = \frac{1}{c_s^2}\frac{\omega^2-c_s^2k_H^2}{\omega(m+i\nu)}\hat{p} = \frac{\omega(m-i\nu)}{\omega^2-N^2}\hat{p}, \qquad (4.3.59)$$

$$\hat{\rho} = \frac{k_H^2+\left(m+\frac{i}{2H_\rho}\right)^2}{\omega^2-i\left(m+\frac{i}{2H_\rho}\right)g}\hat{p} = \frac{1}{c_s^2}\frac{\omega^2+i\left(m+\frac{i}{2H_\rho}\right)\frac{c_s^2N^2}{g}}{\omega^2-N^2}\hat{p}. \qquad (4.3.60)$$

The spatial structure of the perturbations (ρ', p', u', v', w') can be constructed from (4.3.10) with (4.3.18). For $\omega = 0$, we have $\hat{p} = \hat{w} = 0$. This is a vortical mode with horizontal motions. For $\omega^2 = N^2$, we have $\hat{p} = \hat{u} = \hat{v} = 0$ and $m = -i\nu$: the wave is external. For Lamb waves, in particular, we can see that the vertical motion is always zero $\hat{w} = 0$ by substituting $\omega^2 = c_s k_H^2$ (4.3.30) into (4.3.59).

The phase relations for the waves in the Boussinesq system are given by $c_s^2 \to \infty$ and $H_\rho \to \infty$ in (4.3.53)–(4.3.57) and neglecting ω^2 on the left-hand side of (4.3.56):

$$\hat{u} = \frac{k}{\omega}\hat{p}, \qquad \hat{v} = \frac{l}{\omega}\hat{p}, \qquad (4.3.61)$$

$$\hat{w} = -\frac{k_H^2}{\omega m}\hat{p} = \frac{\omega m}{\omega^2-N^2}\hat{p}, \qquad (4.3.62)$$

$$\hat{\rho} = i\frac{\boldsymbol{k}^2}{m}\frac{\hat{p}}{g} = i\frac{mN^2}{\omega^2-N^2}\frac{\hat{p}}{g}. \qquad (4.3.63)$$

Fig. 4.4 shows the phase relations of a gravity wave in the xz section for $l = 0$, $k > 0$, and $m < 0$. Note $\omega < N$ in this case. Fig. 4.5 also shows the relation between the direction of group velocity and the phase line for four gravity waves generated from a point source. The phase lines are parallel to the directions of the group velocities and their vertical propagations are opposite to those of the group velocities.

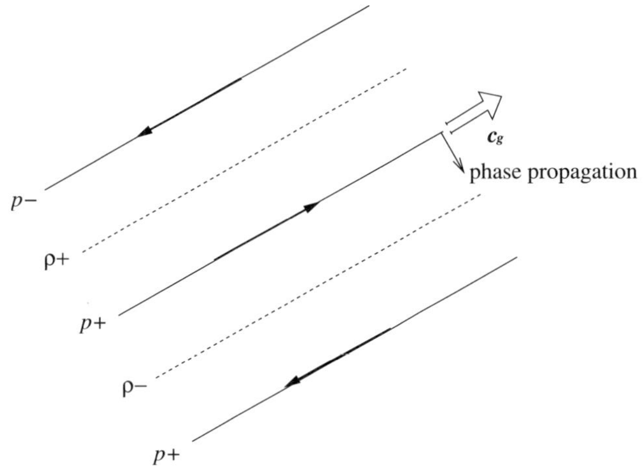

FIGURE 4.4: Structure of a gravity wave for $l = 0$, $k > 0$, and $m < 0$ in the xz section. The directions of group velocity $\boldsymbol{c}_g$ and phase propagation are indicated by arrows in the upper-right of the figure. Bold arrows indicate the directions of fluid motions.

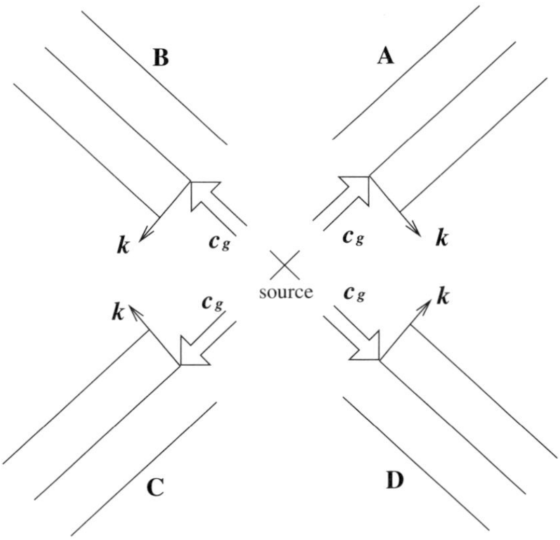

FIGURE 4.5: The relation between the wave number vectors of the gravity waves and the group velocities in the height-horizontal two-dimensional section. The symbols A, B, C, and D denote four propagating waves generated from the wave source designated by the cross. Arrows with $\boldsymbol{k}$ indicate wave number vectors and open arrows with $\boldsymbol{c}_g$ are group velocities.

4.4 Inertial waves

Waves also occur in the fluids in the rotating frame without gravity, in which the Coriolis force works as a restoring force. The governing equations are given from (4.2.1)–(4.2.3) by setting $\Phi = 0$ as

$$\frac{d\rho}{dt} + \rho\nabla\cdot\boldsymbol{v} = 0, \tag{4.4.1}$$

$$\frac{d\boldsymbol{v}}{dt} + 2\boldsymbol{\Omega}\times\boldsymbol{v} = -\frac{1}{\rho}\nabla p, \tag{4.4.2}$$

$$\frac{ds}{dt} = 0. \tag{4.4.3}$$

We can set the angular velocity $\boldsymbol{\Omega} = (0, 0, f/2)$ without loss of generality. We assume that the basic state is at rest relative to the rotating frame and has uniform density and pressure: $\overline{\rho} =$ const. and $\overline{p} =$ const. The linearized equations for the perturbation field are written as

$$\frac{\partial\rho'}{\partial t} + \overline{\rho}\nabla\cdot\boldsymbol{v}' = 0, \tag{4.4.4}$$

$$\frac{\partial u'}{\partial t} - fv' = -\frac{1}{\overline{\rho}}\frac{\partial p'}{\partial x}, \tag{4.4.5}$$

$$\frac{\partial v'}{\partial t} + fu' = -\frac{1}{\overline{\rho}}\frac{\partial p'}{\partial y}, \tag{4.4.6}$$

$$\frac{\partial w'}{\partial t} = -\frac{1}{\overline{\rho}}\frac{\partial p'}{\partial z}, \tag{4.4.7}$$

$$\frac{1}{c_s^2}\frac{\partial p'}{\partial t} - \frac{\partial\rho'}{\partial t} = 0. \tag{4.4.8}$$

Assuming that the variables (ρ', p', u', v', w') have the sinusoidal form proportional to $e^{i(kx+ly+mz-\omega t)}$ as (4.2.12), we have the coefficient matrix as

$$\begin{pmatrix} -i\omega & 0 & ik & il & im \\ ic_s^2\omega & -i\omega & 0 & 0 & 0 \\ 0 & ik & -i\omega & -f & 0 \\ 0 & il & f & -i\omega & 0 \\ 0 & im & 0 & 0 & -i\omega \end{pmatrix} \begin{pmatrix} \hat{\rho} \\ \hat{p} \\ \overline{\rho}\hat{u} \\ \overline{\rho}\hat{v} \\ \overline{\rho}\hat{w} \end{pmatrix} = \begin{pmatrix} 0 \\ 0 \\ 0 \\ 0 \\ 0 \end{pmatrix}. \tag{4.4.9}$$

The dispersion relation is given by setting the determinant of this matrix to zero:

$$\omega\left[\omega^4 - \left(c_s^2\boldsymbol{k}^2 + f^2\right)\omega^2 + c_s^2 f^2 m^2\right] = 0. \tag{4.4.10}$$

This is analogous to the dispersion relation in a stratified fluid without rotation (4.3.20). Eq. (4.4.10) has two categories of solutions for ω^2 other than vertical mode $\omega = 0$. The solution for the larger ω^2 corresponds to sound waves affected by the rotation. The solution for the smaller ω^2 is a new category of waves where the Coriolis force works as a restoring force. They are called *inertial waves*.

The similar dispersion relation can be derived in the case of incompressible fluids. The linearized equations of the incompressible fluids in the rotating frame are written as

$$\nabla \cdot \boldsymbol{v}' = 0, \tag{4.4.11}$$

$$\frac{\partial u'}{\partial t} - fv' = -\frac{1}{\bar{\rho}}\frac{\partial p'}{\partial x}, \tag{4.4.12}$$

$$\frac{\partial v'}{\partial t} + fu' = -\frac{1}{\bar{\rho}}\frac{\partial p'}{\partial y}, \tag{4.4.13}$$

$$\frac{\partial w'}{\partial t} = -\frac{1}{\bar{\rho}}\frac{\partial p'}{\partial z}, \tag{4.4.14}$$

$$\frac{\partial \rho'}{\partial t} = 0. \tag{4.4.15}$$

Substituting the sinusoidal form into these equations, we obtain the dispersion relation by $\omega = 0$ and

$$\omega^2 = \frac{f^2 m^2}{\boldsymbol{k}^2}. \tag{4.4.16}$$

This corresponds to the limit $c_s^2 \to \infty$ in (4.4.10).

We can rewrite the dispersion relation of inertial waves (4.4.10) in a non-dimensional form by setting $k_H^* = c_s k_H/f$, $m^* = c_s m/f$, and $\omega^* = \omega/f$ as

$$\omega^{*4} - (k_H^{*2} + m^{*2} + 1)\omega^{*2} + m^{*2} = 0. \tag{4.4.17}$$

Figs. 4.6 (a) and (b) display contours of m^* as a function of k_H^* and ω^* and those of k_H^* as a function of m^* and ω^*. (c) and (d) show the relations between k_H^* and m^* for sound wave mode and inertial wave mode, respectively. These figures are very similar to Fig. 4.3 for gravity waves if the x and z axes are exchanged.

4.5 Inertio-gravity waves

4.5.1 Dispersion relation

In the rotating system with stratification, both gravity waves and inertial waves exist in the perturbation fields. In general, the direction of the rotating axis is different from that of acceleration due to gravity. In such a case, we can consider shear flows in the geostrophic balance as a steady state. As was described in Section 2.4, such a steady state may be unstable to perturbations and various types of unstable waves will develop. Here, we do not consider the general perturbation fields in shear flows. We simply examine the effect of rotation on the gravity waves in a stratified fluid. We assume that the direction of the rotation axis is parallel to that of gravity and the basic state has no motion. First, we will derive the dispersion relation in a compressible stratified fluid in the rotating frame. Next, we will see how the dispersion relation is modified in the Boussinesq approximation.

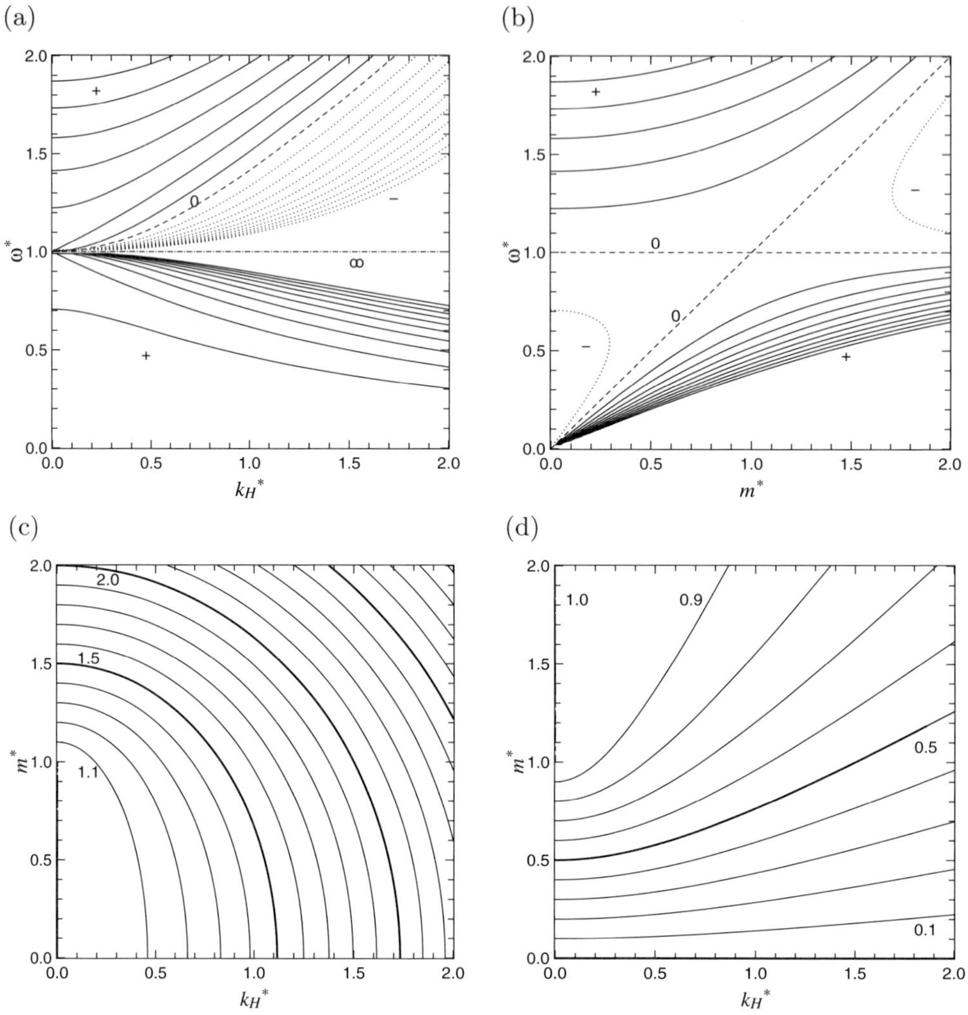

FIGURE 4.6: The dispersion relation of sound waves and inertial waves for the fluid with no gravity. (a) Contours of $m^{*2} = n/2$ (n is integer) as a function of k_H^* and ω^*. (b) Contours of $k_H^{*2} = n/2$ as a function of m^* and ω^*. (c) Contours of $\omega^* = n/10$ as a function of k_H^* and m^* for the sound wave mode. (d) The same as (c) but for the inertial wave mode. In (a) and (b), positive values are shown by solid curves, negative values by dotted curves, and zero by dashed curves. In (c) and (d), only positive values of ω^* are shown.

The governing equations in the rotating frame with gravity are given by (4.2.1)–(4.2.3). As in the previous section, we set the rotation angular vector as $\mathbf{\Omega} = (0, 0, f/2)$ and the acceleration vector of gravity as $\nabla\Phi = (0, 0, g)$: both vectors are parallel. In the same way as in Section 4.3.1, the basic state is at rest with respect to the rotating frame and is in the hydrostatic balance (4.3.4). The linearized equations for the perturbation field are written as

$$\frac{\partial \rho'}{\partial t} + w'\frac{\partial \overline{\rho}}{\partial z} + \overline{\rho}\nabla \cdot \boldsymbol{v}' = 0, \tag{4.5.1}$$

$$\frac{\partial u'}{\partial t} - fv' = -\frac{1}{\overline{\rho}}\frac{\partial p'}{\partial x}, \tag{4.5.2}$$

$$\frac{\partial v'}{\partial t} + fu' = -\frac{1}{\overline{\rho}}\frac{\partial p'}{\partial y}, \tag{4.5.3}$$

$$\frac{\partial w'}{\partial t} = -\frac{1}{\overline{\rho}}\frac{\partial p'}{\partial z} - \frac{\rho'}{\overline{\rho}}g, \tag{4.5.4}$$

$$\frac{1}{c_s^2}\frac{\partial p'}{\partial t} - \frac{\partial \rho'}{\partial t} + \frac{N^2}{g}\overline{\rho}w' = 0. \tag{4.5.5}$$

Substituting (4.3.10) into this equation set, we obtain

$$\frac{\partial \tilde{\rho}}{\partial t} + \nabla \cdot \tilde{\boldsymbol{v}} - \frac{1}{2H_\rho}\tilde{w} = 0, \tag{4.5.6}$$

$$\frac{\partial \tilde{u}}{\partial t} - f\tilde{v} = -\frac{\partial \tilde{p}}{\partial x}, \tag{4.5.7}$$

$$\frac{\partial \tilde{v}}{\partial t} + f\tilde{u} = -\frac{\partial \tilde{p}}{\partial y}, \tag{4.5.8}$$

$$\frac{\partial \tilde{w}}{\partial t} = -\frac{\partial \tilde{p}}{\partial z} + \frac{1}{2H_\rho}\tilde{p} - \tilde{\rho}g, \tag{4.5.9}$$

$$\frac{\partial \tilde{p}}{\partial t} - c_s^2\frac{\partial \tilde{\rho}}{\partial t} + \frac{c_s^2 N^2}{g}\tilde{w} = 0. \tag{4.5.10}$$

We assume that the basic state is isothermal such that all the coefficients are constant. Let us seek solutions proportional to $e^{i(\boldsymbol{k}\cdot\boldsymbol{x}-\omega t)}$ as (4.3.18). In this case, the coefficient matrix is given by

$$\begin{pmatrix} -i\omega & 0 & ik & il & im - \frac{1}{2H_\rho} \\ ic_s^2\omega & -i\omega & 0 & 0 & \frac{c_s^2 N^2}{g} \\ 0 & ik & -i\omega & -f & 0 \\ 0 & il & f & -i\omega & 0 \\ g & im - \frac{1}{2H_\rho} & 0 & 0 & -i\omega \end{pmatrix} \begin{pmatrix} \hat{\rho} \\ \hat{p} \\ \hat{u} \\ \hat{v} \\ \hat{w} \end{pmatrix} = \begin{pmatrix} 0 \\ 0 \\ 0 \\ 0 \\ 0 \end{pmatrix}. \tag{4.5.11}$$

Setting the determinant of this matrix to zero, we obtain the dispersion relation:

$$\omega\left\{\omega^4 - c_s^2\left(\boldsymbol{k}^2 + \frac{1}{4H_\rho^2} + \frac{f^2}{c_s^2}\right)\omega^2 + c_s^2\left[N^2 k_H^2 + f^2\left(m^2 + \frac{1}{4H_\rho^2}\right)\right]\right\} = 0. \tag{4.5.12}$$

In the special case with no rotation $f = 0$, this relation reduces to (4.3.20). The solutions to (4.5.12) are vortical mode $\omega = 0$ and

$$\omega^2 = \frac{c_s^2}{2}\left(\boldsymbol{k}^2 + \frac{1}{4H_\rho^2} + \frac{f^2}{c_s^2}\right) \pm\sqrt{\frac{c_s^4}{4}\left(\boldsymbol{k}^2 + \frac{1}{4H_\rho^2} + \frac{f^2}{c_s^2}\right)^2 - c_s^2\left[N^2k_H^2 + f^2\left(m^2 + \frac{1}{4H_\rho^2}\right)\right]} \tag{4.5.13}$$

$$= \frac{c_s^2}{2}\left(\boldsymbol{k}^2 + \frac{1}{4H_\rho^2} + \frac{f^2}{c_s^2}\right)(1 \pm A^{\frac{1}{2}}), \tag{4.5.14}$$

where

$$A \equiv \sqrt{1 - \frac{4\left[N^2k_H^2 + f^2\left(m^2 + \frac{1}{4H_\rho^2}\right)\right]}{c_s^2\left(\boldsymbol{k}^2 + \frac{1}{4H_\rho^2} + \frac{f^2}{c_s^2}\right)^2}}. \tag{4.5.15}$$

The solution for the larger ω^2 corresponds to sound wave mode and the solution for the smaller ω^2 corresponds to gravity wave mode. In the latter case, the gravity waves are affected by the rotation; they are called *inertio-gravity waves.*

The relation between the frequencies of the two modes in the case of $N^2 < f^2$ is given as follows. Let ω_s denote the frequency of sound wave mode and ω_g denote that of inertio-gravity wave mode. In the limit of $\boldsymbol{k}^2 \to 0$ in (4.5.14), we have

$$\omega_s^2 \to \frac{c_s^2}{4H_\rho^2}, \tag{4.5.16}$$

$$\omega_g^2 \to f^2, \tag{4.5.17}$$

whereas, in the limit of $k_H^2 \to \infty$, we have

$$\omega_s^2 \to c_s^2\left(\boldsymbol{k}^2 + \frac{1}{4H_\rho^2} + \frac{f^2}{c_s^2}\right) \approx c_s^2\boldsymbol{k}^2, \tag{4.5.18}$$

$$\omega_g^2 \to \frac{N^2k_H^2 + f^2\left(m^2 + \frac{1}{4H_\rho^2}\right)}{\boldsymbol{k}^2 + \frac{1}{4H_\rho^2} + \frac{f^2}{c_s^2}} \approx N^2. \tag{4.5.19}$$

For internal waves with $m^2 > 0$, it can be found from (4.5.14) that

$$\omega_s^2 \geq \frac{c_s^2}{4H_\rho^2} > N^2 \geq \omega_g^2 \geq f^2 \tag{4.5.20}$$

(i.e., the frequency of sound wave mode is always larger than that of gravity wave mode for internal waves).

The inside of the square of (4.5.13) is rewritten as

$$D = \frac{c_s^4}{4}\left[\left(k_H^2 + m^2 + \nu^2 - \frac{N^2 - f^2}{c_s^2}\right)^2 + 4(N^2 - f^2)(m^2 + \nu^2)\right], \tag{4.5.21}$$

where ν is given by (4.3.28). In the special case of $m^2 = -\nu^2$, the solutions are

$$\omega^2 = c_s k_H^2 + f^2, \quad (4.5.22)$$
$$\omega^2 = N^2. \quad (4.5.23)$$

The former corresponds to Lamb waves affected by the rotation. These waves are external waves and their vertical structure is the same as that of Lamb waves without rotation. In particular, in the case of $k_H = (N^2 - f^2)/c_s^2$, we have $D = 0$; sound wave mode degenerates to gravity wave mode.

The dispersion relation (4.5.12) can be rewritten by nondimensionalizing with $k_H^* = c_s k_H/N$, $m^* = c_s m/N$, $\omega^* = \omega/N$, $\delta = c_s/2H_\rho N$, and $\varepsilon = f/N$ as

$$\omega^{*4} - (k_H^{*2} + m^{*2} + \delta^2 + \varepsilon^2)\omega^{*2} + k_H^{*2} + \varepsilon^2 m^{*2} + \delta^2\varepsilon^2 = 0. \quad (4.5.24)$$

Figure 4.7 shows this nondimensionalized dispersion relation in the case $\varepsilon = 0.3$ and $\delta = 1.107$. In (a), Lamb waves are shown by the dashed-dotted contour, which is designated by $m^{*2} = 1 - \delta^2$ starting from $\omega^* = \varepsilon$ at $k_H^{*2} = 0$. The contours in (c) and (d) are the dispersion relations of sound waves and inertio-gravity waves; these are very similar to those of the no-rotation field shown in Fig. 4.3 (c) and (d), respectively. This is because the effect of rotation is not important in this case for $\varepsilon = 0.3$.

4.5.2 Inertio-gravity waves in the hydrostatic Boussinesq approximation

As in Section 4.3.2, we examine how the disturbance of inertio-gravity waves is modified in the hydrostatic balance and in the Boussinesq approximation. First, we assume the perturbation field is in the hydrostatic balance; using (4.3.33) in place of (4.5.4) in the equation set (4.5.1)–(4.5.5), we derive the dispersion relation as

$$\omega\left\{\left(m^2 + \frac{1}{4H_\rho^2}\right)\omega^2 - \left[N^2 k_H^2 + f^2\left(m^2 + \frac{1}{4H_\rho^2}\right)\right]\right\} = 0. \quad (4.5.25)$$

Thus, the dispersion relation of inertio-gravity waves is given by

$$\omega^2 = \frac{N^2 k_H^2}{m^2 + \frac{1}{4H_\rho^2}} + f^2. \quad (4.5.26)$$

Second, if the Boussinesq approximation is introduced, we only have a solution for inertio-gravity waves; sound waves are not included in the system. By introducing Coriolis forces into (4.3.35)–(4.3.37), we have the linearized Boussinesq equations in the rotating frame as

$$\nabla \cdot \boldsymbol{v}' = 0, \quad (4.5.27)$$
$$\frac{\partial u'}{\partial t} - fv' = -\frac{1}{\overline{\rho}}\frac{\partial p'}{\partial x}, \quad (4.5.28)$$
$$\frac{\partial v'}{\partial t} + fu' = -\frac{1}{\overline{\rho}}\frac{\partial p'}{\partial y}, \quad (4.5.29)$$

$$\frac{\partial w'}{\partial t} = -\frac{1}{\overline{\rho}}\frac{\partial p'}{\partial z} - \frac{\rho'}{\overline{\rho}}g, \tag{4.5.30}$$

$$\frac{\partial \rho'}{\partial t} - \frac{N^2}{g}\overline{\rho}w' = 0. \tag{4.5.31}$$

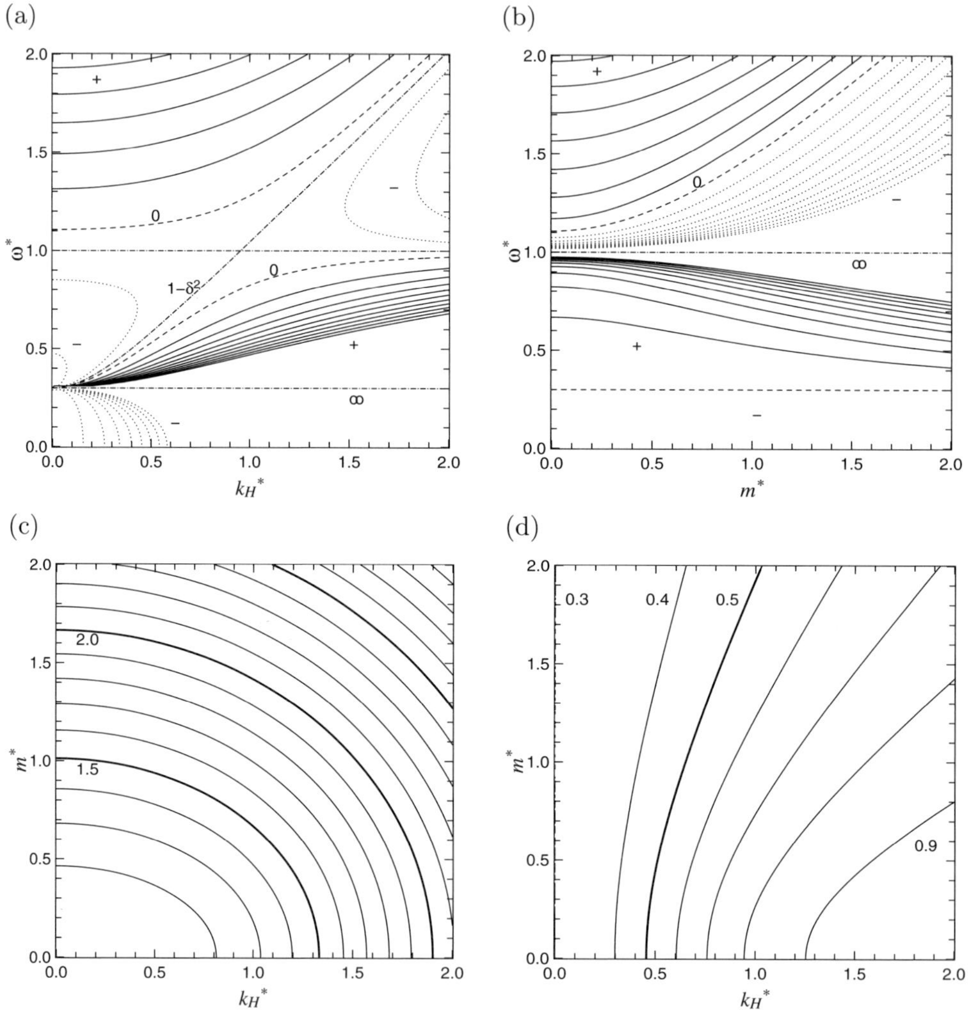

FIGURE 4.7: The dispersion relation of inertio-gravity waves and sound waves for the isothermal atmosphere in the rotating frame for $\varepsilon = 0.3$ and $\delta = 1.107$. (a) Contours of $m^{*2} = n/2$ (n is an integer) as a function of k_H^* and ω^*. (b) Contours of $k_H^{*2} = n/2$ as a function of m^* and ω^*. (c) Contours of $\omega^* = n/10$ as a function of k_H^* and m^* for sound wave mode. (d) The same as (c) but for inertio-gravity wave mode. In (a) and (b), positive values are shown by solid curves, negative values by dotted curves, and zero by dashed curves. In (c) and (d), only positive values of ω^* are shown.

Substituting the sinusoidal form into these equations, we obtain the dispersion relation as

$$\omega \left[\boldsymbol{k}^2 \omega^2 - (N^2 k_H^2 + f^2 m^2) \right] = 0. \tag{4.5.32}$$

Sound waves are not included due to the Boussinesq approximation. Only vortical mode $\omega = 0$ and inertio-gravity wave mode exist. Formally, we may derive (4.5.32) from (4.5.12) by taking the limits $c_s^2 \to \infty$ and $H_\rho \to \infty$. The dispersion relation of inertio-gravity waves is

$$\omega^2 = \frac{N^2 k_H^2 + f^2 m^2}{\boldsymbol{k}^2}. \tag{4.5.33}$$

From this, we generally have $N^2 \geq \omega^2 \geq f^2$.

4.5.3 Group velocity and phase speed

Differentiating the dispersion relation (4.5.12) with respect to k and m, we have group velocity in the horizontal and vertical directions:

$$\begin{aligned} c_{gx} &= \frac{\partial \omega}{\partial k} = \mp \frac{k}{\omega} \frac{N^2 - \omega^2}{\boldsymbol{k}^2 + \frac{1}{4H_\rho^2} + \frac{f^2}{c_s^2}} A^{-\frac{1}{2}} \\ &= \mp c_{px} \frac{\omega^2}{k^2} \frac{N^2 - \omega^2}{\boldsymbol{k}^2 + \frac{1}{4H_\rho^2} + \frac{f^2}{c_s^2}} A^{-\frac{1}{2}}, \end{aligned} \tag{4.5.34}$$

$$\begin{aligned} c_{gz} &= \frac{\partial \omega}{\partial m} = \pm \frac{m}{\omega} \frac{\omega^2 - f^2}{\boldsymbol{k}^2 + \frac{1}{4H_\rho^2} + \frac{f^2}{c_s^2}} A^{-\frac{1}{2}} \\ &= \pm c_{pz} \frac{\omega^2}{m^2} \frac{\omega^2 - f^2}{\boldsymbol{k}^2 + \frac{1}{4H_\rho^2} + \frac{f^2}{c_s^2}} A^{-\frac{1}{2}}, \end{aligned} \tag{4.5.35}$$

where A is given by (4.5.15). Note that $c_{px} = \omega/k$ and $c_{pz} = \omega/m$ are the phase speeds. The lower signs of these equations represent the group velocity of inertio-gravity waves. In the case $f^2 < \omega^2 < N^2$, in particular, we have $c_{gx}c_{px} > 0$ and $c_{gz}c_{pz} < 0$. The latter means that the direction of the vertical component of the group velocity is opposite to that of the phase velocity.

Group velocity in the Boussinesq approximation is given in the limits $c_s \to \infty$ and $H_\rho \to \infty$ of the above equations. Using $A = 1$ and (4.5.33) in this case, group velocity is written as

$$c_{gx} = \frac{k}{\omega} \frac{N^2 - \omega^2}{\boldsymbol{k}^2} = \frac{(N^2 - f^2) k m^2}{(N^2 k_H^2 + f^2 m^2)^{1/2} |\boldsymbol{k}|^3}, \tag{4.5.36}$$

$$c_{gz} = -\frac{m}{\omega} \frac{\omega^2 - f^2}{\boldsymbol{k}^2} = -\frac{(N^2 - f^2) m k_H^2}{(N^2 k_H^2 + f^2 m^2)^{1/2} |\boldsymbol{k}|^3}. \tag{4.5.37}$$

To this approximation, we have $\boldsymbol{c_g} \cdot \boldsymbol{k} = 0$; the direction of group velocity is perpendicular to that of phase velocity, similar to (4.3.44). The phase speeds are written as

$$c_{px} = \frac{\omega}{k} = \frac{(N^2k_H^2 + f^2m^2)^{1/2}}{k|\boldsymbol{k}|}, \tag{4.5.38}$$

$$c_{pz} = \frac{\omega}{m} = \frac{(N^2k_H^2 + f^2m^2)^{1/2}}{m|\boldsymbol{k}|}. \tag{4.5.39}$$

In the case when $f \ll N$ and vertical wave number is much larger than horizontal wave number $m^2 \gg k_H^2$, we have approximation expressions

$$c_{gx} = \frac{N^2k}{m(N^2k^2 + f^2m^2)^{1/2}} = c_{px}\frac{N^2k^2}{N^2k^2 + f^2m^2}, \tag{4.5.40}$$

$$c_{gz} = -\frac{N^2k^2}{m^2(N^2k^2 + f^2m^2)^{1/2}} = -c_{pz}\frac{N^2k^2}{N^2k^2 + f^2m^2}, \tag{4.5.41}$$

where we have assumed that the direction of the phase velocity is parallel to the xz plane, $l = 0$, for simplicity. These expressions are frequently used for illustration of inertio-gravity waves.

4.5.4 Structure of inertio-gravity waves

The structure of the inertio-gravity waves is given in a similar way to that in Section 4.3.4. First, we express $\tilde{u}$, $\tilde{v}$, $\tilde{w}$, and $\tilde{\rho}$ in terms of $\tilde{p}$. From the horizontal components of the equations of momentum, (4.5.7) and (4.5.8), we have

$$\left(\frac{\partial^2}{\partial t^2} + f^2\right)\tilde{u} = -\left(\frac{\partial}{\partial x}\frac{\partial}{\partial t} + f\frac{\partial}{\partial y}\right)\tilde{p}, \tag{4.5.42}$$

$$\left(\frac{\partial^2}{\partial t^2} + f^2\right)\tilde{v} = -\left(\frac{\partial}{\partial y}\frac{\partial}{\partial t} - f\frac{\partial}{\partial x}\right)\tilde{p}. \tag{4.5.43}$$

Eliminating $\tilde{\rho}$ from (4.5.6) and (4.5.9), and eliminating $\tilde{u}$ and $\tilde{v}$ using (4.5.42) and (4.5.43), we have a relation between $\tilde{w}$ and $\tilde{p}$ as

$$\left(\frac{\partial^2}{\partial t^2} + f^2\right)\left(\frac{\partial}{\partial z} - \nu\right)\tilde{w} = \frac{\partial}{\partial t}\left(\frac{\partial^2}{\partial x^2} + \frac{\partial^2}{\partial y^2} - \frac{1}{c_s^2}\frac{\partial^2}{\partial t^2} - \frac{f^2}{c_s^2}\right)\tilde{p}. \tag{4.5.44}$$

Eliminating $\tilde{\rho}$ from (4.5.6) and (4.5.10) gives another relation between $\tilde{w}$ and $\tilde{p}$ as

$$\left(\frac{\partial^2}{\partial t^2} + N^2\right)\tilde{w} = -\frac{\partial}{\partial t}\left(\frac{\partial}{\partial z} + \nu\right)\tilde{p}, \tag{4.5.45}$$

which is the same as (4.3.49). We obtain a single equation for $\tilde{p}$ by eliminating $\tilde{w}$ from (4.5.44) and (4.5.45):

$$\frac{\partial}{\partial t}\left[\left(\frac{\partial^2}{\partial t^2} + N^2\right)\left(\frac{\partial^2}{\partial x^2} + \frac{\partial^2}{\partial y^2} - \frac{1}{c_s^2}\frac{\partial^2}{\partial t^2} - \frac{f^2}{c_s^2}\right) + \left(\frac{\partial^2}{\partial t^2} + f^2\right)\left(\frac{\partial^2}{\partial z^2} - \nu^2\right)\right]\tilde{p} = 0. \tag{4.5.46}$$

We can derive the dispersion relation (4.5.12) by substituting the sinusoidal form into this equation.

The relation between $\tilde{\rho}$ and $\tilde{p}$ can be derived from (4.5.6), (4.5.42), (4.5.43), and (4.5.9) by eliminating $\tilde{u}$, $\tilde{v}$, and $\tilde{w}$:

$$\begin{aligned}\left(\frac{\partial^2}{\partial t^2}+f^2\right)&\left[\frac{\partial^2}{\partial t^2}+\left(\frac{\partial}{\partial z}-\frac{1}{2H_\rho}\right)g\right]\tilde{\rho}\\ &=\left[\frac{\partial^2}{\partial t^2}\left(\frac{\partial^2}{\partial x^2}+\frac{\partial^2}{\partial y^2}\right)+\left(\frac{\partial^2}{\partial t^2}+f^2\right)\left(\frac{\partial}{\partial z}-\frac{1}{2H_\rho}\right)^2\right]\tilde{p}. \qquad (4.5.47)\end{aligned}$$

We also have another relation between $\tilde{\rho}$ and $\tilde{p}$ by eliminating $\tilde{w}$ from (4.5.9) and (4.5.10):

$$\left(\frac{\partial^2}{\partial t^2}+N^2\right)\tilde{\rho} = \left[\frac{1}{c_s^2}\frac{\partial^2}{\partial t^2}-\frac{N^2}{g}\left(\frac{\partial}{\partial z}-\frac{1}{2H_\rho}\right)\right]\tilde{p}, \qquad (4.5.48)$$

which is the same as (4.3.52).

Substituting the sinusoidal form (4.3.18) into (4.5.42), (4.5.43), (4.5.44), (4.5.45), (4.5.47), and (4.5.48), we have the phase relations between the variables:

$$(\omega^2-f^2)\hat{u} = (k\omega+ifl)\hat{p}, \qquad (\omega^2-f^2)\hat{v} = (l\omega-ifk)\hat{p}, \qquad (4.5.49)$$

$$(\omega^2-f^2)(m+i\nu)\hat{w} = \frac{1}{c_s^2}\omega(\omega^2-c_s^2k_H^2-f^2)\hat{p}, \qquad (4.5.50)$$

$$(\omega^2-N^2)\hat{w} = \omega(m-i\nu)\hat{p}, \qquad (4.5.51)$$

$$\begin{aligned}(\omega^2-f^2)&\left[\omega^2-i\left(m+\frac{i}{2H_\rho}\right)g\right]\hat{\rho}\\ &= \left[\omega^2k_H^2+(\omega^2-f^2)\left(m\frac{i}{2H_\rho}\right)^2\right]\hat{p}, \qquad (4.5.52)\end{aligned}$$

$$(\omega^2-N^2)\hat{\rho} = \frac{1}{c_s^2}\left[\omega^2+i\left(m+\frac{i}{2H_\rho}\right)\frac{c_s^2N^2}{g}\right]\hat{p}. \qquad (4.5.53)$$

In the case $\omega \neq f$ and $\omega \neq N$, we can solve them in terms of $\hat{p}$:

$$\hat{u} = \frac{k\omega+ifl}{\omega^2-f^2}\hat{p}, \qquad \hat{v} = \frac{l\omega-ifk}{\omega^2-f^2}\hat{p}, \qquad (4.5.54)$$

$$\hat{w} = \frac{1}{c_s^2}\omega\frac{\omega^2-c_s^2k_H^2-f^2}{(\omega^2-f^2)(m+i\nu)}\hat{p} = \frac{\omega(m-i\nu)}{\omega^2-N^2}\hat{p}, \qquad (4.5.55)$$

$$\begin{aligned}\hat{\rho} &= \frac{\omega^2k_H^2+(\omega^2-f^2)\left(m+\frac{i}{2H_\rho}\right)^2}{(\omega^2-N^2)\left[\omega^2-i\left(m+\frac{i}{2H_\rho}\right)g\right]}\hat{p}\\ &= \frac{1}{c_s^2}\frac{\omega^2+i\left(m+\frac{i}{2H_\rho}\right)\frac{c_s^2N^2}{g}}{\omega^2-N^2}\hat{p}. \qquad (4.5.56)\end{aligned}$$

The structure of the perturbation field (ρ', p', u', v', w') can be constructed using these expressions with (4.3.16) and (4.3.10).

We obtain the phase relations in the Boussinesq approximation by taking the limits $c_s^2 \to \infty$ and $H_\rho \to \infty$ in (4.5.49)–(4.5.53), and neglecting ω^2 in the brackets [ ] of the left-hand side of (4.5.52):

$$\hat{u} = \frac{k\omega + ifl}{\omega^2 - f^2}\hat{p}, \qquad \hat{v} = \frac{l\omega - ifk}{\omega^2 - f^2}\hat{p}, \tag{4.5.57}$$

$$\hat{w} = -\frac{\omega k_H^2}{(\omega^2 - f^2)m}\hat{p} = \frac{\omega m}{\omega^2 - N^2}\hat{p}, \tag{4.5.58}$$

$$\hat{\rho} = i\frac{\omega^2 k_H^2 + (\omega^2 - f^2)m^2}{(\omega^2 - N^2)m}\frac{\hat{p}}{g} = i\frac{mN^2}{\omega^2 - N^2}\frac{\hat{p}}{g}. \tag{4.5.59}$$

Fig. 4.8(a) illustrates the structure of an inertio-gravity wave in the special case for $l = 0$; in this case, the velocity components reduce to

$$\hat{u} = \frac{k\omega}{\omega^2 - f^2}\hat{p}, \qquad \hat{v} = -i\frac{fk}{\omega^2 - f^2}\hat{p}, \qquad \hat{w} = -\frac{\omega k^2}{(\omega^2 - f^2)m}\hat{p}. \tag{4.5.60}$$

Fig. 4.8(b) shows a time change in the horizontal velocity vector at point P shown in (a). It is called a *hodograph*. This shows an anticyclonic change of the velocity vector for the northern hemisphere $f > 0$. We have $\omega = f\cos\phi$, where ϕ is the inclination of the phase line.

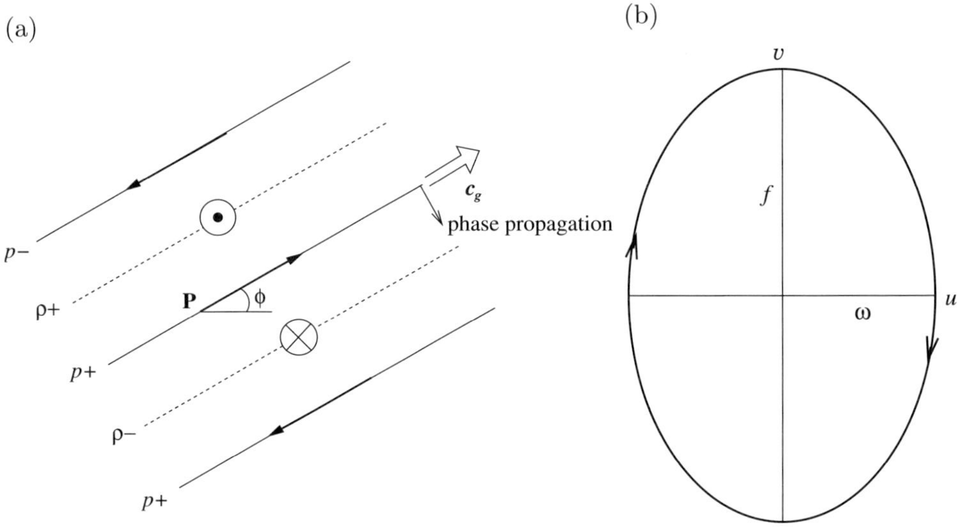

FIGURE 4.8: (a) Structure of an inertio-gravity wave for $l = 0$, $k > 0$, and $m < 0$ in the xz section for $f < \omega < N$, where $\omega = f\cos\phi$. Directions of group velocity and phase velocity are shown by arrows in the upper right of the figure. Solid arrows indicate the directions of fluid motions. Fluid motions in the y-direction are shown at the center of the dotted lines. (b) A hodograph showing the anticyclonic cycle of the velocity vector on a horizontal plane through which an inertial gravity wave passes at P as shown in (a). Note this figure is the case for $f > 0$.

4.6 Rossby waves

In general, the linearized equations of fluids about a certain basic state have five independent solutions. This means that the dispersion relation can be solved for five frequencies. In particular, in the case of a basic state of no motion we have a solution with $\omega = 0$, in which vorticity remains constant irrespective of time. It is appropriately called *vortical mode.* If the basic state has a nonuniform vorticity field, on the other hand, vortical mode becomes propagating wave mode. Such waves are called *Rossby waves.*

First, we characterize Rossby waves in a general form. Rossby waves can be viewed as the propagation of disturbances of potential vorticity. Potential vorticity is conserved under adiabatic and frictionless conditions:

$$\frac{dP}{dt} = 0. \tag{4.6.1}$$

The general form of the equation of potential vorticity is considered in Section 1.3.4 (see (1.3.42)). Let us consider a steady basic state in which potential vorticity $\overline{P}$ and velocity $\overline{\boldsymbol{v}}$ satisfy

$$\overline{\boldsymbol{v}} \cdot \nabla \overline{P} = 0. \tag{4.6.2}$$

From (4.6.1) and (4.6.2), thus, we have the linearized equation of the perturbation of potential vorticity as

$$\left(\frac{\partial}{\partial t} + \overline{\boldsymbol{v}} \cdot \nabla\right) P' + \boldsymbol{v}' \cdot \nabla \overline{P} = 0, \tag{4.6.3}$$

where $P' = P - \overline{P}$ and $\boldsymbol{v}' = \boldsymbol{v} - \overline{\boldsymbol{v}}$. In general, the perturbation of potential vorticity P' is expressed by the perturbation of velocity $\boldsymbol{v}'$ and that of other conservative quantities, such as potential temperature. In the following arguments, we assume that P' can be expressed only by $\boldsymbol{v}'$:

$$P' = \mathcal{L}\{\boldsymbol{v}'\}. \tag{4.6.4}$$

In this case, the two equations (4.6.3) and (4.6.4) can be solved for the two unknown variables P' and $\boldsymbol{v}'$. Substituting the sinusoidal form into these equations, we will obtain a single solution of the frequency ω.

If the gradient of $\overline{P}$ is zero, (4.6.3) becomes

$$\left(\frac{\partial}{\partial t} + \overline{\boldsymbol{v}} \cdot \nabla\right) P' = 0. \tag{4.6.5}$$

The solution is given by $P' =$ const. along the basic flow $\overline{\boldsymbol{v}}$. This corresponds to vortical mode. If $\overline{P}$ has a nonzero gradient, vortical mode becomes a solution of propagating waves. For instance, let us consider the case in which $\overline{P}$ has a gradient in the y-direction (latitude) with $\beta \equiv d\overline{P}/dy$. The velocity field is given by $\overline{\boldsymbol{v}} = (\overline{u}, 0, 0)$ such that (4.6.2) is satisfied. Letting $\boldsymbol{v}' = (u', v', 0)$, we write (4.6.3) as

$$\left(\frac{\partial}{\partial t} + \overline{u}\frac{\partial}{\partial x}\right) P' + \beta v' = 0. \tag{4.6.6}$$

In the following subsections, we use this form of the linearized equation of potential vorticity.

As two special cases for (4.6.4), we consider the two-dimensional nondivergent equations and the three-dimensional quasi-geostrophic equations. The two-dimensional nondivergent equationsprovide a simple example of Rossby waves. Using the three-dimensional quasi-geostrophic equations, the propagation of Rossby waves in a stratified fluid can be examined. In both cases, we introduce the *β-plane approximation* on which the Coriolis parameter varies in the y-direction. In particular, when the basic state is at rest, the gradient of $\overline{P}$ is equal to the change in Coriolis parameter, β.

4.6.1 Two-dimensional Rossby waves

The two-dimensional *nondivergent equations* (or *barotropic equations*) are written as follows:

$$\frac{du}{dt} - fv = -\frac{1}{\rho_0}\frac{\partial p}{\partial x}, \tag{4.6.7}$$

$$\frac{dv}{dt} + fu = -\frac{1}{\rho_0}\frac{\partial p}{\partial y}, \tag{4.6.8}$$

$$\frac{\partial u}{\partial x} + \frac{\partial v}{\partial y} = 0, \tag{4.6.9}$$

where ρ_0 is constant and f is the Coriolis parameter in the β-plane approximation, given by

$$f = f_0 + \beta y, \tag{4.6.10}$$

and the Lagrangian derivative is given by

$$\frac{d}{dt} = \frac{\partial}{\partial t} + u\frac{\partial}{\partial x} + v\frac{\partial}{\partial y}. \tag{4.6.11}$$

From (4.6.7) and (4.6.8), the vorticity equation, or the conservation of absolute vorticity, is given by

$$\frac{d(\zeta + f)}{dt} = 0, \tag{4.6.12}$$

where vorticity is defined as

$$\zeta = \frac{\partial v}{\partial x} - \frac{\partial u}{\partial y}. \tag{4.6.13}$$

If we consider the dispersion of Rossby waves, the conservation of absolute vorticity (4.6.12) in the two-dimensional nondivergent equations plays the same role as the conservation of potential vorticity (4.6.1).†

†Two-dimensional nondivergent equations are equivalent to three-dimensional nondivergent equations with no vertical velocity $w = 0$. In this case, the vertical coordinate z is a conserved quantity since $\frac{dz}{dt} = w = 0$. Therefore, if one chooses $\Psi = \rho_0 z$ in the definition of potential vorticity (1.3.33), potential vorticity is identical to absolute vorticity.

When there is no basic zonal wind,† the linearized equations are written as

$$\frac{\partial u}{\partial t} - fv = -\frac{1}{\rho_0}\frac{\partial p}{\partial x}, \tag{4.6.14}$$

$$\frac{\partial v}{\partial t} + fu = -\frac{1}{\rho_0}\frac{\partial p}{\partial y}. \tag{4.6.15}$$

The corresponding linearized vorticity equation is

$$\frac{\partial \zeta}{\partial t} + \beta v = 0. \tag{4.6.16}$$

From this, we see that vorticity may not be steady due to the β effect. Owing to the nondivergent condition (4.6.9), we can introduce the streamfunction ψ:

$$u = -\frac{\partial \psi}{\partial y}, \qquad v = \frac{\partial \psi}{\partial x}. \tag{4.6.17}$$

Thus, vorticity is expressed as

$$\zeta = \left(\frac{\partial^2}{\partial x^2} + \frac{\partial^2}{\partial y^2}\right)\psi, \tag{4.6.18}$$

and the vorticity equation (4.6.16) is written as

$$\left[\frac{\partial}{\partial t}\left(\frac{\partial^2}{\partial x^2} + \frac{\partial^2}{\partial y^2}\right) + \beta\frac{\partial}{\partial x}\right]\psi = 0. \tag{4.6.19}$$

Substituting a sinusoidal solution

$$\psi = \hat{\psi} e^{i(kx+ly-\omega t)}, \tag{4.6.20}$$

we obtain the dispersion relation:

$$\omega = -\frac{k\beta}{k^2 + l^2}. \tag{4.6.21}$$

The type of waves described by this dispersion relation is called *Rossby waves*, or *nondivergent Rossby waves*. The phase speeds in the x- and y-directions are expressed as

$$c_{px} = \frac{\omega}{k} = -\frac{\beta}{k^2 + l^2}, \qquad c_{py} = \frac{\omega}{l} = -\frac{\beta k}{l(k^2 + l^2)}. \tag{4.6.22}$$

Since $c_{px} < 0$, nondivergent Rossby waves always propagate westward. Group velocity is given by

$$c_{gx} = \frac{\partial \omega}{\partial k} = \frac{\beta(k^2 - l^2)}{(k^2 + l^2)^2}, \qquad c_{gy} = \frac{\partial \omega}{\partial l} = \frac{2\beta kl}{(k^2 + l^2)^2}. \tag{4.6.23}$$

---

†In Section 17.3.3, we will consider the propagation of Rossby waves under the condition that the basic wind $\overline{u}$ has nonzero shear.

Introducing the length scale of a Rossby wave L, we obtain the nondimensional dispersion relation from (4.6.21) with $k^* = kL$, $l^* = lL$, and $\omega^* = \omega/\beta L$:

$$\omega^* = -\frac{k^*}{k^{*2} + l^{*2}}, \tag{4.6.24}$$

which is rewritten as

$$\left(k^* + \frac{1}{2\omega^*}\right)^2 + l^{*2} = \frac{1}{4\omega^{*2}}. \tag{4.6.25}$$

Fig. 4.9 shows this relationship. The contours of constant ω^{*2} are circles with radius $1/2\omega^*$ at the center $(-1/2\omega^*, 0)$. Every circle passes through the origin $(0, 0)$. Fig. 4.10 shows the relation between group velocity and phase lines. Although the direction of the group velocity is outward from the wave source, all the phase speeds have a westward component.

We now consider the structure of Rossby waves. Using (4.6.17) and (4.6.18), we express the variables in terms of the streamfunction ψ. Substituting $(u, v, \zeta) = e^{i(kx+ly-\omega t)}(\hat{u}, \hat{v}, \hat{\zeta})$ into these equations, we have the phase relationships as

$$\hat{u} = -ik\hat{\psi}, \qquad \hat{v} = il\hat{\psi}, \tag{4.6.26}$$

$$\hat{\zeta} = -(k^2 + l^2)\hat{\psi}. \tag{4.6.27}$$

Let us define the geostrophic pressure p_g that is in geostrophic balance with the velocity components:

$$-f_0 v = -\frac{1}{\rho_0}\frac{\partial p_g}{\partial x}, \qquad f_0 u = -\frac{1}{\rho_0}\frac{\partial p_g}{\partial y}. \tag{4.6.28}$$

As shown below, we may have a different choice of geostrophic balance. However, we will obtain an approximate pressure field here. Substituting $p_g = e^{i(kx+ly-\omega t)} \times \hat{p}_g$ gives

$$\hat{p}_g = f_0 \rho_0 \hat{\psi}. \tag{4.6.29}$$

The relation between group velocity and the directions of propagation of a Rossby wave is shown in Fig. 4.11. Geostrophic pressure is also shown here.

In this case, total pressure p is different from geostrophic pressure p_g. Actually, p must satisfy the divergence equation, which is given from (4.6.14) and (4.6.15) as

$$-f\zeta + \beta u = -\frac{1}{\rho_0}\left(\frac{\partial^2}{\partial x^2} + \frac{\partial^2}{\partial y^2}\right)p. \tag{4.6.30}$$

Precisely, p is not a simple sinusoidal form since f depends on y. If we assume that f is constant f_0 in this equation, we obtain an approximate solution of pressure in the form $p = e^{i(kx+ly-\omega t)}\hat{p}$, where

$$\hat{p} = \left(f_0 + i\frac{\beta l}{k^2 + l^2}\right)\rho_0\hat{\psi} = \hat{p}_g + i\frac{\beta l}{k^2 + l^2}\rho_0\hat{\psi}. \tag{4.6.31}$$

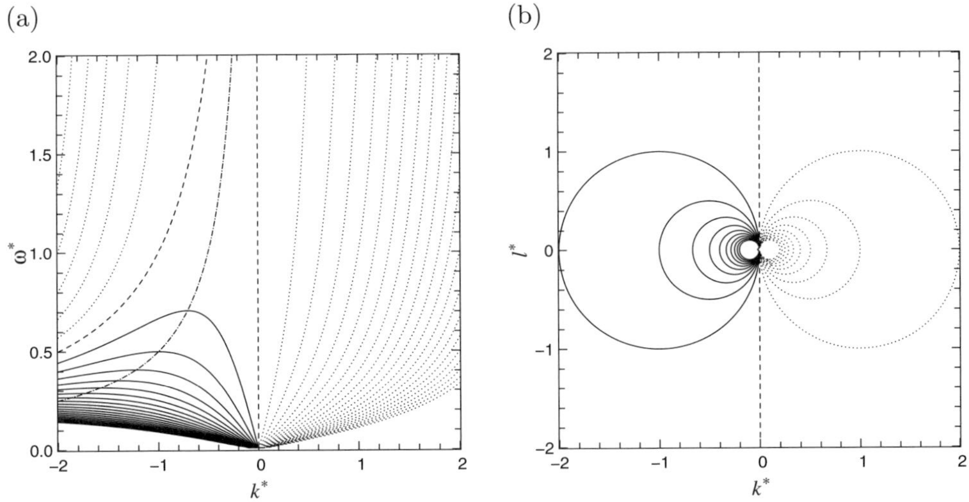

FIGURE 4.9: The dispersion relation of Rossby waves in two-dimensional non-convergence equations. (a) shows the contours of l^{*2} as a function of k^* and ω^*. (b) shows the contours of ω^* as a function of k^* and l^*. The contour interval is 0.5. Positive values are shown by solid curves, negative values by dotted curves, and contours of zero value by dashed curves. The dashed-dotted curves in (a) are positions where $\frac{\partial \omega^*}{\partial k^*} = 0$ is satisfied.

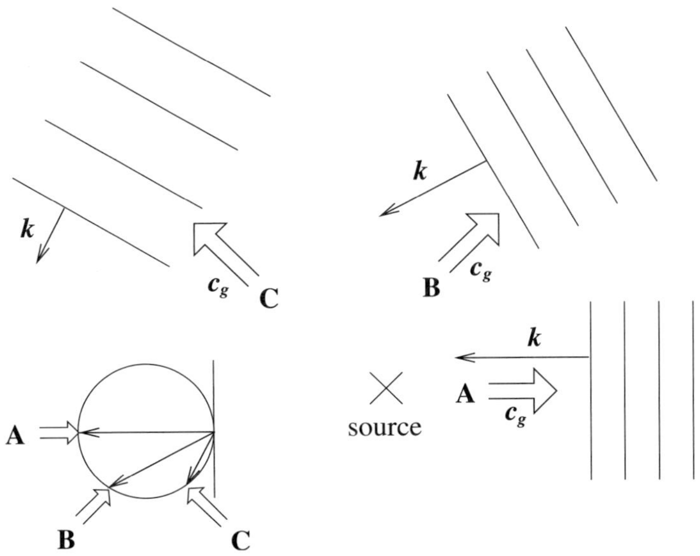

FIGURE 4.10: The relation between the wave number vectors of Rossby waves and the group velocities in two-dimensional nondivergent equations. The symbols A, B, and C denote three propagating waves in three directions relative to the wave source denoted by the cross. The diagram shown at the left bottom represents the contour of constant ω as a function of k and l. Bold arrows are wave number vectors and open arrows are group velocities.

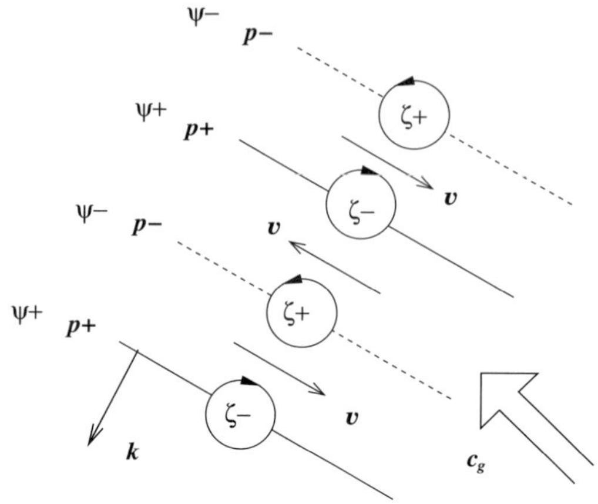

FIGURE 4.11: Structure of a two-dimensional nondivergent Rossby wave for $k < 0$ and $l < 0$. The direction of group velocity is shown by the arrow with $\boldsymbol{c}_g$. The direction of phase velocity is the same as the wave number vector $\boldsymbol{k}$.

The second term on the right-hand side is regarded as an ageostrophic component of the pressure, which will be denoted by p_a.

Generally, geostrophic balance is not uniquely determined. The geostrophically balanced state between u_g, v_g, and p_g is arbitrarily chosen; the only requirement is that the ageostrophic components $u_a = u - u_g$, $v_a = v - v_g$, and $p_a = p - p_g$ must be smaller than the geostrophic components. In the above derivation, the geostrophic pressure p_g is directly defined from the velocity field u and v; in this case, ageostrophic winds are identically zero: $u_a = v_a = 0$. In contrast, we may define the geostrophic winds, u_g and v_g, from the pressure p such that $p_a = 0$.

The general form of geostrophic balance is written as

$$-f_0 v_g = -\frac{1}{\rho_0}\frac{\partial p_g}{\partial x}, \tag{4.6.32}$$

$$f_0 u_g = -\frac{1}{\rho_0}\frac{\partial p_g}{\partial y}, \tag{4.6.33}$$

$$p_g = f_0 \rho_0 \psi_g, \tag{4.6.34}$$

$$\zeta_g = \frac{\partial v_g}{\partial x} - \frac{\partial u_g}{\partial y} = \left(\frac{\partial^2}{\partial x^2} + \frac{\partial^2}{\partial y^2}\right)\psi_g, \tag{4.6.35}$$

where ψ_g and ζ_g denote the streamfunction and vorticity of the geostrophic field, respectively. If the ageostrophic components are smaller than the geostrophic components and βy is smaller than f_0, (4.6.14) and (4.6.15) become

$$\frac{\partial u_g}{\partial t} - f_0 v_a - \beta y v_g = -\frac{1}{\rho_0}\frac{\partial p_a}{\partial x}, \tag{4.6.36}$$

$$\frac{\partial v_g}{\partial t} + f_0 u_a + \beta y u_g = -\frac{1}{\rho_0}\frac{\partial p_a}{\partial y}. \tag{4.6.37}$$

These should be compared with (3.2.24) and (3.2.25) of the quasi-geostrophic equations (Section 3.2.2). For this approximation, the vorticity equation is written in the same form as (4.6.16):

$$\frac{\partial \zeta_g}{\partial t} + \beta v_g = 0. \tag{4.6.38}$$

From this, the dispersion relation of Rossby waves can be given in the same way as (4.6.21).

When the geostrophic pressure p_g is defined from the velocity field by choosing $u_g = u$ and $v_g = v$, p_g becomes different from the pressure p as shown by (4.6.31). Instead, in the case that $p_g = p$ and $p_a = 0$, the velocity field becomes different from the geostrophic winds. Substituting $p_a = 0$ in (4.6.36) and (4.6.37), the ageostrophic components are expressed as

$$u_a = -\frac{1}{f_0}\frac{\partial v_g}{\partial t} - \frac{\beta y}{f_0} u_g = -\frac{1}{f_0}\frac{\partial^2 \psi_g}{\partial t \partial x} + \frac{\beta y}{f_0}\frac{\partial \psi_g}{\partial y}, \tag{4.6.39}$$

$$v_a = \frac{1}{f_0}\frac{\partial u_g}{\partial t} - \frac{\beta y}{f_0} v_g = -\frac{1}{f_0}\frac{\partial^2 \psi_g}{\partial t \partial y} - \frac{\beta y}{f_0}\frac{\partial \psi_g}{\partial x}. \tag{4.6.40}$$

It can be shown from these equations that the ageostrophic winds are also nondivergent; this is a peculiar characteristic of two-dimensional nondivergent equations.

4.6.2 Rossby waves in a stratified fluid

We next consider Rossby waves in a three-dimensional fluid. Unlike the two-dimensional nondivergent equations, velocity components in a three-dimensional fluid are not expressed by a streamfunction, since there exists divergence or convergence in general. In the quasi-geostrophic equation system, however, the zero-th order velocity in Rossby number expansion is a geostrophic wind, so that the streamfunction can be introduced. Rossby waves are clearly defined using the streamfunction.

Let us consider the linear equations for a perturbation field from the basic state at rest. The quasi-geostrophic equations are given by (3.2.24)–(3.2.28), and the corresponding potential vorticity equation is given by (3.2.32). Here, the Coriolis parameter is denoted by $f = f_0 + \beta y$. From (3.2.38), the quasi-geostrophic potential vorticity of the basic state and that of the perturbation are written respectively as

$$\overline{\Pi} = f_0 + \beta y, \tag{4.6.41}$$

$$\Pi' = \left(\frac{\partial^2}{\partial x^2} + \frac{\partial^2}{\partial y^2}\right)\psi + \frac{1}{\rho_s}\frac{\partial}{\partial z}\left(\rho_s \frac{f^{(0)2}}{N^2}\frac{\partial}{\partial z}\psi\right). \tag{4.6.42}$$

From (3.2.32), the linearized potential vorticity equation becomes

$$\frac{\partial}{\partial t}\Pi' + v^{(0)}\frac{\partial}{\partial y}\overline{\Pi} = 0. \tag{4.6.43}$$

The geostrophic velocity components $u^{(0)}$ and $v^{(0)}$ can be expressed by the streamfunction ψ:

$$u^{(0)} = -\frac{\partial \psi}{\partial y}, \qquad v^{(0)} = \frac{\partial \psi}{\partial x}. \tag{4.6.44}$$

Using the above relations, (4.6.43) can be written as a single equation for ψ:

$$\frac{\partial}{\partial t}\left[\left(\frac{\partial^2}{\partial x^2}+\frac{\partial^2}{\partial y^2}\right)\psi+\frac{1}{\rho_s}\frac{\partial}{\partial z}\left(\rho_s\frac{f_0^2}{N^2}\frac{\partial}{\partial z}\psi\right)\right]+\beta\frac{\partial}{\partial x}\psi \;=\; 0. \qquad (4.6.45)$$

For simplicity, we consider the case when ρ_s and N^2 are constant. Substituting

$$\psi \;=\; \hat{\psi}e^{i(kx+ly+mz-\omega t)} \qquad (4.6.46)$$

in (4.6.45), we obtain the dispersion relation of divergent Rossby waves:

$$\omega(k^2+l^2+\varepsilon^2 m^2)+\beta k \;=\; 0, \qquad (4.6.47)$$

where

$$\varepsilon \;\equiv\; \frac{f_0}{N}. \qquad (4.6.48)$$

Thus, the frequency of a Rossby wave is given by

$$\omega \;=\; -\frac{\beta k}{k^2+l^2+\varepsilon^2 m^2}, \qquad (4.6.49)$$

the phase speeds are

$$c_{px} \;=\; \frac{\omega}{k} \;=\; -\frac{\beta}{k^2+l^2+\varepsilon^2 m^2}, \qquad (4.6.50)$$

$$c_{py} \;=\; \frac{\omega}{l} \;=\; -\frac{\beta k}{l(k^2+l^2+\varepsilon^2 m^2)}, \qquad (4.6.51)$$

$$c_{pz} \;=\; \frac{\omega}{m} \;=\; -\frac{\beta k}{m(k^2+l^2+\varepsilon^2 m^2)}, \qquad (4.6.52)$$

and the group velocity is

$$c_{gx} \;=\; \frac{\partial\omega}{\partial k} \;=\; -\frac{2k\omega+\beta}{k^2+l^2+\varepsilon^2 m^2} \;=\; \frac{\beta(k^2-l^2-\varepsilon^2 m^2)}{(k^2+l^2+\varepsilon^2 m^2)^2}, \qquad (4.6.53)$$

$$c_{gy} \;=\; \frac{\partial\omega}{\partial l} \;=\; -\frac{2l\omega}{k^2+l^2+\varepsilon^2 m^2} \;=\; \frac{2\beta kl}{(k^2+l^2+\varepsilon^2 m^2)^2}, \qquad (4.6.54)$$

$$c_{gz} \;=\; \frac{\partial\omega}{\partial m} \;=\; -\frac{2\varepsilon^2 m\omega}{k^2+l^2+\varepsilon^2 m^2} \;=\; \frac{2\beta\varepsilon^2 km}{(k^2+l^2+\varepsilon^2 m^2)^2}. \qquad (4.6.55)$$

Let L be the horizontal length scale of a Rossby wave. Setting $k^* = kL$, $l^* = lL$, and $m^* = mL$, we rewrite the dispersion relation in nondimensional form

$$\omega^* \;=\; -\frac{k^*}{k^{*2}+l^{*2}+\varepsilon^2 m^{*2}}, \qquad (4.6.56)$$

which can be rewritten as

$$\left(k^*+\frac{1}{2\omega^*}\right)^2+l^{*2} \;=\; \frac{1}{4\omega^{*2}}-\varepsilon^2 m^{*2}. \qquad (4.6.57)$$

This indicates that contours of ω^{*2} are circles with their center at $(-1/2\omega^*, 0)$ and radius $\sqrt{1/4\omega^{*2} - \varepsilon^2 m^{*2}}$. Fig. 4.12 shows the dispersion relation in the case $\varepsilon m^* = 0.3$.

The structure of a divergent Rossby wave can be expressed by using the streamfunction ψ. From (3.2.30), (3.2.36), and (3.2.40), we obtain

$$\begin{aligned} \zeta^{(0)} &= \left(\frac{\partial^2}{\partial x^2} + \frac{\partial^2}{\partial y^2}\right)\psi, \\ \pi^{(0)} &= \frac{f_0}{\theta_0}\psi, \qquad \theta^{(0)} = -\frac{\theta_0}{g} f_0 \frac{\partial}{\partial z}\psi, \\ w^{(1)} &= -\frac{g}{N^2\theta_0}\frac{\partial}{\partial t}\theta^{(0)} = -\frac{f_0}{N^2}\frac{\partial^2}{\partial t \partial z}\psi. \end{aligned}$$

Substituting the sinusoidal form

$$(u^{(0)}, v^{(0)}, w^{(0)}, \theta^{(0)}, \pi^{(0)}, \psi^{(0)}, \zeta^{(0)}) = (\hat{u}, \hat{v}, \hat{w}, \hat{\theta}, \hat{\pi}, \hat{\psi}, \hat{\zeta})e^{i(kx+ly+mz-\omega t)},$$

into the above relations, we obtain

$$\begin{aligned} \hat{u} &= -il\hat{\psi}, \qquad \hat{v} = ik\hat{\psi}, \qquad \hat{\pi} = \frac{f_0}{\theta_0}\hat{\psi}, \\ \hat{\zeta} &= -(k^2 + l^2)\hat{\psi}, \qquad \hat{\theta} = -im\frac{\theta_0}{g}f_0\hat{\psi}, \qquad \hat{w} = -\omega m\frac{f_0}{N^2}\hat{\psi}. \end{aligned}$$

The structure of a Rossby wave given by the above phase relations for the case $k < 0$, $l < 0$, and $m < 0$ is shown in Fig. 4.13 ((a) shows the phase relation in the horizontal plane, while (b) shows that in the vertical section).

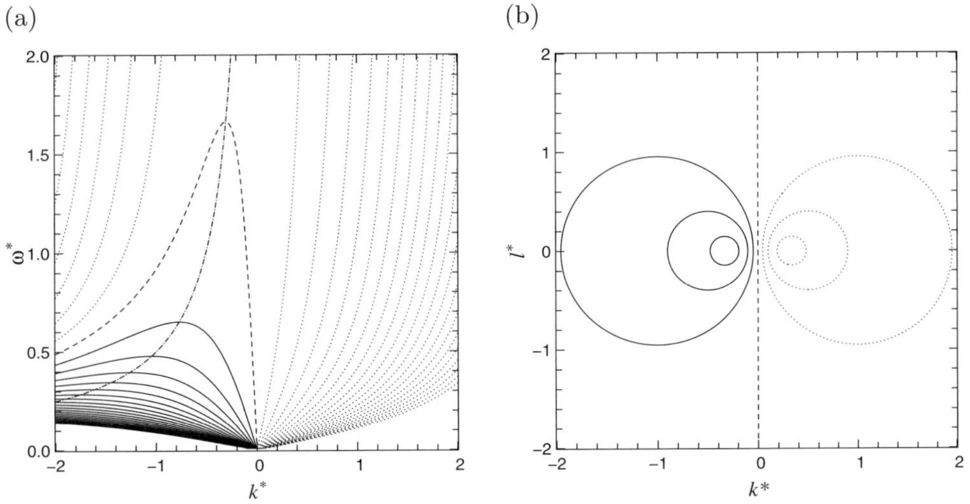

FIGURE 4.12: The dispersion relation of a Rossby wave in a three-dimensional stratified fluid. (a) Contours of l^{*2} as a function of k^* and ω^*. (b) Contours of ω^* as a function of k^* and l^*. The contour interval is 0.5. Positive values are drawn by solid curves, negative values by dotted curves, and the contours of zero value by dashed curves. The dashed-dotted curves are positions where $\frac{\partial \omega^*}{\partial k^*} = 0$ is satisfied.

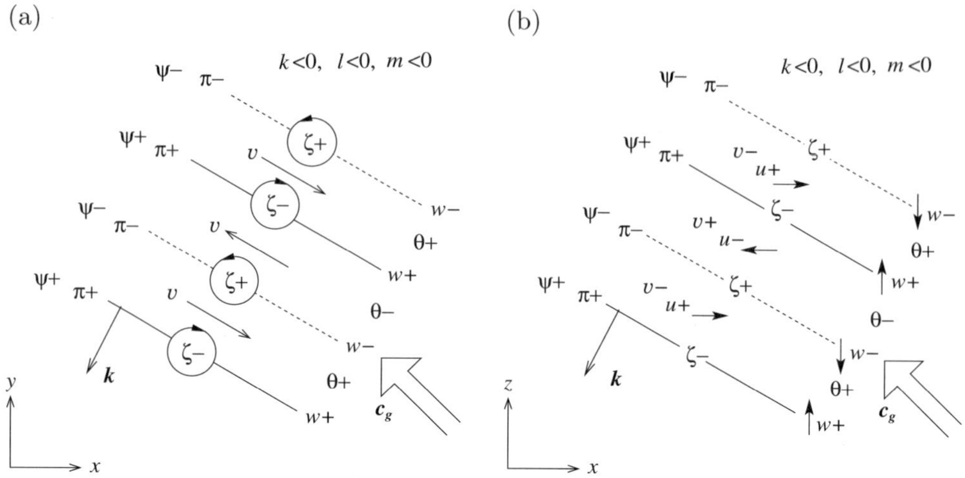

FIGURE 4.13: Structure of a Rossby wave in the quasi-geostrophic equations for $k < 0$, $l < 0$, and $m < 0$. (a) The structure in the xy section, and (b) the structure in the xz section.

4.6.3 Propagation of Rossby waves

The mechanism of propagation of Rossby waves can be interpreted using the schematic figure depicted in Fig. 4.14. The right-hand side of the figure is the profile of the potential vorticity (PV) of the basic field; the value of PV increases from the bottom toward the top of the figure. Imagine a streamline depicted by a wavy curve; the positive vortices with clock-wise flow are located at the ridges of the wave, while the negative vortices with counterclock-wise flow are located at the troughs of the wave. In this case, a secondary flow is induced in the direction depicted by the filled and open arrows between the ridges and the troughs. If the induced flow is in the same direction as the gradient of PV (upward), the smaller PV is advected from the lower side. On the other hand, if the induced flow is in the opposite direction to the gradient of PV (downward), the larger PV is advected from the upper side. As a result, the phase of vorticity propagates to the left side (westward), and the wavy streamline also propagates westward.

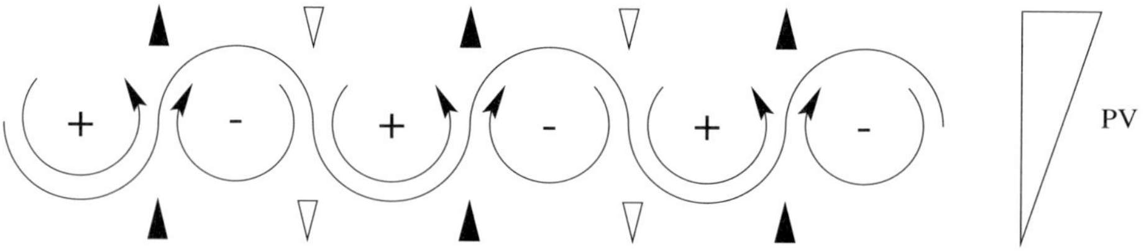

FIGURE 4.14: Schematic figure of propagation of a Rossby wave. The right-hand side is a distribution of the basic PV, the wavy curve is a streamline, + and − are the signs of vortices, and the filled and open arrows indicate induced secondary flows by the vortices. The wave propagates westward (left) by the secondary flows. (Courtesy of Dr. Takayabu.)

Let the gradient of the basic state PV be denoted by $\overline{P}_y$, where the y-axis is taken in the direction of the gradient of PV. (The y-axis is regarded as the northward direction.) A perturbation of PV induced by a displacement η in the y-direction is given by

$$P' \approx \overline{P}_y \eta. \tag{4.6.58}$$

If the effect of stratification is negligible, the change in PV is expressed by the change in relative vorticity ζ. In this case, northward velocity induced by the perturbation of vorticity is given by

$$v' \approx \zeta\lambda \approx \overline{P}_y \eta \lambda, \tag{4.6.59}$$

where λ is the wavelength. During the time interval v'/η, the wave propagates westward at a distance of one wavelength λ; then the phase speed is given by

$$c \approx -\lambda \frac{v'}{\eta} \approx -\overline{P}_y \lambda^2. \tag{4.6.60}$$

This corresponds to the phase speed of a two-dimensional nondivergent Rossby wave c_{px} in (4.6.22), if $\overline{P}_y$ is replaced by β.

The mechanism of meridional and vertical propagations of Rossby waves can also be inferred from Fig. 4.14. This is schematically shown in Fig. 4.15. As in Fig. 4.14, we assume that $\overline{P}$ increases with y. At first, a packet of Rossby waves is confined to region A and no disturbances exist in region B that is adjacent to A. Secondary flows induced by Rossby waves propagating westward in region A have a meridional extension λ in the y-direction and have a vertical depth $H = (f/N)\lambda$ in the z-direction. As a result, wavy disturbances are generated in region B. In turn,

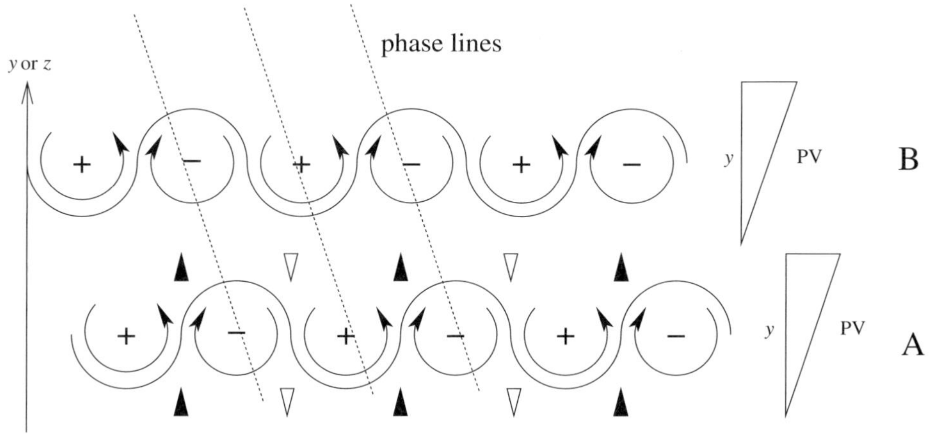

FIGURE 4.15: Schematic figure of the latitudinal and vertical propagations of Rossby waves. Region B is located in either a northern or upper adjacent region to A. When the Rossby wave propagates from region A to region B, the secondary flows in A induce displacements in B and the phase lines are tilted to the west.

the waves in B also induce secondary flows in A, which reduces the amplitude of the displacements in A. Thus, the wave packet propagates from the original region to the adjacent regions in both y- and z-directions. In Fig. 4.15, since the wave propagates westward, the phase lines are tilted westward in the upward and northward directions.

4.7 Waves on a sphere

Linearized primitive equations can be divided into equations of horizontal and vertical structures. The disturbances in horizontal directions are described by shallow water equations. The vertical structure equation determines the vertical profiles of the disturbance if upper and lower boundary conditions are prescribed. In this section, we first derive a general form of the governing equations of waves by linearizing the primitive equations on a sphere. Then, we discuss the approximate forms in the mid- and low latitudes.

4.7.1 Shallow water equations and the equation of vertical structure

We consider primitive equations on the sphere (3.3.15)–(3.3.18) and (3.3.5). Linearized equations about a state at rest are written as

$$\frac{\partial \rho'}{\partial t} + w'\frac{\partial \overline{\rho}}{\partial z} + \overline{\rho}\left[\frac{1}{R\cos\varphi}\frac{\partial u'}{\partial \lambda} + \frac{1}{R\cos\varphi}\frac{\partial}{\partial \varphi}(v'\cos\varphi) + \frac{\partial w'}{\partial z}\right] = 0, \tag{4.7.1}$$

$$\frac{\partial u'}{\partial t} - fv' = -\frac{1}{\overline{\rho}}\frac{1}{R\cos\varphi}\frac{\partial p'}{\partial \lambda}, \tag{4.7.2}$$

$$\frac{\partial v'}{\partial t} + fu' = -\frac{1}{\overline{\rho}}\frac{1}{R}\frac{\partial p'}{\partial \varphi}, \tag{4.7.3}$$

$$0 = -\frac{1}{\overline{\rho}}\frac{\partial p'}{\partial z} - \frac{\rho'}{\overline{\rho}}g, \tag{4.7.4}$$

$$\frac{1}{c_s^2}\frac{\partial p'}{\partial t} - \frac{\partial \rho'}{\partial t} + \frac{N^2}{g}\overline{\rho}w' = 0, \tag{4.7.5}$$

where the Coriolis parameter is defined as $f = 2\Omega\sin\varphi$ and the linearized equation of entropy (4.3.8) is used. Substituting from (4.3.10) into the above equations yields

$$\frac{\partial \tilde{p}}{\partial t} + \frac{1}{R\cos\varphi}\frac{\partial \tilde{u}}{\partial \lambda} + \frac{1}{R\cos\varphi}\frac{\partial}{\partial \varphi}(\tilde{v}\cos\varphi) + \left(\frac{\partial}{\partial z} - \frac{1}{2H_\rho}\right)\tilde{w} = 0, \tag{4.7.6}$$

$$\frac{\partial \tilde{u}}{\partial t} - f\tilde{v} = -\frac{1}{R\cos\varphi}\frac{\partial \tilde{p}}{\partial \lambda}, \tag{4.7.7}$$

$$\frac{\partial \tilde{v}}{\partial t} + f\tilde{u} = -\frac{1}{R}\frac{\partial \tilde{p}}{\partial \varphi}, \tag{4.7.8}$$

$$0 = -\left(\frac{\partial}{\partial z} - \frac{1}{2H_\rho}\right)\tilde{p} - \tilde{\rho}g, \tag{4.7.9}$$

$$\frac{1}{c_s^2}\frac{\partial \tilde{p}}{\partial t} - \frac{\partial \tilde{\rho}}{\partial t} + \frac{N^2}{g}\tilde{w} = 0. \tag{4.7.10}$$

Eliminating $\tilde{\rho}$ from (4.7.6), (4.7.9), and (4.7.10), we have the following two equations:

$$\frac{\partial}{\partial t}\left(\frac{\partial}{\partial z} + \nu\right)\tilde{p} + N^2\tilde{w} = 0, \tag{4.7.11}$$

$$\frac{1}{c_s^2}\frac{\partial \tilde{p}}{\partial t} + \frac{1}{R\cos\varphi}\frac{\partial \tilde{u}}{\partial \lambda} + \frac{1}{R\cos\varphi}\frac{\partial}{\partial \varphi}(\tilde{v}\cos\varphi) + \left(\frac{\partial}{\partial z} - \nu\right)\tilde{w} = 0, \tag{4.7.12}$$

where (4.3.16) and (4.3.28) are used. If c_s is independent of height (i.e., in the case of the isothermal atmosphere), (4.7.7), (4.7.8), and (4.7.12) have solutions in which $\tilde{u}$, $\tilde{v}$, and $\tilde{p}$ are all proportional to a single function of z. In addition, from (4.7.9), $\tilde{p}$ and $\tilde{\rho}$ are proportional to the same function of x, y, and t. Thus, these variables can be expressed in the following forms:

$$\begin{aligned} \tilde{p}(x,y,z,t) &= \check{p}(x,y,t)P(z), & \tilde{\rho}(x,y,z,t) &= \check{p}(x,y,t)Z(z), \\ \tilde{u}(x,y,z,t) &= \check{u}(x,y,t)P(z), & \tilde{v}(x,y,z,t) &= \check{v}(x,y,t)P(z), \\ \tilde{w}(x,y,z,t) &= \check{w}(x,y,t)W(z). \end{aligned} \tag{4.7.13}$$

This kind of expression is not available when c_s depends on the height (i.e., the basic state temperature is not uniform). If c_s is formally regarded as constant, or at the limit of nondivergence $c_s \to \infty$, we can similarly separate the variables.

Substituting (4.7.13) into (4.7.11) and (4.7.12) and separating the variables, we obtain

$$\frac{1}{\check{w}}\frac{\partial \check{p}}{\partial t} = -N^2 W\left[\left(\frac{d}{dz} + \nu\right)P\right]^{-1} = C_1, \tag{4.7.14}$$

$$\begin{aligned} &-\frac{1}{\check{w}}\left[\frac{1}{c_s^2}\frac{\partial \check{p}}{\partial t} + \frac{1}{R\cos\varphi}\frac{\partial \check{u}}{\partial \lambda} + \frac{1}{R\cos\varphi}\frac{\partial}{\partial \varphi}(\check{v}\cos\varphi)\right] \\ &= \frac{1}{P}\left(\frac{d}{dz} - \nu\right)W = C_2, \end{aligned} \tag{4.7.15}$$

where C_1 and C_2 are constants. Without loss of generality, we can choose

$$C_1 = gh, \qquad C_2 = 1. \tag{4.7.16}$$

Here, h is a constant whose dimension is length or height. Note that h can be either positive or negative. Thus, we obtain the equations for vertical structure:

$$\left(\frac{d}{dz} + \nu\right)P = -\frac{N^2}{gh}W, \tag{4.7.17}$$

$$\left(\frac{d}{dz} - \nu\right)W = P. \tag{4.7.18}$$

These equations are rewritten as the equations for P and W:

$$\left(\frac{d^2}{dz^2}+\frac{N^2}{gh}-\nu^2\right)P = 0, \tag{4.7.19}$$

$$\left(\frac{d^2}{dz^2}+\frac{N^2}{gh}-\nu^2\right)W = 0. \tag{4.7.20}$$

We obtain an infinite set of eigenvalues $\frac{N^2}{gh}$ and the eigenfunctions W if appropriate boundary conditions are given to W. In the case of the atmosphere, the boundary conditions can be given as†

$$W = 0, \qquad \text{at } z=0, \tag{4.7.21}$$

$$|W| \text{ is finite}, \qquad \text{as } z\to\infty. \tag{4.7.22}$$

If ν is constant, a solution that satisfies the boundary conditions is given by

$$W(z) = W_0 \sin mz, \tag{4.7.23}$$

where W_0 is constant and m is a real number. In this case, we have from (4.7.20)

$$gh = \frac{N^2}{m^2+\nu^2}. \tag{4.7.24}$$

Substituting (4.7.23) into (4.7.18), we obtain the structure of pressure as

$$P(z) = W_0(m\cos mz - \nu\sin mz). \tag{4.7.25}$$

In a similar way, we obtain the vertical structure of density from (4.7.9)

$$\begin{aligned} Z(z) &= -\frac{1}{g}\left(\frac{d}{dz}-\frac{1}{2H_\rho}\right)P \\ &= \frac{W_0}{g}\left[\left(m^2-\frac{\nu}{2H_\rho}\right)\sin mz - m\left(\nu+\frac{1}{2H_\rho}\right)\cos mz\right]. \end{aligned} \tag{4.7.26}$$

The equations for horizontal structure, on the other hand, are given from (4.7.7), (4.7.8), (4.7.14), and (4.7.15):

$$\frac{\partial \breve{u}}{\partial t} - f\breve{v} = -\frac{1}{R\cos\varphi}\frac{\partial \breve{p}}{\partial \lambda}, \tag{4.7.27}$$

$$\frac{\partial \breve{v}}{\partial t} + f\breve{u} = -\frac{1}{R}\frac{\partial \breve{p}}{\partial \varphi}, \tag{4.7.28}$$

$$\left(\frac{1}{c_s^2}+\frac{1}{gh}\right)\frac{\partial \breve{p}}{\partial t} + \frac{1}{R\cos\varphi}\frac{\partial \breve{u}}{\partial \lambda} + \frac{1}{R\cos\varphi}\frac{\partial}{\partial \varphi}(\breve{v}\cos\varphi) = 0, \tag{4.7.29}$$

$$\frac{\partial \breve{p}}{\partial t} = gh\breve{w}, \tag{4.7.30}$$

†The condition (4.7.22) is a little too restrictive. Since $w \propto \sqrt{\overline{\rho}}W$ and $\overline{\rho}\to 0$ as $z\to\infty$, W need not remain finite in order for w to be finite. In the special case that $\tilde{w}\equiv 0$, no boundary condition is required for W. In this case, we have $h\to\infty$ and $P = P_0 e^{-\nu z}$ from (4.7.17).

where (4.7.16) is used. Substituting (4.7.24) and (4.3.29) into the coefficients of $\check{p}$ in (4.7.29), we obtain

$$\frac{1}{gH} \equiv \frac{1}{c_s^2} + \frac{1}{gh} = \frac{1}{N^2}\left(m^2 + \frac{1}{4H_\rho^2}\right). \tag{4.7.31}$$

We have defined a new constant H on the left-hand side. Linearized shallow water equations on the sphere can be derived from (4.7.27), (4.7.28), and (4.7.29) with the further definition of η by

$$\check{p} = g\eta. \tag{4.7.32}$$

If we omit the symbol ˇ, we finally have the following set of equations:

$$\frac{\partial u}{\partial t} - fv = -\frac{g}{R\cos\varphi}\frac{\partial \eta}{\partial \lambda}, \tag{4.7.33}$$

$$\frac{\partial v}{\partial t} + fu = -\frac{g}{R}\frac{\partial \eta}{\partial \varphi}, \tag{4.7.34}$$

$$\frac{\partial \eta}{\partial t} = -H\left[\frac{1}{R\cos\varphi}\frac{\partial u}{\partial \lambda} + \frac{1}{R\cos\varphi}\frac{\partial}{\partial \varphi}(v\cos\varphi)\right], \tag{4.7.35}$$

$$\frac{\partial \eta}{\partial t} = hw. \tag{4.7.36}$$

These are equivalent to linearized shallow water equations. Solutions to these equations are considered in the next subsection.

4.7.2 Spherical waves

Linearized shallow water equations (4.7.33)–(4.7.35) have solutions for spherical waves. Using $\mu = \sin\varphi$, we can rewrite the equations as

$$\frac{\partial u}{\partial t} - fv = -\frac{g}{R\sqrt{1-\mu^2}}\frac{\partial \eta}{\partial \lambda}, \tag{4.7.37}$$

$$\frac{\partial v}{\partial t} + fu = -\frac{g\sqrt{1-\mu^2}}{R}\frac{\partial \eta}{\partial \mu}, \tag{4.7.38}$$

$$\frac{\partial \eta}{\partial t} = -H\left[\frac{1}{R\sqrt{1-\mu^2}}\frac{\partial u}{\partial \lambda} + \frac{1}{R}\frac{\partial}{\partial \mu}\left(v\sqrt{1-\mu^2}\right)\right]. \tag{4.7.39}$$

From (4.7.37) and (4.7.38), we obtain

$$\left(\frac{\partial^2}{\partial t^2} + f^2\right)u = -\frac{g}{R\sqrt{1-\mu^2}}\left[\frac{\partial^2}{\partial\lambda\partial t} + 2\Omega\mu(1-\mu^2)\frac{\partial}{\partial \mu}\right]\eta, \tag{4.7.40}$$

$$\left(\frac{\partial^2}{\partial t^2} + f^2\right)v = -\frac{g}{R\sqrt{1-\mu^2}}\left[(1-\mu^2)\frac{\partial^2}{\partial\mu\partial t} - 2\Omega\mu\frac{\partial}{\partial \lambda}\right]\eta. \tag{4.7.41}$$

Since the coefficients of these equations depend on the latitudinal coordinate μ, we cannot assume a simple sinusoidal solution in the latitudinal direction. To obtain functions on μ, we seek a solution in the form

$$(u, v, \eta) = (\tilde{u}(\mu), \tilde{v}(\mu), \tilde{\eta}(\mu))\, e^{i(s\lambda - 2\Omega\sigma t)}. \tag{4.7.42}$$

Frequency is written as $\omega = 2\Omega\sigma$ and the Coriolis parameter is written as $f = 2\Omega\mu$. Thus, (4.7.40) and (4.7.41) can be solved for $\tilde{u}$ and $\tilde{v}$ as

$$\tilde{u} = \frac{1}{\sigma^2 - \mu^2} \frac{g}{R\sqrt{1-\mu^2}} \left[s\sigma + \mu(1-\mu^2)\frac{d}{d\mu} \right] \tilde{\eta}, \tag{4.7.43}$$

$$\tilde{v} = -i\frac{1}{\sigma^2 - \mu^2} \frac{g}{R\sqrt{1-\mu^2}} \left[\sigma(1-\mu^2)\frac{d}{d\mu} + \mu s \right] \tilde{\eta}. \tag{4.7.44}$$

Substituting (4.7.42) into (4.7.39) and substituting (4.7.43) and (4.7.44) into the result, we obtain the *Laplace tidal equation*:

$$\left\{ \frac{d}{d\mu} \frac{1-\mu^2}{\sigma^2 - \mu^2} \frac{d}{d\mu} - \frac{1}{\sigma^2 - \mu^2} \left[-\frac{s(\sigma^2 + \mu^2)}{\sigma(\sigma^2 + \mu^2)} + \frac{\sigma}{1-\mu^2} \right] + \varepsilon \right\} \tilde{\eta} = 0, \tag{4.7.45}$$

where

$$\varepsilon \equiv \frac{4\Omega^2 R^2}{gH}, \tag{4.7.46}$$

is called the *Lamb parameter.*

The Laplace tidal equation (4.7.45) is not easy to solve in this form. We derive the equations of vorticity and divergence to obtain the solutions for spherical waves. Let us define the streamfunction Ψ and the velocity potential Φ so as to satisfy

$$u = \frac{1}{R\cos\varphi} \frac{\partial \Phi}{\partial \lambda} - \frac{1}{R} \frac{\partial \Psi}{\partial \varphi}, \tag{4.7.47}$$

$$v = \frac{1}{R} \frac{\partial \Phi}{\partial \varphi} + \frac{1}{R\cos\varphi} \frac{\partial \Psi}{\partial \lambda}. \tag{4.7.48}$$

Vorticity and divergence are written as

$$\zeta = \frac{1}{R\cos\varphi} \left[\frac{\partial v}{\partial \lambda} - \frac{\partial}{\partial \varphi}(u\cos\varphi) \right] = \nabla_H^2 \Psi, \tag{4.7.49}$$

$$\delta = \frac{1}{R\cos\varphi} \left[\frac{\partial u}{\partial \lambda} + \frac{\partial}{\partial \varphi}(v\cos\varphi) \right] = \nabla_H^2 \Phi, \tag{4.7.50}$$

where

$$\nabla_H^2 = \frac{1}{R^2\cos\varphi} \left[\frac{\partial}{\partial \varphi} \left(\cos\varphi \frac{\partial}{\partial \varphi} + \frac{1}{\cos\varphi} \frac{\partial^2}{\partial \lambda^2} \right) \right]. \tag{4.7.51}$$

The shallow water equations (4.7.33), (4.7.34), and (4.7.39) are rewritten as the following set of equations of vorticity, divergence, and vertical displacement:

$$\frac{\partial \zeta}{\partial t} + 2\Omega\sin\varphi\,\delta + \frac{2\Omega}{R}\cos\varphi\, v = 0, \tag{4.7.52}$$

$$\frac{\partial \delta}{\partial t} - 2\Omega\sin\varphi\,\zeta + \frac{2\Omega}{R}\cos\varphi\, u = -g\nabla_H^2 \eta, \tag{4.7.53}$$

$$\frac{\partial \eta}{\partial t} + H\delta = 0. \tag{4.7.54}$$

These are expressed by using Φ and Ψ as

$$\left(\frac{\partial}{\partial t}\nabla_H^2 + \frac{2\Omega}{R^2}\frac{\partial}{\partial\lambda}\right)\Psi + 2\Omega\left(\sin\varphi\nabla_H^2 + \frac{\cos\varphi}{R^2}\frac{\partial}{\partial\varphi}\right)\Phi = 0, \tag{4.7.55}$$

$$\left(\frac{\partial}{\partial t}\nabla_H^2 + \frac{2\Omega}{R^2}\frac{\partial}{\partial\lambda}\right)\Phi - 2\Omega\left(\sin\varphi\nabla_H^2 + \frac{\cos\varphi}{R^2}\frac{\partial}{\partial\varphi}\right)\Psi = -g\nabla_H^2\eta, \tag{4.7.56}$$

$$\frac{\partial\eta}{\partial t} + H\nabla_H^2\Phi = 0. \tag{4.7.57}$$

We then seek a solution in the form

$$(\Psi, \Phi, \eta) = (\tilde{\Psi}(\mu), \tilde{\Phi}(\mu), \tilde{\eta}(\mu))\, e^{i(s\lambda - 2\Omega\sigma t)}. \tag{4.7.58}$$

Substituting this into (4.7.55)–(4.7.57) and eliminating η, we obtain

$$(\sigma\nabla_H^{2*} - s)i\tilde{\Psi} - (\mu\nabla_H^{2*} + D)\tilde{\Phi} = 0, \tag{4.7.59}$$

$$\left(\sigma\nabla_H^{2*} - s + \frac{\nabla_H^{4*}}{\varepsilon\sigma}\right)\tilde{\Phi} - (\mu\nabla_H^{2*} + D)i\tilde{\Psi} = 0, \tag{4.7.60}$$

where

$$\nabla_H^{2*} = \frac{d}{d\mu}\left[(1-\mu^2)\frac{d}{d\mu}\right] - \frac{s^2}{1-\mu^2}, \tag{4.7.61}$$

$$D = (1-\mu^2)\frac{d}{d\mu}. \tag{4.7.62}$$

Here, we expand $\tilde{\Phi}$ and $\tilde{\Psi}$ in a series as

$$\tilde{\Phi} = \sum_{n=s}^{\infty} A_n^s P_n^s(\mu), \qquad \tilde{\Psi} = \sum_{n=s}^{\infty} iB_n^s P_n^s(\mu), \tag{4.7.63}$$

where P_n^s is the normalized associated Legendre function, defined by (21.2.16).[†] We use the following recurrence relation

$$\mu P_n^s(\mu) = \varepsilon_{n+1}^s P_{n+1}^s + \varepsilon_n^s P_{n-1}^s, \tag{4.7.64}$$

where

$$\varepsilon_n^s = \sqrt{\frac{n^2 - s^2}{4n^2 - 1}}. \tag{4.7.65}$$

Substituting (4.7.63) into (4.7.59) and (4.7.60) and using (4.7.64), we obtain the relations between the coefficients A_n^s and B_n^s as

$$L_n B_n^s + q_{n-1}A_{n-1}^s + p_{n+1}A_{n+1}^s = 0, \tag{4.7.66}$$

$$K_n A_n^s + q_{n-1}B_{n-1}^s + p_{n+1}B_{n+1}^s = 0, \tag{4.7.67}$$

[†] Characteristics of the associated Legendre functions are summarized in more detail in Section 21.8.

where for $n \geq s$

$$
\begin{aligned}
K_n &= \sigma + \frac{s}{n(n+1)} - \frac{n(n+1)}{\varepsilon\sigma}, & L_n &= \sigma + \frac{s}{n(n+1)}, \\
p_n &= \frac{n+1}{n}\varepsilon_n^s, & q_n &= \frac{n}{n+1}\varepsilon_{n+1}^s,
\end{aligned}
$$

and $q_n = 0$ for $n = s - 1$.

Eqs. (4.7.66) and (4.7.67) can be divided into two groups:

$$\left(\sigma I + C - \frac{1}{\varepsilon\sigma}J_{sym}\right)\boldsymbol{X}_{sym} = 0, \tag{4.7.68}$$

$$\left(\sigma I + C - \frac{1}{\varepsilon\sigma}J_{asym}\right)\boldsymbol{X}_{asym} = 0, \tag{4.7.69}$$

where I is the identity matrix, and

$$
\begin{aligned}
\boldsymbol{X}_{sym} &= (A_s^s, B_{s+1}^s, A_{s+2}^s, B_{s+3}^s, \cdots)^t, \\
\boldsymbol{X}_{asym} &= (B_s^s, A_{s+1}^s, B_{s+2}^s, A_{s+3}^s, \cdots)^t, \\
J_{sym,ii} &= \begin{cases} (s+i-1)(s+i), & \text{for} \quad i = \text{odd}, \\ 0, & \text{for} \quad i = \text{even}, \end{cases} \\
J_{asym,ii} &= \begin{cases} (s+i-1)(s+i), & \text{for} \quad i = \text{even}, \\ 0, & \text{for} \quad i = \text{odd}, \end{cases} \\
C_{ij} &= \begin{cases} \dfrac{s}{(s+i-1)(s+i)}, & \text{for} \quad i = j, \\ \dfrac{s+i+1}{s+i}\varepsilon_{s+1}^s, & \text{for} \quad i+1 = j, \\ \dfrac{s+i-1}{s+i}\varepsilon_{s+1}^s, & \text{for} \quad i-1 = j. \end{cases}
\end{aligned}
$$

The subscript *sym* represents the symmetric mode about the equator, whereas the subscript *asym* represents the antisymmetric mode. J_{sym} and J_{asym} are diagonal matrices, and C is a tri-diagonal matrix. Eqs. (4.7.68) and (4.7.69) are regarded as characteristic equations for the eigenvalue σ for a given constant $\eta \equiv 1/\varepsilon\sigma$ and can be solved numerically.

Fig. 4.16 shows the dependency of frequency σ on ε for $s = 1$. Eastward waves $\sigma > 0$ and westward waves $\sigma < 0$ are shown for both cases of $\varepsilon > 0$ and $\varepsilon < 0$. For $\varepsilon > 0$, the eastward waves are gravity waves, and the westward waves are either gravity waves or Rossby waves. We can see the intermediate waves between the Rossby and gravity waves; these are called *Kelvin waves* and *mixed Rossby-gravity waves*. Fig. 4.17 particularly shows the dependency on zonal wave number for $\varepsilon = 10$.

In the special case, $\varepsilon \to 0$, we have an approximation:

$$0 = \left|\sigma I + C - \frac{1}{\varepsilon\sigma}J\right| \approx \prod_{i=1}^{\infty}\left|\sigma + C_{ii} - \frac{1}{\varepsilon\sigma}J_{ii}\right|, \tag{4.7.70}$$

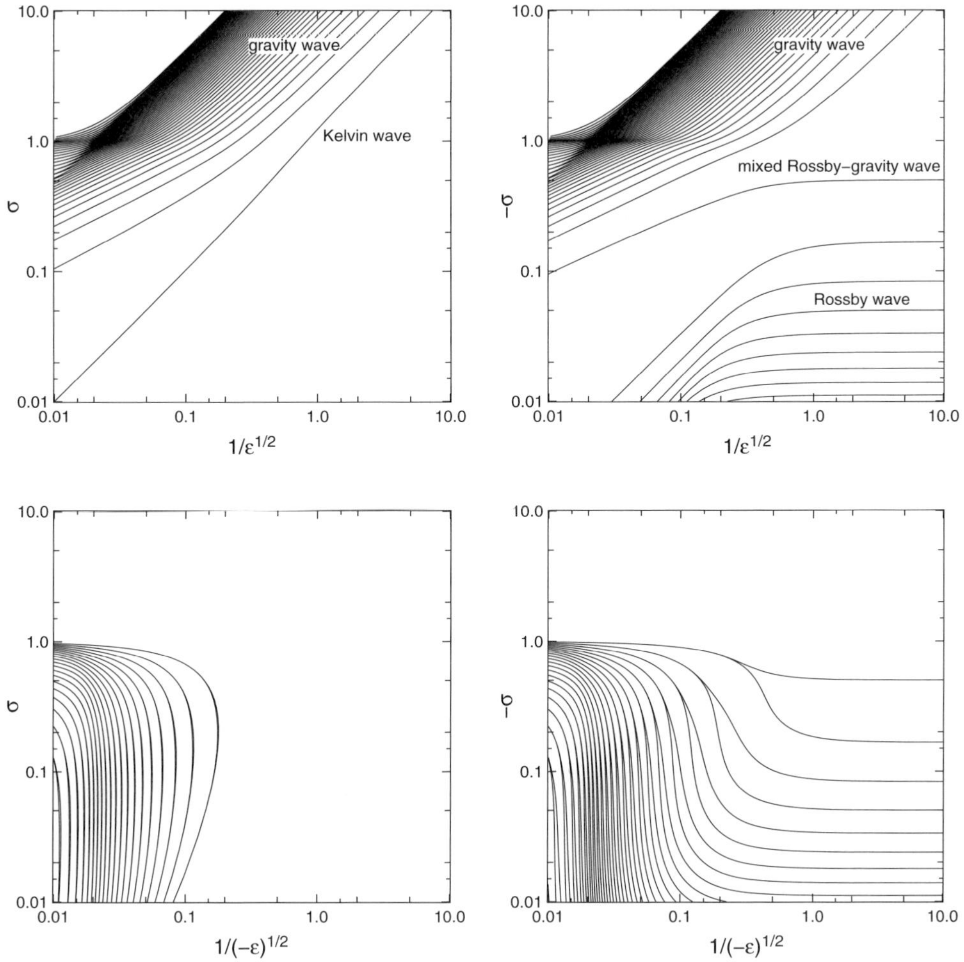

FIGURE 4.16: Examples of the eigenvalues of spherical waves for $s = 1$. The ordinate is $1/\sqrt{|\varepsilon|}$ and the abscissa is the frequency σ; upper left: eastward waves for $\varepsilon > 0$; upper right: westward waves for $\varepsilon > 0$; lower left: eastward waves for $\varepsilon < 0$; and lower right: westward waves for $\varepsilon < 0$.

where J represents either J_{sym} or J_{asym}. For the nonzero component of J_{ii} (the odd i for the symmetric mode, and the even i for the antisymmetric mode), letting $n = s + i - 1$, we obtain

$$\sigma + \frac{s}{n(n+1)} - \frac{n(n+1)}{\varepsilon\sigma} \quad = \quad 0. \tag{4.7.71}$$

From this, we obtain the frequency in the case of $\varepsilon > 0$ as

$$\sigma \quad = \quad -\frac{s}{2n(n+1)} \pm \sqrt{\frac{s^2}{4n^2(n+1)^2} + \frac{n(n+1)}{\varepsilon}} \quad \approx \quad \pm\sqrt{\frac{n(n+1)}{\varepsilon}}, \tag{4.7.72}$$

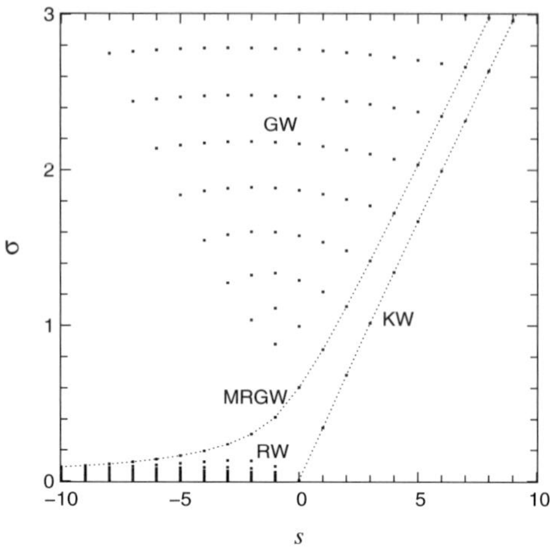

FIGURE 4.17: Dependency of frequency σ on zonal wave number s of equatorial waves for $\varepsilon = 10$. The ordinate is zonal wave number s and the abscissa is frequency σ. Note that for $\sigma < 0$ signs of s and σ are converted. The dotted line denoted by KW represents Kelvin waves and that with MRGW the mixed Rossby-gravity waves. GW represents gravity waves and RW Rossby waves.

or

$$\omega \quad = \quad 2\Omega\sigma \quad \approx \quad \pm\sqrt{\frac{n(n+1)gh}{R}}. \tag{4.7.73}$$

The corresponding eigenfunctions are $\Psi = 0$ and

$$\Phi \quad = \quad AP_n^s(\mu)e^{i(s\lambda-2\Omega\sigma t)}, \tag{4.7.74}$$

where A is constant. This mode corresponds to gravity waves of spherical shallow water with depth h in the nonrotating system. The positive and negative signs of the frequency correspond to eastward and westward propagating gravity waves, respectively.

On the other hand, corresponding to the zero components of J_{ii} in (4.7.70), we obtain

$$\sigma \quad = \quad -\frac{s}{n(n+1)}, \tag{4.7.75}$$

or

$$\omega \quad = \quad 2\Omega\sigma \quad = \quad -\frac{2\Omega s}{n(n+1)}. \tag{4.7.76}$$

The eigenfunctions are $\Phi = 0$ and

$$\Psi \quad = \quad BP_n^s(\mu)e^{i(s\lambda-2\Omega\sigma t)}, \tag{4.7.77}$$

where B is constant. This mode has just a vorticity component with no divergence; it corresponds to westward propagating Rossby waves. Specifically, in the case of $n = s = 1$ in (4.7.76), we have $\omega = -\Omega$. This wave rotates with angular velocity Ω in the direction opposite to the rotation of the frame (i.e., this wave is stationary with respect to the inertial frame).

4.7.3 Equatorial waves

In this subsection, we derive the approximate forms of spherical waves in equatorial latitudes. In particular, we will investigate how Rossby waves (Section 4.6.1) and gravity waves (Section 4.6.2) are related to these approximations. We assume that $|\varphi| \ll 1$, and write $x = R\lambda$, $y = R\varphi$, $\beta = 2\Omega/R$, and $f = 2\Omega \sin\varphi \approx 2\Omega\varphi = \beta y$. Then, using $\cos\varphi \approx 1$, (4.7.33)–(4.7.35) are approximated to

$$\frac{\partial u}{\partial t} - fv = -g\frac{\partial \eta}{\partial x}, \tag{4.7.78}$$

$$\frac{\partial v}{\partial t} + fu = -g\frac{\partial \eta}{\partial y}, \tag{4.7.79}$$

$$\frac{\partial \eta}{\partial t} = -H\left(\frac{\partial u}{\partial x} + \frac{\partial v}{\partial y}\right). \tag{4.7.80}$$

Eliminating u or v from (4.7.78) and (4.7.79) yields

$$\left(\frac{\partial^2}{\partial t^2} + f^2\right) u = -g\left(\frac{\partial^2}{\partial x \partial t} + f\frac{\partial}{\partial y}\right)\eta, \tag{4.7.81}$$

$$\left(\frac{\partial^2}{\partial t^2} + f^2\right) v = -g\left(\frac{\partial^2}{\partial y \partial t} - f\frac{\partial}{\partial x}\right)\eta. \tag{4.7.82}$$

Subtracting a derivative of (4.7.79) with respect to x and a derivative of (4.7.78) with respect to y, and substituting the two into (4.7.80), we obtain

$$\frac{\partial}{\partial t}\left(\zeta - \frac{f}{H}\eta\right) + \beta v = 0, \tag{4.7.83}$$

where ζ is the vorticity:

$$\zeta = \frac{\partial v}{\partial x} - \frac{\partial u}{\partial y}. \tag{4.7.84}$$

Note that (4.7.83) is also given from the equation of potential vorticity.

Eliminating u from (4.7.80) and (4.7.83) yields

$$\frac{\partial}{\partial t}\left(\frac{\partial^2}{\partial y \partial t} - f\frac{\partial}{\partial x}\right)\eta = -H\left[\frac{\partial}{\partial t}\left(\frac{\partial^2}{\partial x^2} + \frac{\partial^2}{\partial y^2}\right) + \beta\frac{\partial}{\partial x}\right] v. \tag{4.7.85}$$

Eliminating η from (4.7.82) and (4.7.85), we obtain a partial differential equation for v as

$$\frac{\partial}{\partial t}\left[\left(\frac{\partial^2}{\partial t^2} + f^2\right) v - c^2\left(\frac{\partial^2}{\partial x^2} + \frac{\partial^2}{\partial y^2}\right) v\right] - c^2\beta\frac{\partial v}{\partial x} = 0, \tag{4.7.86}$$

where $c^2 = gH$. Since the coefficients depend on y through f, we can assume the solution for v in the form

$$v = \tilde{v}(y)e^{i(kx - \omega t)}. \tag{4.7.87}$$

Substituting this into (4.7.86) gives

$$\left[\frac{d^2}{dy^2} + \left(\frac{\omega^2}{c^2} - k^2 - \frac{\beta k}{\omega}\right) - \frac{f^2}{c^2}\right]\tilde{v} \;=\; 0. \tag{4.7.88}$$

Substituting $f = \beta y$ and introducing the nondimensional variables $y^* = \sqrt{\beta/c}y$, $\omega^* = \omega/\sqrt{c\beta}$, and $k^* = \sqrt{c/\beta}k$, (4.7.88) is rewritten as

$$\left[\frac{d^2}{dy^{*2}} + \left(\omega^{*2} - k^{*2} - \frac{k^*}{\omega^*}\right) - y^{*2}\right]\tilde{v} \;=\; 0. \tag{4.7.89}$$

This has the form of the *Weber equation*:

$$\left[\frac{d^2}{dz^2} + \left(n + \frac{1}{2} - \frac{z^2}{4}\right)\right]w \;=\; 0. \tag{4.7.90}$$

It is known that this equation has a solution that is finite at infinity for nonnegative integer n,

$$w \;=\; D_n(z) \;=\; e^{-\frac{z^2}{4}}H_n(z), \tag{4.7.91}$$

where $D_n(z)$ are parabolic functions and $H_n(z)$ are Hermitian polynomials. The same Hermitian polynomials are expressed as $H_0(z) = 1$, $H_1(z) = z$, $H_2(z) = z^2 - 1$, and $H_3(z) = z^3 - 3z$. Therefore, the solution to (4.7.89) can be written in the form

$$\tilde{v} \;=\; v_0 D_n\left(\sqrt{\frac{2\beta}{c}}y\right) \;=\; v_0 \exp\left(-\frac{\beta y^2}{2c}\right)H_n\left(\sqrt{\frac{2\beta}{c}}y\right), \tag{4.7.92}$$

where

$$n \;=\; \frac{1}{2}\left(\omega^{*2} - k^{*2} - \frac{k^*}{\omega^*} - 1\right), \tag{4.7.93}$$

that is,

$$\left(\frac{\omega}{c}\right)^2 - k^2 - \frac{\beta k}{\omega} \;=\; (2n+1)\frac{\beta}{c}, \quad \text{for} \quad n = 0, 1, 2, \cdots . \tag{4.7.94}$$

Fig. 4.18 shows the relation between wave number k^* and frequency ω^* for each n.

The waves for $n \geq 1$ are categorized into either Rossby waves or gravity waves. The wave $n = 0$ has characteristics of both and is called a mixed Rossby-gravity wave.† This figure also shows Kelvin waves for $n = -1$, which will be described later.

We can easily determine the structures of η and u by introducing new variables q and r, defined by

$$q \;=\; u + \frac{g}{c}\eta, \quad r \;=\; -u + \tfrac{g}{c}\eta. \tag{4.7.95}$$

†In the literature, the term, mixed-Rossby wave, is only given to westward propagating waves of mode $n = 0$. See Fig. 4.19.

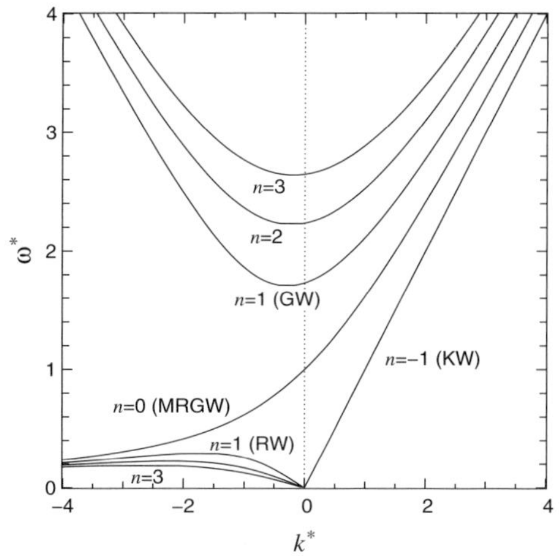

FIGURE 4.18: Dispersion relation for equatorial waves. The abscissa is k^* and the ordinate is ω^*. The dotted lines with KW represent Kelvin waves, those with MRGW represent mixed Rossby-gravity waves, GW represents gravity waves and RW Rossby waves.

From (4.7.78) and (4.7.80), we have

$$\left(\frac{\partial}{\partial t} + c\frac{\partial}{\partial x}\right) q + c\frac{\partial v}{\partial y} - fv = 0, \tag{4.7.96}$$

$$\left(\frac{\partial}{\partial t} - c\frac{\partial}{\partial x}\right) r + c\frac{\partial v}{\partial y} + fv = 0. \tag{4.7.97}$$

Let us assume that the variables have the following form:

$$(u, \eta, q, r) = (\tilde{u}(y), \tilde{\eta}(y), \tilde{q}(y), \tilde{\eta}(y))\, e^{i(kx-\omega t)}. \tag{4.7.98}$$

Substituting this into (4.7.96) and (4.7.97), we have

$$\tilde{q} = -\frac{ic}{\omega - ck}\left(\frac{d}{dy} - \frac{\beta}{c}y\right)\tilde{v}, \tag{4.7.99}$$

$$\tilde{r} = -\frac{ic}{\omega + ck}\left(\frac{d}{dy} + \frac{\beta}{c}y\right)\tilde{v}. \tag{4.7.100}$$

Using (4.7.92) and the recursive relation of the parabolic functions

$$\left(\frac{d}{dz} + \frac{z}{2}\right) D_n = nD_{n-1}, \qquad \left(\frac{d}{dz} - \frac{z}{2}\right) D_n = -D_{n+1}, \tag{4.7.101}$$

we obtain

$$\tilde{q} = -v_0\frac{ic}{\omega - ck}D_{n+1}\left(\sqrt{\frac{2\beta}{c}}y\right), \tag{4.7.102}$$

$$\tilde{r} = -v_0\frac{ic}{\omega + ck}D_{n-1}\left(\sqrt{\frac{2\beta}{c}}y\right). \tag{4.7.103}$$

Therefore, we have

$$\begin{aligned}\tilde{u} &= \frac{q-r}{2} \\ &= iv_0\sqrt{\frac{\beta c}{2}}\left[\frac{1}{\omega - ck}D_{n+1}\left(\sqrt{\frac{2\beta}{c}}y\right) + \frac{n}{\omega + ck}D_{n-1}\left(\sqrt{\frac{2\beta}{c}}y\right)\right],\end{aligned} \tag{4.7.104}$$

$$\begin{aligned}\tilde{\eta} &= \frac{c}{g}\frac{q+r}{2} \\ &= iv_0\frac{c}{g}\sqrt{\frac{\beta c}{2}}\left[\frac{1}{\omega - ck}D_{n+1}\left(\sqrt{\frac{2\beta}{c}}y\right) - \frac{n}{\omega + ck}D_{n-1}\left(\sqrt{\frac{2\beta}{c}}y\right)\right].\end{aligned} \tag{4.7.105}$$

The case $n = 0$ is the exception, since $\tilde{r} = 0$ from (4.7.100) and (4.7.101). From (4.7.94), the dispersion relation for $n = 0$ is given by

$$\frac{\omega}{c} - k - \frac{\beta}{\omega} = 0, \tag{4.7.106}$$

where we have used the relation $\omega + ck \neq 0$ (see Eq. (4.7.115)). The structure of this type of wave is given by

$$\tilde{v} = v_0 D_0\left(\sqrt{\frac{2\beta}{c}}y\right) = v_0 \exp\left(-\frac{\beta y^2}{2c}\right), \tag{4.7.107}$$

$$\tilde{u} = \frac{g}{c}\tilde{\eta} = \frac{\tilde{q}}{2} = iv_0\frac{\beta y}{\omega - ck}\exp\left(-\frac{\beta y^2}{2c}\right) = iv_0\frac{\omega y}{c}\exp\left(-\frac{\beta y^2}{2c}\right). \tag{4.7.108}$$

This corresponds to mixed Rossby-gravity waves. Fig. 4.19 shows the structures of the eastward propagating wave ($k > 0$) and the westward propagating wave ($k < 0$). The westward propagating wave has a geostrophic character in higher latitudes, while the eastward propagating wave is more like a gravity wave.

Eq. (4.7.86) also has a solution $\tilde{v} \equiv 0$, but the dispersion relation in this case cannot be given from (4.7.86). To obtain the dispersion relation, we set $v = 0$ in the original equations (4.7.78)–(4.7.80),

$$\frac{\partial u}{\partial t} = -g\frac{\partial \eta}{\partial x}, \tag{4.7.109}$$

$$fu = -g\frac{\partial \eta}{\partial y}, \tag{4.7.110}$$

$$\frac{\partial \eta}{\partial t} = -H\frac{\partial u}{\partial x}, \tag{4.7.111}$$

which gives an equation for η as

$$\frac{\partial^2 \eta}{\partial t^2} - c^2\frac{\partial^2 \eta}{\partial x^2} = 0. \tag{4.7.112}$$

Substitution from (4.7.98) yields

$$\omega^2 - c^2k^2 = 0. \tag{4.7.113}$$

From (4.7.109) and (4.7.110), the latitudinal structure is described by

$$\frac{1}{\tilde{\eta}}\frac{d\tilde{\eta}}{dy} = -\frac{k\beta}{\omega}y. \tag{4.7.114}$$

From this, η remains finite as $|y| \to \infty$ in the case of $\omega > 0$; that is,

$$\omega = ck. \tag{4.7.115}$$

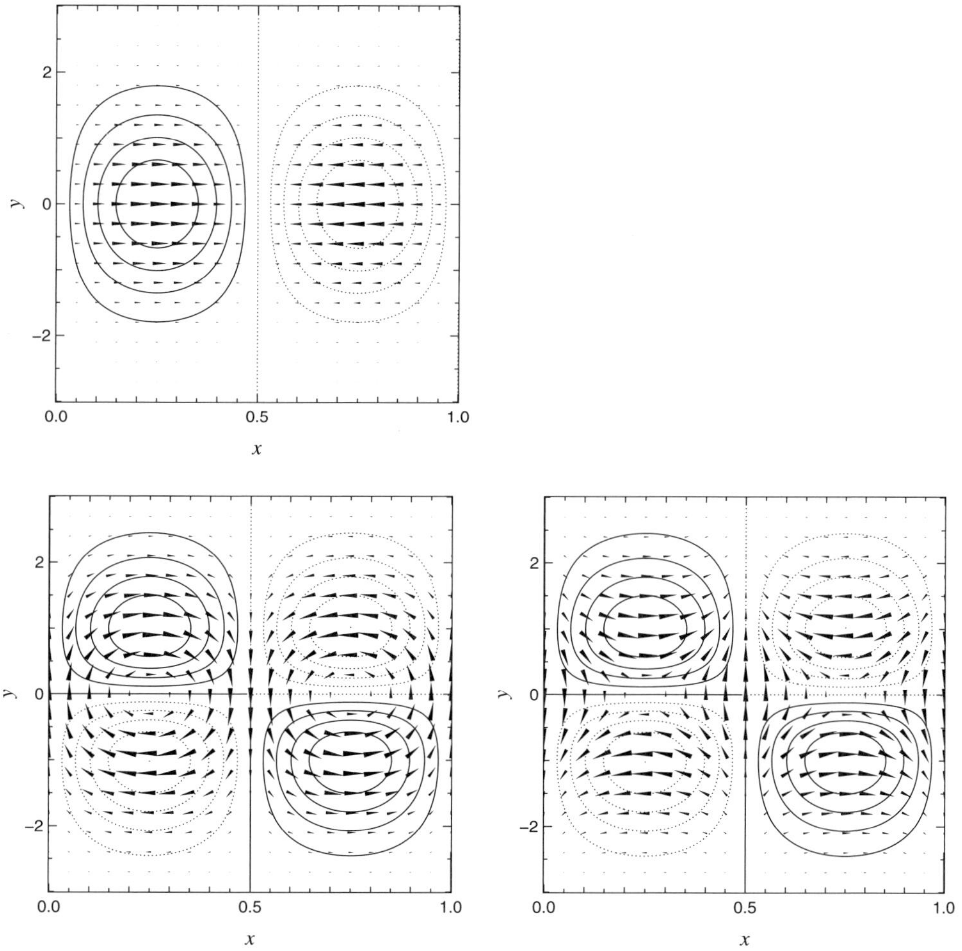

FIGURE 4.19: Horizontal structure of a Kelvin wave (top), a westward propagating mixed Rossby-gravity wave (bottom left), and a eastward propagating mixed Rossby-gravity wave (bottom right). Contours represent surface height (solid: positive; dotted: negative) and vectors show horizontal velocity with appropriate scales.

The structure of this type of wave is given by

$$\tilde{\eta} = \eta_0 \exp\left(-\frac{\beta y^2}{2c}\right), \tag{4.7.116}$$

$$\tilde{u} = \frac{\eta_0}{c} \exp\left(-\frac{\beta y^2}{2c}\right), \tag{4.7.117}$$

$$\tilde{v} = 0, \tag{4.7.118}$$

and is depicted as Fig. 4.19. We note that solution (4.7.115) is included in (4.7.93) with $n = -1$. This corresponds to the Kelvin wave.

References and suggested reading

Atmospheric waves are one of the central points of dynamic meteorology, and are fully described in standard textbooks. Among them, we cite Gill (1982), in which each wave property is extensively examined in separate chapters. For a thorough discussion of wave theory in Section 4.1, the readers should refer to textbooks by Lighthill (1978) and Whitham (1974). The explanation of Rossby wave propagations in Section 4.6.3 is based on Hoskins et al. (1985); that of spherical waves in Section 4.7.2 follows Longuet-Higgins (1968). The description of spherical waves in Swarztrauber and Kasahara (1985) is also useful. The description of equatorial waves in Section 4.7.3, which were first examined by Matsuno (1966), follows Gill (1980, 1982).

Gill, A. E., 1980: Some simple solutions for heat-induced tropical circulation. *Q. J. Roy. Meteorol. Soc.*, **106**, 447–462.

Gill, A. E., 1982: *Atmosphere-Ocean Dynamics.* Academic Press, San Diego, 662 pp.

Hoskins, B. J., McIntyre, M. E., and Robertson, A. W., 1985: On the use and significance of isentropic potential vorticity maps. *Q. J. Roy. Meteorol. Soc.*, **111**, 877–946.

Lighthill, M. J., 1978: *Waves in Fluids.* Cambridge University Press, Cambridge, UK, 504 pp.

Longuet-Higgins, M. S., 1968: The eigenfunctions of Laplace's tidal equations over a sphere. *Phil. Trans. Roy. Soc. Lond.*, **A262**, 511–607.

Matsuno, T., 1966: Quasi-geostrophic motions in the equatorial area. *J. Meteorol. Soc. Japan*, **44**, 25–43.

Swarztrauber, P. N. and Kasahara, A., 1985: The vector harmonic analysis of Laplace's tidal equations. *SIAM J. Sci. and Stat. Comput.*, **6**, 464–491.

Whitham, G. B., 1974: *Linear and Nonlinear Waves.* Wiley-Interscience. New York, 636 pp.

5

Instability

Various types of unstable disturbances are examined in this chapter. We discuss the motions and structures of unstable disturbances superimposed on a balanced state of the atmosphere. These are in contrast to the neutral disturbances considered in the previous chapter. We examined the condition for instability of the balanced states of the atmospheric motions in Chapter 2, in which the main interest revolved around the characteristics of balanced states. In this chapter, however, we mainly consider the properties of unstable disturbances. The structures of unstable disturbances will be investigated using linear stability analysis in this chapter.

First, we briefly introduce the concept of linear stability analysis. Then, we subsequently use it to consider various instabilities related to the atmospheric motions. The main subjects are convective instability, inertial instability, barotropic instability, and baroclinic instability. Specifically, the Rayleigh-Bénard convection is a basic tool for understanding convective motions in the atmosphere. Inertial instability can be described by a formulation similar to that of convective instability. Both barotropic and baroclinic instabilities are commonly described as shear instability problems. In particular, baroclinic instability is set up as the Eady problem and the Charney problem, both of which are fundamental frameworks for mid-latitude circulations, which will be considered in Chapter 18.

5.1 Linear stability analysis

We generally obtain linear equations by linearizing the governing equations of fluids around a specified basic state. We symbolically write linear equations as (4.1.1), where ϕ is a set of variables describing the perturbations. We then assume a sinusoidal form of the solution to the linear equations:

$$\phi(\boldsymbol{x}, t) \quad = \quad \phi_0 e^{i(\boldsymbol{k}\cdot\boldsymbol{x}-\omega t)}, \tag{5.1.1}$$

where ω is a frequency and $\boldsymbol{k}$ is a wave number vector. ω can be a complex number in this case. Substituting this form into the linear equations, we obtain the dispersion relation between ω and $\boldsymbol{k}$ as

$$\Omega(\omega, \boldsymbol{k}) = 0. \tag{5.1.2}$$

In Chapter 4, we examined the cases in which ω is real such that waves do not grow or decay. In general, however, ω is not necessarily a real number as a solution of the dispersion relation. When ω is complex, ω is written with real and imaginary parts as

$$\omega = \omega_r + i\omega_i. \tag{5.1.3}$$

Thus, (5.1.1) is rewritten as

$$\phi(\boldsymbol{x}, t) = \phi_0 e^{i(\boldsymbol{k}\cdot\boldsymbol{x} - \omega_r t)} e^{\omega_i t}. \tag{5.1.4}$$

In the case that

$$\omega_i > 0, \tag{5.1.5}$$

the amplitude of the disturbance increases exponentially. ω_i is called the *growth rate*.

The basic state is regarded as *unstable* if there is a solution with $\omega_i > 0$ for any wave number $\boldsymbol{k}$ in the dispersion relation (5.1.2). In contrast, the basic state is *stable* if $\omega_i < 0$ for all wave numbers. The solutions to linear equations with $\omega_i > 0$, $\omega_i < 0$, and $\omega_i = 0$ are called the unstable, stable, and neutral solutions.

Let us consider the case when the dispersion relation (5.1.2) depends on an external parameter R. It may be probable that stable solutions exist for a certain range of R and that unstable solutions appear if R exceeds a critical value R_c. In such a case, if a value of R varies from the stable range to the unstable range, a marginally unstable wave that is close to the neutral solution emerges first just above the critical value R_c. Generally, wave numbers of the neutral solutions are related to R as

$$\omega_i(R, \boldsymbol{k}) = 0. \tag{5.1.6}$$

We assume that this can be solved as

$$R = R(\boldsymbol{k}). \tag{5.1.7}$$

Neutral solutions exist only in the unstable range of R. If the basic state is unstable in the range $R > R_c$, (5.1.7) has a real solution of $\boldsymbol{k}$ in the range $R \geq R_c$. The *critical wave number* $\boldsymbol{k}_c$ is given as a solution to

$$\frac{\partial R}{\partial \boldsymbol{k}}(\boldsymbol{k}_c) = 0, \tag{5.1.8}$$

and the critical value of R is given as $R_c = R(\boldsymbol{k}_c)$.

Even in the case of the unstable range $R > R_c$, we cannot ascertain which kind of wave number emerges as an unstable wave just from linear stability analysis. Since we generally consider nonlinear fluid systems, the neglected terms in linear equations become important if the amplitude of disturbances grows. However, one may extend application of linear stability analysis to the nonlinear regime. According to linear analysis, the unstable mode that has the fastest growth rate grows most rapidly. Such an unstable mode is called the *most unstable mode.* If initial small disturbances have various wave numbers, the most unstable mode overcomes the other modes. The wave number of the most unstable mode is a function of R and is given as a solution to

$$\frac{\partial \omega_i}{\partial \boldsymbol{k}}(R, \boldsymbol{k}) = 0. \tag{5.1.9}$$

Stability analysis may be extended to include the effects of nonlinear terms. However, we do not describe nonlinear stability analysis in this chapter. When we try to explain phenomena in the real atmosphere in terms of instability, it is important that we define appropriately the basic state and a set of external parameters R.

5.2 Convective instability

If a fluid parcel is heated within the gravitational field, the density becomes lighter and the heated parcel rises through buoyancy. Similarly, if the fluid is stratified such that density decreases with height, a fluid parcel will get more buoyancy when it is displaced upward, and the upward motion will be accelerated. This argument can be more accurately described by the parcel method in Section 2.3. The stability condition is given by (2.3.6):

$$\frac{\partial s}{\partial z} > 0 \tag{5.2.1}$$

(i.e., the fluid is stable if entropy increases with height). This condition is derived on the assumptions of the parcel method that the pressure of the fluid parcel is kept the same as that of the environment, and the effects of friction and diffusion are negligible. In more general cases where the effects of friction and diffusion exist, however, the fluid parcel loses its heat by thermal diffusion, and its upward motion will be suppressed by friction. In such cases, the simple parcel method is not applicable for explanation of the stability condition of the fluid. Instead, we use linear stability analysis to discuss the stability condition by calculating the eigenvalues of the linearized governing equations of perturbation fields.

In this section, we concentrate on the Boussinesq fluid to consider convective instability. We begin with convection in the non-rotating frame. Next, we examine convection in the rotating frame (i.e., on the f-plane). The equations of the Boussinesq fluid on the f-plane are given by adding the Coriolis forces to (3.1.21)–(3.1.25) as

$$\frac{du}{dt} - fv = -\frac{1}{\rho_0}\frac{\partial p}{\partial x} + \nu \nabla^2 u, \tag{5.2.2}$$

$$\frac{dv}{dt} + fu = -\frac{1}{\rho_0}\frac{\partial p}{\partial y} + \nu \nabla^2 v, \tag{5.2.3}$$

$$\frac{dw}{dt} = -\frac{1}{\rho_0}\frac{\partial p}{\partial z} + \alpha g T + \nu \nabla^2 w, \tag{5.2.4}$$

$$\frac{\partial u}{\partial x} + \frac{\partial v}{\partial y} + \frac{\partial w}{\partial z} = 0, \tag{5.2.5}$$

$$\frac{dT}{dt} = \kappa \nabla^2 T. \tag{5.2.6}$$

Here, the acceleration due to gravity g, the coefficient of viscosity ν, and the expansion coefficient α are all constant. The temperature T is a deviation from the constant T_0. Since the density of the Boussinesq fluid depends only on temperature, the stability condition (5.2.1) is written as

$$\frac{\partial T}{\partial z} > 0. \tag{5.2.7}$$

If the effects of viscosity and thermal diffusion exist, however, it will be shown that the fluid is stable unless the lapse rate of the temperature exceeds a critical value. We will see that the critical value becomes larger when the effect of rotation is introduced, thus the convection is further suppressed in the rotating frame.

5.2.1 Rayleigh-Bénard convection

First, we consider convective instability in the non-rotating frame with $f = 0$. The convection in the Boussinesq fluid without rotation is called the *Rayleigh-Bénard convection*, or *Bénard convection.* Let us consider convective instability in the following situation: a fluid is placed between rigid horizontal plates at $z = 0$ and H. The fluid spans infinite horizontal dimensions. The temperatures at the top and bottom boundaries are kept fixed with $T = T_a$ at $z = 0$ and $T = T_b$ at $z = H$. The temperature at the top boundary is lower than that at the bottom boundary: $\Delta T \equiv T_a - T_b > 0$. As for the boundary conditions of the velocity, we assume *free slip conditions* at $z = 0$, H:

$$\frac{\partial u}{\partial z} = \frac{\partial v}{\partial z} = 0, \tag{5.2.8}$$

$$w = 0. \tag{5.2.9}$$

This assumption is introduced only to simplify stability analysis. From the view point of practical application, the *rigid lid condition,*

$$u = v = w = 0, \tag{5.2.10}$$

is more appropriate, though the calculation becomes a little complicated.

We consider a steady basic state at rest. Let overline ($\bar{\ }$) denote a variable of the basic state. From (5.2.6), the temperature profile of the basic state is vertically linear. Thus, the basic state is given by

$$\overline{u} = \overline{v} = \overline{w} = 0, \tag{5.2.11}$$

$$\overline{T} = T_a + (T_b - T_a)\frac{z}{H} = T_a - \Gamma z, \tag{5.2.12}$$

where

$$\Gamma \equiv -\frac{d\overline{T}}{dz} = \frac{T_a - T_b}{H} = \frac{\Delta T}{H}. \tag{5.2.13}$$

The pressure of the basic state must satisfy

$$0 = -\frac{1}{\rho_0}\frac{\partial \overline{p}}{\partial z} + \alpha g \overline{T}. \tag{5.2.14}$$

The perturbation of temperature T' and that of pressure p' is defined as $T = \overline{T} + T'$ and $p = \overline{p} + p'$. For the velocity components, we have $u' = u$, $v' = v$, $w' = w$. The linearized equations of the perturbation field are given from (3.1.21)–(3.1.25) with $f = 0$ as

$$\frac{\partial u'}{\partial t} = -\frac{1}{\rho_0}\frac{\partial p'}{\partial x} + \nu\nabla^2 u', \tag{5.2.15}$$

$$\frac{\partial v'}{\partial t} = -\frac{1}{\rho_0}\frac{\partial p'}{\partial y} + \nu\nabla^2 v', \tag{5.2.16}$$

$$\frac{\partial w'}{\partial t} = -\frac{1}{\rho_0}\frac{\partial p'}{\partial z} + \alpha g T' + \nu\nabla^2 w', \tag{5.2.17}$$

$$\frac{\partial u'}{\partial x} + \frac{\partial v'}{\partial y} + \frac{\partial w'}{\partial z} = 0, \tag{5.2.18}$$

$$\frac{\partial T'}{\partial t} - \Gamma w' = \kappa\nabla^2 T'. \tag{5.2.19}$$

We can rewrite (5.2.15), (5.2.16), and (5.2.17) using (5.2.19) to eliminate T' as

$$\left(\frac{\partial}{\partial t} - \nu\nabla^2\right) u' = -\frac{1}{\rho_0}\frac{\partial p'}{\partial x}, \tag{5.2.20}$$

$$\left(\frac{\partial}{\partial t} - \nu\nabla^2\right) v' = -\frac{1}{\rho_0}\frac{\partial p'}{\partial y}, \tag{5.2.21}$$

$$\left[\left(\frac{\partial}{\partial t} - \kappa\nabla^2\right)\left(\frac{\partial}{\partial t} - \nu\nabla^2\right) - \alpha g\Gamma\right] w' = -\left(\frac{\partial}{\partial t} - \kappa\nabla^2\right)\frac{1}{\rho_0}\frac{\partial p'}{\partial z}. \tag{5.2.22}$$

Applying the divergence operator to these equations and using (5.2.18), we obtain a single equation for p':

$$\left[\left(\frac{\partial}{\partial t} - \kappa\nabla^2\right)\left(\frac{\partial}{\partial t} - \nu\nabla^2\right)\nabla^2 - \alpha g\Gamma\left(\frac{\partial^2}{\partial x^2} + \frac{\partial^2}{\partial y^2}\right)\right] p' = 0. \tag{5.2.23}$$

Now, let us put aside the vertical boundary conditions for a while and assume that the domain is not bounded in the vertical direction. In such a case, p' may be written in the form

$$p' = p_0 e^{i(kx+ly+mz-\omega t)}. \tag{5.2.24}$$

Substituting this into (5.2.23), we obtain the dispersion relation as

$$(-i\omega + \kappa \boldsymbol{k}^2)(-i\omega + \nu \boldsymbol{k}^2)\boldsymbol{k}^2 - \alpha g \Gamma k_H^2 \quad = \quad 0, \tag{5.2.25}$$

where $k_H^2 = k^2 + l^2$ and $\boldsymbol{k}^2 = k^2 + l^2 + m^2$. In particular, in the case that $\kappa = 0$, $\nu = 0$, and $-\Gamma > 0$, we have the frequency of neutral waves: $\omega^2 = \alpha g |\Gamma| k_H^2 / \boldsymbol{k}^2$. This dispersion relation corresponds to that of gravity waves (4.3.38), since the Brunt-Väisälä frequency is given by $N = \sqrt{\alpha g |\Gamma|}$.

Eq. (5.2.25) is solved for ω as

$$\omega \quad = \quad -i \left\{ (\kappa + \nu)\boldsymbol{k}^2 \pm \left[(\kappa + \nu)^2 \boldsymbol{k}^4 - \kappa\nu \boldsymbol{k}^4 + \alpha g \Gamma \frac{k_H^2}{\boldsymbol{k}^2} \right]^{\frac{1}{2}} \right\}. \tag{5.2.26}$$

This equation has an unstable solution with $\omega_i > 0$ if the inequality,

$$-\kappa\nu \boldsymbol{k}^4 + \alpha g \Gamma \frac{k_H^2}{\boldsymbol{k}^2} \quad > \quad 0, \tag{5.2.27}$$

is satisfied. Here, we define the *Rayleigh number* by

$$Ra \quad \equiv \quad \frac{\alpha g \Gamma H^4}{\kappa \nu} = \frac{\alpha g \Delta T H^3}{\kappa \nu}. \tag{5.2.28}$$

Then, the inequality (5.2.27) is rewritten as

$$Ra \quad > \quad \frac{\boldsymbol{k}^6}{k_H^2} H^4. \tag{5.2.29}$$

Here, since we are assuming that the vertical domain is unbounded, H is an appropriately defined height scale. If we specify the vertical wave number m, we find that the right-hand side of (5.2.29) takes the smallest value at

$$k_H^2 \quad = \quad \frac{m^2}{2}. \tag{5.2.30}$$

In this case, the Rayleigh number is

$$Ra_{min}(m) \quad = \quad \frac{27}{4} m^4 H^4, \tag{5.2.31}$$

which can be called the *critical Rayleigh number*. If the domain is not bounded vertically, m can take any small number. As m becomes smaller, Ra_{min} becomes smaller. This means that the fluid tends to be more unstable as the vertical scale of a perturbation is deeper. From this argument, however, we do not have any information on the relation between the horizontal wave numbers k and l in the x- and y-directions, respectively; we only know the magnitude of k_H.

Let us go back to the original boundary conditions, in which the fluid is confined between two walls at $z = 0$ and H. The vertical structure of perturbations depends on the boundary conditions, so that we seek a solution of p' in the form

$$p' \quad = \quad \tilde{p}(z) e^{i(kx + ly - \omega t)}. \tag{5.2.32}$$

Substituting this into (5.2.23) we get

$$\begin{aligned}\Big\{\Big[\kappa\left(\frac{d^2}{dz^2}-k_H^2\right)+i\omega\Big]\Big[\nu\left(\frac{d^2}{dz^2}-k_H^2\right)+i\omega\Big]\left(\frac{d^2}{dz^2}-k_H^2\right)\\ +\alpha g\Gamma k_H^2\Big\}\tilde{p} &= 0. \end{aligned} \quad (5.2.33)$$

In particular, in the case of the neutral wave $\omega = 0$, the above equation becomes

$$\left[\left(\frac{d^2}{dz^2}-k_H^2\right)^3 + Ra\frac{k_H^2}{H^4}\right]\tilde{p} = 0. \quad (5.2.34)$$

Substituting (5.2.8), (5.2.9), and $T' = 0$ into (5.2.15)–(5.2.19), we obtain the boundary condition for $\tilde{p}$ as

$$\frac{d\tilde{p}}{dz} = \frac{d^3\tilde{p}}{dz^3} = \frac{d^5\tilde{p}}{dz^5} = \cdots = 0, \quad \text{at } z = 0,\ H. \quad (5.2.35)$$

This equation has a solution in the form

$$\tilde{p}(z) = \tilde{p}_0 \cos mz, \quad (5.2.36)$$

$$m = \frac{(2n+1)\pi}{H}, \quad n = 0, 1, 2, \ldots, \quad (5.2.37)$$

which is written as a linear combination of (5.2.24). Thus, from (5.2.29), the relation between the Rayleigh number and the wave number of the neutral solutions is given by

$$Ra = \frac{(k_H^2+m^2)^3}{k_H^2}H^4. \quad (5.2.38)$$

For a fixed vertical wave number m, Ra takes its minimum value (5.2.31) at (5.2.30). From (5.2.37), m takes the smallest value $m = \pi/H$ at $n = 0$. In this case, the Rayleigh number and the wave number are given as

$$Ra_c = \frac{27}{4}\pi^4 = 657.51, \quad (5.2.39)$$

$$k_{Hc} = \frac{\pi}{\sqrt{2}H} = \frac{2.22}{H}. \quad (5.2.40)$$

Ra_c is called the critical Rayleigh number, and k_{Hc} is called the critical wave number. The neutral curve is shown in Fig. 5.1.

Next, let us consider the structure of unstable waves. The *growth rate* is defined by

$$\sigma \equiv -i\omega = \omega_i - i\omega_r. \quad (5.2.41)$$

The wave is unstable if the real part of σ is positive (i.e., $\omega_i > 0$). From (5.2.26) and (5.2.27), ω_r becomes zero if ω_i is positive. That is, the unstable waves satisfies

$$\sigma = \omega_i. \quad (5.2.42)$$

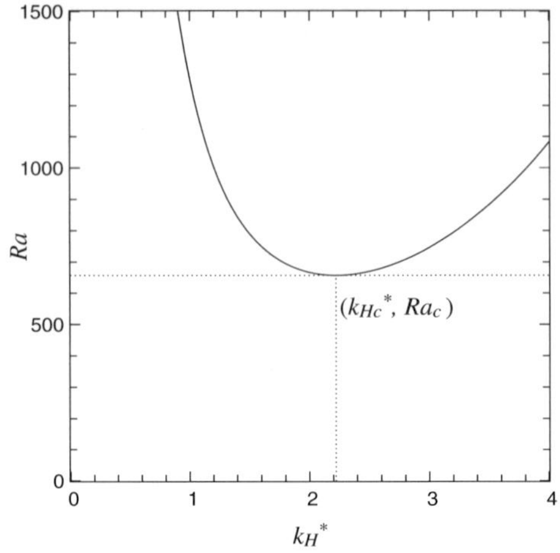

FIGURE 5.1: The neutral curve of convective instability. The abscissa is $k_H^* = k_H H$ and the ordinate is Ra. The Rayleigh number takes the minimum value Ra_c at $k_H^* = k_{Hc}H$.

From (5.2.32) and (5.2.36), p' is expressed as

$$p' = \tilde{p}_0 \cos mz \, e^{i(kx+ly)} \, e^{\sigma t}. \tag{5.2.43}$$

From (5.2.20)–(5.2.22), and (5.2.19), the other variables are expressed as follows:

$$u' = -i\frac{k}{\nu \boldsymbol{k}^2 + \sigma} \frac{\tilde{p}_0}{\rho_0} \cos mz \, e^{i(kx+ly)} \, e^{\sigma t}, \tag{5.2.44}$$

$$v' = -i\frac{l}{\nu \boldsymbol{k}^2 + \sigma} \frac{\tilde{p}_0}{\rho_0} \cos mz \, e^{i(kx+ly)} \, e^{\sigma t}, \tag{5.2.45}$$

$$w' = -\frac{k_H^2}{m(\nu \boldsymbol{k}^2 + \sigma)} \frac{\tilde{p}_0}{\rho_0} \sin mz \, e^{i(kx+ly)} \, e^{\sigma t}, \tag{5.2.46}$$

$$\begin{aligned} T' &= -\frac{k_H^2}{m(\nu \boldsymbol{k}^2 + \sigma)(\kappa \boldsymbol{k}^2 + \sigma)} \frac{\tilde{p}_0 \Gamma}{\rho_0} \sin mz \, e^{i(kx+ly)} \, e^{\sigma t} \\ &= -i\tilde{T}_0 \sin mz \, e^{i(kx+ly)} \, e^{\sigma t}, \end{aligned} \tag{5.2.47}$$

where $\boldsymbol{k}^2 = k_H^2 + m^2$, m is given by (5.2.37), and $\tilde{T}_0$ is the amplitude of perturbation of temperature. Fig. 5.2 shows the phase relationship of an unstable wave.

Vertical transports are second-order quantities. For instance, the horizontal average of vertical heat flux is given by

$$\overline{Re(w')Re(T')} = \frac{(\kappa \boldsymbol{k}^2 + \sigma)\tilde{T}_0^2}{2\Gamma} \sin^2 mz \, e^{2\sigma t}, \tag{5.2.48}$$

where $Re(A)$ represents the real part of A. For the critical mode in which $\sigma = 0$ and $\boldsymbol{k}^2 = 3m^2/2 = 3\pi^2/2H^2$, heat transport is

$$\overline{Re(w')Re(T')} = \frac{3\pi^2 \kappa \tilde{T}_0^2}{4\Gamma H^2} \sin^2 mz = 7.4\frac{\kappa \tilde{T}_0^2}{\Gamma H^2} \sin^2 mz. \tag{5.2.49}$$

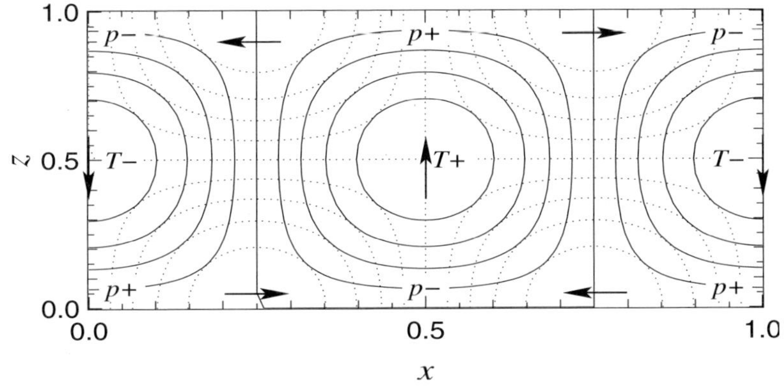

FIGURE 5.2: Phase relationship of the unstable wave of convective instability in the xz section with $l = 0$ and $n = 0$ (i.e., $m = \pi/H$), in the vertical. Solid curves are contours of temperature or vertical velocity, and dotted curves are those of lateral velocity.

The ratio of total heat flux to the heat flux due to thermal conductivity is called the *Nusselt number*. Since the horizontal average of the conductive heat flux is $\kappa\Gamma$, the Nusselt number for the critical mode is

$$Nu = \frac{\overline{Re(w)Re(T')} + \kappa\Gamma}{\kappa\Gamma} = 7.4\frac{\tilde{T}_0^2}{\Gamma^2 H^2}\sin^2 mz + 1. \tag{5.2.50}$$

The amplitude $\tilde{T}_0$ is an infinitesimal quantity on the basis of linear theory. Even in the case of the nonlinear regime, the amplitude of temperature is bounded by half of the temperature difference between the top and bottom boundaries. Therefore, the amplitude of temperature is in the range $\tilde{T}_0 \leq \Gamma H/2$. If we give the maximum bound of the temperature $\tilde{T}_0 = \Gamma H/2$, the Nusselt number estimated by (5.2.50) takes the maximum value 2.85.

If the Rayleigh number is in the unstable range above the critical value, unstable waves will develop and the fluid motion will have a finite amplitude. After a long time, the fluid may be either steady or unsteady; the final state may depend on the Rayleigh number and the initial state of the disturbance. For instance, at slightly larger Rayleigh numbers than the critical value, it is expected that convective cells will be finite and steady. Nonlinear analysis is required to obtain its amplitude and the preferable horizontal scale. We do not intend here to describe the characteristics of the nonlinear range of convective instability. We only mention the dependency of the Nusselt number on the Rayleigh number based on dimensional analysis.

When convection is fully developed in the nonlinear regime, we assume that the *local* Rayleigh number defined in the thermal boundary layer is equal to the critical Rayleigh number. In this case, the temperature difference in the thermal boundary layer is $\Delta T/2$, since the temperature in the inner region is homogeneously mixed. Letting the thickness of the thermal boundary layer be δ_T, we express the critical Rayleigh number as

$$Ra_c = \frac{\alpha g \Delta T \delta_T^3}{2\kappa\nu}. \tag{5.2.51}$$

From the definition of the Rayleigh number Ra, (5.2.28), the thickness of the thermal boundary layer is given by

$$\frac{\delta_T}{H} = \left(\frac{Ra}{2Ra_c}\right)^{-\frac{1}{3}}, \tag{5.2.52}$$

In the statistically steady state, the horizontal average of the vertical heat flux is constant irrespective of height. In particular, the horizontal average of the conductive heat flux in the boundary layer is equal to that of the total heat flux. Therefore, the Nusselt number is given by

$$Nu = \frac{\kappa\Delta T/2\delta_T}{\kappa\Delta T/H} = \frac{H}{2\delta_T} = \frac{1}{2}\left(\frac{Ra}{2Ra_c}\right)^{\frac{1}{3}} \sim Ra^{\frac{1}{3}} \tag{5.2.53}$$

(i.e., the Nusselt number is proportional to the Rayleigh number to the one-third power). This is sometimes called the *one-third power law*.

5.2.2 Convective instability in a rotating frame

We next consider the effects of rotation on convective motion. The discussion of this section follows Chandrasekhar (1961), who fully discussed convection in a rotating frame. The rotation axis is assumed to agree with the direction of gravity and the boundary conditions are the same as in the previous section. The basic state in the previous section is also a steady basic solution in the rotating frame. Linearized equations around the basic state are given by adding the Coriolis terms to (5.2.15)–(5.2.19):

$$\frac{\partial u'}{\partial t} - fv' = -\frac{1}{\rho_0}\frac{\partial p'}{\partial x} + \nu\nabla^2 u', \tag{5.2.54}$$

$$\frac{\partial v'}{\partial t} + fu' = -\frac{1}{\rho_0}\frac{\partial p'}{\partial y} + \nu\nabla^2 v', \tag{5.2.55}$$

$$\frac{\partial w'}{\partial t} = -\frac{1}{\rho_0}\frac{\partial p'}{\partial z} + \alpha g T' + \nu\nabla^2 w', \tag{5.2.56}$$

$$\frac{\partial u'}{\partial x} + \frac{\partial v'}{\partial y} + \frac{\partial w'}{\partial z} = 0, \tag{5.2.57}$$

$$\frac{\partial T'}{\partial t} - \Gamma w' = \kappa\nabla^2 T'. \tag{5.2.58}$$

The vorticity equation is given by subtracting the y-derivative of (5.2.54) from the x-derivative of (5.2.55), and the divergence equation is given by adding the x-derivative of (5.2.54) and the y-derivative of (5.2.55). Eliminating p' from the divergence equation and (5.2.56), we obtain the relation between w', ζ', and T', where $\zeta' = \frac{\partial v'}{\partial x} - \frac{\partial u'}{\partial y}$ is the perturbation vorticity. Hence, with the vorticity equation and (5.2.58), we have three equations for w', ζ', and T':

$$\left(\frac{\partial}{\partial t} - \nu\nabla^2\right)\zeta' - f\frac{\partial}{\partial z}w' = 0, \tag{5.2.59}$$

$$\left(\frac{\partial}{\partial t} - \nu\nabla^2\right)\nabla^2 w' + f\frac{\partial}{\partial z}\zeta' - \alpha g\nabla_H^2 T' = 0, \tag{5.2.60}$$

$$\left(\frac{\partial}{\partial t} - \kappa\nabla^2\right) T' - \Gamma w' \quad = \quad 0, \tag{5.2.61}$$

where $\nabla_H^2 = \frac{\partial^2}{\partial x^2} + \frac{\partial^2}{\partial y^2}$ is the horizontal Laplacian. Eliminating T' and ζ from these three equations, we have a single equation for w':

$$\left(\frac{\partial}{\partial t} - \nu\nabla^2\right)^2 \left(\frac{\partial}{\partial t} - \kappa\nabla^2\right) \nabla^2 w' - \alpha g\Gamma \left(\frac{\partial}{\partial t} - \nu\nabla^2\right) \nabla_H^2 w'$$
$$+f^2 \left(\frac{\partial}{\partial t} - \kappa\nabla^2\right) \frac{\partial^2}{\partial z^2} w' \quad = \quad 0. \tag{5.2.62}$$

We assume a solution that satisfies the boundary conditions in the form,

$$w' \quad = \quad \tilde{w}_0 \sin mz e^{i(kx+ly-\omega t)}, \tag{5.2.63}$$
$$m \quad = \quad \frac{(2n+1)\pi}{H}, \qquad n = 0, 1, 2, \cdots. \tag{5.2.64}$$

Substituting this into (5.2.62) yields

$$(-i\omega + \nu\boldsymbol{k}^2)^2(-i\omega + \kappa\boldsymbol{k}^2)\boldsymbol{k}^2 - \alpha g\Gamma(-i\omega + \nu\boldsymbol{k}^2)k_H^2$$
$$+f^2(-i\omega + \kappa\boldsymbol{k}^2)m^2 \quad = \quad 0. \tag{5.2.65}$$

Eq. (5.2.65) is a cubic polynomial for ω that has three solutions for ω. In the special case that $\kappa = 0$, $\nu = 0$, and $-\Gamma > 0$, a neutral wave exists; Eq. (5.2.65) reduces to the dispersion relation of the inertio-gravity wave (4.5.32) with $N = \sqrt{\alpha T|\Gamma|}$. In the case that $f = 0$, on the other hand, (5.2.65) reduces to the equation for the irrotational system (5.2.25).

If $f = 0$, the unstable regime starts from the neutral state where both $\omega_i = 0$ and $\omega_r = 0$ are satisfied. In the case that $f \neq 0$, however, the starting point of instability $\omega_i = 0$ does not necessarily mean $\omega_r = 0$. If $\omega_r = 0$ at the neutral state $\omega_i = 0$, it is called the *exchange of stability*. On the other hand, if $\omega_r \neq 0$ at $\omega_i = 0$, the propagating wave is unstable. In this case, the system is said to be *overstable*. It is also called the *oscillating instability*. Here, we will obtain solutions only for $\omega_r = \omega_i = 0$. The oscillating instability can be similarly examined by setting $\omega_i = 0$ under the condition $\omega_r \neq 0$. Details of the oscillating instability are described in Chandrasekhar (1961).

In order to obtain the neutral solution with $\omega_r = \omega_i = 0$, we substitute $\omega = 0$ into (5.2.65):

$$\boldsymbol{k}^6 - \frac{\alpha g\Gamma}{\nu\kappa} k_H^2 + \frac{f^2}{\nu^2} m^2 \quad = \quad 0. \tag{5.2.66}$$

Setting $k_H^* = k_H H$ and $m^* = mH$, we have the non-dimensional equation as

$$(k_H^{*2} + m^{*2})^3 - Ra\, k_H^{*2} + Ta\, m^{*2} \quad = \quad 0, \tag{5.2.67}$$

where

$$Ta \quad \equiv \quad \frac{f^2 H^4}{\nu^2}, \tag{5.2.68}$$

is called the *Taylor number*. The Rayleigh number Ra is already given by (5.2.28). Let us obtain the critical Rayleigh number, which is the minimum of Ra that satisfies (5.2.67) under the condition that the Taylor number Ta is kept constant. First, seek the minimum value of Ra by varying k_H^{*2} and keeping m^* constant. Differentiating (5.2.67) with respect to k_H^{*2} and setting $\frac{\partial Ra}{\partial k_H^{*2}} = 0$, we have

$$3(k_H^{*2} + m^{*2})^2 - Ra \quad = \quad 0. \tag{5.2.69}$$

Eliminating Ra from this and (5.2.67), we obtain

$$k_H^{*6} + \frac{3}{2} m^{*2} k_H^{*4} - \frac{1}{2}(Ta + m^{*4}) m^{*2} \quad = \quad 0. \tag{5.2.70}$$

If we introduce

$$X \equiv \frac{k_H^{*2}}{m^{*2}}, \qquad \cosh\phi \equiv 1 + 2\frac{Ta}{m^{*4}}, \tag{5.2.71}$$

Eq. (5.2.70) is written as

$$X^3 + \frac{3}{2}X^2 - \frac{1}{4}(\cosh\phi + 1) \quad = \quad 0. \tag{5.2.72}$$

It is known that this equation has a solution in the form

$$X \quad = \quad \cosh\frac{\phi}{3} - \frac{1}{2}. \tag{5.2.73}$$

From this, it can be shown that the Rayleigh number takes the minimum value if

$$k_H^{*2} \quad = \quad \left(\cosh\frac{\phi}{3} - \frac{1}{2}\right) m^{*2} \tag{5.2.74}$$

is satisfied; from (5.2.69), the minimum value is given as

$$Ra \quad = \quad 3(k_H^{*2} + m^{*2})^2 \quad = \quad 3\left(\cosh\frac{\phi}{3} + \frac{1}{2}\right)^2 m^{*4}. \tag{5.2.75}$$

Ra takes the smallest value if m^{*2} is the smallest at $n = 0$ in (5.2.37) (i.e., $m^{*2} = \pi$). Thus, the critical Rayleigh number is given by

$$Ra_c \quad = \quad 3\pi^4 \left(\cosh\frac{\phi}{3} + \frac{1}{2}\right)^2, \tag{5.2.76}$$

where

$$\cosh\phi \quad = \quad 1 + 2\frac{Ta}{\pi^4}, \tag{5.2.77}$$

and the critical wave number is

$$k_H \quad = \quad \left(\cosh\frac{\phi}{3} - \frac{1}{2}\right)^{\frac{1}{2}} \pi H. \tag{5.2.78}$$

If $Ta \to 0$, then $\phi \to 0$ and (5.2.76) approaches (5.2.39) (i.e., the critical Rayleigh number in the non-rotating system). In contrast, if $Ta \to \infty$, then $e^{\phi} \to 4Ta/\pi^4$ and

$$Ra_c = 3\left(\frac{\pi^2 Ta}{2}\right)^{\frac{2}{3}}, \qquad k_H = \left(\frac{\pi^2 Ta}{2}\right)^{\frac{1}{6}}. \tag{5.2.79}$$

The fact that the critical Rayleigh number increases with rotation rate indicates that the convection is stabilized by the effect of rotation. At the same time, the horizontal wave number of the neutral wave increases and the horizontal scale of convection becomes smaller as the rotation rate increases.

5.3 Inertial instability

In this section, the stability of a zonal flow U in the f-plane is considered. The basic flow U is a function of y and satisfies the geostrophic balance with the basic pressure field P as

$$fU = -\frac{1}{\rho_0}\frac{\partial P}{\partial y}. \tag{5.3.1}$$

For simplicity, we assume $U_y = \text{const.}$, which is consistent with the condition that the frictional force of the x component is zero. The angular momentum in the f-plane is given by (2.5.8):

$$L = U - fy. \tag{5.3.2}$$

We assume that the temperature is uniform and no heating or cooling is given; this means that we do not need to consider the effect of buoyancy. We write perturbations of zonal wind and pressure as u' and p', respectively. The linearized equations of (5.2.2)–(5.2.5) become

$$\left(\frac{\partial}{\partial t} + U\frac{\partial}{\partial x}\right)u' + L_y v' = -\frac{1}{\rho_0}\frac{\partial p'}{\partial x} + \nu\nabla^2 u', \tag{5.3.3}$$

$$\left(\frac{\partial}{\partial t} + U\frac{\partial}{\partial x}\right)v' + fu' = -\frac{1}{\rho_0}\frac{\partial p'}{\partial y} + \nu\nabla^2 v', \tag{5.3.4}$$

$$\left(\frac{\partial}{\partial t} + U\frac{\partial}{\partial x}\right)w' = -\frac{1}{\rho_0}\frac{\partial p'}{\partial z} + \nu\nabla^2 w', \tag{5.3.5}$$

$$\frac{\partial u'}{\partial x} + \frac{\partial v'}{\partial y} + \frac{\partial w'}{\partial z} = 0, \tag{5.3.6}$$

where $L_y = U_y - f$.

Eqs. (5.3.3)–(5.3.6) have solutions that depend on the x-direction. However, here we only consider symmetric disturbances (i.e., the perturbations that are uniform in the x-direction and have a sinusoidal structure in the y- and z-directions). The governing equations for the perturbations are

$$\frac{\partial u'}{\partial t} + L_y v' = \nu\nabla^2 u', \tag{5.3.7}$$

$$\frac{\partial v'}{\partial t} + fu' = -\frac{1}{\rho_0}\frac{\partial p'}{\partial y} + \nu\nabla^2 v', \tag{5.3.8}$$

$$\frac{\partial w'}{\partial t} = -\frac{1}{\rho_0}\frac{\partial p'}{\partial z} + \nu\nabla^2 w', \tag{5.3.9}$$

$$\frac{\partial v'}{\partial y} + \frac{\partial w'}{\partial z} = 0, \tag{5.3.10}$$

where $\nabla^2 = \frac{\partial^2}{\partial y^2} + \frac{\partial^2}{\partial z^2}$. From these equations, a single equation for p' is derived as

$$\left[\left(\frac{\partial}{\partial t} - \nu\nabla^2\right)^2 \nabla^2 - fL_y\frac{\partial^2}{\partial z^2}\right] p' = 0. \tag{5.3.11}$$

Assuming that p' has a sinusoidal form as (5.2.24), we obtain the dispersion relation. In particular, the neutral mode with $\omega = 0$ satisfies

$$\frac{fL_y}{\nu^2} = \frac{(l^2+m^2)^3}{m^2}. \tag{5.3.12}$$

This describes the relation between l, m, and the parameter fL_y/ν^2. Eq. (5.3.12) is analogous to (5.2.38) for convective instability. Letting the fluid width be D in the y-direction and applying the slip condition at the lateral boundaries, we obtain the critical values as

$$Ta_c = \frac{fL_yD^4}{\nu^2} = \frac{27}{4}\pi^4, \tag{5.3.13}$$

$$m_c = \frac{\pi}{\sqrt{2}D}. \tag{5.3.14}$$

The non-dimensional parameter fL_yD^4/ν^2 corresponds to the Taylor number. If the Taylor number is slightly larger than the critical value Ta_c, only a disturbance with wave number m_c in the z-direction grows.

The above mentioned instability is categorized into the *inertial instability* derived by the parcel method in Sections 2.4 and 2.5. At that time, we obtain the necessary condition for the instability in the case of nonfriction as

$$fL_y = f(U_y - f) > 0, \tag{5.3.15}$$

as seen from (2.5.13). If there is an effect of friction, on the other hand, the inequality (5.3.15) is not a sufficient condition for the instability. The instability occurs only if the Taylor number exceeds the critical value (5.3.13).

In reality, inertial instability occurs near the upper troposphere close to the equator. Near the equator, since f becomes smaller, the condition (5.3.15) may be satisfied if horizontal shear exists (Held and Hou, 1980; Hayashi et al., 2002).

5.4 Barotropic instability

We may further consider the gravitational effect on the stability of a shear flow. In this case, the temperature of the basic state is assumed to be in thermal balance with the shear. In such a system, there are unstable perturbations that are uniform in the x-direction; this is called *symmetric instability* (see Section 2.4). On the other hand, there is also an unstable perturbation that is inhomogeneous in the x-direction; this is called *shear instability. Barotropic instability* is a special case of shear instability where the basic flow U depends only on y. If U depends on z, there is another type of shear instability called *baroclinic instability.* We will examine these two kinds of shear instability in this and the next sections, respectively.

5.4.1 Formulation

First, we consider the stability of a two-dimensional parallel flow. We assume that the flow is confined in a channel with $0 \leq y \leq L$ and that the basic flow is expressed as $\overline{u} = U(y)$.[†] The flow is barotropic in the sense that it is uniform in the z-direction. Note that, in general, the basic flow with arbitrary shear $U(y)$ is not a steady solution to the governing equations if friction exists. We also assume that perturbation is uniform in the z-direction (i.e., barotropic). Setting $w' = 0$ in (5.3.3)–(5.3.6), we obtain linearized equations for the perturbation as

$$\left(\frac{\partial}{\partial t} + U\frac{\partial}{\partial x}\right) u' + (U_y - f)v' = -\frac{1}{\rho_0}\frac{\partial p}{\partial x} + \nu\nabla^2 u', \tag{5.4.1}$$

$$\left(\frac{\partial}{\partial t} + U\frac{\partial}{\partial x}\right) v' + fu' = -\frac{1}{\rho_0}\frac{\partial p}{\partial y} + \nu\nabla^2 v', \tag{5.4.2}$$

$$\frac{\partial u'}{\partial x} + \frac{\partial v'}{\partial y} = 0. \tag{5.4.3}$$

Here, we generalize the Coriolis parameter as $f = f_0 + \beta y$ by introducing the β effect. From (5.4.1)–(5.4.3), we obtain the vorticity equation as

$$\left(\frac{\partial}{\partial t} + U\frac{\partial}{\partial x}\right) \nabla^2\psi' + (\beta - U_{yy})\frac{\partial \psi'}{\partial x} = \nu\nabla^2\nabla^2\psi', \tag{5.4.4}$$

where ψ' is the streamfunction, which is related to the velocity components and vorticity ζ' as

$$u' = -\frac{\partial \psi'}{\partial y}, \qquad v' = \frac{\partial \psi'}{\partial x}, \tag{5.4.5}$$

$$\zeta' = \frac{\partial v'}{\partial x} - \frac{\partial u'}{\partial y} = \nabla^2\psi'. \tag{5.4.6}$$

We assume that the perturbation is periodic in the x-direction and specify the lateral boundary conditions as $\psi' = 0$ at $y = 0$ and L; this is equivalent to the conditions that $v' = 0$ at the boundaries and the domain integral of u' is zero. Since the coefficients of (5.4.4) do not depend on x, we can assume that the solution has a sinusoidal form in the x-direction. Assuming exponential dependency in time, we write the streamfunction as

$$\psi' = \phi(y)e^{i(kx-\omega t)} = \phi(y)e^{ik(x-ct)}, \tag{5.4.7}$$

[†]In this section, the symbol L represents the width of the channel, while L was used for angular momentum in the previous section.

where c is a complex phase speed defined as $c = c_r + ic_i$. The boundary condition is $\phi = 0$ at $y = 0$ and L. For unstable waves with $c_i > 0$, we can rewrite (5.4.4) as

$$(U - c)\left(\frac{d^2\phi}{dy^2} - k^2\phi\right) + (\beta - U_{yy})\phi \quad = \quad \frac{\nu}{ik}\left(\frac{d^2}{dy^2} - k^2\right)^2 \phi, \tag{5.4.8}$$

which is called the *Orr-Sommerfeld equation.* In particular, in the limit of no friction, $\nu \to 0$, this equation is reduced to the *Rayleigh equation*:

$$(U - c)\left(\frac{d^2\phi}{dy^2} - k^2\phi\right) + (\beta - U_{yy})\phi \quad = \quad 0. \tag{5.4.9}$$

We consider the basic characteristics of unstable waves in shear flows using the Rayleigh equation.

5.4.2 Integral theorems

The necessary condition for instability of a shear flow can be given from an integral form of the Rayleigh equation. Multiplying ϕ^* to (5.4.9), integrating in the region $0 \le y \le L$ by part, and using the boundary condition we obtain

$$\int_0^L \left(\left|\frac{d\phi}{dy}\right|^2 + k^2|\phi|^2 - \frac{\beta - U_{yy}}{U - c}|\phi|^2\right) dy \quad = \quad 0. \tag{5.4.10}$$

The imaginary part of the integral is written as

$$c_i \int_0^L \frac{\beta - U_{yy}}{|U - c|^2}|\phi|^2 \, dy \quad = \quad 0. \tag{5.4.11}$$

Thus, in order for the basic flow to be unstable with $c_i > 0$, the following integral must vanish:

$$\int_0^L \frac{\beta - U_{yy}}{|U - c|^2}|\phi|^2 \, dy \quad = \quad 0. \tag{5.4.12}$$

To satisfy this constraint, the sign of $\beta - U_{yy}$ must change at least once in the domain. It can be written as a gradient of the zonal-mean absolute vorticity ζ_a:

$$\frac{d\overline{\zeta_a}}{dy} \quad = \quad \frac{d}{dy}\left(f + \overline{\zeta}\right) \quad = \quad \frac{d}{dy}(f - U_y) \quad = \quad \beta - U_{yy}. \tag{5.4.13}$$

Thus, the necessary condition for instability of a basic flow is that the absolute vorticity of the zonal flow has an extremum within the domain. This constraint is called *Kuo's theorem.* In particular for the case $\beta = 0$, the necessary condition for instability reads that there is at least one point at which $U_{yy} = 0$ is satisfied (i.e., the basic flow is unstable if U has an inflection point). This is called *Rayleigh's theorem.*

The real part of (5.4.10) is written as

$$\int_0^L \left(\left|\frac{d\phi}{dy}\right|^2 + k^2|\phi|^2 - (U - c_r)\frac{\beta - U_{yy}}{|U - c|^2}|\phi|^2\right) dy \quad = \quad 0. \tag{5.4.14}$$

Thus, we have

$$\int_0^L (U - c_r)\frac{\beta - U_{yy}}{|U-c|^2}|\phi|^2\,dy \;=\; \int_0^L \left(\left|\frac{d\phi}{dy}\right|^2 + k^2|\phi|^2\right)dy \;>\; 0. \quad (5.4.15)$$

Let us assume that U is monotonic within the domain $0 \leq y \leq L$ and that the absolute vorticity takes its extreme at $y = y_s$ $(0 < y_s < L)$:

$$\frac{d\overline{\zeta_a}}{dy}(y_s) \;=\; \beta - U_{yy}(y_s) \;=\; 0. \quad (5.4.16)$$

Multiplying $c_r - U(y_s)$ to (5.4.12) and adding the result with (5.4.15), we have

$$\int_0^L (U - U(y_s))\frac{d\overline{\zeta_a}}{dy}\frac{|\phi|^2}{|U-c|^2}\,dy \;>\; 0. \quad (5.4.17)$$

From the condition (5.4.16) and the monotonicity of U, the sign of $(U - U(y_s))\frac{d\overline{\zeta_a}}{dy}$ does not change within the range $0 \leq y \leq L$. Therefore, from (5.4.17), we obtain the necessary condition for instability as

$$(U - U(y_s))\frac{d\overline{\zeta_a}}{dy} \;\geq\; 0. \quad (5.4.18)$$

In the case of $\beta = 0$, this condition reads that the absolute vorticity $\overline{\zeta_a}$ takes its minimum at y_s if U is an increasing function, or $\overline{\zeta_a}$ takes its maximum at y_s if U is a decreasing function. This is called *Fjørtoft's theorem.*

Next, we consider the bounds of the phase speed of unstable waves. Letting

$$\phi \;=\; (U - c)\Phi, \quad (5.4.19)$$

we rewrite (5.4.9) as

$$(U-c)\frac{d^2\Phi}{dy^2} + 2U_y\frac{d\Phi}{dy} - k^2(U-c)\Phi + \beta\Phi \;=\; 0. \quad (5.4.20)$$

Multiplying $(U - c)\Phi^*$ and integrating within the domain $0 \leq y \leq L$, we have

$$\int_0^L (U-c)^2\left(\left|\frac{d\Phi}{dy}\right|^2 + k^2|\Phi|^2\right)dy \;=\; \beta\int_0^L (U-c)|\Phi|^2\,dy. \quad (5.4.21)$$

For the unstable waves $c_i > 0$, the imaginary part of this equation becomes

$$\int_0^L (U-c_r)\left(\left|\frac{d\Phi}{dy}\right|^2 + k^2|\Phi|^2\right)dy \;=\; \frac{\beta}{2}\int_0^L |\Phi|^2\,dy. \quad (5.4.22)$$

If $\beta = 0$, the left-hand side of (5.4.22) must vanish for instability (i.e., $U - c_r$ must change its sign at least once within the domain). Thus, letting the minimum and the maximum of U be U_{min} and U_{max}, respectively, we have the bounds for c_r as

$$U_{min} \;<\; c_r \;<\; U_{max}. \quad (5.4.23)$$

This means that the phase speed of unstable waves is within the range of basic zonal velocity.

If $\beta \neq 0$, by expanding

$$\Phi = \sum_{j=1}^{\infty} A_j \sin \frac{\pi j y}{L}, \tag{5.4.24}$$

we have

$$\frac{d\Phi}{dy} = \frac{\pi}{L} \sum_{j=1}^{\infty} A_j j \cos \frac{\pi j y}{L}, \tag{5.4.25}$$

$$\int_0^L \left| \frac{d\Phi}{dy} \right|^2 dy = \frac{\pi^2}{2L} \sum_{j=1}^{\infty} |A_j|^2 j^2 \geq \frac{\pi^2}{2L} \sum_{j=1}^{\infty} |A_j|^2 = \frac{\pi^2}{L^2} \int_0^L |\Phi|^2 \, dy. \tag{5.4.26}$$

From (5.4.22), we write the phase speed as

$$c_r = \frac{\int_{-1}^{1} U \left(\left| \frac{d^2\Phi}{dy^2} \right|^2 + k^2 |\Phi|^2 \right) dy}{\int_{-1}^{1} \left(\left| \frac{d^2\Phi}{dy^2} \right|^2 + k^2 |\Phi|^2 \right) dy} - \frac{\beta}{2} \frac{\int_{-1}^{1} |\Phi|^2 \, dy}{\int_{-1}^{1} \left(\left| \frac{d^2\Phi}{dy^2} \right|^2 + k^2 |\Phi|^2 \right) dy}. \tag{5.4.27}$$

Using (5.4.26), we have

$$U_{min} - \frac{\beta}{2 \left(\frac{\pi^2}{L^2} + k^2 \right)} < c_r < U_{max}. \tag{5.4.28}$$

Although the bounds for c_r depend on k, the basic flow can be unstable if c_r is smaller than U_{min} in the case of $\beta \neq 0$.

5.4.3 Instability of the flow U= tanh y

As an example of barotropic instability, we consider the stability of such a flow on the f-plane ($\beta = 0$) that

$$U(y) = \tanh y, \tag{5.4.29}$$

with $-\infty < y < \infty$. In this case,

$$U_y = -\overline{\zeta} = \operatorname{sech}^2 y, \tag{5.4.30}$$

$$U_{yy} = -\frac{d\overline{\zeta}}{dy} = -2 \operatorname{sech}^2 y \tanh y, \tag{5.4.31}$$

where $\overline{\zeta}$ is the vorticity of the basic state. This flow has an inflection point at $y = 0$ and satisfies the necessary condition for instability stated by Fjørtoft's theorem. The equation for perturbation is given by the Rayleigh equation (5.4.9) with $\beta = 0$:

$$(U - c) \left(\frac{d^2\phi}{dy^2} - k^2 \phi \right) - U_{yy} \phi = 0. \tag{5.4.32}$$

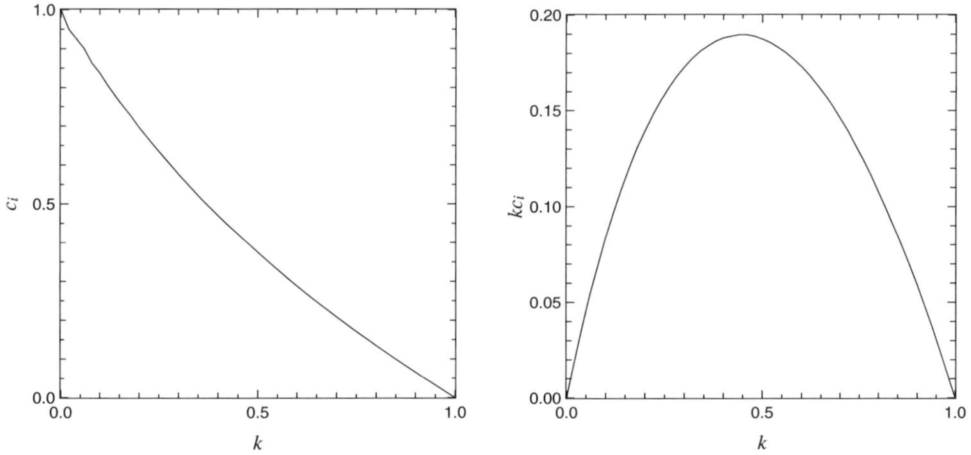

FIGURE 5.3: The phase speed and growth rate of barotropic instability for the basic flow $U = \tanh y$. (Left) Relation between the wave number k and the imaginary part of the phase speed c_i. (Right) Relation between the wave number k and the growth rate kc_i.

First, we seek a solution to a neutral wave whose phase speed is zero. Setting $c = 0$ in (5.4.32) and substituting (5.4.29) and (5.4.31), we have

$$\frac{d^2\phi}{dy^2} - (k^2 - 2\,\mathrm{sech}^2 y)\phi = 0. \tag{5.4.33}$$

This equation has two independent solutions as

$$\phi_1 = k \cosh ky - \sinh ky\ \tanh ky, \tag{5.4.34}$$

$$\phi_2 = k \sinh ky - \cosh ky\ \tanh ky, \tag{5.4.35}$$

where ϕ_1 is symmetric about $y = 0$, and ϕ_2 is antisymmetric about $y = 0$. If we assume that the solution is finite as $y \to \pm\infty$, we must choose $k = 1$; that is,

$$\phi_1 = \mathrm{sech}\ y \tag{5.4.36}$$

and $\phi_2 = 0$.

In general cases with $c \neq 0$, (5.4.32) can be solved with numerical methods (e.g., Michalke, 1968; Tanaka, 1975). Fig. 5.3 shows the dependency of c_i and the growth rate kc_i on wave number. The growth rate takes the maximum value 0.190 at $k = 0.446$. Fig. 5.4 shows structures of the streamfunction and vorticity for the neutral wave (5.4.36) and Fig. 5.5 those for the most unstable wave.

5.4.4 Interpretation of barotropic instability

Barotropic instability can be viewed as the interaction between Rossby waves. If absolute vorticity has a gradient, there exist Rossby waves whose propagation depends on the direction of the gradient. Let us assume that the basic flow U is a monotonically increasing function of y and Fjørtoft's theorem is applicable. In this

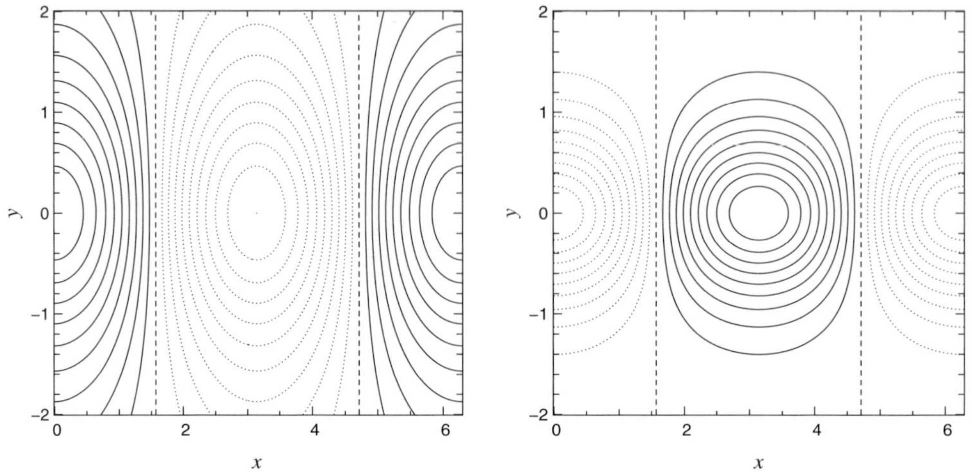

FIGURE 5.4: Structures of the neutral wave for the flow $U = \tanh y$. (Left) streamfunction and (right) vorticity. The maximum value is 1 and the contour interval is 0.1. Positive values are represented by solid lines and negative values by dotted.

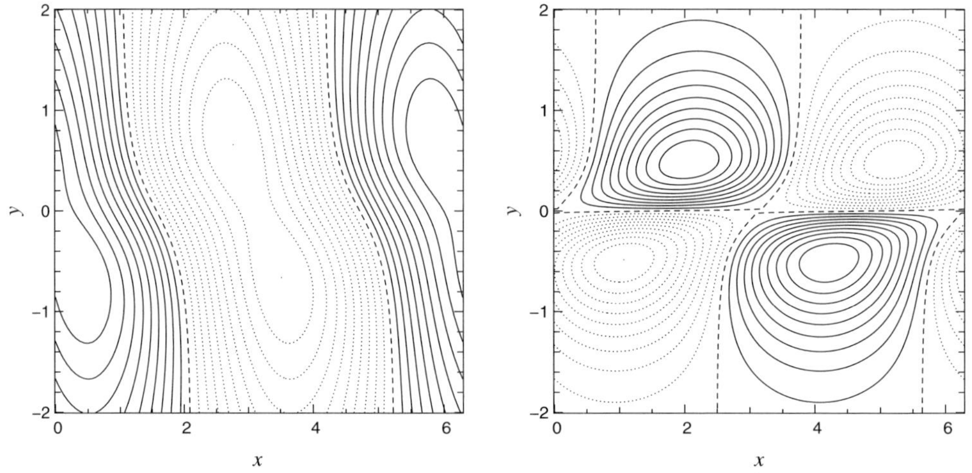

FIGURE 5.5: Same as Fig. 5.3 but for the most unstable wave $k = 0.446$.

case, the zonal-mean absolute vorticity $\overline{\zeta_a}$ takes its minimum at some point $y = y_s$ with $U > U(y_s)$, $\frac{d\overline{\zeta_a}}{dy} > 0$ in $y > y_s$, and $U < U(y_s)$, $\frac{d\overline{\zeta_a}}{dy} < 0$ in $y < y_s$. In this case, the phase speed of Rossby waves is westward (the negative x-direction) in $y > y_s$ and is eastward (the positive x-direction) in $y < y_s$. Thus, the directions of Rossby wave propagation in the two sides of y_s are opposite and the two Rossby waves can be in such a phase relationship that they are reinforced by each other. Therefore, it can be said that the basic flow is unstable.

Fig. 5.6 schematically shows the phase relation of two Rossby waves in the case

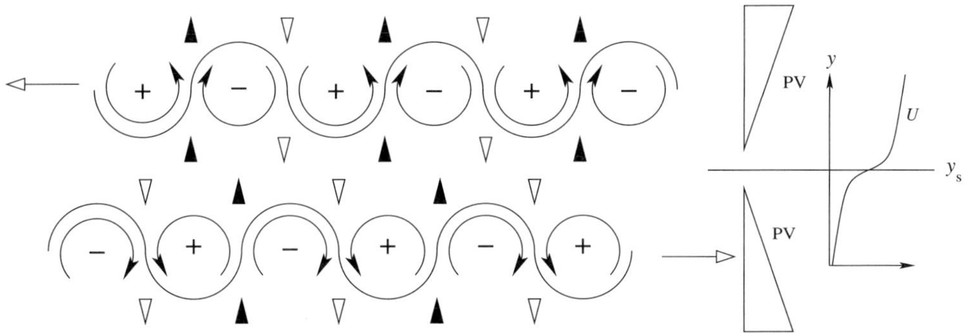

FIGURE 5.6: Schematic figure of the interaction between Rossby waves for barotropic instability. The profiles of absolute vorticity (or potential vorticity; PV) and zonal velocity U are shown at the right; PV is minimum at the center y_S. In $y > y_S$, Rossby waves propagate toward the left (westward), whereas, in $y < y_S$, Rossby waves propagate toward the right (eastward). The basic flow is unstable if the two waves are in phase relation such that the vorticity of each Rossby wave is intensified by the y-component of velocity associated with the other Rossby wave.

of unstable waves. In order for the necessary condition for instability to be satisfied, the gradient of the basic state absolute vorticity should be opposite in two adjacent regions. In the two regions, Rossby waves propagate toward opposite directions. If the phase relation is preferable, the two Rossby waves are amplified.

In the case of an unstable wave in the flow $U = \tanh y$ in the previous subsection, the phase lines of the streamfunction are inclined westward so that $\overline{u'v'} < 0$ as shown in Fig. 5.5. Since $U_y > 0$, energy is converted from the basic flow to the perturbation (i.e., the perturbation is amplified).† In contrast, the perturbation will decay if the phase lines are inclined to the opposite direction and $\overline{u'v'} > 0$. In this case, energy is converted from the perturbation to the basic flow. The center of positive vorticity in $y > 0$ is located at almost the same x-coordinate as the maximum of the southward component of velocity $-v$ associated with the vorticity in $y < 0$. In $y > 0$, since the gradient of the basic vorticity is positive, a relatively larger vorticity is advected by the southward flow $v < 0$. Thus, the perturbation of the positive vorticity in $y > 0$ grows. This applies to both the positive and negative vorticities in $y > 0$ and $y < 0$. Therefore, the wave fields in $y > 0$ and $y < 0$ are mutually intensified.

5.5 Baroclinic instability

While barotropic instability occurs in the case when the basic flow has lateral shear, the baroclinic instability considered in this section occurs in the case when the basic flow has vertical shear. We consider the two typical problems of baroclinic instability: the *Eady problem* and the *Charney problem.* The Eady problem is the baroclinic instability in the system on a uniform rotating frame (the f-plane) with rigid top and bottom boundaries. The Charney problem, in contrast, is that on the

†Energy conversions of barotropic and baroclinic instabilities will be described in Section 5.5.3.

β plane with a rigid bottom boundary and a continuously infinite upper boundary. After examination of unstable solutions, the thermal and momentum transports associated with baroclinic instability and secondary circulation are considered in the framework of the Eady problem. At the end of section, baroclinic instability is interpreted based on the integral theorem and interaction between Rossby waves.

5.5.1 Eady problem

5.5.1.1 Formulation

We consider the stability of a zonal flow in a channel with $0 \leq y \leq L$ and $0 \leq z \leq H$ on the f-plane. Periodicity is assumed in the x-direction, and the channel is surrounded by rigid walls at $y = 0,\ L$ and $z = 0,\ H$. The basic flow has linear zonal shear in the z-direction with no meridional flow.

The Eady problem is formulated with the quasi-geostrophic equations of the Boussinesq fluid. First, we write down the quasi-geostrophic equations with no friction and no diabatic heating on the f-plane. Following Section 3.2, we use potential temperature θ instead of temperature T for the Boussinesq fluid. The quasi-geostrophic vorticity equation is given by

$$\left(\frac{\partial}{\partial t} + \boldsymbol{u_g} \cdot \nabla_H\right) \Pi \quad = \quad 0, \tag{5.5.1}$$

where Π is the quasi-geostrophic potential vorticity:

$$\Pi \quad = \quad \frac{1}{f}\left(\frac{\partial^2}{\partial x^2} + \frac{\partial^2}{\partial y^2} + \frac{f^2}{N^2}\frac{\partial^2}{\partial z^2}\right)\psi. \tag{5.5.2}$$

The streamfunction ψ is related to the geostrophic winds $\boldsymbol{u_g} = (u_g, v_g)$, vorticity ζ, potential temperature θ, and pressure p as

$$u_g \quad = \quad -\frac{1}{f}\frac{\partial \psi}{\partial y}, \qquad v_g \quad = \quad \frac{1}{f}\frac{\partial \psi}{\partial x}, \tag{5.5.3}$$

$$\zeta \quad = \quad \frac{\partial v_g}{\partial x} - \frac{\partial u_g}{\partial y} \quad = \quad \left(\frac{\partial^2}{\partial x^2} + \frac{\partial^2}{\partial y^2}\right)\psi, \tag{5.5.4}$$

$$\theta \quad = \quad \frac{\theta_0}{g}\frac{\partial \psi}{\partial z}, \qquad p \quad = \quad \rho_0 \psi. \tag{5.5.5}$$

In these equations, θ and p are deviations from the basic fields θ_s and p_s. The basic profile of the potential temperature is written as

$$\theta_s(z) \quad = \quad \theta_s(0) + \frac{N^2\theta_0}{g} z. \tag{5.5.6}$$

The equations of the geostrophic winds, the hydrostatic balance, the equation of potential temperature, and the continuity equation are given by

$$\left(\frac{\partial}{\partial t} + \boldsymbol{u_g} \cdot \nabla_H\right) u_g = f v_a, \tag{5.5.7}$$

$$\left(\frac{\partial}{\partial t} + \boldsymbol{u_g} \cdot \nabla_H\right) v_g = -f u_a, \tag{5.5.8}$$

$$0 = -\frac{1}{\rho_0}\frac{\partial p}{\partial z} + \frac{\theta}{\theta_0} g, \tag{5.5.9}$$

$$\left(\frac{\partial}{\partial t} + \boldsymbol{u_g} \cdot \nabla_H\right) \theta + \frac{N^2 \theta_0}{g} w = 0, \tag{5.5.10}$$

$$\frac{\partial u_a}{\partial x} + \frac{\partial v_a}{\partial y} + \frac{\partial w}{\partial z} = 0, \tag{5.5.11}$$

where u_a, v_a are the ageostrophic components of the winds. The vorticity equation is

$$\left(\frac{\partial}{\partial t} + \boldsymbol{u_g} \cdot \nabla_H\right) \zeta - f\frac{\partial w}{\partial z} = 0. \tag{5.5.12}$$

The boundary conditions are $v_g = 0$ at $y = 0,\ L$, and $w = 0$ at $z = 0,\ H$. From (5.5.10) and (5.5.5), the condition $w = 0$ is expressed using the streamfunction as

$$\left(\frac{\partial}{\partial t} + \boldsymbol{u_g} \cdot \nabla_H\right) \frac{\partial}{\partial z}\psi = 0. \tag{5.5.13}$$

Let us consider uniform shear flow as the basic state:

$$\overline{u}_g = U(z) = \Lambda z, \qquad \overline{v}_g = 0. \tag{5.5.14}$$

In this case, the other variables are written as

$$\overline{\psi} = -f\int_0^y \overline{u}_g dy = -f\Lambda yz, \tag{5.5.15}$$

$$\overline{\Pi} = f, \tag{5.5.16}$$

$$\overline{\theta} = \frac{\theta_0}{g}\frac{\partial \overline{\psi}}{\partial z} = -\frac{f\theta_0}{g}\Lambda y. \tag{5.5.17}$$

Let the perturbation of the streamfunction be ψ'. The linearized equation of the quasi-geostrophic potential vorticity for perturbation is written as

$$\left(\frac{\partial}{\partial t} + \Lambda z\frac{\partial}{\partial x}\right)\left(\frac{\partial^2}{\partial x^2} + \frac{\partial^2}{\partial y^2} + \frac{f^2}{N^2}\frac{\partial^2}{\partial z^2}\right)\psi' = 0. \tag{5.5.18}$$

The boundary conditions are given as

$$\frac{\partial}{\partial t}\frac{\partial \psi'}{\partial z} - \Lambda\frac{\partial \psi'}{\partial x} = 0, \quad \text{at } z = 0, \tag{5.5.19}$$

$$\left(\frac{\partial}{\partial t} + \Lambda H\frac{\partial}{\partial x}\right)\psi' - \Lambda\frac{\partial \psi'}{\partial x} = 0, \quad \text{at } z = H, \tag{5.5.20}$$

$$\frac{\partial \psi'}{\partial x} = 0, \quad \text{at } y = 0, L. \tag{5.5.21}$$

From (5.5.21), we particularly specify

$$\psi' = 0, \quad \text{at} \quad y = 0, L. \tag{5.5.22}$$

We assume the following sinusoidal form of the solution that satisfies the boundary condition:

$$\psi' = \tilde{\psi}(z) \sin ly \, e^{i(kx-\omega t)} = \tilde{\psi}(z) \sin ly \, e^{ik(x-ct)}, \tag{5.5.23}$$

where

$$l = \frac{n\pi}{L}, \quad \text{for} \quad n = 1, 2, \cdots, \tag{5.5.24}$$

and $c = \omega/k$ is a complex phase speed. Substituting (5.5.23) into the linearized equations (5.5.18), (5.5.19), and (5.5.20), we obtain

$$(\Lambda z - c)\left(\frac{d^2}{dz^2} - \mu^2\right)\tilde{\psi} = 0, \tag{5.5.25}$$

$$c\frac{d\tilde{\psi}}{dz} + \Lambda\tilde{\psi} = 0, \quad \text{at} \quad z = 0, \tag{5.5.26}$$

$$(c - \Lambda H)\frac{d\tilde{\psi}}{dz} + \Lambda\tilde{\psi} = 0, \quad \text{at} \quad z = H, \tag{5.5.27}$$

where

$$\mu^2 \equiv \frac{N^2}{f^2}(k^2 + l^2). \tag{5.5.28}$$

The following two types of solutions satisfy (5.5.25). The first type of solution has no singularity, satisfying

$$\left(\frac{d^2}{dz^2} - \mu^2\right)\tilde{\psi} = 0. \tag{5.5.29}$$

The second type of solution has singularity at z_c and satisfies

$$\left(\frac{d^2}{dz^2} - \mu^2\right)\tilde{\psi} = A\delta(z - z_c), \tag{5.5.30}$$

where $\delta(z)$ is Dirac's delta function,†

$$z_c \equiv \frac{c}{\Lambda}, \quad 0 < z_c < H, \tag{5.5.31}$$

and A is constant. In this case, the phase speed c is a real value in the range

$$0 < c < \Lambda H. \tag{5.5.32}$$

In the following two subsections we will obtain each type of solution.

---

†Dirac's delta function $\delta(z)$ satisfies the following relations:

$$\delta(z) = \begin{cases} 0, & \text{for} \quad z \neq 0, \\ \infty, & \text{for} \quad z = 0; \end{cases} \qquad \int_{-\infty}^{\infty} \delta(z)\, dz = 1.$$

5.5.1.2 Nonsingular solutions

First, as a nonsingular solution to (5.5.29), we can set the following form

$$\tilde{\psi}(z) \quad = \quad a'e^{\mu z} + b'e^{-\mu z} \quad = \quad a\cosh\mu z + b\sinh\mu z, \tag{5.5.33}$$

where a and b, or a' and b', are constant determined from the boundary conditions. Substituting (5.5.33) into (5.5.26) and (5.5.27), we have the relations

$$c\mu b + \Lambda a \quad = \quad 0, \tag{5.5.34}$$

$$\begin{aligned}(c - \Lambda H)\mu(a\sinh\mu H + b\cosh\mu H)& \\ +\Lambda(a\cosh\mu H + b\sinh\mu H) \quad &= \quad 0.\end{aligned} \tag{5.5.35}$$

A solution with $(a, b) \neq (0, 0)$ exists if the determinant of the coefficient matrix of a and b is equal to zero. Thus, the phase speed c is given by solution of

$$c^2 - c\Lambda H + \frac{\Lambda^2 H}{\mu}\coth\mu H - \frac{\Lambda^2}{\mu^2} \quad = \quad 0. \tag{5.5.36}$$

From this, we obtain

$$c \quad = \quad \frac{\Lambda H}{2} \pm \frac{\Lambda H}{2}\left[\left(1 - \frac{2}{\mu H}\coth\frac{\mu H}{2}\right)\left(1 - \frac{2}{\mu H}\tanh\frac{\mu H}{2}\right)\right]^{\frac{1}{2}}. \tag{5.5.37}$$

To examine the properties of solutions, we introduce a constant x_c that satisfies

$$\frac{\coth x_c}{x_c} \quad = \quad 1 \tag{5.5.38}$$

(i.e., $x_c = 1.1997$), and a parameter

$$\mu_c \quad = \quad \frac{2x_c}{H} \quad = \quad \frac{2.3994}{H}. \tag{5.5.39}$$

From (5.5.37), we can see that c is real if $\mu > \mu_c$, whereas c has an imaginary part if $0 < \mu < \mu_c$. Thus, using (5.5.28), the basic state is unstable if

$$\mu \quad = \quad \frac{N}{f}(k^2 + l^2)^{\frac{1}{2}} \quad < \quad \mu_c, \tag{5.5.40}$$

or

$$L \quad > \quad \pi\frac{N}{f\mu_c} \quad = \quad \frac{\pi}{2.3994}\frac{NH}{f}, \tag{5.5.41}$$

where (5.5.24) is used. This means that only in the case that the channel width is sufficiently large, the basic state is unstable. The largest wave number of the unstable wave is

$$k_c \quad = \quad \left(\frac{f^2}{N^2}\mu_c^2 - \frac{\pi^2}{L^2}\right)^{\frac{1}{2}}. \tag{5.5.42}$$

If the wavelength is shorter than $2\pi/k_c$, perturbations do not grow. This limitation to the wavelength is called *Eady's short-wave cut-off.* In the case $\mu < \mu_c$, the growth rate of the instability is given from (5.5.37) as

$$\begin{aligned}\omega_i &= kc_i \\ &= \frac{\Lambda H}{2}k\left[\left(\frac{2}{\mu H}\coth\frac{\mu H}{2}-1\right)\left(1-\frac{2}{\mu H}\tanh\frac{\mu H}{2}\right)\right]^{\frac{1}{2}}. \qquad (5.5.43)\end{aligned}$$

Fig. 5.7 displays the dependency of phase speed on wave number. The dispersion relation (5.5.37) is nondimensionalized with $\mu^* = \mu H$ and $c^* = c/\Lambda H$ as

$$c^* = \frac{1}{2} \pm \frac{1}{2}\left[\left(1-\frac{2}{\mu^*}\coth\frac{\mu^*}{2}\right)\left(1-\frac{2}{\mu^*}\tanh\frac{\mu^*}{2}\right)\right]^{\frac{1}{2}}. \qquad (5.5.44)$$

The figure shows the imaginary part and the real part of phase speed where $c^* = c_r^* + ic_i^*$. We have $c^* = \frac{1}{2} \pm i\frac{1}{2\sqrt{3}}$ in the limit $\mu^* \to 0$, whereas we have $c^* = 0$, 1 in the limit $\mu^* \to \infty$. Fig. 5.8 shows the dependency of growth rate on wave number. Using $k^* = (NH/f)k$, $l^* = (NH/f)l$, $\mu^* = \mu H$, $c^* = c/\Lambda H$, and $\omega^* = (N/f\Lambda)\omega$, the expression of the growth rate (5.5.43) is rewritten as

$$\omega_i^* = k^*c_i^* = \frac{1}{2}k^*\left[\left(1-\frac{2}{\mu^*}\coth\frac{\mu^*}{2}\right)\left(1-\frac{2}{\mu^*}\tanh\frac{\mu^*}{2}\right)\right]^{\frac{1}{2}}, \qquad (5.5.45)$$

where $\mu^{*2} = k^{*2} + l^{*2}$. The maximum value of the growth rate is given as $\omega_{i*} = 0.3098$ at $k^* = 1.606$ and $l^* = 0$.

Now, we examine the structure of unstable waves. From (5.5.34), we can set $(a, b) = \psi_0(1, -\Lambda/\mu c)$. Then, (5.5.33) becomes

$$\tilde{\psi}(z) = \psi_0\left(\cosh\mu z - \frac{\Lambda}{\mu c}\sinh\mu z\right). \qquad (5.5.46)$$

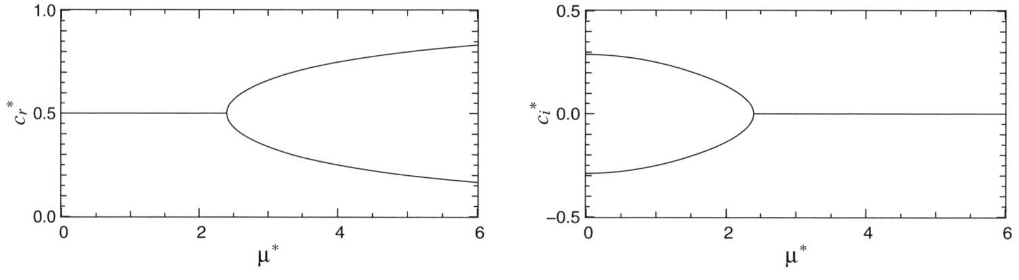

FIGURE 5.7: The dependency of phase speed on wave number μ^* for Eady solutions. Left: the real part c_r^*; right: the imaginary part c_i^*.

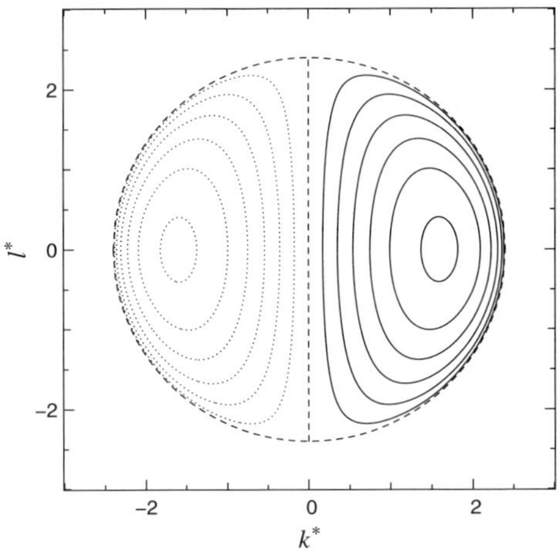

FIGURE 5.8: The dependency of growth rate ω_i^* on wave numbers k^* and l^* for Eady solutions. The contour interval of ω_i^* is 0.05.

For the unstable wave with $\mu < \mu_c$, substituting this equation into (5.5.23), we obtain for the real part

$$\begin{aligned} \psi' &= \psi_0 Re\left[\left(\cosh\mu z - \frac{c_r\Lambda}{\mu|c|^2}\sinh\mu z + i\frac{c_i\Lambda}{\mu|c|^2}\sinh\mu z\right)e^{ik(x-c_rt)}\right] \\ &\quad \times \sin ly\; e^{kc_it} \\ &= \Psi(z)\sin ly\;\cos[k(x-c_rt)+\alpha(z)]\;e^{kc_it}. \end{aligned} \tag{5.5.47}$$

Here, $\Psi(z)$ is the amplitude and $\alpha(z)$ is the phase, given as

$$\Psi(z) \equiv \psi_0\left[\left(\cosh\mu z - \frac{c_r\Lambda}{\mu|c|^2}\sinh\mu z\right)^2 + \frac{c_i^2\Lambda^2}{\mu^2|c|^4}\sinh^2\mu z\right]^{\frac{1}{2}}, \tag{5.5.48}$$

$$\alpha(z) = \tan^{-1}\frac{\Lambda c_i\sinh\mu z}{\mu|c|^2\cosh\mu z - \Lambda c_r\sinh\mu z}. \tag{5.5.49}$$

The direction of the phase line can be found from the sign of $d\alpha/dz$; if we write (5.5.49) as $\tan\alpha = Y/X$, we have

$$\frac{d\alpha}{dz} = \frac{XY' - X'Y}{X^2+Y^2} = \frac{\psi_0^2}{\Psi(z)^2}\frac{\Lambda c_i}{|c|^2}. \tag{5.5.50}$$

Thus, we have $\frac{d\alpha}{dz} > 0$ for the unstable waves $c_i > 0$. In this case, phase lines incline toward the negative x-direction (westward) with height. On the other hand, for neutral waves with $c_i = 0$, the phase is independent of height.

The other variables can be expressed with respect to $\tilde{\psi}$ from (5.5.3)–(5.5.5) and

(5.5.10) as

$$
\begin{aligned}
u_g' &= -\frac{l}{f}\tilde{\psi}(z)\cos ly\; e^{ik(x-ct)}, \\
v_g' &= i\frac{k}{f}\tilde{\psi}(z)\sin ly\; e^{ik(x-ct)}, \\
\theta' &= \frac{\theta_0}{g}\frac{d\tilde{\psi}(z)}{dz}\sin ly\; e^{ik(x-ct)}, \\
p' &= \rho_0\tilde{\psi}(z)\sin ly\; e^{ik(x-ct)}, \\
w' &= i\frac{k}{N^2}\left[(c-\Lambda z)\frac{d\tilde{\psi}(z)}{dz}+\Lambda\tilde{\psi}(z)\right]\sin ly\; e^{ik(x-ct)}.
\end{aligned} \tag{5.5.51}
$$

For unstable waves ($\mu < \mu_c$) the real part of these equations is given as

$$
\begin{aligned}
u_g' &= -\frac{l}{f}\Psi(z)\cos ly\;\cos[k(x-c_r t)+\alpha(z)]\; e^{kc_i t}, \\
v_g' &= -\frac{k}{f}\Psi(z)\sin ly\;\sin[k(x-c_r t)+\alpha(z)]\; e^{kc_i t}, \\
\theta' &= \frac{\theta_0}{g}\Psi_\theta(z)\sin ly\;\cos[k(x-c_r t)+\alpha_\theta(z)]\; e^{kc_i t}, \\
p' &= \rho_0\Psi(z)\sin ly\;\cos[k(x-c_r t)+\alpha(z)]\; e^{kc_i t}, \\
w' &= -\frac{k}{N^2}\Psi_w(z)\sin ly\;\sin[k(x-c_r t)+\alpha_w(z)]\; e^{kc_i t},
\end{aligned}
$$

where $(\Psi_\theta, \alpha_\theta)$ and (Ψ_w, α_w) are a pair of the amplitude and phase of the following functions, respectively,

$$
\begin{aligned}
\tilde{\psi}_\theta(z) &\equiv \psi_0\left(\mu\sinh\mu z - \frac{c_r\Lambda}{|c|^2}\cosh\mu z + i\frac{c_i^2\Lambda^2}{|c|^4}\cosh^2\mu z\right), \\
\tilde{\psi}_w(z) &\equiv \psi_0\left\{c_r\left[\left(\mu-\frac{\Lambda z}{c_r}\mu-\frac{\Lambda^2}{\mu|c|^2}\right)\sinh\mu z+\frac{\Lambda^2 z}{|c|^2}\cosh\mu z\right]\right. \\
&\quad \left. +ic_i\left[\left(\mu+\frac{\Lambda^2}{\mu|c|}\right)\sinh\mu z-\frac{\Lambda^2 z}{|c|^2}\cosh\mu z\right]\right\}.
\end{aligned}
$$

Fig. 5.9 shows the vertical profiles of the amplitude of the most unstable wave. The amplitudes of ψ' and θ' take a maximum value at the top and bottom boundaries, while the maximum of the amplitude of w' occurs at the middle layer. Fig. 5.10 shows the zonal-height section of the structure of the unstable wave at the middle of the channel $y = L/2$. The phase lines of ψ and w are inclined westward with height, while those of θ are inclined eastward with height. At the middle layer, the phases of w', θ', and v_g' agree with each other. These figures are for the parameter $NH/fL = 0.3$. In this case, the growth rate is maximum at the wave numbers $k^* = kL = 4.98$ and $l^* = lL = \pi$. The phase speed is $c_r^* = c_r/\Lambda H = 0.5$ and $c_i^* = c_i/\Lambda H = 0.172$, and the growth rate is $\omega_i^* = k^* c_i^* = 0.858$.

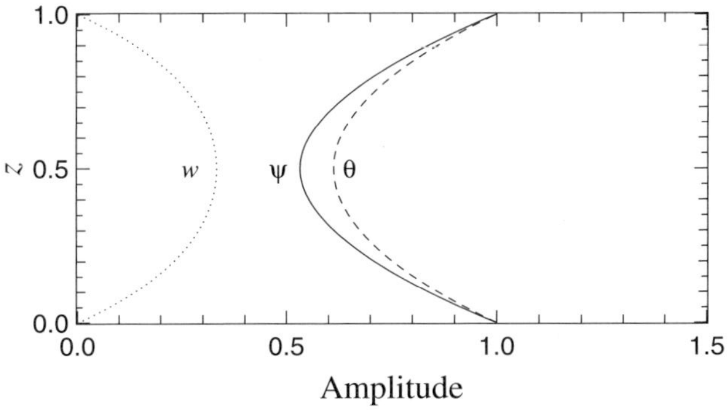

FIGURE 5.9: Vertical profiles of the amplitudes of the most unstable wave for $NH/fL = 0.3$. The solid curve is the amplitude of Ψ, the dashed curve is that of potential temperature Ψ_θ, and the dotted curve is that of vertical velocity Ψ_w. The scales of these amplitudes are appropriately normalized.

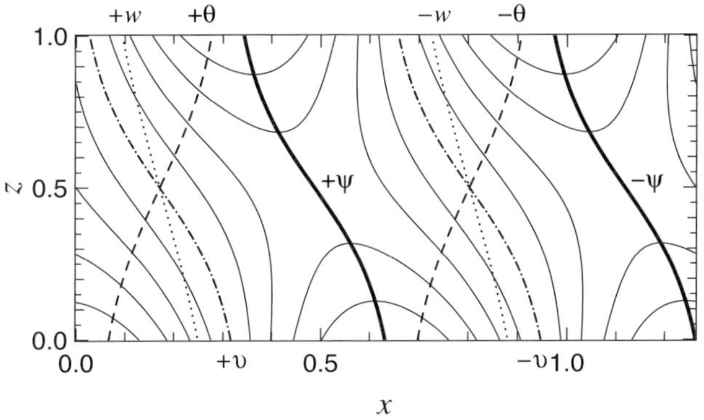

FIGURE 5.10: Zonal-height cross section of the most unstable wave at the middle of the channel $y = L/2$. The solid curves depict the streamfunction ψ. The thick solid curves are the phase lines of the maximum and minimum values of ψ. The dashed, dotted, and dashed-dotted curves are those of θ, w, and v_g, respectively. The parameters are the same as in Fig. 5.9. The wavelength in the x-direction is $2\pi/k^* = 1.26$.

At the end of this section, we consider transport due to unstable waves. The product of two complex values A and B is denoted by $\overline{Re(A)Re(B)}$, where $\overline{(\ )}$ denotes the average over phase X. If $A = |A|e^{iX}$ and $B = |B|e^{i(X+\alpha)}$ are given, we have the formula

$$\frac{1}{2}Re(AB^*) \quad = \quad \overline{Re(A)Re(B)} \quad = \quad \frac{1}{2}|A||B|\cos\alpha. \tag{5.5.52}$$

Substituting (5.5.51) into this, we obtain the transports of various quantities. First, the meridional transports of momentum and heat are given by

$$\frac{1}{2}Re(u'v'^*) = \frac{kl}{2f^2}Re(i|\tilde{\psi}|^2)\cos ly\ \sin ly\ e^{2kc_it} = 0, \tag{5.5.53}$$

$$\begin{aligned}\frac{1}{2}Re(v'\theta'^*) &= \frac{k\theta_0}{2gf}Re\left(i\tilde{\psi}\frac{d\tilde{\psi}^*}{dz}\right)\sin^2 ly\ e^{2kc_it} \\ &= \psi_0^2\frac{kc_i\Lambda\theta_0}{4gf|c|^2}(1-\cos 2ly)\ e^{2kc_it}.\end{aligned} \tag{5.5.54}$$

It is clear that there exists no meridional momentum transport due to the Eady baroclinic wave. Note that we have used the following relation with (5.5.50) to derive the heat transport (5.5.54):

$$Re\left(i\tilde{\psi}\frac{d\tilde{\psi}^*}{dz}\right) = Re\left[i\Psi e^{i\alpha}\cdot\frac{d}{dz}(\Psi e^{-i\alpha})\right] = \frac{d\alpha}{dz}\Psi^2 = \frac{\Lambda c_i}{|c|^2}\psi_0^2.$$

Since $\frac{d\alpha}{dz} > 0$ for $c_i > 0$, heat transport is positive: $\overline{v'\theta'} > 0$. It can be seen that heat transport is vertically constant.

Second, the vertical transport of momentum and heat are given by

$$\begin{aligned}\frac{1}{2}Re(u'w'^*) &= -\frac{kl}{2fN^2}Re\left\{-i\left[(c^*-\Lambda z)\frac{d\tilde{\psi}^*}{dz}+\Lambda\tilde{\psi}^*\right]\tilde{\psi}\right\} \\ &\quad\times\sin ly\ \cos ly\ e^{2kc_it} \\ &= \psi_0^2\frac{klc_i}{2fN^2}\left[\frac{1}{2\mu}\left(\mu^2+\frac{\Lambda^2}{|c|^2}\right)\sinh 2\mu z - \frac{c_r\Lambda}{|c|^2}\cosh 2\mu z\right. \\ &\quad\left.+(c_r-\Lambda z)\frac{\Lambda}{|c|^2}\right]\sin 2ly\ e^{2kc_it}, \\ \frac{1}{2}Re(\theta'w'^*) &= \frac{k\theta_0}{2gN^2}Re\left\{-i\left[(c^*-\Lambda z)\frac{d\tilde{\psi}^*}{dz}+\Lambda\tilde{\psi}^*\right]\frac{d\tilde{\psi}}{dz}\right\}\sin^2 ly\ e^{2kc_it} \\ &= \psi_0^2\frac{k\theta_0 c_i}{4gN^2}\left[-\left(\mu^2+\frac{\Lambda^2}{|c|^2}\right)\sinh^2\mu z+\frac{c_r\Lambda\mu}{|c|^2}\sinh 2\mu z\right] \\ &\quad\times(1-\cos 2ly)\ e^{2kc_it}.\end{aligned}$$

If these are averaged over the channel width in y, the domain-averaged vertical momentum transport becomes zero, while the domain-averaged vertical heat transport is upward for unstable waves ($c_i > 0$).

5.5.1.3 Singular solutions

Returning to (5.5.30), we consider singular solutions whose phase speed is a real value in the range (5.5.32):

$$c = \Lambda z_c, \quad (0 < z_c < H), \tag{5.5.55}$$

The solution to (5.5.30) can be written in the form

$$\begin{aligned}\tilde{\psi}(z) &= a_1 \cosh \mu z + b_1 \sinh \mu z, \quad (0 < z < z_c), & (5.5.56)\\ \tilde{\psi}(z) &= a_2 \cosh \mu z + b_2 \sinh \mu z, \quad (z_c < z < H). & (5.5.57)\end{aligned}$$

From the boundary conditions at $z = 0$ and H, (5.5.26) and (5.5.27), we have the relations between the coefficients:

$$\begin{aligned}\Lambda a_1 + c\mu b_1 &= 0, & (5.5.58)\\ [(c - \Lambda H)\mu \sinh \mu H + \Lambda \cosh \mu H]\, a_2 & \\ + [(c - \Lambda H)\mu \cosh \mu H + \Lambda \sinh \mu H]\, b_2 &= 0. & (5.5.59)\end{aligned}$$

Thus, the solution can be written as

$$\begin{aligned}\tilde{\psi}(z) &= C_1(\mu z_c \cosh \mu z - \sinh \mu z)\\ &\equiv C_1 U(z), \qquad (0 < z < z_c), & (5.5.60)\\ \tilde{\psi}(z) &= C_2 \left[\frac{(z_c - H)\mu \cosh \mu H + \sinh \mu H}{(z_c - H)\mu \sinh \mu H + \cosh \mu H} \cosh \mu z - \sinh \mu z\right]\\ &\equiv C_2 V(z), \qquad (z_c < z < H), & (5.5.61)\end{aligned}$$

where C_1 and C_2 are new coefficients. If we assume that both the streamfunction and its first derivative with respect to height (i.e., the potential temperature) are continuous at $z = z_c$, we also have

$$\begin{aligned}\tilde{\psi}(z_c + 0) &= \tilde{\psi}(z_c - 0), & (5.5.62)\\ \frac{d\tilde{\psi}}{dz}(z_c + 0) - \frac{d\tilde{\psi}}{dz}(z_c - 0) &= A, & (5.5.63)\end{aligned}$$

where A is the coefficient defined in (5.5.30). From these relations, we obtain

$$\begin{aligned}C_1 &= A \frac{V(z_c)}{U'(z_c)V(z_c) - U(z_c)V'(z_c)}, & (5.5.64)\\ C_2 &= A \frac{U(z_c)}{U'(z_c)V(z_c) - U(z_c)V'(z_c)}. & (5.5.65)\end{aligned}$$

We can express the denominator as

$$\begin{aligned}&U'(z_c)V(z_c) - U(z_c)V'(z_c)\\ &= \frac{\mu^3 \sinh \mu H}{(z_c - H)\mu \sinh \mu H + \cosh \mu H} \left(z_c - \frac{c_+}{\Lambda}\right)\left(z_c - \frac{c_-}{\Lambda}\right), \qquad (5.5.66)\end{aligned}$$

where c_+ and c_- are two phase speeds of the nonsingular solution given by (5.5.37).

5.5.1.4 Edge waves

There exist neutral waves in a basic shear flow. Neutral waves are most clearly formulated under the top boundary condition given by

$$\tilde{\psi} \to 0, \quad \text{as} \quad z \to \infty, \qquad (5.5.67)$$

instead of the rigid boundary (5.5.27). The solutions to (5.5.25) under the lower boundary condition (5.5.26) and the upper boundary condition (5.5.67) are given by

$$\tilde{\psi}(z) \quad = \quad \Psi_0 e^{-\mu z}, \tag{5.5.68}$$

with the phase speed

$$c \quad = \quad \frac{\Lambda}{\mu}. \tag{5.5.69}$$

Since c is real, this wave is neutral. This kind of neutral wave is called the *edge wave*. Frequency is given by

$$\omega \quad = \quad kc \quad = \quad \frac{k\Lambda}{\mu} \quad = \quad \frac{\Lambda f}{N}\frac{k}{\sqrt{k^2+l^2}}, \tag{5.5.70}$$

where (5.5.28) is used. The edge wave has a vertical scale of μ^{-1}. It is trapped near the lower boundary and propagates eastward. If the top boundary condition is rigid, similar neutral waves are trapped near the upper boundary. These edge waves play a central role if baroclinic instability is viewed as the interaction between two waves (see Section 5.5.5).

5.5.1.5 Secondary circulation

A zonal-mean meridional circulation is associated with Eady unstable waves. We define the zonal mean by $\overline{(\ )}$ to derive meridional circulation. First, from (5.5.3), the zonal-mean geostrophic component is

$$\overline{v_g} \quad = \quad \frac{1}{f}\overline{\frac{\partial \psi}{\partial x}} \quad = \quad 0 \tag{5.5.71}$$

(i.e., the meridional circulation of the geostrophic component is identically zero). As for the ageostrophic component, to the first order approximation of linear theory the zonal average of (5.5.52) gives

$$\overline{w} \quad = \quad 0 \tag{5.5.72}$$

(i.e., the meridional circulation of the linear part of the ageostrophic component is also identically zero). Thus, we need to consider the nonlinear ageostrophic component to obtain the meridional circulation. Here, we take the second-order nonlinearity of the perturbation that is given by the linear theory. In this sense, the meridional circulation given by the following procedure is called the *secondary circulation*.

The equations of zonal velocity and temperature can be divided into linear and nonlinear parts. Substituting the basic field (5.5.14) and (5.5.17) into (5.5.7), (5.5.10), we have the equations for the perturbation

$$\left(\frac{\partial}{\partial t} + \Lambda z\frac{\partial}{\partial x}\right)u_g - fv_a \quad = \quad -\boldsymbol{u_g}\cdot\nabla_H u_g, \tag{5.5.73}$$

$$\left(\frac{\partial}{\partial t} + \Lambda z\frac{\partial}{\partial x}\right)\frac{\partial \psi'}{\partial z} + \Lambda\frac{\partial \psi'}{\partial x} + N^2 w \quad = \quad -\boldsymbol{u_g}\cdot\nabla_H\frac{\partial \psi'}{\partial z}. \tag{5.5.74}$$

We expand the perturbation in a series as

$$\psi' = \varepsilon\psi^{(1)} + \varepsilon^2\psi^{(2)} + \cdots , \tag{5.5.75}$$

where ε is the magnitude of the perturbation and satisfies $\varepsilon \ll 1$. To the first order $O(\varepsilon^1)$, we obtain a set of linear equations: $\psi^{(1)}$ is the linear solution that is already given by (5.5.47). Next, the second-order $O(\varepsilon^2)$ zonally averaged equations are given by

$$\frac{\partial}{\partial t}\overline{u_g^{(2)}} - f\overline{v_a^{(2)}} = -\overline{\boldsymbol{u_g}^{(1)} \cdot \nabla_H u_g^{(1)}}. \tag{5.5.76}$$

$$\frac{\partial}{\partial t}\frac{\partial \overline{\psi^{(2)}}}{\partial z} + N^2\overline{w^{(2)}} = -\overline{\boldsymbol{u_g}^{(1)} \cdot \nabla_H \frac{\partial \psi^{(1)}}{\partial z}}. \tag{5.5.77}$$

The right-hand sides are rewritten as

$$-\overline{\boldsymbol{u_g}^{(1)} \cdot \nabla_H u_g^{(1)}} = -\overline{\nabla_H \cdot \left(\boldsymbol{u_g}^{(1)} u_g^{(1)}\right)} = -\frac{\partial}{\partial y}\overline{u_g^{(1)} v_g^{(1)}}, \tag{5.5.78}$$

$$-\overline{\boldsymbol{u_g}^{(1)} \cdot \nabla_H \frac{\partial \psi^{(1)}}{\partial z}} = -\frac{\partial}{\partial y}\overline{v_g^{(1)}\frac{\partial \psi^{(1)}}{\partial z}}. \tag{5.5.79}$$

The zonal mean of the continuity equation (5.5.11) is

$$\frac{\partial \overline{v_a^{(2)}}}{\partial y} + \frac{\partial \overline{w^{(2)}}}{\partial z} = 0. \tag{5.5.80}$$

Thus, we can define the meridional streamfunction χ as

$$\overline{v_a^{(2)}} = -\frac{\partial \chi}{\partial z}, \qquad \overline{w^{(2)}} = \frac{\partial \chi}{\partial y}. \tag{5.5.81}$$

Since the boundary conditions are given as

$$w^{(2)} = 0, \quad \text{at } z = 0, H, \tag{5.5.82}$$

$$v_a^{(2)} = 0, \quad \text{at } y = 0, L, \tag{5.5.83}$$

we can set $\chi = 0$ at the boundaries. Now, using

$$u_g^{(2)} = -\frac{1}{f}\frac{\partial \psi^{(2)}}{\partial y}, \tag{5.5.84}$$

we can eliminate the time derivatives in (5.5.76) and (5.5.77), so that we obtain an equation for the meridional streamfunction χ as

$$\left(\frac{\partial^2}{\partial y^2} + \frac{f^2}{N^2}\frac{\partial^2}{\partial z^2}\right)\chi = -\frac{f}{N^2}\frac{\partial^2}{\partial y \partial z}\overline{u_g^{(1)} v_g^{(1)}} - \frac{1}{N^2}\frac{\partial^2}{\partial y^2}\overline{v_g^{(1)}\frac{\partial \psi^{(1)}}{\partial z}}. \tag{5.5.85}$$

where (5.5.81) is used. Substituting the linear solutions (5.5.53), (5.5.54)

$$\overline{u_g^{(1)} v_g^{(1)}} = 0, \tag{5.5.86}$$

$$\overline{v_g^{(1)}\frac{\partial \psi^{(1)}}{\partial z}} = \psi_0^2 \frac{k c_i \Lambda}{4f|c|^2}(1 - \cos 2ly)\, e^{2kc_i t} \tag{5.5.87}$$

into the right-hand side of (5.5.85), we obtain

$$\left(\frac{\partial^2}{\partial y^2}+\frac{f^2}{N^2}\frac{\partial^2}{\partial z^2}\right)\chi = B\cos 2ly, \tag{5.5.88}$$

where

$$B \equiv -\psi_0^2\frac{kl^2\Lambda}{fN^2}\frac{c_i}{|c|^2}\,e^{2kc_it}. \tag{5.5.89}$$

Now, let us seek a solution to (5.5.88) in the form

$$\chi = \chi_1(z)\cos 2ly+\chi_2(y,z), \tag{5.5.90}$$

where χ_1 and χ_2 respectively satisfy

$$\frac{f^2}{N^2}\frac{d^2}{dz^2}\chi_1-4l^2\chi_1 = B, \tag{5.5.91}$$

$$\left(\frac{\partial^2}{\partial y^2}+\frac{f^2}{N^2}\frac{\partial^2}{\partial z^2}\right)\chi_2 = 0. \tag{5.5.92}$$

The boundary conditions at $z=0,\ H$ are given as $\chi_1=\chi_2=0$. First, we have

$$\chi_1 = -\frac{B}{4l^2}\left(1-\cosh\eta z+\frac{\cosh\eta H-1}{\sinh\eta H}\sinh\eta z\right), \tag{5.5.93}$$

where

$$\eta \equiv \frac{2lN}{f}. \tag{5.5.94}$$

Then, to satisfy the boundary condition $\chi=0$ at $y=0,\ L$, we have

$$\begin{aligned}\chi_2 = &\sum_{m=1}^{\infty}\frac{B}{2l^2}\frac{\eta^2H^2}{m\pi(\eta^2H^2+m^2\pi^2)}\frac{1-\cos m\pi}{\cosh(\nu_mL/2)}\\ &\times\cosh\nu_m\left(y-\frac{L}{2}\right)\sin\left(m\pi\frac{z}{H}\right),\end{aligned} \tag{5.5.95}$$

where

$$\nu_m \equiv \frac{m\pi f}{NH},\quad (m=1,2,\cdots). \tag{5.5.96}$$

From (5.5.81), therefore, secondary circulation is given as

$$\begin{aligned}\overline{v_a^{(2)}} = &-\frac{B\eta}{4l^2}\left(-\sinh\eta z+\frac{\cosh\eta H-1}{\sinh\eta H}\cosh\eta z\right)\cos 2ly\\ &+\frac{B}{2l^2}\sum_{m=1}^{\infty}\frac{\eta^2H}{\eta^2H^2+m^2\pi^2}\frac{1-\cos m\pi}{\cosh(\nu_mL/2)}\\ &\times\cosh\nu_m\left(y-\frac{L}{2}\right)\cos\left(m\pi\frac{z}{H}\right),\end{aligned} \tag{5.5.97}$$

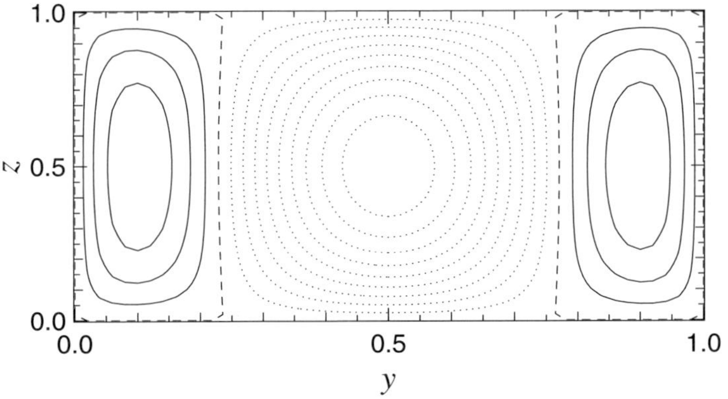

FIGURE 5.11: Distribution of the secondary circulation χ of the Eady solution in meridional cross section. Both the abscissa y and the ordinate z are normalized to 1. The contour interval is 25 $\mathrm{m^2\ s^{-1}}$. Dotted curves are negative values (indirect circulation) and solid values are positive values (direct circulation).

$$
\begin{aligned}
\overline{w^{(2)}} &= -\frac{B\eta}{2l}\left(1-\cosh\eta z+\frac{\cosh\eta H-1}{\sinh\eta H}\sinh\eta z\right)\sin 2ly \\
&\quad +\frac{B}{2l^2}\sum_{m=1}^{\infty}\frac{\nu_m\eta^2H^2}{m\pi(\eta^2H^2+m^2\pi^2)}\frac{1-\cos m\pi}{\cosh(\nu_m L/2)} \\
&\quad \times\sinh\nu_m\left(y-\frac{L}{2}\right)\sin\left(m\pi\frac{z}{H}\right). \qquad (5.5.98)
\end{aligned}
$$

Figure 5.11 shows an example of secondary circulation. The values of the parameters are $L = 4{,}000$ km, $H = 10$ km, $U = 50\ \mathrm{m\ s^{-1}}$, $\Lambda = U/H = 5\times10^{-3}\ \mathrm{s^{-1}}$, $N = 10^{-2}\ \mathrm{s^{-1}}$, $f = 2\Omega\sin 30° = 7.27\times10^{-5}\ \mathrm{s^{-1}}$, $c_i = c_r = U$, $k = l = f/U$, and $\psi_0 e^{2kc_it} = U^2$. In this case, $B = -1.542\times10^{-9}\ \mathrm{s^{-1}}$ and $\eta = 2.16\times10^{-4}\ \mathrm{m^{-1}}$. The maximum value of the streamfunction of meridional circulation is 249.7 $\mathrm{m^2\ s^{-1}}$. This figure shows that secondary circulation is $\chi < 0$ in the middle of the channel (i.e., the meridional flow is poleward in the lower layer while it is equatorward in the upper layer). This kind of circulation is called the *indirect circulation.* The secondary circulation of baroclinic instability is characterized by indirect circulation. There are also two cells with $\chi > 0$ in both sides of the indirect circulation. The meridional circulation $\chi > 0$ is called *direct circulation.* These direct circulations are associated with the normal mode of baroclinic instability.

This three-cell structure is seemingly similar to that of meridional circulations observed in the real atmosphere. Nonlinearity of angular momentum transport is important in fact, however, so that the dynamics of the three-cell structure in Fig. 5.11 are very different from those of the real atmosphere.

5.5.2 Charney problem

We next consider the second type of baroclinic instability, called the *Charney problem.* This type of baroclinic instability occurs on the β plane between a rigid boundary at the bottom and an infinite upper boundary. In this section, we briefly summarize the properties of the solutions following Pedlosky (1987).

The Charney problem is formulated with the quasi-geostrophic equations on the β plane in a stratified atmosphere. Under adiabatic and frictionless conditions the quasi-geostrophic potential vorticity equation is given by

$$\left(\frac{\partial}{\partial t} + \boldsymbol{u_g} \cdot \nabla_H\right) \Pi \quad = \quad 0, \tag{5.5.99}$$

where

$$\Pi \quad = \quad \frac{1}{f_0}\left[\frac{\partial^2 \psi}{\partial x^2} + \frac{\partial^2 \psi}{\partial y^2} + \frac{1}{\rho_s}\frac{\partial}{\partial z}\left(\rho_s \frac{f_0^2}{N_s^2}\frac{\partial \psi}{\partial z}\right)\right] + f_0 + \beta y \tag{5.5.100}$$

is the quasi-geostrophic potential vorticity. Here, ρ_s and N_s^2 are functions of altitude z. The thermodynamic equation is given by

$$\left(\frac{\partial}{\partial t} + \boldsymbol{u_g} \cdot \nabla_H\right) \frac{\partial \psi}{\partial z} + N_s^2 w \quad = \quad 0. \tag{5.5.101}$$

From the geostrophic balance, we have

$$u_g \quad = \quad -\frac{1}{f}\frac{\partial \psi}{\partial y}, \qquad v_g \quad = \quad \frac{1}{f}\frac{\partial \psi}{\partial x}, \tag{5.5.102}$$

$$\theta \quad = \quad \frac{\theta_{s0}}{g}\frac{\partial \psi}{\partial z}, \qquad \pi' \quad = \quad \frac{1}{c_p \theta_{s0}}\psi. \tag{5.5.103}$$

θ is the deviation of potential temperature from the basic state potential temperature θ_s, which is also a function of z, and θ_{s0} is the vertical average of θ_s. We assume that the basic state has constant temperature $T(z) = \theta_{s0}$ and has linear shear. Thus, we have

$$\rho_s \quad = \quad \rho_{s0} e^{-\frac{z}{H}}, \tag{5.5.104}$$

$$N_s^2 \quad = \quad \frac{\theta_{s0}}{g}\frac{d\theta_s}{dz} \quad = \quad \frac{g^2}{c_p \theta_{s0}} \quad \equiv \quad N^2, \tag{5.5.105}$$

$$H \quad = \quad \frac{R_d \theta_{s0}}{g}, \tag{5.5.106}$$

where H is scale height. Eqs. (5.5.14), (5.5.15), and (5.5.17) in the Eady model are also used in this case:

$$\overline{u}_g \quad = \quad U(z) \quad = \quad \Lambda z, \tag{5.5.107}$$

$$\overline{v}_g \quad = \quad 0, \tag{5.5.108}$$

$$\overline{\psi} \quad = \quad -f\int_0^y \overline{u}_g dy \quad = \quad -f\Lambda yz, \tag{5.5.109}$$

$$\overline{\theta} \quad = \quad \frac{\theta_{s0}}{g}\frac{\partial \overline{\psi}}{\partial z} \quad = \quad -\frac{f_0 \theta_{s0}}{g}\Lambda y. \tag{5.5.110}$$

Substituting (5.5.104), (5.5.105), and (5.5.109) into (5.5.100), we obtain the basic state quasi-geostrophic potential vorticity as

$$\overline{\Pi} = \frac{1}{f_0}\frac{1}{\rho_s}\frac{\partial}{\partial z}\left(\rho_s \frac{f_0^2}{N^2}\frac{\partial \overline{\psi}}{\partial z}\right) + f_0 + \beta y = \frac{f_0^2}{N^2}\frac{\Lambda}{H}y + f_0 + \beta y, \tag{5.5.111}$$

and the meridional gradient of potential vorticity as

$$\overline{\Pi}_y = \frac{\partial \overline{\Pi}}{\partial y} = \frac{f_0^2}{N^2}\frac{\Lambda}{H} + \beta = \beta\left(1 + \frac{h}{H}\right), \tag{5.5.112}$$

where h is a typical vertical scale defined by

$$h \equiv \frac{\Lambda f_0^2}{\beta N^2}. \tag{5.5.113}$$

The strength of shear and static stability greatly affect the value of h. A typical value for $\Lambda = 3 \times 10^{-3}$ m s^{-1} and $N^2 = 10^{-4}$ s^{-2} with $f_0 = 2\Omega \sin(45°)$ s^{-1} and $\beta = 2\Omega \cos(45°)/R$ s^{-1} m^{-1} gives $h = 2.5$ km. In the case of $H = 8$ km, this also gives $h/H = 0.3$.

Let us define the perturbation of the streamfunction by ψ'. The linearized equations of potential vorticity and potential temperature are given by

$$\left(\frac{\partial}{\partial t} + \Lambda z \frac{\partial}{\partial x}\right)\left(\frac{\partial^2 \psi'}{\partial x^2} + \frac{\partial^2 \psi'}{\partial y^2} + \frac{f_0^2}{N^2}\frac{\partial^2 \psi'}{\partial z^2} - \frac{f_0^2}{N^2}\frac{1}{H}\frac{\partial \psi'}{\partial z}\right)$$

$$+\overline{\Pi}_y \frac{\partial \psi'}{\partial x} = 0, \tag{5.5.114}$$

$$\left(\frac{\partial}{\partial t} + \Lambda z \frac{\partial}{\partial x}\right)\frac{\partial \psi'}{\partial z} - \Lambda \frac{\partial \psi'}{\partial x} + N^2 w = 0. \tag{5.5.115}$$

The boundary conditions are given by

$$\frac{\partial}{\partial t}\frac{\partial \psi'}{\partial z} - \Lambda \frac{\partial \psi'}{\partial x} = 0, \quad \text{at } z = 0, \tag{5.5.116}$$

$$\rho_s |\psi'|^2 \rightarrow 0, \quad \text{as } z \rightarrow \infty, \tag{5.5.117}$$

$$\frac{\partial \psi'}{\partial x} = 0, \quad \text{at } y = 0, L. \tag{5.5.118}$$

The boundary condition in the limit $z \rightarrow \infty$ comes from the constraint that the energy is finite.† As a solution that satisfies the conditions at $y = 0$, L, we assume the modal solution (5.5.23). Substituting (5.5.23) into (5.5.114), (5.5.116), and (5.5.117), we have

$$(\Lambda z - c)\left[\frac{d^2 \tilde{\psi}}{dz^2} - \frac{1}{H}\frac{\partial \tilde{\psi}}{\partial z} - \frac{N^2}{f_0^2}(k^2 + l^2)\tilde{\psi}\right] + \frac{N^2}{f_0^2}\overline{\Pi}_y \tilde{\psi} = 0, \tag{5.5.120}$$

---

†The boundary condition (5.5.117) can be written using perturbation velocity as

$$\frac{1}{2}\rho_s ({u'_g}^2 + {v'_g}^2) \rightarrow 0, \quad \text{as } z \rightarrow \infty. \tag{5.5.119}$$

The energetics of baroclinic instability are described in Section 5.5.3.

and the boundary conditions

$$c\frac{d\tilde{\psi}}{dz} + \Lambda\tilde{\psi} \quad = \quad 0, \qquad \text{at} \quad z = 0, \tag{5.5.121}$$

$$e^{-\frac{z}{H}}|\tilde{\psi}|^2 \quad \to \quad 0, \qquad \text{as} \quad z \to \infty. \tag{5.5.122}$$

Eq. (5.5.120) can be rewritten as

$$\frac{d^2\tilde{\psi}}{dz^2} - \frac{1}{H}\frac{\partial\tilde{\psi}}{\partial z} + \left(\frac{\alpha}{z - z_c} - \mu^2\right)\tilde{\psi} \quad = \quad 0, \tag{5.5.123}$$

where

$$\mu^2 \quad = \quad \frac{N^2}{f_0^2}(k^2 + l^2), \tag{5.5.124}$$

$$\alpha \quad \equiv \quad \frac{N^2}{f_0^2}\frac{\overline{\Pi}_y}{\Lambda} = \frac{1}{H} + \frac{1}{h}, \tag{5.5.125}$$

$$z_c \quad = \quad \frac{c}{\Lambda}. \tag{5.5.126}$$

At the height z_c, the phase speed c becomes equal to the zonal wind of the basic state; z_c is called the *steering level.* Now, setting

$$\tilde{\psi}(z) \quad = \quad e^{-\nu(z-z_c)}(z - z_c)\Phi(z - z_c), \tag{5.5.127}$$

with parameters

$$\delta \quad \equiv \quad \left(1 + 4\mu^2H^2\right)^{\frac{1}{2}}, \tag{5.5.128}$$

$$\nu \quad \equiv \quad \frac{1}{2H}\left[\left(1 + 4\mu^2H^2\right)^{\frac{1}{2}} - 1\right] \quad = \quad \frac{1}{2H}(\delta - 1), \tag{5.5.129}$$

we can rewrite (5.5.123), (5.5.121), (5.5.122) as

$$x\frac{d^2\Phi}{dx^2} + (2 - x)\frac{d\Phi}{dx} - (1 - r)\Phi \quad = \quad 0. \tag{5.5.130}$$

$$x_0\left(\frac{d\Phi}{dx} - \frac{\delta - 1}{2\delta}\Phi\right) \quad = \quad 0, \qquad \text{at} \quad x = x_0, \tag{5.5.131}$$

$$|\Phi|^2 e^{-x} \quad = \quad 0, \qquad \text{as} \quad x \to \infty, \tag{5.5.132}$$

where

$$r \quad = \quad \frac{1 + \frac{H}{h}}{\delta}, \tag{5.5.133}$$

$$x \quad = \quad \frac{z - z_c}{H}\delta, \tag{5.5.134}$$

$$x_0 \quad = \quad x(z = 0) \quad = \quad -\frac{z_c}{H}\delta. \tag{5.5.135}$$

It is known that the function $\Phi(x)$ can be expressed by confluent hypergeometric functions:

$$\Phi(x) \quad = \quad \begin{cases} c_1\chi_1(1 - r, 2; x), & \text{for} \quad r = n, \\ c_2\chi_2(1 - r, 2; x), & \text{for} \quad r \neq n, \end{cases} \tag{5.5.136}$$

where $n = 1, 2, \cdots$, and

$$\chi_1(\alpha, 2; x) = \sum_{n=0}^{\infty} \frac{\Gamma(\alpha+n)}{\Gamma(\alpha)(n+1)!n!} x^n, \tag{5.5.137}$$

$$\begin{aligned}\chi_2(\alpha, 2; x) &= \frac{1}{\alpha-1}\frac{1}{x} + \sum_{n=0}^{\infty} \frac{\Gamma(\alpha+n)}{\Gamma(\alpha)(n+1)!n!} \\ &\times[\ln x + \phi(\alpha+n) - \phi(1+n) - \phi(2+n)]x^n.\end{aligned} \tag{5.5.138}$$

$\Gamma(x)$ is called the Gamma function, which satisfies

$$\Gamma(x+1) = x\Gamma(x). \tag{5.5.139}$$

In particular, $\Gamma(n) = (n-1)!$ for integer n. Using the Gamma function, we have

$$\phi(x) = \frac{d}{dx} \log \Gamma(x), \tag{5.5.140}$$

$$\phi(n+a) = \sum_{m=0}^{n-1} \frac{1}{a+m} + \phi(a). \tag{5.5.141}$$

In the case $r = n$, since $\Phi(x)$ is a polynomial with degree $n-1$ as given by (5.5.137), we have n solutions for x_0 from the boundary condition (5.5.131). Corresponding phase velocities c can be calculated from (5.5.144). It can be shown that all phase speeds c are real with $c \leq 0$. This means that there is no unstable wave for positive integer r. In the case $r \neq n$, substituting the solution of Φ given by (5.5.138) into the boundary condition (5.5.131), we obtain a number of phase velocities c. In general, there exists a complex solution among them (i.e., an unstable wave). From further investigation, the properties of the solution are summarized as follows: If $0 < r < 1$, only one unstable solution exists. If $r = n$ ($n = 1, 2, \cdots$), only n neutral solutions exist. If $n < r < n+1$, n neutral solutions and one unstable solution exist.

The typical parameters of unstable waves are given as follows. For given r and h, the horizontal scale is given from (5.5.124) and (5.5.133) by

$$k^2 + l^2 = \frac{f_0^2}{4N^2H^2}\left[\frac{(1+H/h)^2}{r^2} - 1\right]. \tag{5.5.142}$$

From (5.5.129), the vertical scale is given by

$$\nu = \frac{1}{2H}\left(\frac{1+H/h}{r} - 1\right), \tag{5.5.143}$$

The value of x_0 can be determined from the boundary condition (5.5.131). From this, the dispersion relation between phase speed c and wave numbers k and l are related to x_0. In particular, if r is an integer, c is real and given by

$$c = -\frac{\Lambda H x_0}{\left[1 + 4\frac{N^2}{f_0^2}(k^2+l^2)H^2\right]^{\frac{1}{2}}}. \tag{5.5.144}$$

This implies that $c \leq 0$.

Fig. 5.12 shows the dependency of x_0 on r given by the boundary condition (5.5.131) for specified values of δ: $\delta \to \infty$ and $\delta = 2$. Both cases indicate that $x_0 = 0$ for any positive integer $r = n$. Figs. 5.13 and 5.14 show the dependencies of phase speed and growth rate on wave numbers. Here, we define the nondimensional phase speed by $c^* = c/\Lambda H$, nondimensional wave numbers by $k^* = kHN/f_0$, l^*HN/f_0, and $\mu^* = \mu H$, and the nondimensional growth rate by $\omega^* = k^*c^* = \omega N/(\Lambda f_0)$. Fig. 5.13 is the relation between the total wave number μ^* and the phase speed c^*.

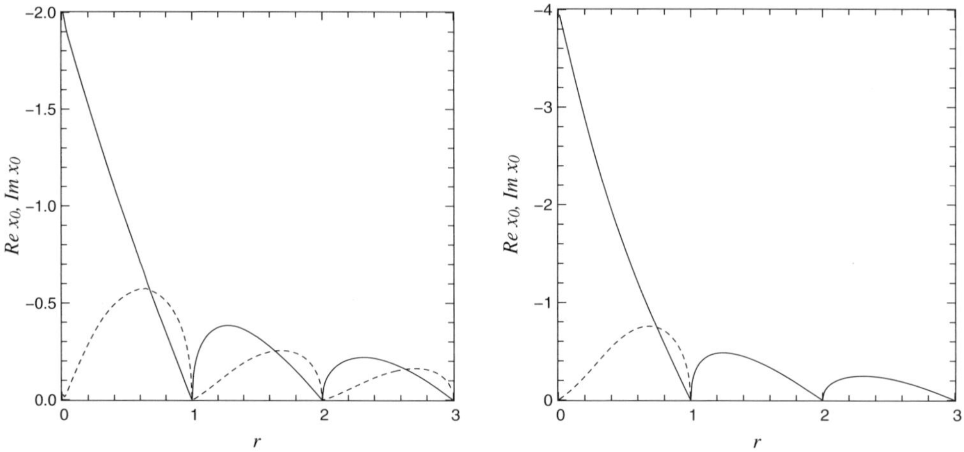

FIGURE 5.12: The dependency of x_0 on r for (left) $\delta \to \infty$ and (right) $\delta = 2$. The solid curve is the real part $Re\ x_0$, and the dashed curve is the imaginary part $Im\ x_0$.

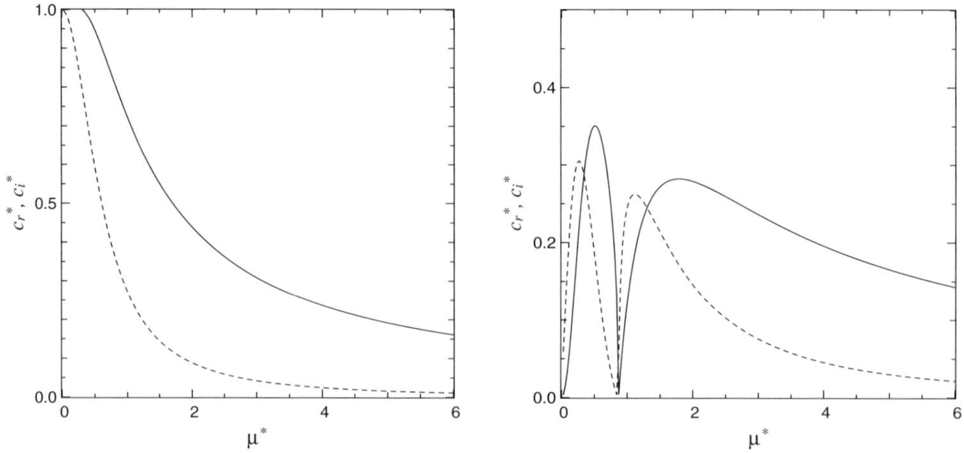

FIGURE 5.13: The relation between phase speed c^* and total wave number μ^* for $H/h = 0$ and 1. The solid curve is the real part c_r^*, and the dashed curve is the imaginary part c_i^*.

Fig. 5.14 is the growth rate ω^* as a function of the wave number k^* for $l^* = 0$, and Fig. 5.15 shows the dependency of the growth rate ω_i^* on k^* and H/h. In the case of $H/h = 0$, growth rates are always positive for all wave numbers: $\omega_i^* > 0$. In the case of $H/h > 0$, in turn, there exists a neutral mode $\omega_i^* = 0$ at $r = 1$ (i.e., from (5.5.133) and (5.5.128)),

$$\mu^* \quad = \quad \mu_c^* \quad \equiv \quad \frac{\sqrt{(1 + H/h)^2 - 1}}{2}. \tag{5.5.145}$$

For $H/h = 1$, we have $\mu_c^* = 0.866$. The wave that grows largest exists at the wave number in the region $\mu^* > \mu_c^*$ (i.e., $0 < r < 1$).

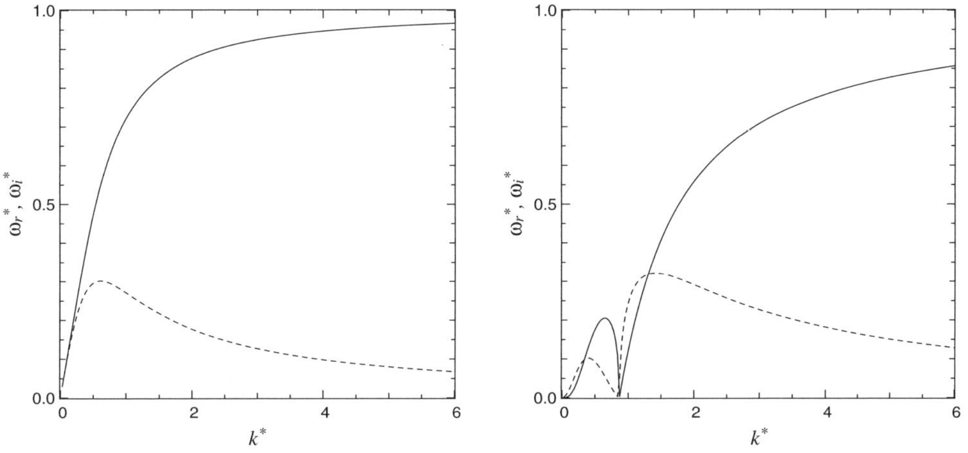

FIGURE 5.14: The relation between growth rate ω^* and zonal wave number k^* for $H/h = 0$ and 1. The solid curve is the real part ω_r^*, and the dashed curve is the imaginary part ω_i^*.

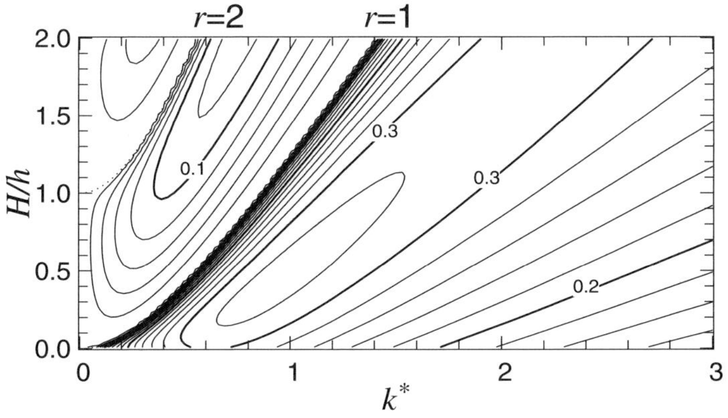

FIGURE 5.15: The dependency of growth rate ω_i^* on zonal wave number k^* and the parameter H/h for $l^* - 0$. The counter interval is 0.02 for thin curves and 0.1 for thick curves. The dotted curves correspond to $r = 1$ and 2 where the growth rate is zero.

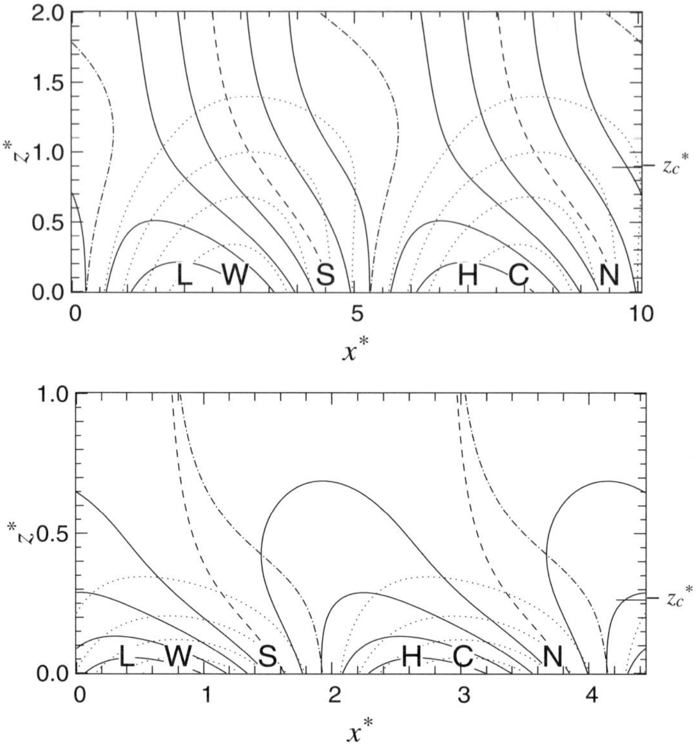

FIGURE 5.16: Structure of the most unstable waves of the Charney problem for $H/h = 0$ (top) and 1 (bottom). The streamfunction ψ' (solid) and the potential temperature θ' (dotted) in the zonal and height section are shown; the contour interval is one-fifth of the maximum amplitude of each quantity. The zero values are denoted by the dashed and dotted-dashed curves, respectively. L/H denotes the centers of low/high pressures; N/S denotes the maxima of northerly/southerly winds; W/C denotes the warmest/coldest positions of the perturbations of potential temperature. The steering level z_c^* is also shown by a tick on the right ordinate. The growth rate and zonal wave number are $\omega_i^* = 0.302$, $k^* = 0.625$ for $H/h = 0$, and $\omega_i^* = 0.322$, $k^* = 1.416$ for $H/h = 1$.

Figure 5.16 shows the structure of most unstable waves in the zonal and height cross section for $H/h = 0$ and 1. The amplitude of the streamfunction becomes larger as height lowers, and becomes smaller in the upper layers, which is different from the Eady problem. Near the surface, the warmest air is found at the front of a low-level cyclone, while the coldest air is found at the front of a low-level anticyclone. This means that the heat is transported poleward. The magnitude of heat transport is also larger near the surface.

5.5.3 Energetics

We have examined the disturbances of baroclinic instability based on the Eady problem and the Charney problem. The energetics of baroclinic instability can be studied in the same framework as that of barotropic instability. We now consider the energetics of shear instability in general for a basic flow on the β plane. We assume that zonal velocity has an arbitrary distribution $\overline{u}(y,z)$. The linearized quasi-geostrophic potential vorticity equation for the perturbation field is given by (5.5.114):

$$\left(\frac{\partial}{\partial t}+\overline{u}\frac{\partial}{\partial x}\right)\Pi'+\overline{\Pi}_y\frac{1}{f_0}\frac{\partial\psi'}{\partial x} = 0, \tag{5.5.146}$$

where

$$\Pi' = \frac{1}{f_0}\left[\frac{\partial^2\psi'}{\partial x^2}+\frac{\partial^2\psi'}{\partial y^2}+\frac{1}{\rho_s}\frac{\partial}{\partial z}\left(\rho_s\frac{f_0^2}{N^2}\frac{\partial\psi'}{\partial z}\right)\right], \tag{5.5.147}$$

$$\overline{\Pi}_y = \beta-\frac{\partial^2\overline{u}}{\partial y^2}-\frac{1}{\rho_s}\frac{\partial}{\partial z}\left(\rho_s\frac{f_0^2}{N^2}\frac{\partial\overline{u}}{\partial z}\right). \tag{5.5.148}$$

We assume that the flow is confined in the region with $0\le z\le H$ and $0\le y\le L$. From (5.5.115), the boundary condition at $z=0,\ H$ is given by

$$\left(\frac{\partial}{\partial t}+\overline{u}\frac{\partial}{\partial x}\right)\frac{\partial\psi'}{\partial z}-\frac{\partial\overline{u}}{\partial z}\frac{\partial\psi'}{\partial x} = 0. \tag{5.5.149}$$

The meridional boundary condition is $v'=0$ at $y=0,\ L$. From this and using (5.5.7), we obtain,

$$\left(\frac{\partial}{\partial t}+\overline{u}\frac{\partial}{\partial x}\right)\frac{\partial\psi'}{\partial y} = 0. \tag{5.5.150}$$

Multiplying $\rho_s\psi'$ by (5.5.146) and averaging over the x-direction, we have the zonal-mean energy equation:

$$\begin{aligned}\frac{\partial}{\partial t}&\left\{\frac{\rho_s}{2}\left[\overline{\left(\frac{\partial\psi'}{\partial x}\right)^2}+\overline{\left(\frac{\partial\psi'}{\partial y}\right)^2}+\frac{f_0^2}{N^2}\overline{\left(\frac{\partial\psi'}{\partial z}\right)^2}\right]\right\}\\ &= \frac{\partial}{\partial y}\left[\overline{\rho_s\psi'\cdot\left(\frac{\partial}{\partial t}+\overline{u}\frac{\partial}{\partial x}\right)\frac{\partial\psi'}{\partial y}}\right]\\ &\quad+\frac{\partial}{\partial z}\left[\overline{\rho_s\frac{f_0^2}{N^2}\psi'\cdot\left(\frac{\partial}{\partial t}+\overline{u}\frac{\partial}{\partial x}\right)\frac{\partial\psi'}{\partial z}}\right]\\ &\quad+\rho_s\frac{\partial\overline{u}}{\partial y}\overline{\frac{\partial\psi'}{\partial x}\frac{\partial\psi'}{\partial y}}+\rho_s\frac{f_0^2}{N^2}\frac{\partial\overline{u}}{\partial z}\overline{\frac{\partial\psi'}{\partial x}\frac{\partial\psi'}{\partial z}}.\end{aligned} \tag{5.5.151}$$

Integrating in the y-z cross section, and using the boundary conditions (5.5.149) and (5.5.150), we have

$$\begin{aligned}\frac{dE}{dt} &= \frac{1}{f_0^2}\int_0^L\int_0^H\left(\rho_s\frac{\partial\overline{u}}{\partial y}\overline{\frac{\partial\psi'}{\partial x}\frac{\partial\psi'}{\partial y}}+\rho_s\frac{f_0^2}{N^2}\frac{\partial\overline{u}}{\partial z}\overline{\frac{\partial\psi'}{\partial x}\frac{\partial\psi'}{\partial z}}\right)dy\,dz\\ &= \int_0^L\int_0^H\left(-\rho_s\frac{\partial\overline{u}}{\partial y}\overline{u'v'}+\rho_s\frac{g}{\theta_0 f_0}\frac{f_0^2}{N^2}\frac{\partial\overline{u}}{\partial z}\overline{v'\theta'}\right)dy\,dz,\end{aligned} \tag{5.5.152}$$

where

$$E = \frac{1}{f_0^2}\int_0^L\int_0^H \left\{ \frac{\rho_s}{2}\left[\overline{\left(\frac{\partial\psi'}{\partial x}\right)^2} + \overline{\left(\frac{\partial\psi'}{\partial y}\right)^2} + \frac{f_0^2}{N^2}\overline{\left(\frac{\partial\psi'}{\partial z}\right)^2}\right]\right\} dz\,dy$$

$$= \int_0^L\int_0^H \frac{\rho_s}{2}\left(\overline{u'^2} + \overline{v'^2} + \frac{g^2}{N^2}\frac{\overline{\theta'^2}}{\theta_0^2}\right) dz\,dy \qquad (5.5.153)$$

is the total energy. In the case with an infinite upper boundary, the domain of the above integral is taken from the ground to $H = \infty$. Eq. (5.5.152) states that, in the case that $\frac{\partial \overline{u}}{\partial z} = 0$ and $\frac{\partial \overline{u}}{\partial y} > 0$, the disturbance will grow if $\overline{u'v'} < 0$; this is the case for barotropic instability. On the other hand, in the case that $\frac{\partial \overline{u}}{\partial y} = 0$ and $\frac{\partial \overline{u}}{\partial z} > 0$, the disturbance will grow if $\overline{v'\theta'} > 0$; this is the case for baroclinic instability. In general cases when both $\frac{\partial \overline{u}}{\partial y}$ and $\frac{\partial \overline{u}}{\partial z}$ are different from zero, the stability of the flow depends on the structure of the disturbance.

5.5.4 Necessary condition for instability

5.5.4.1 Energy equation

The energy equation can be used to discuss the necessary condition for shear instability. We assume that the perturbation has a sinusoidal form in the quasi-geostrophic potential vorticity equation (5.5.146),

$$\psi' = \phi(y,z)e^{ik(x-ct)}, \qquad (5.5.154)$$

where c is a complex phase speed

$$c = c_r + ic_i. \qquad (5.5.155)$$

Substituting (5.5.154) into (5.5.147), we rewrite (5.5.146) as

$$(\overline{u} - c)\left[-k^2\phi + \frac{\partial^2\phi}{\partial y^2} + \frac{1}{\rho_s}\frac{\partial}{\partial z}\left(\rho_s\frac{f_0^2}{N^2}\frac{\partial\phi}{\partial z}\right)\right] + \overline{\Pi}_y\phi = 0. \qquad (5.5.156)$$

The boundary conditions (5.5.149) and (5.5.150) become

$$(\overline{u} - c)\frac{\partial\phi}{\partial z} - \frac{\partial\overline{u}}{\partial z}\phi = 0, \quad \text{at } z = 0, H, \qquad (5.5.157)$$

$$\phi = 0, \quad \text{at } y = 0, L. \qquad (5.5.158)$$

Multiplying $\rho_s\phi^*/(\overline{u} - c)$ by (5.5.156) and integrating in the y-z cross section using the above boundary conditions, (5.5.157) and (5.5.158), we obtain the energy equation as

$$\int_0^L\int_0^H \rho_s\left(k^2|\phi|^2 + \left|\frac{\partial\phi}{\partial y}\right|^2 + \frac{f_0^2}{N^2}\left|\frac{\partial\phi}{\partial z}\right|^2\right) dz\,dy$$

$$= -\int_0^L \left[\rho_s\frac{f_0^2}{N^2}\frac{|\phi|^2}{\overline{u} - c}\frac{\partial\overline{u}}{\partial z}\right]_0^H dy + \int_0^L\int_0^H \rho_s\frac{|\phi|^2}{\overline{u} - c}\overline{\Pi}_y\, dz\,dy, \qquad (5.5.159)$$

which corresponds to (5.5.152).

Let us consider the real and imaginary parts of this equation to consider the condition for instability. Using

$$\frac{1}{\overline{u}-c} = \frac{1}{|\overline{u}-c|^2}(\overline{u}-c_r+ic_i), \tag{5.5.160}$$

we obtain the imaginary part of (5.5.159) as

$$c_i\left\{\int_0^L\left[\rho_s\frac{f_0^2}{N^2}\frac{|\phi|^2}{|\overline{u}-c|^2}\frac{\partial\overline{u}}{\partial z}\right]_0^H dy - \int_0^L\int_0^H \rho_s\frac{|\phi|^2}{|\overline{u}-c|^2}\overline{\Pi}_y\,dz\,dy\right\} = 0. \tag{5.5.161}$$

In order to have an unstable wave with $c_i > 0$, we need the inside of the brace in the above equation to vanish. Thus, replacing

$$P = \rho_s\frac{|\phi|^2}{|\overline{u}-c|^2}, \tag{5.5.162}$$

we can rewrite the condition as

$$0 = -\int_0^L\left[P\frac{f_0^2}{N^2}\frac{\partial\overline{u}}{\partial z}\right]_0^H dy + \int_0^L\int_0^H P\overline{\Pi}_y\,dz\,dy \tag{5.5.163}$$

$$= \int_0^L\left[P\frac{f_0 g}{N^2\theta_0}\frac{\partial\overline{\theta}}{\partial y}\right]_0^H dy + \int_0^L\int_0^H P\overline{\Pi}_y\,dz\,dy, \tag{5.5.164}$$

where we have used the thermal wind balance relation

$$\frac{\partial\overline{u}}{\partial z} = -\frac{g}{\theta_0 f_0}\frac{\partial\overline{\theta}}{\partial y}. \tag{5.5.165}$$

This gives the necessary condition for instability. In order to satisfy this condition, the signs of the three quantities,

$$\left.\frac{\partial\overline{\theta}}{\partial y}\right|_{z=0}, \quad -\left.\frac{\partial\overline{\theta}}{\partial y}\right|_{z=H}, \quad \text{and} \quad \overline{\Pi}_y, \tag{5.5.166}$$

or

$$-\left.\frac{\partial\overline{u}}{\partial z}\right|_{z=0}, \quad \left.\frac{\partial\overline{u}}{\partial z}\right|_{z=H}, \quad \text{and} \quad \overline{\Pi}_y, \tag{5.5.167}$$

must not be identical.

In particular, let us examine the case when $\frac{\partial\overline{u}}{\partial y} = 0$: the pure baroclinic instability. First, we assume that there is no meridional gradient of potential temperature at the top and bottom boundaries. In this situation, (5.5.164) becomes

$$\int_0^L\int_0^H P\overline{\Pi}_y\,dz\,dy = 0. \tag{5.5.168}$$

In order to satisfy this equation, it is required that the sign of

$$\overline{\Pi}_y = \beta - \frac{1}{\rho_s}\frac{\partial}{\partial z}\left(\rho_s \frac{f_0^2}{N^2}\frac{\partial \overline{u}}{\partial z}\right) \tag{5.5.169}$$

must change at some z; this statement corresponds to Kuo's or Rayleigh's theorem for barotropic instability.

Second, if we assume that there is no meridional gradient of potential vorticity within the inner domain, $\overline{\Pi}_y = 0$, (5.5.164) becomes

$$\int_0^L \left[P\frac{f_0 g}{N^2\theta_0}\frac{\partial \overline{\theta}}{\partial y}\right]_0^H dy = 0 \tag{5.5.170}$$

(i.e., the instability depends on the meridional gradient of potential temperature at the top and bottom boundaries). The Eady problem is categorized into this case.

Third, if we assume that the upper boundary condition plays no role in (5.5.164) (i.e., $\frac{\partial \overline{\theta}}{\partial y} = 0$ at $z = H$, or the upper domain is infinite as $z \to \infty$), the necessary condition becomes

$$\int_0^L \left[P\frac{f_0 g}{N^2\theta_0}\frac{\partial \overline{\theta}}{\partial y}\right]_{z=0} dy + \int_0^L \int_0^H P\overline{\Pi}_y \, dz \, dy = 0. \tag{5.5.171}$$

The Charney problem is categorized into this case. From this, it can be seen that if there is a meridional gradient of potential temperature at the lower boundary, the basic flow is not unstable as long as the potential vorticity is constant in the inner domain. This is the case of the modified Eady problem without the upper boundary.

Next, returning to (5.5.159), let us consider the real part of the energy equation

$$\begin{aligned}\int_0^L \int_0^H \rho_s \left(k^2|\phi|^2 + \left|\frac{\partial \phi}{\partial y}\right|^2 + \frac{f_0^2}{N^2}\left|\frac{\partial \phi}{\partial z}\right|^2\right) dz\, dy \\ = -\int_0^L \left[P\frac{f_0^2}{N^2}(\overline{u}-c_r)\frac{\partial \overline{u}}{\partial z}\right]_0^H dy + \int_0^L \int_0^H P(\overline{u}-c_r)\overline{\Pi}_y \, dz\, dy.\end{aligned} \tag{5.5.172}$$

In order to consider the conditions for instability, we use (5.5.163) which is satisfied if the perturbation is unstable. Introducing an arbitrary constant c_0, multiplying (5.5.163) by $(c_0 - c_r)$, and summing up the result with (5.5.172) we have a relation

$$\begin{aligned}\int_0^L \int_0^H \rho_s \left(k^2|\phi|^2 + \left|\frac{\partial \phi}{\partial y}\right|^2 + \frac{f_0^2}{N^2}\left|\frac{\partial \phi}{\partial z}\right|^2\right) dz\, dy \\ = -\int_0^L \left[\frac{f_0^2}{N^2}P(\overline{u}-c_0)\frac{\partial \overline{u}}{\partial z}\right]_0^H dy + \int_0^L \int_0^H P(\overline{u}-c_0)\overline{\Pi}_y \, dz\, dy.\end{aligned} \tag{5.5.173}$$

The left-hand side is positive for unstable waves. In the special case that $\frac{\partial \overline{u}}{\partial z} = 0$ and the top and bottom boundary conditions are negligible, $\overline{\Pi}_y = 0$ holds at some altitude z according to (5.5.168). We set c_0 equal to the wind speed at this height: $c_0 = \overline{u}(z_c) \equiv \overline{u}_s$ where $\overline{\Pi}_y(z_c) = 0$. The positive condition of the second term on the right-hand side of (5.5.173) is written as

$$\int_0^H P(\overline{u} - \overline{u}_s)\overline{\Pi}_y \, dz \quad > \quad 0. \tag{5.5.174}$$

In particular, if $\beta = 0$ and ρ_s, N^2 are constant, it is rewritten as

$$\int_0^H P(\overline{u} - \overline{u}_s)\frac{\partial^2 \overline{u}}{\partial z^2} \, dz \quad < \quad 0. \tag{5.5.175}$$

This means that the absolute value of $\left|\frac{\partial \overline{u}}{\partial z}\right|$ is maximum at the inflection point of $\overline{u}$: this corresponds to Fjørtoft's theorem of barotropic instability.

5.5.4.2 Potential vorticity flux

The above necessary condition for instability (5.5.163) is also derived from the relation of potential vorticity flux. The zonal average of quasi-geostrophic potential vorticity flux is written as

$$\begin{aligned}
\rho_s \overline{v'\Pi'} &= \overline{\frac{\rho_s}{f_0^2}\frac{\partial \psi'}{\partial x}\left[\frac{\partial^2 \psi'}{\partial x^2} + \frac{\partial^2 \psi'}{\partial y^2} + \frac{1}{\rho_s}\frac{\partial}{\partial z}\left(\rho_s \frac{f_0^2}{N^2}\frac{\partial \psi'}{\partial z}\right)\right]} \\
&= \frac{\partial}{\partial y}\left(\frac{\rho_s}{f_0^2}\overline{\frac{\partial \psi'}{\partial x}\frac{\partial \psi'}{\partial y}}\right) + \frac{\partial}{\partial z}\left(\frac{\rho_s}{N^2}\overline{\frac{\partial \psi'}{\partial x}\frac{\partial \psi'}{\partial z}}\right) \\
&= -\frac{\partial}{\partial y}\left(\rho_s \overline{u'v'}\right) + \frac{\partial}{\partial z}\left(\rho_s \frac{f_0 g}{\theta_0 N^2}\overline{v'\theta'}\right).
\end{aligned} \tag{5.5.176}$$

Using $v' = 0$ at $y = 0$ and L, the domain integral of the above equation is written as

$$\int_0^L \int_0^H \rho_s \overline{v'\Pi'} \, dz \, dy - \int_0^L \left[\rho_s \frac{f_0 g}{\theta_0 N^2}\overline{v'\theta'}\right]_0^H dy \quad = \quad 0. \tag{5.5.177}$$

Now, we can show how (5.5.161) can be derived from this. The quasi-geostrophic potential vorticity equation and the boundary conditions at $z = 0$ and H are given as (5.5.146)–(5.5.149)

$$\left(\frac{\partial}{\partial t} + \overline{u}\frac{\partial}{\partial x}\right)\Pi' + v'\overline{\Pi}_y \quad = \quad 0, \tag{5.5.178}$$

$$\left(\frac{\partial}{\partial t} + \overline{u}\frac{\partial}{\partial x}\right)\theta' + v'\overline{\theta}_y \quad = \quad 0, \quad \text{at} \quad z = 0,\, H \tag{5.5.179}$$

Introducing meridional displacement η by

$$\left(\frac{\partial}{\partial t} + \overline{u}\frac{\partial}{\partial x}\right)\eta \quad = \quad v', \tag{5.5.180}$$

we rewrite (5.5.178) as

$$\left(\frac{\partial}{\partial t}+\overline{u}\frac{\partial}{\partial x}\right)\Pi' \quad = \quad -\left(\frac{\partial}{\partial t}+\overline{u}\frac{\partial}{\partial x}\right)\eta\cdot\overline{\Pi}_y. \tag{5.5.181}$$

Thus, the perturbation of potential vorticity is expressed as

$$\Pi' \quad = \quad -\eta\overline{\Pi}_y. \tag{5.5.182}$$

Using this and (5.5.180), we obtain

$$\overline{v'\Pi'} \quad = \quad -\left(\frac{\partial}{\partial t}\frac{\overline{\eta^2}}{2}\right)\overline{\Pi}_y. \tag{5.5.183}$$

At the boundaries $z=0$ and H, (5.5.179) is also rewritten as

$$\left(\frac{\partial}{\partial t}+\overline{u}\frac{\partial}{\partial x}\right)\theta' \quad = \quad -\left(\frac{\partial}{\partial t}+\overline{u}\frac{\partial}{\partial x}\right)\eta\cdot\overline{\theta}_y, \tag{5.5.184}$$

which gives

$$\theta' \quad = \quad -\eta\overline{\theta}_y. \tag{5.5.185}$$

Using this and (5.5.180), we also have

$$\overline{v'\theta'} \quad = \quad -\left(\frac{\partial}{\partial t}\frac{\overline{\eta^2}}{2}\right)\overline{\theta}_y = \frac{\theta_0 f_0}{g}\left(\frac{\partial}{\partial t}\frac{\overline{\eta^2}}{2}\right)\frac{\partial\overline{u}}{\partial z}. \tag{5.5.186}$$

Substituting (5.5.183) and (5.5.186) into (5.5.177) yields

$$\int_0^L\int_0^H \rho_s\left(\frac{\partial}{\partial t}\frac{\overline{\eta^2}}{2}\right)\overline{\Pi}_y\,dz\,dy+\int_0^L\left[\rho_s\frac{f_0^2}{N^2}\left(\frac{\partial}{\partial t}\frac{\overline{\eta^2}}{2}\right)\frac{\partial\overline{u}}{\partial z}\right]_0^H dy \quad = \quad 0. \tag{5.5.187}$$

For the sinusoidal perturbation (5.5.154), the displacement is written from (5.5.180) as

$$\eta \quad = \quad Re\frac{\phi}{\overline{u}-c}e^{ik(x-ct)}. \tag{5.5.188}$$

Then, we have

$$\frac{\partial}{\partial t}\frac{\overline{\eta^2}}{2} \quad = \quad \frac{kc_i|\phi|^2}{|\overline{u}-c|^2}e^{2kc_it}. \tag{5.5.189}$$

Substituting this into (5.5.187), we obtain the identical equation to (5.5.161).

The relation (5.5.176) is derived from quasi-geostrophic potential vorticity flux and always holds for both stable and unstable waves. This relation is equivalent to (5.5.187) which is expressed using the amplitude η. Furthermore, (5.5.161) can be derived if the sinusoidal form is assumed to be η. These relations imply that quasi-geostrophic potential vorticity flux grows with time for unstable waves.

5.5.4.3 Extension of potential vorticity

The above relation is compactly rewritten by extending quasi-geostrophic potential vorticity using potential temperature at the top and bottom boundaries:

$$\Pi^* = \Pi + \frac{f_0 g}{\theta_0 N^2}\theta[\delta(z-0) - \delta(z-H+0)], \tag{5.5.190}$$

where $\delta(z)$ is Dirac's delta function. In this equation, it is assumed that the contributions of potential temperature are confined to just above the bottom boundary $z = +0$ and to just below the top boundary $z = H - 0$ in the inner domain. In this case, (5.5.178) and (5.5.179) are combined to the equation of extended quasi-geostrophic potential vorticity:

$$\left(\frac{\partial}{\partial t} + \overline{u}\frac{\partial}{\partial x}\right)\Pi^{*\prime} + v'\overline{\Pi^*}_y = 0, \tag{5.5.191}$$

where

$$\Pi^{*\prime} = \Pi' + \frac{f_0 g}{\theta_0 N^2}\theta' \, [\delta(z-0) - \delta(z-H+0)], \tag{5.5.192}$$

$$\overline{\Pi^*}_y = \frac{\partial \overline{\Pi^*}}{\partial y} = \overline{\Pi}_y + \frac{f_0 g}{\theta_0 N^2}\overline{\theta}_y \, [\delta(z-0) - \delta(z-H+0)]. \tag{5.5.193}$$

In this case, the integral of the quasi-geostrophic potential vorticity flux, (5.5.177), is written as

$$\int_0^L \int_0^H \rho_s \overline{v'\Pi^{*\prime}} \, dz \, dy = 0. \tag{5.5.194}$$

This indicates that we can introduce the contribution of potential temperature at the boundaries to quasi-geostrophic potential vorticity in the inner region. Although potential temperature has a gradient at the boundaries in general, this effect can be included in quasi-geostrophic potential vorticity, and thus potential temperature can be thought to be constant at the boundaries. Fig. 5.17 schematically shows this situation where potential temperature surfaces are sharply bent so as to be parallel to the boundaries.

By rewriting $\Pi^{*\prime}$ using the displacement η analogous to (5.5.182), we obtain the necessary condition for instability $c_i > 0$ using (5.5.183) and (5.5.189):

$$\int_0^L \int_0^H P \, \overline{\Pi^*}_y \, dz \, dy = 0, \tag{5.5.195}$$

where P is defined by (5.5.162) and always positive. This condition does not contain the contribution of the boundaries and is formally similar to (5.5.168). This indicates that it is necessary for instability that the sign of $\overline{\Pi^*}_y$ changes within the inner domain. In the case of the Eady problem, we have $\overline{\Pi^*}_y < 0$ at $z = +0$ and $\overline{\Pi^*}_y > 0$ at $z = H - 0$. In the case of the Charney problem, we have $\overline{\Pi^*}_y < 0$ at $z = +0$ and $\overline{\Pi^*}_y > 0$ in $z > 0$ (see Fig. 5.18).

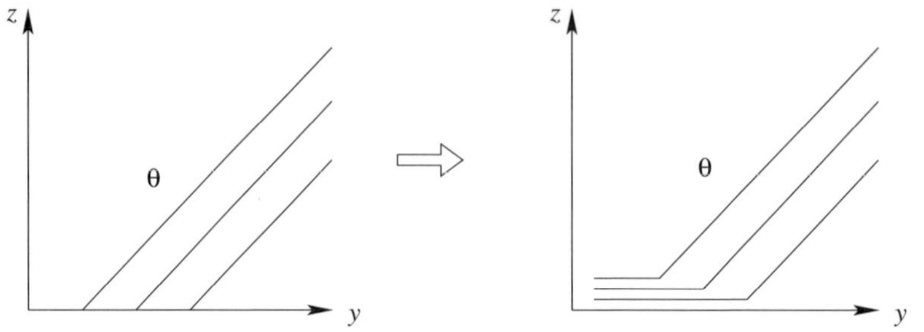

FIGURE 5.17: The contours of potential temperature in meridional cross section near the bottom boundary (left) and those of potential temperature used for extended quasi-geostrophic potential vorticity Π^* (right).

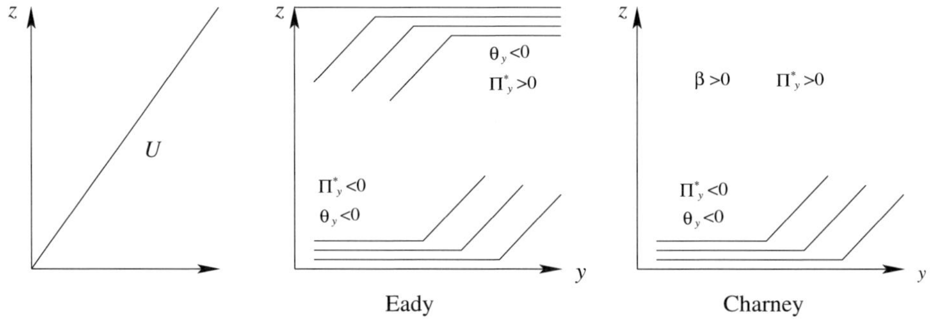

FIGURE 5.18: Interpretation of the Eady and Charney problems by extended quasi-geostrophic potential vorticity Π^*. The meridional gradients of Π^* are opposite between the lower and upper layers.

5.5.5 Interpretation of baroclinic instability

Analogous to the interpretation of barotropic instability described in Section 5.4.4, baroclinic instability can also be interpreted from the interaction of Rossby waves. Baroclinic instability occurs if the meridional gradients of potential vorticity of the basic state are opposite in the upper and lower layers. In this case, Rossby waves propagate in opposite x-directions in each layer, and they can be in such a phase relation that each wave amplifies the amplitude of the other wave. As depicted in Fig. 5.19, such a phase relation is realized if the phase line inclines westward with height.

In the case of the Eady problem (the middle panel of Fig. 5.18), edge waves at the boundaries play the roles of Rossby waves. If the basic field has a westerly shear in a confined region with top and bottom boundaries, the lower edge wave propagates eastward along the lower boundary and the upper edge wave propagates westward along the upper boundary. Since zonal winds are westward in the upper layers, these two waves have a phase relation with no relative motion. In such a phase-locked situation, the poleward flow associated with the lower edge wave is located at about the same longitude of the trough of the upper edge wave and the

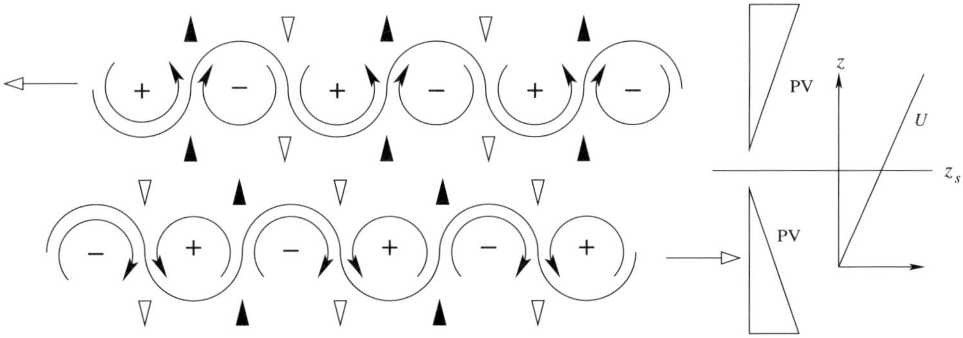

FIGURE 5.19: Schematic figure of the interaction of Rossby waves for interpretation of baroclinic instability. The vertical profiles of the potential vorticity (PV) and zonal velocity U are shown at the right; PV is minimum at the middle z_s. The Rossby wave propagates westward in the upper layer $z > z_s$, while it propagates eastward in the lower layer $z < z_s$. The basic flow is unstable if the upper and lower waves are in such a phase relation that each vorticity is amplified by the meridional flow of the other wave.

equatorward flow is located at about the same longitude of the ridge of the upper edge wave. The poleward/equatorward flow in the upper layer also corresponds to the ridge/trough in the lower layer. Thus, each wave amplifies the amplitude of the counterpart wave, and the amplitude of the two waves grows as a whole.

Eady's short wave cut-off is estimated by this interaction of the Rossby waves. If the wavelength of the edge wave is given by L, the vertical scale is given by $(f/N)L$. If the distance between the top and bottom boundaries H is smaller than $(f/N)L$, the two waves interact with each other

$$H < \frac{f}{N}L. \tag{5.5.196}$$

Thus, instability occurs if the wavelength is long enough that

$$L > \frac{N}{f}H. \tag{5.5.197}$$

This condition corresponds to (5.5.40).

In the case of the Charney problem (the right panel of Fig. 5.18), Rossby waves in the free atmosphere interact with the edge wave at the lower boundary. In the free atmosphere, the Rossby wave propagates westward due to the β effect, while the edge wave is trapped near the lower boundary and propagates eastward. The structure of the unstable wave of the Charney problem is characterized by the phase relations of the two waves where each wave amplifies the amplitude of the other wave.

References and suggested reading

As stated in the references of Chapter 2, the theories of instability are detailed in Chandrasekhar (1961) and Drazin and Reid (1981). In particular, a more complete discussion on convective instability in the rotating frame is given by Chandrasekhar (1961). Shear instability, or barotropic instability, is more fully described in Lin (1955) and Drazin and Reid (1981). The original works on baroclinic instability are Eady (1949) and Charney (1947). The mathematical formulation of baroclinic instability in this chapter almost follows Pedlosky (1987). The interpretation of barotropic and baroclinic instability in Sections 5.4.4 and 5.5.5 is based on Hoskins et al. (1985). Secondary circulation due to baroclinic instability in Section 5.5.1.5 is an extension of the two-layer model by Phillips (1954).

Chandrasekhar, S., 1961: *Hydrodynamic and Hydromagnetic Stability.* Oxford University Press, New York, 654 pp.

Charney, J. G., 1947: The dynamics of long waves in a baroclinic westerly current. *J. Meteor.*, **4**, 135–162.

Drazin, P. G. and Reid, W. H., 1981: *Hydrodynamic Stability.* Cambridge University Press, Cambridge, UK, 527 pp.

Eady, E. T., 1949: Long waves and cyclone waves. *Tellus*, **1**, 35–52.

Hayashi, H., Shiotani, M., and Gille, J. C., 2002: Horizontal wind disturbances induced by inertial instability in the equatorial middle atmosphere as seen in rocketsonde observations. *J. Geophys. Res.*, **107**, 10.1029/2001JD000922.

Held, I. M. and Hou, A. Y., 1980: Nonlinear axially symmetric circulations in a nearly inviscid atmosphere. *J. Atmos. Sci.*, **37**, 515–533.

Hoskins, B. J., McIntyre, M. E., and Robertson, A. W., 1985: On the use and significance of isentropic potential vorticity maps. *Q. J. Roy. Meteorol. Soc.*, **111**, 877–946.

Lin, C. C., 1955: *The Theory of Hydrodynamic Stability.* Cambridge University Press, Cambridge, UK, 155 pp.

Michalke, A., 1968: On the inviscid instability of the hyperbolic-tangent velocity profile. *J. Fluid Mech.*, **19**, 543–556.

Pedlosky, J., 1987: *Geophysical Fluid Dynamics*, 2nd ed. Springer-Verlag, New York, 710 pp.

Phillips, N. A., 1954: Energy transformations and meridional circulations associated with simple baroclinic waves in a two-level, quasi-geostrophic model. *Tellus*, **4**, 273–286.

Tanaka, H., 1975: Quasi-linear and non-linear interactions of finite amplitude perturbation in a stably stratified fluid with hyperbolic tangent *J. Meteor. Soc. Japan*, **53**, 1–31.

6

Forced motions

In this chapter, thermally and mechanically forced motions of the atmosphere or the ocean are examined. The underlying concepts are propagation of disturbances generated by forcing and their adjustment process. These are discussed using the knowledge of waves described in Chapter 4. We begin with geostrophic adjustment as an initial value problem. Although it is not categorized as a forced motion, it gives a key notion for understanding the relation between the propagation of inertial gravity waves and the adjustment process, through which geostrophic winds are established. We subsequently describe thermal responses on the f- and β-planes. The response to thermal and mechanical forcing of axisymmetric flows is also described. At the end of this chapter, we consider the effects of frictional forcing on circulation and explain Ekman transport.

6.1 Geostrophic adjustment

If a fluid in the rotating frame is disturbed, a part of the disturbance propagates in the form of waves and the other part remains as geostrophic motions at the location where the disturbance is applied. The process involved when geostrophic balance is established is called *geostrophic adjustment.*

We use linearized shallow water equations to illustrate geostrophic adjustment. Linearizing (3.4.1) and (3.4.2) around a basic state of no motion with depth H, we obtain

$$\frac{\partial u}{\partial t} - fv = -g\frac{\partial \eta}{\partial x}, \tag{6.1.1}$$

$$\frac{\partial v}{\partial t} + fu = -g\frac{\partial \eta}{\partial y}, \tag{6.1.2}$$

$$\frac{\partial \eta}{\partial t} + H\left(\frac{\partial u}{\partial x} + \frac{\partial v}{\partial y}\right) = 0, \tag{6.1.3}$$

in which friction terms are neglected. The corresponding linearized potential vorticity equation is given by

$$\frac{\partial}{\partial t}\left[\frac{1}{H}\left(\frac{\partial v}{\partial x} - \frac{\partial u}{\partial y} - f\frac{\eta}{H}\right)\right] = 0, \tag{6.1.4}$$

from which it is found that perturbation potential vorticity,

$$q' = \frac{\partial v}{\partial x} - \frac{\partial u}{\partial y} - f\frac{\eta}{H}, \tag{6.1.5}$$

is constant irrespective of time.

If we assume that the flow field approaches a steady state as $t \to \infty$, the final state is described by (6.1.1)–(6.1.3) by omitting the tendency terms:

$$-fv = -g\frac{\partial \eta}{\partial x}, \qquad fu = -g\frac{\partial \eta}{\partial y}, \qquad \frac{\partial u}{\partial x} + \frac{\partial v}{\partial y} = 0 \tag{6.1.6}$$

(i.e., geostrophic balance is expected as the steady final state). In this case, perturbation potential vorticity is written as

$$q' = \frac{g}{f}\left(\frac{\partial^2 \eta}{\partial x^2} + \frac{\partial^2 \eta}{\partial y^2}\right) - f\frac{\eta}{H} = \frac{g}{f}\left(\frac{\partial^2 \eta}{\partial x^2} + \frac{\partial^2 \eta}{\partial y^2} - \frac{\eta}{\lambda^2}\right), \tag{6.1.7}$$

where

$$\lambda = \frac{\sqrt{gH}}{f} \tag{6.1.8}$$

is called the *Rossby radius of deformation.*

Let us consider evolutions of the shallow water system giving a perturbation in the region $0 \le x \le L$ at the initial state $t = 0$. We consider two types of initial states: a perturbation is given to either surface height field or velocity field. First, we assume $u = v = 0$ and a perturbation of the surface height given by

$$\eta = \begin{cases} 0, & (x < 0, L < x), \\ \eta_0, & (0 \le x \le L). \end{cases} \tag{6.1.9}$$

In this case, potential vorticity is given by

$$q' = \begin{cases} 0, & (x < 0, L < x), \\ -f\dfrac{\eta_0}{H}, & (0 \le x \le L). \end{cases} \tag{6.1.10}$$

If we assume that the flow field approaches a steady state in the limit $t \to \infty$, the surface height has a steady profile at the final state. Since potential vorticity is constant irrespective of time, we find from (6.1.4) that the surface height at the final state satisfies

$$\frac{\partial^2 \eta}{\partial x^2} - \frac{\eta}{\lambda^2} = \begin{cases} 0, & (x < 0, L < x), \\ -\dfrac{\eta_0}{\lambda^2}, & (0 \le x \le L), \end{cases} \tag{6.1.11}$$

Assuming that η and $\frac{\partial \eta}{\partial x}$ are continuous at $x = 0$, L, and $\eta \to 0$ as $|x| \to \infty$, we obtain the steady solution as

$$\eta = \begin{cases} \frac{\eta_0}{2}\left(1 - e^{-\frac{L}{\lambda}}\right) e^{\frac{x}{\lambda}}, & (x < 0), \\ \eta_0 - \frac{\eta_0}{2}\left(e^{\frac{x-L}{\lambda}} + e^{-\frac{x}{\lambda}}\right), & (0 \le x \le L), \\ \frac{\eta_0}{2}\left(e^{\frac{L}{\lambda}} - 1\right) e^{-\frac{x}{\lambda}}, & (x > L). \end{cases} \quad (6.1.12)$$

The corresponding velocity field v is given from (6.1.6) and $u = 0$. The maximum surface height occurs at $x = \frac{L}{2}$:

$$\eta = \eta_0 \left(1 - e^{-\frac{L}{2\lambda}}\right). \quad (6.1.13)$$

This implies that surface height gets very small if $L \gg \lambda$. On the other hand, initial surface elevation almost disappears if $L \ll \lambda$. This is the case when rotation is small. Fig. 6.1 shows the profiles of η for $\lambda/L = 0.1$, $1/3$, and 1.

Second, we consider a perturbation of the initial velocity field. At $t = 0$, we assume $u = \eta = 0$ and

$$v = \begin{cases} 0, & (x < 0, L < x), \\ v_0, & (0 \le x \le L). \end{cases} \quad (6.1.14)$$

Potential vorticity is given by

$$q' = \begin{cases} 0, & (x < 0, L < x), \\ v_0 \left[\delta(x) - \delta(x - L)\right], & (0 \le x \le L), \end{cases} \quad (6.1.15)$$

where $\delta(x)$ is the delta function. If we assume a steady solution in the limit $t \to \infty$, the surface height at the final state satisfies

$$\frac{\partial^2 \eta}{\partial x^2} - \frac{\eta}{\lambda^2} = \begin{cases} 0, & (x < 0, L < x), \\ \frac{f v_0}{g}\left[\delta(x) - \delta(x - L)\right], & (0 \le x \le L). \end{cases} \quad (6.1.16)$$

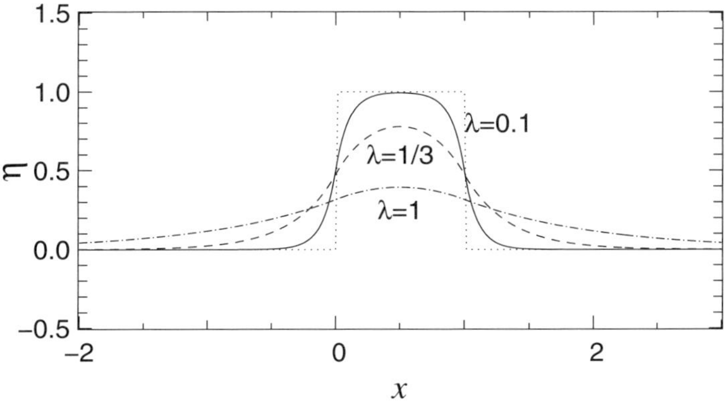

FIGURE 6.1: The profiles of surface height η at the final state due to geostrophic adjustment. The initial perturbation of surface height is $\eta = 1$ in $0 \le x \le 1$ $(L = 1)$ (dotted line). Solid: $\lambda = 0.1$; dashed: $\lambda = 1/3$; and dashed-dotted: $\lambda = 1$.

There are discontinuities in $\frac{\partial \eta}{\partial x}$ at $x = 0$, L by $\pm\frac{fv_0}{g}$. Thus, the solution is given by

$$\eta = \begin{cases} \frac{Hv_0}{2f\lambda}\left(e^{-\frac{L}{\lambda}} - 1\right)e^{\frac{x}{\lambda}}, & (x < 0), \\ \frac{Hv_0}{2f\lambda}\left(e^{\frac{x-L}{\lambda}} - e^{-\frac{x}{\lambda}}\right), & (0 \le x \le L), \\ \frac{Hv_0}{2f\lambda}\left(e^{\frac{L}{\lambda}} - 1\right)e^{-\frac{x}{\lambda}}, & (x > L). \end{cases} \tag{6.1.17}$$

In this case, v is discontinuous at $x = 0$, L, but η is continuous and $u = 0$. The surface height takes the maximum and minimum values at $x = 0$ and L, respectively. In particular, the surface height at $x = L$ is

$$\eta = \frac{Hv_0}{2f\lambda}\left(1 - e^{-\frac{L}{\lambda}}\right). \tag{6.1.18}$$

Thus, $\eta > 0$ is always satisfied. If $L \gg \lambda$, surface height is close to zero everywhere. Fig. 6.2 shows the steady solutions of η for $\lambda/L = 0.1$, $1/3$, and 1.

The profiles of the steady solution (6.1.18) are different from those for the case when the perturbation of surface height is given initially. In the case $L \gg \lambda$, the change in surface height is very small in (6.1.13), while surface height becomes very flat in (6.1.18). This difference comes from perturbations in potential vorticity. If a perturbation of surface height is given initially, potential vorticity q' has nonzero values in the range $0 \le x \le L$, whereas if a perturbation of momentum is given initially, q' is different from zero only at the edges $x = 0$ and L. The effect of the perturbation of potential vorticity on surface height η is confined in the horizontal distance about λ.

As an extreme case, if a constant momentum v_0 is given in the whole region, q' is everywhere zero. This implies that no change in surface height occurs at the final solution. In this case, however, (6.1.1) and (6.1.2) are written as

$$\frac{\partial u}{\partial t} - fv = 0, \qquad \frac{\partial v}{\partial t} + fu = 0, \tag{6.1.19}$$

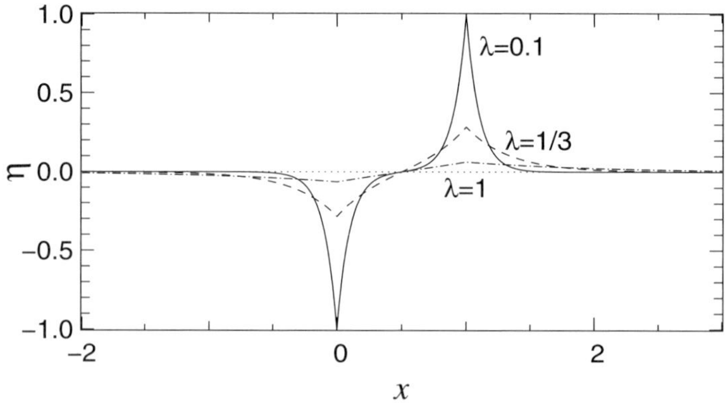

FIGURE 6.2: Same as Fig. 6.1 but the initial perturbation is given to the momentum.

from which a single equation for v is given as

$$\frac{\partial^2 v}{\partial t^2} = -f^2 v \tag{6.1.20}$$

(i.e., v has an oscillatory motion with frequency f; this is an inertial oscillation and no steady solution exists). In general, waves are generated by an initial perturbation as a restoring process. If a perturbation is given in a limited region initially, waves propagate to remote regions and only geostrophic motions remain near the region where the perturbation is given initially.

The energy change due to geostrophic adjustment can be estimated as follows. The energy equation is constructed from (6.1.1)–(6.1.3) as a second-order equation

$$\frac{\partial}{\partial t}\left(g\frac{\eta^2}{2} + H\frac{u^2+v^2}{2}\right) = gH\left[\frac{\partial}{\partial x}(u\eta) + \frac{\partial}{\partial y}(v\eta)\right]. \tag{6.1.21}$$

Integrating this in the entire domain and assuming that no flux exists at the infinite boundaries, we obtain the conservation of energy as

$$\frac{\partial}{\partial t}\int\left(g\frac{\eta^2}{2} + H\frac{u^2+v^2}{2}\right) dxdy = 0. \tag{6.1.22}$$

Let us consider the energy change between the initial state $t = 0$ and the final state $t \to \infty$ for the case when a perturbation is given only to surface height initially. The initial surface height is (6.1.9), and the final state is given by the steady geostrophic solution (6.1.12). We define total energy by

$$E \equiv \int_{-\infty}^{\infty}\left(g\frac{\eta^2}{2} + H\frac{u^2+v^2}{2}\right) dx, \tag{6.1.23}$$

where since all the quantities are uniform in the y-direction, integration is taken only in the x-direction here. At $t = 0$, the initial energy is given by

$$E = \int_0^L g\frac{\eta_0^2}{2}\,dx = gL\frac{\eta_0^2}{2}. \tag{6.1.24}$$

On the other hand, as $t \to \infty$, using (6.1.1), (6.1.11), and (6.1.12), the energy at the final state is

$$\begin{aligned} E &= \int_{-\infty}^{\infty}\left(g\frac{\eta^2}{2} + H\frac{v^2}{2}\right) dx = \int_{-\infty}^{\infty}\left[g\frac{\eta^2}{2} + \frac{g\lambda^2}{2}\left(\frac{\partial\eta}{\partial x}\right)^2\right] dx \\ &= \int_{-\infty}^{\infty}\left(g\frac{\eta^2}{2} - \frac{g\lambda^2}{2}\eta\frac{\partial^2\eta}{\partial x^2}\right) dx = \frac{g\eta_0}{2}\int_0^L \eta\,dx \\ &= gL\frac{\eta_0^2}{2}\left[1 - \frac{\lambda}{L}\left(1 - e^{-\frac{L}{\lambda}}\right)\right]. \end{aligned} \tag{6.1.25}$$

Hence, there exists a difference in the energies between the initial state and the steady state in geostrophic balance. The difference is given by

$$\Delta E = g\lambda\frac{\eta_0^2}{2}\left(1 - e^{-\frac{L}{\lambda}}\right). \tag{6.1.26}$$

Thus, it contradicts the conservation of energy (6.1.22). It can be proved that this contradiction comes from the assumption that the steady geostrophic state is the final solution. We need to take account of the nongeostrophic component even in the limit $t \to \infty$. This is a part of propagating inertial gravity waves.

Similarly, if a perturbation of momentum is given initially as (6.1.14), the energy at $t = 0$ is expressed by

$$E = LH\frac{v_0^2}{2}, \tag{6.1.27}$$

whereas the energy in the limit $t \to \infty$ is

$$E = \lambda H\frac{v_0^2}{2}\left(1 - e^{-\frac{L}{\lambda}}\right). \tag{6.1.28}$$

Thus, in this case, again, there is a difference between the initial energy and the final steady energy. In particular, as λ becomes smaller, the final energy of the part made up by the steady geostrophic balance becomes smaller.

6.2 Forced motions on the f-plane

In the following sections, we consider forced motions in the atmosphere or fluids. First, let us examine thermally forced motions on the f-plane. We particularly consider large-scale motions in the hydrostatic balance; in this case, horizontal structure and vertical structure can be separately considered as described in Section 4.7.1. The separation of variables is also applicable if there is a forcing in the system. Introducing the Boussinesq approximation to (4.7.6)–(4.7.10) and neglecting the metric terms on the f-plane, we obtain the basic equations:

$$\frac{\partial u}{\partial t} - fv = -\frac{1}{\rho_0}\frac{\partial p}{\partial x}, \tag{6.2.1}$$

$$\frac{\partial v}{\partial t} + fu = -\frac{1}{\rho_0}\frac{\partial p}{\partial y}, \tag{6.2.2}$$

$$0 = -\frac{\partial p}{\partial z} - \rho g, \tag{6.2.3}$$

$$\frac{\partial u}{\partial x} + \frac{\partial v}{\partial y} + \frac{\partial w}{\partial z} = 0, \tag{6.2.4}$$

$$\frac{\partial \rho}{\partial t} - \frac{N^2}{g}w = -\alpha\rho_0 Q. \tag{6.2.5}$$

Note that the forcing term is given as the right-hand side of (6.2.5), using (3.1.12) for the Boussinesq approximation. In the following arguments, we assume that the buoyancy frequency N^2 is constant for simplicity. Each of the variables u, v, w, ρ, and p is expressed as a product of a function of (x, y, t) and a function of z. The former describes the horizontal structure, while the latter is the vertical structure. The vertical profiles of w, ρ, and p are denoted by W, Z, and P, respectively. If W is set proportional to a sine function as (4.7.23), Z and P are expressed using

(4.7.25) and (4.7.26) as follows:

$$W(z) = W_0 \sin mz, \tag{6.2.6}$$
$$P(z) = W_0 m \cos mz, \tag{6.2.7}$$
$$Z(z) = -\frac{W_0 m^2}{g} \sin mz. \tag{6.2.8}$$

In addition, we also expand the thermal forcing Q by sine waves proportional to W and Z in the vertical direction:

$$\frac{\alpha \rho_0 g}{N^2} Q(x, y, z, t) = \int_0^\infty \tilde{Q}(m; x, y, t) W_0 \sin mz \, dm. \tag{6.2.9}$$

Thus, similar to (4.7.33)–(4.7.35), we obtain shallow water equations for the horizontal structure:

$$\frac{\partial u}{\partial t} - fv = -g\frac{\partial \eta}{\partial x}, \tag{6.2.10}$$
$$\frac{\partial v}{\partial t} + fu = -g\frac{\partial \eta}{\partial y}, \tag{6.2.11}$$
$$\frac{\partial \eta}{\partial t} + H\left(\frac{\partial u}{\partial x} + \frac{\partial v}{\partial y}\right) = -Q, \tag{6.2.12}$$

where the symbol ˜ is omitted from the variables of horizontal structure, and we define H and c by

$$c^2 \equiv gH = \frac{N^2}{m^2}. \tag{6.2.13}$$

One may note that the thermal forcing Q corresponds to the source/sink of mass in shallow water equations. Eliminating u and v from (6.2.10)–(6.2.12), we obtain a single equation for η as

$$\frac{\partial}{\partial t}\left[\frac{\partial^2}{\partial t^2} + f^2 - c^2\left(\frac{\partial^2}{\partial x^2} + \frac{\partial^2}{\partial y^2}\right)\right]\eta = -\left(\frac{\partial^2}{\partial t^2} + f^2\right) Q. \tag{6.2.14}$$

To consider the thermal response in shallow water equations, we further assume that motions are uniform in the y-direction. Letting τ denote the time scale of motion, we can characterize forced motions by relative magnitude between τ and f^{-1}. In the case $\tau \ll f^{-1}$, (6.2.14) is approximated to

$$\left(\frac{\partial^2}{\partial t^2} - c^2\frac{\partial^2}{\partial x^2}\right)\eta = -\frac{\partial}{\partial t}Q, \tag{6.2.15}$$

while in the case $\tau \gg f^{-1}$, on the other hand, (6.2.14) becomes

$$\frac{\partial}{\partial t}\left(f^2 - c^2\frac{\partial^2}{\partial x^2}\right)\eta = -f^2 Q. \tag{6.2.16}$$

The corresponding equations for $\tau \ll f^{-1}$ are

$$\frac{\partial u}{\partial t} = -g\frac{\partial \eta}{\partial x}, \tag{6.2.17}$$

$$\frac{\partial \eta}{\partial t} + H\frac{\partial u}{\partial x} = -Q. \tag{6.2.18}$$

The set of these equations describes the propagation of pure gravity waves. On the other hand, the equations for $\tau \gg f^{-1}$ are

$$-fv = -g\frac{\partial \eta}{\partial x}, \tag{6.2.19}$$

$$\frac{\partial v}{\partial t} + fu = 0, \tag{6.2.20}$$

$$\frac{\partial \eta}{\partial t} + H\frac{\partial u}{\partial x} = -Q. \tag{6.2.21}$$

In this case, velocity v always satisfies geostrophic balance.

We apply a thermal forcing at the point $x = 0$ and the time $t = 0$ to the initial state at rest. This type of forcing is expressed by a function

$$Q(m; x, t) = Q_0(m)\delta(x)H(t), \tag{6.2.22}$$

where $\delta(x)$ is the delta function and $H(t)$ is the step function: $H(t) = 0$ for $t < 0$ and $H(t) = 1$ for $t \geq 0$. If $\tau \ll f^{-1}$, it can be shown that the solution to (6.2.15) is given by

$$\eta = -\frac{Q_0}{4c}\left[\,\mathrm{sgn}(x+ct) - \mathrm{sgn}(x-ct)\right], \tag{6.2.23}$$

$$u = -\frac{Q_0}{4c^2}g\left[\,\mathrm{sgn}(x) - \frac{1}{2}\,\mathrm{sgn}(x+ct) - \frac{1}{2}\,\mathrm{sgn}(x-ct)\right], \tag{6.2.24}$$

where $\mathrm{sgn}(x) = [H(x) - H(-x)]/2$: $\mathrm{sgn}(x) = 1$ for $x > 0$ and $\mathrm{sgn}(x) = -1$ for $x < 0$. If $\tau \gg f^{-1}$, on the other hand, the solution to (6.2.16) is given by

$$\eta = -\frac{fQ_0}{2c}e^{-\frac{f}{c}|x|}\,tH(t), \tag{6.2.25}$$

$$u = -\frac{fQ_0}{2c^2}ge^{-\frac{f}{c}|x|}\,\mathrm{sgn}(x)\,tH(t), \tag{6.2.26}$$

$$v = \frac{fQ_0}{2c^2}ge^{-\frac{f}{c}|x|}\,\mathrm{sgn}(x)\,tH(t), \tag{6.2.27}$$

where $\lambda = c/f$ is the Rossby radius of deformation.

Examples of forced motions are shown in Figs. 6.3 and 6.4. In these cases, we apply a point mass sink at the origin $q_0 > 0$. In the case $\tau \ll f^{-1}$, the forced motions are driven in the region that expands in time with the speed of c. Velocity u is convergent toward the mass sink and its amplitude is constant irrespective of time. In the case of $\tau \gg f^{-1}$, on the other hand, the responses of η and v to the forcing are confined only in the range with the Rossby radius of deformation λ

from the mass sink. In this region, η and v grow in proportion to time, while the distribution of u is invariant.

The thermal response in the hydrostatic atmosphere is expressed as the products of the solutions to shallow water equations and vertical structure. In the case that the vertical domain is limited by rigid boundaries at $z = 0, 1$, the function $\sin \pi z$ satisfies this boundary condition for w. In this case, the corresponding vertical profile of u is given by $\cos \pi z$. If we write a solution to shallow water equations as $\tilde{u}$, solutions to (6.2.1)–(6.2.5) are expressed as $u = \tilde{u} \cos \pi z$. In addition, if the motions are uniform in the y-direction, the streamfunction ψ can be defined in the

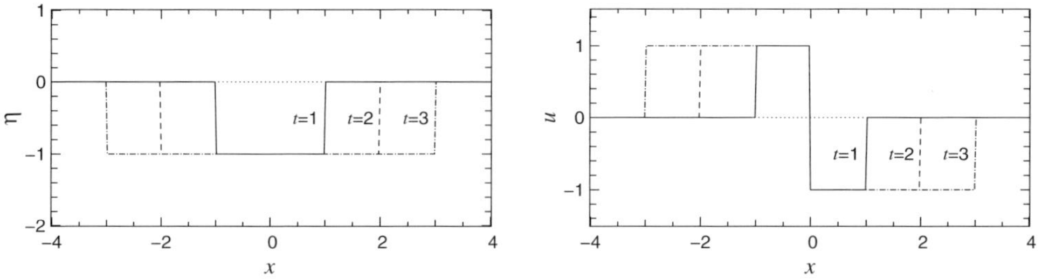

FIGURE 6.3: Forced motions in shallow water equations for $\tau \ll f^{-1}$. The distributions of (left) surface height η and (right) velocity u at $t = 1$ (solid curve), $t = 2$ (dashed curve), and $t = 3$ (dotted-dashed curve). The abscissa is x/c, and the ordinates are (left) $\eta/(Q_0/2c)$ and (right) $u/(Q_0 g/2c^2)$.

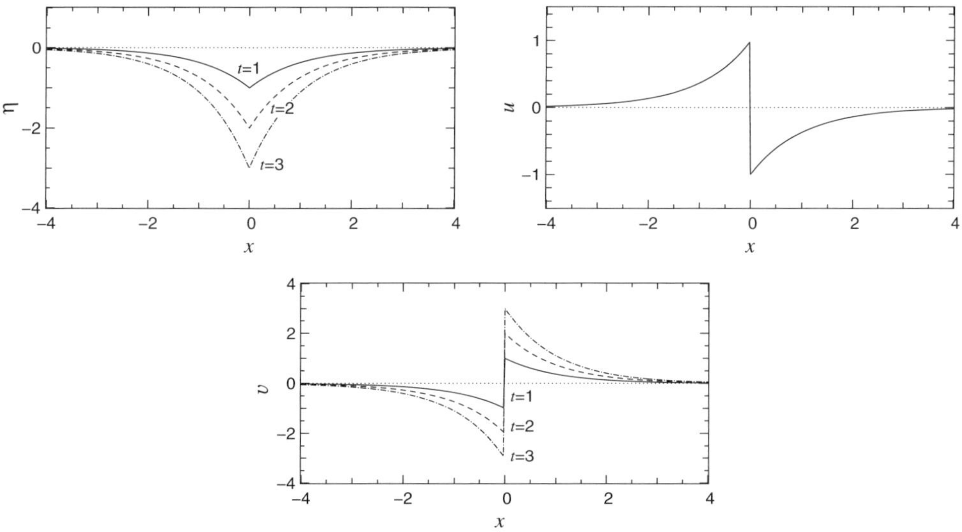

FIGURE 6.4: Forced motions in shallow water equations for $\tau \gg f^{-1}$ and $\lambda = c/f = 1/3$. The distributions of (top left) surface hight η, (top right) lateral velocity u, and (bottom) normal velocity v at $t = 1$ (solid), $t = 2$ (dashed), and $t = 3$ (dotted-dashed). The ordinates are $\eta/(fQ_0/2c)$, $u/(Q_0 g/2c^2)$, and $v/(fQ_0 g/2c^2)$, respectively. The abscissa is x/λ with $\lambda = c/f = 1/3$.

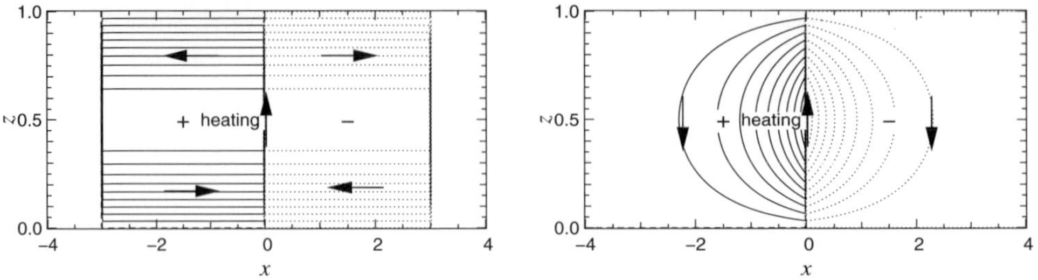

FIGURE 6.5: Distributions of the streamfunctions of forced motions for (left) $\tau \ll f^{-1}$ and (right) $\tau \gg f^{-1}$. These figures correspond to $t = 3$ of Figs. 6.3 and 6.4, respectively. Thermal forcing has a maximum at $x = 0$, $z = 0.5$.

xz-section such that $u = \frac{\partial \psi}{\partial z}$ and $w = -\frac{\partial \psi}{\partial x}$. The two panels in Fig. 6.5 show the streamfunctions at $t = 3$ for the cases of Figs. 6.3 and 6.4, respectively. In particular, the two distributions of horizontal velocity u along $z = 0$ in Fig. 6.5 agree with those shown in Figs. 6.3 and 6.4, respectively.

As shown in Fig. 6.5, the streamfunction in the xz-section expands with time in the case $\tau \ll f^{-1}$. On the other hand, in the case $\tau \gg f^{-1}$, the extent of the streamfunction is confined within the distance of the Rossby radius of deformation λ. However, its strength increases with time. In order to keep the strength of the streamfunction constant, we need additional damping to the equations of velocity and mass. Giving the damping terms proportional to velocity or mass in (6.2.10)–(6.2.12), we obtain balance in the steady state:

$$su - fv = -g\frac{\partial \eta}{\partial x}, \tag{6.2.28}$$

$$sv + fu = -g\frac{\partial \eta}{\partial y}, \tag{6.2.29}$$

$$s'\eta + H\left(\frac{\partial u}{\partial x} + \frac{\partial v}{\partial y}\right) = -Q, \tag{6.2.30}$$

where s and s' are coefficients of the damping term. Assuming uniformity in the y-direction, we combine these equations to

$$\left[s'(s^2 + f^2) - sc^2\frac{\partial^2}{\partial x^2}\right]\eta = -(s^2 + f^2)Q. \tag{6.2.31}$$

If forcing is given by $Q = Q_0\delta(x)$, the solution to this equation is expressed as

$$\eta = -(s^2 + f^2)Q_0 \exp\left(-\sqrt{\frac{s'(s^2 + f^2)}{sc^2}}|x|\right). \tag{6.2.32}$$

Therefore, the extent of the forced motions is given by

$$L = \sqrt{\frac{sc^2}{s'(s^2 + f^2)}}. \tag{6.2.33}$$

In the limit $s' \to 0$ or $s \to 0$ under $f = 0$, we have $L \to \infty$. As shown in Fig. 6.3, the region of thermally forced motions expands infinitely if $s \to 0$ and $f = 0$. In the case $s' = 0$, the amplitude of η grows infinitely, since the response of surface height η cannot be balanced by Q.

6.3 Axisymmetric flows†

We have seen in previous sections that the geostrophic balance (6.2.19) holds if $\tau \gg f^{-1}$ in hydrostatic equations or in shallow water equations. In that case, v or ρ grows linearly with time, whereas the strength of the streamfunction in the vertical section remains the same. In this section, we generalize this situation to thermally or mechanically forced axisymmetric flows.

We use the symbols introduced in Section 2.4 for the equations of axisymmetric flows. We further introduce the Boussinesq approximation and assume that the geopotential Φ is a function of z. Acceleration due to gravity is $\frac{\partial \Phi}{\partial z} = g$. We express the changes in angular momentum and entropy as

$$\frac{dl}{dt} = rF_\varphi, \tag{6.3.1}$$

$$\frac{ds}{dt} = \frac{C_p Q}{T}, \tag{6.3.2}$$

where F_φ is a forcing term and Q is a diabatic term. Using (3.1.12) for the Boussinesq approximation, (6.3.2) is reduced to

$$\frac{d\rho}{dt} = -\rho_0 \alpha Q. \tag{6.3.3}$$

Neglecting density variation in the continuity equation (2.4.21), we have

$$\frac{1}{r}\frac{\partial}{\partial r}(rv_r) + \frac{\partial}{\partial z}v_z = 0. \tag{6.3.4}$$

From this, the streamfunction ψ can be introduced as

$$rv_r = \frac{\partial \psi}{\partial z}, \qquad rv_z = -\frac{\partial \psi}{\partial r}. \tag{6.3.5}$$

Balances in the r- and z-directions, (2.4.26) and (2.4.27), are rewritten as

$$-\frac{l^2}{r^3} = -\frac{1}{\rho_0}\frac{\partial p}{\partial r}, \tag{6.3.6}$$

$$0 = -\frac{1}{\rho}\frac{\partial p}{\partial z} - g. \tag{6.3.7}$$

Hence, the thermal wind balance is written as

$$\frac{1}{r^3}\frac{\partial l^2}{\partial z} = -\frac{g}{\rho_0}\frac{\partial \rho}{\partial r}. \tag{6.3.8}$$

†This section follows Eliassen (1951).

The angular momentum equation (6.3.1) and the density equation (6.3.3) become

$$\frac{\partial l^2}{\partial t} + v_r \frac{\partial l^2}{\partial r} + v_z \frac{\partial l^2}{\partial z} = 2lrF_\varphi, \tag{6.3.9}$$

$$\frac{\partial \rho}{\partial t} + v_r \frac{\partial \rho}{\partial r} + v_z \frac{\partial \rho}{\partial z} = -\rho_0 \alpha Q. \tag{6.3.10}$$

Eliminating time derivatives from (6.3.9) and (6.3.10) using (6.3.8), we obtain the equation for ψ:

$$\frac{\partial}{\partial r}\left(A\frac{\partial \psi}{\partial r} + B\frac{\partial \psi}{\partial z}\right) + \frac{\partial}{\partial z}\left(B\frac{\partial \psi}{\partial r} + C\frac{\partial \psi}{\partial z}\right) = \frac{\partial E}{\partial r} + \frac{\partial F}{\partial z}, \tag{6.3.11}$$

where

$$A = -\frac{g}{\rho_0 r}\frac{\partial \rho}{\partial z}, \qquad B = \frac{g}{\rho_0 r}\frac{\partial \rho}{\partial r} = -\frac{1}{r^4}\frac{\partial l^2}{\partial z},$$

$$C = \frac{1}{r^4}\frac{\partial l^2}{\partial r}, \qquad E = -\alpha g Q, \qquad F = \frac{2l}{r^2}F_\varphi. \tag{6.3.12}$$

Eq. (6.3.11) is an elliptic equation if the inequality,

$$D \equiv AC - B^2 = -\frac{g}{\rho_0 r^5}\left(\frac{\partial \rho}{\partial z}\frac{\partial l^2}{\partial r} - \frac{\partial \rho}{\partial r}\frac{\partial l^2}{\partial z}\right) > 0, \tag{6.3.13}$$

is satisfied. We note here that, using $\frac{d\rho}{\rho_0} = -\frac{d\theta}{\theta}$ in (2.4.55), the condition for the stability (2.4.53) is rewritten as

$$\det M = -\frac{g}{\rho_0 r^3}\left(\frac{\partial \rho}{\partial z}\frac{\partial l^2}{\partial r} - \frac{\partial \rho}{\partial r}\frac{\partial l^2}{\partial z}\right) > 0. \tag{6.3.14}$$

Therefore, if the axisymmetric flow is stable, the response to thermal or mechanical forcing is described by an elliptic equation. If forcing is localized, the induced flows are confined to a limited region.

If A, B, and C are constant, it can be shown that the solution to (6.3.11) is given by

$$\psi(r,z) = \int G(r,z;r_0,z_0)\left(\frac{\partial E}{\partial r_0} + \frac{\partial F}{\partial z_0}\right)\bigg|_{(r_0,z_0)} dr_0\, dz_0, \tag{6.3.15}$$

where G represents the Green function, given by

$$G(r,z;r_0,z_0) = \frac{1}{2\pi D} \times \ln\left[C(r-r_0)^2 - 2B(r-r_0)(z-z_0) + A(z-z_0)^2\right]^{\frac{1}{2}}. \tag{6.3.16}$$

If just the heat source E exists and $F = 0$, the solution becomes

$$\psi(r,z) = \int G(r,z;r_0,z_0)\frac{\partial E(r_0,z_0)}{\partial r_0} dr_0\, dz_0 = -\int \frac{\partial G(r,z;r_0,z_0)}{\partial r_0} E(r_0,z_0)\, dr_0\, dz_0. \tag{6.3.17}$$

In addition, if E is a point source localized at (r_0, z_0) and normalized as $\int E dr dz = 1$, we have

$$\begin{aligned} \psi(r,z) &= -\frac{\partial G(r,z;r_0,z_0)}{\partial r_0} \\ &= \frac{1}{2\pi D}\frac{C(r-r_0)-B(z-z_0)}{C(r-r_0)^2-2B(r-r_0)(z-z_0)+A(z-z_0)^2}. \end{aligned} \tag{6.3.18}$$

Thus, configuration of the streamlines is given by

$$\begin{aligned} C(r-r_0)^2 - 2B(r-r_0)(z-z_0) + A(z-z_0)^2 & \\ -\alpha[C(r-r_0)-B(z-z_0)] &= 0, \end{aligned} \tag{6.3.19}$$

where α is constant. This equation indicates that streamlines consist of ellipses that cross the point (r_0, z_0). The direction of flow at this point is, from (6.3.12),

$$\left.\frac{dz}{dr}\right|_{(r_0,z_0)} = \frac{C}{B} = -\frac{\frac{\partial l^2}{\partial r}}{\frac{\partial l^2}{\partial z}} = \left.\frac{dz}{dr}\right|_l \tag{6.3.20}$$

(i.e., the direction of flow is tangential to the contour of angular momentum; a point heat source drives a circulation along the contour of angular momentum near the heat source). The left panel of Fig. 6.6 schematically shows this case.

On the other hand, if only a localized torque F exists at (r_0, z_0) and $E = 0$, the streamlines are also ellipses. The contours of ψ are described by

$$\begin{aligned} C(r-r_0)^2 - 2B(r-r_0)(z-z_0) + A(z-z_0)^2 & \\ -\gamma[A(z-z_0)-B(r-r_0)] &= 0, \end{aligned} \tag{6.3.21}$$

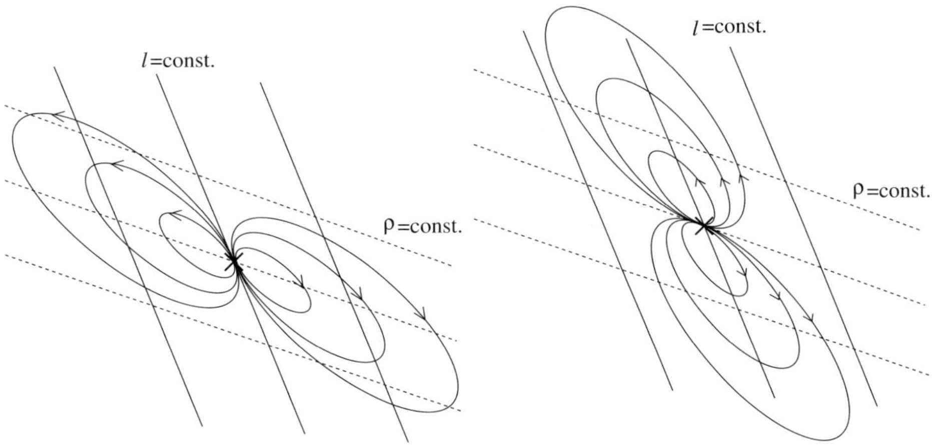

FIGURE 6.6: Response of streamlines to a point-source of (left) heat and (right) angular momentum in the axisymmetric flow. The point source is applied at (r_0, z_0) denoted by a thick cross. Contours of angular momentum are denoted by $l =$ const. (solid), and those of density or entropy by $\rho =$ const. (dashed). The ellipses are streamlines.

where γ is constant. The inclination of the streamfunction near the torque is

$$\left.\frac{dz}{dr}\right|_{(r_0,z_0)} = \frac{B}{A} = -\frac{\frac{\partial\rho}{\partial r}}{\frac{\partial\rho}{\partial z}} = \left.\frac{dz}{dr}\right|_{\rho} \tag{6.3.22}$$

(i.e., the direction of flow is tangential to the contour of the density (entropy) at the point where the torque is applied). The right panel of Fig. 6.6 schematically shows this case.

6.4 Forced motions on the β-plane

We consider the thermally forced motions of the geostrophic regime on the β-plane using shallow water equations. In the case of $f = f_0 + \beta y$, the vorticity equation is given from (6.2.10)–(6.2.12) as

$$\frac{\partial\zeta}{\partial t} + \beta v + fD = 0, \tag{6.4.1}$$

where $\zeta = \frac{\partial v}{\partial x} - \frac{\partial u}{\partial y}$ and $D = \frac{\partial u}{\partial x} + \frac{\partial v}{\partial y}$. Using this equation and (6.2.12), the potential vorticity equation is written as

$$\frac{\partial}{\partial t}\left(\zeta - \frac{f}{H}\eta\right) + \beta v = \frac{f}{H}Q. \tag{6.4.2}$$

Assuming that geostrophic balance holds between u, v, and η for simplicity, we rewrite the potential vorticity equation as

$$\frac{\partial}{\partial t}\left(\frac{\partial^2}{\partial x^2} + \frac{\partial^2}{\partial y^2} - \lambda^{-2}\right)\eta + \beta\frac{\partial}{\partial x}\eta = \frac{Q}{\lambda^2}, \tag{6.4.3}$$

where $\lambda = gH/f$ is the Rossby radius of deformation. Assuming that η is proportional to $\exp(ily)$ and introducing a damping term with a coefficient s, we obtain the equation for the steady solution as

$$\left(\frac{d^2}{dx^2} + \frac{\beta}{s}\frac{d}{dx} - l^2 - \lambda^{-2}\right)\eta = \frac{Q}{s\lambda^2}. \tag{6.4.4}$$

Let us consider the response to a point source at $x = 0$. If the response of η has the form of $e^{-\nu|x|}$, we have solutions of ν as

$$\nu_+ = \frac{\beta}{2s} + \sqrt{\frac{\beta^2}{4s^2} + l^2 + \lambda^{-2}}, \quad \text{for} \quad x > 0, \tag{6.4.5}$$

$$\nu_- = -\frac{\beta}{2s} + \sqrt{\frac{\beta^2}{4s^2} + l^2 + \lambda^{-2}}, \quad \text{for} \quad x < 0. \tag{6.4.6}$$

This means that the response reaches farther on the negative side $x < 0$ than on the positive side $x > 0$. In particular, in the inviscid limit $s \to 0$, we have $\nu_+ \to \infty$.

In the special case of steady solution in the potential vorticity equation (6.4.2), we have a balance

$$\beta v = \frac{f}{H}Q. \tag{6.4.7}$$

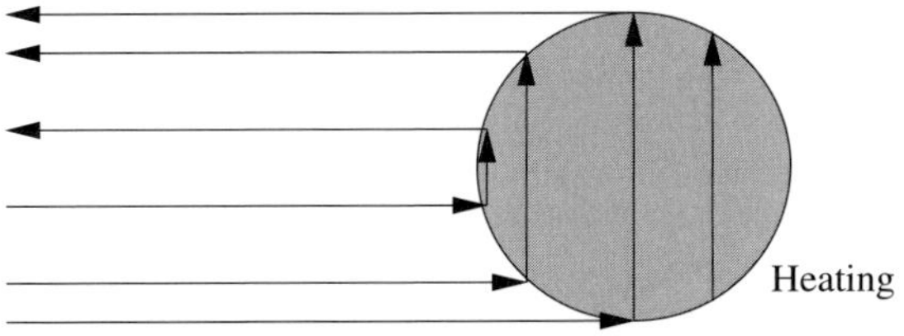

FIGURE 6.7: Schematic figure of thermally forced motions on the mid-latitude β-plane. The shaded circle is the heating region. If the circulation is steady, the response in the heating region is $v > 0$, whereas that of the outer region is $v = 0$. The influence of heating spreads only on the western side.

This indicates that a poleward flow $v > 0$ exists in the heating region $Q > 0$, whereas no meridional flow $v = 0$ exists if there is no heating. Thus, the response to heating spreads only in the x-direction and its meridional width is confined to that of the heating. The thermal response spreads in the negative x-direction, since the group velocity of Rossby waves is westward as shown below: the dispersion relation of Rossby waves is given by (4.6.21). Since $\omega = 0$ for the steady motion, we have $k = 0$. From (4.6.23), therefore, the group velocity is

$$c_{gx} = \frac{\partial \omega}{\partial k} = -\beta < 0, \tag{6.4.8}$$

Fig. 6.7 shows a schematic response to heating on the β-plane.

6.5 Forced motions on the equatorial β-plane

As the heating region gets closer to the equator, the balance in (6.4.7) breaks down in the limit $f \to 0$. This implies that the geostrophic balance used in the previous section becomes inappropriate and that gravity waves show important roles in the equatorial region. We use the shallow water equations on the equatorial β-plane to consider the thermal response in the equatorial region, which is given by substituting $f = \beta y$ into (6.2.28)–(6.2.30). We simply consider the case when the damping coefficients of heat and momentum are equal: $s' = s$. Thus, the balance equations in the equatorial β-plane are written as

$$su - \beta y v = -g\frac{\partial \eta}{\partial x}, \tag{6.5.1}$$

$$sv + \beta y u = -g\frac{\partial \eta}{\partial y}, \tag{6.5.2}$$

$$s\eta + H\left(\frac{\partial u}{\partial x} + \frac{\partial v}{\partial y}\right) = -Q. \tag{6.5.3}$$

Introducing the variables q and r as defined by (4.7.95), we rewrite the above equations as

$$\left(s + c\frac{\partial}{\partial x}\right) q + \left(c\frac{\partial}{\partial y} - \beta y\right) v = -\frac{g}{c}Q, \tag{6.5.4}$$

$$\left(s+c\frac{\partial}{\partial x}\right)r+\left(c\frac{\partial}{\partial y}+\beta y\right)v \quad = \quad -\frac{g}{c}Q, \tag{6.5.5}$$

$$\left(c\frac{\partial}{\partial y}+\beta y\right)q+\left(c\frac{\partial}{\partial y}-\beta y\right)r+sv \quad = \quad 0. \tag{6.5.6}$$

Let us expand these variables with the parabolic functions (4.7.91); for instance,

$$q(x,y) \quad = \quad \sum_{n=0}^{\infty} q_n(x) D_n\left(\sqrt{\frac{2\beta}{c}}y\right). \tag{6.5.7}$$

We also expand r, v, and Q in a similar way. Using the recursive relations (4.7.101), we have a set of equations:

$$\left(s+c\frac{d}{dx}\right)q_0 \quad = \quad -\frac{g}{c}Q_0, \tag{6.5.8}$$

$$\left(s+c\frac{d}{dx}\right)q_{n+1}-\sqrt{2\beta c}\,v_n \quad = \quad -\frac{g}{c}Q_{n+1}, \quad \text{for } n\geq 0, \tag{6.5.9}$$

$$\left(s-c\frac{d}{dx}\right)r_{n-1}+\sqrt{2\beta c}\,nv_n \quad = \quad -\frac{g}{c}Q_{n-1}, \quad \text{for } n\geq 1, \tag{6.5.10}$$

$$(n+1)q_1+\frac{s}{\sqrt{2\beta c}}v_0 \quad = \quad 0, \tag{6.5.11}$$

$$(n+1)q_{n+1}-r_{n-1}+\frac{s}{\sqrt{2\beta c}}v_n \quad = \quad 0, \quad \text{for } n\geq 1. \tag{6.5.12}$$

Specifically, we consider heating that is symmetric about the equator:

$$Q \quad = \quad Q_0 e^{ikx} D_0\left(\sqrt{\frac{2\beta}{c}}y\right) \quad = \quad Q_0 e^{ikx}\exp\left(-\frac{\beta y^2}{2c}\right). \tag{6.5.13}$$

From the symmetry, we promptly obtain $q_1 = v_0 = 0$. It can be found that there are two types of solutions, which are respectively expressed by q_0 and (q_2, r_0, v_1). First, q_0 is determined by (6.5.8):

$$q_0 \quad = \quad -\frac{1}{s+ikc}\frac{g}{c}Q_0. \tag{6.5.14}$$

This solution corresponds to the response due to the Kelvin wave. The response emerges on the eastern side of heating, and the length scale of the response is c/s. Next, (6.5.9), (6.5.10), and (6.5.12) give

$$q_2 \quad = \quad -\frac{1}{(3+\epsilon)s-ikc}\frac{g}{c}Q_0, \tag{6.5.15}$$

where

$$\epsilon \quad = \quad \frac{s^2+k^2c^2}{2\beta c}. \tag{6.5.16}$$

The solution (6.5.15) corresponds to the response due to the Rossby wave. The response emerges on the western side of heating. If the wavelength is long enough and the damping small enough such that $\epsilon \ll 1$, then the scale of the response is $c/3s$, which is one-third of the length scale of the Kelvin response. For the Kelvin response, the structure of velocity and surface height is written as

$$u = \frac{g}{c}\eta = \frac{q_0}{2}\exp\left(-\frac{\beta y^2}{2c}\right), \qquad v = 0, \tag{6.5.17}$$

and, for the Rossby response,

$$u = \frac{q_2}{2}\left[\frac{2\beta}{c}y^2 - 3 - \frac{s(s+ikc)}{2\beta c}\right]\exp\left(-\frac{\beta y^2}{2c}\right), \tag{6.5.18}$$

$$v = q_2\frac{s+ikc}{c}y\exp\left(-\frac{\beta y^2}{2c}\right), \tag{6.5.19}$$

$$\frac{g}{c}\eta = \frac{q_2}{2}\left[\frac{2\beta}{c}y^2 + 1 + \frac{s(s+ikc)}{2\beta c}\right]\exp\left(-\frac{\beta y^2}{2c}\right). \tag{6.5.20}$$

The above shallow water model corresponds to the lower layer of stratified fluids. If we introduce vertical structure, we obtain three-dimensional circulation. Let the depth of the atmosphere be H and the vertical structure of the vertical velocity be proportional to $\sin(\pi z/H)$. From the nondivergent condition, the amplitude of the vertical velocity w is given by

$$w = -H\left(\frac{\partial u}{\partial x} + \frac{\partial v}{\partial y}\right) = s\eta + Q, \tag{6.5.21}$$

where (6.5.3) is used.

Fig. 6.8 shows an example of thermal response to longitudinally cyclic heating in the case of the long wave limit $\epsilon \ll 1$. The center of the forcing is located at $x = 0$. The Kelvin response is seen along the equator $y = 0$ on the eastern side of the center of heating. On the western side, the Rossby response is seen with the maximum in surface height around $y = \pm 2$. There is a confluent wind along the equator near the heating region, which diverges in the meridional direction. The maximum of upward motion is located at the convergent area of horizontal winds at the equator (i.e., the position of maximum heating).

6.6 Ekman transport

As an example of the response to mechanical forcing, we consider Ekman transport. This can be illustrated by using shallow water equations. By introducing the source terms of momentum (F_x, F_y), linearized shallow water equations are given as

$$\frac{\partial u}{\partial t} - fv = -g\frac{\partial \eta}{\partial x} + F_x, \tag{6.6.1}$$

$$\frac{\partial v}{\partial t} + fu = -g\frac{\partial \eta}{\partial y} + F_y, \tag{6.6.2}$$

$$\frac{\partial \eta}{\partial t} + H\left(\frac{\partial u}{\partial x} + \frac{\partial v}{\partial y}\right) = 0. \tag{6.6.3}$$

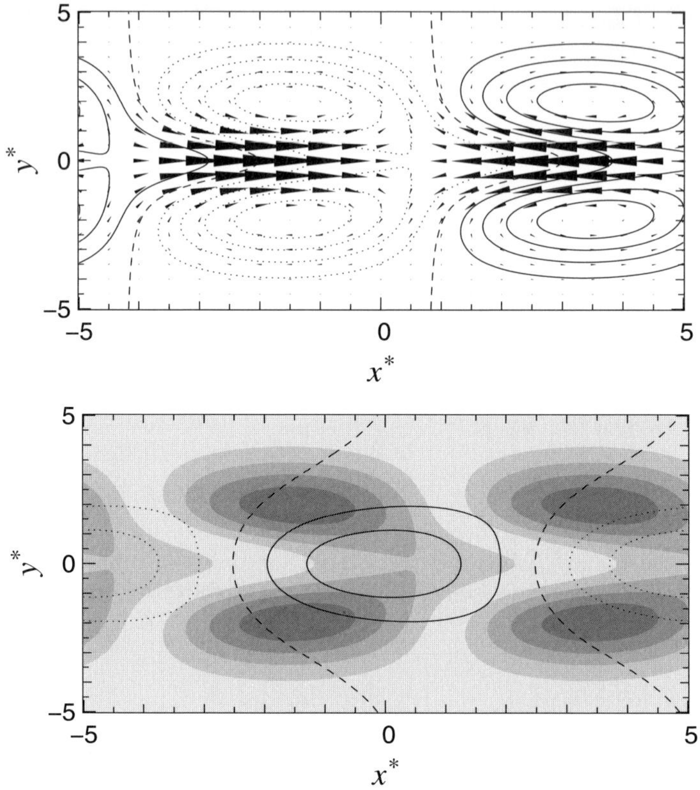

FIGURE 6.8: Forced motions on the equatorial β-plane. (Top) The contour represents η; positive values are shown by solid curves and negative values by dashed curves. The arrows are velocity vectors. (Bottom) The contour represents vertical velocity w. The positive values correspond to upward motions and the negative values correspond to downward motions. The gray scale is η. The parameters are $s/\sqrt{2\beta c} = 0.1$, $k\sqrt{c/2\beta} = 0.2\pi$, and $\epsilon \ll 1$. The maximum in the heat sink is located at $x = 0$, and the heating has a profile with $Q = Q_0 \cos kx\, D_0(y^*)$. The ordinates are normalized as $x^* = x\sqrt{2\beta/c}$ and $y^* = y\sqrt{2\beta/c}$.

These are combined with the vorticity and potential vorticity equation as

$$\frac{\partial \zeta}{\partial t} + \beta v + fD = \frac{\partial F_y}{\partial x} - \frac{\partial F_x}{\partial y}, \tag{6.6.4}$$

$$\frac{\partial}{\partial t}\left(\zeta - \frac{f}{H}\eta\right) + \beta v = \frac{\partial F_y}{\partial x} - \frac{\partial F_x}{\partial y}. \tag{6.6.5}$$

We consider a frictional force as a momentum source that works in the opposite direction to the velocity of the fluid. In this case, we may set the frictional force to $F_x = -ku$ and $F_y = -kv$, where k is a coefficient. The vorticity equation (6.6.4) becomes

$$\frac{\partial \zeta}{\partial t} + \beta v + fD = -k\zeta. \tag{6.6.6}$$

In a special case of a steady vorticity field on the f-plane with $\beta = 0$, we have

$$fD = -k\zeta \qquad (6.6.7)$$

(i.e., the flow is convergent where the vorticity is positive). This characteristic is called *Ekman convergence.* Although vorticity is steady in this case, height η is not steady since the flow is convergent. In the region of cyclonic vorticity ($\zeta > 0$), mass tends to be transported toward the center of the cyclone because of convergent flow, thus the pressure of the cyclone increases. For instance, if the field is uniform in the x-direction, (6.6.1) becomes

$$fv = ku. \qquad (6.6.8)$$

Meridional flow is poleward $v > 0$ in the region of eastward flow $u > 0$ and is equatorward $v < 0$ in the region of westward flow $u < 0$. Since $\frac{\partial \eta}{\partial y} < 0$ corresponds to zonal flow $u > 0$ in the geostrophic balance, the pressure gradient tends to be reduced by these meridional flows. This kind of flow orthogonal to the geostrophic flow is called *Ekman transport.*

In the case $\beta \neq 0$, there is a solution where potential vorticity is time-independent. By omitting the tendency term in (6.6.5), we obtain

$$\beta v = \frac{\partial F_y}{\partial x} - \frac{\partial F_x}{\partial y}, \qquad (6.6.9)$$

which is called the *Sverdrup balance.* This balance is used to explain the planetary-scale circulation of the ocean. The ocean current is partly driven by wind stress exerted by atmospheric surface wind. If surface wind is uniform in the x-direction $U(y)$, and wind stress is approximately given by $F_x = \alpha U$, the Sverdrup balance of the ocean current is given by

$$\beta v = -\alpha \frac{\partial U}{\partial y}. \qquad (6.6.10)$$

In an approximate sense, atmospheric surface winds are easterly in low latitudes $U < 0$ and westerly in the mid- and higher latitudes $U > 0$, so that we generally have $\frac{\partial U}{\partial y} > 0$ in the northern hemisphere. Thus, the ocean current is everywhere equatorward $v < 0$ from (6.6.10). To compensate for this equatorward current, the counter poleward current is driven in the confined western boundary of the ocean. This is the *western boundary current.* Fig. 6.9 is a schematic distribution of the planetary-scale ocean current and the profile of wind stress. If zonal wind in the atmosphere has a latitudinal profile $U = -U_0 \cos \pi y$ $(0 \leq y \leq 1)$, meridional flow in the ocean is $v = U_0(\alpha\pi/\beta) \sin \pi y$ from the Sverdrup balance (6.6.10). The streamfunction is calculated as $\psi = U_0(\alpha\pi/\beta) \sin \pi y \cdot (x - 1)$ using the boundary condition $\psi = 0$ at the eastern end $x = 1$. Thus, $\psi \neq 0$ at the western end $x = 0$; this implies a poleward returning flow within a very thin western boundary layer in the ocean.

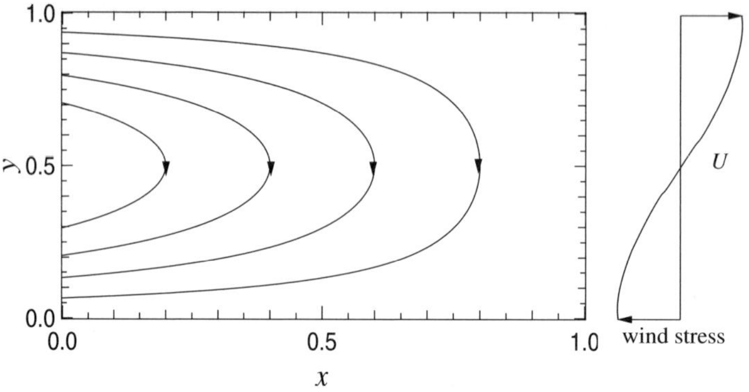

FIGURE 6.9: Schematic distribution of the ocean current: the Sverdrup balance and the western boundary current. Left: contours of the streamfunction and direction of flow; right: meridional distribution of surface winds in the atmosphere (wind stress). The poleward returning flow is confined near the western boundary layer.

Returning to the geostrophic flow on the f-plane, we examine the height dependency of the flow near the surface modified by friction in the surface boundary layer. Let us consider a horizontally uniform flow over a flat surface. Here, we use incompressible equations with constant density ρ and a diffusion-type friction:

$$\frac{\partial u}{\partial t} - fv = -\frac{1}{\rho}\frac{\partial p}{\partial x} + \nu\nabla^2 u, \tag{6.6.11}$$

$$\frac{\partial v}{\partial t} + fu = -\frac{1}{\rho}\frac{\partial p}{\partial y} + \nu\nabla^2 v, \tag{6.6.12}$$

$$-\frac{\partial p}{\partial z} - \rho g = 0, \tag{6.6.13}$$

$$\frac{\partial u}{\partial x} + \frac{\partial v}{\partial x} = 0. \tag{6.6.14}$$

We assume a steady and uniform flow in the x- and y-directions and no vertical velocity: $w = 0$. The boundary conditions are $u = v = 0$ at $z = 0$ and $u = u_g$, $v = 0$ as $z \to \infty$, where u_g is the geostrophic wind that satisfies

$$fu_g = -\frac{1}{\rho}\frac{\partial p}{\partial y}. \tag{6.6.15}$$

Since the horizontal gradient of p is independent of z, we obtain the relations for the ageostrophic components of horizontal winds as

$$-fv_a = \nu\frac{\partial^2 u_a}{\partial z^2}, \qquad fu_a = \nu\frac{\partial^2 v_a}{\partial z^2}, \tag{6.6.16}$$

where $u_a = u - u_g$ and $v_a = v$. These two equations are combined to

$$f^2 v_a = \nu^2\frac{\partial^4 v_a}{\partial z^4}. \tag{6.6.17}$$

We obtain a general solution to this as

$$
\begin{aligned}
v_a &= C_1 \exp\left[\sqrt{\frac{f}{2\nu}}(1+i)z\right] + C_2 \exp\left[\sqrt{\frac{f}{2\nu}}(1-i)z\right] \\
&\quad + C_3 \exp\left[-\sqrt{\frac{f}{2\nu}}(1+i)z\right] + C_4 \exp\left[-\sqrt{\frac{f}{2\nu}}(1-i)z\right], \qquad (6.6.18)
\end{aligned}
$$

where we assume $f > 0$. From the boundary conditions, we have $u_g + u_a = v_a = 0$ at $z = 0$, and $u_a = v_a = 0$ as $z \to \infty$. Thus, we obtain the vertical profiles of ageostrophic components:

$$
u_a = -u_g \exp\left(-\sqrt{\frac{f}{2\nu}}z\right) \cos\left(\sqrt{\frac{f}{2\nu}}z\right), \qquad (6.6.19)
$$

$$
v_a = u_g \exp\left(-\sqrt{\frac{f}{2\nu}}z\right) \sin\left(\sqrt{\frac{f}{2\nu}}z\right). \qquad (6.6.20)
$$

From this solution, the height scale of the effect of friction is estimated as

$$
d = \sqrt{\frac{2\nu}{f}}. \qquad (6.6.21)
$$

The directions of velocity change spirally as depicted in Fig. 6.10; this is called the *Ekman spiral.* In the case $u_g > 0$, meridional flow is almost poleward: $v_a > 0$. Integrating v_a in the z-direction, we have the total transport:

$$
\int_0^\infty v_a \, dz = \frac{d}{2} u_g. \qquad (6.6.22)
$$

This is the *Ekman transport,* which corresponds to (6.6.8) in the case of shallow water equations.

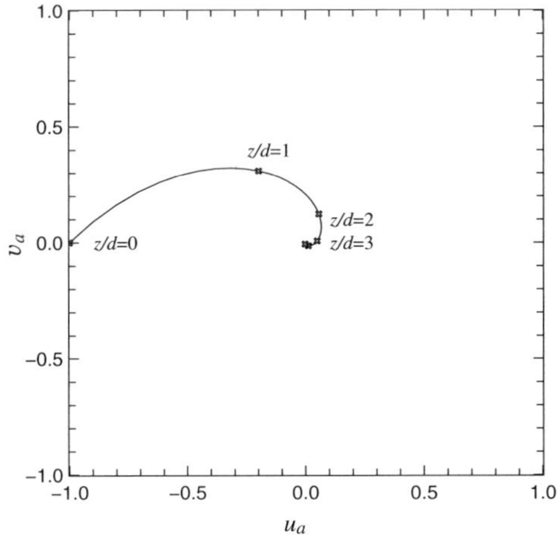

FIGURE 6.10: Ekman spiral: the dependency of (u_a, v_a) on height for $u_g = 1$ and $v_g = 0$. The flows at height $z/d = 1$, 2, 3, 4, and 5 are marked.

References and suggested reading

A systematic analysis of the thermal response of the atmosphere is given by Gill (1980). Emanuel (1983) also summarizes elementary aspects of the thermal response in stratified fluids. The discussion on axisymmetric flow in Section 6.3 follows Eliassen (1951). The thermal response on the β-plane in Section 6.4 is inherently related to the theory of Rossby wave propagation. For the Rossby wave response to heating on the sphere, we refer to Hoskins and Karoly (1981), for instance. The equatorial response in Section 6.5 is based on Matsuno (1966) and Gill (1982). A more detailed discussion on oceanic wind-driven circulation described in Section 6.6 is given in chapter 5 of Pedlosky (1987).

Emanuel, K., 1983: Elementary aspects of the interaction between cumulus convection and the large-scale environment. In: Lilly, D. K. and Gal-Chen, T. (eds.), *Mesoscale Meteorology – Theories, observations and models.* NATO ASI Series: Mathematical and Physical Sciences, **C114**, Reidel, Dordrecht, Netherlands, pp. 551–575.

Eliassen, A., 1951: Slow thermally or frictionally controlled meridional circulation in a circular vortex. *Astrophys. Norv.*, **5**, 19–60.

Gill, A. E., 1980: Some simple solutions for heat-induced tropical circulation. *Q. J. Roy. Meteorol. Soc.*, **106**, 447–462.

Gill, A. E., 1982: *Atmosphere-Ocean Dynamics.* Academic Press, San Diego, 662 pp.

Hoskins, B. J. and Karoly, D., 1981: The steady linear response of a spherical atmosphere to thermal and orographic forcing. *J. Atmos. Sci.*, **38**, 1179–1196.

Matsuno, T., 1966: Quasi-geostrophic motions in the equatorial area. *J. Meteorol. Soc. Japan*, **44**, 25–43.

Pedlosky, J., 1987: *Geophysical Fluid Dynamics*, 2nd ed. Springer-Verlag, New York, 710 pp.

7

Eddy transport

To study transport in the atmosphere, we normally use temporally and spatially averaged quantities associated with waves or disturbances. For instance, the meridional transports of the general circulation of the atmosphere are described with the zonal average along a latitudinal circle. In general, the characteristics of mean transport depend on the averaging procedure. A spatial mean along one direction is called the *Eulerian mean*, while a mean over a set of fluid parcels is called the *Lagrangian mean*. For purely wavy oscillating disturbances, for instance, the positive and negative phases are canceled out to zero by the Eulerian mean regardless of its amplitude. For the Lagrangian mean, however, the mean position of fluid parcels generally has a motion, and a net transport of energy or momentum occurs if the amplitude is finite. This concept is connected to the formulation of the generalized Lagrangian mean.

We first define the generalized Lagrangian mean in this chapter and explain its relation to the Eulerian mean. Some examples are shown to distinguish the two mean transports. Then, a diffusion coefficient tensor is introduced to formulate the residual mean and the transformed Eulerian mean equations. At the end of this chapter, we describe the relation between eddy transport and transport in isentropic coordinates.

The topics of the present chapter are closely related to the meridional transport of the general circulation of the atmosphere. In particular, both the Eulerian and Lagrangian mean are used to study the mean transport in the mid-latitudes in Chapter 18.

7.1 Transport due to finite amplitude waves

7.1.1 Generalized Lagrangian mean

We introduce the concept of the generalized Lagrangian mean to consider transport due to finite amplitude waves. Let a velocity vector at position $\boldsymbol{x}$ and time t be denoted by $\boldsymbol{u}(\boldsymbol{x}, t)$. To simplify the argument, we use Cartesian coordinates with

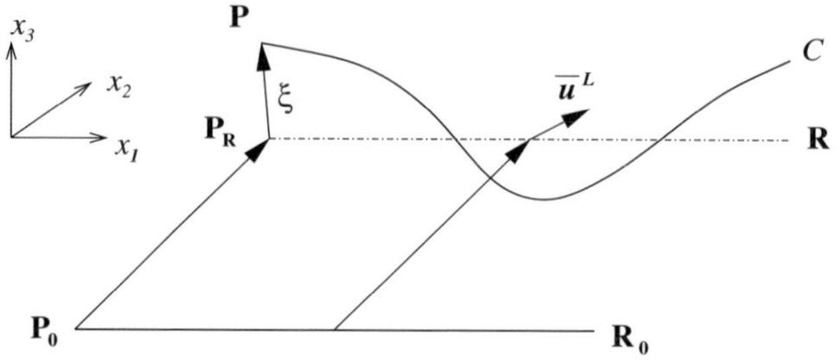

FIGURE 7.1: Schematic figure for the explanation of the GLM velocity. The GLM velocity $\overline{\boldsymbol{u}}^L$ is given by the velocity of the mass center of the curve C, and is equal to the average motion of the fluid parcels composing the rod R initially.

$\boldsymbol{x} = (x_1, x_2, x_3)$, and consider the average along the x_1-axis.† The *Eulerian mean* is defined by

$$\overline{\boldsymbol{u}}(\boldsymbol{x}, t) = \lim_{L\to\infty} \frac{1}{2L} \int_{-L}^{L} \boldsymbol{u}(\boldsymbol{x}, t)\, dx_1. \tag{7.1.1}$$

We imply $\overline{\boldsymbol{u}}(\boldsymbol{x}, t) = \overline{\boldsymbol{u}}(x_2, x_3, t)$ since the left-hand side is independent of x_1.

In contrast, the *generalized Lagrangian mean* (GLM) is introduced as follows. Let us consider fluid parcels located along a rod R_0 parallel to the x_1-axis at $t = t_0$. We assume that this rod moves to a curve $\mathcal{C}$ at time t. Fig. 7.1 schematically represents this situation; the parcel at the point P_0 at $t = t_0$ moves to the point P at t on the curve $\mathcal{C}$. Let the mass center of the curve $\mathcal{C}$ move with velocity $\boldsymbol{v}$. If $\boldsymbol{v}$ is known, the point P_R is defined for each point P_0 on the rod R_0 as

$$\overrightarrow{P_0 P_R} = \int_{t_0}^{t} \boldsymbol{v}\, dt. \tag{7.1.2}$$

Let the position vector at the point P_R be $\boldsymbol{x}$ and define

$$\boldsymbol{\xi} = \overrightarrow{P_R P}. \tag{7.1.3}$$

From the definition of the mass center of the curve $\mathcal{C}$, we have

$$\overline{\boldsymbol{\xi}}(\boldsymbol{x}, t) = \boldsymbol{0}. \tag{7.1.4}$$

Since the position vector at the point P is $\boldsymbol{x} + \boldsymbol{\xi}(\boldsymbol{x}, t)$, the velocity of P, denoted by $\boldsymbol{u}^{\boldsymbol{\xi}}$, is given by

$$\begin{aligned} \boldsymbol{u}^{\boldsymbol{\xi}}(\boldsymbol{x}, t) &\equiv \boldsymbol{u}(\boldsymbol{x} + \boldsymbol{\xi}(\boldsymbol{x}, t), t) = \frac{d}{dt}(\boldsymbol{x} + \boldsymbol{\xi}(\boldsymbol{x}, t)) \\ &= \boldsymbol{v} + \left(\frac{\partial}{\partial t} + \boldsymbol{v} \cdot \nabla\right) \boldsymbol{\xi}. \end{aligned} \tag{7.1.5}$$

†Here we assumed that (x_1, x_2, x_3) are components of Cartesian coordinates. A similar formulation is possible for the spherical coordinate (λ, φ, r). By taking x_1 as the longitude λ, we obtain the average along the zonal direction in the meridional cross section.

Let us define

$$\boldsymbol{u}^l(\boldsymbol{x},t) \equiv \left(\frac{\partial}{\partial t} + \boldsymbol{v}\cdot\nabla\right)\boldsymbol{\xi}, \tag{7.1.6}$$

then we have from (7.1.4)

$$\overline{\boldsymbol{u}^l}(\boldsymbol{x},t) = \boldsymbol{0}. \tag{7.1.7}$$

We define the GLM velocity $\overline{\boldsymbol{u}}^L$ as the Eulerian average of $\boldsymbol{u}^{\boldsymbol{\xi}}$:

$$\overline{\boldsymbol{u}}^L(\boldsymbol{x},t) \equiv \overline{\boldsymbol{u}^{\boldsymbol{\xi}}}(\boldsymbol{x},t) = \overline{\boldsymbol{u}}(\boldsymbol{x}+\boldsymbol{\xi}(\boldsymbol{x},t),t) = \boldsymbol{v} \tag{7.1.8}$$

(i.e., the GLM velocity is the velocity of the mass center (Fig. 7.1)). We also define the *Stokes correction* as

$$\overline{\boldsymbol{u}}^S \equiv \overline{\boldsymbol{u}}^L - \overline{\boldsymbol{u}}, \tag{7.1.9}$$

and the deviation from the Eulerian mean, or the Eulerian perturbation velocity, as

$$\boldsymbol{u}' = \boldsymbol{u} - \overline{\boldsymbol{u}}. \tag{7.1.10}$$

The following relations are satisfied if the amplitude of disturbance is small enough (i.e., the amplitude $|\boldsymbol{u}'|$ is smaller than the mean flow $|\overline{\boldsymbol{u}}|$: $|\boldsymbol{u}'| \ll |\overline{\boldsymbol{u}}|$). Letting a denote the magnitude of the wave amplitude, we have $\boldsymbol{\xi} = O(a)$ and

$$u_i^{\boldsymbol{\xi}} = u_i(\boldsymbol{x}+\boldsymbol{\xi},t) = u_i(\boldsymbol{x},t) + \xi_j\frac{\partial u_i}{\partial x_j} + \frac{\xi_j\xi_l}{2}\frac{\partial^2 \overline{u}_i}{\partial x_j \partial x_k} + O(a^3), \tag{7.1.11}$$

$$\overline{u}_i^L = \overline{u}_i(\boldsymbol{x}+\boldsymbol{\xi},t) = \overline{u}_i + \overline{\xi_j\frac{\partial u_i'}{\partial x_j}} + \frac{\overline{\xi_j\xi_l}}{2}\frac{\partial^2 \overline{u}_i}{\partial x_j \partial x_k} + O(a^3). \tag{7.1.12}$$

Therefore,

$$\overline{u_i}^S \equiv \overline{u_i}^L - \overline{u}_i = \overline{\xi_j\frac{\partial u_i'}{\partial x_j}} + \frac{\overline{\xi_j\xi_l}}{2}\frac{\partial^2 \overline{u}_i}{\partial x_j \partial x_k} + O(a^3) = O(a^2), \tag{7.1.13}$$

$$u_i^l \equiv u_i^{\boldsymbol{\xi}} - \overline{u_i}^L = u_i' + \xi_j\frac{\partial \overline{u_i}}{\partial x_j} + O(a^2). \tag{7.1.14}$$

Eq. (7.1.13) defines the Stokes correction for waves. The magnitude of the Stokes correction is a second-order quantity of a.

7.1.2 Examples of finite amplitude waves

7.1.2.1 Square wave

In this section, we show some examples of finite amplitude waves to get a better understanding of the GLM and the Stokes correction. First, we consider the square wave illustrated in Fig. 7.2; this is a longitudinal wave, in which the motion of fluid particles is parallel to the direction of wave propagation. The amplitude of velocity is denoted by u_0 and the wavelength by λ. The wave propagates toward the

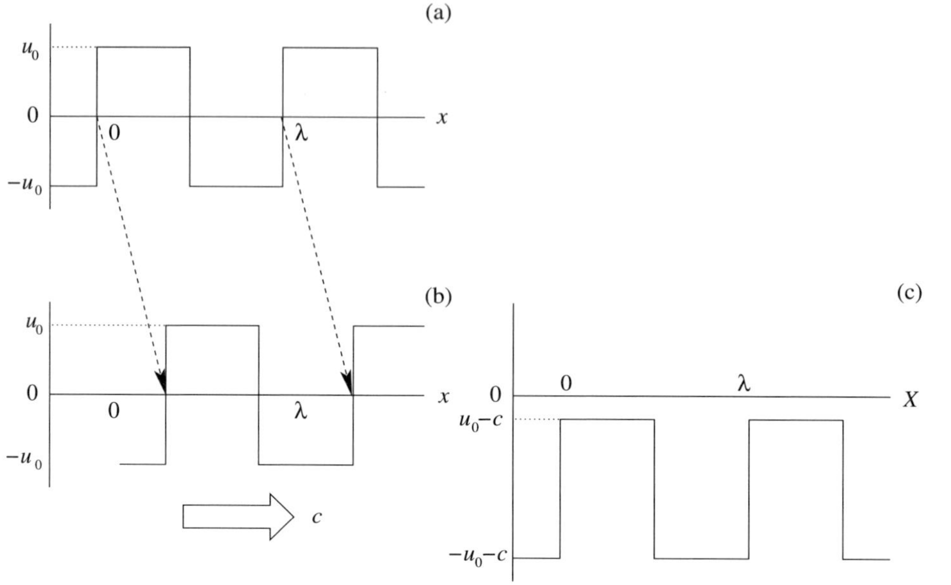

FIGURE 7.2: A square wave. (a) and (b) show velocity viewed from the stationary frame, and (c) shows velocity viewed from the moving frame with the speed of the wave c. In this moving frame, the wave is stationary. $X = x - ct$ and $u_0 < c$.

positive x-direction at a phase speed c. Fig. 7.2 shows the longitudinal velocity of fluid particles; (a) and (b) show the velocity viewed from the stationary frame, whereas (c) is that viewed from the moving frame with the speed c. In this moving frame, the wave is stationary. We define the coordinate in the moving frame as $X = x - ct$ and assume that $u_0 < c$.

The Eulerian mean velocity is given by averaging the velocity in the stationary frame in the x-direction. It is simply given as

$$\overline{u} = 0. \tag{7.1.15}$$

The Eulerian period of the wave is given by

$$T_E = \frac{\lambda}{c}. \tag{7.1.16}$$

On the other hand, by considering the time required for a fluid particle to move from $X = 0$ to $X = -\lambda$ as shown in Fig. 7.2(c), we obtain the Lagrangian period T_L as

$$T_L = \frac{\lambda/2}{c - u_0} + \frac{\lambda/2}{c + u_0} = \frac{\lambda c}{c^2 - u_0^2} > T_E. \tag{7.1.17}$$

Thus, the Lagrangian period is longer than the Eulerian period. If viewed from the stationary frame, the displacement of a particle during the Lagrangian period T_L is

$$d = -\lambda + cT_L = c(T_L - T_E) > 0 \tag{7.1.18}$$

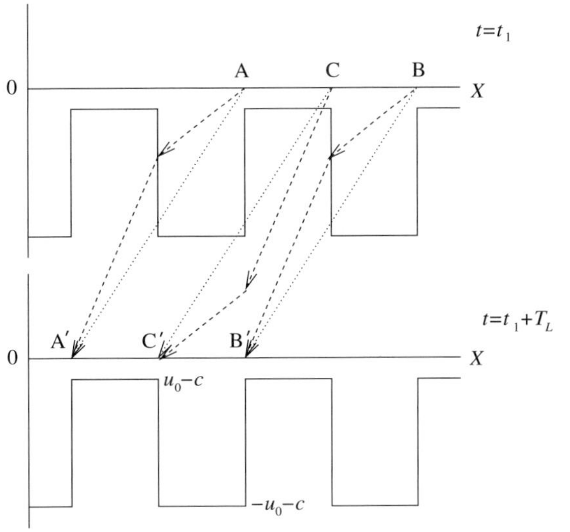

FIGURE 7.3: The motions of fluid particles (dashed lines) and those of the mass center (dotted lines) viewed from the moving frame with the speed c for the square wave between time t_1 to $t_1 + T_L$.

(i.e., the displacement is positive). Hence, we obtain the Lagrangian mean velocity $\overline{u}^{(L)}$ by

$$\overline{u}^{(L)} = \frac{d}{T_L} = c\left(1 - \frac{T_E}{T_L}\right) > 0. \tag{7.1.19}$$

Velocity $\overline{u}^{(L)}$ is the Lagrangian velocity averaged for one period of a fluid particle and is identical to the GLM velocity of all fluid particles. As shown below, this relation can be understood using Fig. 7.3. If viewed from the moving frame with the speed c, it takes one period T_L for a fluid particle to move one wavelength. The fluids in the segment AB of Fig. 7.3 move to the segment A′B′ after the period T_L. Thus, the mean velocity of the mass center of the fluids in AB is $-\lambda/T_L = -cT_E/T_L$. This velocity is the GLM velocity, $\overline{u}^L$, if viewed from the stationary frame, and is identical to the Lagrangian mean velocity $\overline{u}^{(L)}$ of a fluid particle, (7.1.19). This relation remains the same if the region for averaging is extended from the segment AB to a wider range. Thus, we obtain

$$\overline{u}^L = \overline{u}^{(L)}. \tag{7.1.20}$$

Here, it should be recalled that the square wave is a one-dimensional divergent flow; we must be careful about calculation of the position of the mass center. To obtain the position of the mass center of AB, we note the difference of density between segments AC and CB. Denoting the ratio of the density of AB to that of CB by $a : (1 - a)$, we have the relation

$$a(u_0 - c) + (1 - a)(-u_0 - c) = -\frac{\lambda}{T_L}, \tag{7.1.21}$$

since the fluid velocity in the segment AC is $u_0 - c$ and that in the segment CB is $-u_0 - c$. From this equation, the ratio of the density is given by

$$a = \frac{1}{2}\left(1 + \frac{u_0}{c}\right). \tag{7.1.22}$$

The mass center of AB is the interior division point by $(1-a) : a$ between the half points of AC and CB; therefore, it is located at $\lambda u_0/4c$ to the left of C.

7.1.2.2 Sinusoidal wave

The second example is the finite amplitude sinusoidal wave shown in Fig. 7.4. It is a one-dimensional longitudinal wave with amplitude u_0 and wavelength $\lambda = 2\pi/k$. The velocity of a fluid particle is given by

$$u = u_0 \cos k(x - ct). \tag{7.1.23}$$

If viewed from the moving frame with the speed c, the velocity of a fluid particle is

$$\dot{X} = u - c = u_0 \cos k(x - ct) - c. \tag{7.1.24}$$

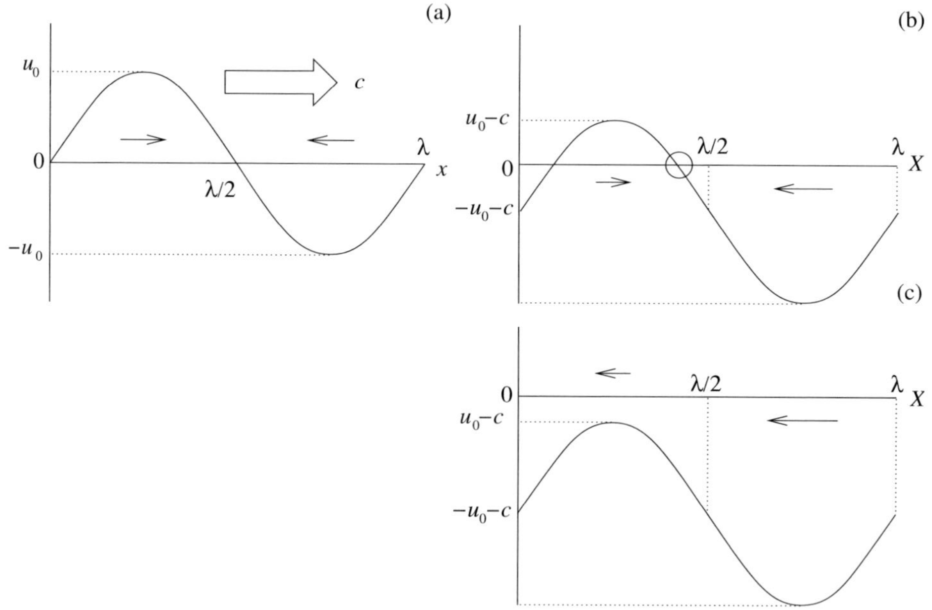

FIGURE 7.4: A sinusoidal wave. (a) shows velocity viewed from the stationary frame. The sinusoidal wave has wavelength λ and phase speed c. (b) and (c) show velocity viewed from the moving frame with the speed c: for the case (b) $u_0 < c$ and (c) $u_0 > c$. Here, $X = x - ct$. In (b), fluid particles are trapped at the position designated by the circle.

The Eulerian period is $T_E = \lambda/c$, whereas the Lagrangian period is

$$T_L = \int_0^\lambda \frac{dX}{c - u_0 \cos k(x - ct)} = \begin{cases} \dfrac{T_E}{\sqrt{1-(u_0/c)^2}}, & (u_0 < c), \\ \infty, & (u_0 > c) \end{cases} \tag{7.1.25}$$

(i.e., there are two cases: the finite period and the infinite period). The generalized Lagrangian mean velocity is respectively given by

$$\overline{u}^L = \frac{-\lambda + cT_L}{T_L} = \begin{cases} c(1 - \sqrt{1-(u_0/c)^2}), & (u_0 < c) \\ c, & (u_0 > c). \end{cases} \tag{7.1.26}$$

If the wave amplitude is large, $u_0 > c$, fluid particles are trapped by the wave and carried with the propagation of the wave as shown in Fig. 7.4(b).

In the case of an infinitesimally small amplitude ($\varepsilon = u_0/c \ll 1$), the GLM velocity (7.1.26) is approximated to

$$\overline{u}^L = \frac{u_0^2}{2c} = \frac{c}{2}\left(\frac{u_0}{c}\right)^2 = O(\varepsilon^2). \tag{7.1.27}$$

Although the Eulerian mean is $\overline{u} = 0$, the Lagrangian velocity is different from zero as a second-order quantity in the amplitude. The difference between Eulerian velocity and Lagrangian velocity is the Stokes correction. Using (7.1.6), (7.1.13), and (7.1.14), we obtain

$$u^l = u' = \frac{\partial \xi}{\partial t}, \qquad \overline{u}^S = \overline{\xi \frac{\partial u'}{\partial x}}. \tag{7.1.28}$$

Therefore, we also have

$$\xi = -\frac{u_0}{kc}\sin k(x - ct), \tag{7.1.29}$$

$$\overline{u}^S = \frac{u_0}{kc}\overline{\sin k(x-ct)\cdot k u_0 \sin k(x-ct)} = \frac{u_0^2}{2c}. \tag{7.1.30}$$

Thus, the Stokes correction is equal to the GLM velocity given by (7.1.27).

7.1.2.3 Two-dimensional nondivergent wave

As the third example, we consider waves in a two-dimensional nondivergent fluid. We can define the streamfunction of wave motion that has a phase speed c by

$$\Psi = \Psi(x - ct, y). \tag{7.1.31}$$

In the moving coordinates $(X, Y) = (x - ct, y)$, the streamfunction and velocity are written as

$$\Phi = \Psi(X, Y) + cY, \tag{7.1.32}$$

$$\dot{X} = -\Phi_Y = -\Psi_Y(X, Y) - c, \tag{7.1.33}$$

$$\dot{Y} = \Phi_X = \Psi_X(X, Y). \tag{7.1.34}$$

Since Φ is stationary, fluid particles on a contour of Φ remain on the same contour. Let a contour of Φ be denoted by $Y = Y(X)$. If this contour passes through a point (X_0, Y_0), the function $Y = Y(X)$ can be given by solving

$$\Phi(X, Y(X)) = \Phi(X_0, Y_0). \tag{7.1.35}$$

Note that $Y(X)$ may be a multivalued function. If Φ is periodic with wavelength λ, the Lagrangian period and the GLM velocity are expressed in terms of $Y(X)$ as

$$T_L = \int_0^{-\lambda} \frac{dX}{-\Phi_Y(X, Y(X))}, \tag{7.1.36}$$

$$\overline{u}^L = c\left(1 - \frac{T_E}{T_L}\right), \tag{7.1.37}$$

where $T_E = \lambda/c$ is the Eulerian period.

As an example, we consider

$$\Psi = \frac{u_0}{k}\cos k(x - ct)\cos ky, \tag{7.1.38}$$

where we assume $|y| \le \lambda/4$ where $\lambda = 2\pi/k$. The streamfunction in the moving frame with the speed c is

$$\Phi = \frac{u_0}{k}\cos kX \cos kY + cY. \tag{7.1.39}$$

Fig. 7.5 shows the distribution of this streamfunction. Fig. 7.5(b) is the case for $u_0 > c$, where closed contours exist near $y = \lambda/4$. In this case, fluid particles are trapped on the closed contours and propagate with the wave. Therefore, these fluid particles have a positive Lagrangian velocity. Fig. 7.5(c) is the case for $u_0 < c$. In this case, no closed contour exists such that all the particles are left behind the wave.

In the neighborhood of $y = \lambda/4$, in particular, we have approximations as

$$\Psi = \frac{u_0}{k}\cos k(x - ct) \cdot k\left(y - \frac{\lambda}{4}\right), \tag{7.1.40}$$

$$u = -\Psi_y = u_0 \cos k(x - ct). \tag{7.1.41}$$

This distribution has a positive Lagrangian velocity as shown by the previous example, (7.1.27). In the neighborhood of $y = 0$, on the other hand, we have

$$\Phi = \frac{u_0}{k}\cos kX \cdot \left(1 - \frac{k^2}{2}Y^2\right) + cY. \tag{7.1.42}$$

By setting $\Phi(X_0, Y_0) = 0$, we can approximately solve it for Y with $Y^2 \ll 1$:

$$Y \approx -\frac{u_0}{kc}\cos kX. \tag{7.1.43}$$

Thus, velocity is given by

$$u = -\Psi_y = \frac{u_0}{2}\cos k(x - ct) \cdot ky \approx -\frac{u_0^2}{2c}\cos^2 k(x - ct) < 0. \tag{7.1.44}$$

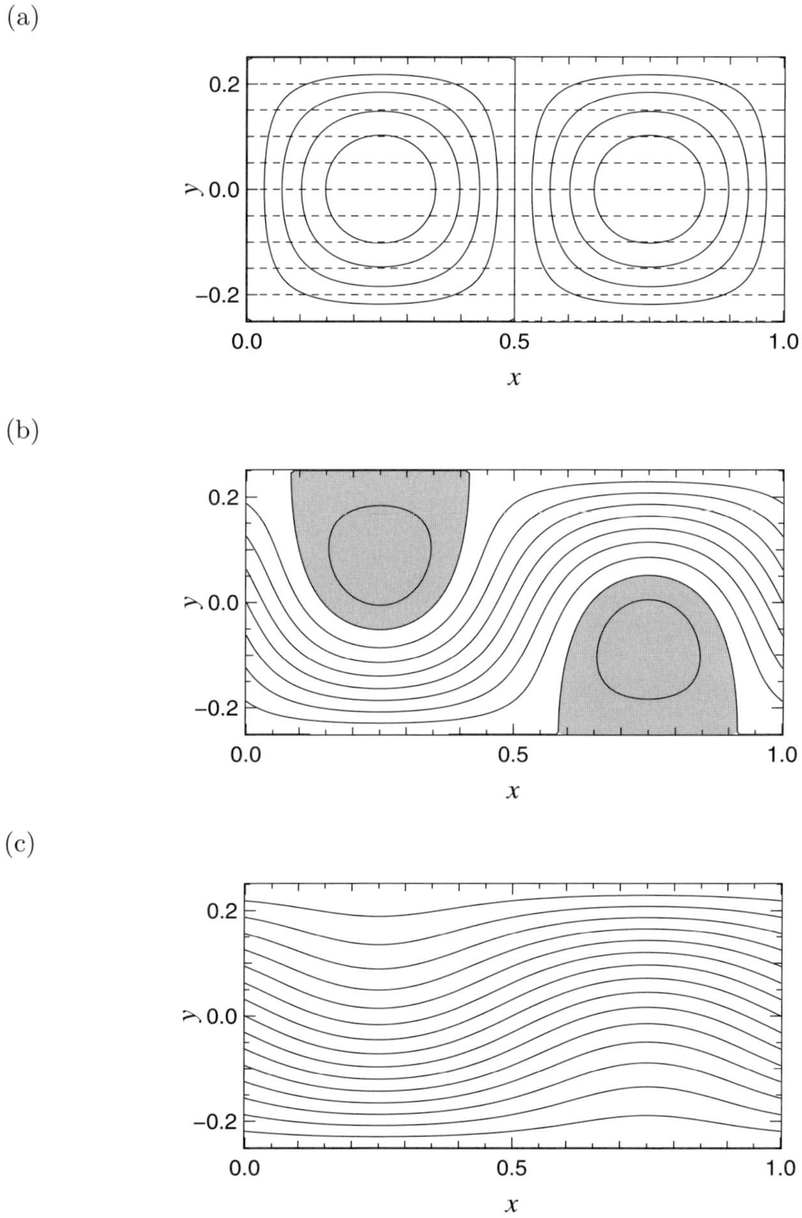

FIGURE 7.5: Waves in a two-dimensional nondivergent fluid. (a) Streamfunctions of a vorticity-like wave propagating with the speed c (solid) and the mean velocity with c (dashed) in the case $k = 2\pi$ and $u_0 = 1$. (b) and (c) are streamfunctions viewed from the moving frame with the speed c for (b) $c = 0.5$ ($u_0 > c$) and (c) $c = 2.0$ ($u_0 < c$), respectively. In (b), the fluid particles in the hatched region are trapped by the wave. In (c), all the fluid particles are left behind the wave. The contour intervals are arbitrarily chosen.

Hence, from (7.1.36) and (7.1.37), the GLM velocity is given as

$$\overline{u}^L = c\left[1-\sqrt{1+\left(\frac{u_0}{c}\right)^2}\right] < 0 \tag{7.1.45}$$

(i.e., the fluid particles propagate in the opposite direction to wave propagation irrespective of wave speed). This means that fluid particles on the contours near $y = 0$ are not carried by the wave. This comes from the fact that the contours are not closed in this vicinity.

7.2 Diffusion in the meridional section

7.2.1 Governing equations and the Eulerian mean

In the following sections, we consider the zonally averaged transport of materials in the meridional cross section for the case when the amplitude of disturbances is sufficiently small. To clarify the argument, we concentrate on primitive equations in the Boussinesq approximations and in the hydrostatic balance given by (3.1.6)–(3.1.8):

$$\frac{\partial u}{\partial t} + u\frac{\partial u}{\partial x} + v\frac{\partial u}{\partial y} + w\frac{\partial u}{\partial z} - fv = -\theta_s\frac{\partial \pi}{\partial x} + G_x, \tag{7.2.1}$$

$$\frac{\partial v}{\partial t} + u\frac{\partial v}{\partial x} + v\frac{\partial v}{\partial y} + w\frac{\partial v}{\partial z} + fu = -\theta_s\frac{\partial \pi}{\partial y} + G_y, \tag{7.2.2}$$

$$0 = -\theta_s\frac{\partial \pi}{\partial z} - \frac{\theta}{\theta_s}g, \tag{7.2.3}$$

$$\frac{\partial \theta}{\partial t} + u\frac{\partial \theta}{\partial x} + v\frac{\partial \theta}{\partial y} + w\frac{\partial \Theta}{\partial z} = Q, \tag{7.2.4}$$

$$\nabla\cdot\boldsymbol{u} = 0, \tag{7.2.5}$$

where $f = f_0 + \beta y$ is the Coriolis parameter, G_x and G_y are frictional forces, and Q is heating. Potential temperature is written as $\Theta = \theta_s + \theta$, where θ_s is the basic state potential temperature and θ is the perturbation. The Exner function is also partitioned as $\Pi = \pi_s + \pi$, where π_s is the basic state and π is the perturbation. In the continuity equation, we neglect the variation of density based on the Boussinesq approximation.

First, we consider the Eulerian mean along the x-direction of (7.2.1)–(7.2.5). Denoting the zonal mean of a quantity A by $\overline{A}$, and the departure from it by $A' = A - \overline{A}$, we have the following set of equations for the Eulerian mean:

$$\frac{\partial \overline{u}}{\partial t} + \overline{v}\left(\frac{\partial u}{\partial y} - y\right) + \overline{w}\frac{\partial \overline{u}}{\partial z} = -\frac{\partial \overline{u'v'}}{\partial y} - \frac{\partial \overline{u'w'}}{\partial z} + \overline{G_x}, \tag{7.2.6}$$

$$\frac{\partial \overline{v}}{\partial t} + \overline{v}\frac{\partial \overline{v}}{\partial y} + \overline{w}\frac{\partial \overline{v}}{\partial z} + f\overline{u} = -\frac{1}{\theta_s}\frac{\partial \overline{\pi}}{\partial y} - \frac{\partial \overline{v'^2}}{\partial y} - \frac{\partial \overline{v'w'}}{\partial z} + \overline{G_y}, \tag{7.2.7}$$

$$0 = -\theta_s\frac{\partial \overline{\pi}}{\partial z} + \frac{\overline{\theta}}{\theta_s}g, \tag{7.2.8}$$

$$\frac{\partial \overline{\theta}}{\partial t} + \overline{v}\frac{\partial \overline{\theta}}{\partial y} + \overline{w}\frac{\partial \Theta}{\partial z} = -\frac{\partial \overline{\theta' v'}}{\partial y} - \frac{\partial \overline{\theta' w'}}{\partial z} + \overline{Q}, \tag{7.2.9}$$

$$\frac{\partial \overline{v}}{\partial y} + \frac{\partial \overline{w}}{\partial z} = 0. \tag{7.2.10}$$

We consider a state of no meridional motion as a stationary reference state. Neglecting all the second-order terms $\overline{A'B'}$, the frictions $\overline{G_x}$, $\overline{G_y}$, and the heating $\overline{Q}$ we have the balance equations for the steady zonal flow:

$$f\overline{u} = -\frac{1}{\theta_s}\frac{\partial \overline{\pi}}{\partial y}, \tag{7.2.11}$$

$$0 = -\theta_s\frac{\partial \overline{\pi}}{\partial z} + \frac{\overline{\theta}}{\theta_s}g, \tag{7.2.12}$$

and $\overline{v} = \overline{w} = 0$. From (7.2.11) and (7.2.12), the thermal wind balance is given by

$$f\frac{\partial \overline{u}}{\partial z} = -\frac{g}{\theta_s}\frac{\partial \overline{\theta}}{\partial y}. \tag{7.2.13}$$

Hereafter, the variables of this stationary state will be denoted by a subscript 0 (e.g., $\overline{u}_0$).

We consider a disturbance as the departure from the steady state and assume that its magnitude is given by $O(a)$. If the amplitude of the disturbance is small enough, it can be shown that the departure of the zonal-mean zonal wind $\overline{u}$ from the steady state is $O(a^2)$. Thus, $\overline{v}$ and $\overline{w}$ are also $O(a^2)$, and the advective term of v, $\overline{v}\frac{\partial \overline{v}}{\partial y} + \overline{w}\frac{\partial \overline{v}}{\partial z}$, is $O(a^4)$.

7.2.2 Tracer transport

We then consider tracer transport in the meridional cross section associated with eddy disturbances. Let a change of mass concentration of a tracer q be written as

$$\frac{dq}{dt} = S, \tag{7.2.14}$$

where S is a source term (or a sink term if $S < 0$). To consider the meridional transport of q, we take the Eulerian mean of (7.2.14); the zonal-mean equation of q is given as

$$\frac{\partial \overline{q}}{\partial t} + \overline{u}_i\frac{\partial \overline{q}_0}{\partial x_i} = -\frac{\partial \overline{u_i' q'}}{\partial x_i} + \overline{S}, \tag{7.2.15}$$

in which the summation convention is used for the repeated indices of i. From the difference between (7.2.15) and (7.2.14), the change in the eddy component of q is given by

$$\frac{Dq'}{Dt} + u_i'\frac{\partial \overline{q}}{\partial x_i} = S' + O(a^2), \tag{7.2.16}$$

where

$$\frac{D}{Dt} = \frac{\partial}{\partial t} + \overline{u}_0 \frac{\partial}{\partial x}, \tag{7.2.17}$$

$$\frac{\partial}{\partial t} + \overline{u}_i \frac{\partial}{\partial x_i} = \frac{\partial}{\partial t} + \overline{u}_0 \frac{\partial}{\partial x} + O(a^2) = \frac{D}{Dt} + O(a^2). \tag{7.2.18}$$

We will investigate the basic characteristics of the eddy flux $\overline{u_i'q'}$ in (7.2.15). To find the direction of the eddy flux $\overline{u_i'q'}$, we multiply q' by (7.2.16) and average in the x-direction. Thus, we have

$$\overline{u_i'q'}\frac{\partial \overline{q}_0}{\partial x_i} = \overline{q'S'} - \frac{\partial}{\partial t}\frac{\overline{q'^2}}{2}. \tag{7.2.19}$$

In the case of no eddy source $S' = 0$, this becomes

$$\overline{u_i'q'}\frac{\partial \overline{q}_0}{\partial x_i} = -\frac{\partial}{\partial t}\frac{\overline{q'^2}}{2} \equiv -\sigma_t. \tag{7.2.20}$$

If the disturbance is a neutral wave, the right-hand side is identically zero: $\sigma_t = 0$. In this case, the direction of the flux $\overline{u_i'q'}$ is perpendicular to the gradient of the basic state $\frac{\partial \overline{q}_0}{\partial x_i}$. If the wave is growing $\sigma_t > 0$, the direction of the flux $\overline{u_i'q'}$ is down-gradient of $\overline{q}$. If the wave is decaying $\sigma_t < 0$, on the other hand, the direction of the flux $\overline{u_i'q'}$ is up-gradient. These relations between the gradient of $\overline{q}$ and the eddy flux $\overline{u_i'q'}$ are depicted in Fig. 7.6.

The relation between the eddy flux $\overline{\boldsymbol{u}'q'}$ and the gradient of $\overline{q}$ is expressed by using the parcel displacement of the disturbance. From (7.1.6), the displacement vector ξ_i is defined as

$$\left(\frac{\partial}{\partial t} + \overline{u}_j^L \frac{\partial}{\partial x_j}\right)\xi_i = u_i^l. \tag{7.2.21}$$

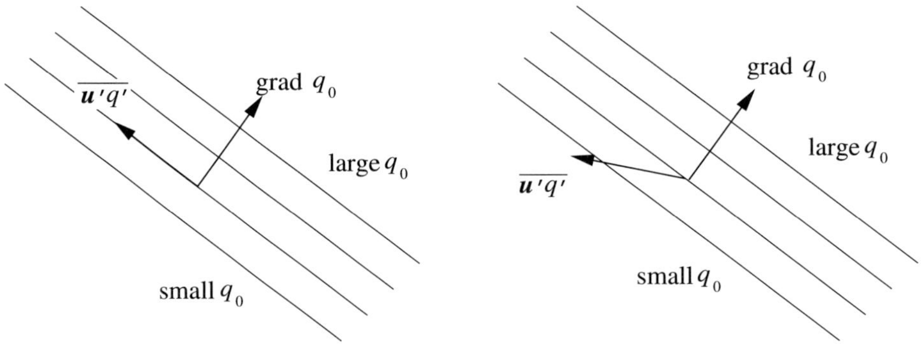

FIGURE 7.6: Relations between the eddy flux $\overline{\boldsymbol{u}'q'}$ and the gradient of the basic state $\nabla\overline{q}_0$. (Left) the case of the neutral wave $\sigma_t = 0$; the directions of the two vectors are perpendicular. (Right) the case of a growing wave $\sigma_t > 0$; the flux $\overline{\boldsymbol{u}'q'}$ has an up-gradient component of $\nabla\overline{q}_0$.

Using (7.1.12)–(7.1.14), this can be written as

$$\left[\frac{\partial}{\partial t} + \left(\overline{u}_j + \overline{u}_j^S\right)\frac{\partial}{\partial x_j}\right]\xi_i \quad = \quad u_i' + \xi_j\frac{\partial \overline{u}_i}{\partial x_j} + O(a^2). \tag{7.2.22}$$

Since $\overline{u}_j^S = O(a^2)$ and $\overline{u}_i = \overline{u}_0\delta_{ix} + O(a^2)$, the relation between the eddy component of velocity $\boldsymbol{u}' = (u', v', w')$ and the displacement vector $\boldsymbol{\xi} = (\eta, \zeta, \xi)$ is given as

$$\frac{D\eta}{Dt} \quad = \quad u' + \zeta\frac{\partial \overline{u}_0}{\partial y} + \xi\frac{\partial \overline{u}_0}{\partial z} \quad = \quad u^l, \tag{7.2.23}$$

$$\frac{D\zeta}{Dt} \quad = \quad v' \quad = \quad v^l, \tag{7.2.24}$$

$$\frac{D\xi}{Dt} \quad = \quad w' \quad = \quad w^l. \tag{7.2.25}$$

We also define a quantity s which satisfies

$$\frac{Ds}{Dt} \quad = \quad S', \tag{7.2.26}$$

where S' is the source term for the eddy component. Since (7.2.16) is written as

$$\frac{Dq'}{Dt} + \left(\frac{D\xi_i}{Dt}\right)\frac{\partial \overline{q}_0}{\partial x_i} \quad = \quad \frac{Ds}{Dt}, \tag{7.2.27}$$

thus we have

$$q' \quad = \quad -\xi_i\frac{\partial \overline{q}_0}{\partial x_i} + s. \tag{7.2.28}$$

Multiplying u_i' by this and averaging in the x-direction, we obtain

$$\overline{u_i'q'} \quad = \quad -\overline{u_i'\xi_j}\frac{\partial \overline{q}_0}{\partial x_j} + \overline{u_i's}. \tag{7.2.29}$$

This relates the eddy flux $\overline{u_i'q'}$ to the gradient of the basic state $\overline{q}_0$. The tensor $\overline{u_i'\xi_j}$ is regarded as a generalized diffusion coefficient and is called the *diffusion tensor*. Here, subscripts i and j are used for representing the y- and z-components, respectively.

To investigate the roles of eddy on transport, we divide the diffusion tensor $\overline{u_i'\xi_j}$ into symmetric and antisymmetric parts:

$$\overline{u_i'\xi_j} \quad = \quad K_{ij} + L_{ij}, \tag{7.2.30}$$

where

$$K_{ij} \quad = \quad \frac{1}{2}\left(\overline{u_i'\xi_j} + \overline{u_j'\xi_i}\right) \quad = \quad \frac{\partial}{\partial t}\frac{\overline{\xi_i\xi_j}}{2}, \tag{7.2.31}$$

$$L_{ij} \quad = \quad \frac{1}{2}\left(\overline{u_i'\xi_j} - \overline{u_j'\xi_i}\right), \tag{7.2.32}$$

or

$$K = \begin{pmatrix} \frac{\partial}{\partial t}\frac{\overline{\zeta^2}}{2} & \frac{\partial}{\partial t}\frac{\overline{\zeta\xi}}{2} \\ \frac{\partial}{\partial t}\frac{\overline{\zeta\xi}}{2} & \frac{\partial}{\partial t}\frac{\overline{\zeta^2}}{2} \end{pmatrix}, \tag{7.2.33}$$

$$L = \begin{pmatrix} 0 & \frac{1}{2}(\overline{v'\xi} - \overline{w'\zeta}) \\ -\frac{1}{2}(\overline{v'\xi} - \overline{w'\zeta}) & 0 \end{pmatrix}. \tag{7.2.34}$$

K is a symmetric tensor and L is an antisymmetric tensor. Substituting (7.2.30) into (7.2.29), we rewrite (7.2.15) as

$$\frac{\partial \overline{q}}{\partial t} = -\overline{u}_i \frac{\partial \overline{q}_0}{\partial x_i} - \frac{\partial}{\partial x_i}\left[-(K_{ij} + L_{ij})\frac{\partial \overline{q}_0}{\partial x_j} + \overline{u_i' s}\right] + \overline{S} \tag{7.2.35}$$

$$= -\left(\overline{u}_i + \frac{\partial L_{ij}}{\partial x_j}\right)\frac{\partial \overline{q}_0}{\partial x_i} + \frac{\partial}{\partial x_i}\left(K_{ij}\frac{\partial \overline{q}_0}{\partial x_j}\right) + \left(\overline{S} - \overline{u_i' s}\right). \tag{7.2.36}$$

The first term on the right-hand side of (7.2.36) plays the role of advective transport, and $\frac{\partial L_{ij}}{\partial x_j}$ is regarded as an additional advection due to eddy. This can be related to the Stokes correction. From (7.1.13) for $i = y$ and z, we have

$$\overline{u}_i^S = \overline{\xi_j \frac{\partial u_i'}{\partial x_j}} + O(a^3). \tag{7.2.37}$$

Using the continuity equation (7.2.29), we have

$$\frac{D}{Dt}\left(\frac{\partial \xi_i}{\partial x_i}\right) = \frac{\partial}{\partial x_i}\left(\frac{D\xi_i}{Dt}\right) - \frac{\partial \overline{u}_0}{\partial x_i}\frac{\partial \xi_i}{\partial x} = \frac{\partial u_i^l}{\partial x_i} - \frac{\partial \overline{u}_0}{\partial x_i}\frac{\partial \xi_i}{\partial x}$$

$$= \frac{\partial}{\partial x_i}\left(u_i' + \xi_j \frac{\partial \overline{u}_0}{\partial x_i}\delta_{ix}\right) - \frac{\partial \overline{u}_0}{\partial x_i}\frac{\partial \xi_i}{\partial x} = 0. \tag{7.2.38}$$

Thus, if $\frac{\partial \xi_i}{\partial x_i} = 0$ is satisfied at any initial time $t = t_0$, we always have

$$\frac{\partial \xi_i}{\partial x_i} = 0. \tag{7.2.39}$$

From (7.2.37) and (7.2.39), the Stokes correction is rewritten as

$$\overline{u}_i^S = \frac{\partial \overline{\xi_j u_i'}}{\partial x_j}. \tag{7.2.40}$$

From (7.2.32), therefore, we have the relation:

$$\frac{\partial L_{ij}}{\partial x_j} = \frac{\partial}{\partial x_j}\frac{1}{2}\left(\overline{u_i'\xi_j} - \overline{u_j'\xi_i}\right)$$

$$= \frac{\partial}{\partial x_j}\left(\overline{u_i'\xi_j}\right) - \frac{1}{2}\frac{\partial}{\partial x_j}\left(\overline{u_i'\xi_j} + \overline{u_j'\xi_i}\right)$$

$$= \overline{u}_i^S - \frac{\partial K_{ij}}{\partial x_j}. \tag{7.2.41}$$

This implies that the additional advection due to eddy consists of two parts. The first is the Stokes correction, which is the effect of the finite amplitude of eddy. The second is an apparent advection which arises when the eddy grows or decays inhomogeneously. Adding mean advection $\overline{u}_i$ to the above relation, we rewrite total advective velocity in the first term on the right-hand side of (7.2.36) as

$$\overline{u}_i + \frac{\partial L_{ij}}{\partial x_j} = \overline{u}_i^L + \frac{\partial K_{ij}}{\partial x_j}. \tag{7.2.42}$$

In the case of a neutral wave, we have $K_{ij} = 0$ so that (7.2.42) reduces to the GLM velocity $\overline{u}_i^L$. If the wave amplitude changes inhomogeneously, on the other hand, an apparent advection associated with K_{ij} occurs.

The second term on the right-hand side of (7.2.36) is regarded as diffusion induced by the growth or decay of disturbances. For a neutral wave, the contribution of this term vanishes since $K = 0$ from (7.2.31). The role of this diffusion term can be examined using (7.2.35) with no source term $S = s = 0$:

$$\frac{\partial \overline{q}}{\partial t} = -\overline{u}_i \frac{\partial \overline{q}_0}{\partial x_i} + \frac{\partial}{\partial x_i}\left[(K_{ij} + L_{ij})\frac{\partial \overline{q}_0}{\partial x_j}\right]. \tag{7.2.43}$$

Multiplying $\overline{q}$ and averaging in the whole domain, we have

$$\begin{aligned} \frac{\partial}{\partial t}\frac{\langle \overline{q}^2 \rangle}{2} &= \left\langle \overline{q}\frac{\partial}{\partial x_i}\left[(K_{ij} + L_{ij})\frac{\partial \overline{q}_0}{\partial x_j}\right]\right\rangle \\ &= -\left\langle K_{ij}\frac{\partial \overline{q}_0}{\partial x_i}\frac{\partial \overline{q}_0}{\partial x_j}\right\rangle + O(a^2), \end{aligned} \tag{7.2.44}$$

where $\langle A \rangle$ represents the domain average of A. In the case $K_{yy} > 0$, $K_{zz} > 0$, and $K_{yy}K_{zz} > K_{yz}^2$, the right-hand side is always positive for any distribution of $\overline{q}_0$. These inequalities hold if the diffusion tensor is isotropic. In this case, we have

$$\frac{\partial \langle \overline{q}^2 \rangle}{\partial t} < 0. \tag{7.2.45}$$

Since we have assumed that $\langle \overline{q} \rangle$ is conservative, the tendency of the variance of $\overline{q}$ becomes

$$\frac{\partial}{\partial t}\langle \overline{q}^2 \langle \overline{q} \rangle^2 \rangle = \frac{\partial}{\partial t}\left(\langle \overline{q}^2 \rangle - \langle \overline{q} \rangle^2\right) = \frac{\partial \langle \overline{q}^2 \rangle}{\partial t} 0. \tag{7.2.46}$$

This indicates that $\overline{q}$ tends to be homogenized in the region considered. In particular, the tracer spreads in the whole region even though the amplitude of disturbance is statistically steady.

7.3 Residual circulation

It has been found from (7.2.29) that the eddy flux $\overline{u_i' q'}$ receives additional advection from tracer q. As seen in (7.2.42), the sum of advection due to eddy and Eulerian mean velocity is the GLM velocity in the case $K = 0$. This indicates that Eulerian mean transport contains a spurious transport induced by eddy. Thus, the transport due to eddy can be interpreted as a correction to the Eulerian mean. The *residual circulation* described below is given by a systematic procedure to remove the spurious transport induced by eddy from Eulerian mean transport.

Transport due to eddy comes from a component of $\overline{u_i'q'}$ tangential to isolines of $\overline{q}_0$. To extract this component of $\overline{u_i'q'}$, we introduce two unit vectors, $\boldsymbol{n}$, $\boldsymbol{t}$, which are normal and tangential to isolines of $\overline{q}_0$, respectively:

$$\boldsymbol{n} = \frac{\nabla\overline{q}_0}{|\nabla\overline{q}_0|} = \frac{\boldsymbol{j}\frac{\partial\overline{q}_0}{\partial y} + \boldsymbol{k}\frac{\partial\overline{q}_0}{\partial z}}{|\nabla\overline{q}_0|}, \tag{7.3.1}$$

$$\boldsymbol{t} = \boldsymbol{i}\times\boldsymbol{n} = \frac{\boldsymbol{k}\frac{\partial\overline{q}_0}{\partial y} - \boldsymbol{j}\frac{\partial\overline{q}_0}{\partial z}}{|\nabla\overline{q}_0|}. \tag{7.3.2}$$

$\boldsymbol{i}$, $\boldsymbol{j}$, and $\boldsymbol{k}$ are the unit vectors in the x, y, and z directions, respectively. These are rewritten in the tensor form as

$$n_i = \frac{\frac{\partial\overline{q}_0}{\partial x_i}}{|\nabla\overline{q}_0|}, \qquad t_i = -\frac{\varepsilon_{1ij}\frac{\partial\overline{q}_0}{\partial x_j}}{|\nabla\overline{q}_0|}, \tag{7.3.3}$$

where ε_{1ij} is the antisymmetric tensor. Using these vectors, we write the eddy flux as

$$\overline{u_i'q'} = \overline{u_j'q'}n_jn_i + \overline{u_j'q'}t_jt_i = \chi\frac{\partial\overline{q}_0}{\partial x_i} + \psi\varepsilon_{1ij}\frac{\partial\overline{q}_0}{\partial x_j}, \tag{7.3.4}$$

where

$$\chi = \frac{\overline{u_j'q'}\frac{\partial\overline{q}_0}{\partial x_i}}{|\nabla\overline{q}_0|^2} = \frac{\overline{v'q'}\frac{\partial\overline{q}_0}{\partial y} + \overline{w'q'}\frac{\partial\overline{q}_0}{\partial z}}{|\nabla\overline{q}_0|^2}, \tag{7.3.5}$$

$$\psi = \frac{\varepsilon_{1ij}\overline{u_i'q'}\frac{\partial\overline{q}_0}{\partial x_j}}{|\nabla\overline{q}_0|^2} = \frac{\overline{v'q'}\frac{\partial\overline{q}_0}{\partial z} - \overline{w'q'}\frac{\partial\overline{q}_0}{\partial y}}{|\nabla\overline{q}_0|^2}. \tag{7.3.6}$$

Substitution of (7.3.4) into (7.2.15) yields

$$\begin{aligned}\frac{\partial\overline{q}}{\partial t} &= -\overline{u}_i\frac{\partial\overline{q}_0}{\partial x_i} - \frac{\partial}{\partial x_i}\left(\chi\frac{\partial\overline{q}_0}{\partial x_i} + \psi\varepsilon_{1ij}\frac{\partial\overline{q}_0}{\partial x_j}\right) + \overline{S}\\ &= -\left(\overline{u}_i - \varepsilon_{1ij}\frac{\partial\psi}{\partial x_j}\right)\frac{\partial\overline{q}_0}{\partial x_i} - \frac{\partial}{\partial x_i}\left(\chi\frac{\partial\overline{q}_0}{\partial x_i}\right) + \overline{S}\\ &= -\overline{u}_i^*\frac{\partial\overline{q}_0}{\partial x_i} - \frac{\partial}{\partial x_i}\left(\chi\frac{\partial\overline{q}_0}{\partial x_i}\right) + \overline{S},\end{aligned} \tag{7.3.7}$$

where

$$\overline{u}_i^* = \overline{u}_i - \varepsilon_{1ij}\frac{\partial\psi}{\partial x_j}, \tag{7.3.8}$$

that is

$$\overline{v}^* = \overline{v} - \frac{\partial\psi}{\partial z}, \qquad \overline{w}^* = \overline{w} + \frac{\partial\psi}{\partial y}. \tag{7.3.9}$$

The scalar ψ plays the role of a meridional streamfunction whose associated flow is induced by eddy. The flow $\overline{u}_i^*$ is called *residual circulation*, defined as a residual by subtracting the flow associated with ψ from Eulerian mean circulation. The scalar χ is, on the other hand, a coefficient for the meridional diffusive transport of the tracer due to eddy.

By comparison between (7.3.4) and (7.2.29), we can relate χ and ψ to the diffusion tensors K and L using the displacement vector ξ_i. Substituting (7.2.19) into (7.3.5), we have

$$\chi \quad = \quad \frac{1}{|\nabla \overline{q}_0|^2}\left(-\frac{\partial}{\partial t}\frac{\overline{q'^2}}{2} + \overline{q'S'}\right), \tag{7.3.10}$$

or substituting (7.2.29) into (7.3.5) and using (7.2.30), we obtain

$$\begin{aligned} \chi &= -K_{ij}n_i n_j + \frac{\overline{u_i's}}{|\nabla \overline{q}_0|^2}\frac{\partial \overline{q}_0}{\partial x_i} \\ &= -\frac{\partial}{\partial t}\left[\frac{1}{2}\overline{(\xi_i' n_i)^2}\right] + \frac{\overline{u_i's}\, n_i}{|\nabla \overline{q}_0|}. \end{aligned} \tag{7.3.11}$$

If there is no source or sink of the tracer, $S' = 0$, $s = 0$, we have

$$\chi \quad = \quad -\frac{1}{|\overline{q}_0|^2}\frac{\partial}{\partial t}\frac{\overline{q'^2}}{2} \quad = \quad -K_{ij}n_i n_j \quad = \quad -\frac{\partial}{\partial t}\left[\frac{1}{2}\overline{(\xi_i' n_i)^2}\right]. \tag{7.3.12}$$

Thus, we obtain $\chi = 0$ in the special case that the amplitude of eddy is steady. We also have $\chi < 0$ if the eddy is growing and $\chi > 0$ if the eddy is decaying. This indicates that the tracer is diffusive if $\chi < 0$, while it tends to concentrate in a smaller region if $\chi > 0$.

The expression of ψ can be given by substituting (7.2.29) into (7.3.6):

$$\begin{aligned} \psi &= -\varepsilon_{1ij}(K_{il} + L_{il})n_l n_j + \frac{\varepsilon_{1ij}}{|\nabla \overline{q}_0|}\overline{u_i's}\, n_j \\ &= -K_{il}n_l t_i - L_{23} + \frac{\overline{u_i's}\, t_i}{|\nabla \overline{q}_0|} \\ &= -\frac{\partial}{\partial t}\left[\frac{1}{2}\overline{(\xi_l' n_l)(\xi_i' t_i)}\right] - L_{23} + \frac{\overline{u_i's}\, t_i}{|\nabla \overline{q}_0|}, \end{aligned} \tag{7.3.13}$$

where we have used the relations $L_{il} = \varepsilon_{1il}L_{23}$ and $\varepsilon_{1ij}\varepsilon_{1il} = \delta_{jl}$. In the case $s = 0$, we have

$$\psi \quad = \quad -K_{il}n_l t_i - L_{23} \quad = \quad -\frac{\partial}{\partial t}\left[\frac{1}{2}\overline{(\xi_l' n_l)(\xi_i' t_i)}\right] - L_{23}. \tag{7.3.14}$$

In addition, if the amplitude of eddy is steady or the displacement of eddy is either normal or parallel to isolines of $\overline{q}_0$, we have

$$\psi \quad = \quad -L_{23}, \tag{7.3.15}$$

or

$$L_{ij} = -\varepsilon_{1ij}\psi. \tag{7.3.16}$$

The total advective transport including the effect of eddy is expressed by the first term on the right-hand side of (7.2.36). Using (7.3.8) and (7.3.13), we can rewrite total advection as

$$\begin{aligned}\overline{u}_i + \frac{\partial L_{ij}}{\partial x_j} &= \overline{u}_i^* + \varepsilon_{1ij}\frac{\partial \psi}{\partial x_j} + \frac{\partial L_{ij}}{\partial x_j} = \overline{u}_i^* + \varepsilon_{1ij}\frac{\partial}{\partial x_j}(\psi + L_{23}) \\ &= \overline{u}_i^* + \varepsilon_{1ij}\frac{\partial}{\partial x_j}\left(-K_{kl}n_l t_k + \frac{\varepsilon_{1kl}}{|\nabla \overline{q}_0|}\overline{u_k' s}\, t_l\right). \end{aligned} \tag{7.3.17}$$

By comparing this with (7.2.42), we can relate residual circulation to GLM circulation as

$$\overline{u}_i^* = \overline{u}_i^L - \frac{\partial K_{ij}}{\partial x_j} - \varepsilon_{1ij}\frac{\partial}{\partial x_j}\left(-K_{kl}n_l t_k + \frac{\varepsilon_{1kl}}{|\nabla \overline{q}_0|}\overline{u_k' s}\, t_l\right). \tag{7.3.18}$$

In particular, in the case $s = 0$ and $K = 0$, we have

$$\overline{u}_i^* = \overline{u}_i^L. \tag{7.3.19}$$

This means that the residual circulation is equal to GLM circulation if the tracer is conservative and the disturbance is neutral.

7.4 Transformed Eulerian mean equations

7.4.1 Generalized transformed Eulerian mean equations

We can obtain a set of zonal-mean equations using the residual circulation from the equation set given by (7.2.1)–(7.2.5). Residual circulation can be defined by using any tracer constituent q. Here, we choose potential temperature Θ as a tracer q for residual circulation. In this case, the equation for Θ is given by (7.2.4). From (7.3.5) and (7.3.6), we have expressions for χ and ψ as

$$\chi = \frac{\overline{v'\theta'}\frac{\partial\overline{\Theta}_0}{\partial v} + \overline{w'\theta'}\frac{\partial\overline{\Theta}_0}{\partial w}}{|\nabla\overline{\Theta}_0|^2}, \tag{7.4.1}$$

$$\psi = \frac{\overline{v'\theta'}\frac{\partial\overline{\Theta}_0}{\partial z} - \overline{w'\theta'}\frac{\partial\overline{\Theta}_0}{\partial y}}{|\nabla\overline{\Theta}_0|^2}, \tag{7.4.2}$$

where $\overline{\Theta}_0 = \theta_s + \overline{\theta}_0$ is the potential temperature of the time mean field. Residual circulation is given by substituting (7.4.2) into (7.3.9).

Using residual circulation, the zonal-mean equations (7.2.6)–(7.2.10) are rewritten as

$$\frac{\partial \overline{u}}{\partial t} + \overline{v}^* \frac{\partial M}{\partial y} + \overline{w}^* \frac{\partial M}{\partial z} = \nabla \cdot \boldsymbol{F} + \overline{G_x}, \tag{7.4.3}$$

$$\frac{\partial \overline{v}}{\partial t} + f\overline{u} = -\theta_s \frac{\partial \overline{\pi}}{\partial y} - \frac{\partial \overline{v'^2}}{\partial y} - \frac{\partial \overline{v'w'}}{\partial z} + \overline{G_y} + O(a^4), \tag{7.4.4}$$

$$0 = -\theta_s \frac{\partial \overline{\pi}}{\partial z} + \frac{\overline{\theta}}{\theta_s} g, \tag{7.4.5}$$

$$\frac{\partial \overline{\theta}}{\partial t} + \overline{v}^* \frac{\partial \overline{\theta}}{\partial y} + \overline{w}^* \frac{\partial \overline{\Theta}}{\partial z} = -\nabla \cdot (\chi \nabla \overline{\Theta}_0) + \overline{Q}, \tag{7.4.6}$$

$$\frac{\partial \overline{v}^*}{\partial y} + \frac{\partial \overline{w}^*}{\partial z} = 0, \tag{7.4.7}$$

where

$$M = \overline{u} - fy, \tag{7.4.8}$$

$$\nabla \cdot \boldsymbol{F} = -\frac{\partial}{\partial y}\left(\overline{u'v'} - \psi \frac{\partial M}{\partial z}\right) - \frac{\partial}{\partial z}\left(\overline{u'w'} + \psi \frac{\partial M}{\partial y}\right), \tag{7.4.9}$$

$$\boldsymbol{F} = \left(0, -\overline{u'v'} + \psi \frac{\partial M}{\partial z}, -\overline{u'w'} - \psi \frac{\partial M}{\partial y}\right). \tag{7.4.10}$$

$\boldsymbol{F}$ is called the generalized *Eliassen-Palm flux*. The zonal-mean equations using residual circulation (7.4.3)–(7.4.7) are called *transformed Eulerian mean equations* (hereafter, TEM equations). From (7.4.7), residual circulation can be written by using a meridional streamfunction ψ^* as

$$\overline{v}^* = -\frac{\partial \psi^*}{\partial z}, \qquad \overline{w}^* = \frac{\partial \psi^*}{\partial y}. \tag{7.4.11}$$

Subtracting (7.4.3)–(7.4.7) from (7.2.1)–(7.2.5) and neglecting second-order terms, we obtain the linearized perturbation equations:

$$\frac{Du'}{Dt} + v' \frac{\partial M}{\partial y} + w' \frac{\partial M}{\partial z} = -\theta_s \frac{\partial \pi'}{\partial x} + G'_x, \tag{7.4.12}$$

$$\frac{Dv'}{Dt} + fu' = -\theta_s \frac{\partial \pi'}{\partial y} + G'_y, \tag{7.4.13}$$

$$0 = -\theta_s \frac{\partial \pi'}{\partial z} + \frac{\theta'}{\theta_s} g, \tag{7.4.14}$$

$$\frac{D\theta'}{Dt} + v' \frac{\partial \theta}{\partial y} + w' \frac{\partial \Theta}{\partial z} = Q', \tag{7.4.15}$$

$$\nabla \cdot \boldsymbol{u}' = 0. \tag{7.4.16}$$

Making use of these equations and the equations of displacements, (7.2.23)–(7.2.25) as shown in Section 7.6, we obtain divergence of the Eliassen-Palm flux, which is the right-hand side of (7.4.3):

$$\begin{aligned}
\nabla \cdot \boldsymbol{F} &= \frac{\partial \overline{\zeta G'_x}}{\partial y} + \frac{\partial \overline{\xi G'_y}}{\partial y} + \overline{\frac{\partial \eta}{\partial x} G'_x} + \overline{\frac{\partial \zeta}{\partial x} G'_y} + \frac{g}{\theta_s} \overline{\frac{\partial \xi}{\partial x} h} \\
&\quad - \frac{\partial}{\partial x_i} \left\{ \left[\varepsilon_{1ij} \left(K_{km} n_m t_k + \frac{\overline{u'_k h t_k}}{|\nabla \overline{\Theta}_0|} \right) + K_{ij} \right] \frac{\partial M}{\partial x_j} \right\} \\
&\quad - \frac{\partial}{\partial t} \left[\frac{\partial}{\partial x_i} \overline{\xi_i u'} + \overline{\frac{\partial \eta}{\partial x} (u^l - f\zeta)} + \overline{\frac{\partial \zeta}{\partial x} v'} \right] \\
&\quad + \left(f \frac{\partial M}{\partial z} + \frac{g}{\theta_s} \frac{\partial \overline{\Theta}}{\partial y} \right) \overline{\frac{\partial \xi}{\partial x} \zeta},
\end{aligned} \tag{7.4.17}$$

where we introduced h that satisfies $Dh/Dt = Q'$. This is a generalized form of the *Eliassen-Palm relation.* The last term on the right-hand side of (7.4.17) is the deviation from thermal wind balance, which is $O(a^4)$ because of (7.2.13). In the case of the non-diffusive ($G'_x = 0$, $G'_y = 0$), diabatic ($h = 0$), and steady ($\frac{\partial}{\partial t} = 0$) condition, (7.4.17) becomes zero for this approximation. This means that there is no acceleration in zonal-mean zonal wind $\overline{u}$ due to eddy flux. This statement is referred to as the *nonacceleration theorem.*

Under the condition where the nonacceleration theorem is satisfied, (7.4.3)–(7.4.7) become

$$\frac{\partial \overline{u}}{\partial t} + \overline{v}^* \frac{\partial M}{\partial y} + \overline{w}^* \frac{\partial M}{\partial z} = \overline{G_x}, \tag{7.4.18}$$

$$f\overline{u} = -\theta_s \frac{\partial \overline{\pi}}{\partial y}, \tag{7.4.19}$$

$$0 = -\theta_s \frac{\partial \overline{\pi}}{\partial z} + \frac{\overline{\theta}}{\theta_s} g, \tag{7.4.20}$$

$$\frac{\partial \overline{\theta}}{\partial t} + \overline{v}^* \frac{\partial \overline{\theta}}{\partial y} + \overline{w}^* \frac{\partial \overline{\Theta}}{\partial z} = \overline{Q}, \tag{7.4.21}$$

$$\frac{\partial \overline{v}^*}{\partial y} + \frac{\partial \overline{w}^*}{\partial z} = 0. \tag{7.4.22}$$

We also have $(\overline{v}^*, \overline{w}^*) = (\overline{v}^L, \overline{w}^L)$ from (7.3.19).

If residual circulation is defined by using potential temperature, the equation for other tracers is written as

$$\frac{\partial \overline{q}}{\partial t} + \overline{u}_i^* \frac{\partial \overline{q}}{\partial x_i} = -\frac{\partial}{\partial x_i} \left(\overline{u'_i q'} - \varepsilon_{1ij} \psi \frac{\partial \overline{q}}{\partial x_j} \right) + \overline{S}. \tag{7.4.23}$$

If the eddy component of the tracer is conservative $S' = 0$ and its amplitude is steady, the first term on the right-hand side vanishes. Thus, the equation of the tracer simplifies to

$$\frac{\partial \overline{q}}{\partial t} + \overline{v}^* \frac{\partial \overline{q}}{\partial y} + \overline{w}^* \frac{\partial \overline{q}}{\partial z} = \overline{S}. \tag{7.4.24}$$

If $\overline{Q}$ is independent of $\overline{S}$, residual circulation can be viewed as thermally driven circulation by $\overline{Q}$: $(\overline{v}^*, \overline{w}^*)$ are calculated from (7.4.18)–(7.4.22) by giving the distribution of $\overline{Q}$. Substituting this circulation into (7.4.24) and giving an appropriate source term $\overline{S}$, we obtain the change in $\overline{q}$.

7.4.2 Quasi-geostrophic TEM equations

Residual circulation has a simplified form for quasi-geostrophic approximation. To derive the equation set for this approximation, we use the following quasi-geostrophic equations:

$$\left(\frac{\partial}{\partial t}+u_g\frac{\partial}{\partial x}+v_g\frac{\partial}{\partial y}\right)u_g - fv_a = -\theta_s\frac{\partial \pi_a}{\partial x}+G_x, \tag{7.4.25}$$

$$\left(\frac{\partial}{\partial t}+u_g\frac{\partial}{\partial x}+v_g\frac{\partial}{\partial y}\right)v_g + fu_a = -\theta_s\frac{\partial \pi_a}{\partial y}+G_y, \tag{7.4.26}$$

$$0 = -\theta_s\frac{\partial \pi_a}{\partial z}+\frac{\theta_a}{\theta_s}g, \tag{7.4.27}$$

$$\left(\frac{\partial}{\partial t}+u_g\frac{\partial}{\partial x}+v_g\frac{\partial}{\partial y}\right)\theta_g + w_a\frac{\partial \theta_s}{\partial z} = Q, \tag{7.4.28}$$

$$\nabla\cdot\boldsymbol{u_a} = 0, \tag{7.4.29}$$

where subscript g denotes the zero-th order quantities of the Rossby number expansion, while subscript a denotes first-order quantities. The zero-th order quantities satisfy the geostrophic balance as

$$u_g = -\frac{\theta_s}{f_0}\frac{\partial \pi_g}{\partial y}, \qquad v_g = \frac{\theta_s}{f_0}\frac{\partial \pi_g}{\partial x}, \qquad \frac{\theta_g}{\theta_s} = \frac{\theta_s}{g}\frac{\partial \pi_g}{\partial z}. \tag{7.4.30}$$

The quasi-geostrophic potential vorticity is given by

$$P_g = f+\frac{\theta_s}{f_0}\left(\frac{\partial^2}{\partial x^2}+\frac{\partial^2}{\partial y^2}+\frac{f_0^2}{N^2}\frac{\partial^2}{\partial z^2}\right)\pi_g, \tag{7.4.31}$$

and its evolution equation is given by

$$\left(\frac{\partial}{\partial t}+u_g\frac{\partial}{\partial x}+v_g\frac{\partial}{\partial y}\right)P_g = S, \tag{7.4.32}$$

where S can be written by using the frictional terms G_x, G_y and the diabatic term Q. We use the β-plane approximation $f = f_0+\beta y$ and N is the Brunt-Väisälä frequency.

Using $\overline{v_g}=0$, the Eulerian mean equations of (7.4.25)–(7.4.29) are given as

$$\frac{\partial \overline{u_g}}{\partial t} - f\overline{v_a} = -\frac{\partial}{\partial y}\overline{u_g'v_g'}+\overline{G_x}, \tag{7.4.33}$$

$$f\overline{u_a} = -\frac{\partial}{\partial y}\overline{v_g'^2}-\theta_s\frac{\partial\overline{\pi_a}}{\partial y}+\overline{G_y}, \tag{7.4.34}$$

$$0 = -\theta_s \frac{\partial \overline{\pi_a}}{\partial z} + \frac{\overline{\theta_a}}{\theta_s} g, \tag{7.4.35}$$

$$\frac{\partial \overline{\theta_g}}{\partial t} + \overline{w_a}\frac{\partial \theta_s}{\partial z} = -\frac{\partial}{\partial y}\overline{v_g'\theta_g'} + \overline{Q}, \tag{7.4.36}$$

$$\frac{\partial \overline{v_a}}{\partial y} + \frac{\partial \overline{w_a}}{\partial z} = 0. \tag{7.4.37}$$

From (7.4.32), the zonal-mean quasi-geostrophic potential vorticity equation is

$$\frac{\partial \overline{P_g}}{\partial t} = \frac{\partial}{\partial y}\overline{v_g' P_g'} - \overline{S}, \tag{7.4.38}$$

where

$$\overline{P_g} = f + \frac{\theta_s}{f_0}\left(\frac{\partial^2}{\partial y^2} + \frac{f_0^2}{N^2}\frac{\partial^2}{\partial z^2}\right)\overline{\pi_g}. \tag{7.4.39}$$

As can be seen from (7.4.36), the eddy heat flux has only the horizontal component for quasi-geostrophic approximation. In this case, the appropriate choice of residual circulation is

$$\overline{v}^* = \overline{v_a} - \frac{\partial}{\partial z}\left(\frac{\overline{v_g'\theta_g'}}{\frac{\partial \theta_s}{\partial z}}\right), \qquad \overline{w}^* = \overline{w_a} + \frac{\partial}{\partial y}\left(\frac{\overline{v_g'\theta_g'}}{\frac{\partial \theta_s}{\partial z}}\right), \tag{7.4.40}$$

in which the streamfunction is defined as

$$\psi = \frac{\overline{v_g'\theta_g'}}{\frac{\partial \theta_s}{\partial z}}. \tag{7.4.41}$$

Thus, we obtain the transformed Eulerian mean equations:

$$\frac{\partial \overline{u_g}}{\partial t} - f\overline{v}^* = \nabla \cdot \boldsymbol{F} + \overline{G_x}, \tag{7.4.42}$$

$$f\overline{u_a} = -\theta_s \frac{\partial \overline{\pi_a}}{\partial y}, \tag{7.4.43}$$

$$0 = -\theta_s \frac{\partial \overline{\pi_a}}{\partial z} + \frac{\overline{\theta_a}}{\theta_s} g, \tag{7.4.44}$$

$$\frac{\partial \overline{\theta_g}}{\partial t} + \overline{w}^*\frac{\partial \theta_s}{\partial z} = \overline{Q}, \tag{7.4.45}$$

$$\frac{\partial \overline{v}^*}{\partial y} + \frac{\partial \overline{w}^*}{\partial z} = 0. \tag{7.4.46}$$

We have assumed geostrophic balance in (7.4.43) by neglecting the $O(a^2)$ terms. In (7.4.42), $\boldsymbol{F}$ is the quasi-geostrophic *Eliassen-Palm flux*, given by

$$\boldsymbol{F} = \left(0, \quad \overline{u_g'v_g'}, \quad f_0\frac{\overline{v_g'\theta_g'}}{\frac{\partial \theta_s}{\partial z}}\right). \tag{7.4.47}$$

The divergence of the Eliassen-Palm flux is written as

$$\nabla \cdot \boldsymbol{F} \quad = \quad -\frac{\partial}{\partial y}\overline{u'_g v'_g} + \frac{\partial}{\partial z}\left(f_0 \frac{\overline{v'_g \theta'_g}}{\frac{\partial \theta_s}{\partial z}}\right). \tag{7.4.48}$$

In the general form, the divergence of the Eliassen-Palm flux is more complicated as shown by (7.4.17). For quasi-geostrophic approximation, however, $\nabla \cdot \boldsymbol{F}$ has a simplified form and can be related to transport of quasi-geostrophic potential vorticity, as shown below. Subtracting (7.4.38) from (7.4.32) gives the perturbation equation of quasi-geostrophic potential vorticity,

$$\frac{\partial P'_g}{\partial t} + v'_g \frac{\partial \overline{P_g}}{\partial y} \quad = \quad S' + O(a^2), \tag{7.4.49}$$

where

$$P'_g \quad = \quad \frac{\theta_s}{f_0}\left(\frac{\partial^2}{\partial x^2} + \frac{\partial^2}{\partial y^2} + \frac{f_0^2}{N^2}\frac{\partial^2}{\partial z^2}\right)\pi'_g. \tag{7.4.50}$$

Hence, we have

$$\begin{aligned} \overline{v'_g P'_g} &= \overline{\frac{\theta_s^2}{f_0^2}\frac{\partial \pi'_g}{\partial x}\left(\frac{\partial^2}{\partial x^2} + \frac{\partial^2}{\partial y^2} + \frac{f_0^2}{N^2}\frac{\partial^2}{\partial z^2}\right)\pi'_g} \\ &= -\frac{\partial}{\partial y}\overline{u'_g v'_g} + \frac{\partial}{\partial z}\left(f_0 \frac{\overline{v'_g \theta'_g}}{\frac{\partial \theta_s}{\partial z}}\right) \quad = \quad \nabla \cdot \boldsymbol{F}. \end{aligned} \tag{7.4.51}$$

If the β-effect is nonzero with $\frac{\partial \overline{P_g}}{\partial y} \neq 0$, multiplying P'_g by (7.4.49), averaging in the zonal direction, and dividing by $\frac{\partial \overline{P_g}}{\partial y}$ we obtain

$$\frac{\partial}{\partial t}\left(\frac{1}{2}\frac{P_g'^2}{\frac{\partial \overline{P_g}}{\partial y}}\right) \quad = \quad -\overline{P'_g v'_g} + \overline{P'_g S'} \quad = \quad -\nabla \cdot \boldsymbol{F} + \overline{P'_g S'}. \tag{7.4.52}$$

This equation is formally written as

$$\frac{\partial A}{\partial t} \quad = \quad -\nabla \cdot \boldsymbol{F} + \overline{P'_g S'}, \tag{7.4.53}$$

where

$$A \quad = \quad \frac{1}{2}\frac{P_g'^2}{\frac{\partial \overline{P_g}}{\partial y}}, \tag{7.4.54}$$

is called the *wave activity*. Eq. (7.4.52), or (7.4.53), is called the *Eliassen-Palm relation*. If the wave activity is time-independent and there is no dissipation, the divergence of the Eliassen-Palm flux is equal to zero. In this case, according to (7.4.42), there exists no acceleration of the zonal wind due to eddy flux. This is the original form of the *nonacceleration theorem* for quasi-geostrophic approximation.

Using (7.4.42)–(7.4.46), we have the equation for the streamfunction of residual circulation ψ^*. From (7.4.46), we can define ψ^* by

$$\overline{v}^* = -\frac{\partial \psi^*}{\partial z}, \qquad \overline{w}^* = \frac{\partial \psi^*}{\partial y}, \tag{7.4.55}$$

and similarly the streamfunction of the Eulerian mean ψ_E by

$$\overline{v_a} = -\frac{\partial \psi_E}{\partial z}, \qquad \overline{w_a} = \frac{\partial \psi_E}{\partial y}. \tag{7.4.56}$$

From (7.4.41) and (7.4.40), we have a relation

$$\psi^* = \psi_E - \psi = \psi_E - \frac{\overline{v_g'\theta_g'}}{\frac{\partial \theta_s}{\partial z}}. \tag{7.4.57}$$

The equations of $\overline{u_g}$ and ψ^* are given from (7.4.42)–(7.4.45) by replacing f with f_0 and assuming θ_s and N^2 are constant. That is,

$$\left(\frac{\partial^2}{\partial y^2} + \frac{f_0^2}{N^2}\frac{\partial^2}{\partial z^2}\right)\frac{\partial \overline{u_g}}{\partial t} = -\frac{\partial^2}{\partial y^2}\left(\nabla \cdot \boldsymbol{F} + \overline{G_x}\right) - \frac{\partial}{\partial z}\left(\frac{f_0}{\frac{\partial \theta_s}{\partial z}}\frac{\partial \overline{Q}}{\partial y}\right), \tag{7.4.58}$$

$$\left(\frac{\partial^2}{\partial y^2} + \frac{f_0^2}{N^2}\frac{\partial^2}{\partial z^2}\right)\psi^* = \frac{\partial}{\partial z}\left[\frac{f_0}{N^2}\left(\nabla \cdot \boldsymbol{F} + \overline{G_x}\right)\right] + \frac{1}{\frac{\partial \theta_s}{\partial z}}\frac{\partial \overline{Q}}{\partial y}. \tag{7.4.59}$$

Eq. (7.4.59) determines the residual circulation ψ^*.

7.5 Eulerian mean equations in isentropic coordinates

Eulerian mean equations in isentropic coordinates have many similarities to the generalized transformed Eulerian mean equations that are described in Section 7.4.1. To show the relation, we use the equations in the isentropic coordinates in the hydrostatic balance on the f-plane. The flux-form momentum equation in the x-direction, (3.3.69), the advective-form momentum equation in the y-direction, (3.3.63), the hydrostatic balance (3.3.64), the continuity equation (3.3.65), and the equation of potential temperature (3.3.66) are written as

$$\frac{\partial}{\partial t}(\rho_\theta u) + \frac{\partial}{\partial x}(\rho_\theta u^2) + \frac{\partial}{\partial y}(\rho_\theta uv) + \frac{\partial}{\partial \theta}(\rho_\theta u\dot{\theta}) - f\rho_\theta v$$
$$= -\frac{\partial}{\partial x}\left(\frac{\kappa}{\kappa+1}p\pi\right) + \frac{\partial}{\partial \theta}\left(\frac{p}{g}\frac{\partial \Psi}{\partial x}\right) + \rho_\theta G_x, \tag{7.5.1}$$

$$\frac{dv}{dt} + fu = -\frac{\partial \Psi}{\partial y} + G_y, \tag{7.5.2}$$

$$\frac{\partial \Psi}{\partial \theta} = \pi, \tag{7.5.3}$$

$$\frac{\partial \rho_\theta}{\partial t} + \frac{\partial}{\partial x}(\rho_\theta u) + \frac{\partial}{\partial y}(\rho_\theta v) + \frac{\partial}{\partial \theta}(\rho_\theta \dot{\theta}) = 0, \tag{7.5.4}$$

$$\dot{\theta} = Q. \tag{7.5.5}$$

The derivatives with respect to t, x, and y are taken along isentropic surfaces.

We define the zonal average of a quantity A along an isentrope by $\overline{A}$, and the deviation from it by A'. We also define the mass-weighted zonal average by $\overline{A}^*$, and the deviation from it by $\hat{A}$; that is,

$$\overline{A}^* = \frac{\overline{\rho_\theta A}}{\overline{\rho_\theta}}, \qquad \hat{A} = A - \overline{A}^*. \tag{7.5.6}$$

Thus, we have

$$\overline{\rho_\theta A} = \overline{\rho_\theta}\,\overline{A} + \overline{\rho'_\theta A'} = \overline{\rho_\theta}\,\overline{A}^*, \tag{7.5.7}$$

$$\overline{A}^* = \overline{A} + \frac{\overline{\rho'_\theta A'}}{\overline{\rho_\theta}}. \tag{7.5.8}$$

If the isentropes intersect with the ground, we assume $\rho_\theta = 0$ in the region where the potential temperature is lower than the potential temperature at the ground.

Using (7.5.7), the zonal average of the continuity equation (7.5.4) becomes

$$\frac{\partial \overline{\rho_\theta}}{\partial t} + \frac{\partial}{\partial y}(\overline{\rho_\theta}\,\overline{v}^*) + \frac{\partial}{\partial \theta}(\overline{\rho_\theta}\,\overline{\dot{\theta}}^*) = 0. \tag{7.5.9}$$

The zonal average of the zonal momentum equation (7.5.1) becomes

$$\frac{\partial}{\partial t}(\overline{\rho_\theta u}) + \frac{\partial}{\partial y}(\overline{\rho_\theta u v}) + \frac{\partial}{\partial \theta}(\overline{\rho_\theta u \dot{\theta}}) - f\overline{\rho_\theta v}$$
$$= \frac{\partial}{\partial \theta}\left(\overline{\frac{p}{g}\frac{\partial \Psi}{\partial x}}\right) + \overline{\rho_\theta G_x}. \tag{7.5.10}$$

Using (7.5.7) and (7.5.9) and dividing by $\overline{\rho_\theta}$, the momentum equation is rewritten as

$$\frac{\partial \overline{u}}{\partial t} + \overline{v}^* \frac{\partial M}{\partial y} + \overline{\dot{\theta}}^* \frac{\partial M}{\partial \theta} = -\frac{1}{\overline{\rho_\theta}}\frac{\partial}{\partial t}(\overline{\rho'_\theta u'}) + \frac{1}{\overline{\rho_\theta}}\nabla_\theta \cdot \boldsymbol{F}_\theta + \overline{G_x}^*, \tag{7.5.11}$$

where $M = \overline{u} - fy$ is the angular momentum and

$$\nabla_\theta \cdot \boldsymbol{F}_\theta = -\frac{\partial}{\partial y}\overline{(\rho_\theta v)' u'} + \frac{\partial}{\partial \theta}\left(\overline{\frac{p'}{g}\frac{\partial \Psi'}{\partial x}} - \overline{(\rho_\theta \dot{\theta})' u'}\right), \tag{7.5.12}$$

$$\boldsymbol{F}_\theta = \left(0,\ -\overline{(\rho_\theta v)' u'},\ \overline{\frac{p'}{g}\frac{\partial \Psi'}{\partial x}} - \overline{(\rho_\theta \dot{\theta})' u'}\right). \tag{7.5.13}$$

We should point out the similarity between (7.5.11) and the momentum equation of the transformed Eulerian mean equation (7.4.3). Eq. (7.5.13) corresponds to the Eliassen-Palm flux (7.4.10).

The zonal averages of (7.5.2), (7.5.3), and (7.5.5) are written as

$$f\overline{u} + \frac{\partial \overline{\Psi}}{\partial y} = -\overline{\frac{dv}{dt}} + \overline{G_y}, \tag{7.5.14}$$

$$\frac{\partial \overline{\Psi}}{\partial \theta} = \overline{\pi}, \tag{7.5.15}$$

$$\overline{\dot{\theta}}^* = \overline{Q}^*. \tag{7.5.16}$$

Eq. (7.5.14) reduces to the geostrophic balance if the right-hand side is negligible. For statistically steady states, the meridional streamfunction ψ_θ can be defined from (7.5.9) as

$$\overline{\rho_\theta}\,\overline{v}^* = -\frac{\partial \psi_\theta}{\partial \theta}, \qquad \overline{\rho_\theta}\,\overline{\dot{\theta}}^* = \frac{\partial \psi_\theta}{\partial y}. \tag{7.5.17}$$

In this case, the momentum balance (7.5.11) is written as

$$-\frac{\partial \psi_\theta}{\partial \theta}\frac{\partial M}{\partial y} + \frac{\partial \psi_\theta}{\partial y}\frac{\partial M}{\partial \theta} = \nabla_\theta \cdot \boldsymbol{F}_\theta + \overline{\rho_\theta}\overline{G_x}^*. \tag{7.5.18}$$

If $\overline{Q}^*$ is specified, the vertical velocity $\overline{\dot{\theta}}^*$ is calculated from (7.5.16), and hence ψ_θ is determined from (7.5.17) using appropriate boundary conditions. If the Eliassen-Palm flux (7.5.13) is known from the properties of disturbances, the distribution of angular momentum M is calculable from (7.5.18).

A useful relation between the Eliassen-Palm flux and potential vorticity can be driven in the isentropic coordinates. The zonal average of the zonal momentum equation in the form of (3.3.74) gives

$$\frac{\partial \overline{u}}{\partial t} - \overline{v\omega_{a\theta}} + \overline{\dot{\theta}\frac{\partial u}{\partial \theta}} = \overline{G_x}, \tag{7.5.19}$$

where $\omega_{a\theta}$ is absolute vorticity; the friction term is denoted by G_x in this section. Absolute vorticity is expressed as

$$\omega_{a\theta} = \frac{\partial v}{\partial x} - \frac{\partial u}{\partial y} + f = \rho_\theta P, \tag{7.5.20}$$

where P is potential vorticity given by (3.3.70). Using (7.5.6), we have

$$\begin{aligned} \overline{v\omega_{a\theta}} &= \overline{v\rho_\theta P} = \overline{\rho_\theta}\overline{vP}^* = \overline{\rho_\theta}(\overline{v}^*\overline{P}^* + \overline{\hat{v}\hat{P}}^*) = \overline{v}^*\overline{\omega_\theta} + \overline{\rho_\theta}\overline{\hat{v}\hat{P}}^* \\ &= -\overline{v}^*\frac{\partial M}{\partial y} + \overline{\rho_\theta}\overline{\hat{v}\hat{P}}^*. \end{aligned} \tag{7.5.21}$$

Therefore, (7.5.19) becomes

$$\frac{\partial \overline{u}}{\partial t} + \overline{v}^*\frac{\partial M}{\partial y} + \overline{\dot{\theta}}^*\frac{\partial M}{\partial \theta} = \overline{\rho_\theta}\overline{\hat{v}\hat{P}}^* - \overline{\rho_\theta}\overline{\hat{\dot{\theta}}\widehat{\left(\frac{1}{\rho_\theta}\frac{\partial M}{\partial \theta}\right)}}^* + \overline{G_x}, \tag{7.5.22}$$

From comparison between (7.5.11) and (7.5.22), we have a relation:

$$\overline{\rho_\theta}^2 \overline{\hat{v}\hat{P}}^* = \nabla_\theta \cdot \boldsymbol{F}_\theta - \frac{\partial}{\partial t}(\overline{\rho'_\theta u'}) + \overline{\rho'_\theta G'_x} + \overline{\rho_\theta}^2 \overline{\hat{\dot\theta}\left(\widehat{\frac{1}{\rho_\theta}\frac{\partial M}{\partial \theta}}\right)}^* . \tag{7.5.23}$$

This corresponds to (7.4.51) in the case of quasi-geostrophic equations, but in the isentropic case it is no longer a simple relation between potential vorticity flux and the Eliassen-Palm flux.

7.6 Appendix: Derivation of the generalized Eliassen-Palm relation

We derive the generalized Eliassen-Palm relation (7.4.17) in this appendix. Substituting (7.3.13) into the generalized Eliassen-Palm flux (7.4.10) gives

$$\begin{aligned}
F_i &= -\overline{u'u'_i} + \varepsilon_{1ij}\psi\frac{\partial M}{\partial x_i} \\
&= -\overline{u'u'_i} + \varepsilon_{1ij}\left[-(K_{km}+L_{km})n_m t_k + \frac{\overline{u'_k h t_k}}{|\nabla\overline{\Theta}_0|}\right]\frac{\partial M}{\partial x_j} \\
&= -\overline{u'u'_i} - L_{ij}\frac{\partial M}{\partial x_j} - \varepsilon_{1ij}(K_{km}n_m t_k + \mathcal{H})\frac{\partial M}{\partial x_j}.
\end{aligned} \tag{7.6.1}$$

We have used potential temperature Θ for the definition of ψ (see Section 7.3) and introduced h and $\mathcal{H}$ by

$$\frac{Dh}{Dt} = Q', \qquad \mathcal{H} = \frac{\overline{u'_k h t_k}}{|\nabla\overline{\Theta}_0|}.$$

First, we use (7.4.12):

$$\frac{Du'}{Dt} + u'_j\frac{\partial M}{\partial x_j} = -\theta_s\frac{\partial \pi'}{\partial x} + G'_x.$$

Multiplying ξ_i by this and taking the zonal average using the relation,

$$\overline{\xi_i\frac{Du'}{Dt}} = \frac{\partial\overline{\xi_i u'}}{\partial t} - \overline{u'u^l_i},$$

we have

$$\frac{\partial\overline{\xi_i u'}}{\partial t} - \overline{u'u^l_i} + \overline{\xi_i u'_j}\frac{\partial M}{\partial x_j} = -\theta_s\overline{\xi_i\frac{\partial \pi'}{\partial x}} + \overline{\xi_i G'_x}. \tag{7.6.2}$$

Next, noting that

$$\begin{aligned}
\overline{u'u^l_i} &= \overline{u'u'_i}, \quad (i = y, z), \\
\overline{\xi_i u'_i} &= K_{ij} + L_{ji} = K_{ij} - L_{ij}, \\
\overline{\frac{\partial \pi'}{\partial x}\xi_i} &= -\overline{\pi'\frac{\partial \xi}{\partial x}},
\end{aligned}$$

we rewrite (7.6.2) as

$$\frac{\partial\overline{\xi_i u'}}{\partial t} - \overline{u'u^l_i} + (K_{ij} - L_{ij})\frac{\partial M}{\partial x_j} = \theta_s\overline{\pi'\frac{\partial \xi}{\partial x}} + \overline{\xi_i G'_x}.$$

Thus, (7.6.1) becomes

$$
\begin{aligned}
F_i &= \theta_s \overline{\pi' \frac{\partial \xi}{\partial x}} - \frac{\partial \overline{\xi_i u'}}{\partial t} - K_{ij}\frac{\partial M}{\partial x_j} \\
&+ \overline{\xi_i G'_x} - \varepsilon_{1ij}(K_{km} n_m t_k + \mathcal{H})\frac{\partial M}{\partial x_j}.
\end{aligned} \tag{7.6.3}
$$

Finally, from the zonal average of the inner product of (7.4.12), (7.4.13), and (7.4.14) with $\frac{\partial \eta}{\partial x}$, $\frac{\partial \zeta}{\partial x}$, and $\frac{\partial \xi}{\partial x}$, we have a relation:

$$
\begin{aligned}
\theta_s \frac{\partial}{\partial x_i}\left(\overline{\pi' \frac{\partial \xi_i}{\partial x}}\right) &= \overline{\frac{\partial \eta}{\partial x} G'_x} + \overline{\frac{\partial \zeta}{\partial x} G'_y} + \overline{\frac{\partial \xi}{\partial x} h}\frac{g}{\theta_s} \\
-\frac{\partial}{\partial t}&\left[\overline{\frac{\partial \eta}{\partial x}(u^l - f\zeta)} + \overline{\frac{\partial \zeta}{\partial x} v'}\right] + \left(f\frac{\partial M}{\partial z} + \frac{g}{\theta_s}\frac{\partial \overline{\Theta}}{\partial y}\right)\overline{\frac{\partial \xi}{\partial x}\zeta}.
\end{aligned} \tag{7.6.4}
$$

In this derivation, we have used the following relations:

$$
\begin{aligned}
u' + \xi_i \frac{\partial \overline{u}}{\partial x_i} &= u^l, \\
\theta' + \xi_i \frac{\partial \overline{\Theta}}{\partial x_i} &= h, \\
\frac{Du'}{Dt} + u'_i \frac{\partial M}{\partial x_i} &= \frac{Du^l}{Dt} - fv' = \frac{D}{Dt}(u^l - f\zeta), \\
\overline{\frac{\partial \eta}{\partial x}\left(\frac{Du'}{Dt} + u'_i\frac{\partial M}{\partial x_i}\right)} &= \frac{\partial}{\partial t}\left[\overline{\frac{\partial \eta}{\partial x}(u^l - f\zeta)}\right] + f\overline{\frac{\partial u'}{\partial x}\zeta} + f\frac{\partial M}{\partial z}\overline{\frac{\partial \xi}{\partial x}\zeta}, \\
\overline{\frac{\partial \zeta}{\partial x}\left(\frac{Dv'}{Dt} + fu'\right)} &= \frac{\partial}{\partial t}\left(\overline{\frac{\partial \zeta}{\partial x} v'}\right) - f\overline{\zeta\frac{\partial u'}{\partial x}}, \\
\overline{\frac{\partial \xi}{\partial x}\theta'}\frac{g}{\theta_s} &= -\overline{\frac{\partial \xi}{\partial x}\zeta}\frac{\partial \overline{\Theta}}{\partial y}\frac{g}{\theta_s} - \overline{\frac{\partial \xi}{\partial x} h}\frac{g}{\theta_s},
\end{aligned}
$$

and, since $\frac{\partial \xi_i}{\partial x_i} = 0$,

$$
\overline{\frac{\partial \xi_i}{\partial x}\frac{\partial \pi'}{\partial x_i}} = \frac{\partial}{\partial x_i}\left(\overline{\pi' \frac{\partial \xi_i}{\partial x}}\right).
$$

Using (7.6.3) and (7.6.4), therefore, we have

$$
\begin{aligned}
\frac{\partial F_i}{\partial x_i} &= -\frac{\partial}{\partial x_i}\left[\frac{\partial \overline{\xi_i u'}}{\partial t} + K_{ij}\frac{\partial M}{\partial x_j} - \overline{\xi_i G'_x} + \varepsilon_{1ij}(K_{km} n_m t_k + \mathcal{H})\frac{\partial M}{\partial x_j}\right] \\
&+ \overline{\frac{\partial \eta}{\partial x} G'_x} + \overline{\frac{\partial \zeta}{\partial x} G'_y} + \overline{\frac{\partial \xi}{\partial x} h}\frac{g}{\theta_s} - \frac{\partial}{\partial t}\left[\overline{\frac{\partial \eta}{\partial x}(u^l - f\zeta)} + \overline{\frac{\partial \zeta}{\partial x} v'}\right] \\
&+ \left(f\frac{\partial M}{\partial z} + \frac{g}{\theta_s}\frac{\partial \overline{\Theta}}{\partial y}\right)\overline{\frac{\partial \xi}{\partial x}\zeta}.
\end{aligned}
$$

From this, it is easy to derive (7.4.17).

References and suggested reading

Most of the material in this chapter refer to the textbook by Andrews et al. (1987). The original work on the generalized Eulerian mean of Section 7.1.1 was done by Andrews and McIntyre (1978b), and that on the transformed Eulerian mean equations of Section 7.4 was done by Andrews and McIntyre (1976,1978a). The finite amplitude wave theory in Section 7.1.2 is based on Flierl (1981). The diffusion tensor and the residual circulation described in Sections 7.2 and 7.3 follows Plumb (1979). Note that since generalized Eulerian mean equations are not uniquely constructed, the formulation in Section 7.4 is just one approach and is slightly different from Andrews et al. (1987).

Andrews, D. G. and McIntyre, M. E., 1976: Planetary waves in horizontal and vertical shear: The generalized Eliassen-Palm relation and the mean zonal acceleration. *J. Atmos. Sci.*, **33**, 2031–2048.

Andrews, D. G. and McIntyre, M. E., 1978a: Generalized Eliassen-Palm and Charney-Drazin theorems for waves on axisymmetric flows in compressible atmospheres. *J. Atmos. Sci.*, **35**, 175–185.

Andrews, D. G. and McIntyre, M. E., 1978b: An exact theory of nonlinear waves on a Lagrangian mean flow. *J. Fluid Mech.*, **89**, 609–646.

Andrews, D. G., Holton, J. R., and Leovy, C. B., 1987: *Middle Atmosphere Dynamics.* Academic Press, San Diego, 489 pp.

Flierl, G. R., 1981: Particle motions in large-amplitude wave fields. *Geophys. Astrophys. Fluid Dyn.*, **18**, 39–74.

Plumb, R. A., 1979: Eddy flux of conserved quantities by small-amplitude waves. *J. Atmos. Sci.*, **36**, 1699–1704.

8

Thermodynamics of moist air

Up to now, we have described the dynamics of a dry atmosphere, which is a gas with well-mixed components but without water vapor. In this and the next chapters, we will explain the basic properties and the governing equations of moist air. Moist air is a mixture of dry air and water in which water experiences phase changes between vapor, liquid water, and ice. Interactions between various phases of water play fundamental roles in meso-scale moist circulation (i.e., 1–10 km horizontal scale convection). In this book, however, we are mainly concerned with the large-scale dynamics of the atmosphere, which can be described using a simplified formulation of moist thermodynamics. We mainly consider a mixture of vapor and liquid water and assume that the two phases are in phase equilibrium. The ice phase and interactions between different phases will not be explicitly considered.

In this chapter, we describe the thermodynamic properties of moist air. First, we derive the thermodynamic expressions of moist air in a general form. Then, we simplify the thermodynamic expressions of moist air and introduce the assumptions of ideal gas heat and constant specific heat. In addition, with the further assumption that the water content is low in air, we introduce the thermodynamic expressions of moist air conventionally used in the literature.

8.1 Formulation

8.1.1 Definition of moist air

When we consider hydrological circulation in the atmosphere, the atmosphere should be viewed as the *moist atmosphere*. The moist atmosphere consists of *moist air*, which includes the various phases of water substance. Note that moist air is not a pure gas, since water substance experiences phase changes between gas, liquid, and solid phases (i.e., vapor, water, and ice). In the framework of thermodynamic equilibrium, the water phase solely depends on the thermodynamic state (i.e., temperature and pressure). In reality, however, water may be supersaturated or supercooled, so that different phases of water may exist in the same thermodynamic state. Thus, it is complicated to determine the exact composition of each water phase. For simplicity, we do not consider such supersaturation and supercooling in the formulation of moist thermodynamics in this chapter; these effects are secondary to understanding the large-scale motions of the atmosphere.

When moist air does not include a liquid or ice phase, the air is a pure gas and behaves like dry air. Since the composition of vapor is generally inhomogeneous in the moist atmosphere, vapor has a diffusive process. Energy transport is also associated with the diffusion of vapor. When moist air contains a liquid or ice phase, the governing equations of the moist atmosphere become very complicated because it becomes a multi phase flow. Water substance exists in the atmosphere in the form of liquid or ice particles. In conventional terminology, if liquid particles move with the gas part of the air, they are called cloud particles, whereas if they have relative motions to the gas, they are called raindrops. Similar categories are used for ice particles, but they are categorized as snow, graupel, or hail according to their sizes. Strictly, we must consider the surface processes of individual liquid or ice particles and the collision processes between the particles. In practice, however, if we are interested in the mean properties of a sufficiently large volume of air, the interactions between different phases need to be formulated for the collection of a large number of liquid or ice particles in the gas.

The volume of moist air to be considered depends on many factors. In numerical models, moist air is treated differently according to the resolvable scale used. For example, the horizontal resolutions of large-scale circulations used in general circulation models are normally more than 100 km. If one explicitly calculates cloud motions by using cumulus-resolving models, it is thought that the horizontal resolution should be less than a few kilometers. If the inner motions of clouds are considered, air parcels with a diameter of about 100 m would be required. In fact, moist air can have different phases and must be very inhomogeneous in the 100 m to 100 km scale.

In the following formulation, we neglect all the inhomogeneity of moist air and consider thermodynamic state variables by assuming that moist air is described as a local thermodynamic state of a homogeneously mixed fluid.

8.1.2 Basic thermodynamic equations

We assume that moist air has the following thermodynamic properties. For simplicity, we only consider the gas and liquid phases of water, but not the ice phase.

(1) The gas component consists of dry air and vapor and is the ideal gas.

(2) The gas component is in phase equilibrium with the liquid component.

(3) The volume of liquid component is negligible compared with the fractional volume of vapor.

(4) The entropy of moist air is the sum of the entropy of the gas component and that of the liquid component. This implies that the entropy of mixing between the gas and liquid particles is negligible.

(5) Liquid particles consist of a single component (i.e., water), and the gas phase is insoluble in liquid particles.

The assumptions of the ideal gas (1) and neglect of the volume of the liquid component (3) are normally introduced. However, in extreme situations, such as in the initial evolution of the proto-atmosphere, these assumptions break down (e.g., Abe and Matsui, 1988). In this chapter, we first derive general expressions without these two assumptions. Later, we introduce them to obtain the familiar relationship used for the present atmosphere.

In general, the expressions of thermodynamic variables can be derived if the specific heat at constant pressure and the equation of state are known. The definition of moist air enables us to construct the specific heat at constant pressure and the equation of state for the moist air from those of each component. We use subscript k for each of the gas or the liquid part, g for the gas part, and c for the liquid part of moist air. Among the gas components, subscript d is used for dry air and v is used for vapor. Although dry air consists of many gas components, we do not distinguish them in this chapter. We designate the molar specific heat at constant pressure and the molar specific volume of the k-th component by c_{pk} and v_k, respectively.† The equation of state for the k-th component is generally written in the form

$$v_k = v_k(p, T). \tag{8.1.1}$$

In particular, the equation of state for gas components is given by that of the ideal gas:

$$v_k = \frac{R^* T}{p}, \tag{8.1.2}$$

where R^* is the universal gas constant. The equation of state for the ideal gas has the same function for any gas components. Although the liquid part has its own equation of state, we do not specify a specific form of the equation here; we simply use the general expression (8.1.1). According to the above assumption (3), the equation of state for the liquid component does not make any contribution to the moist air in the actual atmosphere. Specific heats generally depend on both pressure and temperature. However, if the dependency of the specific heats on temperature at arbitrary pressure p_0 is known, all the expressions of thermodynamic variables can be determined.

In moist air, water substance can be in gas phase (vapor) or liquid phase (water). If the air parcel is not saturated, all the water in the air parcel is in the form of vapor. If saturated, on the other hand, both vapor and water are contained in the air parcel. In this case, the liquid water is airborne as cloud particles. We use quantities per unit mole for the thermodynamic formulations in this section. The number of moles of the k-th component per unit mass of air, or *molar concentration*,

† In this section, subscripts k, g, d, v, and c are used for the quantities per unit mole of each component. Quantities per unit mass of moist air are denoted by symbols without subscripts.

is denoted by n_k.[†] The gas phase is the sum of dry air and vapor, and the water substance is the sum of vapor and cloud particles; that is,

$$n_g = n_d + n_v, \qquad n_w = n_v + n_c. \tag{8.1.3}$$

Let v denote the specific volume of moist air. Density is given by $\rho = 1/v$. The internal energy and the entropy per unit mass of moist air are denoted by u and s, respectively. We assume that the specific volume, the internal energy, and the entropy of moist air are given by the sum of those of gas and cloud particles:

$$v = n_g v_g + n_c v_c, \qquad u = n_g u_g + n_c u_c, \qquad s = n_g s_g + n_c s_c, \tag{8.1.4}$$

where v_g and v_c are molar volumes, u_g and u_c are molar internal energies, and s_g and s_c are the molar entropies of gas and liquid components, respectively. Assumption (4) is used for the expression of entropy. Enthalpy h, Helmholtz's free energy f, and Gibbs free energy g are defined by

$$h = u + pv, \tag{8.1.5}$$

$$f = u - Ts, \tag{8.1.6}$$

$$g = h - Ts = u + pv - Ts. \tag{8.1.7}$$

These are also quantities per unit mass. The specific heat of moist air at constant pressure per unit mass C_p is defined by the derivative of the entropy with respect to temperature:

$$\left(\frac{\partial s}{\partial T}\right)_p = \frac{C_p}{T}, \tag{8.1.8}$$

which is the same relation as (1.1.18). Hence, from (8.1.5)–(8.1.8), specific heat, enthalpy, and free energies are given by the sum of those of gas and cloud particles:

$$C_p = n_g c_{pg} + n_c c_{pc}, \qquad h = n_g h_g + n_c h_c, \tag{8.1.9}$$

$$f = n_g f_g + n_c f_c, \qquad g = n_g g_g + n_c g_c. \tag{8.1.10}$$

In the following arguments, we calculate the thermodynamic variables of gas and cloud components respectively and obtain the thermodynamic expressions of moist air as a mixture of multiphase fluids using the above formulas. The gas component is regarded as a mixture of dry air and vapor. Then, we first obtain the thermodynamic functions of a single component of gas and next consider the effect of mixing to obtain the thermodynamic functions of the gas component of moist air.

We summarize useful thermodynamic relations in the rest of this subsection. If the thermodynamic state of moist air changes but keeps thermodynamic equilibrium, the first law of thermodynamics is written by

$$du = Tds - pdv + \sum_k \mu_k dn_k, \tag{8.1.11}$$

[†]In the next section, we will use the mass concentration of the k-th component q_k which is defined as the mass of the k-th component per unit mass of air. Molar concentration is related as $n_k = q_k m_k$, where m_k is molecular weight.

where μ_k is the molar chemical potential of the k-th component. Using this equation, the total derivatives of thermodynamic energies are expressed as

$$dh = Tds + vdp + \sum_k \mu_k dn_k, \tag{8.1.12}$$

$$df = -sdT - pdv + \sum_k \mu_k dn_k, \tag{8.1.13}$$

$$dg = -sdT + vdp + \sum_k \mu_k dn_k. \tag{8.1.14}$$

From these equations, the partial derivatives of thermodynamic functions can be related to quantities of state as (1.1.9)–(1.1.12). In particular, the partial derivatives with respect to the molar concentration of each component are given as

$$\begin{aligned} \left(\frac{\partial u}{\partial n_k}\right)_{s,v,n_{l,l\neq k}} &= \left(\frac{\partial h}{\partial n_k}\right)_{s,p,n_{l,l\neq k}} \\ = \left(\frac{\partial f}{\partial n_k}\right)_{T,v,n_{l,l\neq k}} &= \left(\frac{\partial g}{\partial n_k}\right)_{T,p,n_{l,l\neq k}} = \mu_k, \end{aligned} \tag{8.1.15}$$

$$\left(\frac{\partial s}{\partial n_k}\right)_{u,v,n_{l,l\neq k}} = -\frac{\mu_k}{T}. \tag{8.1.16}$$

The Maxwell relations are given by

$$\left(\frac{\partial s}{\partial p}\right)_{T,n_k} = -\left(\frac{\partial v}{\partial T}\right)_{p,n_k}, \quad \left(\frac{\partial s}{\partial v}\right)_{p,n_k} = \left(\frac{\partial p}{\partial T}\right)_{s,n_k}, \tag{8.1.17}$$

$$\left(\frac{\partial s}{\partial v}\right)_{T,n_k} = \left(\frac{\partial p}{\partial T}\right)_{v,n_k}, \quad \left(\frac{\partial s}{\partial p}\right)_{v,n_k} = -\left(\frac{\partial v}{\partial T}\right)_{s,n_k}, \tag{8.1.18}$$

which corresponds to (1.1.13)–(1.1.16).

8.1.3 Thermodynamic functions of a single component

We first obtain expressions for the thermodynamic functions of a single component k using the equation of state $v_k(p,T)$ and the specific heat at constant pressure $c_{pk}(p_0,T)$ at a specified pressure p_0. Thermodynamic functions are expressed by pressure p, temperature T, and these two functions. As in Chapter 1, molar specific heat c_{pk}, entropy s_k, enthalpy h_k, and the internal energy u_k of a single component are given by (1.1.21), (1.1.25), (1.1.26), and (1.1.27). Similarly, (8.1.7) also holds for each component g_k. Thus,

$$\begin{aligned} c_{pk}(p,T) &= c_{pk}(p_0,T) + \int_{p_0}^{p} \left(\frac{\partial c_{pk}}{\partial p}\right)_T dp \\ &= c_{pk}(p_0,T) - T\int_{p_0}^{p} \left(\frac{\partial^2 v_k(p',T)}{\partial T^2}\right)_{p'} dp', \end{aligned} \tag{8.1.19}$$

$$
\begin{aligned}
h_k(p,T) &= h_0 + \int_{T_0}^{T} c_{pk}(p_0,T')dT' \\
&\quad - T^2 \int_{p_0}^{p} \left[\frac{\partial}{\partial T}\left(\frac{v_k(p',T)}{T}\right)\right]_{p'} dp', \qquad (8.1.20)\\
u_k(p,T) &= h_k(p,T) - p v_k(p,T) \\
&= h_0 + \int_{T_0}^{T} c_{pk}(p_0,T')dT' \\
&\quad - T^2 \int_{p_0}^{p} \left[\frac{\partial}{\partial T}\left(\frac{v_k(p',T)}{T}\right)\right]_{p'} dp' - p v_k(p,T), \qquad (8.1.21)\\
s_k(p,T) &= s_0 + \int_{T_0}^{T} \frac{c_{pk}(p_0,T')}{T'}dT' - \int_{p_0}^{p}\left(\frac{\partial v_k(p',T)}{\partial T}\right)_{p'} dp', \qquad (8.1.22)\\
g_k(p,T) &= h_k(p,T) - T s_k \\
&= h_0 - T s_0 + \int_{T_0}^{T} c_{pk}(p_0,T')\left(1 - \frac{T}{T'}\right)dT' \\
&\quad + \int_{p_0}^{p} v_k(p',T)dp'. \qquad (8.1.23)
\end{aligned}
$$

where the entropy and enthalpy at (p_0, T_0) are denoted by s_0 and h_0, respectively. For an ideal gas, substituting the equation of state (8.1.2) into v_k gives

$$
\begin{aligned}
c_{pk}(p,T) &= c_{pk}(p_0,T), \qquad (8.1.24)\\
h_k(p,T) &= h_0 + \int_{T_0}^{T} c_{pk}(p_0,T')dT', \qquad (8.1.25)\\
u_k(p,T) &= h_k - p v_k = h_0 + \int_{T_0}^{T} c_{pk}(p_0,T')dT' - R^*T, \qquad (8.1.26)\\
s_k(p,T) &= s_0 + \int_{T_0}^{T} \frac{c_{pk}(p_0,T')}{T'}dT' - R^* \ln\frac{p}{p_0}, \qquad (8.1.27)\\
g_k(p,T) &= h_k - T s_k \\
&= h_0 - T s_0 + \int_{T_0}^{T} c_{pk}(p_0,T')\left(1 - \frac{T}{T'}\right)dT' + R^*T\ln\frac{p}{p_0}. \\
&\qquad (8.1.28)
\end{aligned}
$$

8.1.4 Mixing of ideal gases

Next we consider the mixing of dry air and vapor as ideal gases. Subscript j denotes each component of gas and subscript g denotes the mixed gas.† Since any kind of ideal gas has the same specific volume v_j, we can define the universal specific volume

†Subscript k is used for any phase of components, while j is used only for the gas phase.

v^I of ideal gas as

$$v_j(p,T) = v^I(p,T) \equiv \frac{R^*T}{p}. \tag{8.1.29}$$

In (8.1.24)–(8.1.28), we change the independent variables from pressure and temperature (p, T) to volume and temperature (v^I, T) in order to consider the mixing of gases. Let the values at pressure p_0 be denoted by superscript 0. The above equations are written as

$$c_{pj}(v^I,T) = c_{pj}^0(T), \quad h_j(v^I,T) = h_j^0(T), \quad u_j(v^I,T) = u_j^0(T), \tag{8.1.30}$$

$$s_j(v^I,T) = s_j^0(T) + R^* \ln \frac{v_j}{v_j^0}, \tag{8.1.31}$$

$$g_j(v^I,T) = g_j^0(T) - R^*T \ln \frac{v_j}{v_j^0}, \tag{8.1.32}$$

where $c_{pj}^0(T) = c_{pj}(p_0,T)$ and

$$h_j^0(T) = h_0 + \int_{T_0}^{T} c_{pj}^0(T')dT', \tag{8.1.33}$$

$$u_j^0(T) = h_j^0(T) - R^*T = h_0 + \int_{T_0}^{T} c_{pj}^0(T')dT' - R^*T, \tag{8.1.34}$$

$$s_j^0(T) = s_0 + \int_{T_0}^{T} \frac{c_{pj}^0(T')}{T'} dT', \tag{8.1.35}$$

$$\begin{aligned} g_j^0(T) &= h_j^0(T) - Ts_j^0(T) \\ &= h_0 - Ts_0 + \int_{T_0}^{T} c_{pj}^0(T')\left(1 - \frac{T}{T'}\right) dT'. \end{aligned} \tag{8.1.36}$$

Then, we mix the ideal gases. The thermodynamic states before the mixing, denoted by prime $'$, are simply given by the sum of thermodynamic expressions of each gas, (8.1.30)–(8.1.32), times each molar concentration:

$$\begin{aligned} n_g c'_{pg}(v_g,T) &= \sum_j n_j c_{pj}^0(T), \quad n_g h'_g(v_g,T) = \sum_j n_j h_j^0(T), \\ n_g u'_g(v_g,T) &= \sum_j n_j u_j^0(T), \end{aligned} \tag{8.1.37}$$

$$n_g s'_g(v_g,T) = \sum_j n_j s_j^0(T) + R^* \sum_j n_j \ln \frac{v_j}{v_j^0}, \tag{8.1.38}$$

$$n_g g'_g(v_g,T) = \sum_j n_j g_j^0(T) - R^*T \sum_j n_j \ln \frac{v_j}{v_j^0}, \tag{8.1.39}$$

where $v_j = v_g = v^I$; subscripts j and g are recovered for later convenience.

During the mixing process, the volume occupied by gas j is changed from $n_j v_j$ ($= n_j v^I$) to $n_g v_g$ ($= n_g v^I$). After mixing, the above thermodynamic variables become

$$\begin{aligned} n_g c_{pg}(v_g, T) &= \sum_j n_j c^0_{pj}(T), \quad n_g h_g(v_g, T) = \sum_j n_j h^0_j(T), \\ n_g u_g(v_g, T) &= \sum_j n_j u^0_j(T), \end{aligned} \tag{8.1.40}$$

$$n_g s_g(v_g, T) = \sum_j n_j s^0_j(T) + R^* \sum_j n_j \ln \frac{n_g v_g}{n_j v^0_j}, \tag{8.1.41}$$

$$n_g g_g(v_g, T) = \sum_j n_j g^0_j(T) - R^* T \sum_j n_j \ln \frac{n_g v_g}{n_j v^0_j}. \tag{8.1.42}$$

Hence, the differences between variables before and after mixing are given by

$$n_g \Delta c_{pg} = 0, \quad n_g \Delta h_g = 0, \quad n_g \Delta u_g = 0, \tag{8.1.43}$$

$$n_g \Delta s_g = -R^* \sum_j n_j \ln \frac{n_j}{n_g}, \tag{8.1.44}$$

$$n_g \Delta g_g = R^* T \sum_j n_j \ln \frac{n_j}{n_g} = -n_g T \Delta s_g. \tag{8.1.45}$$

Since these differences are independent of volume and are functions only of temperature T, (8.1.40)–(8.1.42) can be rewritten with respect to pressure. Therefore, we obtain expressions for the thermodynamic variables of mixed gas:

$$\begin{aligned} n_g c_{pg}(p, T) &= n_g c'_{pg}(p, T) = \sum_j n_j c^0_{pj}(T), \\ n_g h_g(p, T) &= n_g h'_g(p, T) = \sum_j n_j h^0_j(T), \\ n_g u_g(p, T) &= n_g u'_g(p, T) = \sum_j n_j u^0_j(T), \end{aligned} \tag{8.1.46}$$

$$\begin{aligned} n_g s_g(p, T) &= n_g s'_g(p, T) + n_g \Delta s_g \\ &= \sum_j n_j \left(s^0_j(T) - R^* \ln \frac{n_j p}{n_g p_0} \right), \end{aligned} \tag{8.1.47}$$

$$\begin{aligned} n_g g_g(p, T) &= n_g g'_g(p, T) - n_g T \Delta s_g \\ &= \sum_j n_j \left(g^0_j(T) + R^* T \ln \frac{n_j p}{n_g p_0} \right), \end{aligned} \tag{8.1.48}$$

which are written as

$$n_g c_{pg}(p,T) = \sum_j n_j c_{pj}, \quad n_g h_g(p,T) = \sum_j n_j h_j,$$
$$n_g u_g(p,T) = \sum_j n_j u_j, \tag{8.1.49}$$
$$n_g s_g(p,T) = \sum_j n_j s_j, \quad n_g g_g(p,T) = \sum_j n_j g_j. \tag{8.1.50}$$

Since specific heat, enthalpy, and internal energy, given by (8.1.46), are independent of pressure p, we have $c_{pj} = c^0_{pj}(T)$, $h_{pj} = h^0_{pj}(T)$, and $u_{pj} = u^0_{pj}(T)$. The entropy and the Gibbs free energy of the j-th component are defined by

$$s_j(p_j,T) = s^0_j(T) - R^* \ln \frac{n_j p}{n_g p_0} = s^0_j(T) - R^* \ln \frac{p_j}{p_0}, \tag{8.1.51}$$
$$g_j(p_j,T) = g^0_j(T) + R^* T \ln \frac{n_j p}{n_g p_0} = g^0_j(T) + R^* T \ln \frac{p_j}{p_0}, \tag{8.1.52}$$

where

$$p_j \equiv \frac{n_j}{n_g} p \tag{8.1.53}$$

is the *partial pressure* of the j-the component.

8.1.5 Thermodynamic variables of liquid phase

To derive the thermodynamic variables of cloud particles, we consider a fluid with a single component with two phases: vapor and liquid water. The thermodynamic variables of such a pure substance are denoted by superscript $*$ to distinguish them from other components. The equation of state of the gas phase (vapor) is that for the ideal gas:

$$v^*_v = \frac{R^* T}{p}, \tag{8.1.54}$$

while the equation of state of the liquid phase is given as a general expression:

$$v^*_c = v^*_c(p,T). \tag{8.1.55}$$

We assume that the volume of liquid phase is negligible:

$$v^*_v \gg v^*_c. \tag{8.1.56}$$

This corresponds to assumption (3) described in Section 8.1.1.

We define the saturation vapor pressure at temperature T by $p^*(T)$. In the case of liquid-vapor equilibrium, the chemical potential of vapor is equal to that of the liquid phase:

$$\mu^*_c(p^*,T) = \mu^*_v(p^*,T). \tag{8.1.57}$$

If a slightly different state $(p^* + dp^*, T + dT)$ is also in phase equilibrium, we have

$$\mu_v^*(p^* + dp^*, T + dT) = \mu_c^*(p^* + dp^*, T + dT). \tag{8.1.58}$$

The differences between (8.1.58) and (8.1.57) are written as

$$\left(\frac{\partial \mu_v^*}{\partial p^*}\right)_T dp^* + \left(\frac{\partial \mu_v^*}{\partial T}\right)_{p^*} dT = \left(\frac{\partial \mu_c^*}{\partial p^*}\right)_T dp^* + \left(\frac{\partial \mu_c^*}{\partial T}\right)_{p^*} dT. \tag{8.1.59}$$

Making use of (1.1.12), we obtain

$$v_v^* dp^* - s_v^* dT = v_c^* dp^* - s_s^* dT, \tag{8.1.60}$$

that is,

$$\frac{dp^*}{dT} = \frac{s_v^* - s_c^*}{v_v^* - v_c^*}. \tag{8.1.61}$$

The latent heat l per unit mole is defined by

$$s_v^* - s_c^* = \frac{l}{T}. \tag{8.1.62}$$

Thus, (8.1.61) is rewritten as

$$\frac{dp^*}{dT} = \frac{l}{T(v_v^* - v_c^*)}. \tag{8.1.63}$$

This is the *Clausius-Clapeyron equation.* Using the equation of state of the ideal gas (8.1.54) and neglecting the volume of the liquid phase $v_v^* \gg v_c^*$, the Clausius-Clapeyron equation becomes

$$\frac{dp^*}{dT} = \frac{lp^*}{R^* T^2}. \tag{8.1.64}$$

If $l(T)$ is known, the function $p^*(T)$ is calculated using this equation.

The thermodynamic variables of condensable gas are expressed as follows. Now we are considering a single component system with a condensable gas, the chemical potential μ^* of which is equal to g in (8.1.23). Using (8.1.14), the chemical potential at (p, T) is expressed as

$$\begin{aligned} \mu_c^*(p, T) &= \mu_v^*(p^*, T) + \int_{p^*}^{p} v_c^* dp' \\ &= \mu_v^*(p, T) - \int_{p^*}^{p} (v_v^* - v_c^*) dp'. \end{aligned} \tag{8.1.65}$$

From (1.1.12), entropy is expressed as

$$\begin{aligned} s_c^*(p, T) &= -\left(\frac{\partial \mu_c^*}{\partial T}\right)_p = s_v^*(p, T) + \frac{\partial}{\partial T} \int_{p^*}^{p} (v_v^* - v_c^*) dp' \\ &= s_v^*(p, T) - \frac{dp^*}{dT}(v_v^* - v_c^*)_{p^*} + \int_{p^*}^{p} \left(\frac{\partial v_v^*}{\partial T} - \frac{\partial v_c^*}{\partial T}\right) dp'. \end{aligned} \tag{8.1.66}$$

Note that (8.1.61) is directly given from this equation by setting $p = p^*$. Next, enthalpy and internal energy are given by

$$\begin{aligned} h_c^*(p,T) &= \mu_c^*(p,T) + Ts_c^*(p,T) \\ &= h_v^*(p,T) - \int_{p^*}^{p} (v_v^* - v_c^*)dp' \\ &\quad -T\frac{dp^*}{dT}(v_v^* - v_c^*)_{p^*} + T\int_{p^*}^{p} \left(\frac{\partial v_v^*}{\partial T} - \frac{\partial v_c^*}{\partial T}\right) dp', \end{aligned} \tag{8.1.67}$$

$$\begin{aligned} u_c^*(p,T) &= h_c^*(p,T) - pv_c^*(p,T) \\ &= u_v^*(p,T) + p(v_v^* - v_c^*) - \int_{p^*}^{p} (v_v^* - v_c^*)dp' \\ &\quad -T\frac{dp^*}{dT}(v_v^* - v_c^*)_{p^*} + T\int_{p^*}^{p} \left(\frac{\partial v_v^*}{\partial T} - \frac{\partial v_c^*}{\partial T}\right) dp'. \end{aligned} \tag{8.1.68}$$

From (1.1.18), specific heat is given by

$$\begin{aligned} c_{pc}^*(p,T) &= T\left(\frac{\partial s_c^*}{\partial T}\right)_p = c_{pv}^*(p,T) + T\frac{\partial^2}{\partial T^2}\int_{p^*}^{p} (v_v^* - v_c^*)dp' \\ &= c_{pv}^*(p,T) + T\left[-\frac{d^2p^*}{dT^2}(v_v^* - v_c^*)_{p^*} - 2\frac{dp^*}{dT}\left(\frac{\partial v_v^*}{\partial T} - \frac{\partial v_c^*}{\partial T}\right)_{p^*}\right. \\ &\quad \left. -\left(\frac{dp^*}{dT}\right)^2\left(\frac{\partial v_v^*}{\partial p} - \frac{\partial v_c^*}{\partial p}\right)_{p^*} + \int_{p^*}^{p}\left(\frac{\partial^2 v_v^*}{\partial T^2} - \frac{\partial^2 v_c^*}{\partial T^2}\right) dp'\right]. \end{aligned} \tag{8.1.69}$$

Up to now, no assumptions have been made about the equation of state for gas and liquid phases. Using the equation of state for the ideal gas for v_v^* (8.1.54) and neglecting the volume of the liquid phase v_c^*, (8.1.66)–(8.1.69) become

$$c_{pc}^*(p,T) = c_{pv}^{*0}(T) - \frac{R^*T^2}{p^*}\frac{d^2p^*}{dT^2} - 2\frac{R^*T}{p^*}\frac{dp^*}{dT} + \frac{R^*T^2}{p^{*2}}\left(\frac{dp^*}{dT}\right)^2, \tag{8.1.70}$$

$$h_c^*(p,T) = h_v^{*0}(T) - \frac{R^*T^2}{p^*}\frac{dp^*}{dT}, \tag{8.1.71}$$

$$u_c^*(p,T) = u_v^{*0}(T) + R^*T - \frac{R^*T^2}{p^*}\frac{dp^*}{dT}, \tag{8.1.72}$$

$$s_c^*(p,T) = s_v^{*0}(T) - R^*\ln\frac{p^*}{p_0} - \frac{R^*T}{p^*}\frac{dp^*}{dT}, \tag{8.1.73}$$

$$\mu_c^*(p,T) = \mu_v^{*0}(T) + R^*T\ln\frac{p^*}{p_0}. \tag{8.1.74}$$

Variables with superscript 0 are given by (8.1.33)–(8.1.36). These expressions are rewritten with latent heat l using (8.1.64):

$$\begin{aligned}
c_{pc}^*(T) &= c_{pv}^{*0}(T) - \frac{dl}{dT}, &(8.1.75)\\
h_c^*(T) &= h_v^{*0}(T) - l, &(8.1.76)\\
u_c^*(T) &= u_v^{*0}(T) + R^*T - l, &(8.1.77)\\
s_c^*(T) &= s_v^{*0}(T) - R^* \ln \frac{p^*}{p_0} - \frac{l}{T}, &(8.1.78)\\
\mu_c^*(T) &= \mu_v^{*0}(T) + R^*T \ln \frac{p^*}{p_0}. &(8.1.79)
\end{aligned}$$

We can see that the above expressions are independent of pressure p. Eq. (8.1.75) is called *Kirchhoff's equation.* Furthermore, using (8.1.33)–(8.1.36), these are also rewritten with specific heats as

$$\begin{aligned}
h_c^*(T) &= h_{v0} + \int_{T_0}^{T} c_{pv}^{*0}(T')dT' - l\\
&= h_{v0} + \int_{T_0}^{T} c_{pc}^{*0}(T')dT', &(8.1.80)\\
u_c^*(T) &= h_{v0} + \int_{T_0}^{T} c_{pv}^{*0}(T')dT' + R^*T - l\\
&= h_{v0} + \int_{T_0}^{T} c_{pc}^{*0}(T')dT' + R^*T, &(8.1.81)\\
s_c^*(T) &= s_{v0} + \int_{T_0}^{T} \frac{c_{pv}^{*0}(T')}{T'}dT' - R^* \ln \frac{p^*}{p_0} - \frac{l}{T}\\
&= s_{v0} + \int_{T_0}^{T} \frac{c_{pc}^{*0}(T')}{T'}dT', &(8.1.82)\\
g_c^*(T) &= h_{v0} - Ts_{v0} + \int_{T_0}^{T} c_{pv}^{*0}(T')\left(1 - \frac{T}{T'}\right)dT' + R^*T \ln \frac{p^*}{p_0}\\
&= h_{v0} - Ts_{v0} + \int_{T_0}^{T} c_{pc}^{*0}(T')\left(1 - \frac{T}{T'}\right)dT'. &(8.1.83)
\end{aligned}$$

8.1.6 Thermodynamic variables of moist air

The thermodynamic variables of moist air are written as the sum of those of gas and liquid phases: specific heat and enthalpy by (8.1.9); entropy and internal energy by (8.1.4). If the mixture of dry air and vapor is ideal gas, its thermodynamic expressions are given by the sum of the corresponding variables of dry air and vapor as (8.1.46)–(8.1.47). Thus, the thermodynamic expressions for moist air are given by

$$\begin{aligned}
C_p &= n_d c_{pd} + n_v c_{pv} + n_c c_{pc}, & h &= n_d h_d + n_v h_v + n_c h_c,\\
u &= n_d u_d + n_v u_v + n_c u_c, & s &= n_d s_d + n_v s_v + n_c s_c. &(8.1.84)
\end{aligned}$$

The variables of the liquid phase are given by (8.1.70)–(8.1.74), or by (8.1.75)–(8.1.79) if latent heat l is used.

When gas and liquid phases are in phase equilibrium, we get

$$\Delta g(p, T, n_v, n_g) \equiv \mu_c(p, T) - \mu_v(p, T, n_v, n_g) = 0, \tag{8.1.85}$$

where, from (8.1.52) and (8.1.74),

$$\mu_v(p, T, n_v, n_g) = \mu_v^{*0}(T) + R^* T \ln \frac{n_v p}{n_g p_0}, \tag{8.1.86}$$

$$\mu_c(p, T) = \mu_v^{*0}(T) + R^* T \ln \frac{p^*}{p_0}. \tag{8.1.87}$$

Hence, the condition for phase equilibrium is given by

$$p^* = \frac{n_v}{n_g} p. \tag{8.1.88}$$

The right-hand side is the *partial pressure* of vapor. Using $n_g = n_d + n_v$, we also have

$$n_v = \frac{p^*}{p - p^*} n_d. \tag{8.1.89}$$

In the case of phase equilibrium, substituting (8.1.46)–(8.1.47) and (8.1.75)–(8.1.77) into (8.1.84) and using (8.1.88), we have

$$\begin{aligned} C_p &= n_d c_{pd}^0(T) + n_v c_{pv}^0(T) + n_c \left(c_{pv}^{*0}(T) - \frac{dl}{dT} \right) \\ &= n_d c_{pd}^0(T) + n_w c_{pv}^0(T) - n_c \frac{dl}{dT}, \end{aligned} \tag{8.1.90}$$

$$\begin{aligned} h &= n_d h_d^0(T) + n_v h_v^0(T) + n_c (h_v^0(T) - l) \\ &= n_d h_d^0(T) + n_w h_v^0(T) - n_c l, \end{aligned} \tag{8.1.91}$$

$$\begin{aligned} u &= n_d u_d^0(T) + n_v u_v^0(T) + n_c (u_v^0(T) + R^* T - l) \\ &= n_d u_d^0(T) + n_w u_v^0(T) + n_c (R^* T - l). \end{aligned} \tag{8.1.92}$$

$$\begin{aligned} s &= n_d \left(s_d^0(T) - R^* \ln \frac{n_d p}{n_g p_0} \right) + n_v \left(s_v^0(T) - R^* \ln \frac{n_v p}{n_g p_0} \right) \\ &\quad + n_c \left(s_v^0(T) - R^* \ln \frac{p^*}{p_0} - \frac{l}{T} \right) \\ &= n_d \left(s_d^0(T) - R^* \ln \frac{n_d p}{n_g p_0} \right) + n_w \left(s_v^0(T) - R^* \ln \frac{p^*}{p_0} \right) - n_c \frac{l}{T}, \end{aligned} \tag{8.1.93}$$

These thermodynamic variables are measured from the base values $c_{pv}^0(T)$, $s_v^0(T)$, $h_v^0(T)$, and $u_v^0(T)$. We may use different bases for the thermodynamic variables of the liquid phase. Since the variables of the liquid phase are independent of vapor pressure, we can write them as

$$\begin{aligned} C_p &= n_d c_{pd}^0(T) + n_v \left(c_{pc}^0(T) + \frac{dl}{dT} \right) + n_c c_{pc}^0(T) \\ &= n_d c_{pd}^0(T) + n_w c_{pc}^0(T) + n_v \frac{dl}{dT}, \end{aligned} \tag{8.1.94}$$

$$\begin{aligned} h &= n_d h_d^0(T) + n_v (h_c^0(T) + l) + n_c h_c^0(T) \\ &= n_d h_d^0(T) + n_w h_c^0(T) + n_v l, \end{aligned} \tag{8.1.95}$$

$$\begin{aligned} u &= n_d u_d^0(T) + n_v (u_c^0(T) - R^* T + l) + n_c u_c^0(T) \\ &= n_d u_d^0(T) + n_w u_c^0(T) + n_v (l - R^* T), \end{aligned} \tag{8.1.96}$$

$$\begin{aligned} s &= n_d \left(s_d^0(T) - R^* \ln \frac{n_d p}{n_g p_0} \right) + n_v \left(s_c^0(T) + \frac{l}{T} \right) + n_c s_c^0(T), \\ &= n_d \left(s_d^0(T) - R^* \ln \frac{n_d p}{n_g p_0} \right) + n_w s_c^0(T) + n_v \frac{l}{T}, \end{aligned} \tag{8.1.97}$$

where $c_{pc}^0(T)$, $h_c^0(T)$, $u_c^0(T)$, and $s_c^0(T)$ are given by (8.1.75), (8.1.80), (8.1.81), and (8.1.82), respectively.

For general use in meteorology, further assumptions are introduced to the above expressions. We derive approximate expressions in Section 8.2. But before doing so, we obtain the adiabat of moist air in the next subsection.

8.1.7 Moist adiabat

The adiabat, or the adiabatic lapse rate, is the temperature change experienced by an air parcel in an isentropic process. In the case of moist air, temperature change in saturated air is different from that of unsaturated air because latent heat release is associated. Precisely, the adiabat also depends on the amount of water vapor. We present a general formula of the adiabat of moist air in this subsection. In Section 8.2.4, we derive the approximate expression of moist adiabat conventionally used in meteorology.

If the amount of vapor contained in a moist air parcel is sufficient, vapor is condensed and latent heat is released as the air parcel ascends. If the air parcel does not exchange any heat with the environment during this ascent, the air parcel undergoes an adiabatic process. The temperature change in the adiabatic process of moist air is smaller than that of dry air due to latent heat release. The rate of the moist adiabatic temperature change is called the *moist adiabatic lapse rate.* In the real atmosphere, some condensates fall out as precipitation, while others remain as cloud particles and move with the air. Although the ratio between precipitable water and cloud particles depends on circumstances, we may consider two extreme cases. The first is the case when no precipitation occurs, while the second is the case when no cloud particles remain. In the first case, the total amount of water substance is conserved by moist air. In the second case, all the condensates are removed from the moist air. These two extremes are called the *reversible moist adiabat* and the *pseudo-moist adiabat*, respectively. They are collectively called the *moist adiabat.*

We examine changes in thermodynamic variables in the moist adiabat under two conditions: (1) entropy is constant and (2) the gas and liquid phases are in equilibrium. In particular, the dependencies of temperature and humidity on pressure are derived in general forms. The rate of temperature change with pressure has an equivalent meaning to the rate of temperature change with altitude if pressure is related to height through the hydrostatic balance.

First, we consider the reversible moist adiabat where all the water substance in an air parcel remains with the air parcel. The condition that the components of dry air and water substance are conserved is written as

$$dn_d = 0, \qquad dn_w = d(n_v + n_c) = 0. \tag{8.1.98}$$

The adiabatic condition and the phase equilibrium are written as

$$0 = ds = n_g S_T dT + n_g S_p dp + S_n dn_v, \tag{8.1.99}$$

$$0 = d\Delta g = G_T dT + G_p dp + \frac{1}{n_g} G_n dn_v, \tag{8.1.100}$$

where the following symbols are defined:

$$S_T \equiv \frac{1}{n_g}\left(\frac{\partial s}{\partial T}\right)_{p,n}, \qquad S_p \equiv \frac{1}{n_g}\left(\frac{\partial s}{\partial p}\right)_{p,n}, \tag{8.1.101}$$

$$S_n \equiv \left(\frac{\partial s}{\partial n_v}\right)_{p,T,n_d,n_c} - \left(\frac{\partial s}{\partial n_c}\right)_{p,T,n_d,n_v}, \tag{8.1.102}$$

$$G_T \equiv \left(\frac{\partial \Delta g}{\partial T}\right)_{p,n}, \qquad G_p \equiv \left(\frac{\partial \Delta g}{\partial p}\right)_{p,n}, \tag{8.1.103}$$

$$G_n \equiv n_g\left(\frac{\partial \Delta g}{\partial n_v}\right)_{p,T,n_d,n_c}. \tag{8.1.104}$$

Thus, the dependencies of temperature and vapor content on pressure are written as

$$\left(\frac{\partial T}{\partial p}\right)_s = \frac{S_n G_p - S_p G_n}{S_T G_n - S_n G_T}, \tag{8.1.105}$$

$$\left(\frac{\partial n_v}{\partial p}\right)_s = n_g \frac{S_p G_T - S_T G_p}{S_T G_n - S_n G_T}. \tag{8.1.106}$$

On the assumptions that the gas phase is the ideal gas and that the volume of liquid phase is negligibly small, the entropy of moist air per unit mass is expressed using (8.1.84), (8.1.51), and (8.1.73) as

$$\begin{aligned} s &= n_d s_d + n_v s_v + n_c s_c \\ &= n_d\left(s_d^0(T) - R^* \ln\frac{n_d p}{n_g p_0}\right) + n_v\left(s_v^0(T) - R^* \ln\frac{n_v p}{n_g p_0}\right) \\ &\quad + n_c\left(s_v^{*0}(T) - R^* \ln\frac{p^*}{p_0} - \frac{R^* T}{p^*}\frac{dp^*}{dT}\right). \end{aligned} \tag{8.1.107}$$

From (8.1.85), (8.1.86), and (8.1.87), the condition of the phase equilibrium is written as

$$\Delta g(p, T, n_d, n_v) \equiv \mu_c - \mu_v = R^*T \ln \frac{p^*}{p_0} - R^*T \ln \frac{n_v p}{n_g p_0} = 0. \quad (8.1.108)$$

Making use of (8.1.107) and (8.1.108), we have the expressions for (8.1.101)–(8.1.104) as

$$S_T = \frac{1}{n_g T}(n_d c_{pd} + n_v c_{pv} + n_c c_{pc}), \quad (8.1.109)$$

$$S_p = -\frac{R^*}{p}, \qquad S_n = G_T = \frac{l}{T}, \quad (8.1.110)$$

$$G_p = -\frac{R^*T}{p}, \qquad G_n = -R^*T\frac{n_d}{n_v}. \quad (8.1.111)$$

Thus, substituting (8.1.109)–(8.1.111) into (8.1.105) gives the *moist adiabatic lapse rate* or the *reversible moist adiabatic lapse rate*:

$$\begin{aligned} \gamma_m \equiv \left(\frac{\partial T}{\partial p}\right)_s &= \frac{-\frac{Rl}{T} - \frac{n_d}{n_v}\frac{R^2T}{p}}{-R^*\frac{n_d}{n_v}\frac{1}{n_g}(n_d c_{pd} + n_v c_{pv} + n_c c_{pc}) - \left(\frac{l}{T}\right)^2} \\ &= \gamma_d \frac{n_g}{n_d} \frac{1 + \frac{n_v}{n_d}\frac{l}{R^*T}}{1 + \frac{n_v c_{pv} + n_c c_{pc}}{n_d c_{pd}} + \frac{l^2}{c_{pd} R^* T^2}\frac{n_v n_g}{n_d^2}}, \end{aligned} \quad (8.1.112)$$

where γ_d is the dry adiabatic lapse rate defined by (1.1.57), or

$$\gamma_d \equiv \frac{R^*T}{c_{pd}p}. \quad (8.1.113)$$

Similarly, the change in the molar concentration of vapor is given by substituting (8.1.109)–(8.1.111) into (8.1.106):

$$\left(\frac{\partial n_v}{\partial p}\right)_s = \frac{n_g n_d}{p n_v} \frac{1 + \frac{n_v c_{pv} + n_c c_{pc}}{n_d c_{pd}} + \frac{n_g l}{n_d c_{pd} T}}{1 + \frac{n_v c_{pv} + n_c c_{pc}}{n_d c_{pd}} + \frac{l^2}{c_{pd} R^* T^2}\frac{n_v n_g}{n_d^2}}. \quad (8.1.114)$$

In this equation, the value of n_v is given by solving (8.1.108) or (8.1.89).

Next, we consider the other extreme case of the moist process, the pseudo-moist adiabat, in which all the liquid particles are completely removed from an air parcel when condensation occurs; this is a very simple model of the precipitation process in a moist atmosphere. The dependencies of temperature and moisture on pressure are given by substituting $n_c = 0$ into (8.1.105) and (8.1.106):

$$\left(\frac{\partial T}{\partial p}\right)_{s'} = \frac{S_n G_p - S'_p G_n}{S'_T G_n - S_n G_T}, \quad (8.1.115)$$

$$\left(\frac{\partial n_v}{\partial p}\right)_{s'} = n_g \frac{S'_p G_T - S'_T G_p}{S'_T G_n - S_n G_T}, \quad (8.1.116)$$

where subscript s' represents the pseudo-moist adiabat process. S'_T and S'_p are given from (8.1.101) with $n_c = 0$:

$$S'_T = \frac{1}{n_g T}(n_d c_{pd} + n_v c_{pv}), \qquad S'_p = S_p = -\frac{R^*}{p}. \tag{8.1.117}$$

Thus, from (8.1.115), the pseudo-moist adiabat is given by

$$\gamma'_m \equiv \left(\frac{\partial T}{\partial p}\right)_{s'} = \gamma_d \frac{n_g}{n_d} \frac{1 + \frac{n_v}{n_d}\frac{l}{R^* T}}{1 + \frac{n_v c_{pv}}{n_d c_{pd}} + \frac{l^2}{c_{pd} R^* T^2}\frac{n_v n_g}{n_d^2}}. \tag{8.1.118}$$

If $\Delta g > 0$ is satisfied, moist air is undersaturated. The adiabatic lapse rate of unsaturated air is given by

$$\gamma_m = \left(\frac{\partial T}{\partial p}\right)_s = -\frac{S'_p}{S'_T}. \tag{8.1.119}$$

Substituting (8.1.117) into S'_p and S'_T yields

$$\gamma_m = \frac{n_g R^* T/p}{n_d c_{pd} + n_v c_{pv}} = \gamma_d \frac{n_g}{n_d} \frac{1}{1 + \frac{n_v c_{pv}}{n_d c_{pd}}} = \frac{1}{\rho C_p}, \tag{8.1.120}$$

where the equation of state $\rho^{-1} = n_g R^* T/p$ and that of specific heat $C_p = n_d c_{pd} + n_v c_{pv}$ are used. When there is no humidity, this lapse rate reduces to the dry adiabat γ_d, (8.1.113).

8.2 Expressions of thermodynamic variables

8.2.1 Constants

Here, we derive the conventionally used expressions of the thermodynamic variables of moist air by introducing an additional assumption that specific heats are independent of temperature. First, we summarize the values of thermodynamic constants. The molecular weights of dry air and water are respectively given by

$$m_d = 28.966 \times 10^{-3} \text{ kg mol}^{-1},$$
$$m_w = 18.0160 \times 10^{-3} \text{ kg mol}^{-1}.$$

The universal gas constant is $R^* = 8.31436 \text{ J mol}^{-1} \text{ K}^{-1}$. The gas constant for dry air R_d and that for vapor R_v are respectively defined as

$$R_d \equiv \frac{R^*}{m_d} = 287.04 \text{ J kg}^{-1} \text{ K}^{-1}, \tag{8.2.1}$$

$$R_v \equiv \frac{R^*}{m_v} = 461.50 \text{ J kg}^{-1} \text{ K}^{-1}. \tag{8.2.2}$$

The ratio of the two gas constants is denoted by

$$\varepsilon \equiv \frac{m_w}{m_d} = \frac{R_d}{R_v} = 0.62197. \tag{8.2.3}$$

We express thermodynamic variables using those values at a temperature of 0°C (i.e., at $T_0 = 273.15$ K). Specific heats generally depend on temperature and their numerical values are given as a table (List, 1951; Iribarne and Godson, 1981). We use the following representative values for the specific heats at constant pressure of dry air, vapor, and liquid water per unit mass (Appendix A2):

$$\begin{aligned} C_{pd} &= \frac{7R_d}{2} = 1004.6 \ \text{J kg}^{-1} \text{ K}^{-1}, \\ C_{pv} &= 4R_v = 1846 \ \text{J kg}^{-1} \text{ K}^{-1}, \\ C_{pc} &= 4218 \ \text{J kg}^{-1} \text{ K}^{-1}. \end{aligned}$$

We assume that these specific heats are constant irrespective of temperature. Specific heats per unit mass are related to those per unit mole as $c_{pd} = C_{pd}m_d$, $c_{pv} = C_{pv}m_w$, and $c_{pc} = C_{pc}m_w$. Specific heats at the constant volume of dry air and vapor are given by

$$C_{vd} = C_{pd} - R_d, \tag{8.2.4}$$

$$C_{vv} = C_{pv} - R_v. \tag{8.2.5}$$

Specific heats at the constant volume of liquid water are the same as C_{pc}. Similarly, the specific heat of condensed water is

$$C_{pi} = 2106 \ \text{J kg}^{-1} \text{ K}^{-1}.$$

We get the thermodynamic expressions for ice by replacing C_{pc} by C_{pi} in the following formulas.

Latent heat per unit mass L and the saturation vapor pressure p^* are related to specific heats C_{pv} and C_{pc} through Kirchhoff's equation and the Clapeyron-Clausius equation. Latent heat and the saturation vapor pressure at temperature 0°C are given by

$$\begin{aligned} L_0 &= 2.501 \times 10^6 \ \text{J kg}^{-1}, \\ p_0^* &= 6.1078 \times 10^2 \ \text{Pa}. \end{aligned}$$

Kirchhoff's equation (8.1.75) is

$$C_{pv} - C_{pc} = \frac{dL}{dT}, \tag{8.2.6}$$

where the latent heat per unit mole $l = Lm_w$ is used. Thus, the temperature dependence of latent heat is given by

$$L = L_0 + (C_{pv} - C_{pc})(T - T_0) = L_{00} + (C_{pv} - C_{pc})T, \tag{8.2.7}$$

where

$$L_{00} = L_0 - (C_{pv} - C_{pc})T_0 \tag{8.2.8}$$

is constant and is thought to be the latent heat at 0 K. From the Clapeyron-Clausius equation (8.1.64), the temperature dependence of the saturation vapor pressure is written as

$$\frac{1}{p^*}\frac{dp^*}{dT} = \frac{L}{R_v T^2}. \tag{8.2.9}$$

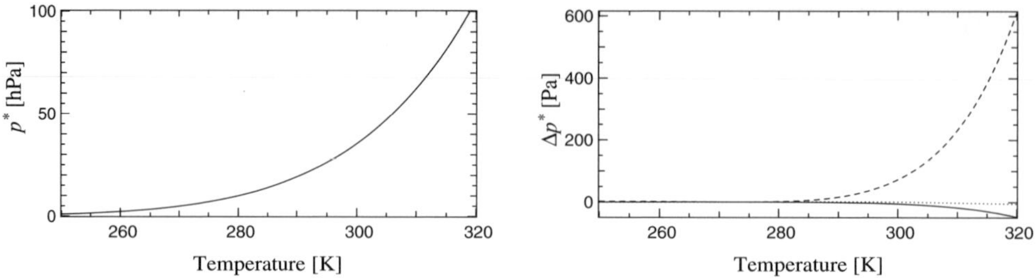

FIGURE 8.1: Temperature dependence of the saturation vapor pressure by the Goff-Gratch formula (left), and differences from that given by the Goff-Gratch formula (right) for formula; solid (8.2.10), dashed (8.2.11), and dotted (8.2.13).

Substituting (8.2.7) to this and integrating with respect to temperature, we obtain

$$p^*(T) = p_0^* \left(\frac{T}{T_0}\right)^{\frac{C_{pv}-C_{pc}}{R_v}} \exp\left[\frac{L_{00}}{R_v}\left(\frac{1}{T_0}-\frac{1}{T}\right)\right]. \tag{8.2.10}$$

This is approximated to the first order of temperature as

$$p^*(T) = p_0^* \exp\left[\frac{L_0}{R_v}\left(\frac{1}{T_0}-\frac{1}{T}\right)\right]. \tag{8.2.11}$$

A more accurate function of the saturation vapor pressure is given by the Goff-Gratch formula (List, 1951):

$$\begin{aligned} \log_{10}\frac{p^*}{p_s^*} &= -7.90298(t^{-1}-1) - 5.02808\log_{10} t \\ &\quad -1.3816\times 10^{-7}[10^{11.344(1-t)}-1] \\ &\quad +8.1328\times 10^{-3}[10^{-3.19149(t^{-1}-1)}-1], \end{aligned} \tag{8.2.12}$$

where $t = T/T_s$, $T_s = 373.16$ K is the steam-point temperature, and $p_s^* = 101324.6$ Pa is the standard pressure. The following simplified formula (Teten's formula) is also frequently used:

$$\log_{10}\frac{p^*(T)}{p_0^*} = \frac{A(T-T_0)}{T-T_0+B}, \tag{8.2.13}$$

where $A = 7.5$ and $B = 237.3$ on liquid water and $A = 9.5$ and $B = 265.5$ on ice. Fig. 8.1 shows the dependence of the saturation vapor pressure on temperature, $p^*(T)$. The above three formulas are compared in this figure.

8.2.2 Mass concentrations and saturation condition

There are several ways to express the mass of dry air and that of water substance. It is straightforward to use the *molar mixing ratio* r, which is the number of moles of water substance per unit mole of dry air. Any thermodynamic variable of moist air f is expressed in terms of pressure p, temperature T, and the mixing ratio r, such that $f = f(T, p, r)$. On the other hand, one can use the mass of water substance per unit mass of moist air (i.e, the mixture of dry air and water substance including liquid water). Let the mass of dry air and that of water substance per unit mass of moist air, or *mass concentration*, be denoted by q_d and q_w, respectively; the mass of vapor and liquid water per unit mass of moist air by q_v and q_c, respectively. q_v is called *specific humidity*, which will be simply referred to as q.

The molar mixing ratio of water substance is defined by

$$r \equiv \frac{n_w}{n_d}. \tag{8.2.14}$$

Mass concentrations of dry air and water substance are written as

$$q_d = n_d m_d, \qquad (q_w, q_v, q_c) = (n_w, n_v, n_c) m_w, \tag{8.2.15}$$

which satisfies

$$q_d + q_w = q_d + q_v + q_c = 1. \tag{8.2.16}$$

From (8.2.15), and (8.2.16), using (8.2.3), mass concentrations are related to the molar mixing ratio as

$$q_d = \frac{1}{1+\varepsilon r}, \qquad q_w = \frac{\varepsilon r}{1+\varepsilon r}. \tag{8.2.17}$$

or r is expressed by q_w:

$$r = \varepsilon^{-1} \frac{q_w}{1-q_w}. \tag{8.2.18}$$

Expression of the specific humidity of saturated air is different from that of unsaturated air. We use n_v^* to express the molar concentration of vapor corresponding to the saturation vapor pressure of moist air at pressure p, temperature T, and mixing ratio r. From (8.1.89), the saturation molar mixing ratio is defined by

$$r^*(p, T) \equiv \frac{n_v^*}{n_d} = \frac{p^*(T)}{p - p^*(T)}. \tag{8.2.19}$$

Using this quantity, then, we can judge whether the moist air with mixing ratio r is saturated or not. In the case $r < r^*$, the air is unsaturated with

$$q_v = \frac{\varepsilon r}{1+\varepsilon r}, \qquad q_c = 0, \tag{8.2.20}$$

while in the case $r \geq r^*$, the air is saturated with

$$q_v = \frac{\varepsilon r^*}{1+\varepsilon r}, \qquad q_c = \frac{\varepsilon(r - r^*)}{1+\varepsilon r}. \tag{8.2.21}$$

From (8.2.17) and (8.2.19), the condition of saturation is rewritten in terms of the mass concentration:

$$q_w \geq q^*(p, T) \equiv \frac{\varepsilon r^*(p, T)}{1+\varepsilon r^*(p, T)} = \frac{\varepsilon p^*(T)}{p - (1-\varepsilon) p^*(T)}. \tag{8.2.22}$$

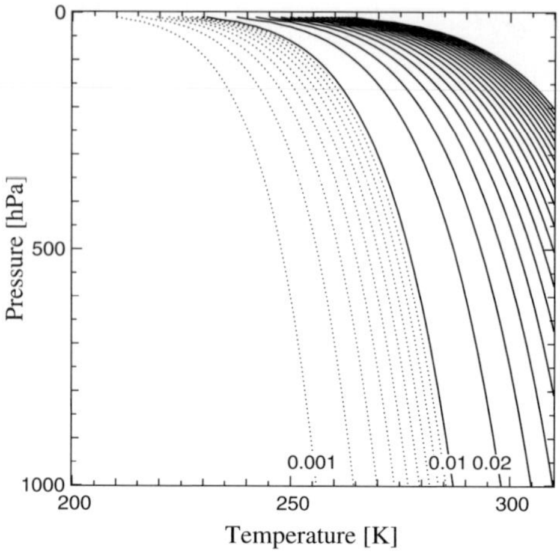

FIGURE 8.2: Dependency of saturated specific humidity q^* on pressure and temperature. The ordinate is pressure, which is positive downward, and the abscissa is temperature. The contour interval of solid lines is 0.01 kg kg^{-1}, and that of dashed lines is 0.001 kg kg^{-1}.

$q^*(p,T)$ may be called *saturation specific humidity.* Note that, even if the air is saturated, q^* is generally different from the specific humidity q_v of saturated moist air since the air may have liquid water. Only in the case $r = r^*$, is q_v equal to q^*. Fig. 8.2 shows the dependency of q^* on pressure and temperature.

8.2.3 Thermodynamic functions

Let us express the thermodynamic functions of moist air with pressure p, temperature T, and mass concentrations q_d, q_v, and q_c. Note that in this section, the quantities per unit mole used in Section 8.1 will be denoted by ($\tilde{}$) to distinguish them from those per unit mass. For instance, the entropy of dry air per unit mass is s_d while that per unit mole is $\tilde{s}_d$; these are related as $\tilde{s}_d = s_d m_d$.

First, the equation of state for the ideal gas is expressed as

$$p = (n_d + n_v)\rho R^* T = (q_d R_d + q_v R_v)\rho T = p_d + p_v, \tag{8.2.23}$$

where p_d and p_v are the partial pressures of dry air and vapor, respectively:

$$p_d = \frac{n_d}{n_d + n_v} p = \frac{q_d}{q_d + \varepsilon^{-1} q_v} p, \tag{8.2.24}$$

$$p_v = \frac{n_v}{n_d + n_v} p = \frac{\varepsilon^{-1} q_v}{q_d + \varepsilon^{-1} q_v} p. \tag{8.2.25}$$

From (8.1.84), specific heat, enthalpy, internal energy, and entropy are given by

$$C_p = q_d C_{pd} + q_v C_{pv} + q_c C_{pc}, \tag{8.2.26}$$

$$h = q_d h_d + q_v h_v + q_c h_c, \tag{8.2.27}$$

$$u = q_d u_d + q_v u_v + q_c u_c, \tag{8.2.28}$$

$$s = q_d s_d + q_v s_v + q_c s_c, \tag{8.2.29}$$

where the quantities of each component are given from (8.1.94)–(8.1.97) as

$$\begin{aligned} h_d &= C_{pd}T, \quad h_v = C_{pc}T + L = C_{pv}T + L_{00}, \\ h_c &= C_{pc}T, \end{aligned} \tag{8.2.30}$$

$$\begin{aligned} u_d &= C_{vd}T, \quad u_v = C_{vv}T + L - R_vT = C_{vv}T + L_{00}, \\ u_c &= C_{pc}T, \end{aligned} \tag{8.2.31}$$

$$\begin{aligned} s_d &= C_{pd}\ln\frac{T}{T_0} - R_d\ln\frac{p_d}{p_0}, \quad s_v = C_{pv}\ln\frac{T}{T_0} - R_v\ln\frac{p_v}{p_0^*} + \frac{L_0}{T_0}, \\ s_c &= C_{pc}\ln\frac{T}{T_0}. \end{aligned} \tag{8.2.32}$$

We have chosen the temperature 0 K as the origin of energy and the temperature T_0 as the origin of entropy. The definitions of entropy and energy of vapor are based on those of liquid water. The above expressions are satisfied even when vapor is supersaturated. The entropy, the enthalpy, and the internal energy of saturated vapor, s_v^*, h_v^*, and u_v^*, are given by

$$s_v^* - s_c = \frac{L}{T}, \tag{8.2.33}$$

$$h_v^* - h_c = L, \tag{8.2.34}$$

$$u_v^* - u_c = L - R_vT. \tag{8.2.35}$$

The density of moist air is written from (8.2.23) as

$$\rho = \frac{1}{q_d + \varepsilon^{-1}q_v}\frac{p}{R_dT} = \frac{1}{1 + (\varepsilon^{-1} - 1)q_v - q_c}\frac{p}{R_dT}, \tag{8.2.36}$$

which is rewritten as

$$\rho = \frac{p}{R_dT_v}, \tag{8.2.37}$$

where T_v is called the *virtual temperature*:

$$\begin{aligned} T_v &= (q_d + \varepsilon^{-1}q_v)T = [1 + (\varepsilon^{-1} - 1)q_v - q_c]T \\ &= (1 + 0.608q_v - q_c)T, \end{aligned} \tag{8.2.38}$$

where $\varepsilon = 0.622$ is used. The virtual temperature is used as a proxy of density (see Section 9.2).

Using (8.2.27), (8.2.30), and (8.2.8), enthalpy is written as

$$\begin{aligned} h &= q_dC_{pd}T + q_v(C_{pv}T + L_{00}) + q_cC_{pc}T \\ &= (q_dC_{pd} + q_wC_{pc})T + q_vL. \end{aligned} \tag{8.2.39}$$

From (8.2.29), (8.2.32), and (8.2.9), entropy is given as

$$\begin{aligned} s &= q_d\left(C_{pd}\ln\frac{T}{T_0} - R_d\ln\frac{p_d}{p_0}\right) + q_v\left(C_{pv}\ln\frac{T}{T_0} - R_v\ln\frac{p_v}{p_0^*} + \frac{L_0}{T_0}\right) \\ &\quad + q_cC_{pc}\ln\frac{T}{T_0} \\ &= (q_dC_{pd} + q_wC_{pc})\ln\frac{T}{T_0} - R_d\ln\frac{p_d}{p_0} + q_v\frac{L}{T} - q_vR_v\ln\frac{p_v}{p^*}. \end{aligned} \tag{8.2.40}$$

If we use further approximations $q_w C_{pc} \ll C_{pd}$ and $q_d \approx 1$, $L \approx L_0$, the enthalpy is reduced to

$$h = C_{pd}T + L_0 q_v. \tag{8.2.41}$$

This expression is conventionally used in meteorology for simplicity. Using the saturation condition $p_v = p^*$ and $p_d \approx p$, entropy is approximated to

$$\begin{aligned} s &= C_{pd} \ln \frac{T}{T_0} - R_d \ln \frac{p}{p_0} + \frac{L_0 q_v}{T} \\ &= C_{pd} \ln \frac{\theta}{T_0} + \frac{L_0 q_v}{T} \equiv C_{pd} \ln \frac{\theta_e}{T_0}, \end{aligned} \tag{8.2.42}$$

where θ is the potential temperature and θ_e is called the *equivalent potential temperature*: these two are defined by

$$\theta = T \left(\frac{p_0}{p} \right)^{\frac{R_d}{C_{pd}}}, \tag{8.2.43}$$

$$\theta_e = \theta \exp \left(\frac{L_0 q_v}{C_{pd} T} \right). \tag{8.2.44}$$

The adiabatic condition of saturated moist air is approximately expressed as $\theta_e =$ const. Similar to the virtual temperature (8.2.38), the *virtual potential temperature* can be introduced by

$$\theta_v = (q_d + \varepsilon^{-1} q_v)\theta = [1 + (\varepsilon^{-1} - 1)q_v - q_c]\theta = T_v \left(\frac{p_0}{p} \right)^{\frac{R_d}{C_{pd}}}. \tag{8.2.45}$$

The virtual potential temperature is used for consideration of the buoyancy effect of moist air.

8.2.4 Moist adiabat

In Section 8.1.7, we derived a general form of moist adiabat. Here, we write the approximate expressions of the moist adiabat conventionally used in meteorology. Eq. (8.1.112) can be rewritten by using the actual mixing ratio $r = n_w/n_d$ and the saturation mixing ratio $r^* = n_v/n_d$ as

$$\gamma_m = \gamma_d (1 + r^*) \frac{1 + r^* \frac{l}{R^* T}}{1 + \frac{r^* c_{pv} + (r - r^*) c_{pc}}{c_{pd}} + \frac{l^2}{c_{pd} R^* T^2} r^* (1 + r^*)}. \tag{8.2.46}$$

Normally, we neglect the contribution of liquid water and assume that the contribution of vapor is small to obtain an approximate form of moist adiabat. Assuming $r = r^* \ll 1$ in (8.2.46), we have an expression

$$\gamma_m = \gamma_d \frac{1 + \frac{l}{R^* T} r}{1 + \frac{c_{pv}}{c_{pd}} \left(1 + \frac{l^2}{c_{pv} R^* T^2} \right) r}. \tag{8.2.47}$$

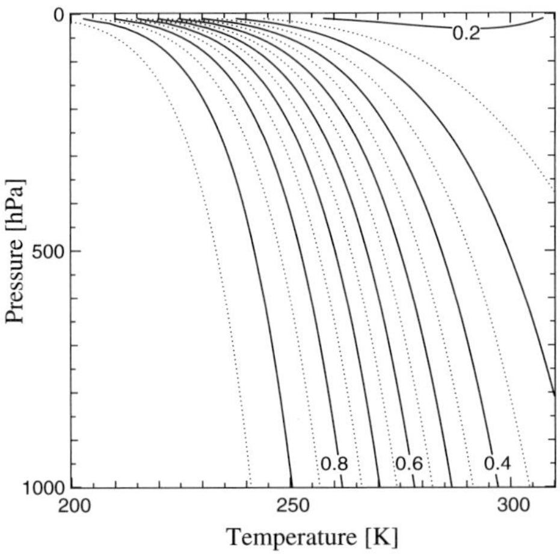

FIGURE 8.3: The ratio of moist adiabat to dry adiabat. The moist adiabat is calculated using the approximate formula (8.2.49). The contour interval of solid lines is 0.1.

In this approximation, the pseudo-moist adiabat agrees with the moist adiabat. Since the second term in parentheses in the denominator is estimated as

$$\frac{l^2}{c_{pv}R^*T^2} = \frac{\varepsilon L^2}{C_{pv}R_dT^2} \approx 98,$$

where $T = 273$ K is used, then the first term (i.e., 1) in parentheses can be neglected. If the temperature dependence of latent heat is neglected such that l is replaced by a constant l_0, and the specific humidity is approximated as $q_v = \varepsilon r/(1+\varepsilon r) \approx \varepsilon r$, we have

$$\gamma_m \approx \gamma_d \frac{1+\frac{l_0}{R^*T}r}{1+\frac{l_0^2}{c_{pd}R^*T^2}r} \approx \gamma_d \frac{1+\frac{L_0q_v}{R_dT}}{1+\frac{\varepsilon L_0^2 q_v}{C_{pd}R_dT^2}}. \tag{8.2.48}$$

Furthermore, if we use an approximation $q_v \approx \varepsilon p^*(T)/p$, we obtain

$$\gamma_m \approx \gamma_d \frac{1+\frac{\varepsilon L_0}{R_dT}\frac{p^*(T)}{p}}{1+\frac{\varepsilon^2 L_0^2}{C_{pd}R_dT^2}\frac{p^*(T)}{p}}. \tag{8.2.49}$$

The right-hand side is a function of (p, T). Fig. 8.3 shows the ratio of γ_m to γ_d using this expression.

From the approximate form of entropy (8.2.42), the condition that the equivalent potential temperature is constant $d\theta_e = 0$ is written as

$$ds = \frac{C_{pd}}{T}dT - \frac{R_d}{p}dp + \frac{L_0}{T}dq_v - \frac{L_0q_v}{T^2}dT. \tag{8.2.50}$$

Under the conditions $ds = 0$ and $q_v \approx \varepsilon p^*/p$, using (8.2.9), the moist adiabat is rewritten as

$$\gamma_m \quad \approx \quad \gamma_d \frac{1+\frac{L_0 q_v}{R_d T}}{1+\left(\frac{L_0}{R_v T}-1\right)\frac{L_0 q_v}{C_{pd} T}}. \tag{8.2.51}$$

We obtain the same expression as (8.2.48) by assuming $\frac{L_0}{R_v T} \gg 1$ in the parentheses.

Up to now, we have described changes in the thermodynamic state of an air parcel under the adiabatic process. These thermodynamic changes can be used as vertical distributions of thermodynamic functions if the atmosphere has an isentropic vertical structure. In this case, the atmosphere is neutrally stable as described in Section 2.3.

In the case that the atmosphere has an isentropic stratification and that the total amount of water is uniform with phase equilibrium, an infinitesimal difference between the thermodynamic quantities at two adjacent levels is written as

$$ds \;=\; 0, \quad dn_d \;=\; 0, \quad d(n_v+n_c) \;=\; 0, \quad \mu_v \;=\; \mu_c. \tag{8.2.52}$$

Using the thermodynamic relation and the hydrostatic balance

$$dh \quad = \quad Tds + \frac{1}{\rho}dp + \sum_k \mu_k dn_k, \tag{8.2.53}$$

$$0 \quad = \quad -\frac{1}{\rho}dp - d\Phi, \tag{8.2.54}$$

and (8.2.52), we have

$$d\sigma \quad \equiv \quad d(h+\Phi) \quad = \quad 0, \tag{8.2.55}$$

where

$$\sigma \quad \equiv \quad h + \Phi, \tag{8.2.56}$$

is *static energy*; (8.2.55) means that static energy is uniform in the saturated homogeneously mixed isentropic atmosphere. Using the expression of enthalpy (8.2.41) and geopotential $\Phi = gz$, static energy is written as

$$\sigma \quad = \quad C_{pd}T + L_0 q_v + gz \quad = \quad \sigma_d + L_0 q_v, \tag{8.2.57}$$

$$\sigma_d \quad \equiv \quad C_{pd}T + gz, \tag{8.2.58}$$

σ_d is called *dry static energy*, and (8.2.57) is specifically called *moist static energy*. The temperature lapse rate is given by

$$\frac{\partial T}{\partial z} \quad = \quad \left(\frac{\partial T}{\partial p}\right)_{s,n} \frac{\partial p}{\partial z} \quad = \quad -\rho g \gamma_m. \tag{8.2.59}$$

If the air is unsaturated, from (8.1.120), the lapse rate becomes that of the dry adiabat:

$$\frac{\partial T}{\partial z} \quad = \quad -\frac{g}{C_p}, \tag{8.2.60}$$

where C_p is the specific heat of moist air and gets contributions of water substance.

References and suggested reading

Iribarne and Godson (1981) is the most frequently cited textbook for the moist thermodynamics of meteorology. See also chapter 4 of Emanuel (1994). The thermodynamic expressions of moist air generally used in meteorology are based on a number of assumptions that are specific to the present terrestrial atmosphere. In the proto-atmosphere, for instance, the assumption of the ideal gas may not be applicable. Formulation of the thermodynamic expressions in this chapter follows the concept of Abe and Matsui (1988). The values of the thermodynamic variables are tabulated in List (1951).

Abe, Y. and Matsui, T., 1988: Evolution of an impact-generated H_2O-CO_2 atmosphere and formation of a hot proto-ocean on earth. *J. Atmos. Sci.*, **45**, 3081–3101.

Emanuel, K., 1994: *Atmospheric Convection.* Oxford University Press, New York, 580 pp.

Iribarne, J. V. and Godson, W. L., 1981: *Atmospheric Thermodynamics*, 2nd ed. D. Reidel, Dordrecht, 259 pp.

List, R. J., 1951: *Smithsonian Meteorological Tables*, 6th rev. ed. Smithsonian Institution Press, Washington D. C., 527 pp.

9

Basic equations of moist air

Following the description of the thermodynamic variables of moist air in the previous chapter, the governing equations of a moist atmosphere are derived in this chapter. The same assumptions as in Section 8.1.1 for moist air are used here (i.e., air parcels in moist air consist of dry air, vapor, and cloud particles).

The governing equations of dry air are described in Chapter 1. The present chapter is its counterpart for moist air. Moist air is described by the conservation laws of mass quantities, momentum, and energy, similar to dry air. In the case of moist air, since vapor is variable, the mass conservations of various components must be considered. This introduces the diffusion of mass. Energy transport is also associated with diffusion. The form of the equations of motion is, however, unchanged if we assume that all the constituents of water substance including the liquid phase have no relative motions to the air. In this chapter, only the diffusion of vapor is considered, though the treatment of diffusion is generally applicable to other incondensable gases such as ozone.

It is necessary to use the equations of a moist atmosphere to study atmospheric general circulation. In practice, however, the equations of a dry atmosphere can be used if the moist effect is introduced only as a diabatic heating due to latent heat release. In the final section, the equations of a moist atmosphere are approximated for conventional use; we will have two equivalent equation sets (i.e., dry equations with latent heat release and moist equations). In Chapters 12 and 15, for instance, we will use these two equation sets for consideration of the energetics and circulation of a moist atmosphere.

9.1 Conservation of mass variables

We assume as in the previous chapter that air parcels consist of dry air, vapor, and cloud particles, and that vapor and cloud particles are in phase equilibrium if the liquid phase exists. Dry air is composed of well-mixed ideal gases and is denoted by subscript d. Vapor and cloud particles are denoted by subscripts v

and c, respectively. The gas component of moist air is a mixture of dry air and vapor and is denoted by subscript g. Subscript k is used for general components irrespective of phases and components. The liquid phase that has relative motions with respect to the gas phase is regarded as rain. However, we do not consider any motions of rain in this chapter. The motions of rain and their interaction have important roles in meso-scale moist circulation. The treatment of the rain process can be seen in Ooyama (2001), for instance.

Let mass per unit volume, or density, of dry air, vapor, and cloud particles be denoted by ρ_d, ρ_v and ρ_c, respectively, and the velocity of the center of mass of each component by $\boldsymbol{v}_d$, $\boldsymbol{v}_v$, and $\boldsymbol{v}_c$, respectively. The conservation of mass of the k-th component is generally written as

$$\frac{\partial \rho_k}{\partial t} + \nabla \cdot \rho_k \boldsymbol{v}_k = S[\rho_k], \tag{9.1.1}$$

where $S[\rho_k]$ is a source term of the k-th component. Specifically, the conservations of the mass of dry air, vapor, and cloud particles are respectively written as

$$\frac{\partial \rho_d}{\partial t} + \nabla \cdot \rho_d \boldsymbol{v}_d = 0, \tag{9.1.2}$$

$$\frac{\partial \rho_v}{\partial t} + \nabla \cdot \rho_v \boldsymbol{v}_v = S[\rho_v], \tag{9.1.3}$$

$$\frac{\partial \rho_c}{\partial t} + \nabla \cdot \rho_c \boldsymbol{v}_c = -S[\rho_v], \tag{9.1.4}$$

where the source (sink) of vapor is equal to the sink (source) of cloud particles. If vapor is condensed to cloud particles, $S[\rho_v]$ is negative; if cloud particles evaporate, $S[\rho_v]$ is positive. The total mass per unit volume, or density of moist air, ρ, and velocity of center of total mass, or barycentric velocity, $\boldsymbol{v}$ are defined as

$$\rho \equiv \sum_k \rho_k = \rho_d + \rho_v + \rho_c, \tag{9.1.5}$$

$$\rho \boldsymbol{v} \equiv \sum_k \rho_k \boldsymbol{v}_k = \rho_d \boldsymbol{v}_d + \rho_v \boldsymbol{v}_v + \rho_c \boldsymbol{v}_v. \tag{9.1.6}$$

Therefore, the sum of (9.1.2)–(9.1.4) gives the conservation of total mass:

$$\frac{\partial \rho}{\partial t} + \nabla \cdot \rho \boldsymbol{v} = 0. \tag{9.1.7}$$

This is the familiar form of the continuity equation.

Using molecular weight m_k [kg mol^{-1}], density ρ_k [kg m^{-3}] is related to molar concentration n_k [mol kg^{-1}] and mass concentration q_k [kg kg^{-1}] as

$$\rho_k = \rho q_k = \rho n_k m_k, \qquad q_k = n_k m_k. \tag{9.1.8}$$

These quantities are expressed by the molar mixing ratio of total water r using (8.2.17)–(8.2.20). From (9.1.5), we have

$$\sum_k q_k = \sum_k n_k m_k = 1, \tag{9.1.9}$$

which is already given as (8.2.16). Let us define the difference of the velocities of each center of mass from the barycentric velocity of moist air by

$$\Delta_k \equiv \boldsymbol{v}_k - \boldsymbol{v}. \tag{9.1.10}$$

In the case of a gas component, this difference is due to diffusion. In the case of cloud particles, however, there is no diffusion in general, but we may have a difference of velocity due to gravity or inertia. Thus, from (9.1.6), we may write

$$\rho \boldsymbol{v} = \sum_k \rho_k (\boldsymbol{v} + \Delta_k), \tag{9.1.11}$$

and, from (9.1.5),

$$\sum_k \rho_k \Delta_k = 0. \tag{9.1.12}$$

If we define

$$\boldsymbol{i}_k \equiv \rho_k \Delta_k, \tag{9.1.13}$$

we also have

$$\sum_k \boldsymbol{i}_k = 0. \tag{9.1.14}$$

In the case of a gas component, $\boldsymbol{i}_k$ is called the density of *diffusion flux*. Using $\boldsymbol{i}_k$, (9.1.1) is rewritten as

$$\frac{\partial \rho_k}{\partial t} + \nabla \cdot (\rho_k \boldsymbol{v} + \boldsymbol{i}_k) = S[\rho_k]. \tag{9.1.15}$$

The equation of mass concentration q_k is given by substituting ρ_k (9.1.8) into (9.1.15) as

$$\frac{\partial (\rho q_k)}{\partial t} + \nabla \cdot (\rho q_k \boldsymbol{v} + \boldsymbol{i}_k) = S[\rho_k]. \tag{9.1.16}$$

This is rewritten in advective form by using the equation of mass (9.1.7) as

$$\rho \frac{dq_k}{dt} = -\nabla \cdot \boldsymbol{i}_k + S[\rho_k], \tag{9.1.17}$$

that is, for each component,

$$\rho \frac{dq_d}{dt} = -\nabla \cdot \boldsymbol{i}_d, \tag{9.1.18}$$

$$\rho \frac{dq_v}{dt} = -\nabla \cdot \boldsymbol{i}_v + S[\rho_v], \tag{9.1.19}$$

$$\rho \frac{dq_c}{dt} = -\nabla \cdot \boldsymbol{i}_c - S[\rho_v]. \tag{9.1.20}$$

The change in the mass of total water substance is given by the sum of (9.1.19) and (9.1.20):

$$\rho \frac{d(q_v + q_c)}{dt} = -\nabla \cdot (\boldsymbol{i}_v + \boldsymbol{i}_c) = \nabla \cdot \boldsymbol{i}_d. \qquad (9.1.21)$$

In the case $\boldsymbol{i}_c = 0$, cloud particles completely follow the motion of the gas component, while in the case $\boldsymbol{i}_c \neq 0$, cloud particles have relative velocity to the gas component. In the latter case, liquid water can be called rain and $\boldsymbol{i}_c$ corresponds to the precipitation flux. For simplicity, we only consider the case $\boldsymbol{i}_c = 0$.

9.2 Conservation of momentum

The momentum equation of moist air is apparently the same as that of dry air as long as liquid water moves with the gas component. The equation of motion is written as

$$\rho \frac{dv_i}{dt} = \frac{\partial \sigma_{ij}}{\partial x_j} - \rho g_i, \qquad (9.2.1)$$

where subscript i denotes the i-th component of the Cartesian coordinate, σ_{ij} is the stress tensor given by the sum of pressure tensor $-p\delta_{ij}$ and the residual σ'_{ij}:

$$\sigma_{ij} = -p\delta_{ij} + \sigma'_{ij}. \qquad (9.2.2)$$

In the case of moist air including cloud particles, σ'_{ij} can be different from the viscous stress tensor used for dry air, (1.2.17), since cloud particles may have contributions to spin angular momentum (see arguments in Section 1.3.1). In practice, however, no special treatment of the stress tensor is introduced for moist air and the expression of the stress tensor for dry air is used. The external force g_i is derived only from a gravitational potential field Φ as

$$g_i = \frac{\partial \Phi}{\partial x_i}. \qquad (9.2.3)$$

Eq. (9.2.1) can be rewritten to the flux-form conservation of momentum as

$$\frac{\partial}{\partial t}\rho v_i + \frac{\partial}{\partial x_j}(\rho v_i v_j + p\delta_{ij} - \sigma'_{ij}) = -\rho \frac{\partial \Phi}{\partial x_j}, \qquad (9.2.4)$$

which is the same as (1.2.18).

We can introduce a buoyancy force as the sum of the pressure gradient force and the gravity force. First, we need to define an arbitrary static state in the hydrostatic balance:

$$0 = -\frac{\partial p_s}{\partial x_i} - \rho_s g_i, \qquad (9.2.5)$$

where p_s and ρ_s are the pressure and the density of a reference state. Introducing perturbations from the reference state as $p = p_s + p'$ and $\rho = \rho_s + \rho'$, we rewrite the equation of motion as

$$\frac{dv_i}{dt} = -\frac{1}{\rho}\frac{\partial p'}{\partial x_i} - \frac{\rho'}{\rho}g_i + \frac{1}{\rho}\frac{\partial \sigma'_{ij}}{\partial x_j}, \tag{9.2.6}$$

The second term on the right-hand side is the buoyancy term. Using (8.2.37) under the condition $p \approx p_s$, the buoyancy term is approximately rewritten as

$$-\frac{\rho'}{\rho}g_i \approx \frac{T'_v}{T_{vs}}g_i, \tag{9.2.7}$$

where T_{vs} is the virtual temperature of the reference state and T'_v is the deviation from it. Thus, the virtual temperature plays a role of buoyancy when the effect of water is included. Since from (8.2.38)

$$T_v = (1 + 0.608q_v - q_c)T, \tag{9.2.8}$$

it can be seen that air is more buoyant as vapor q_v is more abundant. On the other hand, air is less buoyant as cloud particles q_c are more abundant; this is called the loading effect.

If one considers the rainwater that has a relative velocity with respect to the gas component, the motion of rainwater must be included in the equation set. However, the precise form of the equation of rainwater is very complicated since the motion of rainwater depends on its size. In practice, many assumptions have been introduced to treat the bulk motion of rainwater. In the numerical modeling of meso-scale convection, for instance, one may assume that some rainwater has a singular relative vertical velocity that is generally given by terminal velocity, while having the same horizontal velocity as the gas component. Ooyama (2001) formulates the conservative equation set of moist air which includes the relative motions of rainwater.

9.3 Conservation of energy and entropy

The conservation equation of the energy of moist air is the same as that of dry air, (1.2.42), if the internal energies of vapor and liquid water are included in the internal energy of moist air. Let u denote the internal energy of moist air per unit mass. We obtain the equation of internal energy as

$$\frac{\partial(\rho u)}{\partial t} + \nabla \cdot (\rho u \boldsymbol{v} + \boldsymbol{F}^{ene}) = -p\nabla \cdot \boldsymbol{v} + \varepsilon, \tag{9.3.1}$$

where ε is the dissipation rate and $\boldsymbol{F}^{ene}$ is the energy flux, but excluding the advective flux. In practice, $\boldsymbol{F}^{ene}$ consists of the energy flux due to local gradients of thermodynamic variables $\boldsymbol{F}^{therm}$ and radiative flux $\boldsymbol{F}^{rad}$:

$$\boldsymbol{F}^{ene} = \boldsymbol{F}^{rad} + \boldsymbol{F}^{therm}. \tag{9.3.2}$$

Using the equation of internal energy (9.3.1) and the continuity equation (9.1.7), we have

$$\rho\left(\frac{du}{dt} + p\frac{dv_s}{dt}\right) = \varepsilon - \nabla \cdot \boldsymbol{F}^{ene}, \tag{9.3.3}$$

where $v_s = 1/\rho$ is the specific volume. Using the thermodynamic relation,

$$du = Tds - pdv_s + \sum_k \mu_k dn_k, \tag{9.3.4}$$

we obtain

$$\rho\left(T\frac{ds}{dt} + \sum_k \mu_k \frac{dn_k}{dt}\right) = \varepsilon - \nabla \cdot \boldsymbol{F}^{ene}. \tag{9.3.5}$$

Since $\mu_v = \mu_c$ is satisfied based on the assumption that vapor and cloud particles are in phase equilibrium, using $n_k = q_k/m_k$ and (9.1.18)–(9.1.20), we have

$$\sum_k \mu_k \frac{dn_k}{dt} = -\sum_k \frac{\mu_k}{m_k}\nabla \cdot \boldsymbol{i}_k. \tag{9.3.6}$$

Using (9.1.14), the right-hand side can bc written as

$$-\sum_k \frac{\mu_k}{m_k}\nabla \cdot \boldsymbol{i}_k = -\left(\frac{\mu_d}{m_d} - \frac{\mu_v}{m_v}\right)\nabla \cdot \boldsymbol{i}_d. \tag{9.3.7}$$

So we can rewrite (9.3.5) by using (9.3.6) as

$$\rho T\frac{ds}{dt} = \sum_k \frac{\mu_k}{m_k}\nabla \cdot \boldsymbol{i}_k + \varepsilon - \nabla \cdot \boldsymbol{F}^{ene}. \tag{9.3.8}$$

Substituting (9.3.2) into $\boldsymbol{F}^{ene}$, we have

$$\rho T\frac{ds}{dt} = \varepsilon - \nabla \cdot \boldsymbol{F}^{rad} - \nabla \cdot \left(\boldsymbol{F}^{therm} - \sum_k \frac{\mu_k}{m_k}\boldsymbol{i}_k\right) - \sum_k \boldsymbol{i}_k \cdot \nabla \frac{\mu_k}{m_k}. \tag{9.3.9}$$

This is further rewritten as

$$\begin{aligned} \rho\frac{ds}{dt} = & -\frac{1}{T^2}\left(\boldsymbol{F}^{therm} - \sum_k \frac{\mu_k}{m_k}\boldsymbol{i}_k\right)\cdot \nabla T - \frac{1}{T}\sum_k \boldsymbol{i}_k \cdot \nabla \frac{\mu_k}{m_k} \\ & -\nabla \cdot \left[\frac{1}{T}\left(\boldsymbol{F}^{therm} - \sum_k \frac{\mu_k}{m_k}\boldsymbol{i}_k\right)\right] + \frac{\varepsilon}{T} - \frac{\nabla \cdot \boldsymbol{F}^{rad}}{T}. \end{aligned} \tag{9.3.10}$$

If there are no cloud particles, the first and the second terms on the right-hand side represent the production of entropy due to thermal and material diffusions, respectively. These terms must be positive according to the second law of thermodynamics. The third term represents the transport of entropy. The fourth term is the production of entropy due to dissipation and is positive. The entropy change due to radiation is represented by the last term, whose sign is not definite.

9.4 Transport process

For a pure gas without cloud particles, the requirement that the production of entropy due to thermal and material diffusions must be positive in (9.3.10) determines the expressions of fluxes duc to thermal and material diffusions (Landau and Lifshitz, 1987). These fluxes can be written in the form

$$\frac{\boldsymbol{i}_k}{m_k} = -\alpha_k \nabla \mu_k - \beta_k \nabla T, \tag{9.4.1}$$

$$\boldsymbol{F}^{therm} - \sum_k \mu_k \frac{\boldsymbol{i}_k}{m_k} = -\sum_k \beta_k T \nabla \mu_k - \gamma \nabla T, \tag{9.4.2}$$

where α_k, β_k, γ, and δ_k are the coefficients of transport. Eliminating $\nabla \mu_k$ from (9.4.1) and (9.4.2) and letting the coefficient of ∇T in $\boldsymbol{F}^{therm}$ be denoted by $-\kappa_T$, we obtain

$$\kappa_T = \gamma - \sum_k \frac{\beta_k^2}{\alpha_k} T. \tag{9.4.3}$$

Thus, the first and the second terms on the right-hand side of (9.3.10) are written as

$$-\frac{1}{T^2}\left(\boldsymbol{F}^{therm} - \sum_k \frac{\mu_k}{m_k}\boldsymbol{i}_k\right) \cdot \nabla T - \frac{1}{T}\sum_k \boldsymbol{i}_k \cdot \nabla \frac{\mu_k}{m_k}$$
$$= \frac{\kappa_T |\nabla T|^2}{T^2} + \sum_k \frac{1}{\alpha_k T} \frac{\boldsymbol{i}_k^2}{m_k^2}. \tag{9.4.4}$$

From this, we have the requirements that $\kappa_T > 0$ and $\alpha_k > 0$.

In (9.4.1), using

$$\nabla \mu_k = \left(\frac{\partial \mu_k}{\partial p}\right)_{T,n_k} \nabla p + \left(\frac{\partial \mu_k}{\partial T}\right)_{p,n_k} \nabla T + \sum_l \left(\frac{\partial \mu_k}{\partial n_l}\right)_{p,T} \nabla n_l, \tag{9.4.5}$$

the diffusion flux $\boldsymbol{i}_k$ is expressed as

$$\boldsymbol{i}_k = -\rho D_k \left[\nabla q_k + \frac{k_{T,k}}{T}\nabla T + \frac{k_{p,k}}{p}\nabla p\right], \tag{9.4.6}$$

where D_k is called the *diffusion coefficient*, and $k_{T,k}$ and $k_{p,k}$ are non-dimensional coefficients. These coefficients satisfy the following relations:

$$D_k = \frac{\alpha_k}{\rho}\left(\frac{\partial \mu_k}{\partial n_k}\right)_{p,T}, \tag{9.4.7}$$

$$\rho D_k \frac{k_{T,k}}{T} = \alpha_k \left(\frac{\partial \mu_k}{\partial T}\right)_{p,n_k} + \beta_k, \tag{9.4.8}$$

$$\rho D_k \frac{k_{p,k}}{p} = \alpha_k \left(\frac{\partial \mu_k}{\partial p}\right)_{T,n_k}. \tag{9.4.9}$$

In particular, we can see that diffusion coefficients are always positive. $k_{T,k}D_k$ and $k_{p,k}D_k$ are called the *thermal diffusion coefficient* and the *barodiffusion coefficient*, respectively. These two coefficients imply that material diffusion exists if there is a gradient of temperature or pressure. In the limits of a small or large mass concentration, however, these terms should vanish. In the case of $q_v \ll 1$ and $q_d \approx 1$, therefore, we may set

$$\boldsymbol{i}_k = -\rho D_k \nabla q_k. \tag{9.4.10}$$

As for the thermal diffusion flux $\boldsymbol{F}^{therm}$, eliminating $\nabla \mu_k$ from (9.4.1) and (9.4.2) and using (9.4.7) and (9.4.8), we obtain

$$\boldsymbol{F}^{therm} = \sum_k \left[k_{T,k} \left(\frac{\partial \mu_k}{\partial n_k} \right)_{T,p} - T \left(\frac{\partial \mu_k}{\partial T} \right)_{p,n_k} + \mu_k \right] \frac{\boldsymbol{i}_k}{m_k} - \kappa_T \nabla T. \tag{9.4.11}$$

From (1.1.12) and (1.1.5), we generally have

$$-T \left(\frac{\partial \mu_k}{\partial T} \right)_{p,n_k} + \mu_k = -T s_k + \mu_k = h_k, \tag{9.4.12}$$

Thus, (9.4.11) is rewritten as

$$\boldsymbol{F}^{therm} = \sum_k \left[\frac{k_{T,k}}{m_k} \left(\frac{\partial \mu_k}{\partial n_k} \right)_{T,p} + \frac{h_k}{m_k} \right] \boldsymbol{i}_k - \kappa_T \nabla T. \tag{9.4.13}$$

In the case of $q_v \ll 1$ and $q_d \approx 1$, we may set $k_{T,k} = 0$ as in (9.4.10); thus we have

$$\boldsymbol{F}^{therm} = \sum_k \frac{h_k}{m_k} \boldsymbol{i}_k - \kappa_T \nabla T, \tag{9.4.14}$$

where h_k/m_k is the enthalpy per unit mass of the k-th component.

The relative velocity of the mass center of cloud particles $\boldsymbol{i}_c$ cannot be written in the form of the diffusion flux. The expression of $\boldsymbol{i}_c$ depends on the definitions of cloud particles. Cloud particles have a size distribution in a fluid parcel, and individual cloud particles have their own falling speed depending on their size due to gravity and drag forces from environmental air. $\boldsymbol{i}_c$ is given as the sum of the falling speeds of all the cloud particles. In the case $\boldsymbol{i}_c \neq 0$, there is a contribution to the transport of energy from the motion of cloud particles. In a formal expression, the contribution of cloud particles, designated by subscript c, is added to the sum of the first term on the right-hand side of (9.4.14).

9.5 Approximate equations of moist air

The governing equations of moist air are made suitably approximate for application to the situations considered. For large-scale dynamics or general circulation modeling, one may use a simplified approximation in which vapor does not appear in the energy budget as long as phase change does not occur. The energy budget is formulated in the form of the change in dry enthalpy, and the moist effect is introduced only when phase change occurs. The effects of the mass of water substance are also taken into account in the buoyancy term in the equation of vertical motion.

In this subsection, the mass concentration of vapor is denoted by q instead of q_v, the diffusion flux by $\boldsymbol{i}$ instead of $\boldsymbol{i}_v$, and the condensation term by S_q instead of $S[\rho_v]$. The equation of vapor (9.1.19) is written as

$$\rho \frac{dq}{dt} = -\nabla \cdot \boldsymbol{i} + S_q. \tag{9.5.1}$$

When latent heat is large enough, the enthalpy of cloud particles is negligible compared with that of vapor. For simplicity, we also write the specific heat at the constant pressure of dry air as C_p instead of C_{pd}, and latent heat as L instead of L_0. The enthalpies of these components are approximated as

$$\frac{h_d}{m_d} = C_p T, \quad \frac{h_v}{m_v} = L, \quad \frac{h_c}{m_v} = 0. \tag{9.5.2}$$

The enthalpy per unit mass of moist air is given by

$$h = C_p T + Lq, \tag{9.5.3}$$

which is equivalent to (8.2.41). This indicates that we implicitly assume $q_d \approx 1$ and $q \ll 1$. To be consistent with this approximation, we assume that dry air does not diffuse; that is, $\boldsymbol{i}_d$ is neglected in the equation of thermal diffusion flux (9.4.14), which is rewritten as

$$\boldsymbol{F}^{therm} = L\boldsymbol{i} - \kappa_T \nabla T = \boldsymbol{F}^{lh} + \boldsymbol{F}^{sh}, \tag{9.5.4}$$

where we have defined

$$\boldsymbol{F}^{sh} = -\kappa_T \nabla T, \tag{9.5.5}$$
$$\boldsymbol{F}^{lh} = L\boldsymbol{i} = -\rho L \kappa_q \nabla q, \tag{9.5.6}$$

in which κ_q is the diffusion coefficient of vapor, $\boldsymbol{F}^{sh}$ is called *sensible heat flux*, and $\boldsymbol{F}^{lh}$ is called *latent heat flux*. From (9.3.3), the equation of enthalpy is derived as

$$\rho \frac{dh}{dt} = \frac{dp}{dt} + \varepsilon - \nabla \cdot \boldsymbol{F}^{rad} - \nabla \cdot \boldsymbol{F}^{therm}. \tag{9.5.7}$$

This can be rewritten as

$$\rho \frac{d}{dt}(C_p T + Lq) = \frac{dp}{dt} + \varepsilon - \nabla \cdot \boldsymbol{F}^{rad} - \nabla \cdot (L\boldsymbol{i} + \boldsymbol{F}^{sh}), \tag{9.5.8}$$

Multiplying L by (9.5.1) and substituting it into the above equation, we obtain

$$\rho \frac{d}{dt}(C_p T) = \frac{dp}{dt} + \varepsilon - \nabla \cdot \boldsymbol{F}^{rad} - \nabla \cdot \boldsymbol{F}^{sh} - LS_q, \tag{9.5.9}$$

where $-LS_q$ is the release of the latent heat of vapor. Let us write the diabatic terms as

$$Q_m = -\frac{LS_q}{\rho C_p}, \tag{9.5.10}$$

$$Q_d = \frac{1}{\rho C_p}[\varepsilon - \nabla \cdot (\boldsymbol{F}^{rad} + \boldsymbol{F}^{sh})]. \tag{9.5.11}$$

Q_d is the diabatic term of dry air defined by (1.2.54). Then, (9.5.9) is rewritten as

$$\rho \frac{d}{dt}(C_p T) = \frac{dp}{dt} + \rho C_p (Q_d + Q_m) = \frac{dp}{dt} + \rho C_p Q. \tag{9.5.12}$$

Thus, the equation of the energy of moist air is simply given by adding latent heat release Q_m to diabatic heating in the equation of dry air. In this case, the total diabatic term is written as

$$Q = Q_d + Q_m = \frac{1}{\rho C_p}[\varepsilon - \nabla \cdot (\boldsymbol{F}^{rad} + \boldsymbol{F}^{sh}) - LS_q]. \tag{9.5.13}$$

Mass exchange between the air and the ground surface takes place by means of evaporation and precipitation. Evaporation is the upward diffusion flux at the surface, while precipitation is the downward flux of raindrops at the surface. If we assume that all condensed water falls out from the atmosphere, precipitation is given by the vertical integral of the phase transform from vapor to cloud particles; this assumption is used when the pseudo-moist adiabatic process is considered. In this case, evaporation and precipitation are given, respectively, as

$$E_v \equiv i_z(z_0), \tag{9.5.14}$$

$$P_r \equiv -\int_{z_0}^{\infty} S_q \, dz. \tag{9.5.15}$$

Evaporation E_r is the mass inflow per unit area from the surface to the atmosphere, while precipitation P_r is the mass outflow per unit area from the atmosphere to the ground. In this expression, the contribution of cloud particles is neglected; it is assumed that all condensed water falls out to the surface as soon as vapor is converted to the liquid phase. Note that this is a highly idealistic assumption since the effects of the re-evaporation of raindrops is important in reality. A more precise approach is given by using the equations of cloud particles and raindrops with the transformation terms from cloud particles to raindrops. The downward flux of raindrops at the surface corresponds to precipitation.

The equation of moist internal energy $u = h - p/\rho$ is given from the equation of moist enthalpy (9.5.7). Adding the equation of moist internal energy to the equations of kinetic energy and potential energy, we obtain the equation of the total energy of moist air. Formally, the equation of moist internal energy is the same as (1.2.43), and the equation of total energy is the same as (1.2.47). In the moist case, the energy flux is given by

$$\boldsymbol{F}^{ene} = \boldsymbol{F}^{rad} + \boldsymbol{F}^{therm} = \boldsymbol{F}^{rad} + \boldsymbol{F}^{sh} + \boldsymbol{F}^{lh}. \tag{9.5.16}$$

Using this with (1.2.49), the equation of total energy (1.2.47) can be converted to

$$\frac{\partial}{\partial t}\left\{\rho\left(\frac{v^2}{2}+\Phi+u\right)\right\}$$
$$+\nabla\cdot\left\{\rho\boldsymbol{v}\left(\frac{v^2}{2}+\sigma\right)-v_j\sigma'_{ij}+\boldsymbol{F}^{rad}+\boldsymbol{F}^{sh}+\boldsymbol{F}^{lh}\right\} = 0, \quad (9.5.17)$$

where $\sigma = h + \Phi$ is the moist static energy defined by (8.2.56).

References and suggested reading

Detailed discussion on the diffusion process and its associated energy transport in Sections 9.1–9.4 can be seen in de Groot and Mazur (1984) and Landau and Lifshitz (1987). The discussion on the entropy budget of a moist atmosphere is more extensible if precipitation occurs (Pauluis and Held, 2002). For the formulation of the equation set of the moist atmosphere with rain, Ooyama (2001) and Bannon(2002) are referred to. As stated in Section 9.5, there are two methods for analyzing the energy budget of a moist atmosphere: dry enthalpy formulation and moist enthalpy formulation. This difference is described in Lorenz (1967) for the energy budget of the global atmosphere.

Bannon, P. R., 2002: Theoretical foundations for models of moist convection. *J. Atmos. Sci.*, **59**, 1967–1982.

de Groot, S.R. and Mazur, P., 1984: *Non-equilibrium Thermodynamics.* Dover, New York, 510 pp.

Landau, L. and Lifshitz, E. M., 1987: *Fluid Mechanics*, 2nd ed. Butterworth-Heinemann, 539 pp.

Lorenz, E. N., 1967: *The Nature and Theory of the General Circulation of the Atmosphere.* World Meteorological Organization, Geneva, 161 pp.

Ooyama, K. V., 2001: A dynamic and thermodynamic foundation for modeling the moist atmosphere with parameterized microphysics. *J. Atmos. Sci.*, **58**, 2073–2102.

Pauluis, O. and Held, I., 2002: Entropy budget of an atmosphere in radiative-convective equilibrium. Part II: Latent heat transport and moist processes. *J. Atmos. Sci.*, **59**, 140–149.

10

Radiation process

The theoretical bases of the *physical processes* for atmospheric modeling are described in this and the next chapters. Generally, the moist process, the radiation process, and the turbulent process are considered as the major physical processes for atmospheric modeling. These processes have their own deep and detailed theoretical backgrounds. In this book, however, from the view point of the atmospheric modeling of large-scale dynamics, we only describe the minimum theoretical requirements for these processes. First, the theory of the radiation process is briefly described in this chapter. The turbulent process will be described in the next chapter. Although the formulation of the moist process is given in the previous two chapters, it will be further explained in Chapter 15, where the roles of moist circulations on atmospheric structure are considered. We will calculate the radiative flux for the one-dimensional vertical model of the atmosphere in Chapter 14 using the formulation described in this chapter. However, we only use a simplified radiation process (i.e., a gray radiation model) to study the basic properties of atmospheric structure. Detailed calculation procedure and its application to realistic cases are not touched on in this book.

10.1 Blackbody radiation

Radiation inside a constant temperature enclosure in thermodynamic equilibrium with matter is called *blackbody radiation*. The state of blackbody radiation is described by the statistical equilibrium of photon gas. The statistical state of photon gas (hereafter referred to as light for simplicity) is specified by temperature T and frequency ν. Frequency ν is related to angular frequency ω, wave number k, and wavelength λ as

$$\omega = ck = 2\pi\nu = \frac{2\pi c}{\lambda}, \qquad (10.1.1)$$

where c is the speed of light. If light is in statistical equilibrium at temperature T, the distribution of the energy density of light per unit volume is given by Bose statistics:

$$dE_\nu = \frac{8\pi h}{c^3}\frac{\nu^3}{e^{\frac{h\nu}{k_B T}}-1}d\nu, \tag{10.1.2}$$

where h is the Planck constant and k_B is the Boltzmann constant (see Appendix A2). The energy flux of blackbody radiation is isotropic. The energy flux per unit solid angle within the frequency range $d\nu$ is given by

$$B_\nu d\nu = \frac{c}{4\pi}dE_\nu = \frac{2h\nu^3}{c^2}\frac{d\nu}{e^{\frac{h\nu}{k_B T}}-1}, \tag{10.1.3}$$

where B_ν is called the *Planck function.* It is related to the energy flux per unit wavelength B_λ as $B_\nu d\nu = B_\lambda d\lambda$. Then, since $\nu = c/\lambda$, we have

$$B_\lambda d\lambda = \frac{2hc^2}{\lambda^5}\frac{d\lambda}{e^{\frac{hc}{k_B T\lambda}}-1}. \tag{10.1.4}$$

It can be found that B_ν is maximum at the frequency

$$\nu_m = 2.82144\frac{k_B T}{h} = 5.87896\times 10^{10}T, \tag{10.1.5}$$

while B_λ is maximum at the wavelength

$$\lambda_m = \frac{1}{4.9651}\frac{hc}{k_B T} = \frac{2.89776\times 10^{-3}}{T} \tag{10.1.6}$$

(i.e., the wavelength of maximum radiation intensity for a blackbody is inversely proportional to temperature). This relation is called *Wien's displacement law.*

By integrating (10.1.3) over all wave numbers, we obtain the energy flux per unit solid angle of blackbody radiation as

$$\begin{aligned} B(T) &\equiv \int_0^\infty B_\nu d\nu = \frac{2h}{c^2}\int_0^\infty \frac{\nu^3 d\nu}{e^{\frac{h\nu}{k_B T}}-1} = \frac{2k_B^4T^4}{h^3c^2}\int_0^\infty \frac{x^3dx}{e^x-1} \\ &= \frac{\sigma_B}{\pi}T^4, \end{aligned} \tag{10.1.7}$$

where σ_B is the Stefan-Boltzmann constant:

$$\sigma_B = \frac{2\pi^5 k_B^4}{15c^2h^3} = 5.67051\times 10^{-8}\ \mathrm{W\ m^{-1}\ K^{-4}}. \tag{10.1.8}$$

Using this constant, we can express the energy flux per unit area on a plane radiated by a blackbody with temperature T as

$$F(T) = \int_0^{\pi/2} B(T)2\pi\cos\zeta\sin\zeta d\zeta = \pi B(T) = \sigma_B T^4, \tag{10.1.9}$$

where ζ is the angle between the direction of radiative flux and the direction normal to the plane.

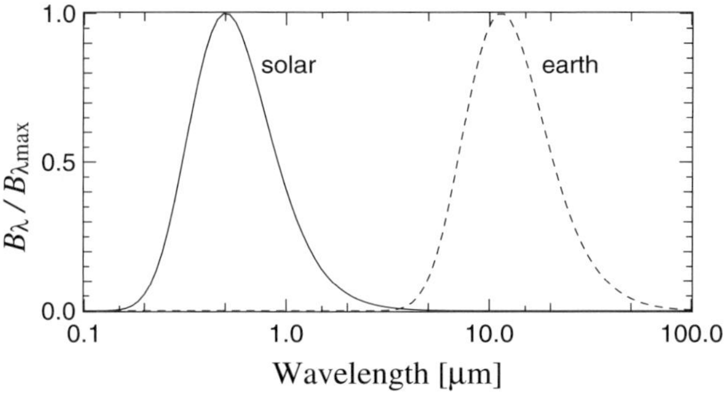

FIGURE 10.1: Schematic relation of the spectra of solar radiation (solid) and terrestrial radiation (dash) as functions of wavelength λ. The solar radiation is for the blackbody temperature at 5778 K, while the terrestrial radiation is for 255 K. The radiations are normalized by the respective maximum intensities.

10.2 Solar radiation and planetary radiation

The total radiative flux of the sun is $L_\odot = 3.846 \times 10^{26}$ W. The corresponding blackbody temperature is

$$T_\odot = \left(\frac{L_\odot}{4\pi R_\odot^2 \sigma_B}\right)^{\frac{1}{4}} = 5778 \text{ K}, \tag{10.2.1}$$

where $R_\odot = 6.960 \times 10^8$ m is the radius of the sun. $T_\odot$ is thought to be the representative temperature of the outermost surface of the sun, and is called the *effective temperature* of the sun (see Section 12.1.1). From Wien's displacement law, (10.1.6), the wavelength of maximum energy flux is about 0.50 μm. On the other hand, if one assumes the representative effective temperature of the earth to be 255 K as described in Section 12.1.1, the wavelength of the maximum energy flux for the earth is about 11.36 μm. The relation between the radiation spectra of the sun and the earth is shown in Fig. 10.1. Since the two spectra have little overlap in the wavelength, the two radiative fluxes are almost separable. The radiation of the earth is called *planetary radiation*. It is also called *infrared radiation* or *long-wave radiation*.

10.3 Absorption bands

A portion of solar radiative flux is absorbed or scattered in the atmosphere, and the remainder reaches the ground. As for radiative flux radiated at the terrestrial surface, some is also absorbed or scattered in the atmosphere, and the remainder goes out to space. The atmosphere itself also radiates in all directions according to its local temperature. The characteristics of absorption and scattering of solar and terrestrial radiation are summarized in Figs. 10.2 and 10.3. Absorption depends on the types of gas in the atmosphere. Each gas has its own characteristic absorbing

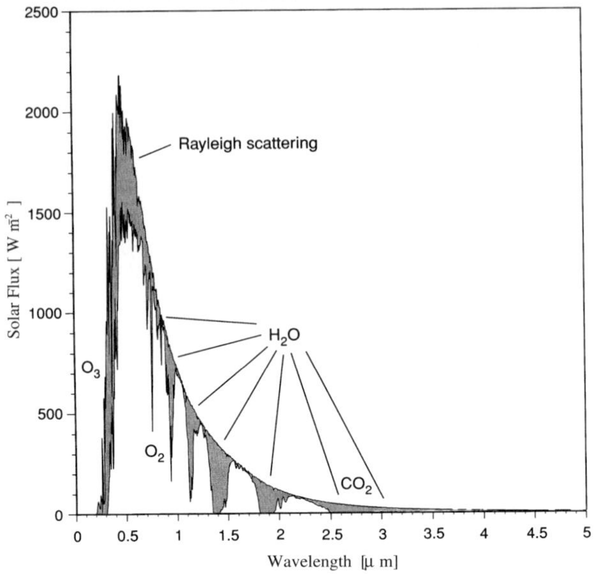

FIGURE 10.2: The absorption and scattering of solar radiation. The upper curve is the solar radiation at the top of the atmosphere, and the lower curve is that at the surface for a solar zenith angle of 60° in a clear sky atmosphere. Absorption and scattering regions are denoted in the figure. Reprinted from Liou (2002) by permission of Elsevier (copyright, 2003).

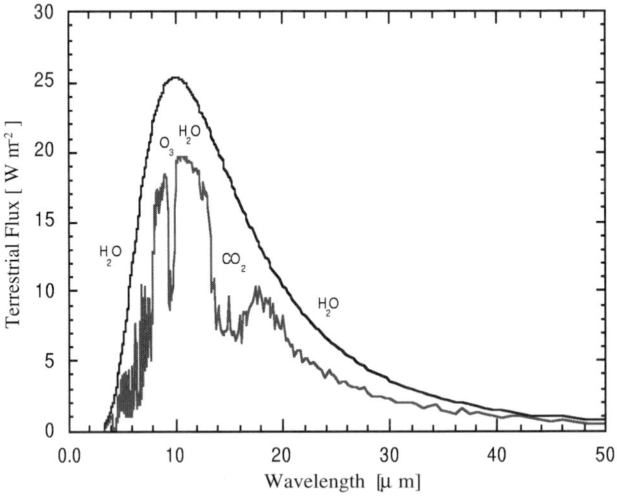

FIGURE 10.3: The absorption of terrestrial radiation. The upper curve is blackbody radiation at the surface. The lower curve is the long-wave radiation at the top of the atmosphere for globally cloudy condition. Major absorption gases are denoted in the figure. After Kiehl and Trenberth (1997) by permission of the American Meteorological Society.

bands depending on the wavelength of light. Scattering in the atmosphere also depends on the wavelength of light. As solar radiation penetrates through the atmosphere, therefore, the dependency on wavelength deviates from that of the Planck function. Terrestrial radiation likewise deviates in its dependency on wavelength from that of the Planck function, especially at the top of the atmosphere.

As shown in Figs. 10.2 and 10.3, the absorbing bands of solar radiation and terrestrial radiation are summarized as follows. Among the atmospheric gases, H_2O, CO_2, O_2, O_3, N, and N_2 have contributions for the absorption of solar radiation. In the ultraviolet radiation spectrum where the wavelength is smaller than 0.3 μm, O_3, O_2, N_2, O, and N have absorption bands. In the near-infrared radiation spectrum where the wavelength is about 0.74–3 μm, H_2O, CO_2, and O_3 have absorption bands; among them H_2O has the largest contribution. The central wavelengths of the main absorption bands in the visible and the near-infrared radiation spectrum are 0.94, 1.1, 1.38, 1.87, 2.7, 3.2, and 6.3 μm for H_2O, 1.4, 1.6, 2.0, 2.7, and 4.3 μm for CO_2, and 0.7 μm for O_2.

For the absorption of terrestrial radiation in which the wavelength is longer than 5 μm, H_2O, CO_2, and O_3 have primary contributions. The main absorption band of CO_2 is the vibration-rotation band centered at 15 μm. H_2O has a broad absorption band longer than 12 μm due to the rotation spectrum. The main absorption band of O_3 is the vibration-rotation spectrum centered at 9.6 μm.

10.4 Radiative transfer equation

If the absorption and scattering properties of gases and their distributions are known together with appropriate boundary conditions, one can calculate the energy flux and diabatic heating due to radiation in the atmosphere. In the following two sections, we formulate the radiative transfer equation for the calculation of the absorption and scattering of radiation.

The energy flux due to radiation per unit area per unit solid angle is called *radiance*, which is a function of spatial coordinates, direction, and frequency. The radiance at a point P in the direction of $\boldsymbol{s}$ with frequency ν is the radiative energy flux on a plane normal to the direction $\boldsymbol{s}$ per unit time, unit solid angle, and unit frequency; it is written as $I_\nu(P, \boldsymbol{s})$. Hereafter, the symbol P is omitted from the arguments. The radiance per unit frequency integrated over solid angles is denoted by I_ν, the radiance per unit solid angle integrated over frequency is denoted by $I(\boldsymbol{s})$, and the radiance integrated over solid angles o and frequency is denoted by I; we then have the relations:

$$I_\nu = \int I_\nu(\boldsymbol{s})\, do, \tag{10.4.1}$$

$$I(\boldsymbol{s}) = \int I_\nu(\boldsymbol{s})\, d\nu, \tag{10.4.2}$$

$$I = \int\int I_\nu(\boldsymbol{s})\, d\nu\, do. \tag{10.4.3}$$

In general, the change in radiance at a distance ds is given by

$$dI_\nu(\boldsymbol{s}) = -k_\nu \rho_i I_\nu(\boldsymbol{s})ds + j_\nu \rho_i ds, \tag{10.4.4}$$

where k_ν is called the *extinction coefficient* per unit mass, j_ν is called the *emission coefficient* per unit mass, and ρ_i is the density of absorbing quantities. The first term on the right-hand side of (10.4.4) is the decay of radiance due to the extinction of radiation. The extinction coefficient k_ν is written as the sum of the *absorption coefficient* k_ν^a and the *scattering coefficient* k_ν^s as

$$k_\nu = k_\nu^a + k_\nu^s. \tag{10.4.5}$$

The emission coefficient j_ν is also written as the sum of the emission of absorption j_ν^a and the emission of scattering j_ν^s as

$$j_\nu = j_\nu^a + j_\nu^s. \tag{10.4.6}$$

Generally, both k_ν and j_ν depend on direction $\boldsymbol{s}$. Dividing (10.4.4) by ds, or $-k_\nu \rho_i ds$, we have the *radiative transfer equation*:

$$\frac{dI_\nu(\boldsymbol{s})}{ds} = -k_\nu \rho_i I_\nu(\boldsymbol{s}) + j_\nu \rho_i, \tag{10.4.7}$$

or

$$-\frac{1}{k_\nu \rho_i}\frac{dI_\nu(\boldsymbol{s})}{ds} = I_\nu(\boldsymbol{s}) - J_\nu(\boldsymbol{s}), \tag{10.4.8}$$

where

$$J_\nu = \frac{j_\mu}{k_\nu} \tag{10.4.9}$$

is called the *source function.*

Let us first find a solution to the radiative transfer equation for the case when no source function exists: $J_\nu = 0$. In this case, (10.4.8) becomes

$$-\frac{1}{k_\nu \rho_i}\frac{dI_\nu(\boldsymbol{s})}{ds} = I_\nu(\boldsymbol{s}). \tag{10.4.10}$$

Integrating from $s = 0$ to s, we have

$$\begin{aligned} I_\nu(s;\boldsymbol{s}) &= I_\nu(0,\boldsymbol{s}) \exp\left(-\int_0^s k_\nu \rho_i ds\right) \\ &= I_\nu(0,\boldsymbol{s})e^{-\tau_\nu(0,s)} = I_\nu(0,\boldsymbol{s})\mathcal{T}_\nu(0,s), \end{aligned} \tag{10.4.11}$$

where we have introduced

$$\tau_\nu(s_0, s) = \int_{s_0}^s k_\nu \rho_i ds, \tag{10.4.12}$$

$$\mathcal{T}_\nu(s_0, s) = \exp\left(-\int_{s_0}^s k_\nu \rho_i ds\right) = e^{-\tau_\nu(s_0,s)}. \tag{10.4.13}$$

τ_ν is called the *optical path* and $\mathcal{T}_\nu$ is called the *transmission function.* As the absorption coefficient is larger or the absorbing quantity is more abundant such that the optical path is longer, the intensity of the radiative flux at the destination becomes exponentially weaker. This dependency is called *Beer-Bouger-Lambert's law.*

Next, we consider the case when the source function is given by blackbody radiation in thermodynamic equilibrium. If there is no scattering, the ratio of emitting radiation to absorbing radiation by the medium is independent of the properties of medium or the direction of radiation. In this case, the source function is given by

$$J_\nu = B_\nu(T). \tag{10.4.14}$$

B_ν is the Planck function, (10.1.3). This is called *Kirchhoff's law.* For application to atmospheric radiation, the air is not in thermodynamic equilibrium since temperature is not uniform. For almost every region except for the higher atmosphere, however, it can be said that the air is locally in thermodynamic equilibrium and that the source function is expressed by using a local temperature T. Such a case is called *local thermodynamic equilibrium (LTE)* under which Kirchoff's law is applicable. In a nonscattering atmosphere with LTE, substituting (10.4.14) into (10.4.8), we obtain the radiative transfer equation as

$$-\frac{1}{k_\nu \rho_i}\frac{dI_\nu(s;\boldsymbol{s})}{ds} = I_\nu(s;\boldsymbol{s}) - B_\nu(T(s)). \tag{10.4.15}$$

This equation can be rewritten by using the optical path $\tau_\nu(s', s)$. Since τ_ν is zero at s, and $d\tau_\nu = -k_\nu \rho_i ds$, we have

$$\frac{d}{d\tau_\nu}\left[I_\nu(\tau_\nu;\boldsymbol{s})e^{-\tau_\nu}\right] = -B_\nu(T(\tau_\nu))e^{-\tau_\nu}. \tag{10.4.16}$$

Integrating this relation from $s' = 0$ to s, or from $\tau_\nu = \tau_\nu(0, s)$ to $\tau_\nu(s, s)(= 0)$, we have the intensity,

$$I_\nu(s;\boldsymbol{s}) = I_\nu(0;\boldsymbol{s})e^{-\tau_\nu(0,s)} + \int_0^{\tau_\nu(0,s)} B_\nu(T(s'))e^{-\tau_\nu}d\tau_\nu. \tag{10.4.17}$$

This equation implies that the radiation at the destination point s is determined by two factors: the first is the exponentially decreasing term due to absorption within the path, and the second is the gain due to the emission from the medium in the path. If the second term on the right-hand side is negligible, Beer-Bouger-Lambert's law (10.4.11) is recovered.

As a third example of the radiative transfer equation, we consider the case where the scattering of radiation exists; the source function can be written as

$$\begin{aligned} J_\nu(\boldsymbol{s}) &= \frac{1}{k_\nu}(j_\nu^a + j_\nu^s) \\ &= \frac{1}{k_\nu}\left(k_\nu^a B_\nu(T) + k_\nu^s \int I_\nu(\boldsymbol{s}')P(\boldsymbol{s},\boldsymbol{s}')\frac{do'}{4\pi}\right). \end{aligned} \tag{10.4.18}$$

$P(\boldsymbol{s}, \boldsymbol{s}')$ is called the *phase function*, or the scattering probability distribution function, which represents the ratio of the scattering radiation in the direction $\boldsymbol{s}$ to the incident radiation from the direction $\boldsymbol{s}'$. It is normalized as

$$\int P(\boldsymbol{s}, \boldsymbol{s}')\frac{do'}{4\pi} = 1. \tag{10.4.19}$$

We also define the *albedo for single scattering* by

$$\tilde{\omega}_\nu \equiv \frac{k_\nu^s}{k_\nu}. \tag{10.4.20}$$

In particular, for the source function of solar radiation, we have from (10.4.18) with $B_\nu(T) = 0$ since the emission in the atmosphere is negligible,

$$J_\nu(\boldsymbol{s}) = \frac{\tilde{\omega}_\nu}{4\pi}\int I_\nu(\boldsymbol{s}')P(\boldsymbol{s}, \boldsymbol{s}')do'. \tag{10.4.21}$$

The phase function $P(\boldsymbol{s}, \boldsymbol{s}')$ is expressed in terms of angles (θ, φ) where θ is the angle between $\boldsymbol{s}$ and $\boldsymbol{s}'$, and φ is the azimuthal angle of $\boldsymbol{s}$ measured from an arbitrary base direction. It is further rewritten as $P(\mu, \varphi)$ with $\mu = \cos\theta$. In the case when the phase function has no dependency on azimuthal direction φ, the function $P(\mu)$ completely determines the properties of scattering. The phase function of *Rayleigh scattering* is an example, which is given by

$$P(\mu) = \frac{3}{4}(1 + \mu^2). \tag{10.4.22}$$

The molecular scattering of solar radiation in the atmosphere is described by Rayleigh scattering.

10.5 Infrared radiation in a plane parallel atmosphere

We consider the radiative transfer of a *plane parallel atmosphere* in which the abundances of absorbing quantities ρ_i and absorption coefficients k_ν are functions only of height z. The top of the atmosphere is designated by $z = \infty$ where the pressure is defined as $p = 0$. In this case, it is convenient to use *radiative depth* as a vertical coordinate, which is defined as

$$\tau_\nu(z) = \int_z^\infty k_\nu \rho_i dz. \tag{10.5.1}$$

Radiative depth is the radiative path measured from the top of the atmosphere $z = \infty$ down to height z. Although physical quantities are functions only of z, radiation depends on its direction in addition to z. Let us introduce the polar coordinates at each altitude. We define a unit vector in the direction of radiation by $\boldsymbol{s}$, the zenith angle by θ, and the azimuthal angle by φ. The radiative intensity $I_\nu(s; \boldsymbol{s})$ is written as $I_\nu(\tau_\nu; \mu, \varphi)$, where $\mu = \cos\theta$. The radiative transfer equation (10.4.8) is written as

$$\mu\frac{dI_\nu(\tau_\nu; \mu, \varphi)}{d\tau_\nu} = I_\nu(\tau_\nu; \mu, \varphi) - J_\nu. \tag{10.5.2}$$

In order to consider the radiation in the infrared spectrum region, we assume that the atmosphere has no scattering of radiation and is in local thermodynamic equilibrium. We also assume that radiative properties are independent of the azimuthal angle φ. Substituting (10.4.14) in (10.5.2), we have

$$\mu \frac{dI_\nu(\tau_\nu;\mu)}{d\tau_\nu} = I_\nu(\tau_\nu;\mu) - B_\nu(T(\tau_\mu)). \tag{10.5.3}$$

Integrating (10.5.3) from the bottom of the atmosphere to arbitrary depth τ_ν, we obtain the upward radiative intensity. In a similar manner, integrating from the top of the atmosphere to τ_ν, we obtain the downward radiative intensity. The two intensities are given respectively as

$$\begin{aligned} I_\nu^{\uparrow}(\tau_\nu,\mu) &= B_\nu(T_s)e^{-\frac{\tau_{\nu,s}-\tau_\nu}{\mu}} \\ &\quad + \int_{\tau_\nu}^{\tau_{\nu,s}} B_\nu(t)e^{-\frac{t-\tau_\nu}{\mu}}\frac{dt}{\mu}, \quad \text{for} \quad 0<\mu\le 1, \end{aligned} \tag{10.5.4}$$

$$I_\nu^{\downarrow}(\tau_\nu,\mu) = -\int_0^{\tau_\nu} B_\nu(t)e^{-\frac{t-\tau_\nu}{\mu}}\frac{dt}{\mu}, \quad \text{for} \quad -1\le\mu<0, \tag{10.5.5}$$

where $\tau_{\nu,s}$ is the total radiative depth from the top of the atmosphere to the bottom of the atmosphere. As for boundary conditions, we have assumed that the ground surface radiates as a blackbody with temperature T_s: $I_\nu^{\uparrow}(\tau_{\nu,s},\mu) = B_\nu(T_s)$. We also assumed that there is no incidence of infrared radiation from the top of the atmosphere: $I_\nu^{\downarrow}(0,\mu) = 0$.

Total radiance is given by the integral over frequency ν using (10.5.4) and (10.5.5). However, since the absorption coefficient k_ν generally has a strong dependency on ν, it is not easy to calculate the integration of radiance over ν. Thus, various approximations are used for integration in practice. In general, average transmittance, or *transmission function*, over a frequency range $\Delta\nu$ is introduced:

$$\mathcal{T}_{\Delta\nu}(u,u';\mu) = \frac{1}{\Delta\nu}\int_{\Delta\nu} e^{-\frac{|\tau_\nu(u)-\tau_\nu(u')|}{\mu}} d\nu. \tag{10.5.6}$$

The subscript $\Delta\nu$ denotes the average from ν to $\nu+\Delta\nu$. Here, u is an arbitrary vertical coordinate independent of frequency ν. For instance, the mass integral

$$u = \int_z^\infty \rho_i dz, \tag{10.5.7}$$

may be used. The choice of the width of the frequency range $\Delta\nu$ depends on the purpose; if $\Delta\nu$ represents the width of an absorption line, the averaging method is called "line by line"; if a group of adjacent absorbing bands is considered, $\Delta\nu$ is called the "narrow band"; and if all the frequency region is considered as one band, it is called the "broad band". In the following argument, we assume that B_ν is smooth enough over the bandwidth $\Delta\nu$. In this case, the averages of (10.5.4) and

(10.5.5) over $\Delta\nu$ respectively give

$$
\begin{aligned}
I^{\uparrow}_{\Delta\nu}(u,\mu) &= B_\nu(T_s)\mathcal{T}_{\Delta\nu}(u_s,u;\mu) \\
&\quad + \int_u^{u_s} B_\nu(t)\mathcal{T}_{\Delta\nu}(u',u;\mu)\frac{du'}{\mu}, \quad \text{for} \quad 0<\mu\le 1, \qquad (10.5.8) \\
I^{\downarrow}_{\Delta\nu}(u,\mu) &= \int_0^u B_\nu(t)\mathcal{T}_{\Delta\nu}(u',u;-\mu)\frac{du'}{-\mu}, \quad \text{for} \quad -1\le\mu<0, \qquad (10.5.9)
\end{aligned}
$$

where u_s corresponds to the value at the ground surface.

Next, we take integration over the direction of propagation of light. The integrations over the upper and lower hemispheres give

$$
\begin{aligned}
F^{\uparrow}_{\Delta\nu}(u) &\equiv \int_0^{2\pi}\int_0^{\pi/2} I_{\Delta\nu}(u;\boldsymbol{s})\cos\theta\sin\theta\,d\theta\,d\varphi \\
&= 2\pi\int_0^1 I_{\Delta\nu}(u;\mu)\mu\,d\mu. \qquad (10.5.10) \\
F^{\downarrow}_{\Delta\nu}(u) &\equiv -\int_0^{2\pi}\int_{\pi/2}^{\pi} I_{\Delta\nu}(u;\boldsymbol{s})\cos\theta\sin\theta\,d\theta\,d\varphi \\
&= 2\pi\int_0^{-1} I_{\Delta\nu}(u;\mu)\mu\,d\mu, \qquad (10.5.11)
\end{aligned}
$$

where the signs of $F^{\uparrow}_{\Delta\nu}$ and $F^{\downarrow}_{\Delta\nu}$ are chosen such that they are positive. These are called *upward* and *downward radiative flux density*. Substituting from (10.5.8) and (10.5.9), we obtain the total radiative flux density as

$$
\begin{aligned}
F_{\Delta\nu}(u) &\equiv F^{\uparrow}_{\Delta\nu}(u) - F^{\downarrow}_{\Delta\nu}(u) \\
&= \pi B_\nu(T_s)\mathcal{T}^f_{\Delta\nu}(u_s,u) + \int_{u_s}^{u} \pi B_\nu(u')\frac{d\mathcal{T}^f_{\Delta\nu}(u',u)}{du'}du' \\
&\quad - \int_0^u \pi B_\nu(u')\frac{d\mathcal{T}^f_{\Delta\nu}(u,u')}{du'}du', \qquad (10.5.12)
\end{aligned}
$$

where we have defined the transmission function of an air column as

$$
\begin{aligned}
\mathcal{T}^f_{\Delta\nu}(u,u') &= 2\int_0^1 \mu\mathcal{T}_{\Delta\nu}(u,u';\mu)d\mu \\
&= \frac{1}{\Delta\nu}\int_{\Delta\nu}\int_0^1 e^{-\frac{|\tau_\nu(u)-\tau_\nu(u')|}{\mu}} 2\mu d\mu d\nu. \qquad (10.5.13)
\end{aligned}
$$

Note that $\mathcal{T}^f_{\Delta\nu} = 1$ at $u = u'$. We rewrite (10.5.12) by integrating by parts as

$$
\begin{aligned}
F_{\Delta\nu}(u) &= \int_0^{u_s} \frac{d(\pi B_\nu(u'))}{du'}\,\mathcal{T}^f_{\Delta\nu}(u,u')\,du' \\
&\quad + [\pi B_\nu(T_s) - \pi B_\nu(u_s)]\;\mathcal{T}^f_{\Delta\nu}(u_s,u) \\
&\quad + \pi B_\nu(0)\;\mathcal{T}^f_{\Delta\nu}(u,0). \qquad (10.5.14)
\end{aligned}
$$

The first term on the right-hand side is a contribution from the inner range of an air column. The second term arises if surface temperature is different from the temperature at the bottom of the atmosphere. The third term is a contribution from the top of the atmosphere. Although surface temperature equals the temperature at the bottom of the atmosphere in a physical sense, we will consider the case when these two temperatures are different; if just the radiation process determines the thermal equilibrium of the atmosphere, the two contributions must be differently treated (see Section 14.3.1). The sum of radiative flux densities $F_{\Delta\nu}\Delta\nu$ over each of bands $\Delta\nu$ is total radiative flux density. This is the upward radiative flux per unit area on an arbitrary horizontal plane.

10.6 Gray radiation

As a first step to understanding the property of radiation, the assumption of *gray radiation* is introduced, where the absorbing coefficients k_ν^a are assumed to be independent of frequency ν. Since the absorption coefficients of atmospheric gases have a strong dependency on ν, the assumption of gray radiation does not quantitatively describe real atmospheric radiation. Nevertheless, many of the basic properties of radiation can be learned from gray radiation since the simplicity of the radiative transfer equation enables us to theoretically analyze solutions. In practice, only a part of planetary radiation is considered to be gray. To a similar level of approximation, it is further assumed that solar radiation is not absorbed in the atmosphere and directly reaches the surface where it is reflected or absorbed. Under these assumptions, we can construct a simple model of atmospheric radiation to study the vertical structure of the atmosphere (see Chapter 14).

We simply designate the optical depth as τ by omitting the subscript ν for gray radiation. Upward and downward radiative fluxes can be given by multiplying μ by (10.5.3) and integrating over μ:

$$2\pi\frac{d}{d\tau}\int_0^1 I_\nu(\tau,\mu)\mu^2\,d\mu = F_\nu^\uparrow(\tau) - \pi B_\nu(T(\tau)), \tag{10.6.1}$$

$$2\pi\frac{d}{d\tau}\int_0^{-1} I_\nu(\tau,\mu)\mu^2\,d\mu = F_\nu^\downarrow(\tau) - \pi B_\nu(T(\tau)). \tag{10.6.2}$$

Various approximations can be introduced for the integration over μ. The simplest assumption is that the intensity $I_\nu(\tau,\mu)$ is isotropic; in this case, we trivially have $F_\nu^\uparrow(\tau) = F_\nu^\downarrow(\tau) = \pi I_\nu(\tau)$ and no net radiative flux. The next approximation might be to take account of the asymmetry between the upward and downward directions of $I_\nu(\tau,\mu)$ and to assume that the intensity is isotropic in the upper and lower hemispheres. Then, under the assumption that $I_\nu(\tau,\mu)$ is independent of μ in each hemisphere, we may express

$$F_\nu^\uparrow(\tau) = \pi I_\nu(\tau,\mu=1), \tag{10.6.3}$$

$$F_\nu^\downarrow(\tau) = \pi I_\nu(\tau,\mu=-1), \tag{10.6.4}$$

$$2\pi \int_0^1 I_\nu(\tau,\mu)\mu^2\, d\mu = \frac{2\pi}{3} I_\nu(\tau,\mu=1) = \frac{2}{3} F_\nu^\uparrow(\tau), \tag{10.6.5}$$

$$2\pi \int_0^{-1} I_\nu(\tau,\mu)\mu^2\, d\mu = -\frac{2\pi}{3} I_\nu(\tau,\mu=-1) = -\frac{2}{3} F_\nu^\downarrow(\tau). \tag{10.6.6}$$

Thus, integrating (10.6.1) and (10.6.2) over the frequency ν yields

$$\frac{2}{3}\frac{dF^\uparrow(\tau)}{d\tau} = F^\uparrow(\tau) - \pi B(T(\tau)), \tag{10.6.7}$$

$$-\frac{2}{3}\frac{dF^\downarrow(\tau)}{d\tau} = F^\downarrow(\tau) - \pi B(T(\tau)), \tag{10.6.8}$$

where

$$F^\uparrow(\tau) \equiv \int_0^\infty F_\nu^\uparrow(\tau)\, d\nu, \quad F^\downarrow(\tau) \equiv \int_0^\infty F_\nu^\downarrow(\tau)\, d\nu, \tag{10.6.9}$$

$$\pi B(T) \equiv \int_0^\infty \pi B_\nu\, d\nu = \sigma_B T^4. \tag{10.6.10}$$

In a similar way to the previous section, we can obtain the radiative flux density by integrating (10.6.7) and (10.6.8) along the optical depth. For simplicity, we define $\tau^* = (3/2)\tau$. Using the boundary conditions $F^\uparrow = \pi B(T_s) = \sigma_B T_s^4$ at the surface and $F^\downarrow = 0$ at the top of the atmosphere, the integrations give

$$F^\uparrow(\tau) = \pi B(T_s)\mathcal{T}^f(\tau_s^* - \tau^*) - \int_{\tau^*}^{\tau_s^*} \pi B \frac{d\mathcal{T}^f(\tau^{*\prime} - \tau^*)}{d\tau^{*\prime}} d\tau^{*\prime}, \tag{10.6.11}$$

$$F^\downarrow(\tau) = \int_0^{\tau^*} \pi B \frac{d\mathcal{T}^f(\tau^* - \tau^{*\prime})}{d\tau^{*\prime}} d\tau^{*\prime}. \tag{10.6.12}$$

Here, the transmission function is given by

$$\mathcal{T}^f(\tau^{*\prime} - \tau^*) = e^{-(\tau^{*\prime} - \tau^*)}. \tag{10.6.13}$$

Similar to (10.5.14), net radiative flux is given by the partial integral of the difference between (10.6.11) and (10.6.12):

$$\begin{aligned} F(\tau) &= F^\uparrow(\tau) - F^\downarrow(\tau) \\ &= \int_0^{\tau_s^*} \frac{d\pi B(\tau^{*\prime})}{d\tau^{*\prime}}\, \mathcal{T}^f(|\tau^{*\prime} - \tau^*|)\, d\tau^{*\prime} \\ &\quad + [\pi B(T_s) - \pi B(\tau_s^*)]\, \mathcal{T}^f(\tau_s^* - \tau^*) + \pi B(0)\, \mathcal{T}^f(\tau^*). \end{aligned} \tag{10.6.14}$$

This expression is suitable for discretization of the numerical calculations of radiative flux.

In Section 14.3.1, we will calculate some atmospheric vertical structures using the radiative transfer equations of gray radiation, (10.6.7) and (10.6.8).

10.7 Radiative transfer equation of solar radiation

Calculations of the flux of solar radiation require distinction between direct radiation from the sun and radiation scattered by the atmosphere. The source function of the scattering part of solar radiation is given by (10.4.21):

$$J_\nu(\mu,\varphi) = \frac{\tilde{\omega}_\nu}{4\pi}\int_0^{2\pi}\int_{-1}^{1} I_\nu(\mu',\varphi')P(\mu,\varphi;\mu',\varphi')d\mu'\,d\varphi'. \tag{10.7.1}$$

Designating the incident angles of solar radiation as $(\mu_\odot, \varphi_\odot)$, we express the incident radiative intensity of solar radiation at the top of the atmosphere as

$$I_{\odot\nu}(0;\mu,\varphi) = F_{\odot\nu}\delta(\mu-\mu_\odot)\delta(\varphi-\varphi_\odot), \tag{10.7.2}$$

where $F_{\odot\nu}$ is the radiative flux density of solar radiation at the top of the atmosphere. In the atmosphere, there is no emission of radiative flux in the frequency range of solar radiation. Neglecting the scattered radiation in the direction of incident solar radiation, we have the expression for direct solar radiation using Beer-Bouger-Lambert's law, (10.4.11). Let the optical depth from the top of the atmosphere be denoted by τ. We have

$$I_{\odot\nu}(\tau;\mu,\varphi) = f_{\odot\nu}\delta(\mu-\mu_\odot)\delta(\varphi-\varphi_\odot), \tag{10.7.3}$$

where

$$f_{\odot\nu} = F_{\odot\nu}\mathcal{T}_\nu(0,\tau) = F_{\odot\nu}e^{-\frac{\tau}{\mu_\odot}}. \tag{10.7.4}$$

The intensity of solar radiation at the optical depth τ is described by the radiative transfer equation (10.5.2). We separate direct radiation $f_{\odot\nu}$ and the single scattering of solar radiation from further multiple scattering radiations; the radiation scattered by the atmosphere more than once is called *diffuse radiation*. Under these assumptions, the radiative transfer equation (10.5.2) becomes†

$$\begin{aligned}\mu\frac{dI_\nu(\tau;\mu,\varphi)}{d\tau} &= I_\nu(\tau;\mu,\varphi)\\ &-\frac{\tilde{\omega}_\nu}{4\pi}\int_0^{2\pi}\int_{-1}^{1} I_\nu(\tau;\mu',\varphi')P(\mu,\varphi;\mu',\varphi')d\mu'\,d\varphi'\\ &-\frac{\tilde{\omega}_\nu}{4\pi}P(\mu,\varphi;\mu_\odot,\varphi_\odot)f_{\odot\nu}.\end{aligned} \tag{10.7.5}$$

The last term on the right-hand side is the contribution from the single scattering of direct solar radiation at depth τ. The total solar radiative flux is given by the sum of $I_\nu(\tau;\mu',\varphi')$ and the direct radiation (10.7.3).

In the case of a plane parallel atmosphere, the radiative flux density of solar radiation can be calculated by the following procedure. We need to integrate the radiance over each of the upper and lower hemispheres in the directions. Since

†Strictly, this equation is given by averaging over the frequency region $\Delta\nu$. The optical depth is a mean depth over frequency $\Delta\nu$.

(10.7.5) depends on angles μ and φ, integration with respect to directions is not straightforward. Here, we introduce an approximation to help us derive the differential equations of the total radiative intensity and the upward and downward radiative fluxes. First, integrating (10.7.5) over the entire directions gives

$$\frac{d}{d\tau}F_\nu(\tau) = 4\pi(1-\tilde{\omega}_\nu)\overline{I}_\nu(\tau) - \tilde{\omega}_\nu f_{\odot\nu}, \tag{10.7.6}$$

where (10.4.19) is used and

$$\int_0^{2\pi}\int_{-1}^{1} I_\nu(\tau;\mu,\varphi)d\mu d\varphi = 4\pi\overline{I}_\nu(\tau), \tag{10.7.7}$$

$$\int_0^{2\pi}\int_{-1}^{1} I_\nu(\tau;\mu,\varphi)\mu d\mu d\varphi = F_\nu^\uparrow(\tau) - F_\nu^\downarrow(\tau) \equiv F_\nu(\tau) \tag{10.7.8}$$

(see (10.5.10) and (10.5.11)). Next, multiplying (10.7.5) by μ and integrating over directions, we obtain

$$\frac{d}{d\tau}\int_0^{2\pi}\int_{-1}^{1} I_\nu(\tau;\mu,\varphi)\mu^2 d\mu d\varphi = (1-\tilde{\omega}_\nu g)\,F_\nu(\tau) - \tilde{\omega}_\nu g\mu_\odot f_{\odot\nu}, \tag{10.7.9}$$

where g is called the *asymmetric factor*, given by

$$g \equiv \frac{1}{4\pi}\int P(0,0;\mu,\varphi)\mu do = \frac{1}{2}\int_{-1}^{1} P(\mu)\mu d\mu. \tag{10.7.10}$$

In this equation, we assume that the phase function $P(0,0;\mu,\varphi)$ is independent of φ and write it as $P(\mu) = P(0,0;\mu,\varphi)$. We have $g = 0$ if the scattering is isotropic, and $g = 1$ if just forward scattering exists. In addition, we generally have

$$\frac{1}{2}\int_{-1}^{1} P(\mu,\varphi;\mu',\varphi')\mu d\mu = g\mu'. \tag{10.7.11}$$

In the case of Rayleigh scattering, forward scattering $\mu > 0$ and backward scattering $\mu < 0$ equally contribute so that $g = 0$. In the case of *Mie scattering*, which describes scattering by a charged spherical particle, the contribution of forward scattering is larger than that of backward scattering such that $g > 0$. At the limit of an infinitesimal particle, Mie scattering agrees with Rayleigh scattering.

In order to obtain the radiative flux density $F_\nu(\tau)$, we again assume that $I_\nu(\tau;\mu,\varphi)$ is isotropic in each of the upper and lower hemispheres on the left-hand side of (10.7.9). In this case,

$$\begin{aligned}\int_0^{2\pi}\int_{-1}^{1} I_\nu(\tau;\mu,\varphi)\mu^2 d\mu d\varphi &= \frac{2\pi}{3}[I_\nu(\tau;\mu=1) + I_\nu(\tau;\mu=-1)] \\ &= \frac{4\pi}{3}\overline{I}_\nu(\tau).\end{aligned} \tag{10.7.12}$$

Using this to eliminate $\overline{I}_\nu(\tau)$ in (10.7.6) and (10.7.9), we obtain

$$\frac{d^2}{d\tau^2}F_\nu(\tau) = 3(1-\tilde{\omega}_\nu)(1-\tilde{\omega}_\nu g)F_\nu(\tau) - \tilde{\omega}_\nu f_{\odot\nu}[1 + 3(1-\tilde{\omega}_\nu)g\mu_\odot]. \tag{10.7.13}$$

The net radiative flux density $F_\nu(\tau)$ is given by the solution to this equation. Since direct solar radiation is not included in $F_\nu(\tau)$, the total radiative flux of solar radiation is given by the sum of $F_\nu(\tau)$ and

$$\int_0^{2\pi}\int_{-1}^{1} I_{\odot\nu}(\tau;\mu,\varphi)\mu d\mu d\varphi \quad = \quad f_\odot\mu_\odot. \tag{10.7.14}$$

The differential equations for upward and downward radiative intensity can also be constructed as follows. If the dependency on φ is neglected for simplicity, (10.7.5) can be rewritten as

$$\begin{aligned}\mu\frac{dI_\nu(\tau;\mu)}{d\tau} \quad &= \quad I_\nu(\tau;\mu) - \frac{\tilde{\omega}_\nu}{2}\int_{-1}^{1} I_\nu(\tau;\mu')P(\mu,\mu')d\mu' \\ &\quad -\frac{\tilde{\omega}_\nu}{4\pi}P(\mu,\mu_\odot)f_{\odot\nu}.\end{aligned} \tag{10.7.15}$$

Multiplying this equation by μ and integrating over each of the upper and lower hemispheres, we respectively obtain

$$\begin{aligned}&2\pi\frac{d}{d\tau}\int_0^1 I_\nu(\tau;\mu)\mu^2 d\mu \\ &\quad = \quad F_\nu^{\uparrow}(\tau) - \pi\tilde{\omega}_\nu\int_0^1\int_{-1}^{1} I_\nu(\tau;\mu')P(\mu,\mu')\mu d\mu'\, d\mu \\ &\qquad -\frac{\tilde{\omega}_\nu}{2}f_{\odot\nu}\int_0^1 P(\mu,\mu_\odot)\mu d\mu,\end{aligned} \tag{10.7.16}$$

$$\begin{aligned}&-2\pi\frac{d}{d\tau}\int_0^{-1} I_\nu(\tau;\mu)d\mu \\ &\quad = \quad F_\nu^{\downarrow}(\tau) - \pi\tilde{\omega}_\nu\int_0^{-1}\int_{-1}^{1} I_\nu(\tau;\mu')P(\mu,\mu')\mu d\mu'\, d\mu \\ &\qquad -\frac{\tilde{\omega}_\nu}{2}f_{\odot\nu}\int_0^{-1} P(\mu,\mu_\odot)\mu d\mu.\end{aligned} \tag{10.7.17}$$

These equations might be rewritten into various expressions if we make assumptions about dependency on directions. For instance, if intensities are isotropic in each of the upper and lower hemispheres, these are rewritten as

$$\begin{aligned}\frac{2}{3}\frac{d}{d\tau}F_\nu^{\uparrow}(\tau) \quad &= \quad F_\nu^{\uparrow}(\tau) - \tilde{\omega}_\nu(1-\beta)F_\nu^{\uparrow}(\tau) \\ &\quad -\tilde{\omega}_\nu\beta F_\nu^{\downarrow}(\tau) - \frac{\tilde{\omega}_\nu}{2}(1+S)f_{\odot\nu},\end{aligned} \tag{10.7.18}$$

$$\begin{aligned}-\frac{2}{3}\frac{d}{d\tau}F_\nu^{\downarrow}(\tau) \quad &= \quad F_\nu^{\downarrow}(\tau) - \tilde{\omega}_\nu(1-\beta)F_\nu^{\downarrow}(\tau) \\ &\quad -\tilde{\omega}_\nu\beta F_\nu^{\uparrow}(\tau) - \frac{\tilde{\omega}_\nu}{2}(1-S)f_{\odot\nu}.\end{aligned} \tag{10.7.19}$$

The sum of the two equations corresponds to (10.7.6) except for the factor 2/3 on the left-hand side. It can be shown that $\beta = (1-g)/2$ and $S = g\mu_\odot$ from an appropriate treatment of the scattering probability distribution function (Liou, 1992, section 3.3). In this case, (10.7.9) is given by the difference between (10.7.18) and (10.7.19).

We may have a more intuitive expression for solar flux by incorporating direct solar radiation into diffuse radiation. The following derivation contrasts with the one above in which direct and single scattering radiation are distinguished from diffuse radiation. If direct solar radiation is treated as diffuse radiation, we rewrite (10.7.13) as

$$\frac{d^2}{d\tau^2} F_\nu(\tau) \quad = \quad \alpha^2 F_\nu(\tau), \tag{10.7.20}$$

where $\alpha^2 = 3(1 - \tilde{\omega}_\nu)(1 - \tilde{\omega}_\nu g)$. If intensity is isotropic, (10.7.9) can be written as

$$\frac{3}{2}\frac{d}{d\tau}[F_\nu^\uparrow(\tau) + F_\nu^\downarrow(\tau)] \quad = \quad (1 - \tilde{\omega}_\nu g)\, F_\nu(\tau). \tag{10.7.21}$$

The top and bottom boundary conditions are given by

$$F_\nu^\downarrow(0) \quad = \quad F_\odot \mu_\odot, \qquad F_\nu^\uparrow(\tau_s) \quad = \quad 0, \tag{10.7.22}$$

where τ_s is total optical depth. The solution to (10.7.20) is written as

$$F_\nu(\tau) \quad = \quad F_\nu^\uparrow - F_\nu^\downarrow \quad = \quad Ce^{-\alpha\tau} + De^{-\alpha(\tau_s - \tau)}, \tag{10.7.23}$$

where C and D are constant. Substituting this into (10.7.21) and integrating over τ yields

$$F_\nu^\uparrow(\tau) + F_\nu^\downarrow(\tau) \quad = \quad \frac{3}{2\alpha}(1 - \tilde{\omega}_\nu g)\left[-Ce^{-\alpha\tau} + De^{-\alpha(\tau_s - \tau)}\right]. \tag{10.7.24}$$

Therefore, we have expressions for upward and downward fluxes as

$$F_\nu^\uparrow(\tau) \quad = \quad aCe^{-\alpha\tau} + be^{-\alpha(\tau_s - \tau)}, \tag{10.7.25}$$

$$F_\nu^\downarrow(\tau) \quad = \quad bCe^{-\alpha\tau} + ae^{-\alpha(\tau_s - \tau)}, \tag{10.7.26}$$

where

$$a \;=\; \frac{1}{2}\left[1 - \frac{3}{2\alpha}(1 - \tilde{\omega}_\nu g)\right], \qquad b \;=\; \frac{1}{2}\left[1 + \frac{3}{2\alpha}(1 - \tilde{\omega}_\nu g)\right]. \tag{10.7.27}$$

Using the boundary conditions (10.7.22), the coefficients in (10.7.25) and (10.7.26) are given by

$$C \;=\; \frac{b}{b^2 - a^2 e^{-2\alpha\tau_s}} F_\odot \mu_\odot, \qquad D \;=\; -\frac{ae^{-\alpha\tau_s}}{b^2 - a^2 e^{-2\alpha\tau_s}} F_\odot \mu_\odot. \tag{10.7.28}$$

From the above solution for solar radiation, we can calculate *planetary albedo.* The ratio of reflected solar radiation at the top of the atmosphere to incident solar radiation is denoted by planetary albedo $\mathcal{A}$, while the ratio of solar radiation that reaches the surface to incident solar radiation is denoted by *transmissivity* $\mathcal{T}$. These are expressed as

$$\mathcal{A} = \frac{F_\nu^\uparrow(0)}{F_\nu^\downarrow(0)} = B\frac{1-e^{-2\alpha\tau_s}}{1-B^2e^{-2\alpha\tau_s}}, \tag{10.7.29}$$

$$\mathcal{T} = \frac{F_\nu^\downarrow(\tau_s)}{F_\nu^\downarrow(0)} = (1-B^2)\frac{e^{-\alpha\tau_s}}{1-B^2e^{-2\alpha\tau_s}}, \tag{10.7.30}$$

where

$$B = \frac{a}{b} = \frac{\alpha - \frac{3}{2}(1-\tilde{\omega}_\nu g)}{\alpha + \frac{3}{2}(1-\tilde{\omega}_\nu g)}. \tag{10.7.31}$$

We have $\mathcal{A} \to B$ and $\mathcal{T} \to 0$ as the total optical depth becomes thicker $\tau_s \to \infty$.

References and suggested reading

Goody and Yung (1989) and Liou (1992) are standard textbooks on the radiation process. Important topics such as the details of absorption coefficients and band structures, the exact treatment of radiative transfer equations, and multiple scattering of solar radiation, which are not described in this book, can be found in these books. Goody (1995) summarizes the basics of the radiation process. Houghton (2002) gives not only elementary but also fundamental descriptions of the radiation process. The gray radiation in Section 10.6 and diffuse radiation in Section 10.7 follows Houghton (2002).

Goody, R., 1995: *Principles of Atmospheric Physics and Chemistry.* Oxford University Press, New York, 324 pp.

Goody, R. and Yung, Y., 1989: *Atmospheric Radiation*, 2nd ed. Oxford University Press, New York, 528 pp.

Houghton, J. T., 2002: *The Physics of Atmospheres*, 3rd ed. Cambridge University Press, Cambridge, UK, 320 pp.

Kiehl, J. T. and Trenberth, K. E., 1997: Earth's annual global mean energy budget. *Bull. Am. Meteorol. Soc.*, **78**, 197–208.

Liou, K. N., 1992: *Radiation and Cloud Processes in the Atmosphere.* Oxford University Press, New York, 487 pp.

Liou, K. N., 2002: *An Introduction to Atmospheric Radiation*, 2nd ed., International Geophysics Series, No. 83. Academic Press, New York, 487 pp.

11

Turbulence

Two aspects of the turbulence process in the atmosphere are described in this chapter. First, the statistical properties of turbulence are explained using similarity theory. In particular, the roles of turbulence in the general circulation of the atmosphere are briefly introduced. Similarity theory will be applied to describe a global-scale turbulent motion of the atmosphere in Section 12.4. The statistical properties of turbulence are used for the choice of numerical diffusion of the atmospheric general circulation models (Chapter 24). Second, we describe the turbulence in the mixed layer and the boundary layer of the atmosphere. We introduce turbulence models that are used to describe the subgrid-scale motions of numerical models. According to this second aspect, turbulence models in the mixed and boundary layers are viewed as one of the physical processes of atmospheric numerical models.

11.1 Similarity theory

11.1.1 Three-dimensional turbulence

Turbulence is a fluid state that has temporary and spatial irregular motion. If one wants to describe a larger scale fluid motion than the inner structure of turbulence itself, one needs to model the statistical properties of turbulence. In general, large-scale motion, or the environmental field of turbulence, does not possess both statistically isotopic and steady properties. The environmental field cannot be steady if energy is not supplied to the field, since turbulence loses its kinematic energy due to friction. Only if energy is supplied from the environment is the turbulent field kept statistically steady. In such a case, however, inhomogeneity which depends on external forcing is introduced to the field (i.e., the turbulence loses isotropy). This means that there does not exist a homogeneous isotropic turbulence in statistical equilibrium as a whole. In a sufficiently small-scale relative to the scale of external energy inputs, however, turbulence can be assumed to be statistically isotropic and homogeneous. Such a state is called *local statistical equilibrium.* The kinetic energy of turbulence is supplied from the environment in a large-scale range, transferred to a middle-scale range in the turbulence, and dissipated into heat due to friction in a sufficiently small-scale range. In local statistical equilibrium, energy supply, transfer, and dissipation are balanced by each other.

The local statistical equilibrium of turbulence can be described by characteristic variables which are related together by *similarity theory*. *Kolmogorov's first hypothesis* states that the statistical equilibrium of turbulence is spatially uniform and steady in the larger wave number range such that it only depends on two parameters: dissipation rate of energy ε and viscous coefficient ν. From isotropy, the turbulent kinetic energy spectrum $E(k)$ can be defined, where k is the magnitude of a wave number vector. $E(k)$ is related to the volume average of the kinetic energy of turbulence by

$$\frac{1}{V}\int \frac{1}{2}\boldsymbol{v}^2\,dV = \int_0^\infty E(k)\,dk, \tag{11.1.1}$$

where V is the volume of the domain considered. The physical dimensions of these variables are given respectively as

$$[\varepsilon] = \frac{L^2}{T^3}, \qquad [\nu] = \frac{L^2}{T}, \qquad [k] = \frac{1}{L}, \qquad [E(k)] = \frac{L^3}{T^2},$$

where the symbol [ ] denotes the dimension of the quantity inside the brackets, L is the dimension of length, and T is that of time. According to Kolmogorov's first hypothesis, the energy spectrum $E(k)$ can be expressed by ε and ν. From dimensional analysis, since

$$[k] = \left[\varepsilon^{\frac{1}{4}}\nu^{-\frac{3}{4}}\right], \qquad [E(k)] = \left[\varepsilon^{\frac{1}{4}}\nu^{\frac{5}{4}}\right],$$

we may write

$$E(k) = \varepsilon^{\frac{1}{4}}\nu^{\frac{5}{4}} f\left(\frac{k}{k_d}\right), \tag{11.1.2}$$

where $f(x)$ is an appropriate function of $x = k/k_d$, and

$$k_d = \left(\frac{\varepsilon}{\nu^3}\right)^{\frac{1}{4}} \tag{11.1.3}$$

is called the *Kolmogorov wave number*.

Kolmogorov's second hypothesis states that there is a range of spatial scale in a statistical equilibrium state of turbulence where the effect of viscosity is negligible; this scale is larger than the scale at which viscosity effectively works and is smaller than that of the energy input. The wave number range between the scale of energy input and the scale of viscous dissipation is called the *inertial subrange*. In the inertial subrange, energy input from smaller scale wave numbers is balanced by energy transfer to higher wave numbers. Under this hypothesis, $E(k)$ becomes independent of ν in the inertial subrange. We therefore have $f \propto \nu^{-\frac{5}{4}}$ from (11.1.2) and then

$$f\left(\frac{k}{k_d}\right) \propto \left(\frac{k}{k_d}\right)^{-\frac{5}{3}} = \nu^{-\frac{5}{4}}\varepsilon^{\frac{5}{12}}k^{-\frac{5}{3}}. \tag{11.1.4}$$

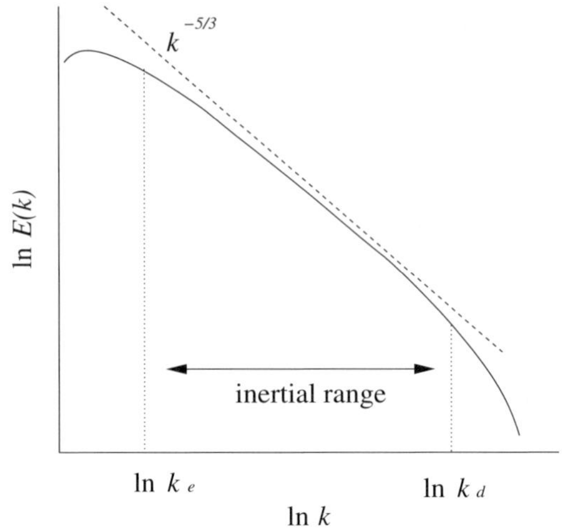

FIGURE 11.1: Schematic figure of the energy spectrum of three-dimensional turbulence. The energy spectrum of the inertial subrange follows Kolmogorov's spectrum (i.e., the five-thirds power law). k_e is the wave number at which energy is input, and k_d is the wave number at which dissipation effectively works (i.e., Kolmogorov wave number).

Thus, we obtain

$$E(k) \quad = \quad A\varepsilon^{\frac{2}{3}}k^{-\frac{5}{3}}, \tag{11.1.5}$$

where A is a universal constant, which has a value in the range 1.3–1.7 according to experimental surveys. This means that the energy spectrum is proportional to the $-5/3$-rd power of the wave number. This relation is referred to as the *five-thirds power law.* This dependency on wave number is a peculiar characteristic of *three-dimensional turbulence.*

The dissipation due to viscosity comes to dominate at the wave number k_d. Let k_e denote the wave number corresponding to the spatial scale of energy input. The inertial subrange exists if

$$Re \quad \equiv \quad \frac{\varepsilon^{\frac{1}{3}}}{\nu k_e^{\frac{4}{3}}} \quad = \quad \left(\frac{k_d}{k_e}\right)^{\frac{4}{3}} \quad \gg \quad 1 \tag{11.1.6}$$

is satisfied, where Re is the *Reynolds number.* Fig. 11.1 shows the schematic energy spectrum of three-dimensional turbulence. Energy input occurs at larger scales than the inertial subrange and the energy is transferred to smaller scales: $k \gg k_e$. At a sufficiently large wave number ($k \gg k_d$), the effect of viscosity becomes significant such that kinetic energy is dissipated into heat.

11.1.2 Two-dimensional turbulence

Next, let us consider the statistical equilibrium state of *two-dimensional turbulence* using similarity theory. The difference between two- and three-dimensional turbulence is that the *enstrophy* is conserved in the limit of small viscosity in the case of two-dimensional turbulence.

For the non-divergent two-dimensional fluid, the vorticity equation is written as

$$\frac{D\omega}{Dt} = \nu\nabla_H^2\omega, \tag{11.1.7}$$

where ω is vorticity and

$$\frac{D}{Dt} = \frac{\partial}{\partial t} + u\frac{\partial}{\partial x} + v\frac{\partial}{\partial y}, \tag{11.1.8}$$

$$\nabla_H^2 = \frac{\partial^2}{\partial x^2} + \frac{\partial^2}{\partial y^2}. \tag{11.1.9}$$

Vorticity ω is expressed by the streamfunction ψ as

$$\omega = -\nabla_H^2\psi. \tag{11.1.10}$$

Multiplying (11.1.7) by ψ or $\nabla_H^2\psi$ and averaging the products over the whole domain S, we obtain the following two equations:

$$\frac{dE}{dt} = -2\nu\Omega, \tag{11.1.11}$$

$$\frac{d\Omega}{dt} = -2\nu P, \tag{11.1.12}$$

where

$$E = \frac{1}{2S}\int (\nabla_H\psi)^2\, dS, \tag{11.1.13}$$

$$\Omega = \frac{1}{2S}\int\int \omega^2\, dS, \tag{11.1.14}$$

$$P = \frac{1}{2S}\int (\nabla_H\omega)^2\, dS. \tag{11.1.15}$$

E is total kinetic energy, or simply called energy. Ω is called *enstrophy*, and P is *palinstrophy*. In the inviscid case, $\nu = 0$, both energy and enstrophy are conserved.

In the case of infinitesimal nonzero viscosity with $\nu \to 0$ and $\nu \neq 0$, it can be shown that energy is still conserved but that enstrophy is not. Actually, enstrophy is finite since $\Omega(t) \leq \Omega(0)$ from (11.1.12) if $\nu > 0$. From (11.1.11), therefore, we have $\frac{dE}{dt} \to 0$ as $\nu \to 0$ (i.e., energy is conserved). On the other hand, from (11.1.7), the change in palinstrophy P is written as

$$\frac{dP}{dt} = -\frac{1}{S}\int \frac{\partial u_j}{\partial x_i}\frac{\partial \omega}{\partial x_i}\frac{\partial \omega}{\partial x_j}\, dS - \frac{\nu}{S}\int \left(\frac{\partial^2\omega}{\partial x_i \partial x_j}\right)^2 dS. \tag{11.1.16}$$

As $\nu \to 0$, the second term on the right-hand side approaches zero, while the first term does not necessarily vanish. Instead, P might diverge to infinity as $\nu \to 0$. If we define the dissipation rate of enstrophy by

$$\eta \equiv -\frac{d\Omega}{dt} = 2\nu P, \tag{11.1.17}$$

there is a possibility that $\eta \neq 0$ as $\nu \to 0$.

The energy spectrum of two-dimensional turbulence is different from that of three-dimensional turbulence. To see this, we study the time dependency of the energy spectrum for the case $\nu = 0$ in the first step. We define the kinetic energy spectrum density $E(k)$ and the enstrophy spectrum density $\Omega(k)$ for the wave number k ($k = |\boldsymbol{k}|$) so as to satisfy

$$E = \int_0^\infty E(k)\,dk, \qquad \Omega = \int_0^\infty \Omega(k)\,dk. \tag{11.1.18}$$

It can be shown from (11.1.13) and (11.1.14) by decomposing into Fourier components that the two spectrum densities are related as

$$\Omega(k) = k^2 E(k). \tag{11.1.19}$$

Let $\Delta E(k)$ denote the difference of the energy spectrum density from its initial value, and $\Delta\Omega(k)$ denote that of the enstrophy spectrum density. The conservation of energy and enstrophy is given by

$$\int_0^\infty \Delta E(k)\,dk = 0, \tag{11.1.20}$$

$$\int_0^\infty \Delta\Omega(k)\,dk = \int_0^\infty k^2 \Delta E(k)\,dk = 0. \tag{11.1.21}$$

These equations describe the conservation for the fully interactive cases between all the wave numbers.

For illustrative purpose, we consider the interaction of three components of waves in a statistical equilibrium state. The wave number vectors of three waves are denoted by $\boldsymbol{k_1}$, $\boldsymbol{k_2}$, and $\boldsymbol{k_3}$. If these three waves interact, the wave numbers must satisfy the relation:

$$\boldsymbol{k_1} + \boldsymbol{k_2} + \boldsymbol{k_3} = 0. \tag{11.1.22}$$

We assume that the magnitudes of wave numbers k_1, k_2, and k_3 are in the order $k_1 < k_2 < k_3$. In this case, the equations of the conservation of energy and enstrophy are written respectively as

$$\Delta E(k_1) + \Delta E(k_2) + \Delta E(k_3) = 0, \tag{11.1.23}$$

$$k_1^2 \Delta E(k_1) + k_2^2 \Delta E(k_2) + k_3^2 \Delta E(k_3) = 0. \tag{11.1.24}$$

Energy changes in the k_1- and k_3-components are solved in terms of the energy change in the k_2-component as

$$\Delta E(k_1) = -\frac{k_2^2 - k_3^2}{k_1^2 - k_3^2}\Delta E(k_2), \qquad \Delta E(k_3) = -\frac{k_2^2 - k_1^2}{k_3^2 - k_1^2}\Delta E(k_2). \tag{11.1.25}$$

This implies that both the energies of k_1- and k_3-components increase as the energy of the k_2-component decreases: $\Delta E(k_2) < 0$. The ratio of energy increases is given by

$$\frac{\Delta E(k_1)}{\Delta E(k_3)} = \frac{k_3^2 - k_2^2}{k_2^2 - k_1^2}. \tag{11.1.26}$$

In general, this ratio may be either greater or smaller than one. It is known, however, that the following inequality holds in many actually interacting cases:

$$\frac{\Delta E(k_1)}{\Delta E(k_3)} > 1, \qquad \frac{\Delta\Omega(k_1)}{\Delta\Omega(k_3)} = \frac{k_1^2 \Delta E(k_1)}{k_3^2 \Delta E(k_3)} < 1. \tag{11.1.27}$$

If energy is injected to the k_2-component, more of it is transferred to the smaller wave number k_1-component than the larger wave number k_3-component. In contrast, more enstrophy is transferred to the larger wave number k_3-component than the smaller wave number k_1-component. This implies that energy is transferred toward larger scales in two-dimensional turbulence. This situation is opposite to the three-dimensional turbulence, where injected energy is transferred toward smaller scales and, as a result, is converted to heat due to molecular viscosity. The energy transfer toward larger scales in two-dimensional turbulence is called the *inverse energy cascade.*

The local equilibrium theory of two-dimensional turbulence in the case of $\nu \neq 0$ can be constructed based on the above consideration. Different similarity theories of the energy spectrum are established for smaller and larger scale regimes relative to the scale of energy input. In the larger scale regime, the inverse energy cascade occurs. In this regime, if we assume that $E(k)$ can be expressed by wave number k and energy production rate ε, the energy spectrum is formulated as the same similarity theory of three-dimensional isotropic turbulence. From (11.1.5), we may have

$$E(k) = C_1 \varepsilon^{\frac{2}{3}} k^{-\frac{5}{3}}, \tag{11.1.28}$$

where C_1 is constant. In contrast, in the smaller scale regime, the dissipation of enstrophy occurs where the dominant parameters are ν and η. From dimensional analysis,

$$[\nu] = \frac{L^2}{T}, \qquad [\eta] = \frac{1}{T^3},$$

then we have

$$E(k) = \eta^{\frac{1}{6}} \nu^{\frac{3}{2}} F\left(\frac{k}{k_d}\right), \tag{11.1.29}$$

$$k_d = \eta^{\frac{1}{6}} \nu^{-\frac{1}{2}}. \tag{11.1.30}$$

As in the case of three-dimensional turbulence, we may assume the existence of an inertial subrange that is independent of viscosity. In this case, it may be referred to as the enstrophy inertial subrange. Thus, we obtain

$$F\left(\frac{k}{k_d}\right) \propto \left(\frac{k}{k_d}\right)^{-3} = \nu^{-\frac{3}{2}} \eta^{\frac{1}{2}} k^{-3}, \tag{11.1.31}$$

and the energy spectrum is given by

$$E(k) = C_2 \eta^{\frac{2}{3}} k^{-3}, \tag{11.1.32}$$

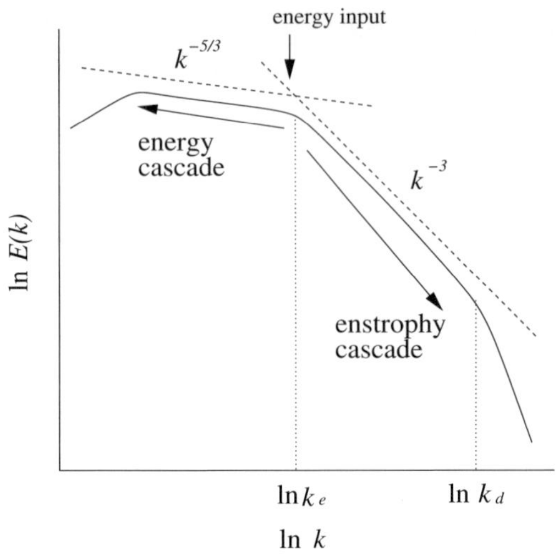

FIGURE 11.2: Schematic energy spectrum of two-dimensional turbulence. The energy spectrum is proportional to $k^{-\frac{3}{5}}$ in the range $k < k_e$, where k_e^{-1} represents the scale of the energy input, and is proportional to k^{-3} in the range $k > k_e$.

where C_2 is a universal constant. From (11.1.19), the enstrophy spectrum $\Omega(k)$ is also given by

$$\Omega(k) = C_2 \eta^{\frac{2}{3}} k^{-1}. \tag{11.1.33}$$

Fig. 11.2 schematically shows the energy spectrum of two-dimensional turbulence. Energy is cascaded to the larger scale relative to the scale of the energy input k_e, whereas enstrophy is cascaded to the smaller scale. Precisely, the assumption of the local equilibrium might be no longer valid in the energy inertial subrange $k < k_e$ of two-dimensional turbulence. This is because the homogeneity of the turbulence is not satisfied since energy is injected at the spatial scale comparable with turbulence. If energy input occurs at random, however, homogeneity can again be assumed in the range with the spatial scale much larger than k_e^{-1} and much smaller than the domain length of the system. The upward energy cascade will continue until the scale of the turbulent eddy becomes the domain size. It is thought that the spectrum of two-dimensional turbulence in the enstrophy inertial subrange corresponds to the energy spectrum observed in the large-scale motion of the atmosphere (see Sections 17.4.2 and 24.2).

11.1.3 Tracer spectrum

Returning back to three-dimensional flow, we consider in this section the spectrum of a tracer which is passively advected by turbulent flow and is also subjected to molecular diffusion. Here, we assume that the tracer does not exert forces on the flow; in this case the tracer is called a *passive tracer*. The mass concentration of a minor quantity can be thought of as such a passive tracer in particular conditions. Suppose that the evolution of the tracer concentration θ is governed by the following advection-diffusion equation:

$$\frac{\partial \theta}{\partial t} + \boldsymbol{u} \cdot \nabla \theta = \kappa \nabla^2 \theta, \tag{11.1.34}$$

where κ is the diffusion coefficient. Let $\overline{(\ )}$ denote the domain average, and $\theta' = \theta - \overline{\theta}$ denote the deviation of the tracer from its average. We assume that the domain averaged tracer $\overline{\theta}$ is conserved. Multiplying this equation by θ, and averaging over the whole domain, we have

$$\frac{\partial \overline{\theta'^2}}{\partial t} = -2\kappa \overline{(\nabla \theta')^2}. \tag{11.1.35}$$

In order to consider the wave number dependency of the spectrum of θ'^2, we introduce the spectrum density $\Gamma(k)$ that satisfies

$$\overline{\theta'^2} = \int_0^\infty \Gamma(k) dk. \tag{11.1.36}$$

It is thought that the tracer spectrum is described by the following two groups of parameters. The first group comprises the parameters of the turbulence of the flow: the dissipation rate of energy ε and the viscous coefficient ν. The second group comprises the parameters that appear in the equation of the tracer: the diffusion coefficient κ and the dissipation rate of the tracer:

$$\chi \equiv 2\kappa \overline{(\nabla \theta')^2}. \tag{11.1.37}$$

The typical wave number at which the diffusion of tracer becomes important may depend on the relative magnitude of ν and κ. In the case $\kappa \gg \nu$, the effect of diffusion becomes greater if the wave number is larger than

$$k_{\kappa 1} \equiv \left(\frac{\varepsilon}{\kappa^3}\right)^{\frac{1}{4}}. \tag{11.1.38}$$

This wave number is derived from dimensional analysis in the same way that the Kolmogorov wave number k_d in (11.1.3) is derived.

In the case $\nu \gg \kappa$, on the other hand, the main process of turbulence governing the spectrum of the tracer is different between the wave number ranges $k \ll k_d$ and $k \gg k_d$. In the range $k \ll k_d$, there exists the inertial subrange of turbulence; advection due to turbulence is important while diffusion is almost negligible. In the range $k \gg k_d$, the turbulence of the flow is suppressed by viscosity and the flow field is characterized by a deformation field.† Since $\nu \gg \kappa$, the effect of diffusion is important at wave numbers sufficiently larger than k_d. Thus, there is another inertial subrange of the tracer spectrum in $k \ll k_d$ which has a different spectrum slope from that of the inertial subrange of energy.

---

†The deformation field is determined by the symmetric part of the deformation tensor $\frac{\partial u_j}{\partial x_i}$. For two-dimensional flow, it is characterized by $\frac{\partial u}{\partial x}$, $\frac{\partial v}{\partial y}$, and $\frac{\partial u}{\partial y} + \frac{\partial v}{\partial x}$.

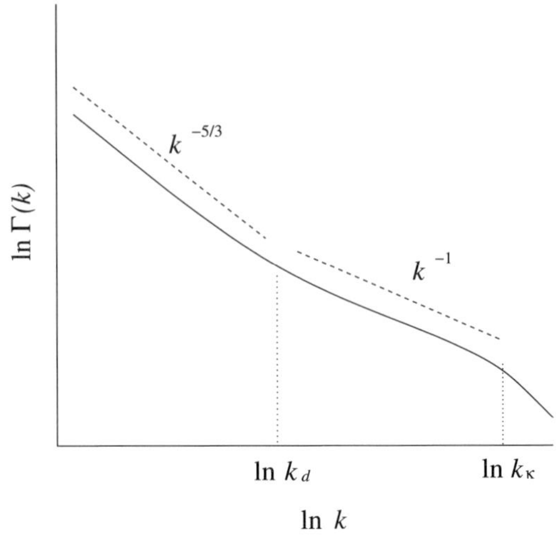

FIGURE 11.3: The schematic distribution of the tracer spectrum in the case $\nu \gg \kappa$. The spectrum is proportional to $k^{-\frac{3}{5}}$ in the smaller wave number range than k_d, whereas the spectrum is proportional to k^{-1} in the range between k_d and k_κ.

Since the transfer of the tracer in the wave number space is constant as given by χ, the dominant parameters of the tracer spectrum in the range $k \ll k_d$ are χ, ε, and k. Dimensional analysis gives

$$\Gamma(k) \quad \propto \quad \chi \varepsilon^{-\frac{1}{3}} k^{-\frac{5}{3}}. \tag{11.1.39}$$

This has a wave number dependency similar to (11.1.5): the five-thirds power law.

In contrast, the strain rate of the flow in the region $k \gg k_d$ is estimated as

$$\left|\frac{\partial u_j}{\partial x_i}\right| \quad \approx \quad \left(\frac{\varepsilon}{\nu}\right)^{\frac{1}{2}} \quad \equiv \quad \gamma. \tag{11.1.40}$$

Thus, the effect of diffusion becomes dominant in the wave number range larger than

$$k_\kappa \quad = \quad \left(\frac{\gamma}{\kappa}\right)^{\frac{1}{2}} \quad = \quad \left(\frac{\varepsilon}{\nu\kappa^2}\right)^{\frac{1}{4}}. \tag{11.1.41}$$

The tracer spectrum in the range $k_d \gg k \gg k_\kappa$ is determined by χ, κ, and γ. From dimensional analysis, we may have

$$\Gamma(k) \quad = \quad \chi \kappa^{\frac{1}{2}} \gamma^{-\frac{3}{2}} g\left(\frac{k}{k_\kappa}\right), \tag{11.1.42}$$

where g is an arbitrary function. In order for Γ to be independent of κ, we need to have $g(n) \propto n^{-1}$. Therefore, we obtain the spectrum of the other type of the inertial subrange of the tracer as

$$\Gamma(k) \quad \propto \quad \chi \gamma^{-1} k^{-1} \tag{11.1.43}$$

(i.e., the spectrum is proportional to the inverse of the wave number). This is called the *Batchelor spectrum*. An example of the spectrum of tracer variance is schematically shown in Fig. 11.3.

11.2 Turbulence models

In the rest of this chapter, the turbulence models used for the numerical models of atmospheric flows are described. Since the resolutions of the numerical models of the atmosphere are generally insufficient to resolve all turbulent flows, smaller scale flows than the resolvable scale of numerical models must be modeled or parameterized to be incorporated into numerical models. Usually, such flows in the unresolvable scale are regarded as turbulence, and are modeled by introducing appropriate closures. We first overview the general theory of the turbulence models used in fluid dynamics. We next describe the *Mellor and Yamada model*, which consists of a hierarchy of the different complexity of turbulent closures. In the next section, we further consider the turbulence models of planetary boundary layers.

11.2.1 Basic equations

In order to numerically consider the evolutions of flows whose spatial scale is larger than turbulence, we need to use filtered equations that are appropriately averaged for the description of mean flows. Let the spatial resolution of a numerical model be denoted by L. If L is within the inertial subrange defined in Section 11.1.1, we can assume that unresolvable flows smaller than L, or subgrid-scale flows, are described by the statistical equilibrium of turbulence. Such a numerical treatment of subgrid-scale flows is called the *large eddy simulation* (LES). For atmospheric flows, in general, it is thought that LES can be applicable if L is about 10 m to 100 m. Thus, LES is only used for simulations of very small-scale flows such as in the boundary layer.

On the other hand, L is about a few kilometers for cloud-resolving models, about 10 km for meso-scale models, and about 100 km for general circulation models. There is no statistical theory for subgrid-scale flows of such large-scale models $L \approx 100$ km. In these cases, the equations for the mean flows are generally based on the statistical average of quantities; the statistical average is defined as an average over a number of experiments under specific conditions. This kind of average is called the *ensemble mean*. In the following argument, we mainly explain filtered equations that are based on the ensemble mean. The ensemble mean of a quantity A is denoted by $\overline{A}$, and the deviation from the average by A': $A = \overline{A} + A'$.

We use the following set of basic equations in the Boussinesq approximation:

$$\frac{\partial}{\partial x_i} u_i = 0, \tag{11.2.1}$$

$$\frac{du_j}{dt} + \varepsilon_{jkl} f_k u_l = -\frac{1}{\rho}\frac{\partial p}{\partial x_j} + \alpha g_j \tilde{\theta}_v + \nu \nabla^2 u_j, \tag{11.2.2}$$

$$\frac{d\theta}{dt} = \kappa \nabla^2 \theta. \tag{11.2.3}$$

ν is the viscous coefficient, κ is the thermal diffusivity, g_j is the vector component of gravity $\boldsymbol{g} = (0, 0, g)$, and f_j is that of the Coriolis force. ε_{jkl} is the antisymmetric tensor. The density ρ is constant and the buoyancy term is approximated as $-\alpha g_j \tilde{\theta}_v$ using the thermal expansion coefficient α. $\tilde{\theta}_v$ is the deviation of virtual potential temperature from its reference value in the hydrostatic balance. The virtual potential temperature is defined by (8.2.45), which is the potential temperature with the buoyancy effect of vapor. the equation of vapor or any other tracer is also governed by equations similar to (11.2.3).

Let the average of velocity components be $U_j = \overline{u_j}$, that of pressure be $P = \overline{p}$, that of potential temperature be $\Theta = \overline{\theta}$, that of the deviation of virtual potential temperature be $\tilde{\Theta}_v = \overline{\tilde{\theta}_v}$. We obtain filtered equations by

$$\frac{\partial}{\partial x_i} U_i = 0, \tag{11.2.4}$$

$$\frac{DU_j}{Dt} + \varepsilon_{jkl} f_k U_l = -\frac{\partial}{\partial x_k}(\overline{u'_k u'_j}) - \frac{1}{\rho}\frac{\partial P}{\partial x_j} + \alpha g_j \tilde{\Theta}_v, \tag{11.2.5}$$

$$\frac{D\Theta}{Dt} = -\frac{\partial}{\partial x_k}(\overline{u'_k \theta'}). \tag{11.2.6}$$

where $\frac{D}{Dt} = \frac{\partial}{\partial t} + U_k \frac{\partial}{\partial x_k}$, and it is assumed that viscous terms and thermal diffusion terms are negligible in filtered equations.

The equations for the average field (11.2.4)–(11.2.6) have the terms of second-order quantities: $\overline{u'_k u'_j}$ and $\overline{u'_k \theta'}$. In particular, $-\rho\ \overline{u'_k u'_j}$ is called the *Reynolds stress*. In order to predict mean field quantities using the above equations, we need to have expressions for these second-order terms. The prognostic equations of second-order quantities can be derived from (11.2.1)–(11.2.3). The equation of the momentum flux is given by the average of $(11.2.2)_i \times u'_j + (11.2.2)_j \times u'_i$, where $()_i$ denotes the i-th component of Eq. (). The equation of heat flux is given by the average of $(11.2.3) \times u'_j + (11.2.2)_j \times \theta'$, and the equation of the square of potential temperature disturbance by the average of $(11.2.3) \times \theta'$:

$$\begin{aligned}
&\frac{D}{Dt}\overline{u'_i u'_j} + \frac{\partial}{\partial x_k}\left(\overline{u'_k u'_i u'_j} - \nu \frac{\partial}{\partial x_k}\overline{u'_i u'_j}\right) + f_k\left(\varepsilon_{jkl}\overline{u'_l u'_i} + \varepsilon_{ikl}\overline{u'_l u'_j}\right)\\
&= \overline{\frac{p'}{\rho}\left(\frac{\partial u'_i}{\partial x_j} + \frac{\partial u'_j}{\partial x_i}\right)} - 2\nu\overline{\frac{\partial u'_i}{\partial x_k}\frac{\partial u'_j}{\partial x_k}} - \overline{u'_k u'_i}\frac{\partial U_j}{\partial x_k} - \overline{u'_k u'_j}\frac{\partial U_i}{\partial x_k}\\
&\quad + \alpha(g_j \overline{\theta'_v u'_i} + g_i \overline{\theta'_v u'_j}) - \frac{1}{\rho}\frac{\partial}{\partial x_j}\overline{p' u'_i} - \frac{1}{\rho}\frac{\partial}{\partial x_i}\overline{p' u'_j},
\end{aligned} \tag{11.2.7}$$

$$\begin{aligned}
&\frac{D}{Dt}\overline{u'_j \theta'} + \frac{\partial}{\partial x_k}\left(\overline{u'_k u'_j \theta'} - \nu\overline{\theta' \frac{\partial u'_j}{\partial x_k}} - \kappa\overline{u'_j \frac{\partial \theta'}{\partial x_k}}\right)\\
&= -\overline{u'_j u'_k}\frac{\partial \Theta}{\partial x_k} - \overline{\theta' u'_k}\frac{\partial U_j}{\partial x_k} + \overline{\frac{p'}{\rho}\frac{\partial \theta'}{\partial x_j}} + \alpha g_j \overline{\theta'_v \theta'}\\
&\quad -(\kappa + \nu)\overline{\frac{\partial u'_j}{\partial x_k}\frac{\partial \theta'}{\partial x_k}} - f_k \varepsilon_{jkl}\overline{u'_l \theta'} - \frac{1}{\rho}\frac{\partial}{\partial x_j}\overline{p' \theta'},
\end{aligned} \tag{11.2.8}$$

$$\frac{D}{Dt}\overline{\theta'^2} + \frac{\partial}{\partial x_k}\left(\overline{u'_k \theta'^2} - \kappa\frac{\partial}{\partial x_k}\overline{\theta'^2}\right) = -2\overline{u'_k \theta'}\frac{\partial \Theta}{\partial x_k} - 2\kappa\overline{\frac{\partial \theta'}{\partial x_k}\frac{\partial \theta'}{\partial x_k}}. \tag{11.2.9}$$

The above second-order equations contain third-order quantities such as $\overline{u_i'u_j'\theta'}$. Therefore, we need to have the expression of third-order quantities to solve these equations. However, the temporal change of third-order quantities contains further higher order quantities and we cannot close the equations. To close the equations, therefore, we need assumptions in which higher order quantities are expressed by lower or same-order quantities. This kind of assumption is called the *closure* of turbulence.

11.2.2 Eddy diffusion coefficients and turbulence models

In order to close the turbulence model, we relate second-order quantities $\overline{u_i'u_j'}$ and $\overline{u_i'\theta'}$ to the gradient of mean field values from the analogy of molecular diffusion:

$$\overline{u_i'u_j'} = -K_M\left(\frac{\partial U_i}{\partial x_j}+\frac{\partial U_j}{\partial x_i}\right)+\frac{2}{3}\delta_{ij}k, \tag{11.2.10}$$

$$\overline{u_i'\theta'} = -K_H\frac{\partial \Theta}{\partial x_i}, \tag{11.2.11}$$

where $k=\overline{u_i'^2}/2$ is turbulent kinetic energy. K_M and K_H are called the *eddy viscous coefficient* and the *eddy diffusion coefficient*, respectively. The former is also called the eddy diffusion coefficient, for simplicity. These coefficients are not necessarily constants and even their signs are not definite. They may not be scalars and can be tensors. These expressions are convenient, however, since we can calculate the evolutions of the basic field if we know the values of K_M and K_H experimentally or theoretically.

Turbulence models are categorized by the closure assumptions used for the expressions for eddy diffusion coefficients. The simplest one is that in which no prognostic equation is used for the evaluation of subgrid values, and is called a *zero-equation model.* Let l be a characteristic length scale of turbulence, and V be a characteristic velocity scale of turbulence. Dimensional analysis gives

$$K_M = lV. \tag{11.2.12}$$

In the case of mixing length theory, V is estimated from shear multiplied by the mixing length l (see (11.3.3)), or the free fall velocity given by buoyancy (see (14.4.4)). In particular, if the eddy viscous coefficient is related to shear,

$$K_M = (C_s l)^2\left[\frac{1}{2}\left(\frac{\partial U_i}{\partial x_j}+\frac{\partial U_j}{\partial x_i}\right)^2\right]^{\frac{1}{2}}, \tag{11.2.13}$$

it is called the *Smagorinsky model*, where C_s is a constant. If the eddy viscous coefficient is related to turbulent kinetic energy k, we may have from dimensional analysis

$$K_M = C_k\, l\, k^{\frac{1}{2}}, \tag{11.2.14}$$

where C_k is a constant. The turbulence model is closed if the values of k and l are determined prognostically or diagnostically from mean field values. We may use the dissipation rate ε instead of l for closure. From dimensional analysis, we can relate the dissipation rate using a constant C_e as

$$\varepsilon = C_e \frac{k^{\frac{3}{2}}}{l}. \tag{11.2.15}$$

Then, (11.2.14) can be rewritten as

$$K_M = C_M \frac{k^2}{\varepsilon}, \tag{11.2.16}$$

where C_M is a constant.

If (11.2.14) or (11.2.16) is used to determine the eddy coefficient, we need to evaluate the turbulent kinetic energy k by introducing further closure assumptions. Here, we derive the prognostic equation of turbulent kinetic energy to obtain k.† First, we assume that the effects of buoyancy and the Coriolis force are negligible, for simplicity. From (11.2.7), by omitting the buoyancy term and the Coriolis terms, and contracting for indices i and j, the equation of turbulent kinetic energy can be derived as

$$\frac{D}{Dt}k = P_s - \varepsilon - \frac{\partial}{\partial x_j}\left(\frac{1}{2}\overline{u_i' u_i' u_j'} + \overline{\frac{p'}{\rho}u_j'}\right) + \nu \frac{\partial^2 k}{\partial {x_j}^2}, \tag{11.2.17}$$

where

$$P_s = -\overline{u_i' u_j'}\frac{\partial U_i}{\partial x_j}, \tag{11.2.18}$$

$$\varepsilon = \nu \overline{\frac{\partial u_i'}{\partial x_l}\frac{\partial u_i'}{\partial x_l}}. \tag{11.2.19}$$

P_s is an energy source term of turbulence due to the shear of the mean field. Substituting (11.2.10) into (11.2.18) gives

$$P_s = \frac{K_M}{2}\left(\frac{\partial U_i}{\partial x_j} + \frac{\partial U_j}{\partial x_i}\right)^2, \tag{11.2.20}$$

from which we have $P_s > 0$ as long as $K_M > 0$ (i.e., the shear of the mean field produces turbulent kinetic energy). Here, we regard the third and following terms on the right-hand side of (11.2.17) as a diffusion of turbulent kinetic energy, and assume that the diffusion flux is written as

$$\frac{1}{2}\overline{u_i' u_i' u_j'} + \overline{\frac{p'}{\rho}u_j'} - \nu \frac{\partial k}{\partial x_j} = -\frac{K_M}{\sigma_k}\frac{\partial k}{\partial x_j}, \tag{11.2.21}$$

where σ_k is an appropriate dimensionless constant. Thus, (11.2.17) can be written as

$$\frac{D}{Dt}k = D_k + P_s - \varepsilon, \tag{11.2.22}$$

† Note that k may be determined by a diagnostic method as will be described in Section 11.3.4.2.

where

$$D_k = \frac{\partial}{\partial x_j}\left(\frac{K_M}{\sigma_k}\frac{\partial k}{\partial x_j}\right) \tag{11.2.23}$$

is the divergence of the diffusion flux of turbulent kinetic energy. If the characteristic length scale of turbulence l is given by a diagnostic method or by an external parameter instead of using a prognostic equation, the dissipation rate ε can be determined from the relation (11.2.15) for a suitable coefficient C_e. Thus, subgrid-scale turbulence is solved using a single prognostic equation for k by (11.2.22). The turbulence model with this approach is called a *one-equation model.*

Thus far, the prognostic equation of k is formulated for neutral stratification, and does not provide the expression for thermal flux. The turbulence model must be extended to include the effect of stratification for application to the atmosphere. In the case that the production term duc to buoyancy exists, the change in turbulent kinetic energy (11.2.22) is rewritten as

$$\frac{D}{Dt}k = D_k + P_s + P_b - \varepsilon, \tag{11.2.24}$$

where

$$P_b = \alpha g \overline{w'\theta'_v} \tag{11.2.25}$$

is the production rate of turbulent kinetic energy due to buoyancy. The *flux Richardson number* R_f is introduced as a ratio between P_s and P_b:

$$R_f = \frac{\alpha g \overline{w'\theta'_v}}{\overline{u'_i u'_j}\frac{\partial U_i}{\partial x_j}} = -\frac{P_b}{P_s}. \tag{11.2.26}$$

In order for the source term to be positive, it requires that $P_s + P_b > 0$. Since $P_s > 0$ in general, this implies that $R_f < 1$. The *Richardson number* R_i is related to the flux Richardson number as

$$R_i = \frac{\alpha g \frac{\partial \Theta}{\partial z}}{\frac{1}{2}\left(\frac{\partial U_i}{\partial x_j} + \frac{\partial U_j}{\partial x_i}\right)^2} = \frac{K_M}{K_H}R_f, \tag{11.2.27}$$

where the eddy diffusion models (11.2.10) and (11.2.11) are used. The flux Richardson number plays the central role among the parameters in the turbulence model of the atmosphere. The *turbulent Prandtl number* P_r is also defined by

$$P_r = \frac{K_M}{K_H}. \tag{11.2.28}$$

If the value of P_r is given using another closure assumption, the eddy diffusion coefficient K_H can be calculated.

The coefficients that appear in the above formula must be determined from comparison with results from laboratory experiments, observations, or numerical simulations. If the grid interval of numerical models Δx is within the range of the inertial subrange, the above filtering procedure can be interpreted as a spatial mean and the turbulence model is used for the LES turbulence model. In this case, l is set equal to Δx. In particular, the Smagorinsky model (11.2.13), which was originally proposed for use in a large-scale atmospheric model, is nowadays widely used as a representative model of the LES turbulence models. As an LES model, $C_s = 0.2$ is usually used. For the coefficients in (11.2.14) and (11.2.15) in the one-equation model, Moeng and Wyngaard (1988) propose values $C_k = 0.1$ and $C_e = 0.93$ from their spectrum analysis, following the original work by Deardorff (1980). In order to determine the thermal eddy diffusion coefficient, the turbulent Prandtl number P_r in (11.2.28) must be specified. As a simple treatment, a constant $P_r = 1/3$ is used, for example (Deardorff, 1972).

In the one-equation model, the length scale l is given by a diagnostic method or by an external parameter. As an extension to the one-equation model, another category of the turbulence model called a *two-equation model* uses an additional prognostic equation. The k-ε model is classified as a two-equation model; the k-ε model internally determines l using two prognostic equations for k and ε, instead of specifying the value of l. By introducing several closure assumptions, the prognostic equation of ε is given as

$$\frac{D}{Dt}\varepsilon = D_\varepsilon + \frac{\varepsilon}{k}(C_{\varepsilon 1}P_s - C_{\varepsilon 2}\varepsilon), \tag{11.2.29}$$

where $C_{\varepsilon 1}$ and $C_{\varepsilon 2}$ are newly introduced constants and

$$D_\varepsilon = \frac{\partial}{\partial x_j}\left(\frac{K_M}{\sigma_\varepsilon}\frac{\partial \varepsilon}{\partial x_j}\right). \tag{11.2.30}$$

For the k-ε model, momentum flux is given by (11.2.10) and the velocity components of the mean field are determined from (11.2.5). Although the k-ϵ model is rarely used in the meteorological context, various values are proposed for the constants for application in engineering. An example is given by Launder and Spalding (1970): $(C_M, \sigma_k, \sigma_\varepsilon, C_{\varepsilon 1}, C_{\varepsilon 2}) = (0.09, 1.0, 1.3, 1.44, 1.92)$.

11.2.3 The Mellor and Yamada model

Next, we turn to describe the turbulence models proposed by Mellor and Yamada (1982) for non-neutrally stratified fluids. The *Mellor and Yamada model* consists of a hierarchy of turbulence closure models, which depend on the number of prognostic equations of second-order quantities of subgrid variables such as (11.2.7)–(11.2.9). These are called Level 4, Level 3, Level $2\frac{1}{2}$, and Level 2 models depending on the approximations.

11.2.3.1 The Level 4 model

The most general Mellor and Yamada model is the Level 4 model. In this model, the following closures are assumed for second-order quantities. From the symmetry of the Reynolds stress tensor, the source terms in the momentum flux equation

(11.2.7) are given as

$$
\begin{aligned}
\overline{\frac{p'}{\rho}\left(\frac{\partial u_i'}{\partial x_j}+\frac{\partial u_j'}{\partial x_i}\right)} &= -\frac{q}{3l_1}\left(\overline{u_i'u_j'}-\frac{\delta_{ij}}{3}q^2\right)+C_1q^2\left(\frac{\partial U_i}{\partial x_j}+\frac{\partial U_j}{\partial x_i}\right), \\
2\nu\overline{\frac{\partial u_i'}{\partial x_k}\frac{\partial u_j'}{\partial x_k}} &= \frac{2}{3}\frac{q^3}{\Lambda_1}\delta_{ij},
\end{aligned}
$$

where $q^2 = \overline{u_i'^2}(=2k)$, l_1 and Λ_1 are constants that have a dimension of length, and C_1 is a nondimensional constant. In the latter equation, the molecular dissipation of momentum is assumed to be isotropic. The source terms of the heat flux equation (11.2.8) and those of the square of potential temperature perturbation (11.2.9) are similarly given by

$$
\begin{aligned}
\overline{\frac{p'}{\rho}\frac{\partial\theta'}{\partial x_j}} &= -\frac{q}{3l_2}\overline{u_i'\theta}, \qquad 2\kappa\overline{\frac{\partial\theta'}{\partial x_k}\frac{\partial\theta'}{\partial x_k}} = 2\frac{q}{\Lambda_2}\overline{\theta'^2}, \\
(\kappa+\nu)\overline{\frac{\partial u_i'}{\partial x_k}\frac{\partial\theta'}{\partial x_k}} &= 0,
\end{aligned}
$$

where l_2 and Λ_2 are length scales. For the closures of third-order quantities $\overline{u_k'u_i'u_j'}$, $\overline{u_i'u_j'\theta'}$, and $\overline{u_k'\theta'^2}$, the following forms are assumed:

$$
\begin{aligned}
\overline{u_k'u_i'u_j'} &= -\frac{3}{5}lqS_q\left(\frac{\partial}{\partial x_k}\overline{u_i'u_j'}+\frac{\partial}{\partial x_j}\overline{u_i'u_k'}+\frac{\partial}{\partial x_i}\overline{u_j'u_k'}\right), \\
\overline{u_i'u_j'\theta'} &= -lqS_{u\theta}\left(\frac{\partial}{\partial x_j}\overline{u_i'\theta'}+\frac{\partial}{\partial x_i}\overline{u_j'\theta'}\right), \\
\overline{u_k'\theta'^2} &= -lqS_\theta\frac{\partial}{\partial x_j}\overline{\theta'^2}.
\end{aligned}
$$

We also assume that the correlation between pressure and velocity or potential temperature is small:

$$
\overline{p'u_i'} = \overline{p'\theta'} = 0,
$$

and that the terms including ν or κ in the fluxes on the left-hand side are negligible. Substituting these expressions into (11.2.7)–(11.2.9), we obtain

$$
\begin{aligned}
&\frac{D}{Dt}\overline{u_i'u_j'}-\frac{\partial}{\partial x_k}\left[\frac{3}{5}lqS_q\left(\frac{\partial}{\partial x_k}\overline{u_i'u_j'}+\frac{\partial}{\partial x_j}\overline{u_i'u_k'}+\frac{\partial}{\partial x_i}\overline{u_j'u_k'}\right)\right] \\
&= -\frac{q}{3l_1}\left(\overline{u_i'u_j'}-\frac{\delta_{ij}}{3}q^2\right)+C_1q^2\left(\frac{\partial U_i}{\partial x_j}+\frac{\partial U_j}{\partial x_i}\right)-\frac{2}{3}\frac{q^3}{\Lambda_1}\delta_{ij} \\
&\quad -\overline{u_k'u_i'}\frac{\partial U_j}{\partial x_k}-\overline{u_k'u_j'}\frac{\partial U_i}{\partial x_k}+\alpha(g_j\overline{\theta_v'u_i'}+g_i\overline{\theta_v'u_j'}) \\
&\quad -f_k\left(\varepsilon_{jkl}\overline{u_l'u_i'}+\varepsilon_{ikl}\overline{u_l'u_j'}\right), \qquad (11.2.31)
\end{aligned}
$$

$$\frac{D}{Dt}\overline{u_j'\theta'} - \frac{\partial}{\partial x_k}\left[lqS_{u\theta}\left(\frac{\partial}{\partial x_j}\overline{u_k'\theta'} + \frac{\partial}{\partial x_k}\overline{u_j'\theta'}\right)\right]$$
$$= -\overline{u_j'u_k'}\frac{\partial \Theta}{\partial x_k} - \overline{\theta' u_k'}\frac{\partial U_j}{\partial x_k} - \frac{q}{3l_2}\overline{u_i'\theta} + \alpha g_j \overline{\theta_v'\theta'} - f_k \varepsilon_{jkl}\overline{u_l'\theta'}, \quad (11.2.32)$$

$$\frac{D}{Dt}\overline{\theta'^2} - \frac{\partial}{\partial x_k}\left[lqS_\theta \frac{\partial}{\partial x_j}\overline{\theta'^2}\right]$$
$$= -2\overline{u_k'\theta'}\frac{\partial \Theta}{\partial x_k} - 2\frac{q}{\Lambda_2}\overline{\theta'^2}. \quad (11.2.33)$$

If the turbulence equations for vapor q_v are similarly constructed, we obtain fifteen equations for second-order quantities. The Level 4 model of turbulence closures is composed of the set of these equations.

11.2.3.2 The Level 3 model

Although the equations of the Level 4 model, (11.2.31)–(11.2.33), can be solved in principle, these equations are quite complicated. For a practical use, further approximations are introduced to simplify the model. Separating the isotropic and anisotropic components of the Reynolds stress tensor, we have

$$\overline{u_i'u_j'} = \left(\frac{\delta_{ij}}{3} + a_{ij}\right) q^2. \quad (11.2.34)$$

If the order of the anisotropy is denoted by $a^2 = O(a_{ij}^2)$, we may assume $a \ll 1$, in general. Then, we expand the equations by a and neglect the $O(a^2)$ and higher order terms. In particular, only temporal changes in q^2 and $\overline{\theta'^2}$ remain as the derivatives with respect to time, and the remaining prognostic equations reduce to diagnostic ones. Thus, the equations of the Level 3 model are given as follows:

$$\frac{D}{Dt}q^2 - \frac{\partial}{\partial x_k}\left[lqS_q \frac{\partial}{\partial x_k}q^2\right] = 2(P_s + P_b - \varepsilon), \quad (11.2.35)$$

$$\frac{D}{Dt}\overline{\theta'^2} - \frac{\partial}{\partial x_k}\left[lqS_\theta \frac{\partial}{\partial x_j}\overline{\theta'^2}\right] = -2\overline{u_k'\theta'}\frac{\partial \Theta}{\partial x_k} - 2\frac{q}{\Lambda_2}\overline{\theta'^2}, \quad (11.2.36)$$

and

$$\overline{u_i'u_j'} = \frac{\delta_{ij}}{3}q^2 - \frac{3l_1}{q}\left[\overline{u_k'u_i'}\frac{\partial U_j}{\partial x_k} + \overline{u_k'u_j'}\frac{\partial U_i}{\partial x_k} + \frac{2}{3}\delta_{ij}P_s\right.$$
$$-C_1 q^2\left(\frac{\partial U_i}{\partial x_j} + \frac{\partial U_j}{\partial x_i}\right) - \alpha(g_j\overline{\theta_v' u_i'} + g_i \overline{\theta_v' u_j'})$$
$$\left. + \frac{2}{3}\delta_{ij}P_b + f_k\left(\varepsilon_{jkl}\overline{u_l'u_i'} + \varepsilon_{ikl}\overline{u_l'u_j'}\right)\right], \quad (11.2.37)$$

$$\overline{u_j'\theta} = -\frac{3l_2}{q}\left[\overline{u_j'u_k'}\frac{\partial \Theta}{\partial x_k} + \overline{\theta' u_k'}\frac{\partial U_j}{\partial x_k} + \alpha g_j \overline{\theta_v'\theta'} + f_k \varepsilon_{jkl}\overline{u_l'\theta'}\right],$$
$$(11.2.38)$$

where

$$P_s = -\overline{u_i'u_j'}\frac{\partial U_i}{\partial x_j}, \qquad P_b = \alpha g\overline{w'\theta_v'}, \tag{11.2.39}$$

$$\varepsilon = \frac{q^3}{\Lambda_1}. \tag{11.2.40}$$

P_s and P_b are the production terms of kinetic energy due to shear and buoyancy, respectively. From (11.2.35), turbulent kinetic energy cannot be maintained unless $P_s + P_b > 0$ is satisfied. The Level 3 model consists of four prognostic equations if the equation for vapor $\overline{q_v'^2}$ and that for covariance $\overline{\theta' q_v'}$ are added.

11.2.3.3 The Level $2\frac{1}{2}$ model and the Level 2 model

For the next approximation to the Level $2\frac{1}{2}$ model, we neglect the time derivative and the diffusion term on the left-hand side of (11.2.36) to obtain the following diagnostic equation:

$$\overline{\theta'^2} = -\frac{\Lambda_2}{q}\overline{u_k'\theta'}\frac{\partial \Theta}{\partial x_k}. \tag{11.2.41}$$

Thus, only the equation for turbulent kinetic energy q^2 remains as the prognostic equation.

Furthermore, for the Level 2 model, all the time derivatives and the diffusion terms are neglected. Therefore, (11.2.35) becomes

$$P_s + P_b = \varepsilon. \tag{11.2.42}$$

These Mellor and Yamada models are rewritten in more concrete forms in the modeling of turbulence in the boundary layer. The boundary layer approximation of Mellor and Yamada models will be described in the subsequent section.

11.3 Boundary layer

11.3.1 Structure of the planetary boundary layer

The atmospheric layer near the earth surface is called the *planetary boundary layer*. The planetary boundary layer has the following layered structure. Just above the surface up to a few millimeters is called the *viscous layer*, where the effects of molecular viscosity and diffusion are predominant. The height of the top of this layer z_0 is called the *roughness length*. Above the viscous layer is the *surface layer*, in which each of the fluxes is regarded to be constant. The depth of this layer is about 30–100 m. The surface layer is also referred to as the *constant flux layer*. If there exists the effect of the Coriolis force on the direction of winds, another layer named the *Ekman layer* emerges above the surface layer (see Section 6.6). In the case of unstable stratification, on the other hand, turbulent motions mix entropy and humidity vertically and makes them vertically uniform. This layer is called the *mixed layer*, which is located above the surface layer. The height of the top of the Ekman layer or the mixed layer is about 500 m to 1 km.

The above classification of the boundary layer is applicable to a very idealistic situation. In reality, the boundary layer is very complicated depending on the surface condition and temporal variation. In some cases, the effects of forests and buildings are taken into account as the effective roughness length in the order of 100 m. In this section, we do not touch upon such a complicated surface condition, and only describe the formulation of turbulent fluxes above a flat surface.

11.3.2 Surface layer

In the surface layer, each of the vertical fluxes is independent of height by definition. In the case of neutral buoyancy, the vertical profiles of velocity, temperature, and vapor can be given based on the *mixing length theory.* For simplicity, we assume that the mean flow is in the x-direction. Let the length scale of turbulence near the surface be denoted by l. According to the mixing length theory, l is set equal to the vertical distance of displacement of an air parcel in the turbulence, or the *mixing length.* In this case, the deviation of a physical quantity s is given by

$$s' = -l\frac{\partial \overline{s}}{\partial z}, \tag{11.3.1}$$

where $\frac{\partial \overline{s}}{\partial z} < 0$ is assumed. Thus, the vertical flux of s is estimated as

$$\overline{s'w'} = -l|w'|\frac{\partial \overline{s}}{\partial z}, \tag{11.3.2}$$

where w' is a turbulent component of vertical velocity. If we take the horizontal wind component u' as a quantity s' and assume the eddy is isotropic, we have

$$|u'| \approx |w'| \approx l\left|\frac{\partial \overline{u}}{\partial z}\right|. \tag{11.3.3}$$

Then, we have

$$\overline{u'w'} = -l^2\left|\frac{\partial \overline{u}}{\partial z}\right|\frac{\partial \overline{u}}{\partial z}. \tag{11.3.4}$$

In the surface layer, the momentum flux $\rho\overline{u'w'}$ is vertically uniform and equal to stress at the surface τ_0. We can define a *friction velocity* u_* by

$$u_*^2 = -\overline{u'w'} = \frac{\tau_0}{\rho}. \tag{11.3.5}$$

u_* is constant irrespective of height within the surface layer. We assume that the mixing length near the surface l is proportional to height from the surface:

$$l = kz, \tag{11.3.6}$$

where k is called the *von Karman constant.* Substituting (11.3.5) and (11.3.6) into (11.3.4), we obtain

$$\frac{\partial \overline{u}}{\partial z} = \frac{u_*}{kz}. \tag{11.3.7}$$

Integrating this over height, we obtain the *logarithmic velocity law*:

$$\overline{u} = \frac{u_*}{k} \ln \frac{z}{z_0}. \tag{11.3.8}$$

The constant z_0 is chosen such that $\overline{u} = 0$ at $z = z_0$. z_0 corresponds to the roughness length, which is the transition level to the viscous layer underneath. It is found by comparison with observation using the logarithmic velocity law that the von Karman constant is given as $k = 0.40$. From (11.3.4) and (11.3.7), eddy viscosity in the neutral case K_M^* is expressed as

$$K_M^* = l^2 \left| \frac{\partial \overline{u}}{\partial z} \right| = kzu_*. \tag{11.3.9}$$

We also have the vertical profiles of temperature and vapor in the case of the neutral stratification. Let F_{H0} denote sensible heat flux at the surface and F_{q0} vapor flux (evaporation) at the surface. In the surface layer, both heat and vapor fluxes are constant irrespective of height and equal to F_{H0}, F_{q0}, respectively. Then, we define the characteristic scales of the deviations of potential temperature θ_* and vapor q_*, which respectively satisfy

$$u_*\theta_* = -\overline{w'\theta'} = -\frac{F_{H0}}{\rho C_p}, \qquad u_*q_* = -\overline{w'q'} = -\frac{F_{q0}}{\rho}. \tag{11.3.10}$$

In the neutral case, constants similar to the von Karman constant in (11.3.7) can be introduced for temperature and vapor profiles. Using the two constants k' and k'', we obtain the temperature and vapor profiles

$$\frac{\partial \overline{\theta}}{\partial z} = \frac{\theta_*}{k'z}, \qquad \frac{\partial \overline{q}}{\partial z} = \frac{q_*}{k''z}. \tag{11.3.11}$$

Thus, we have

$$\overline{\theta} - \overline{\theta}_s = \frac{\theta_*}{k'} \ln \left(\frac{z}{z_{0H}} \right), \qquad \overline{q} - \overline{q}_s = \frac{q_*}{k''} \ln \left(\frac{z}{z_{0q}} \right), \tag{11.3.12}$$

where the roughness length for heat z_{0H} and that for vapor z_{0q} are defined. The eddy diffusion coefficients of heat and vapor K_H^*, K_q^* are therefore given by

$$K_H^* = k'zu_*, \qquad K_q^* = k''zu_*. \tag{11.3.13}$$

These coefficients are for the neutral case and are used for reference to more general cases of stratification.

11.3.3 Bulk method

We next formulate the vertical fluxes between the surface and any level in the surface layer in more general stratified cases. Let z_1 denote an arbitrary level in the surface layer, and $\boldsymbol{u}_1$, θ_1, and q_1 denote the horizontal velocity vector, potential temperature, and vapor at the level z_1, respectively. In the case of neutral stratification, from (11.3.5) and (11.3.8), momentum stress is written as

$$\tau_0 = \rho u_*^2 = \rho \frac{k^2}{\left(\ln \frac{z_1}{z_0}\right)^2} \overline{u}_1^2. \tag{11.3.14}$$

We extend this formula to more general cases for non-neutral stratification as

$$\tau_0 = \rho C_D \overline{u}_1^2, \tag{11.3.15}$$

where the *drag coefficient* C_D is introduced. Decomposing the above equation into each horizontal direction and extending it to heat and vapor fluxes, we have

$$\tau_{x0} = \rho C_D |\overline{\boldsymbol{u}}_1| \overline{u}_1, \qquad \tau_{y0} = \rho C_D |\overline{\boldsymbol{u}}_1| \overline{v}_1, \tag{11.3.16}$$

$$F_{H0} = \rho C_p C_H |\overline{\boldsymbol{u}}_1| (\overline{\theta}_s - \overline{\theta}_1), \qquad F_{q0} = \rho C_q |\overline{\boldsymbol{u}}_1| (\overline{q}_s - \overline{q}_1), \tag{11.3.17}$$

where θ_s and θ_1 are the potential temperatures at the height z_{0H} and z_1, respectively. Similarly, q_s and q_1 are specific humidities at height z_{0q} and z_1, respectively. C_H is called the *Stanton number* and C_q is the *Dalton number*. The coefficients C_D, C_H, and C_q are nondimensional, and are referred to as *bulk coefficients*. Letting C_D^*, C_H^*, and C_q^* denote the bulk coefficients for neutral stratification, we have

$$C_D^* = \frac{k^2}{\left(\ln \frac{z_1}{z_0}\right)^2}, \quad C_H^* = \frac{kk'}{\ln \frac{z_1}{z_0} \ln \frac{z_1}{z_{0H}}}, \quad C_q^* = \frac{kk''}{\ln \frac{z_1}{z_0} \ln \frac{z_1}{z_{0q}}}. \tag{11.3.18}$$

In the case of neutral stratification, (11.3.7) and (11.3.11) hold. These formula are generalized to non-neutral cases using nondimensional vertical profile functions as

$$\phi_M\left(\frac{z}{L}\right) = \frac{kz}{u_*}\frac{\partial \overline{u}}{\partial z}, \quad \phi_H\left(\frac{z}{L}\right) = \frac{kz}{\theta_*}\frac{\partial \overline{\theta}}{\partial z}, \quad \phi_q\left(\frac{z}{L}\right) = \frac{kz}{q_*}\frac{\partial \overline{q}}{\partial z}, \tag{11.3.19}$$

which are called *universal functions*. These are functions of nondimensional height $\zeta = z/L$, where the length scale L is called the *Monin-Obukhov length* defined by

$$L = \frac{u_*^2}{kg\alpha\theta_*}. \tag{11.3.20}$$

Using (11.3.19), (11.3.5), and (11.3.10), we also have

$$L = \frac{z}{\phi_M}\frac{\overline{u'w'}\frac{\partial \overline{u}}{\partial z} + \overline{v'w'}\frac{\partial \overline{v}}{\partial z}}{g\alpha\overline{w'\theta'}} = \frac{z}{\phi_M R_f}. \tag{11.3.21}$$

R_f is the flux Richardson number given by (11.2.26). *Monin-Obukhov's similarity theory* states that the turbulence in the surface layer is characterized by the single parameter L, and the vertical profiles of momentum, temperature, and humidity are expressed by the universal functions ϕ_M, ϕ_H, and ϕ_q, respectively. In neutral case, we have $\phi_M = 1$, $\phi_H = k/k'$, and $\phi_q = k/k''$, and L is independent of height in the surface layer. Using the universal functions, (11.3.9) and (11.3.13), the eddy diffusion coefficients are expressed by

$$K_M = \frac{u_*^2}{\frac{\partial \overline{u}}{\partial z}} = \frac{K_M^*}{\phi_M}, \quad K_H = \frac{u_*\theta_*}{\frac{\partial \overline{\theta}}{\partial z}} = \frac{k}{k'}\frac{K_H^*}{\phi_H}, \quad K_q = \frac{u_* q_*}{\frac{\partial \overline{q}}{\partial z}} = \frac{k}{k''}\frac{K_q^*}{\phi_q}. \tag{11.3.22}$$

From (11.3.15), (11.3.5), (11.3.17), and (11.3.10), the bulk coefficients are expressed as

$$C_D = \left(\frac{u_*}{|\overline{\boldsymbol{u}}_1|}\right)^2 = \frac{k^2}{F_M^2}, \tag{11.3.23}$$

$$C_H = \frac{\theta_*}{\overline{\theta}_1 - \overline{\theta}_s}\frac{u_*}{|\overline{\boldsymbol{u}}_1|} = \frac{k^2}{F_M F_H}, \tag{11.3.24}$$

$$C_q = \frac{q_*}{\overline{q}_1 - \overline{q}_s}\frac{u_*}{|\overline{\boldsymbol{u}}_1|} = \frac{k^2}{F_M F_q}, \tag{11.3.25}$$

where

$$F_M\left(\frac{z_1}{z_0}, \frac{z_1}{L}\right) = \int_{z_0}^{z_1} \frac{\phi_M}{z} dz = \frac{k}{u_*}\int_{z_0}^{z_1} \frac{\partial |\overline{\boldsymbol{u}}|}{\partial z} dz = k\frac{|\overline{\boldsymbol{u}}_1|}{u_*},$$

$$F_H\left(\frac{z_1}{z_{0H}}, \frac{z_1}{L}\right) = \int_{z_{0H}}^{z_1} \frac{\phi_H}{z} dz = \frac{k}{\theta_*}\int_{z_{0H}}^{z_1} \frac{\partial \overline{\theta}}{\partial z} dz = k\frac{\overline{\theta}_1 - \overline{\theta}_s}{\theta_*},$$

$$F_q\left(\frac{z_1}{z_{0q}}, \frac{z_1}{L}\right) = \int_{z_{0q}}^{z_1} \frac{\phi_q}{z} dz = \frac{k}{q_*}\int_{z_{0q}}^{z_1} \frac{\partial \overline{q}}{\partial z} dz = k\frac{\overline{q}_1 - \overline{q}_s}{q_*}.$$

The functions F_M, F_H, and F_q can be calculated from the universal functions. The universal functions ϕ_M, ϕ_H, and ϕ_q can be determined from field experiments or by formulas (given in (11.3.69) and (11.3.70)). If we know ϕ_M, ϕ_H, and ϕ_q and the roughness lengths z_0, z_{0H}, and z_{0q}, the functions F_M, F_H, and F_q can be calculated, thus we obtain the bulk coefficients as functions of z_1 and L. We may use the bulk Richardson number at height z_1 instead of L. The bulk Richardson number is defined by

$$R_{iB} = \frac{g\alpha(\overline{\theta}_1 - \overline{\theta}_s)z_1}{|\overline{\boldsymbol{u}}_1|^2} = \frac{z_1}{L}\frac{F_H}{\phi_M F_M^2}. \tag{11.3.26}$$

In this case, the bulk coefficients are expressed as functions of z_1/z_0 and R_{iB}. See also (11.3.50) and (11.3.21).

11.3.4 Boundary layer models

The Mellor and Yamada model described in Section 11.2.3 is modified for the boundary layer model when the anisotropy between the horizontal and vertical directions near the surface is introduced. We describe the Level $2\frac{1}{2}$ and Level 2 models by introducing the *boundary layer approximation* in this section.

11.3.4.1 The boundary layer Level $2\frac{1}{2}$ model

If one makes the boundary layer approximation, one can assume that the mean field is in the hydrostatic balance and that the derivatives of the mean field with respect to space except for z are negligible. The Coriolis terms are also neglected in the equations of the subgrid quantities. By inserting $\boldsymbol{U} = (U, V, W)$ and $\boldsymbol{g} = (0, 0, g)$

in (11.2.5)–(11.2.6), we rewrite the prognostic equations of the mean field in the hydrostatic approximation as

$$\frac{D}{Dt}U + \frac{\partial}{\partial z}(\overline{u'w'}) = -\frac{1}{\rho}\frac{\partial P}{\partial x} + fV, \tag{11.3.27}$$

$$\frac{D}{Dt}V + \frac{\partial}{\partial z}(\overline{v'w'}) = -\frac{1}{\rho}\frac{\partial P}{\partial y} - fU, \tag{11.3.28}$$

$$0 = -\frac{1}{\rho}\frac{\partial P}{\partial z} + \alpha g\tilde{\Theta}_v, \tag{11.3.29}$$

$$\frac{D}{Dt}\Theta = -\frac{\partial}{\partial z}(\overline{w'\theta'}). \tag{11.3.30}$$

Using the boundary layer approximation, (11.2.35), (11.2.37), and (11.2.38) are explicitly rewritten as

$$\frac{D}{Dt}q^2 - \frac{\partial}{\partial z}\left[lqS_q\frac{\partial}{\partial z}q^2\right] = 2(P_s + P_b - \varepsilon), \tag{11.3.31}$$

$$\overline{u'w'} = \frac{3l_1}{q}\left[-(\overline{w'^2} - C_1q^2)\frac{\partial U}{\partial z} + \alpha g\overline{u'\theta'_v}\right], \tag{11.3.32}$$

$$\overline{v'w'} = \frac{3l_1}{q}\left[-(\overline{w'^2} - C_1q^2)\frac{\partial V}{\partial z} + \alpha g\overline{v'\theta'_v}\right], \tag{11.3.33}$$

$$\overline{w'^2} = \frac{q^2}{3} + \frac{l_1}{q}\left[2\overline{u'w'}\frac{\partial U}{\partial z} + 2\overline{v'w'}\frac{\partial V}{\partial z} + 4P_b\right], \tag{11.3.34}$$

$$\overline{u'\theta'} = \frac{3l_2}{q}\left[-\overline{u'w'}\frac{\partial \Theta}{\partial z} - \overline{w'\theta'}\frac{\partial U}{\partial z}\right], \tag{11.3.35}$$

$$\overline{v'\theta'} = \frac{3l_2}{q}\left[-\overline{v'w'}\frac{\partial \Theta}{\partial z} - \overline{w'\theta'}\frac{\partial V}{\partial z}\right], \tag{11.3.36}$$

$$\overline{w'\theta'} = \frac{3l_2}{q}\left[-\overline{w'^2}\frac{\partial \Theta}{\partial z} + \alpha g\overline{\theta'\theta'_v}\right], \tag{11.3.37}$$

$$\overline{\theta'^2} = -\frac{\Lambda_2}{q}\overline{w'\theta'}\frac{\partial \Theta}{\partial z}, \tag{11.3.38}$$

where

$$P_s = -\overline{u'w'}\frac{\partial U}{\partial z} - \overline{v'w'}\frac{\partial V}{\partial z}, \qquad P_b = \alpha g\overline{w\theta'_v}, \tag{11.3.39}$$

$$\varepsilon = \frac{q^3}{\Lambda_1}. \tag{11.3.40}$$

In (11.3.31), $S_q = 0.2$ is normally used.

Now, we assume that each vertical flux is expressed as follows:

$$\overline{u'w'} = -K_M\frac{\partial U}{\partial z}, \qquad \overline{v'w'} = -K_M\frac{\partial V}{\partial z}, \tag{11.3.41}$$

$$\overline{\theta'w'} = -K_H\frac{\partial \Theta}{\partial z}, \tag{11.3.42}$$

where the eddy diffusion coefficients K_M and K_H are written by using nondimensional parameters S_M and S_H, respectively, as

$$K_M = lqS_M, \qquad K_H = lqS_H. \tag{11.3.43}$$

Furthermore, we introduce nondimensional parameters that depend on the gradient of the mean field:

$$G_M = \frac{l^2}{q^2}\left[\left(\frac{\partial U}{\partial z}\right)^2 + \left(\frac{\partial V}{\partial z}\right)^2\right], \tag{11.3.44}$$

$$G_H = -\frac{l^2}{q^2}\alpha g \frac{\partial \Theta_v}{\partial z}, \tag{11.3.45}$$

and assume that all the length scales are proportional to a single length parameter l as

$$(l_1, \Lambda_1, l_2, \Lambda_2) = (A_1, B_1, A_2, B_2) \times l. \tag{11.3.46}$$

Substituting the above equations into (11.3.32)–(11.3.40), and using $\theta \approx \theta_v$, we obtain

$$\begin{aligned} &S_M(1 + 6A_1^2 G_M - 9A_1A_2G_H) - S_H(12A_1^2G_H + 9A_1A_2G_H) \\ &\quad = A_1(1 - 3C_1), \end{aligned} \tag{11.3.47}$$

$$\begin{aligned} &S_M(6A_1A_2G_M) + S_H(1 - 3A_2B_2G_H - 12A_1A_2G_H) \\ &\quad = A_2. \end{aligned} \tag{11.3.48}$$

The above two equations determine S_M and S_H. If we know the values of the mean field U, V, Θ, and the turbulent kinetic energy q^2, G_M and G_H are determined from (11.3.44) and (11.3.45), where q^2 is given by the prognostic equation (11.3.31). If S_M and S_H are solved from (11.3.47) and (11.3.48), the eddy coefficients K_M, K_H are given from (11.3.43). Thus, the vertical fluxes can be calculated using (11.3.41)–(11.3.42).†

Now that (11.3.39) can be rewritten in the forms

$$P_s = \frac{q^3}{l}S_MG_M, \qquad P_b = \frac{q^3}{l}S_HG_H, \tag{11.3.49}$$

let us define the Richardson number R_i and the flux Richardson number R_f by

$$R_i = \frac{g\alpha\frac{\partial\Theta}{\partial z}}{\left(\frac{\partial U}{\partial z}\right)^2 + \left(\frac{\partial V}{\partial z}\right)^2} = -\frac{G_H}{G_M}, \tag{11.3.50}$$

$$R_f = \frac{g\alpha\overline{w'\theta'}}{\overline{u'w'}\frac{\partial U}{\partial z} + \overline{v'w'}\frac{\partial V}{\partial z}} = -\frac{P_b}{P_s} = -\frac{S_HG_H}{S_MG_M}. \tag{11.3.51}$$

†It is known that the boundary layer Level $2\frac{1}{2}$ model has a pathological behavior for growing turbulence. Some modification is needed for practical application as proposed by Helfand and Labraga (1988), for example.

In order to maintain turbulent kinetic energy, we need $P_s + P_b > 0$ (i.e., $R_f < 1$). We also have

$$\frac{P_s + P_b}{\varepsilon} = B_1(S_M G_M + S_H G_H). \tag{11.3.52}$$

Eliminating G_M in (11.3.47) and (11.3.48) with the use of (11.3.52), we obtain

$$\begin{aligned} & S_M(1 - 9A_1A_2G_H) - S_H(18A_1^2 + 9A_1A_2)G_H \\ &= A_1\left(1 - 3C_1 - \frac{6A_1}{B_1}\frac{P_s + P_b}{\varepsilon}\right), \end{aligned} \tag{11.3.53}$$

$$\begin{aligned} & S_H[1 - (3A_2B_2 + 18A_1A_2)G_H] \\ &= A_2\left(1 - \frac{6A_1}{B_1}\frac{P_s + P_b}{\varepsilon}\right). \end{aligned} \tag{11.3.54}$$

These forms of simultaneous equations are used in the subsequent model.

11.3.4.2 The boundary layer Level 2 model

In the Level 2 model, all the derivatives with respect to time and the diffusion terms in the equations of second-order quantities are neglected. Since the left-hand side of (11.3.31) becomes zero, we have from (11.3.52)

$$1 = \frac{P_s + P_b}{\varepsilon} = B_1(S_M G_M + S_H G_H). \tag{11.3.55}$$

Then, using (11.3.51), we obtain

$$S_H G_H = \frac{1}{B_1(1 - R_f^{-1})}. \tag{11.3.56}$$

By introducing

$$\gamma_1 = \frac{1}{3} - \frac{2A_1}{B_1}, \qquad \gamma_2 = \frac{B_2 + 6A_1}{B_1}, \tag{11.3.57}$$

we can solve (11.3.53) and (11.3.54) for S_H and S_M using (11.3.56) as

$$S_H = \frac{\alpha_1 - \alpha_2 R_f}{1 - R_f}, \qquad \frac{S_M}{S_H} = \frac{\beta_1 - \beta_2 R_f}{\beta_3 - \beta_4 R_f}, \tag{11.3.58}$$

where

$$\begin{aligned} \alpha_1 &= 3A_2\gamma_1, & \alpha_2 &= 3A_2(\gamma_1 + \gamma_2), \\ \beta_1 &= A_1B_1(\gamma_1 - C_1), & \beta_2 &= A_1[B_1(\gamma_1 - C_1) + 6A_1 + 3A_2], \\ \beta_3 &= A_2B_1\gamma_1, & \beta_4 &= A_2[B_1(\gamma_1 + \gamma_2) - 3A_1]. \end{aligned}$$

If the values of the mean field U, V, and Θ are known, the Richardson number R_i is calculated from (11.3.50). From (11.3.50), (11.3.51), and (11.3.58), the relation between R_i and R_f is given by

$$R_f = \frac{S_H}{S_M} R_i = \frac{\beta_3 - \beta_4 R_f}{\beta_1 - \beta_2 R_f} R_i. \tag{11.3.59}$$

Thus, we obtain the flux Richardson number by

$$R_f = \frac{1}{2\beta_2}\left[\beta_1 + \beta_4 R_i - \sqrt{(\beta_1 + \beta_4 R_i)^2 - 4\beta_2\beta_3 R_i}\right], \tag{11.3.60}$$

which is one of the solutions to (11.3.59) that satisfies $R_f = 0$ at $R_i = 0$. Therefore, we obtain S_H and S_M from (11.3.58). Next, from (11.3.49) and (11.3.44), we have

$$P_s = \frac{q^3}{l} S_M G_M = lqS_M\left[\left(\frac{\partial U}{\partial z}\right)^2 + \left(\frac{\partial V}{\partial z}\right)^2\right]. \tag{11.3.61}$$

Using (11.3.40), (11.3.55), (11.3.51), and (11.3.61), the turbulent kinetic energy q^2 is given by

$$\begin{aligned} q^2 &= \Lambda_1 \frac{\varepsilon}{q} = B_1 l(1 - R_f)\frac{P_s}{q} \\ &= B_1 l^2 (1 - R_f) S_M \left[\left(\frac{\partial U}{\partial z}\right)^2 + \left(\frac{\partial V}{\partial z}\right)^2\right]. \end{aligned} \tag{11.3.62}$$

Hence, from (11.3.43), we have the eddy diffusion coefficients by

$$K_M = l^2 S'_M \sqrt{\left(\frac{\partial U}{\partial z}\right)^2 + \left(\frac{\partial V}{\partial z}\right)^2}, \tag{11.3.63}$$

$$K_H = \frac{S_H}{S_M} K_M = l^2 S'_H \sqrt{\left(\frac{\partial U}{\partial z}\right)^2 + \left(\frac{\partial V}{\partial z}\right)^2}, \tag{11.3.64}$$

where

$$S'_M = B_1^{\frac{1}{2}} (1 - R_f)^{\frac{1}{2}} S_M^{\frac{3}{2}}, \qquad S'_H = B_1^{\frac{1}{2}} (1 - R_f)^{\frac{1}{2}} S_M^{\frac{1}{2}} S_H. \tag{11.3.65}$$

Mellor and Yamada propose the following values for the coefficients in (11.3.46):

$$(A_1, B_1, A_2, B_2, C_1) = (0.92, 16.6, 0.74, 10.1, 0.08).$$

In this case, we have

$$(\alpha_1, \alpha_2, \beta_1, \beta_2, \beta_3, \beta_4) = (0.49, 2.58, 2.18, 9.30, 2.73, 12.2).$$

The length scale l of the boundary layer model may be given by the following formula:

$$l = l_0 \frac{kz}{kz + l_0}, \tag{11.3.66}$$

where l_0 is an appropriate length and k is the von Karman constant: $k = 0.40$. l represents the length scale of turbulence. l is proportional to z when z is small, and converges to l_0 for a sufficiently large z. l_0 is sometimes simply specified as a constant value, say $l_0 \approx 200$ m. Eq. (11.3.66) is the more accurate form of (11.3.6) that is used for derivation of the logarithmic law.

Fig. 11.4 shows the relation between R_i and R_f given by (11.3.60). We can see $R_f < \beta_3/\beta_4 = 0.223$ for $R_i > 0$. Fig. 11.5 shows the relation between S_M, S_H and R_f given by (11.3.58), and the relation between S'_M, S'_H and R_f given by (11.3.65).

The results with the Level 2 model can be compared with Monin-Obukhov's similarity theory described in Section 11.3.3. For simplicity, we set $V = 0$. The friction velocity u_* is calculated from (11.3.5), (11.3.41), and (11.3.43) using the expression for q (11.3.62), and the deviation of potential temperature θ_* is from (11.3.10), (11.3.42), and (11.3.43):

$$u_* = \left|\overline{u'w'}\right|^{\frac{1}{2}} = \left(lqS_M\frac{\partial U}{\partial z}\right)^{\frac{1}{2}} = [B_1(1-R_f)S_M^3]^{\frac{1}{4}}l\frac{\partial U}{\partial z}. \tag{11.3.67}$$

$$\theta_* = \frac{\left|\overline{\theta'w'}\right|}{u_*} = \frac{lqS_H}{u_*}\frac{\partial\Theta}{\partial z} = [B_1(1-R_f)S_H^4/S_M]^{\frac{1}{4}}l\frac{\partial\Theta}{\partial z}. \tag{11.3.68}$$

Thus, by assuming $l = kz$ for near-surface flows, the universal functions (11.3.19) are given as

$$\phi_M(\zeta) = \frac{kz}{u_*}\frac{\partial U}{\partial z} = [B_1(1-R_f)S_M^3]^{-\frac{1}{4}}, \tag{11.3.69}$$

$$\phi_H(\zeta) = \frac{kz}{\theta_*}\frac{\partial\Theta}{\partial z} = [B_1(1-R_f)S_H^4/S_M]^{-\frac{1}{4}}, \tag{11.3.70}$$

where, from (11.3.21),

$$\zeta \equiv \frac{z}{L} = \phi_M R_f. \tag{11.3.71}$$

Fig. 11.6 shows the profiles of ϕ_M and ϕ_H with respect to ζ obtained with the above formulas. As $R_f = \alpha_1/\alpha_2 \to 0.19$, ϕ_M, ϕ_H, and ζ asymptotically approach infinity. In the neutral case $R_f = 0$, $\phi_M = [B_1(\beta_1\alpha_1/\beta_3)^3]^{-1/4} = 1.0$ is satisfied. It has been shown that the above profiles are comparable with observational data (Mellor and Yamada 1982).

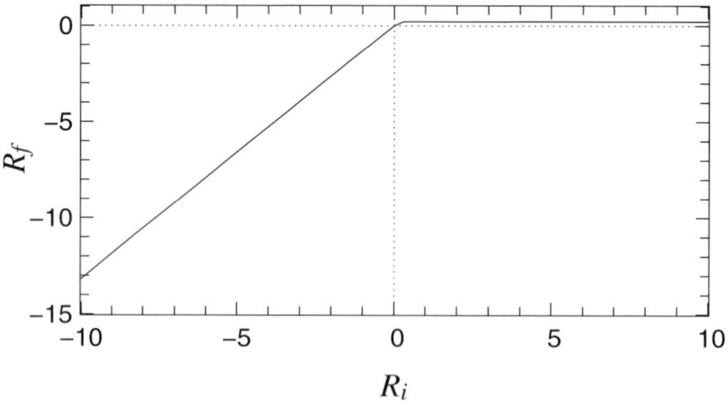

FIGURE 11.4: Relation between the Richardson number R_i and the flux Richardson number R_f.

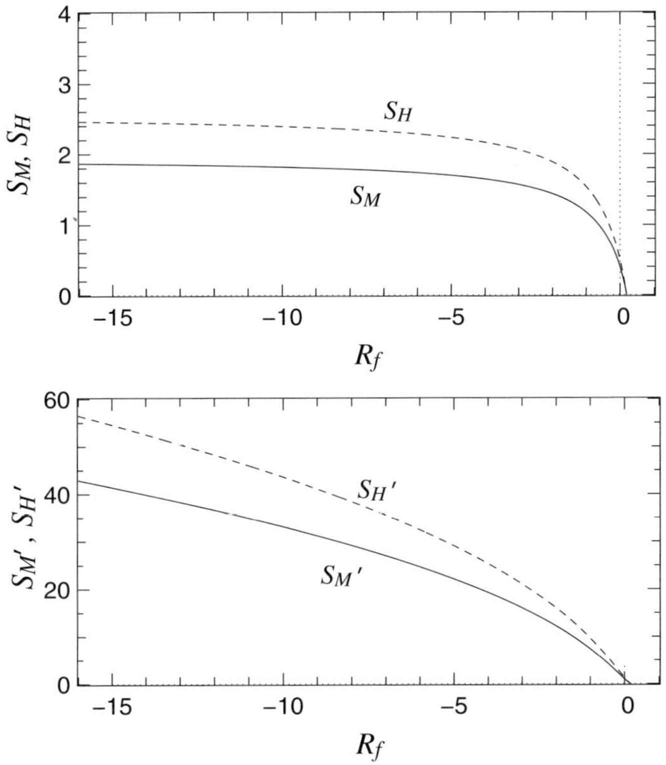

FIGURE 11.5: Relations between the flux Richardson number R_f and the parameters S_M and S_H (top) and S'_M and S'_H (bottom).

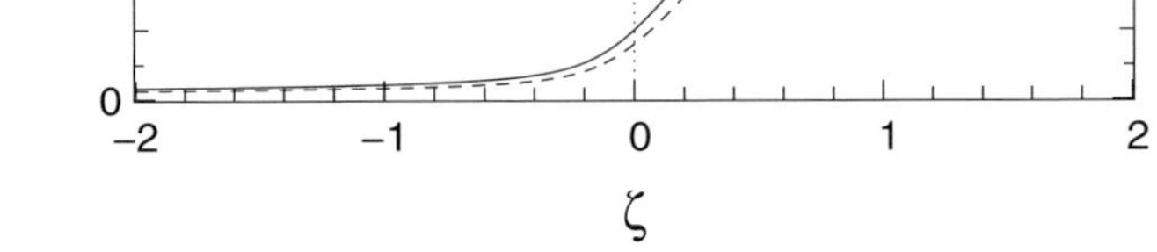

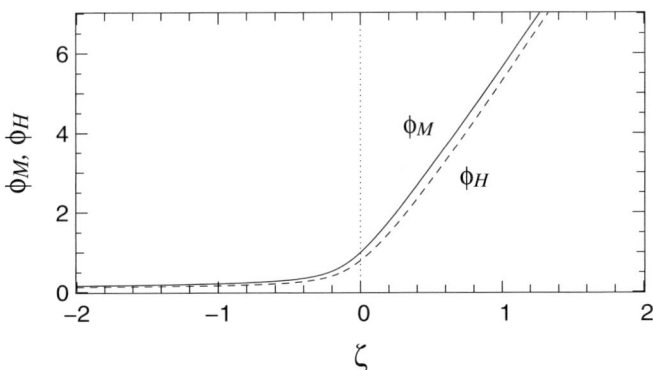

FIGURE 11.6: The profiles of the universal functions ϕ_M and ϕ_H with respect to the non-dimensional height ζ obtained with the Mellor-Yamada Level 2 model.

References and suggested reading

The similarity theory of turbulence in the first half of this chapter can be found in standard textbooks on turbulence: Batchelor (1953) and Monin and Yaglom (1973). The theoretical basis of two-dimensional turbulence in Section 11.1.2 is described in Fjørtoft (1953). Section 11.1.3 on the tracer spectrum is based on Batchelor (1959).

In the second half of this chapter on turbulence models, the classification of the zero-, one-, and two-equation models is described in standard textbooks on engineering. In the meteorological context, the one-equation model by Deardorff (1980) is frequently used. The Smagorinsky model is not only the foundation of the LES model, it is also used for atmospheric numerical models in the evaluation of eddy diffusion in horizontal directions. Section 11.2.3 follows the review by Mellor and Yamada (1982).

Batchelor, G. K., 1953: *The Theory of Homogeneous Turbulence.* Cambridge University Press, Cambridge, UK, 197 pp.

Batchelor, G. K., 1959: Small-scale variation of convected quantities like temperature in turbulent fluid. Part 1: General discussion and the case of small conductivity. *J. Fluid Mech.*, **5**, 113–133.

Deardorff, J. W., 1972: Numerical investigation of neutral and unstable planetary boundary layers. *J. Atmos. Sci.*, **29**, 91–115.

Deardorff, J. W., 1980: Stratocumulus-capped mixed layers derived from a three-dimensional model. *Boundary-Layer Meteorol.*, **18**, 495–527.

Fjørtoft, R., 1953: On the changes in the spectrum distribution of kinetic energy for two dimensional, nondivergent flow. *Tellus*, **5**, 225–230.

Helfand, H. M. and Labraga, J. C., 1988: Design of a nonsingular level 2.5 second-order closure model for the prediction of atmospheric turbulence. *J. Atmos. Sci.*, **45**, 113–132.

Launder, B. E. and Spalding, D. B., 1970: The numerical computation of turbulent flows. *Comp. Meth. Appl. Mech. Eng.*, **3**, 269–289.

Mellor, G. and Yamada, T., 1982: Development of a turbulence closure model for geophysical fluid problems. *Rev. Geophys. Space Phys.*, **20**, 851–875.

Moeng, C.-H. and Wyngaard, J. C., 1988: Spectral analysis of large-eddy simulations of the convective boundary layer. *J. Atmos. Sci.*, **45**, 3573–3587.

Monin, A. and Yaglom, A., 1973: *Statistical Fluid Mechanics.* The MIT Press, Cambridge, MA, 769 pp.

Part II

Atmospheric Structures

In this second part, the basic mechanisms maintaining atmospheric structure are described from various perspectives. The structure of atmospheric general circulation is discussed based on the governing dynamics of the atmospheric general circulation. We consider how atmospheric structure is realized in terms of the latitudinal one-dimensional structure, the vertical one-dimensional structure, the meridional two-dimensional structure, and spherical motions.

In Chapter 12, the energy balance of the entire earth is considered. Following on from this, the energy budget in the latitudinal direction is discussed in Chapter 13, while the vertical structure of the atmosphere is described in Chapter 14.

The atmospheric vertical structure is controlled by moist processes. In Chapter 15, we consider moist convection and its roles in the vertical structure of the atmosphere. The large-scale circulations in the tropical atmosphere can be viewed as moist convective motions. In Chapter 16, as a representative of such moist large-scale circulations in low latitudes, we consider the Walker circulation and the Hadley circulation.

In Chapter 17, the spherical motions on the earth are discussed based on the concepts of wave propagation and the angular momentum budget. In Chapter 18, mid-latitude circulations are considered in terms of meridional circulations. Finally, global-scale mixing motions are reviewed in Chapter 19.

12

Global energy budget

To characterize the global atmosphere, globally averaged quantities of the whole atmosphere can be used. Corresponding to the governing equations of the atmosphere (i.e, the conservations of mass, momentum, and energy), we obtain the global budgets of conserved quantities: mass, angular momentum, and energy.

The global budget of the atmospheric mass plays an important role in the evolution of the atmosphere on the timescale of the earth's history. The change in the total mass of the atmosphere occurs through mass exchange with the solid earth or the ocean at the lowest boundary, or with outer space at the uppermost boundary. On a daily scale, evaporation and precipitation of water at the surface contributes to fluctuation of the total mass of the atmosphere.

The global budget of the total angular momentum of the atmosphere can be given from the global integral of the conservation of angular momentum of the atmosphere around the rotation axis of the earth. The atmospheric total angular momentum changes through the exchange of angular momentum with the ground surface at the lowest boundary. In the case of the earth, in particular, the exchange of angular momentum with the ground has a characteristic latitudinal distribution. This implies that the global angular momentum budget is closely related to the meridional circulation of the atmosphere. Thus, the angular momentum budget will be discussed in Chapters 16 and 18.

In this chapter, we concentrate on the energy budget of the global atmosphere. We review the energy budget of the global atmosphere and explain its theoretical basis. In particular, we show global views of energy budget, energy transformation, and thermal efficiency.

12.1 Energy budget

12.1.1 Effective temperature and global radiative equilibrium

The change in the total energy of the earth (i.e, the atmosphere-ocean-solid earth system) is determined by energy exchange with outer space at the top of the atmosphere. In practice, energy exchange between the earth and outer space takes place in the form of radiation. The energy inflow from outer space to the earth is in the form of solar radiation, and the energy outflow from the earth to outer space is in the form of planetary radiation, which is called *long-wave radiation.* The heat source within the solid earth is negligible compared with these radiative fluxes if the atmosphere-ocean energy budget is considered. Let the total energy of the earth be denoted by E^{tot}, the total inflow of solar radiation by $\mathcal{F}^{isr}$ (isr = incoming shortwave radiation; hereafter referred to as ISR), and the total emission of long-wave radiation by $\mathcal{F}^{olr}$ (olr = outgoing longwave radiation; hereafter referred to as OLR). The equation of the total energy budget is given as

$$\frac{dE^{tot}}{dt} = \mathcal{F}^{isr} - \mathcal{F}^{olr}, \tag{12.1.1}$$

where the unit of $\mathcal{F}^{isr}$ and $\mathcal{F}^{olr}$ is W, and that of E^{tot} is J. In general, since ISR and OLR depend on location, they are referred to as a quantity per unit area with unit of W m^{-2}. However, we use a quantity integrated over the globe to examine the global energy budget in this section (see Section 12.1.2).

We mainly consider the balance of energy over a long time average that statistically could be considered an equilibrium state represented as an annual average, for instance. In this case, the seasonal change of radiative fluxes will be eliminated, and then the time derivative on the left-hand side can be neglected:

$$0 = \mathcal{F}^{isr} - \mathcal{F}^{olr}. \tag{12.1.2}$$

The solar radiation per unit area normal to the direction to the sun at the orbit of the earth is called the *solar constant* $F^{\odot}$. It is inversely proportional to the square of the distance between the earth and the sun. If we assume that the atmosphere is bounded by a sphere with radius R, the total solar radiation radiated to the earth $\mathcal{F}^{sol}$ is written as

$$\mathcal{F}^{sol} = F^{\odot} \cdot \pi R^2, \tag{12.1.3}$$

where πR^2 is the area of the cross section of the earth. Generally, some solar radiation radiated to the earth is reflected back to outer space at the atmosphere or at the ground. The ratio of reflected radiation to total solar flux is called the *planetary albedo* A, which is related as

$$\mathcal{F}^{isr} = (1 - A)\mathcal{F}^{sol}. \tag{12.1.4}$$

If all the atmosphere is regarded as a blackbody with temperature T_e, planetary radiation emitted to outer space $\mathcal{F}^{olr}$ is given by the product of blackbody radiation $\sigma_B T_e^4$ and the area of the surface of the earth:

$$\mathcal{F}^{olr} = \sigma_B T_e^4 \cdot 4\pi R^2, \tag{12.1.5}$$

where σ_B is the Stefan Boltzmann constant (10.1.8). If the outgoing long-wave radiation $\mathcal{F}^{olr}$ is known, on the other hand, the temperature T_e defined by (12.1.5) is called the *effective temperature*. Substitution of (12.1.4) and (12.1.5) into (12.1.2) yields the equation of energy balance:

$$\sigma_B T_e^4 \cdot 4\pi R^2 = (1-A)\, F^{\odot} \cdot \pi R^2. \tag{12.1.6}$$

Thus, effective temperature is expressed in terms of the solar constant and the albedo as

$$T_e = \left[\frac{(1-A)\, F^{\odot}}{4\sigma_B}\right]^{\frac{1}{4}}. \tag{12.1.7}$$

The typical value of the effective temperature of the earth is $T_e = 254.9$ K for $F^{\odot} = 1367.7$ W m^{-2} and $A = 0.3$. This temperature value is considerably lower than the mean surface temperature of the earth, about 288 K. In general, effective temperature T_e is different from surface temperature. The difference can be explained by taking account of the vertical structure of the atmosphere (see Chapter 14). The difference between T_e and the surface temperature is caused by the *greenhouse effect*.

In general, the outward planetary radiation at the top of the atmosphere F^{olr} depends on the vertical profiles of atmospheric temperature and absorbing quantities. For illustrative purposes, let us simply assume that F^{olr} is determined by a representative value of atmospheric temperature $[T]$, which is chosen as a global mean of the surface temperature or the averaged temperature over the whole atmosphere. As a simplest form of the relation between F^{olr} and $[T]$, we use a linear function as

$$\frac{\mathcal{F}^{olr}}{4\pi R^2} = F_0 + B([T] - T_0), \tag{12.1.8}$$

where F_0 and B are empirically determined constants and T_0 is a reference temperature (e.g., $T_0 = 273.15$ K). The effects of the vertical distributions of temperature and absorbing quantities are represented by these constants. If the mean temperature $[T]$ is close to the effective temperature T_e, we obtain the values of F_0 and B by linearizing (12.1.5) about T_0. Using $[T] = T_0 + \Delta T$, we have

$$\sigma_B [T]^4 = \sigma_B (T_0 + \Delta T)^4 \approx \sigma_B T_0^4 + 4\sigma_B T_0^3 \Delta T. \tag{12.1.9}$$

Thus, in this case, we have

$$F_0 = \sigma_B T_0^4 = 315.7 \text{ W m}^{-2}, \tag{12.1.10}$$

$$B = 4\sigma_B T_0^3 = 4.6 \text{ W m}^{-2} \text{ K}^{-1}. \tag{12.1.11}$$

In practice, however, since $[T]$ is different from T_e, empirically determined values of F_0 and B are used to estimate $\mathcal{F}^{olr}$. For instance, Crowley and North (1991) use $F_0 = 210$ W m^{-2} and $B = 2.1$ W m^{-2} K^{-1} for their energy budget analysis. Making use of (12.1.8), we rewrite the equation of the energy balance (12.1.6) as

$$F_0 + B([T] - T_0) = (1-A)\, \frac{F^{\odot}}{4}. \tag{12.1.12}$$

This linearized equation will be used in the following theoretical analysis of the energy budget.

Let us introduce *climate sensitivity*, which is a measure of the sensitivity of the atmospheric mean state to changes in external parameters. For example, if we choose solar radiation $F^{\odot}$ as an external parameter, climate sensitivity can be defined by

$$\beta = \frac{F^{\odot}}{[T]}\frac{d[T]}{dF^{\odot}}, \tag{12.1.13}$$

which represents the change in mean temperature $[T]$ for a given change in the external parameter $F^{\odot}$. Using the equation of energy balance (12.1.12) and assuming that F_0, B, and A are constant, we estimate climate sensitivity with respect to solar radiation as

$$\beta = \frac{(1-A)F^{\odot}}{4B[T]} = \frac{\sigma_B T_e^4}{B[T]}, \tag{12.1.14}$$

where the effective temperature (12.1.7) is used. In the case of $T_0 = T_e = [T]$, we have $\beta = 0.25$ if we use a theoretical value (12.1.11). If we instead use a more realistic value $B = 2.1$ W m^{-2} K^{-1}, we obtain $\beta = 0.40$.

12.1.2 Energy conversion

The total energy of the atmosphere consists of kinetic energy, potential energy, and internal energy. The balance of each energy is given by (1.2.35), (1.2.39), and (1.2.42), respectively, and the conservation of total energy is given by (1.2.47). These equations are rewritten in advective form as

$$\rho\frac{d}{dt}\frac{v^2}{2} = -\frac{\partial}{\partial x_j}(pv_j - \sigma'_{ij}v_i) + p\nabla\cdot\boldsymbol{v} - \rho\boldsymbol{v}\cdot\nabla\Phi - \varepsilon, \tag{12.1.15}$$

$$\rho\frac{d\Phi}{dt} = \rho\boldsymbol{v}\cdot\nabla\Phi, \tag{12.1.16}$$

$$\rho\frac{du}{dt} = -\nabla\cdot\boldsymbol{F}^{ene} - p\nabla\cdot\boldsymbol{v} + \varepsilon, \tag{12.1.17}$$

$$\rho\frac{d}{dt}\left(\frac{v^2}{2} + u + \Phi\right) = -\frac{\partial}{\partial x_j}\left(pv_j - v_i\sigma'_{ij} + F_j\right), \tag{12.1.18}$$

where σ'_{ij} is the viscous stress tensor, ε is the dissipation rate, and $\boldsymbol{F}^{ene}$ is the heat flux given by the sum of the radiative flux $\boldsymbol{F}^{rad}$ and the thermal diffusion flux $\boldsymbol{F}^{therm}$:

$$\boldsymbol{F}^{ene} = \boldsymbol{F}^{rad} + \boldsymbol{F}^{therm}. \tag{12.1.19}$$

If the effect of the hydrological cycle is considered, the transport of internal energy contains a contribution from the transport of water vapor. In this case, the heat flux in (12.1.19) is further divided as

$$\boldsymbol{F}^{therm} = \boldsymbol{F}^{sh} + \boldsymbol{F}^{lh}. \tag{12.1.20}$$

$\boldsymbol{F}^{lh}$ is called the latent heat flux due to water vapor, and the thermal diffusion flux $\boldsymbol{F}^{sh}$ is called the sensible heat flux. The energy budget in the case that the hydrological cycle exists will be further considered in Section 12.2.

The energy budget of the whole atmosphere can be given by integrating (12.1.15)–(12.1.18) over the whole domain of the atmosphere D. Let S denote the boundaries of the atmosphere (i.e., the top and bottom boundaries, S_T and S_B, respectively). The top boundary S_T is set at a sufficiently high altitude, and the bottom boundary S_B is the ground surface of the solid earth or the sea surface. The integrals of the equations of kinetic energy, potential energy, internal energy, and total energy are respectively written as

$$\frac{d}{dt}K = \mathcal{B}_S + \mathcal{W} - \mathcal{C}_G - \mathcal{D}, \tag{12.1.21}$$

$$\frac{d}{dt}G = \mathcal{C}_G, \tag{12.1.22}$$

$$\frac{d}{dt}I = \mathcal{F} - \mathcal{W} + \mathcal{D}, \tag{12.1.23}$$

$$\frac{d}{dt}E^{tot} = \mathcal{B}_S + \mathcal{F}, \tag{12.1.24}$$

where

$$K \equiv \int_D \rho\frac{v^2}{2}\,dV, \qquad G \equiv \int_D \rho\Phi\,dV, \qquad I \equiv \int_D \rho u\,dV,$$

$$E^{tot} = \int_D \rho\left(\frac{v^2}{2} + u + \Phi\right) dV = K + I + G. \tag{12.1.25}$$

K, G, I, and E^{tot} are the domain integrals of kinetic energy, potential energy, internal energy, and total energy of the atmosphere, respectively.† In (12.1.21)–(12.1.24), the source and sink terms and the transformation terms of the energies are given by

$$\mathcal{W} = \int_D p\nabla\cdot\boldsymbol{v}\,dV, \qquad \mathcal{C}_G = \int_D \rho\boldsymbol{v}\cdot\nabla\Phi\,dV, \qquad \mathcal{D} = \int_D \varepsilon\,dV, \tag{12.1.26}$$

$$\mathcal{F} = -\int_D \nabla\cdot\boldsymbol{F}^{ene}\,dV = -\int_S \boldsymbol{F}^{ene}\cdot\boldsymbol{n}\,dS, \tag{12.1.27}$$

$$\mathcal{B}_S = -\int_D \frac{\partial}{\partial x_j}\left(pv_j - v_i\sigma'_{ij}\right) dV = -\int_{S_B} (p\boldsymbol{v}\cdot\boldsymbol{n} - \sigma'_{ij}v_in_j)\,dS, \tag{12.1.28}$$

where $\boldsymbol{n}$ denotes the outward unit vector normal to the boundaries; $\boldsymbol{n}$ is upward at the top boundary S_T, and is approximately downward at the bottom boundary S_B. These transformations of the energies are schematically shown in Fig. 12.1.

---

†Hereafter, E^{tot} is the total energy of the atmosphere, while, in (12.1.1), E^{tot} represents that of the total earth system.

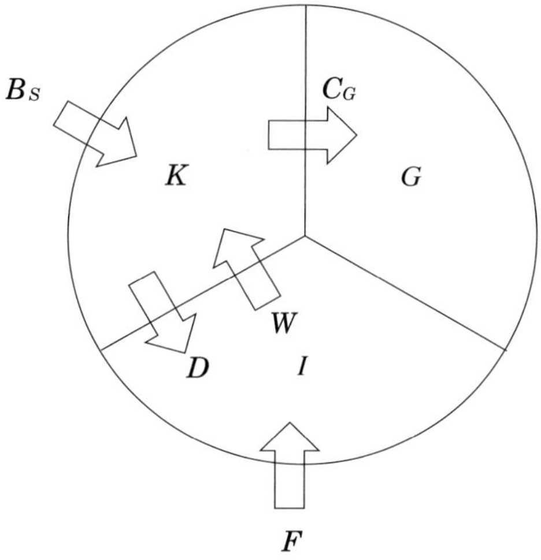

FIGURE 12.1: The transformations of energies. K, G, and I are the domain integrals of kinetic energy, potential energy, and internal energy, respectively. The symbols B_S, C_G, W, D, and F are energy transformation terms.

Over a long time average, the time derivatives on the left-hand side of (12.1.21)–(12.1.24) become negligible. In this case, we have the following balances:

$$\mathcal{B}_S + \mathcal{W} - \mathcal{C}_G - \mathcal{D} = 0, \qquad (12.1.29)$$

$$\mathcal{C}_G = 0, \qquad (12.1.30)$$

$$\mathcal{F} - \mathcal{W} + \mathcal{D} = 0, \qquad (12.1.31)$$

$$\mathcal{B}_S + \mathcal{F} = 0. \qquad (12.1.32)$$

The exchange of energy between the atmosphere and the outside of the atmosphere is in the form of $\mathcal{B}_S$ and $\mathcal{F}$. The term $\mathcal{B}_S$ exists only when the wind velocity of the atmosphere at the boundaries is nonzero (i.e., the interface between the atmosphere and the ocean generally has motion, which might have a contribution to $\mathcal{B}_S$). We neglect this term, however, by assuming that this contribution is small. Then, we have the following balances of energy transformation:

$$\mathcal{F} = 0, \qquad (12.1.33)$$

$$\mathcal{W} = \mathcal{D}. \qquad (12.1.34)$$

The latter is an important consequence of the kinetic energy budget, and is the balance between the generation term of kinetic energy $\mathcal{W}$ and the dissipation term $\mathcal{D}$. $\mathcal{W}$ can also be interpreted as follows. Using hydrostatic pressure p_s and density ρ_s which satisfies

$$0 = -\nabla p_s - \rho_s \nabla \Phi, \qquad (12.1.35)$$

and $p' = p - p_s$, $\rho' = \rho - \rho_s$, the sum of the two source terms of kinetic energy is rewritten from (12.1.26) as

$$\begin{aligned}\mathcal{W}-\mathcal{C}_G &= \int_D (p\nabla\cdot\boldsymbol{v}-\rho\boldsymbol{v}\cdot\nabla\Phi)\,dV \\ &= -\int_D \boldsymbol{v}\cdot(\nabla p+\rho\cdot\nabla\Phi)\,dV = -\int_D \boldsymbol{v}\cdot(\nabla p'+\rho'\nabla\Phi)\,dV \\ &= -\int_D \boldsymbol{v}\cdot\nabla p'\,dV-\int_D \rho\left(-\frac{\rho'}{\rho}wg\right)\,dV, \end{aligned} \tag{12.1.36}$$

where $\Phi = gz$ is used. Although the choice of hydrostatic state is not unique, the second term on the right is regarded as the work done by the buoyancy force, and the first term is generally small.† Over a long time average, making use of (12.1.30), we have

$$\mathcal{W} \approx -\int_D \rho\left(-\frac{\rho'}{\rho}wg\right)\,dV. \tag{12.1.37}$$

This means that $\mathcal{W}$ is interpreted as the work done by the buoyancy force in the whole domain. It may be noted that, since the potential energy G is statistically constant in an equilibrium state, the work done by the buoyancy force has no contribution to the budget of potential energy.

Flux $\mathcal{F}$ gets contributions from radiation, sensible heat, and latent heat fluxes as given by (12.1.19) and (12.1.20). Sensible and latent heat fluxes exist only at the bottom boundary of the atmosphere S_B. As for radiative flux, we decompose it into solar flux and planetary flux according to its wavelength:

$$\boldsymbol{F}^{rad} = \boldsymbol{F}^{radl}+\boldsymbol{F}^{rads}. \tag{12.1.38}$$

$\boldsymbol{F}^{radl}$ is the radiative flux of the longer wavelength region, corresponding to planetary radiation, and $\boldsymbol{F}^{rads}$ is that of the shorter wavelength region, corresponding to solar radiation. We furthermore distinguish the contributions of each flux at the top and bottom boundaries:

$$\begin{aligned}\mathcal{F}_B^{radl} &\equiv -\int_{S_B}\boldsymbol{F}^{radl}\cdot\boldsymbol{n}\,dS, & \mathcal{F}_B^{rads} &\equiv -\int_{S_B}\boldsymbol{F}^{rads}\cdot\boldsymbol{n}\,dS, \\ \mathcal{F}_T^{radl} &\equiv \int_{S_T}\boldsymbol{F}^{radl}\cdot\boldsymbol{n}\,dS, & \mathcal{F}_T^{rads} &\equiv \int_{S_T}\boldsymbol{F}^{rads}\cdot\boldsymbol{n}\,dS, \\ \mathcal{F}^{therm} &\equiv -\int_{S_B}\boldsymbol{F}^{therm}\cdot\boldsymbol{n}\,dS, \end{aligned}$$

and

$$\mathcal{F} = -\mathcal{F}_T^{rads}-\mathcal{F}_T^{radl}+\mathcal{F}_B^{rads}+\mathcal{F}_B^{radl}+\mathcal{F}^{therm}. \tag{12.1.39}$$

Here, the fluxes at the top of the atmosphere S_T and the bottom of the atmosphere S_B are designated by subscripts T and B, respectively. Note that in this convention,

†For geostrophic components, $\boldsymbol{v}_g\cdot\nabla p'_g = 0$ is satisfied. For turbulent motion, we also use the similar assumption (11.2.31).

each of the integrated fluxes on the boundaries is positive if it is upward. The two fluxes at the top boundary, $\mathcal{F}_T^{rads}$ and $\mathcal{F}_T^{radl}$, are related to net incoming solar flux $\mathcal{F}^{isr}$ and the infrared flux emitted from the atmosphere $\mathcal{F}^{olr}$, respectively:

$$\mathcal{F}_T^{radl} = \mathcal{F}^{olr}, \tag{12.1.40}$$
$$\mathcal{F}_T^{rads} = -\mathcal{F}^{isr} = -(1-A)\mathcal{F}^{sol}, \tag{12.1.41}$$

where (12.1.4) is used. The flux defined by (12.1.39) is the sum of the incoming energy fluxes to the atmosphere at the top and bottom boundaries. These fluxes are also grouped into contributions at each of the top and bottom boundaries as

$$\mathcal{F}_T = \mathcal{F}_T^{rads} + \mathcal{F}_T^{radl}, \tag{12.1.42}$$
$$\mathcal{F}_B = \mathcal{F}_B^{rads} + \mathcal{F}_B^{radl} + \mathcal{F}^{therm}, \tag{12.1.43}$$
$$\mathcal{F} = -\mathcal{F}_T + \mathcal{F}_B. \tag{12.1.44}$$

In general, heat transfer from the earth's interior at the ground is negligible for the atmospheric energy budget. In the case of a long time average where the tendency of total energy is negligible, we have

$$\mathcal{F}_T = \mathcal{F}_B = 0 \tag{12.1.45}$$

(i.e., no net flux exists at both the top and bottom boundaries). This relation corresponds to (12.1.2).

The observed annual global mean energy budget is shown in Fig. 12.2 after Kiehl and Trenberth (1997). The values are shown by energy fluxes per unit area. Under the assumption that the depth of the atmosphere is sufficiently shallow compared with the radius of the earth, let $\mathcal{S}$ denote the surface area of the earth:

$$\mathcal{S} = \int_{S_T} dS = \int_{S_B} dS = 4\pi R^2. \tag{12.1.46}$$

The values of the terms in the energy balance (12.1.39) are estimated as

$$\begin{aligned} -\mathcal{F}_T^{rads}/\mathcal{S} &= \mathcal{F}_T^{radl}/\mathcal{S} = 342-107 = 235 \ \text{W m}^{-2}, \\ \mathcal{F}_B^{rads}/\mathcal{S} &= -168 \ \text{W m}^{-2}, \\ \mathcal{F}_B^{radl}/\mathcal{S} &= 390-324 = 66 \ \text{W m}^{-2}, \\ \mathcal{F}_B^{therm}/\mathcal{S} &= 78+24 = 102 \ \text{W m}^{-2}. \end{aligned}$$

Each of the upward and downward radiative fluxes plays important roles in the global budget. This will be discussed in Chapter 14. It is also important to understand the relative magnitude of fluxes. The planetary albedo is the ratio of upward solar flux to downward solar flux at the top of the atmosphere, and is estimated as $A = 107/342 = 31\%$. Thermal flux at the surface $\mathcal{F}^{therm}$ is composed of sensible and latent heat fluxes at the surface, which will be described in Section 12.2.

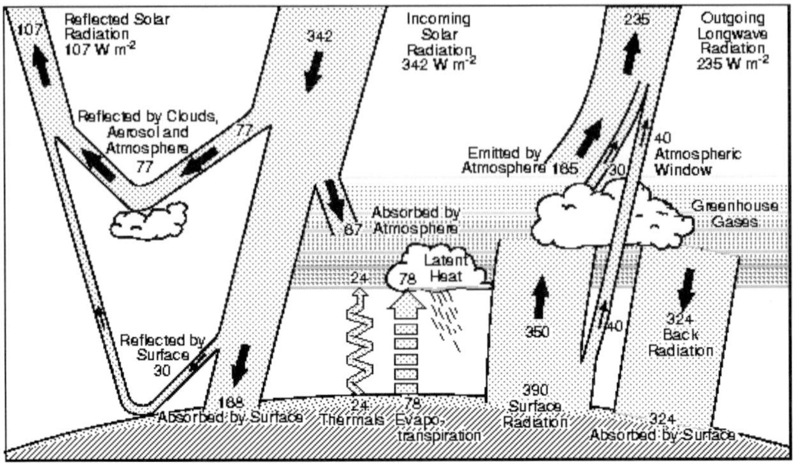

FIGURE 12.2: The annual and global mean energy balance. The energy fluxes per unit area are shown with unit W m^{-2}. After Kiehl and Trenberth (1997) by permission of the American Meteorological Society.

12.1.3 Total potential energy

The sum of internal energy and potential energy is called *total potential energy.* The term, *total* potential energy, is in contrast to *available* potential energy defined in the next section. The global integral of total potential energy is defined as

$$P \equiv I + G = \int_D \rho(u + \Phi)\, dV. \tag{12.1.47}$$

In the case that the atmosphere is in hydrostatic balance, we can derive some useful relations on the total potential energy of an air column per unit square. We use the hydrostatic balance

$$0 = -\frac{\partial p}{\partial z} - \rho g. \tag{12.1.48}$$

Dividing (12.1.25) by the area of the earth surface $\mathcal{S} = 4\pi R^2$, we obtain the energies of an air column as

$$\begin{aligned} K_{col} &\equiv \frac{K}{\mathcal{S}} = \int_{z_0}^{\infty} \rho \frac{v^2}{2} dz = \int_0^{p_0} \frac{v^2}{2} \frac{dp}{g}, \\ I_{col} &\equiv \frac{I}{\mathcal{S}} = \int_0^{p_0} u \frac{dp}{g}, \qquad G_{col} \equiv \frac{G}{\mathcal{S}} = \int_0^{p_0} \Phi \frac{dp}{g}, \\ E_{col}^{tot} &\equiv \frac{E^{tot}}{\mathcal{S}} = \int_0^{p_0} \left(\frac{v^2}{2} + u + \Phi \right) \frac{dp}{g} = K_{col} + I_{col} + G_{col}, \end{aligned} \tag{12.1.49}$$

where z_0 is the height of the earth surface and p_0 is the surface pressure. Note that the right-hand side of these equations represents horizontally averaged quantities; we omit the symbol for the horizontal average $(\bar{\ })$.

A simple relation holds between potential energy and the internal energy under hydrostatic balance. Multiplying (12.1.48) by z and integrating the product over an air column, we have

$$\begin{aligned} 0 &= -\int_{z_0}^{\infty} z\frac{\partial p}{\partial z}dz - \int_{z_0}^{\infty} \rho g z dz \\ &= p_0 z_0 + \int_{z_0}^{\infty} p dz - \int_{z_0}^{\infty} \rho z g dz, \end{aligned} \tag{12.1.50}$$

where we use $p \to 0$ as $z \to \infty$ at the top boundary. This relation is a special case of the *virial theorem*.† In particular, in the case $z_0 = 0$ and $g =$ const. such that $\Phi = gz$, we obtain‡

$$G_{col} = \int_0^{\infty} \rho g z dz = \int_0^{\infty} p dz. \tag{12.1.51}$$

Thus, the total potential energy of a unit column is

$$P_{col} \equiv I_{col} + G_{col} = \int_0^{\infty} (\rho u + p) dz = \int_0^{\infty} \rho h dz = \int_0^{p_0} h \frac{dp}{g}, \tag{12.1.52}$$

where $h = u + p/\rho$ is enthalpy.§ Total potential energy is determined as a thermodynamic quantity and does not depend on fluid motions. In particular, in the case of hydrostatic balance, total potential energy is uniquely determined by temperature distribution on isobaric surfaces. For the ideal gas with constant specific heats, we have

$$P_{col} = \int_0^{p_0} C_p T \frac{dp}{g}. \tag{12.1.53}$$

In this case, we have simple relations

$$\frac{G_{col}}{I_{col}} = \frac{\int_0^{p_0} R_d T dp/g}{\int_0^{p_0} C_v T dp/g} = \frac{R_d}{C_v}, \qquad \frac{G_{col}}{P_{col}} = \frac{R_d}{C_p}, \qquad \frac{I_{col}}{P_{col}} = \frac{C_v}{C_p}. \tag{12.1.54}$$

Using potential temperature θ in (1.1.52), total potential energy is further rewritten as

$$\begin{aligned} P_{col} &= \int_0^{p_0} C_p \theta \left(\frac{p}{p_{00}}\right)^{\kappa} \frac{dp}{g} = \frac{1}{1+\kappa} \frac{C_p}{g} \frac{1}{p_{00}^{\kappa}} \left(\theta p^{\kappa+1}\Big|_0^{p_0} + \int_{\theta_0}^{\infty} p^{\kappa+1} d\theta \right) \\ &= \frac{1}{1+\kappa} \frac{C_p}{g} \frac{1}{p_{00}^{\kappa}} \int_0^{\infty} p^{\kappa+1} d\theta, \end{aligned} \tag{12.1.55}$$

†There are several expressions of the virial theorem in fluid dynamics. See Chandrasekhar (1961) and Lebovitz (1961). In Salmon (1988), the virial theorem is derived from the Lagrangian expression. Derivations of the virial theorem in macroscopic representation of a system of particles can be found in Landau and Lifshitz (1996) and Chandrasekhar (1958).

‡The following expressions can be generalized for the case $z_0 \geq 0$ by using the notation $\rho = 0$ and $p = p_0$ for $z < z_0$.

§It should be noted that total potential energy equals the vertical integral of enthalpy only when the acceleration of gravity g is constant.

where $\kappa = R_d/C_p$ and $p_{00} = 1000$ hPa. In this expression, we formally specify $p = p_0$ for $\theta < \theta_0$, where p_0 is the surface pressure.

The energy budget in the global domain is written as the exchange between total potential energy and kinetic energy. From (12.1.21)–(12.1.23), we have

$$\frac{d}{dt}K = \mathcal{W} - \mathcal{D} - \mathcal{C}_G, \tag{12.1.56}$$

$$\frac{d}{dt}P = \mathcal{F} - \mathcal{W} + \mathcal{D} + \mathcal{C}_G. \tag{12.1.57}$$

Here we have neglected the contributions of $\mathcal{B}_S$. Based on the exchange between the global integral of enthalpy and that of kinetic energy, we have different forms of the transformation from total potential energy to kinetic energy $\mathcal{W} - \mathcal{C}_G$ as shown below. From (1.2.34), the equation of kinetic energy is given as

$$\rho\frac{d}{dt}\frac{v^2}{2} = -\boldsymbol{v}\cdot\nabla p - \rho\boldsymbol{v}\cdot\nabla\Phi + \frac{\partial}{\partial x_j}\sigma'_{ij}v_i - \varepsilon. \tag{12.1.58}$$

The sum of the first and the second terms on the right-hand side is rewritten as

$$\begin{aligned}
-\boldsymbol{v}\cdot\nabla p - \rho\boldsymbol{v}\cdot\nabla\Phi &= -\frac{dp}{dt} + \frac{\partial p}{\partial t} - \rho\left(\frac{d\Phi}{dt} - \frac{\partial\Phi}{\partial t}\right) \\
&= -\frac{dp}{dt} - \rho\left(\frac{d\Phi}{dt} - \left.\frac{\partial\Phi}{\partial t}\right|_p\right) \\
&= -\omega - \rho\left[\nabla_H\cdot(\Phi\boldsymbol{v_H}) - \frac{\partial}{\partial p}(\Phi\omega)\right],
\end{aligned} \tag{12.1.59}$$

where $\omega = dp/dt$ and the symbol $|_p$ denotes a derivative along a constant p-surface. From (1.2.64), the equation of enthalpy is given by

$$\rho\frac{dh}{dt} = \omega + \varepsilon - \nabla\cdot\boldsymbol{F}^{ene}. \tag{12.1.60}$$

Using $\boldsymbol{v} = 0$ at the boundary, the global integrals of (12.1.58) and (12.1.60) give

$$\frac{d}{dt}K = \mathcal{C} - \mathcal{D} + \mathcal{B}_P, \tag{12.1.61}$$

$$\frac{d}{dt}I_h = \mathcal{F} - \mathcal{C} + \mathcal{D}, \tag{12.1.62}$$

and the sum of the two is written as

$$\frac{d}{dt}(K + I_h) = \mathcal{F} + \mathcal{B}_P, \tag{12.1.63}$$

where

$$I_h \equiv \int_D \rho h\, dV = \int_{S_B}\int_{z_0}^{\infty} \rho h dz\, dS = \int_{S_B}\int_0^{p_0} h\frac{dp}{g}\, dS, \tag{12.1.64}$$

$$\mathcal{C} \equiv -\int_D \omega\, dV = -\int_{S_B}\int_{z_0}^{\infty} \omega\, dz\, dS = -\int_{S_B}\int_0^{p_0} \omega\alpha\frac{dp}{g}\, dS, \tag{12.1.65}$$

$$\begin{aligned}\mathcal{B}_P &\equiv -\int_D \rho\left[\nabla_H\cdot(\Phi\boldsymbol{v_H})+\frac{\partial}{\partial p}(\Phi\omega)\right]dV\\ &= -\int_{S_B}\int_0^{p_0}\left[\nabla_H\cdot(\Phi\boldsymbol{v_H})+\frac{\partial}{\partial p}(\Phi\omega)\right]\frac{dp}{g}\,dS\\ &= -\int_{S_B}\omega\big|_{p=p_0} z_0\,dS,\end{aligned} \tag{12.1.66}$$

and $\alpha = 1/\rho$ is the specific volume. In (12.1.66), we have assumed that $\boldsymbol{v}_H = 0$ at the surface and g is constant. These equations are also given directly from the global integral of the energy equations in isobaric coordinates, (3.3.48), (3.3.50), and (3.3.51). By comparing (12.1.56) with (12.1.61), we generally have

$$\mathcal{W} - \mathcal{C}_G = \mathcal{C} + \mathcal{B}_P. \tag{12.1.67}$$

This is just the global integral of (12.1.59). In particular, in the case of $z_0 = 0$, we have $\mathcal{B}_P = 0$ and the global integral of enthalpy agrees with total potential energy: $I_h = P$. We also have $\mathcal{W} - \mathcal{C}_G = \mathcal{C}$ in this case. Over a long time average, $\mathcal{C}_G = 0$ from (12.1.30). Thus, using (12.1.34), we obtain

$$\mathcal{W} = \mathcal{D} = \mathcal{C}. \tag{12.1.68}$$

In Fig. 12.3, the latitudinal distribution of column-integrated energies are compared; these values are calculated for a typical atmospheric state. As shown by (12.1.54), simple relations $G_{col}/I_{col} = R_d/C_v \approx 0.4$ and $I_{col}/P_{col} = C_v/C_p \approx 0.7$ hold. It can be found that the domain integral of kinetic energy of the atmosphere is much smaller than that of total potential energy. Using a typical velocity scale $V \approx 10$ m s^{-1}, average temperature $T \approx 250$ K, and the specific heat at constant pressure $C_p \approx 1000$ J kg^{-1} K^{-1}, the ratio of the two energies can be estimated as

$$\frac{K_{col}}{P_{col}} \approx \frac{V^2}{C_pT} \approx \frac{1}{2500}. \tag{12.1.69}$$

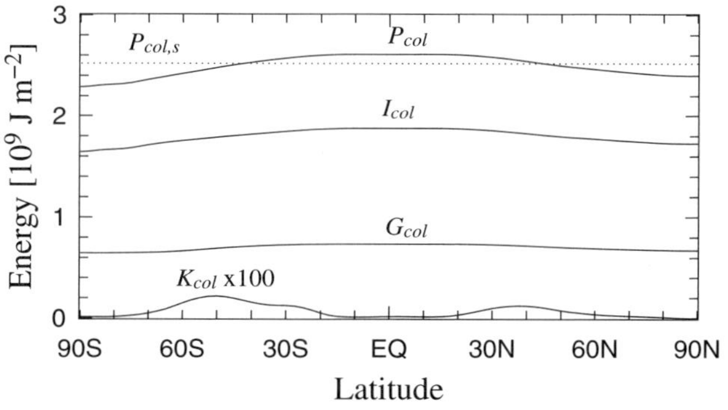

FIGURE 12.3: Latitudinal distributions of the various energies of an air column per unit square. I_{col}: internal energy, G_{col}: potential energy, K_{col}: kinetic energy, P_{col}: enthalpy, and $P_{col,s}$: enthalpy of a reference state used for the calculation of available potential energy (see Fig. 12.4). The values of kinetic energy are multiplied by 100.

12.1.4 Available potential energy

As shown in Fig. 12.3, total potential energy is a thousand times larger than kinetic energy; thus, direct comparison of the two energies is not appropriate for the energy budget of the atmosphere. This fact motivates us to define the available part of total potential energy that can be convertible to kinetic energy so that we can consider the kinetic energy budget. Such a part of potential energy is called *available potential energy.*† The following characteristics are required for available potential energy.

- The sum of available potential energy and kinetic energy is conserved under adiabatic motion.

- Available potential energy is determined solely by the distribution of mass.

- Available potential energy is zero if stratification is horizontally uniform.

- Available potential energy is always positive unless stratification is horizontally uniform.

To fulfill the above requirements, available potential energy can be defined as the difference between the total potential energy of the atmospheric state to be considered and that of a suitably chosen reference state. Here, the choice of reference state introduces arbitrariness, so that available potential energy is not uniquely defined. In general, not all the available potential energy thus defined can be convertible into kinetic energy.

As a reference state, one can choose a state with adiabatic redistribution of mass; all the air parcels of a given state are kinematically redistributed to a horizontally uniform state by keeping their potential temperature. A typical example of the distributions of potential temperature of the two states are shown in Fig. 12.4. We can show that the average pressure over an isentropic surface $\overline{p}(\theta)$ does not change under adiabatic redistribution, where $\overline{p}(\theta)$ is defined as

$$\overline{p}(\theta) \quad \equiv \quad \frac{\int p(\boldsymbol{x_h}, \theta)\, dS}{\mathcal{S}}. \tag{12.1.70}$$

$\boldsymbol{x_h}$ is the horizontal projection of a position vector, and dS that of an area element. In this expression, we formally assume that $p(\boldsymbol{x_h}, \theta)$ is set to the surface pressure p_0 below the ground (i.e., in the region where θ is smaller than the surface value θ_0). Integrating the conservation of mass in isentropic coordinates (3.3.65) along an isentropic surface and integrating in the vertical direction, we have

$$\frac{d}{dt}\int\int_\theta^\infty \rho_\theta(\boldsymbol{x_h}, \theta')d\theta'\, dS + \int \rho_\theta(\boldsymbol{x_h}, \theta')\, \dot{\theta}\, dS\bigg|_\theta^\infty \quad = \quad 0. \tag{12.1.71}$$

†Available potential energy is formulated by Lorenz (1955). There are different ways to define available potential energy. A unified view of available potential energy is presented by Shepherd (1993).

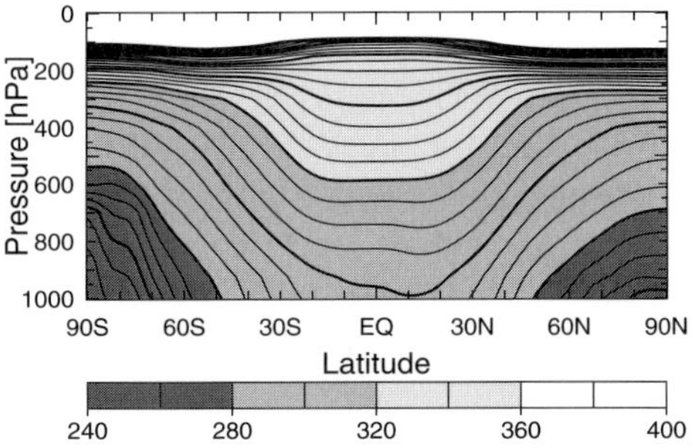

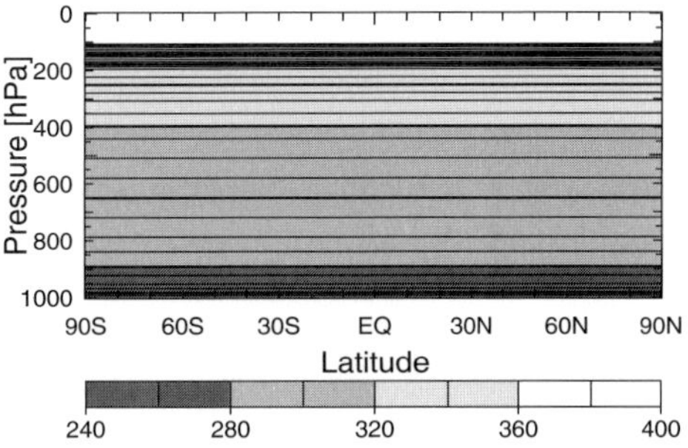

FIGURE 12.4: Meridional distribution of annually averaged potential temperature and its reference state given by adiabatic redistribution. The difference of the total energy between the two states is the available potential energy. The contour interval is 5 K.

Using (12.1.70) and (3.3.62), we have

$$\int \int_{\theta}^{\infty} \rho_{\theta}(\boldsymbol{x_h}, \theta')d\theta'\, dS \quad = \quad -\frac{1}{g}\int dS \cdot \overline{p}(\theta). \tag{12.1.72}$$

For the adiabatic motion $\dot{\theta} = 0$, therefore, (12.1.71) becomes

$$\frac{d}{dt}\overline{p}(\theta) \quad = \quad 0. \tag{12.1.73}$$

Now we add subscript s to the quantities of the reference state. The pressure distribution of the reference state is denoted by $p_s(\theta) = \overline{p}(\theta)$. Available potential energy is defined as

$$A \quad \equiv \quad P - P_s, \tag{12.1.74}$$

and available potential energy per unit air column is

$$A_{col} \quad \equiv \quad P_{col} - P_{col,s}. \tag{12.1.75}$$

Since from (12.1.55)

$$P_{col,s} \quad = \quad \frac{1}{1+\kappa}\frac{C_p}{g}\frac{1}{p_{00}^{\kappa}}\int_0^{\infty} \overline{p}^{\kappa+1} d\theta, \tag{12.1.76}$$

we obtain

$$A_{col} \quad = \quad \frac{1}{1+\kappa}\frac{C_p}{g}\frac{1}{p_{00}^{\kappa}}\int_0^{\infty} \left(\overline{p^{\kappa+1}} - \overline{p}^{\kappa+1}\right) d\theta. \tag{12.1.77}$$

It can be shown that this quantity is positive if the difference between pressure p and $\overline{p}$ is small enough. Letting the average potential temperature on a pressure surface be denoted by $\theta = \overline{\theta}(p)$ and deviations from the reference state by $p' = p - \overline{p}$ and $\theta' = \theta - \overline{\theta}$, we have

$$p' \quad \approx \quad \frac{d\overline{p}}{d\theta}\theta' \quad \approx \quad \left(\frac{d\overline{\theta}}{dp}\right)^{-1} \theta'. \tag{12.1.78}$$

Therefore, available potential energy is approximated as

$$\begin{aligned} A_{col} &\approx \frac{\kappa}{2}\frac{C_p}{g}\frac{1}{p_{00}^{\kappa}}\int_0^{\infty} \overline{p^{\kappa+1}\left(\frac{p'}{\overline{p}}\right)^2}\, d\theta \\ &\approx \frac{\kappa}{2}\frac{C_p}{g}\frac{1}{p_{00}^{\kappa}}\int_0^{p_0} \overline{p}^{\kappa-1}\overline{\theta'^2}\left(-\frac{d\overline{\theta}}{dp}\right)^{-1} dp. \end{aligned} \tag{12.1.79}$$

Since the atmosphere is stably stratified with $-\frac{d\overline{\theta}}{dp} > 0$ in general, available potential energy takes a positive value.

If the reference state is independent of time, the tendency of available potential energy equals that of total potential energy. Using (12.1.67) under the condition $\mathcal{B}_P = 0$ and (12.1.74), Eqs. (12.1.56) and (12.1.57) become

$$\frac{d}{dt}A = \mathcal{F} - \mathcal{W} + \mathcal{D} + \mathcal{C}_G = \mathcal{F} - \mathcal{C} + \mathcal{D}, \tag{12.1.80}$$

$$\frac{d}{dt}(K + A) = \mathcal{F}. \tag{12.1.81}$$

If $\mathcal{F} = 0$, the sum of kinetic energy and available potential energy is conserved.

It should be remarked that all available potential energy is not necessarily *available.* Recall the conservation of angular momentum, another conserved quantity in the system: it is not always possible to transform a given atmospheric state to the reference state by adiabatic redistribution without friction. In this case, one may consider a zonally uniform axisymmetric state in geostrophic balance as a reference state. However, such a state is not uniquely determined.

12.2 Energy budget of a moist atmosphere

The energy budget of a moist atmosphere is formulated in two different ways. The first is by using the thermodynamic quantities of a dry atmosphere. In this case, the dry enthalpy C_pT is used. When water vapor is condensed, the dry atmosphere gains its energy due to the latent heat release of water vapor. The other method is by using the thermodynamic quantities of a moist atmosphere, in which enthalpy is represented by $C_pT + Lq$. A moist atmosphere gains its energy in the form of latent heat when water vapor is supplied by evaporation at the ground surface.

Let us formulate the global budget of a moist atmosphere using the above two methods. Integrating the water vapor equation (9.5.1) over the whole domain gives

$$\frac{d}{dt}I_q = \mathcal{E}_v - \mathcal{P}_r, \tag{12.2.1}$$

where

$$I_q \equiv \int_D \rho q\, dV, \quad \mathcal{E}_v \equiv -\int_{S_B} \boldsymbol{i}\cdot\boldsymbol{n}\, dS, \quad \mathcal{P}_r \equiv -\int_D S_q\, dV. \tag{12.2.2}$$

$\mathcal{E}_v$ is total evaporation from the surface, and $\mathcal{P}_r$ is total precipitation. The contribution of the diffusion of water vapor $\boldsymbol{i}$ is considered only at the bottom boundary S_B (its contribution at the top boundary S_T is thus neglected).

Rewriting the energy equation of a moist atmosphere (9.5.9) in the same form as (12.1.60) with $h = C_pT$, we obtain

$$\rho \frac{d}{dt} h = \omega + \varepsilon - \nabla \cdot \boldsymbol{F}^{ene} - LS_q. \tag{12.2.3}$$

The global integral of this equation gives

$$\frac{d}{dt} I_h = \mathcal{F} - \mathcal{C} + \mathcal{D} + L\mathcal{P}_r. \tag{12.2.4}$$

By comparison with (12.1.62), latent heat release $L\mathcal{P}_r$ is added on the right-hand side. This equation is the budget of the dry enthalpy of a moist atmosphere.

Next, from the sum of (12.2.4) and (12.2.1) multiplied by L, we obtain

$$\frac{d}{dt}(I_h + LI_q) = \mathcal{F} - \mathcal{C} + \mathcal{D} + L\mathcal{E}_v. \tag{12.2.5}$$

$I_h + LI_q$ is the global integral of moist enthalpy $h_m = C_pT + Lq$. Eq. (12.2.5) corresponds to the global integral of (9.5.8). $L\mathcal{E}_v$ is the supply of latent heat from the surface due to evaporation, and is expressed as

$$\mathcal{F}^{lh} \equiv L\mathcal{E}_v = -\int_{S_B} \boldsymbol{F}^{lh} \cdot \boldsymbol{n}\, dS = -L \int_{S_B} \boldsymbol{i} \cdot \boldsymbol{n}\, dS. \tag{12.2.6}$$

The energy budget of a moist atmosphere is formally given by replacing $\mathcal{F}$ by $\mathcal{F} + L\mathcal{P}_r$ in the energy budget of the dry atmosphere that is described in Section 12.1.2. In particular, the conservation of total energy (12.1.24) becomes

$$\frac{dE^{tot}}{dt} = \mathcal{B}_S + \mathcal{F} + L\mathcal{P}_r. \tag{12.2.7}$$

Over a long time average, therefore, the energy balance is given by

$$\mathcal{F} + L\mathcal{P}_r = -\mathcal{F}_T^{rads} - \mathcal{F}_T^{radl} + \mathcal{F}_B^{rads} + \mathcal{F}_B^{radl} + \mathcal{F}^{sh} + L\mathcal{P}_r = 0, \tag{12.2.8}$$

where (12.1.39) is used and the surface stress is neglected: $\mathcal{B}_S = 0$. Since $\mathcal{F}^{lh} = L\mathcal{P}_r$, the energy balance is also written as

$$-\mathcal{F}_T^{rads} - \mathcal{F}_T^{radl} + \mathcal{F}_B^{rads} + \mathcal{F}_B^{radl} + \mathcal{F}^{sh} + \mathcal{F}^{lh} = 0. \tag{12.2.9}$$

According to Fig. 12.2, the observational values of the sensible and latent heat fluxes and the latent heat release are

$$\mathcal{F}^{sh}/\mathcal{S} = 24 \text{ W m}^{-2},$$
$$L\mathcal{P}_r/\mathcal{S} = \mathcal{F}^{lh}/\mathcal{S} = 78 \text{ W m}^{-2}.$$

The ratio of sensible heat flux to latent heat flux is called the *Bowen ratio*, and is estimated as $b = 24/78 = 0.3$ for the global value.

12.3 Entropy budget and thermal efficiency

We can obtain a constraint on the heating in the atmosphere using the balance of entropy over a long time averaged state (i.e., a statistical equilibrium state). Quantities over a long time average are denoted by [ ] in this section. In order to generalize the thermal balance for both cases (i.e., dry and moist atmospheres), we include latent heat release in the diabatic term. Integrating the diabatic term (9.5.13) gives

$$\begin{aligned} \int_D \rho C_p Q \, dV &= \int_D \varepsilon \, dV - \int_D (\nabla \cdot \boldsymbol{F}^{rad} + \nabla \cdot \boldsymbol{F}^{sh}) \, dV - \int_D LS_q \, dV \\ &= \mathcal{D} + \mathcal{F} + L\mathcal{P}_r. \end{aligned} \tag{12.3.1}$$

We have $[\mathcal{F} + L\mathcal{P}_r] = 0$ from (12.2.8). Since the dissipation term is always positive $\mathcal{D} > 0$, we obtain

$$\left[\int_D \rho C_p Q \, dV \right] = [\mathcal{D}] > 0. \tag{12.3.2}$$

Letting diabatic terms other than the dissipation term be denoted by

$$Q_n \equiv -\frac{1}{\rho C_p}(\nabla \cdot \boldsymbol{F}^{rad} + \nabla \cdot \boldsymbol{F}^{sh} + LS_q) = Q - \frac{\varepsilon}{\rho C_p}, \tag{12.3.3}$$

we have

$$\left[\int_D \rho C_p Q_n \, dV \right] = [\mathcal{F} + L\mathcal{P}_r] = 0. \tag{12.3.4}$$

To illustrate the role of the entropy budget, here we divide the whole atmosphere into two domains: one where Q_n is positive and the other where Q_n is negative; the former domain is denoted by D_+ and the latter by D_-. From the energy balance (12.3.4), the total integral of the heating over the domain D_+ is equal to that of cooling over the domain D_-:

$$[\mathcal{F}_+] = [\mathcal{F}_-], \tag{12.3.5}$$

where

$$[\mathcal{F}_+] \equiv \int_{D_+} \rho C_p Q_n \, dV > 0, \tag{12.3.6}$$

$$[\mathcal{F}_-] \equiv -\int_{D_-} \rho C_p Q_n \, dV > 0. \tag{12.3.7}$$

Now let us consider the balance of entropy. In the case of a moist atmosphere, the equation of entropy (1.2.53) is written as

$$\begin{aligned} \rho \frac{ds}{dt} &= \rho \frac{C_p Q}{T} = \frac{\varepsilon}{T} + \rho \frac{C_p Q_n}{T} \\ &= \frac{\varepsilon}{T} - \frac{\nabla \cdot (\boldsymbol{F}^{rad} + \boldsymbol{F}^{sh})}{T} - \frac{LS_q}{T}. \end{aligned} \tag{12.3.8}$$

Integrating over the whole atmosphere gives

$$\frac{d}{dt}\int_D \rho s\, dV = \int_D \rho\frac{C_p Q}{T}\, dV = \int_D \frac{\varepsilon}{T}\, dV + \int_D \rho\frac{C_p Q_n}{T}\, dV. \quad (12.3.9)$$

Since the left-hand side vanishes over a long time average, we obtain

$$\left[\int_D \rho\frac{C_p Q}{T}\, dV\right] = 0, \quad (12.3.10)$$

or

$$\left[\int_D \frac{\varepsilon}{T}\, dV\right] = -\left[\int_D \rho\frac{C_p Q_n}{T}\, dV\right] > 0, \quad (12.3.11)$$

where its sign is determined from $\varepsilon > 0$.

At this point, let us introduce *thermal efficiency*. We need to define the average temperatures in the heating region D_+ and the cooling region D_- respectively by

$$\begin{aligned}\frac{1}{T_+} &\equiv \left[\int_{D_+} \frac{\rho C_p Q_n}{T}\, dV\right] \Big/ \left[\int_{D_+} \rho C_p Q_n\, dV\right] \\ &= \left[\int_{D_+} \frac{\rho C_p Q_n}{T}\, dV\right] \cdot \frac{1}{[\mathcal{F}_+]}, \quad (12.3.12)\end{aligned}$$

$$\begin{aligned}\frac{1}{T_-} &\equiv -\left[\int_{D_-} \frac{\rho C_p Q_n}{T}\, dV\right] \Big/ \left[\int_{D_-} \rho C_p Q_n\, dV\right] \\ &= -\left[\int_{D_-} \frac{\rho C_p Q_n}{T}\, dV\right] \cdot \frac{1}{[\mathcal{F}_-]}. \quad (12.3.13)\end{aligned}$$

In addition, we define an average temperature T_D over the whole domain as a weighted mean by the dissipation rate:

$$\frac{1}{T_D} \equiv \left[\int_V \frac{\varepsilon}{T}\, dV\right] \Big/ \left[\int_V \varepsilon\, dV\right] = \left[\int_V \frac{\varepsilon}{T}\, dV\right] \cdot \frac{1}{[\mathcal{D}]}. \quad (12.3.14)$$

Using these temperatures, (12.3.11) is written as

$$\frac{[\mathcal{D}]}{T_D} = -\frac{[\mathcal{F}_+]}{T_+} + \frac{[\mathcal{F}_-]}{T_-}. \quad (12.3.15)$$

Thus, from (12.3.5), we have an expression,

$$\frac{[\mathcal{D}]}{[\mathcal{F}_+]} = \frac{(T_+ - T_-)T_D}{T_+ T_-} = \frac{\Delta T}{\langle T\rangle}, \quad (12.3.16)$$

where $\Delta T \equiv T_+ - T_-$ and $\langle T\rangle \equiv T_+ T_-/T_D$. ΔT is the temperature difference between the heating and cooling regions, and $\langle T\rangle$ is another average temperature of the atmosphere. Eq. (12.3.16) can be viewed as a definition of the thermal efficiency; $[\mathcal{F}_+]$ is the heating source due to the convergence of thermal flux given to the atmosphere and $[\mathcal{D}]$ is equal to work done by the atmosphere $[\mathcal{W}]$, as seen in (12.1.34). Eq. (12.3.16) states that the ratio of the work done by the atmosphere to the total energy input is given by the ratio of temperature difference between the heating and cooling regions to average temperature.

We can obtain a different type of constraint on heating using the equation of potential temperature instead of that of entropy. The equation of potential temperature (1.2.56) is rewritten as

$$\rho\frac{d\theta}{dt} = \rho\left(\frac{p_0}{p}\right)^{\kappa} Q. \tag{12.3.17}$$

Integrating this over the whole atmosphere gives

$$\frac{d}{dt}\int_D \rho\theta\, dV = \int_D \rho\left(\frac{p_0}{p}\right)^{\kappa} Q\, dV. \tag{12.3.18}$$

Since the left-hand side becomes zero over a long time average, we obtain

$$0 = \left[\int_D \frac{\rho Q}{p^{\kappa}}\, dV\right]. \tag{12.3.19}$$

Thus, since $\varepsilon > 0$, we have another relation of the heating source:

$$\left[\int_D \frac{\varepsilon}{C_p p^{\kappa}}\, dV\right] = -\left[\int_D \rho\frac{Q_n}{p^{\kappa}}\, dV\right] > 0. \tag{12.3.20}$$

We now consider the theorem of entropy production. The equation of entropy (12.3.8) is rewritten as

$$\rho\frac{ds}{dt} = \frac{\varepsilon}{T} + \boldsymbol{F}^{rad}\cdot\nabla\frac{1}{T} - \nabla\cdot\frac{\boldsymbol{F}^{rad}}{T} + \boldsymbol{F}^{sh}\cdot\nabla\frac{1}{T} - \nabla\cdot\frac{\boldsymbol{F}^{sh}}{T} - \frac{LS_q}{T}. \tag{12.3.21}$$

Integrating this over the whole domain gives

$$\begin{aligned}\frac{d}{dt}\int_D \rho s\, dV = {} & \int_D \frac{\varepsilon}{T}\, dV + \int_D \frac{1}{T^2}\boldsymbol{F}^{rad}\cdot\nabla T\, dV - \int_S \frac{\boldsymbol{F}^{rad}}{T}\cdot\boldsymbol{n}\, dS \\ & + \int_D \frac{1}{T^2}\boldsymbol{F}^{sh}\cdot\nabla T\, dV - \int_S \frac{\boldsymbol{F}^{sh}}{T}\cdot\boldsymbol{n}\, dS - \int_D \frac{LS_q}{T}\, dV.\end{aligned} \tag{12.3.22}$$

The first term on the right-hand side is the entropy increase due to dissipation, and the second and fourth terms are the change in entropy due to thermal flux. The third and fifth terms are the inflows and outflows of entropy. The sixth term is the entropy production rate associated with latent heat release. The above equation is written in a similar form as (1.2.58):

$$\frac{dS}{dt} = \frac{d_i S}{dt} + \frac{d_e S}{dt}, \tag{12.3.23}$$

where

$$S = \int_D \rho s \, dV, \tag{12.3.24}$$

$$\frac{d_i S}{dt} = \int_D \frac{\varepsilon}{T} dV + \int_D \frac{1}{T^2} \boldsymbol{F}^{rad} \cdot \nabla T \, dV + \int_D \frac{1}{T^2} \boldsymbol{F}^{sh} \cdot \nabla T \, dV - \int_D \frac{L S_q}{T} dV, \tag{12.3.25}$$

$$\frac{d_e S}{dt} = -\int_S \frac{\boldsymbol{F}^{rad}}{T} \cdot \boldsymbol{n} \, dS - \int_S \frac{\boldsymbol{F}^{sh}}{T} \cdot \boldsymbol{n} \, dS. \tag{12.3.26}$$

$d_e S/dt$ is the entropy change due to the entropy flux at the boundaries of the atmosphere, and is given by two terms: radiative flux and sensible heat flux. $d_i S/dt$ is the rate of increase of entropy due to each process; the first term is the effect of dissipation and is always positive. The third term is the rate of increase due to sensible heat flux and is also always positive as shown by (1.2.62). In general, the fourth term is also positive since we generally have the convergence of water vapor $S_q < 0$. This is the rate of increase due to the diffusion of water vapor. The second term is the change in entropy due to radiative flux, but it does not have a definite sign, in general. When there is no contribution from radiative flux, the production of entropy within the atmosphere must be positive or equal to zero:

$$\frac{d_i S}{dt} \geq 0, \quad \text{(without radiative flux)}. \tag{12.3.27}$$

Neglecting the left-hand side of (12.3.23) over a long time average, we obtain the balance of entropy:

$$-\left[\frac{d_e S}{dt}\right] = \left[\frac{d_i S}{dt}\right] \geq 0, \quad \text{(without radiative flux)}. \tag{12.3.28}$$

This means that the total outflow of entropy through the boundaries from the atmosphere is larger than the total inflow of entropy through the boundaries, since the entropy production rate is always positive within the atmosphere if the contribution of radiative flux is negligible.

The gross entropy production rate of the atmosphere is estimated by assuming that the whole atmosphere is in thermal equilibrium. In this case, the production rate of entropy within the atmosphere is simply given by the radiation budget at the top of the atmosphere. The energy inflow to the atmosphere is by solar flux at the top of the atmosphere: $\mathcal{F}^{isr} = \pi R^2 \cdot (1 - A) F^{\odot}$, which is the same amount of energy flux emitted from the atmosphere to outer space. The production of entropy within the atmosphere is given by the difference between the inward and outward entropy fluxes at the top of the atmosphere. Therefore, the production rate of entropy is given as

$$\frac{d_i S}{dt} = \frac{\mathcal{F}^{olr}}{T_e} - \frac{\mathcal{F}^{isr}}{T_{\odot}} = \pi R^2 \cdot (1 - A) F^{\odot} \left(\frac{1}{T_e} - \frac{1}{T_{\odot}}\right), \tag{12.3.29}$$

where T_e is the effective temperature of the atmosphere and $T_\odot$ is the surface temperature of the sun. Using $(1-A)F^\odot/4 = \sigma_B T_e^4$ given by (12.1.7), the production rate of entropy per unit area is

$$\frac{1}{4\pi R^2}\frac{d_iS}{dt} = \sigma_B T_e^3 \left(1 - \frac{T_e}{T_\odot}\right). \tag{12.3.30}$$

In the case $T_e = 255$ K and $T_\odot = 5760$ K, the production rate of entropy per unit area is about 0.90 W m^{-2} K^{-1}.

12.4 Similarity theory of general circulation

The characteristics of atmospheric general circulation is governed by a set of external parameters. Several nondimensional numbers can be constructed from external parameters. Golitsyn (1970) applied the concept of similarity theory to the general circulations of planetary atmospheres; if the values of nondimensional numbers are equal, atmospheric general circulations are in the same category. In this section, we introduce the nondimensional numbers used by Golitsyn, and describe some of the characteristics of general circulation inferred from similarity theory.

We restrict our consideration to an atmosphere in which the air consists of an ideal gas with constant specific heat and in which the acceleration due to gravity is constant. In this case, we assume that atmospheric circulations are governed by the following external parameters:

F_s	[W m^{-2}]	: average solar radiation per unit area
M	[kg m^{-2}]	: mass of an air column per unit area
C_p	[J kg^{-1} K^{-1}]	: specific heat at constant pressure
R	[m]	: planetary radius
Ω	[s^{-1}]	: angular velocity of rotation
g	[m s^{-2}]	: acceleration due to gravity

Using the solar constant $F^\odot$ and the planetary albedo A, solar radiation is given by

$$F_s = (1-A)\frac{F^\odot}{4}. \tag{12.4.1}$$

Under hydrostatic balance, the mass of an air column is given by

$$M = \frac{1}{4\pi R^2}\int_D \rho dV = p_s g, \tag{12.4.2}$$

where p_s is the average surface pressure. We then construct nondimensional numbers from the above six parameters. Although there are four basic physical dimensions, [kg], [m], [s], and [K], we use the Stefan Boltzmann constant σ_B [W m^{-2} K^{-4}] to eliminate the unit [K]. With this conversion rule, two parameters, C_p and σ_B, appear in the form of $C_p/\sigma_B^{1/4}$ [kg$^{-1/4}$ m^2 s$^{-5/4}$] in nondimensional parameters. Thus, we have three independent nondimensional numbers from the six external parameters. Golitsyn (1970) chose the following three numbers:

$$\Pi_M = \frac{\sigma_B^{3/8} F_s^{5/8}}{C_p^{3/2}} \frac{R}{M}, \tag{12.4.3}$$

$$\Pi_\Omega = \frac{\sigma_B^{1/8} F_s^{-1/8}}{C_p^{1/2}} \Omega R, \tag{12.4.4}$$

$$\Pi_g = \frac{C_p F_s^{1/4}}{\sigma_B^{1/4}} \frac{1}{gR}, \tag{12.4.5}$$

which are thought to characterize M, Ω, and g, respectively. If any of these nondimensional numbers is much larger or much smaller than one, we assume that atmospheric circulations do not depend on that number. This is the assumption of self-similarity.

The meaning of these nondimensional parameters can be given as follows. With the definition of the effective temperature $F_s = \sigma_B T_e^4$, we have

$$\frac{C_p F_s^{1/4}}{\sigma_B^{1/4}} = C_p T_e, \tag{12.4.6}$$

so that

$$\Pi_\Omega = \frac{\Omega a}{(C_p T_e)^{1/2}} = (\gamma - 1)^{1/2} \frac{\Omega a}{c_s}, \tag{12.4.7}$$

$$\Pi_g = \frac{C_p T_e}{gR} = \frac{\gamma}{\gamma - 1} \frac{R_d T}{gR} = \frac{\gamma}{\gamma - 1} \frac{H}{R}, \tag{12.4.8}$$

where

$$c_s = (\gamma R_d T_e)^{1/2}, \quad H = \frac{R_d T_e}{g}, \quad \gamma = \frac{C_p}{C_v}. \tag{12.4.9}$$

c_s is the speed of sound and H is the scale height of the atmosphere. From these expressions, Π_Ω corresponds to the *rotational Mach number*, and Π_g is the aspect ratio of the atmosphere. In general, the aspect ratio of the atmosphere is very small, $\Pi_g \ll 1$. From the assumption of self-similarity, therefore, we conclude that atmospheric circulations are independent of Π_g (i.e., the atmospheric circulations are insensitive to g).

Next, we have

$$\Pi_M = \frac{F_s a}{(C_p T_e)^{3/2} M} = \frac{R}{(C_p T_e)^{1/2} \tau_R} = (\gamma - 1)^{1/2} \frac{R}{c_s \tau_R}, \tag{12.4.10}$$

where

$$\tau_R \equiv \frac{M C_p T_e}{F_s} \tag{12.4.11}$$

is the *radiative relaxation time* of the atmosphere. This means the time required for the atmosphere to be heated up to temperature T_e. R/c_s is the characteristic time of sound waves. Thus, Π_M is given by the ratio of radiation time to time of sound waves.

The aim of similarity theory is to express the characteristic quantities of atmospheric circulations in terms of nondimensional parameters. Here, we particularly consider the expressions of the energy budget. In the first case, we assume that Ω is small enough and g large enough such that

$$\Pi_\Omega \ll 1, \qquad \Pi_g \ll 1.$$

This implies that quantities of atmospheric circulations are independent of Π_Ω and Π_g and can be expressed by the four parameters: F_s, M, $C_p/\sigma_B^{1/4}$, and R. From these parameters, we can construct only one quantity that has the unit of energy [J]:

$$E^* \equiv \frac{\sigma_B^{1/8}}{C_p^{1/2}} F_s^{7/8} R^3. \tag{12.4.12}$$

Note that this is independent of M. Using this quantity, total potential energy is expressed as

$$E_h \equiv 4\pi R^2 M C_p T_e = 4\pi \Pi_M^{-1} E^*. \tag{12.4.13}$$

We express the total kinetic energy of the atmosphere E_k by using a factor B as

$$E_k \equiv BE^*. \tag{12.4.14}$$

The ratio of kinetic energy to total potential energy is given by

$$\frac{E_k}{E_h} = \frac{B\Pi_M}{4\pi}. \tag{12.4.15}$$

The problem is: which parameter does B depend on?

We cannot proceed further with just the estimations of similarity theory. We need to ascertain the structure of atmospheric general circulations to estimate kinetic energy. According to Golitsyn, we can make the following assumptions about the energy budget. First, the convergence of latitudinal energy flux is balanced by the emission of planetary radiation:

$$\frac{1}{R} M C_p U \Delta T \approx F_s, \tag{12.4.16}$$

where U is a characteristic velocity scale, and ΔT is a characteristic temperature difference. Second, the atmosphere is assumed to be turbulent such that U is related to the dissipation rate ϵ as

$$U \approx (\epsilon L)^{1/3}, \tag{12.4.17}$$

where L is a characteristic length scale of atmospheric motion. Using thermal efficiency, ΔT is related to ϵ as

$$\eta = \frac{\epsilon}{F_s/M} = k\frac{\Delta T}{T_+}, \tag{12.4.18}$$

where T_+ is the characteristic temperature of the heating region and η is the thermal efficiency. k is the ratio of η to maximum thermal efficiency $\Delta T/T_+$, and may be called the *effective thermal efficiency.* We hereafter assume $T_+ \approx T_e$. From (12.4.16), (12.4.17), and (12.4.18), we have the relations:

$$\Delta T = \frac{F_s^{9/16}}{k^{1/4}\sigma_B^{1/16}C_p^{3/4}}\left(\frac{R}{M}\right)^{1/2}\left(\frac{R}{L}\right)^{1/4}, \tag{12.4.19}$$

$$\epsilon = \frac{k^{3/4}\sigma_B^{3/16}F_s^{12/16}}{C_p^{3/4}}\left(\frac{R}{M^3}\right)^{1/2}\left(\frac{R}{L}\right)^{1/4}, \tag{12.4.20}$$

$$U = \frac{k^{1/4}\sigma_B^{1/16}F_s^{7/16}}{C_p^{1/4}}\left(\frac{R}{M}\right)^{1/2}\left(\frac{R}{L}\right)^{-1/4}. \tag{12.4.21}$$

Thus, kinetic energy is given by

$$E_k = 4\pi R^2 M\frac{U^2}{2} = 2\pi k^{1/2}\left(\frac{R}{L}\right)^{-1/2} E^*. \tag{12.4.22}$$

Note that this quantity is also independent of the mass M. From (12.4.15), (12.4.14), and (12.4.18), we also have

$$B = 2\pi k^{1/2}\left(\frac{R}{L}\right)^{-1/2}, \tag{12.4.23}$$

$$\frac{E_k}{E_h} = \frac{B\Pi_M}{4\pi} = \frac{k^{1/2}}{2}\left(\frac{R}{L}\right)^{-1/2}\Pi_M, \tag{12.4.24}$$

$$\eta = k^{3/4}\left(\frac{R}{L}\right)^{1/4}\Pi_M^{1/2}. \tag{12.4.25}$$

The length scale L might have different dependencies according to rotation rate; we assume that

$$L \approx \begin{cases} R, & \text{for } \Pi_\Omega \ll 1, \\ \dfrac{\sqrt{gH}}{\Omega}, & \text{for } \Pi_\Omega \gg 1, \end{cases} \tag{12.4.26}$$

where H is the scale height given by (12.4.9). Thus, we have

$$\frac{R}{L} = \begin{cases} 1, & \text{for } \Pi_\Omega \ll 1, \\ \left(\dfrac{\gamma}{\gamma-1}\right)^{1/2}\Pi_\Omega, & \text{for } \Pi_\Omega \gg 1. \end{cases} \tag{12.4.27}$$

We summarize the values of Π_M, Π_Ω, and Π_g of the planetary atmospheres of the solar system in Tables 12.1 and 12.2. For application of the above discussion to real planetary atmospheres, we must note that we do not know the values of U, ΔT, L, and k in (12.4.16), (12.4.17), and (12.4.18). We nevertheless think that the above crude estimations are useful to clarify the relations between the various quantities that characterize atmospheric general circulations.

	T_e [K]	$p_s = Mg$ [Pa]	C_p [J kg^{-1} K^{-1}]	γ	R [km]	$2\pi/\Omega$ [day]	g [m s^{-2}]
Venus	215	9×10^6	850	1.31	6052	243.01	8.93
Earth	255	10^5	1000	1.40	6378	0.9973	9.81
Mars	202	7×10^2	850	1.31	3397	1.0260	3.73
Jupiter	130	(7×10^4)	13000	1.42	71398	0.414	23.3
Saturn	80	(10^5)	13000	1.42	60000	0.444	9.32
Uranus	57	(10^5)	13000	1.42	25400	0.649	8.73
Neptune	45	(10^5)	13000	1.42	24300	0.768	11.7

TABLE 12.1: The external parameters of planetary atmospheres. Note that the values of surface pressure p_s in parentheses are not definite since the atmosphere is very deep.

	Π_M	Π_Ω	Π_g
Venus	9.3×10^{-6}	0.004	3.4×10^{-3}
Earth	1.2×10^{-3}	0.92	4.1×10^{-3}
Mars	2.4×10^{-2}	0.58	1.4×10^{-2}
Jupiter	1.8×10^{-4}	9.6	1.0×10^{-3}
Saturn	1.2×10^{-5}	9.6	1.9×10^{-3}
Uranus	2.1×10^{-6}	3.3	2.1×10^{-3}
Neptune	1.5×10^{-6}	3.0	5×10^{-4}

TABLE 12.2: The nondimensional numbers of planetary atmospheres.

References and suggested reading

Lorenz (1967) is an excellent textbook on the basis of the energy budget of the atmosphere. The recent observational data of the global energy budget is reviewed in IPCC (1996, 2001). Held and Soden (2000) discussed climate sensitivity particularly in terms of water feedback in global warming. Available potential energy was first introduced by Lorenz (1957) and its mathematical foundation is given by Shepherd (1993). The entropy budget of the atmosphere is analyzed by Peixoto et al. (1991) and theoretical consideration of the entropy budget of a moist atmosphere is given by Pauluis and Held (2002). The similarity theory of atmospheric general circulation is discussed by Golitsyn (1970), and a good review is given by Matsuda (2000). A similar dimensional analysis is given by Gierasch et al. (1970).

Chandrasekhar, S., 1958: *An Introduction to the Study of Stellar Structure.* Dover, New York, 507 pp.

Chandrasekhar, S., 1961: A theorem on rotating polytropes. *Astrophys. J.*, **134**, 662–664.

Crowley, T. J. and North, G. R., 1991: *Paleoclimatology.* Oxford University Press, Oxford, UK, 339 pp.

Gierasch, P., Goody, R., and Stone, P., 1970: The energy balance of planetary atmosphere. *Geophys. Fluid Dynamics*, **1**, 1–18.

Golitsyn, G. S., 1970: A similarity approach to the general circulation of planetary atmospheres. *ICARUS*, **13**, 1–24.

Held, I. M. and Soden, B. J., 2000: Water feedback and global warming. *Ann. Rev. Energy Environ.*, **25**, 441–475.

IPCC, 1996: *Climate Change 1995. The Science of Climate Change.* Intergovernmental Panel on Climate Change, J. Houghton, L. G. Meira Filho, B. A. Callander, N. Harris, A. Kattenberg, K. Maskell, eds. Cambridge University Press, Cambridge, UK, 572 pp.

IPCC, 2001: *Climate Change 2001. The Science Basis.* Intergovernmental Panel on Climate Change, J. Houghton, Y. Ding, D. J. Griggs, M. Noguer, P. J. van der Linden, X. Dai, K. Maxwell, and C. A. Johnson, eds. Cambridge University Press, Cambridge, UK, 881 pp.

Kiehl, J. T. and Trenberth, K. E., 1997: Earth's annual global mean energy budget. *Bull. Am. Meteorol. Soc.*, **78**, 197–208.

Landau, L. and Lifshitz, E. M., 1996: *Statistical Physics*, 3rd ed. Butterworth-Heinemann, Oxford, UK, 544 pp.

Lebovitz, N. R., 1961: The virial tensor and its application to self-gravitational fluid. *Astrophys. J.*, **134**, 500–536.

Lorenz, E. N., 1957: Available potential energy and the maintenance of the general circulation. *Tellus*, **7**, 157–167.

Lorenz, E. N., 1967: *The Nature and Theory of the General Circulation of the Atmosphere.* World Meteorological Organization, Geneva, 161 pp.

Matsuda, Y., 2000: *Planetary Meteorology.* University of Tokyo Press, Tokyo, 205 pp. (in Japanese).

Peixoto, J. P., Oort, A. H., De Almeida, M., and Tome, A., 1991: Entropy budget of the atmosphere. *J. Geophys. Res.*, **96**, 10981–10988.

Pauluis, O. and Held, I., 2002: Entropy budget of an atmosphere in radiative-convective equilibrium. Part II: Latent heat transport and moist processes. *J. Atmos. Sci.*, **59**, 140–149.

Salmon, R., 1988: Hamiltonian Fluid Mechanics. *Ann. Rev. Fluid Mech.*, **20**, 225–256.

Shepherd, T. G., 1993: A unified theory of available potential energy. *Atmosphere-Ocean*, **31**, 1–26.

13

Latitudinal energy balance

Following on from the global energy balance described in the previous chapter, the latitudinal energy balance is considered here. The energy balance of the zonally and vertically averaged atmosphere is described using the latitude-dependent energy balance model. The pole-to-equator temperature difference is evaluated with the energy balance model in which meridional heat transport is modeled using a diffusion coefficient. The latitudinal profiles of the radiative fluxes and their balance are also described using the energy balance model. Furthermore, the possibility of multiple equilibrium states is investigated using the relation between solar flux and albedo. In the final section, a constraint on the latitudinal distribution of water vapor is argued using the balance of the latitudinal transport of water vapor.

13.1 Energy balance model

One can study the gross properties of temperature distribution, such as pole-to-equator temperature difference, using a model in which only the energy balance in the latitudinal direction is considered. This type of model is called an *energy balance model* or *EBM*. EBM relates energy fluxes to a temperature that is defined at each latitude as a zonal mean using a differential equation with respect to latitude.

As shown below, EBM is thought to be based on the balance of total potential energy. If the representative temperature of an air column is related to its total potential energy, one obtains the latitudinal distribution of temperature from the balance of total potential energy. Total potential energy is given by the sum of internal energy u and potential energy Φ, and is generally much larger than kinetic energy. This implies that the contribution of kinetic energy is negligible to the first approximation when the temperature structure is considered using the energy balance.

First, we formulate the energy balance of a dry atmosphere. Neglecting kinetic energy and the contribution of viscous stress to total energy balance (1.2.47), we have

$$\frac{\partial}{\partial t}\left[\rho\left(u+\Phi\right)\right]+\nabla\cdot\left(\rho\boldsymbol{v}\sigma+\boldsymbol{F}^{ene}\right) \quad = \quad 0, \tag{13.1.1}$$

where $\sigma = h+\Phi$ is static energy. Rewriting this equation using spherical coordinates with longitude λ, latitude φ, and altitude z and integrating in the zonal and vertical directions, we obtain the energy budget in the latitudinal direction. Here, the zonal average and the vertical integral are denoted by†

$$\overline{X} \quad = \quad \frac{1}{2\pi}\int_0^{2\pi} X\,d\lambda, \tag{13.1.2}$$

$$\langle X\rangle \quad = \quad \frac{1}{2\pi}\int_0^{\infty}\int_0^{2\pi} X\,d\lambda\,dz \quad = \quad \int_0^{\infty}\overline{X}\,dz. \tag{13.1.3}$$

We generally have (12.1.52), which states that total potential energy is equal to the integral of enthalpy:

$$\langle\rho(u+\Phi)\rangle \quad = \quad \langle\rho h\rangle\,. \tag{13.1.4}$$

Although heat flux has three components $\boldsymbol{F}^{ene} = (F_\lambda, F_\varphi, F_z)$, we neglect the latitudinal flux F_φ in favor of the advective flux in the latitudinal direction. In this case, the zonally averaged and vertically integrated energy balance equation is given from (13.1.1) as

$$\frac{\partial}{\partial t}\langle\rho h\rangle+\frac{1}{R\cos\varphi}\frac{\partial}{\partial\varphi}\left(\cos\varphi\,\langle\rho v\sigma\rangle\right)+\overline{F_{zT}}-\overline{F_{zB}} \quad = \quad 0. \tag{13.1.5}$$

where F_{zT} and F_{zB} are the vertical energy flux at the top and bottom of the atmosphere, respectively. Vertical advection is neglected at the top and bottom of the atmosphere.

Enthalpy is expressed as $h = C_pT$ where C_p is the specific heat at constant pressure and assumed to be constant. Then we can define the zonal-mean temperature in a latitude belt as

$$\tilde{T}(\varphi,t) \quad \equiv \quad \frac{\langle\rho C_pT\rangle}{\langle\rho C_p\rangle} \quad = \quad \frac{\langle\rho C_pT\rangle}{C}, \tag{13.1.6}$$

where

$$C \quad \equiv \quad \langle\rho C_p\rangle \quad = \quad C_p\frac{\overline{p_0}}{g} \quad = \quad C_pM \tag{13.1.7}$$

is the zonal average of the heat capacity of an air column per unit area, p_0 is surface pressure, and $M = \overline{p_0}/g$ is the mass of an air column. We assume that C is constant for simplicity, though C is generally a function of time and latitude since surface pressure is variable. For the earth's atmosphere, we have a typical value $C \approx 10^7$ J K^{-1} m^{-2}.

---

†With a given topography, we set $X = 0$ in the region $z < z_0$, where z_0 is the height of the surface.

In a similar way to (12.1.42) and (12.1.43), vertical energy flux is divided into contributions from short-wave radiation, long-wave radiation, and sensible heat fluxes:

$$\overline{F_{zT}} = \overline{F_T^{radl}} + \overline{F_T^{rads}}, \tag{13.1.8}$$

$$\overline{F_{zB}} = \overline{F_B^{radl}} + \overline{F_B^{rads}} + \overline{F^{sh}}, \tag{13.1.9}$$

where F^{sh} is the sensible heat flux between the ground surface and the atmosphere. Note that latent heat flux is ignored at this moment since a dry atmosphere is assumed. Substituting these expressions into (13.1.5), we have the temperature equation as

$$C\frac{\partial}{\partial t}\tilde{T} + \frac{1}{R\cos\varphi}\frac{\partial}{\partial\varphi}(\cos\varphi\,\langle\rho v\sigma\rangle) + (\overline{F_T^{radl}} + \overline{F_T^{rads}}) - (\overline{F_B^{radl}} + \overline{F_B^{rads}} + \overline{F^{sh}}) = 0. \tag{13.1.10}$$

The terms on the left-hand side are the tendency term, the convergence of horizontal advection, and the energy fluxes at the top and the bottom of the atmosphere, respectively. If we know the appropriate forms of the energy fluxes, we obtain the equation for the latitudinal distribution of temperature.

Over a long time average, the tendency term is negligible. The global average of the vertical energy flux at the bottom of the atmosphere becomes zero if the heat source within the solid earth is negligible. However, its latitudinal values are different from zero if there is latitudinal energy transport in the ocean below the atmosphere. Let C_{atm} denote the energy convergence of the atmosphere and C_{ocean} that of the ocean. In a statistical equilibrium state, we have the balance

$$\overline{F_T^{radl}} + \overline{F_T^{rads}} = C_{atm} + C_{ocean}, \tag{13.1.11}$$

$$\overline{F_B^{radl}} + \overline{F_B^{rads}} + \overline{F^{sh}} = C_{ocean}, \tag{13.1.12}$$

where

$$C_{atm} = -\frac{1}{R\cos\varphi}\frac{\partial}{\partial\varphi}(\cos\varphi\,\langle\rho v\sigma\rangle) \tag{13.1.13}$$

is the convergence of latitudinal energy transport in the atmosphere.

13.2 Latitudinal distribution of radiative balance

According to (13.1.11), the energy budget at the top of the atmosphere is expressed by the balance of the radiative fluxes $\overline{F_T^{rads}}$ and $\overline{F_T^{radl}}$. The former is expressed as

$$\overline{F_T^{rads}} = -(1-A)\,\overline{F^{sol}}, \tag{13.2.1}$$

where F^{sol} is the incident solar radiation flux per unit area at the top of the atmosphere and A is the zonal-mean albedo.† The global integral of F^{sol} is equal

†In fact, A is an effective zonal-mean albedo, and is given by a function of latitude and time: $A(\varphi, t) = \overline{A(\lambda, \varphi, t)F^{sol}(\lambda, \varphi, t)}/\overline{F^{sol}}(\varphi, t)$.

to (12.1.3). Dividing the global integral of F^{sol} by $S = 4\pi R^2$, we obtain

$$\frac{1}{S}\int_S F^{sol}\,dS = \frac{1}{2}\int_{-\frac{\pi}{2}}^{\frac{\pi}{2}} \overline{F^{sol}}(\varphi)\cos\varphi\,d\varphi = \frac{F^{\odot}}{4}. \tag{13.2.2}$$

Let us introduce the latitudinal distribution function of solar radiation $f(\varphi)$ and write

$$\overline{F^{sol}}(\varphi) = \frac{F^{\odot}}{4}\,f(\varphi), \tag{13.2.3}$$

where f is normalized as

$$\frac{1}{2}\int_{-\frac{\pi}{2}}^{\frac{\pi}{2}} f(\varphi)\cos\varphi\,d\varphi = 1. \tag{13.2.4}$$

For instance, at the equinox condition, the distribution function is given by

$$f(\varphi) = \frac{4}{\pi}\cos\varphi, \tag{13.2.5}$$

or at the annually averaged condition, it is approximated as

$$f(\varphi) = 1 - cP_2(\varphi) = 1 - \frac{c}{2}(3\sin^2\varphi - 1), \tag{13.2.6}$$

where

$$P_2(\varphi) = \frac{1}{2}(3\sin^2\varphi - 1) \tag{13.2.7}$$

is a second-order Legendre function. An empirical value is given to the constant c, such as 0.477 (North, 1975). Solar radiation actually has large seasonal variation. The left panel of Fig. 13.1 shows the seasonal variation of the latitudinal profiles of solar radiation at the top of the atmosphere.

Using $f(\varphi)$ with (13.2.1) and (13.2.3), the latitudinal distribution of net solar radiation at the top of the atmosphere is expressed as

$$\overline{F_T^{rads}} = -(1 - A)\,\frac{F^{\odot}}{4}\,f(\varphi). \tag{13.2.8}$$

The outgoing long-wave radiation at the top of the atmosphere $\overline{F_T^{radl}}$ is determined by the vertical distributions of temperature and absorbing quantities in an air column (see Chapters 10 and 14). The right panel of Fig. 13.1 also shows the observed seasonal variation of the latitudinal profiles of outgoing long-wave radiation. Here, we assume a simplified form of this dependency as

$$\overline{F_T^{radl}} = F_0 + B(\tilde{T} - T_0), \tag{13.2.9}$$

where $\tilde{T}$ is the zonal-mean temperature at each latitude given by (13.1.6). A similar relationship is also introduced for the global budget (12.1.8). The latitudinal

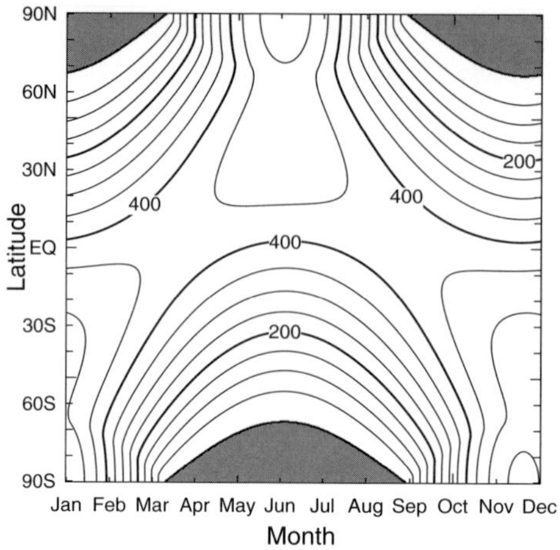

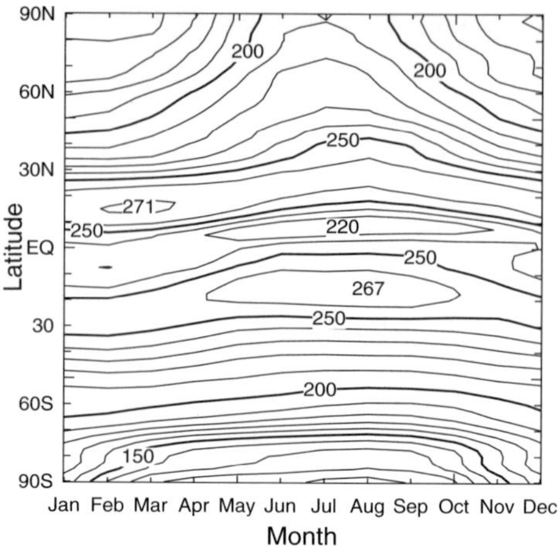

FIGURE 13.1: (Left) The calculated values of the seasonal and latitudinal distribution of daily mean incident solar radiation at the top of the atmosphere. The contour interval is 50 W m^{-2}. Shaded areas represent regions of no insolation. (Right) The seasonal and latitudinal distribution of outgoing long-wave radiation (OLR) for the period between 1979 and 1994. The contour interval is 10 W m^{-2}. The OLR is obtained from satellite data (Gruber and Krueger 1984).

distribution of the effective temperature T_e, on the other hand, can be defined from $\overline{F_T^{radl}}$ by

$$\overline{F_T^{radl}} = \sigma_B T_e^4(\varphi). \tag{13.2.10}$$

In general, T_e is different from the average temperature of an air column $\tilde{T}$. The difference is derived from the greenhouse effect.

The global radiation balance is given by the latitudinal average of (13.1.11):

$$\frac{1}{2}\int_{-\frac{\pi}{2}}^{\frac{\pi}{2}} \overline{F_T^{radl}} \cos\varphi \, d\varphi + \frac{1}{2}\int_{-\frac{\pi}{2}}^{\frac{\pi}{2}} \overline{F_T^{rads}} \cos\varphi \, d\varphi = 0, \tag{13.2.11}$$

which corresponds to (12.1.2) or (12.1.45). Substituting (13.2.8) and (13.2.9) into this equation yields

$$F_0 + B([\tilde{T}] - T_0) = (1 - [A])\frac{F^{\odot}}{4}, \tag{13.2.12}$$

where $[\tilde{T}]$ is the global-mean temperature and $[A]$ is the global-mean albedo (see (12.1.12)), defined by

$$[\tilde{T}] = \frac{1}{2}\int_{-\frac{\pi}{2}}^{\frac{\pi}{2}} \tilde{T} \cos\varphi \, d\varphi, \qquad [A] = \frac{1}{2}\int_{-\frac{\pi}{2}}^{\frac{\pi}{2}} A(\varphi)\, f(\varphi) \cos\varphi \, d\varphi. \tag{13.2.13}$$

From the radiation balance (13.2.12), therefore, the global-mean temperature is expressed as

$$[\tilde{T}] = T_0 - \frac{F_0}{B} + (1 - [A])\frac{F^{\odot}}{4B}. \tag{13.2.14}$$

Fig. 13.2 shows the annually averaged latitudinal profiles of net incoming solar radiation $\overline{F_T^{rads}}$ (ISR) and outgoing long-wave radiation $\overline{F_T^{radl}}$ (OLR) at the top of the atmosphere. The abscissa shows the sine of latitude. Since the energy balance is approximately established, the areas enclosed by the two curves are almost equal. Fig. 13.3 shows the northward energy transport in the atmosphere and the ocean. At latitude about 35°, the energy transport of the atmosphere-ocean system is maximum for the annual mean. This indicates that $\overline{F_T^{rads}} = \overline{F_T^{radl}}$ holds at this latitude.

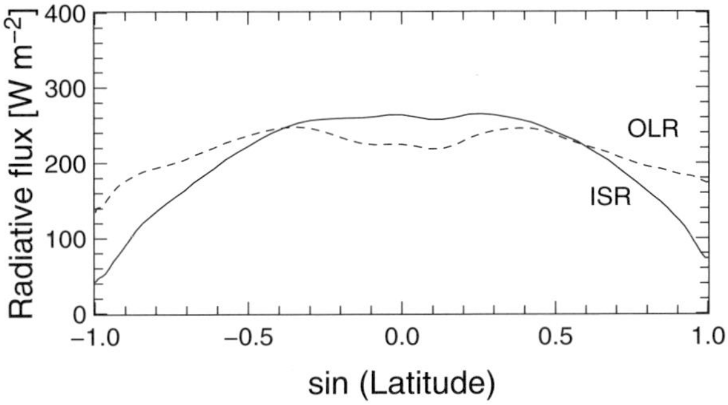

FIGURE 13.2: The annually averaged latitudinal profiles of net incoming solar radiation (ISR: solid curve) and outgoing long-wave radiation (OLR: dashed curve) at the top of the atmosphere. The unit of the ordinate is W m^{-2}. The abscissa is the sine of latitude. NCEP/NCAR reanalysis data is used (see Appendix A3).

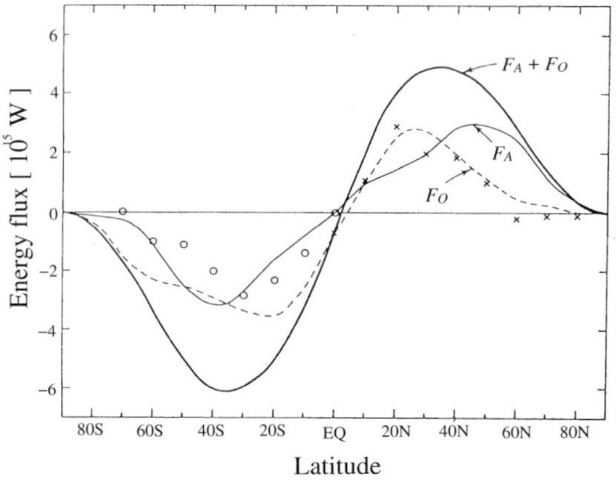

FIGURE 13.3: The latitudinal distribution of the annual-mean northward energy transport for the atmosphere (F_A), the ocean F_O, and the coupled atmosphere-ocean system $F_A + F_O$. The unit is 10^{15} W. After Oort and Peixoto (1983) by permission of Elsevier (copyright, 2003). Analyses by Oort and Vonder Haar (1976) and Trenberth (1979) are indicated by the symbols × and ∘, respectively.

13.3 Latitudinal temperature difference

If there is no latitudinal energy transport in the atmosphere and ocean system, the right-hand side of (13.1.11) is equal to zero. In this case, atmospheric temperature is completely determined by the local net solar radiation, and the latitudinal difference of temperature in the atmosphere solely depends on that of the corresponding solar radiation. Here, we define latitudinal temperature difference by the difference of the temperature between the equator and the pole, ΔT. If the atmosphere and ocean system has a latitudinal energy transport, on the other hand, ΔT is reduced compared with the case of no energy transport. Thus, ΔT is maximum if no energy transport occurs in the atmosphere and ocean system. In this section, we investigate the relation between latitudinal temperature difference ΔT and latitudinal energy transport by assuming that the albedo A is constant irrespective of latitude. In particular, we introduce the diffusion model to EBM in which energy transport is proportional to the latitudinal temperature gradient.

First, if there is no latitudinal energy transport, the radiation balance at each latitude is given by setting the right-hand side of (13.1.11) to zero:

$$\overline{F_T^{radl}} + \overline{F_T^{rads}} \quad = \quad 0. \tag{13.3.1}$$

Using (13.2.1), (13.2.3), and the effective temperature (13.2.10), we have

$$\sigma_B T_e^4(\varphi) \quad = \quad (1 - A)\frac{F^{\odot}}{4} f(\varphi). \tag{13.3.2}$$

Thus, the difference in effective temperature between the equator and the pole is expressed by

$$\begin{aligned} \Delta T_e &= T_e(0) - T_e\left(\frac{\pi}{2}\right) \\ &= \left(f(0)^{\frac{1}{4}} - f\left(\frac{\pi}{2}\right)^{\frac{1}{4}}\right)[T_e] \quad \equiv \quad \Delta f^{\frac{1}{4}} \cdot [T_e], \end{aligned} \tag{13.3.3}$$

where $[T_e]$ is the global-mean effective temperature defined by (12.1.7). The temperature difference is maximized if the equinox distribution function (13.2.4) is used for $f(\varphi)$. In this case, the factor in (13.3.3) is given by $\Delta f^{1/4} = (4/\pi)^{1/4} = 1.06$. If the annually averaged condition (13.2.6) is used for $f(\varphi)$ with $c = 0.477$, we have $\Delta f^{1/4} = (3c/2)^{1/4} = 0.920$. That is, the maximum difference of the effective temperature in the case of no latitudinal energy transport is almost comparable with the magnitude of the global-mean effective temperature: $\Delta T_e \approx [T_e]$.

When the linear formula (13.2.9) is used for the relation between $\overline{F_T^{radl}}$ and zonal-mean temperature $\tilde{T}$, the radiation balance (13.3.1) is written as

$$F_0 + B(\tilde{T}(\varphi) - T_0) \quad = \quad (1 - A)\frac{F^{\odot}}{4} f(\varphi). \tag{13.3.4}$$

In this case, latitudinal temperature distribution is given by

$$\begin{aligned} \tilde{T}(\varphi) &= T_0 - \frac{F_0}{B} + (1 - A)\frac{F^{\odot}}{4B} f(\varphi) \\ &= [\tilde{T}] + (f(\varphi) - 1) \cdot \Delta T_{max} \quad \equiv \quad \tilde{T}_R(\varphi), \end{aligned} \tag{13.3.5}$$

where $[\tilde{T}]$ is the global-mean temperature defined by (13.2.14) and

$$\Delta T_{max} \equiv \frac{(1-A)F^{\odot}}{4B} \approx 114.0\text{ K}. \tag{13.3.6}$$

$\tilde{T}_R(\varphi)$ is determined only by the radiation process and can be viewed as the radiative equilibrium temperature of the energy balance model. Thus, latitudinal temperature difference is given by

$$\Delta T = \left(f(0) - f\left(\frac{\pi}{2}\right)\right) \cdot \Delta T_{max} \equiv \Delta f \cdot \Delta T_{max}. \tag{13.3.7}$$

For the equinox condition (13.2.4), we have $\Delta f = 4/\pi = 1.27$, while for the annually averaged condition (13.2.6), we have $\Delta f = 3c/2 = 0.716$. From this, the upper bound of latitudinal temperature difference is comparable with ΔT_{max} when the linear formula is used.

Second, we consider the case when the atmosphere and ocean system has latitudinal energy transport. If we assume that the atmosphere and ocean circulation is driven by latitudinal temperature difference, latitudinal energy transport becomes larger as ΔT becomes larger. This expectation motivates us to introduce a diffusion-type parameterization in which latitudinal energy transport is proportional to latitudinal temperature difference in order to reconcile the energy balance (13.1.11) with (13.1.13). Using a diffusion coefficient D, we assume that latitudinal energy transport is proportional to the gradient of temperature as

$$\langle \rho v \sigma \rangle = -\frac{D}{R}\frac{\partial \tilde{T}}{\partial \varphi}. \tag{13.3.8}$$

Alternatively, extending this type of diffusion flux to the latitudinal energy transport of the atmosphere and ocean system, we can set total energy convergence on the right-hand side of (13.1.11) as

$$C_{atm} + C_{ocean} = \frac{1}{R\cos\varphi}\frac{\partial}{\partial \varphi}\left(\cos\varphi \frac{D}{R}\frac{\partial \tilde{T}}{\partial \varphi}\right). \tag{13.3.9}$$

The value of the diffusion coefficient D is determined such that the resultant latitudinal temperature difference gets close to that of the real atmosphere. Using this with (13.2.8), (13.2.9), and (13.3.5), the energy balance (13.1.11) is rewritten as

$$B(\tilde{T} - \tilde{T}_R) = \frac{1}{R\cos\varphi}\frac{\partial}{\partial \varphi}\left(\cos\varphi \frac{D}{R}\frac{\partial \tilde{T}}{\partial \varphi}\right). \tag{13.3.10}$$

If B and D are constant and $f(\varphi)$ is expressed by the Legendre function as (13.2.6), we can analytically solve this equation for latitudinal distribution $\tilde{T}(\varphi)$. In general, the n-th order Legendre function P_n satisfies†

$$\frac{1}{\cos\varphi}\frac{\partial}{\partial \varphi}\left(\cos\varphi \frac{\partial P_n(\varphi)}{\partial \varphi}\right) = -n(n+1)P_n(\varphi). \tag{13.3.11}$$

†The characteristics of Legendre functions are summarized in Section 21.8.

Then, from (13.2.6), (13.3.5), and (13.3.10), we have

$$\tilde{T}(\varphi) = [\tilde{T}] - c\frac{B}{B+6D/R^2}\Delta T_{max}P_2(\varphi). \quad (13.3.12)$$

In this case, latitudinal temperature difference is expressed as

$$\Delta T = \frac{3c}{2}\frac{B}{B+6D/R^2}\Delta T_{max}. \quad (13.3.13)$$

This means that latitudinal temperature difference becomes smaller as the diffusion coefficient D becomes larger; ΔT is equal to the maximum value (13.3.7) in the case $D = 0$, while it becomes zero as $D \to \infty$.

Let us examine the meaning of the diffusion coefficient D by comparing it with the energy transport of Golitsyn's model given by (12.4.16). We estimate the energy transport of the atmosphere in (13.3.8) as

$$\langle \rho v \sigma \rangle \approx C\overline{v'T'} \approx C_p M U \Delta T, \quad (13.3.14)$$

where C is defined by (13.1.7), M is the mass of an air column, and $v' = v - \overline{v}$, $T' = T - \overline{T}$ are deviations from the zonal-mean latitudinal velocity and temperature, respectively. The magnitude of v' is estimated as the velocity scale U, and that of T' is tentatively estimated as latitudinal temperature difference ΔT. In general, however, T' is much smaller than ΔT in the earth's atmosphere, so that the difference between T' and ΔT will cause an error. From (13.1.13) and (13.3.14), we have

$$C_{atm} = -\frac{1}{R\cos\varphi}\frac{\partial}{\partial\varphi}\left(\cos\varphi \langle \rho v \sigma \rangle\right) \approx \frac{1}{R} M C_p U \Delta T. \quad (13.3.15)$$

On the other hand, the radiation balance at the top of the atmosphere is estimated as

$$|\overline{F_T^{radl}} + \overline{F_T^{rads}}| < |\overline{F_T^{rads}}| \approx (1-A)\frac{F^{\odot}}{4} f(\varphi) = F_s f(\varphi), \quad (13.3.16)$$

where F_s is defined by (12.4.1). In order to obtain the energy balance of Golitsyn's model (12.4.16), we need to assume $f \approx 1$ and neglect the contributions of long-wave radiation and oceanic transport C_{ocean}. Using these assumptions, (13.3.9) is reduced to

$$\frac{1}{R} M C_p U \Delta T \approx F_s, \quad (13.3.17)$$

which is equivalent to (12.4.16). The right-hand side of this equation was originally the difference between solar radiation and planetary radiation, and oceanic energy transport is added to the left-hand side. Thus, it is thought that this equation gives the upper limit of the latitudinal energy transport of the atmosphere.

If energy transport is expressed as the diffusion-type formula (13.3.8), we have

$$C_{atm} \approx \frac{D}{R^2}\Delta T, \quad (13.3.18)$$

where the oceanic energy transport is neglected once again. By comparison with (13.3.15), the diffusion coefficient is estimated as

$$D \approx RMC_pU. \tag{13.3.19}$$

This shows that D becomes larger as the atmospheric mass M becomes larger. This in turn suggests that latitudinal temperature difference becomes smaller. We cannot reach any conclusion based on just this estimation, however, since the dependence of U is not known.

Returning back to (13.1.10) where the tendency term remains, we rewrite it using (13.3.10) as

$$\frac{\partial}{\partial t}\tilde{T} = -\frac{1}{\tau_R}(\tilde{T}-\tilde{T}_R) + \frac{1}{\tau_D}\frac{1}{\cos\varphi}\frac{\partial}{\partial\varphi}\left(\cos\varphi\frac{\partial}{\partial\varphi}\tilde{T}\right), \tag{13.3.20}$$

where we have defined

$$\tau_R \equiv \frac{C}{B}, \qquad \tau_D \equiv \frac{R^2C}{D}. \tag{13.3.21}$$

These are called the *radiation relaxation time* and the *diffusion time*, respectively. Using (13.3.19) and $C \approx MC_p$, we obtain

$$\tau_D \approx \frac{R}{U}. \tag{13.3.22}$$

This means that the diffusion time is the advection time required to travel the latitudinal distance R at velocity U. From the ratio between the two timescales, a nondimensional number can be defined as

$$\delta \equiv \frac{\tau_R}{\tau_D} = \frac{D}{R^2B}. \tag{13.3.23}$$

As δ becomes larger, the atmospheric temperature becomes more homogenized, while as δ becomes smaller, the atmospheric temperature gets closer to the radiative equilibrium temperature. According to North et al. (1981), a realistic temperature difference can be obtained at $\delta = 0.31$. Fig. 13.4 shows temperature distributions for three values of δ.

13.4 Ice albedo feedback

If albedo depends on temperature, there may exist multiple equilibrium states of the atmosphere for a given solar flux. Let us first consider this situation with globally averaged states. We assume that albedo abruptly increases when temperature becomes colder than freezing point T_i,

$$[A] = A([\tilde{T}]) = \begin{cases} A_f, & \text{for } [\tilde{T}] \geq T_i, \\ A_i, & \text{for } [\tilde{T}] < T_i, \end{cases} \tag{13.4.1}$$

where $A_f < A_i$. Typical values are $A_f = 0.3$ and $A_i = 0.62$ (North et al., 1981).

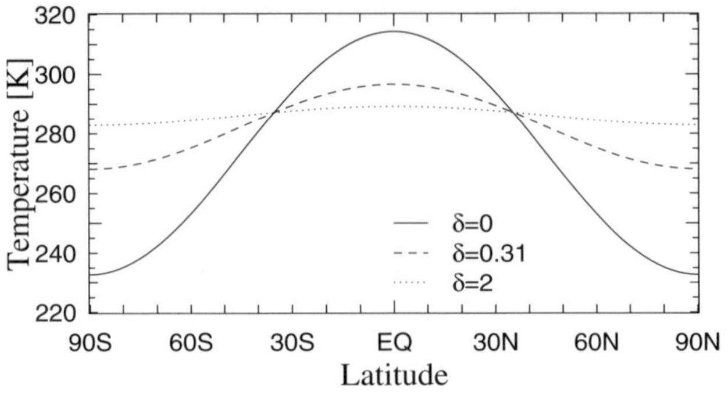

FIGURE 13.4: The latitudinal distribution of the temperature with the energy balance model. Solid line: $\delta = 0$ (radiative equilibrium temperature), dashed line: $\delta = 0.31$, and dotted line: $\delta = 2$. The earth condition is $\delta = 0.31$. The other parameters are $[\tilde{T}] = 287.1$ K, $\Delta T_{max} = 114.0$ K, and $c = 0.477$.

From (13.2.14), radiative equilibrium temperature is given by

$$[\tilde{T}] = T_0 - \frac{F_0}{B} + (1 - A([\tilde{T}]))\frac{F^{\odot}}{4B}. \tag{13.4.2}$$

The relation between $F^{\odot}$ and $[\tilde{T}]$ is shown in Fig. 13.5, where the abscissa denotes the ratio of solar radiation to its present value $F_0^{\odot}$ and $T_i = -10$ °C is used. If solar radiation is within the range

$$4\frac{F_0 + B(T_i - T_0)}{1 - A_f} < F^{\odot} < 4\frac{F_0 + B(T_i - T_0)}{1 - A_i}, \tag{13.4.3}$$

there are two solutions for $[\tilde{T}]$ in (13.4.2). For example, in the case $T_i = -10$ °C and $F_0 = 210$ W m^{-2}, two equilibrium states exist in the range of solar radiation between 1080 and 1989 W m^{-2}. The present value of solar radiation $F_0^{\odot}$ is included in this range. If solar radiation decreases from the state with no ice, the global-mean temperature becomes colder (following the solid line in Fig. 13.5). When solar radiation becomes smaller than 1080 W m^{-2}, temperature abruptly changes to the state on the dashed line. In contrast, if solar radiation increases from the state in which all the globe is frozen, the temperature rises following the dashed line in Fig. 13.5. At the present value of solar radiation, the temperature is still as cold as -40 °C (i.e., the earth is frozen). If solar radiation exceeds the value 1989 W m^{-2}, the temperature abruptly rises from the state on the dashed line to that on the solid line.

Next, we consider *ice albedo feedback* by allowing latitudinal dependence using the energy balance model with diffusion-type energy transport. For simplicity, we assume that all conditions are symmetric about the equator and consider only the northern hemisphere. The albedo at each latitude has the same dependency on local temperature as (13.4.1). If solar radiation is sufficiently small, the earth surface is everywhere frozen, whereas if solar radiation is sufficiently large, all the earth surface is free from ice. In the intermediate case, there is a state in which the polar region is frozen and the equatorial region has no ice. The boundary between the

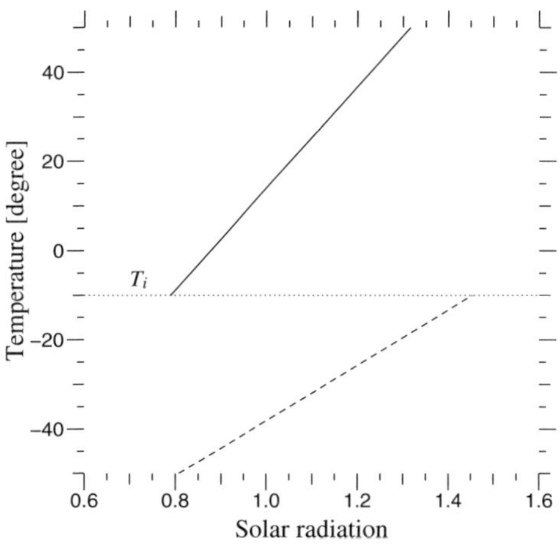

FIGURE 13.5: Ice albedo feedback. The solid line is the equilibrium state for $[\tilde{T}] \geq T_i$, and the dashed line is that for $[\tilde{T}] < T_i$. The abscissa is the ratio of solar radiation to its present value $F^{\odot}/F_0^{\odot}$ where $F_0^{\odot} = 1367.7$ W m^{-2}, and the ordinate is the global-mean temperature in degree [°C]. The dotted line is the freezing point and $T_i = -10$ °C is specified.

frozen region and the ice-free region is referred to as the *ice line.* What we want to know is the location of the ice line for a given amount of solar radiation. To investigate this problem, we try to find the relation between solar radiation and the latitudinal temperature profile assuming that the ice line is located at latitude φ_i. In this case, from (13.4.1), the albedo is given as

$$A(\varphi;\varphi_i) = \begin{cases} A_f, & \text{for } \varphi \leq \varphi_i, \\ A_i, & \text{for } \varphi > \varphi_i. \end{cases} \tag{13.4.4}$$

The energy balance (13.3.10) is written as

$$\tilde{T} - T_0 + \frac{F_0}{B} - (1 - A(\varphi;\varphi_i))\frac{F^{\odot}}{4B} f(\varphi) = \delta\frac{1}{\cos\varphi}\frac{\partial}{\partial\varphi}\left(\cos\varphi\frac{\partial}{\partial\varphi}\tilde{T}\right), \tag{13.4.5}$$

where δ is the nondimensional diffusion coefficient defined by (13.3.23). We expand the temperature using Legendre functions:

$$\tilde{T} = \sum_n t_n P_n(\varphi), \tag{13.4.6}$$

where t_n's are expansion coefficients. Since $\tilde{T}$ is symmetric about $\varphi = 0$, the summation is taken only for even numbers of n. P_n satisfies the orthogonality

$$\int_0^1 P_n(\varphi)P_m(\varphi)\cos\varphi\, d\varphi = \frac{1}{2n+1}\delta_{nm}, \tag{13.4.7}$$

where δ_{nm} equals one only when $n = m$ and is zero otherwise. P_n also satisfies

(13.3.11). From (13.4.5), therefore, we have

$$t_0 = T_0 - \frac{F_0}{B} + \frac{F^{\odot}}{4B} a_0, \tag{13.4.8}$$

$$t_n = \frac{F^{\odot}}{4B} \frac{a_n}{n(n+1)\delta + 1}, \quad \text{for} \quad n = \text{even} \geq 2, \tag{13.4.9}$$

where

$$a_n = (2n+1) \int_0^1 (1 - A(\varphi; \varphi_i))\, f(\varphi)\, P_n(\varphi) \cos\varphi \, d\varphi. \tag{13.4.10}$$

In particular, we have $a_0 = 1 - [A]$, where $[A]$ is given by (13.2.13). Thus, the temperature distribution is given by

$$\tilde{T}(\varphi) = T_0 - \frac{F_0}{B} + \frac{F^{\odot}}{4B} \sum_n \frac{a_n}{n(n+1)\delta + 1} P_n(\varphi). \tag{13.4.11}$$

If the temperature at the ice line is equal to T_i, the relation between solar radiation and the latitude of the ice line is given by

$$F^{\odot}(\varphi_i) = 4[F_0 + B(T_i - T_0)] \left(\sum_n \frac{a_n}{n(n+1)\delta + 1} P_n(\varphi_i) \right)^{-1}. \tag{13.4.12}$$

The left panel of Fig. 13.6 shows this relation for the value $\delta = 0.25$. The global-mean temperature, on the other hand, is simply given by (13.2.14); this relation is shown in the right panel of Fig. 13.6, in which the globally frozen condition and

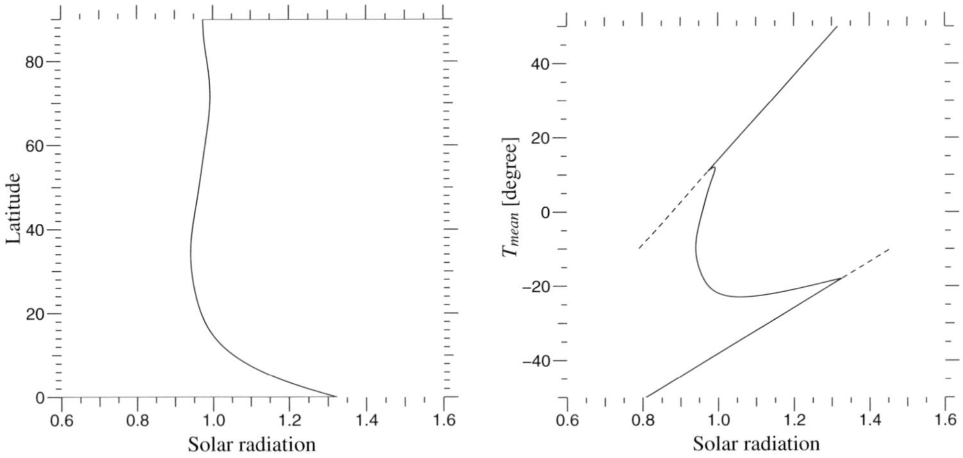

FIGURE 13.6: Ice albedo feedback using the energy balance model with diffusion-type energy transport. The left panel shows the relation between solar radiation $F^{\odot}/F_0^{\odot}$ ($F_0^{\odot} = 1367.7$ W m^{-2}) (abscissa) and latitude of the ice line (ordinate). The right panel shows the relation between solar radiation (abscissa) and global-mean temperature (ordinate). The parameter values are $T_i = -10$ °C, $F_0 = 210$ W m^{-2}, $B = 2.1$ W m^{-2} °C^{-1}, $A_f = 0.3$, $A_i = 0.62$, and $\delta = 0.25$.

the globally ice-free condition are shown by the same lines as in Fig. 13.5. The curve connecting these two lines corresponds to the states in which the ice line exists. For the value of this parameter, there are three equilibrium solutions that are close to the present value of solar radiation. Among the multiple solutions, the intermediate solution for the ice line is thought to be stable and realizable. The stability of the solution provided by EBM is discussed in the review by Crowley and North (1991).

13.5 Water budget

In this final section on latitudinal balance, we continue to consider the transport of water vapor in a moist atmosphere, in which the phase change of water is included. The equation of water vapor (9.5.1) is written using spherical coordinates as

$$\begin{aligned}\frac{\partial}{\partial t}\rho q + \frac{1}{R\cos\varphi}\frac{\partial}{\partial\lambda}(\rho u q + i_\lambda) + \frac{1}{R\cos\varphi}\frac{\partial}{\partial\varphi}[\cos\varphi\,(\rho v q + i_\varphi)] & \\ + \frac{\partial}{\partial z}(\rho w q + i_z) \quad = \quad - S_q, & \qquad (13.5.1)\end{aligned}$$

where the diffusion flux vector is denoted by $\boldsymbol{i} = (i_\lambda, i_\varphi, i_z)$. Integrating this equation in the zonal and vertical directions, and neglecting the latitudinal component of diffusion i_φ and the vertical component of the flux at the top of the atmosphere, we obtain the latitudinally one-dimensional water balance equation as

$$\frac{\partial}{\partial t}\langle\rho q\rangle + \frac{1}{R\cos\varphi}\frac{\partial}{\partial\varphi}(\cos\varphi\,\langle\rho v q\rangle) \quad = \quad \overline{E_v} - \overline{P_r}, \qquad (13.5.2)$$

where E_v is evaporation from the ground surface to the atmosphere and P_r is precipitation, defined by

$$\overline{E_v} \quad \equiv \quad \overline{i_z}(z_0), \qquad (13.5.3)$$
$$\overline{P_r} \quad \equiv \quad -\langle S_q\rangle, \qquad (13.5.4)$$

where z_0 is the height of the surface. Over a long time average of (13.5.2), the global integral of the evaporation of water vapor is equal to that of precipitation:

$$[E_v] \quad = \quad [P_r]. \qquad (13.5.5)$$

We can formulate the energy balance model of a moist atmosphere using the water budget. The zonal and vertical average of the energy equation of a moist atmosphere (9.5.12) gives

$$\begin{aligned}& C\frac{\partial}{\partial t}\tilde{T} + \frac{1}{R\cos\varphi}\frac{\partial}{\partial\varphi}(\cos\varphi\,\langle\rho v\sigma\rangle) \\ & = \quad -(\overline{F_T^{radl}} + \overline{F_T^{rads}}) + (\overline{F_B^{radl}} + \overline{F_B^{rads}} + \overline{F^{sh}}) + L\overline{P_r}, \qquad (13.5.6)\end{aligned}$$

where we have used the same approximation as in Section 13.1 and the relation

$$\langle\rho C_p Q_m\rangle \quad = \quad -L\,\langle S_q\rangle \quad = \quad L\overline{P_r}, \qquad (13.5.7)$$

which is given from (9.5.10) and (13.5.4). An alternative form of the energy balance model is given by using moist enthalpy. Summing up (13.5.6) and (13.5.2) multiplied by L, we obtain

$$\begin{aligned} &\frac{\partial}{\partial t}\langle\rho(C_pT+Lq)\rangle+\frac{1}{R\cos\varphi}\frac{\partial}{\partial\varphi}\left(\cos\varphi\,\langle\rho v(\sigma+Lq)\rangle\right) \\ &\quad= -(\overline{F_T^{radl}}+\overline{F_T^{rads}})+(\overline{F_B^{radl}}+\overline{F_B^{rads}}+\overline{F^{sh}})+L\overline{E_v}, \end{aligned} \quad (13.5.8)$$

where C_pT+Lq is moist enthalpy and $\sigma+Lq$ is moist static energy. This equation corresponds to the zonal and vertical average of (9.5.8).

The first form of the energy balance model (13.5.6) can be written using diffusion-type energy transport and the assumptions that are used to derive (13.3.20) as

$$\frac{\partial}{\partial t}\tilde{T} = -\frac{1}{\tau_R}(\tilde{T}-\tilde{T}_R)+\frac{1}{\tau_D}\frac{1}{\cos\varphi}\frac{\partial}{\partial\varphi}\left(\cos\varphi\frac{\partial}{\partial\varphi}\tilde{T}\right)+\tilde{Q}_m, \quad (13.5.9)$$

where

$$\tilde{Q}_m \equiv \frac{\langle\rho C_pQ_m\rangle}{\langle\rho C_p\rangle} = \frac{L\overline{P_r}}{C} \quad (13.5.10)$$

is latent heat release. It is essential to know the latitudinal distribution of $\tilde{Q}_m$ or $\overline{P_r}$ to use the above energy balance model. This is determined through the water budget (13.5.2). The latitudinal distribution of evaporation and precipitation is to be solved if the latitudinal transport of water $\langle\rho vq\rangle$ is given. In the real atmosphere, water vapor is more abundant in the lower layer where temperature is warmer and the direction of the latitudinal transport of water vapor is greatly dependent on that of the meridional wind in the lower layer. In particular, $\langle\rho vq\rangle$ is equatorward in low latitudes because of the equatorward flow of the Hadley circulation (see Chapter 16). The atmospheric structure in the tropics can be examined by taking account of the lateral water transport associated with the overturning of atmospheric flows.†

As stated above, the diffusion-type model used for temperature (13.3.8) is not a good approximation for water vapor. Nevertheless, a rough idea of the effect of water vapor transport can be captured by assuming water vapor transport occurs in diffusion-type formulas; we assume

$$\langle\rho vq\rangle = -D_q\frac{1}{R}\frac{\partial}{\partial\varphi}\tilde{q}, \quad (13.5.11)$$

where D_q is a diffusion coefficient of water vapor. In this case, a long time average of the balance of (13.5.2) becomes

$$-\frac{1}{R\cos\varphi}\frac{\partial}{\partial\varphi}\left(\cos\varphi\cdot D_q\frac{1}{R}\frac{\partial}{\partial\varphi}\tilde{q}\right) = \overline{E_v}-\overline{P_r}. \quad (13.5.12)$$

† This argument is closely related to that of the low-latitude circulations described in Chapter 16. See Satoh (1994) for the Hadley circulation and Bretherton and Sobel (2002) for the Walker circulation, for example.

In general, evaporation from the surface is proportional to the difference between the water vapor at the surface and that of the lowest layer of the atmosphere (see Section 11.3.3). Precisely, we need to take account of the difference between the temperature of the atmosphere and the surface temperature. We simplify the formula of evaporation as

$$\overline{E_v} \quad = \quad K(q_s - \tilde{q}) \quad \approx \quad K(1-r)q^*(\tilde{T}), \tag{13.5.13}$$

where K is a coefficient of evaporation, r is a characteristic value of relative humidity of the atmospheric boundary layer, q_s is the saturation water vapor content at the surface, and $q^*(\tilde{T})$ is the saturation water vapor content at temperature $\tilde{T}$. Temperature must be obtained by simultaneously solving the equation of water balance and the equation of temperature (13.5.9). Here, in order to understand the qualitative characteristics of water vapor transport, we furthermore assume that temperature and water vapor are given independently. Since temperature is warmer in lower latitudes, water vapor is more abundant in lower latitudes in a global sense. This dependency is modeled as

$$q^*(\tilde{T}) \quad = \quad q_0 - q_2 P_2(\sin\varphi), \tag{13.5.14}$$

where q_0 and q_2 are appropriate constants. The global-mean evaporation is given by $K(1-r)q_0$ from (13.5.13). Thus, from (13.5.12), precipitation is given by

$$\overline{P_r} \quad = \quad K(1-r)q_0 - \left[K(1-r) - \frac{D_q}{R^2}\right] q_2 P_2(\sin\varphi). \tag{13.5.15}$$

This shows that the coefficient of P_2 of precipitation is smaller than that of evaporation in the case $D_q > 0$. This means that precipitation is larger than evaporation in higher latitudes. Therefore, the atmosphere has a net poleward transport of water vapor. If the earth was covered by land and no sea, and if precipitated water did not flow out on the land, the diffusion process of water vapor would result in the concentration of water vapor in the polar region. This process may be applied to Mars, where ice is observed only in the polar region.

References and suggested reading

The ice albedo feedback was first introduced using the energy balance model by Budyko (1969), Sellers (1969), and North (1975). The properties of the energy balance model are summarized in Chapter 1 of Crowley and North (1991), Chapter 2 of Lindzen (1990), or Chapter 9 of Hartmann (1994). The diagnostics of the latitudinal transports of energy and water in the atmosphere and ocean system are summarized by Peixoto and Oort (1992).

Budyko, M. I., 1969: The effect of solar radiation variations on the climate of the earth. *Tellus*, **21**, 611–619.

Bretherton, C. S. and Sobel, A. H., 2002: A simple model of a convectively coupled Walker circulation using the weak temperature gradient approximation. *J. Climate*, **15**, 2907–2920.

Crowley, T. J. and North, G. R., 1991: *Paleoclimatology.* Oxford University Press, New York, 339 pp.

Gruber, A. and Krueger, A. F., 1984: The status of the NOAA outgoing longwave radiation data set. *Bull. Amer. Meteorol. Soc.*, **65**, 958–962.

Hartmann, D. G., 1994: *Global Physical Climatology.* Academic Press, San Diego, 408 pp.

Lindzen, R. S., 1990: *Dynamics in Atmospheric Physics.* Cambridge University Press, Cambridge, UK, 310 pp.

North, G. R., 1975: Theory of energy-balance climate models. *J. Atmos. Sci.*, **32**, 2033–2043.

North, G. R., Chahalan, R. F., and Coakley, J. A. Jr., 1981: Energy balance climate models. *Rev. Geophys. Space Phys.*, **19**, 91–121.

Oort, A. H. and Peixoto, J. P., 1983: Global angular momentum and energy balance requirements from observations. *Advances in Geophys.*, **25**, 355–490.

Oort, A. H. and Vonder Haar, T. H., 1976: On the observed annual cycle in the ocean-atmosphere heat balance over the Northern Hemisphere. *J. Phys. Oceanogr.*, **6**, 781–800.

Peixoto, P. J. and Oort, A. H., 1992: *Physics of Climate.* American Institute of Physics, New York, 520 pp.

Satoh, M., 1994: Hadley circulations in radiative-convective equilibrium in an axially symmetric atmosphere. *J. Atmos. Sci.*, **51**, 1947–1968.

Sellers, W. D., 1969: A global climate model based on the energy balance of the earth-atmosphere system. *J. Appl. Meteorol.*, **8**, 392–400.

Trenberth, K. E., 1979: Mean annual poleward energy transports by the oceans in the southern hemisphere. *Dyn. Atmos. Oceans*, **4**, 57–64.

14

Vertical structure

In this chapter, the vertically one-dimensional structure of the atmosphere is investigated based on energy budget analysis in the vertical direction. The vertical thermal structure of the atmosphere is studied by introducing two key concepts: radiative equilibrium and radiative-convective equilibrium. First, as an introduction of the concept of radiative equilibrium, the greenhouse effect is described using a one-layer glass model. Next, the thermal structure in radiative equilibrium is calculated using a simplified radiation scheme. Then, the discussion on radiative-convective equilibrium follows where the thermal structures of the troposphere and stratosphere are described with the definition of the tropopause. Finally, radiative-convective equilibrium in the case of a moist atmosphere is argued for.

The radiative process plays a fundamental role in the determination of the vertical thermal structure of the atmosphere. In this chapter, the gray radiation model described in Section 10.6 is used to show how the radiative process determines thermal structure. The thermal structure constrained solely by the radiative process is called *radiative equilibrium.* It will be found that the state in radiative equilibrium is in general statically unstable. As studied in Chapter 2, convective motion will occur in such a statically unstable atmosphere.

If convection occurs, both the radiative process and the convective process interact to establish a new thermal state, called *radiative-convective equilibrium.* In a dry atmosphere, convection will be in a form similar to the turbulence in the boundary layer, described in Section 11.3 using different types of turbulence models. In this chapter, instead of using complex turbulence models, convective flux is estimated using the mixing length theory, which is the simplest type of turbulence model. It will be shown that the resultant thermal structure due to convective motion is close to the dry adiabat. This result implies that the effect of convection on thermal structure is equivalent to that of *convective adjustment.*

In the case of a moist atmosphere, since the latent heat release of water vapor is associated, convective motion is much more complicated. In the next chapter, the basic properties of moist convection will be described; it will be shown that, even when moist convection occurs, the thermal structure of radiative-convective equilibrium can be determined by convective adjustment using the moist adiabat.

14.1 Vertically one-dimensional energy balance

In order to examine the vertical structure of the atmosphere, we begin by considering the vertical energy balance of a dry atmosphere. The equation of total energy is approximated as (13.1.1) by neglecting the contributions of kinetic energy and fluxes due to viscous stress. The horizontal average of (13.1.1) gives

$$\frac{\partial}{\partial t}\overline{\rho(u+\Phi)} + \frac{\partial}{\partial z}\left(\overline{F^{conv}} + \overline{F^{rad}} + \overline{F^{sh}}\right) \quad = \quad 0, \tag{14.1.1}$$

where

$$F^{conv} \quad = \quad \rho w \sigma, \tag{14.1.2}$$

is convective energy flux and σ is static energy. The overline ($\overline{\ }$) is used to denote either the global average or zonal average on a horizontal surface. The global average of a quantity X is defined by

$$\overline{X} \quad = \quad \frac{1}{S}\int_S X\,dS \quad = \quad \frac{1}{4\pi}\int_{-\frac{\pi}{2}}^{\frac{\pi}{2}}\int_0^{2\pi} X\,\cos\varphi\,d\lambda\,d\varphi, \tag{14.1.3}$$

and the zonal average is defined by (13.1.2). Eq. (14.1.1) can also be regarded as a zonal-mean equation in the special case when latitudinal energy flux is assumed to be neglected. By omitting the tendency term in (14.1.1) for a long time average, the energy balance is written as

$$\frac{\partial}{\partial z}\left(\overline{F^{conv}} + \overline{F^{rad}} + \overline{F^{sh}}\right) \quad = \quad 0, \tag{14.1.4}$$

from which we obtain

$$\overline{F^{conv}} + \overline{F^{rad}} + \overline{F^{sh}} \quad = \quad \overline{F^{ene}} \quad = \quad \text{const.} \tag{14.1.5}$$

In general, sensible heat flux $\overline{F^{sh}}$ is largest near the surface, and is almost negligible outside the boundary layer.[†] In the free atmosphere, the sum of the convective flux $\overline{F^{conv}}$ and the radiative flux $\overline{F^{rad}}$ is constant irrespective of height. This state of the energy balance is called *radiative-convective equilibrium.* Furthermore, in the case when there is no contribution from convective flux $\overline{F^{conv}}$, the equilibrium state is called *radiative equilibrium.* In reality, convective flux is not at all negligible when compared with radiative flux particularly in the layer close to the surface. The concept of radiative equilibrium is useful, however, since it provides a first approximation to the thermal structure of the atmosphere.

[†]In this formulation, it is physically obvious that $\overline{F^{sh}}$ defines the heat flux due to molecular thermal diffusion within a thin layer just above the surface, since turbulent motion in the boundary layer cannot be distinguishable from convective motion in the free atmosphere, in principle.

In general, it will be found that the thermal structure in radiative equilibrium is statically unstable in the lowest layers of the atmosphere. This suggests that convective flux is required in the lower layers of the atmosphere, and thus that the atmospheric structure is described by radiative-convective equilibrium. In this case, the balance of radiative flux still holds at a sufficiently high altitude. The lower layer in the radiative-convective equilibrium is called the *troposphere* and the upper layer in radiative equilibrium is called the *stratosphere*. The boundary between the two layers is called the *tropopause*.

Fig. 14.1 shows the temperature profile of a standard atmosphere that is a mean state in the mid-latitudes together with the profile of pressure.† The left panel is the temperature structure, which indicates the atmospheric layered structure. In the stratosphere, temperature increases with height and the temperature maximum is located at around the altitude 50 km. This level is the top of the stratosphere and is called the *stratopause*. The region above the stratopause up to about 80 km is called the *mesosphere*, in which temperature decreases with height. Above the mesosphere is the *thermosphere*. In this book, the middle and upper atmosphere above the stratosphere is not further described.

According to the pressure distribution in Fig. 14.1, the pressure at the tropopause level (11 km) is about 200 hPa, which is one-fifth of surface pressure. This means that about 80% of atmospheric mass is contained in the troposphere. The latitudinal dependency of the tropopause will be further described in Section 18.2.3.

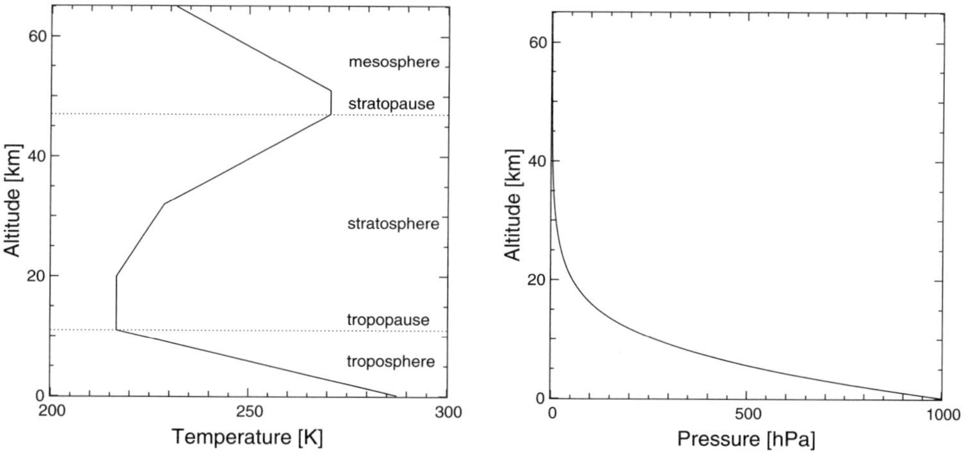

FIGURE 14.1: Temperature and pressure profiles of the standard atmosphere and the regions of the atmospheric layers (from the U.S. Standard Atmosphere, 1976).

†The U.S. standard atmosphere is defined as follows (The U.S. Standard Atmosphere, 1976): with a sea level temperature 288.15 K and sea level pressure 101325 Pa, the lapse rate of temperature $\frac{dT}{dz}$ is given at seven atmospheric layers. The lapse rate is -6.5 K km^{-1} between sea level and 11 km, 0.0 K km^{-1} between 11 km and 20 km, 1.0 K km^{-1} between 20 km and 32 km, 2.8 K km^{-1} between 32 km and 47 km, 0.0 K km^{-1} between 47 km and 51 km, -2.8 K km^{-1} between 51 km and 71 km, and -2.0 K km^{-1} between 71 km and 84.852 km.

14.2 Greenhouse effect

The effective temperature introduced in Section 12.1.1 is different from surface temperature. The difference between effective temperature and surface temperature is caused by the *greenhouse effect.* The simplest model to illustrate the greenhouse effect is a one-layer glass model in which the atmosphere is represented as a sheet of glass. Only radiative transport is allowed as the energy transport in this model. Glass is transparent to solar radiation, while it is completely opaque to planetary radiation. Glass is a blackbody and emits blackbody radiation in both upward and downward directions depending on the temperature of the glass. Here, we consider the radiation balance in a unit square of the surface (Fig. 14.2), assuming that the global atmosphere is horizontally uniform.

In the equilibrium state, total radiative flux is zero (i.e., solar radiation is balanced by planetary radiation). Solar radiation is assumed to pass through the glass and be absorbed or reflected at the ground surface. The ratio of this reflected portion to incident solar radiation is the planetary albedo A. Let the temperature at the ground surface be T_s, and that of the glass T_e. Since outward radiation from the glass corresponds to outgoing planetary radiation at the top of the atmosphere, T_e is equivalent to the effective temperature (12.1.7). Glass emits long-wave radiation $\sigma_B T_e^4$ in both upward and downward directions. The ground surface, on the other hand, emits upward long-wave radiation $\sigma_B T_s^4$, which is completely absorbed by the glass surface. The radiation balance at the top of the atmosphere is given by dividing (12.1.6) by $S = 4\pi R^2$:

$$F^{isr} - F^{olr} = (1 - A)\frac{F^{\odot}}{4} - \sigma_B T_e^4 = 0, \tag{14.2.1}$$

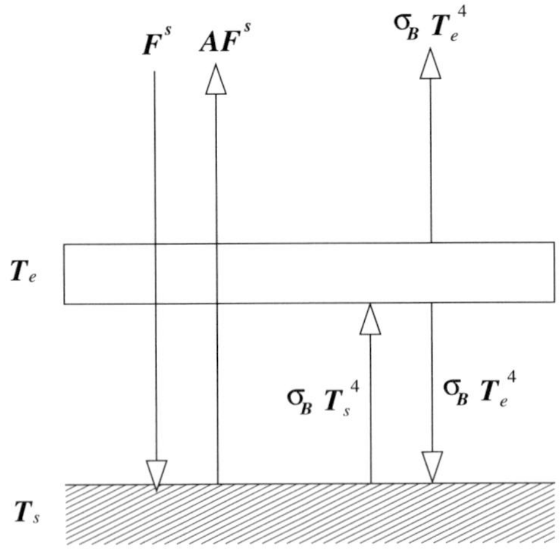

FIGURE 14.2: The radiation balance of the one-layer glass model. The white rectangle is the glass which represents the atmosphere, and the lower shaded area is the ground. The temperature of the glass is T_e, and that of the ground is T_s; both emit blackbody radiation of the corresponding temperature. Solar radiation is $F^s = F^{\odot}/4$ and A is planetary albedo.

where

$$F^{isr} = \frac{\mathcal{F}^{isr}}{4\pi R^2} = (1-A)\,\frac{F^{\odot}}{4}, \tag{14.2.2}$$

$$F^{olr} = \frac{\mathcal{F}^{olr}}{4\pi R^2} = \sigma_B T_e^4. \tag{14.2.3}$$

The radiation balance at the glass and the ground surface is respectively given by

$$\sigma_B T_s^4 - 2\sigma_B T_e^4 = 0, \tag{14.2.4}$$

$$(1-A)\,\frac{F^{\odot}}{4} + \sigma_B T_e^4 - \sigma_B T_s^4 = 0. \tag{14.2.5}$$

Thus, we obtain from (14.2.1) and (14.2.4) or (14.2.5),

$$T_e = \left[\frac{(1-A)\,F^{\odot}}{4\sigma_B}\right]^{\frac{1}{4}}, \tag{14.2.6}$$

$$T_s = 2^{\frac{1}{4}} T_e = \left[\frac{(1-A)\,F^{\odot}}{2\sigma_B}\right]^{\frac{1}{4}}. \tag{14.2.7}$$

In the case $T_e = 254.9$ K, we have $T_s = 303.1$ K (i.e., surface temperature is warmer than effective temperature). This warming effect on the surface temperature due to the existence of the atmosphere, in this case represented by the glass, is called the *greenhouse effect.* This difference comes from the asymmetric property of atmospheric radiation, where the atmosphere is relatively transparent to solar radiation and opaque to planetary radiation. As shown below, the greenhouse effect depends on the distributions of absorbing gases and convective motions in the atmosphere.

The above argument is based on the assumption that the whole atmosphere is regarded as one sheet of glass. This argument can be extensible to multiple sheets of glass; it can be shown that if the number of sheets of glass is increased, the equilibrium surface temperature increases. In the next section, we further investigate the greenhouse effect by using the radiation model that considers the vertical structure of the atmosphere, and show that total optical depth corresponds to the number of sheets of the glass model.

14.3 Radiative equilibrium

14.3.1 Gray radiative equilibrium

In order to study the greenhouse effect using more realistic processes, we use the radiative transfer equation to obtain the relation between radiative flux and the vertical structure of temperature. We consider the vertically one-dimensional radiative equilibrium in which atmospheric temperature is a function of altitude z, and energy transfer is only in the form of radiative flux. We assume the gray atmosphere which enables analytic discussion of the radiative property. In a state of radiative equilibrium, the conservation of energy (14.1.4) is written as

$$\frac{\partial}{\partial z}\overline{F^{rad}} = 0, \tag{14.3.1}$$

where $\overline{F^{rad}}$ is the horizontal mean upward radiative flux.

Consistent with the assumption used for the glass model, solar radiation $(1-A) \times F^{\odot}/4$ is absorbed at the ground without being absorbed in the atmosphere. As for planetary radiation, the upward and downward fluxes satisfy (10.6.7) and (10.6.8), respectively:

$$\frac{2}{3}\frac{dF^{\uparrow}(\tau)}{d\tau} = F^{\uparrow}(\tau) - \pi B(\tau), \tag{14.3.2}$$

$$-\frac{2}{3}\frac{dF^{\downarrow}(\tau)}{d\tau} = F^{\downarrow}(\tau) - \pi B(\tau), \tag{14.3.3}$$

where $\pi B = \sigma_B T^4$. These fluxes are functions of the optical depth τ, which is zero at the top of the atmosphere and τ_s at the ground. The net radiative flux of planetary radiation is given by

$$F(\tau) = F^{\uparrow}(\tau) - F^{\downarrow}(\tau). \tag{14.3.4}$$

The boundary conditions for (14.3.2) and (14.3.3) are given such that downward planetary radiative flux is zero at the top of the atmosphere:

$$F^{\downarrow}(0) = 0, \tag{14.3.5}$$

and that upward flux at the ground is

$$F^{\uparrow}(\tau_s) = \pi B(\tau_s) = \sigma_B T_s^4, \tag{14.3.6}$$

where T_s is surface temperature. Total radiative flux is given by the sum of solar radiation and planetary radiation. If solar radiation is not absorbed or reflected within the atmosphere, total radiative flux is written as

$$\overline{F^{rad}} = -(1-A)\frac{F^{\odot}}{4} + F. \tag{14.3.7}$$

If this flux satisfies the equilibrium condition (14.3.1), the atmosphere is in a state of *radiative equilibrium*. In this case, $\overline{F^{rad}}$ is independent of height. In particular, if there is no heat source in the atmosphere and solid earth system, $\overline{F^{rad}} = 0$ must be satisfied. Therefore, the equilibrium condition reads

$$F = (1-A)\frac{F^{\odot}}{4}. \tag{14.3.8}$$

The solution to (14.3.2) and (14.3.3) can be easily found using the sum and the difference of the two equations:

$$\frac{2}{3}\frac{d}{d\tau}(F^{\uparrow} - F^{\downarrow}) = F^{\uparrow} + F^{\downarrow} - 2\pi B, \tag{14.3.9}$$

$$\frac{2}{3}\frac{d}{d\tau}(F^{\uparrow} + F^{\downarrow}) = F^{\uparrow} - F^{\downarrow}. \tag{14.3.10}$$

In the radiative equilibrium state, the left-hand side of (14.3.9) must vanish, then we have

$$F^{\uparrow} + F^{\downarrow} = 2\pi B. \tag{14.3.11}$$

Substituting this into (14.3.10) and using (14.3.4), we obtain

$$\frac{4\pi}{3}\frac{dB}{d\tau} = F^{\uparrow} - F^{\downarrow} = F. \tag{14.3.12}$$

Integrating this along the optical path, we have

$$B = \frac{3}{4\pi}F\tau + C, \tag{14.3.13}$$

where C is an integral constant. Substituting (14.3.13) into (14.3.11) by using (14.3.4) yields

$$F^{\uparrow} = \frac{1}{4}(3F\tau + 4\pi C) + \frac{F}{2}, \qquad F^{\downarrow} = \frac{1}{4}(3F\tau + 4\pi C) - \frac{F}{2}. \tag{14.3.14}$$

From the boundary condition (14.3.5), the integral constant is given as $C = F/2\pi$. Thus, (14.3.14) reduces to

$$F^{\uparrow} = \frac{F}{2}\left(\frac{3}{2}\tau + 2\right), \qquad F^{\downarrow} = \frac{F}{2}\cdot\frac{3}{2}\tau, \tag{14.3.15}$$

and (14.3.13) becomes

$$\pi B = \frac{F}{2}\left(\frac{3}{2}\tau + 1\right). \tag{14.3.16}$$

The vertical profiles of these radiative fluxes are shown in Fig. 14.3. All these fluxes linearly depend on optical thickness τ. The vertical profile of temperature is also given by a function of optical depth τ. Substituting $\pi B = \sigma_B T^4$ (10.6.10) into (14.3.16), and using (14.3.8) and the effective temperature (14.2.6), we obtain the

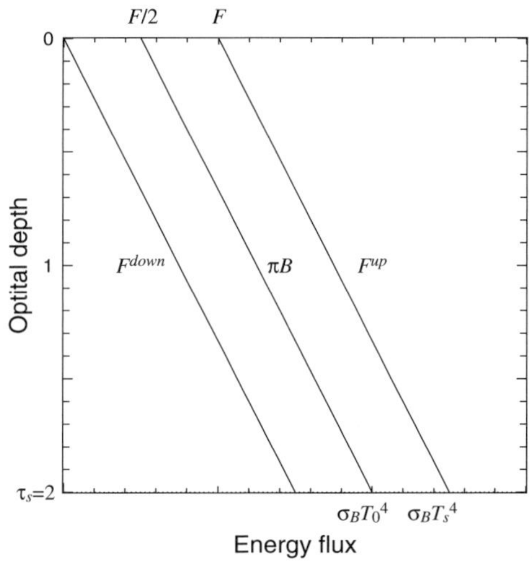

FIGURE 14.3: The vertical profiles of radiative fluxes in radiative equilibrium of the gray atmosphere; F^{up} and F^{down} are the upward and downward long-wave radiative fluxes, respectively. πB is equal to $\sigma_B T^4$. The vertical ordinate is optical depth. Total optical depth $\tau_s = 2$ is used for this figure.

temperature distribution with respect to optical thickness τ

$$\begin{aligned} T(\tau) &= \left[\frac{F}{2\sigma_B}\left(\frac{3}{2}\tau+1\right)\right]^{\frac{1}{4}} = \left[\frac{(1-A)F^{\odot}}{8\sigma_B}\left(\frac{3}{2}\tau+1\right)\right]^{\frac{1}{4}} \\ &= \left[\frac{1}{2}\left(\frac{3}{2}\tau+1\right)\right]^{\frac{1}{4}} T_e. \end{aligned} \tag{14.3.17}$$

From this, we can see that temperature at the altitude $\tau = 2/3$ is equal to effective temperature: $T = T_e$. The temperature at the bottom of the atmosphere at $\tau = \tau_s$ is

$$T(\tau_s) = \left[\frac{F}{2\sigma_B}\left(\frac{3}{2}\tau_s+1\right)\right]^{\frac{1}{4}}. \tag{14.3.18}$$

This is different from surface temperature, which is given from (14.3.6) and (14.3.15):

$$T_s = \left[\frac{F}{2\sigma_B}\left(\frac{3}{2}\tau_s+2\right)\right]^{\frac{1}{4}}. \tag{14.3.19}$$

This means that temperature has a gap between the ground surface and the bottom of the atmosphere. This unrealistic result is derived from the assumption that no energy flux other than radiative flux is allowed as the energy transport. If thermal diffusion is allowed to occur, the temperature gap disappears.

We need to specify the vertical distributions of absorbing constituents to obtain the dependence of temperature on height z. Here, for illustrative purposes, we assume that only one component of the absorbing constituents contributes to radiative flux. Let q_i denote the mass concentration of absorbing constituents and assume that its vertical profile is expressed as the following function of pressure p:

$$q_i = q_{i0}\left(\frac{p}{p_s}\right)^{\alpha}, \tag{14.3.20}$$

where α is constant. If the atmosphere is isothermal, this relation is written as

$$q_i = q_{i0}\exp\left(-\frac{z}{h}\right), \tag{14.3.21}$$

where h is the scale height of mass concentration q_i, which is relatable to the pressure scale height H as $h = H/\alpha$. Substituting $\rho_i = \rho q_i$ in (10.5.1) and using hydrostatic balance, we express optical depth as

$$\begin{aligned} \tau &= \int_z^{\infty} k\rho q_i dz = \int_z^{\infty} k\rho q_{i0}\left(\frac{p}{p_s}\right)^{\alpha} dz \\ &= \frac{kq_{i0}}{(\alpha+1)gp_s^{\alpha}}p^{\alpha+1} = \tau_s\left(\frac{p}{p_s}\right)^{\alpha+1}, \end{aligned} \tag{14.3.22}$$

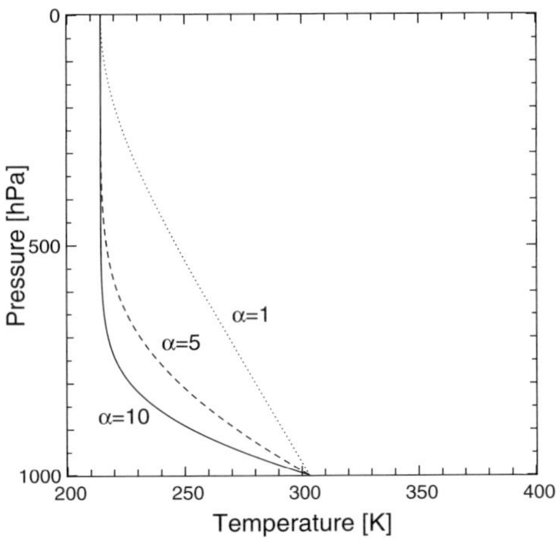

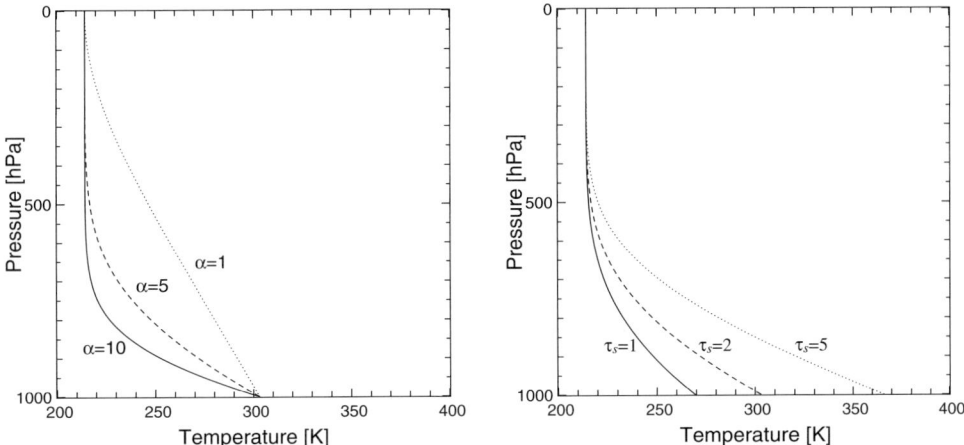

FIGURE 14.4: The vertical profiles of temperature in radiative equilibrium of the gray atmosphere. The vertical ordinate is pressure. $F = 239.3$ [W m^{-2}] is used. The left panel is for $\tau_s = 2$ with $\alpha = 10$ (solid), $\alpha = 5$ (dashed), and $\alpha = 1$ (dotted). The right panel is for $\alpha = 5$ with $\tau_s = 1$ (solid), $\tau_s = 2$ (dashed), and $\tau_s = 5$ (dotted).

where total optical depth is given as

$$\tau_s = \frac{kq_{i0}}{(\alpha+1)} \frac{p_s}{g}. \quad (14.3.23)$$

The vertical distribution of temperature is given from the relation between (14.3.17) and (14.3.22). Examples of temperature profiles are shown in Fig. 14.4 for some values of α and τ_s. For application to the real atmosphere, it is thought that q_i represents water vapor as the most important absorbing constituent. Using typical values of the scale height of specific humidity $h = 2$ km and of the pressure scale height $H = 8$ km, we have $\alpha = 4$. Instead of using (14.3.23), however, the value of τ_s should be chosen so that the resultant temperature is close to a realistic value. Based on the temperature profile of radiative-convective equilibrium described later, the typical value of total optical depth is estimated as $\tau_s = 4$.

14.3.2 Static stability of gray radiative equilibrium

Whether the state in radiative equilibrium is realized depends on the static stability of the thermal structure. In general, the temperature of the gray radiative equilibrium decreases with height. According to Fig. 14.4, as α becomes larger, the lapse rate of the temperature $\frac{dT}{dp}$ in the lower layer of the atmosphere becomes larger. The lapse rate of gray radiative equilibrium can be calculated from (14.3.17) and (14.3.22)

$$\frac{dT}{dp} = \left(\frac{F}{2\sigma_B}\right)^{\frac{1}{4}} \cdot \frac{1}{4}\left(\frac{3}{2}\tau + 1\right)^{-\frac{3}{4}} \frac{3}{2}\frac{d\tau}{dp}$$

$$= \frac{\frac{3}{2}\tau}{\frac{3}{2}\tau + 1} \frac{(\alpha+1)T}{4p} = \frac{\frac{3}{2}\tau}{\frac{3}{2}\tau + 1} \frac{(\alpha+1)}{4\rho R_d}, \tag{14.3.24}$$

where the equation of state $p = \rho R_d T$ is used. The dry adiabatic lapse rate of the ideal gas is given by (1.1.57):

$$\left(\frac{\partial T}{\partial p}\right)_s = \frac{1}{\rho C_p}, \tag{14.3.25}$$

The atmosphere is statically stable if the lapse rate is smaller than that of the dry adiabat. Thus, the condition for stability is given by comparison between (14.3.24) and (14.3.25); the stability condition of radiative equilibrium is given by

$$\frac{dT}{dp} \Big/ \left(\frac{\partial T}{\partial p}\right)_s = \frac{\frac{3}{2}\tau}{\frac{3}{2}\tau + 1} \frac{(\alpha+1)C_p}{4R_d} < 1. \tag{14.3.26}$$

Since $\frac{3}{2}\tau/(\frac{3}{2}\tau + 1) < 1$, the radiative equilibrium of the gray atmosphere is everywhere stable if

$$\alpha < 4\frac{R_d}{C_p} - 1 \tag{14.3.27}$$

is satisfied. This stability criterion is written as $\alpha < 1/7 = 0.14$ for the case $C_p/R_d = 7/2$. If this condition is not satisfied, there exists an unstable layer below the height where the optical depth is given by

$$\tau = \frac{2}{3}\left[\frac{(\alpha+1)C_p}{4R_d} - 1\right]^{-1}. \tag{14.3.28}$$

For example, for the values $\alpha = 4$, $\tau_s = 4$, and $C_p/R_d = 7/2$, the top of the unstable layer is given by $\tau = 0.20$ or $p = 550$ Pa.

Even when the structure of radiative equilibrium is stable, surface temperature is generally warmer than the temperature at the bottom of the atmosphere, as can be seen from the difference between (14.3.18) and (14.3.19). This indicates that if thermal diffusion is allowed between the surface and the atmosphere, the lowest layer of the atmosphere has a large temperature lapse rate and is thought to be statically unstable. Thus, radiative equilibrium is not realizable in fact; instead, we need to consider a different equilibrium state that considers the effect of convection (i.e., radiative-convective equilibrium).

14.3.3 Approximations of radiative transfer

If the radiative equilibrium breaks down, the convergence of radiative flux deviates from zero. In this case, the radiative transfer equations (14.3.2) and (14.3.3) are to be solved to obtain the heating rate due to radiation. Under certain circumstances, however, the heating rate due to radiation can be given by following approximate calculation methods.

If the total optical depth of the atmosphere is sufficiently thin, the magnitude of the radiative flux at a given height is not affected by the surrounding temperature field and is determined by direct interaction with outer space. In this case, the heating rate is a function of the local value of temperature. From (14.3.4), (14.3.7), and (14.3.9), the convergence of radiative flux with respect to optical thickness is given as

$$-\frac{2}{3}\frac{d}{d\tau}\overline{F^{rad}} = -(F^{\uparrow} + F^{\downarrow} - 2\pi B). \tag{14.3.29}$$

Since from (10.5.1)

$$d\tau = -\kappa\rho q_i dz, \tag{14.3.30}$$

the convergence of radiative flux is expressed as

$$-\frac{d}{dz}\overline{F^{rad}} = \frac{3}{2}\kappa\rho q_i\,(F^{\uparrow} + F^{\downarrow} - 2\pi B). \tag{14.3.31}$$

If optical depth τ is thin, we may have approximations $F^{\downarrow} \ll F^{\uparrow}$ and $F^{\uparrow} \approx$ const. In this case, $F^{\uparrow}$ is determined by surface temperature T_s:

$$F^{\uparrow} = \sigma_B T_s^4. \tag{14.3.32}$$

Expanding B with respect to T around the radiative equilibrium temperature $T_R(z)$, we have

$$\begin{aligned} 2\pi B &= 2\sigma_B T^4 = 2\sigma_B T_R^4\left(1 + \frac{T - T_R}{T_R}\right)^4 \\ &\approx 2\sigma_B T_R^4\left(1 + 4\frac{T - T_R}{T_R}\right). \end{aligned} \tag{14.3.33}$$

Thus, (14.3.31) is approximated as

$$\begin{aligned} -\frac{d}{dz}\overline{F^{rad}} &= \frac{3}{2}\kappa\rho q_i\left[\sigma_B T_s^4 - 2\sigma_B T_R^4\left(1 + 4\frac{T - T_R}{T_R}\right)\right] \\ &= -\rho C_p\frac{T - T_e}{\tau_R}, \end{aligned} \tag{14.3.34}$$

where

$$T_e = \frac{T_s^4 + 6T_R^4}{8T_R^3}, \qquad \tau_R = \frac{C_p}{12\kappa q_i\sigma_B T_R^3}. \tag{14.3.35}$$

T_e is the equilibrium temperature, and τ_R is the radiative relaxation time. Note that T_e is different from the radiative equilibrium temperature T_R and that τ_R is slightly different from (12.4.11) and (13.3.21). For the above approximations, the convergence of radiative flux (14.3.34) is in the form of radiative relaxation. This form of radiation is called *Newtonian cooling.* Although these assumptions are not always applicable to the real atmosphere, Newtonian cooling is frequently used to represent the radiative effect for the theoretical consideration of atmospheric structure. In these cases, the profile of the equilibrium temperature T_e is arbitrarily given depending on problems to take account of many factors other than those in (14.3.35).

In the other extreme case that optical depth is sufficiently thick, radiative flux is determined only by the surrounding temperature profiles. To zero-th order approximation, photons and the air are in local thermal equilibrium:

$$F^{\uparrow} \approx F^{\downarrow} \approx \pi B. \tag{14.3.36}$$

To the next order of approximation, from (14.3.2) and (14.3.3), we have

$$F^{\uparrow} = \frac{2}{3}\frac{dF^{\uparrow}}{d\tau} + \pi B \approx \frac{2}{3}\frac{d(\pi B)}{d\tau} + \pi B, \tag{14.3.37}$$

$$F^{\downarrow} = -\frac{2}{3}\frac{dF^{\downarrow}}{d\tau} + \pi B \approx -\frac{2}{3}\frac{d(\pi B)}{d\tau} + \pi B. \tag{14.3.38}$$

From the difference between the two equations and using (14.3.30), we obtain

$$\begin{aligned} F &= F^{\uparrow} - F^{\downarrow} \approx \frac{4}{3}\frac{d(\pi B)}{d\tau} = -\frac{4}{3\kappa\rho q_i}\frac{d}{dz}(\sigma_B T^4) \\ &= -\frac{16\sigma_B T^3}{3\kappa\rho q_i}\frac{dT}{dz}. \end{aligned} \tag{14.3.39}$$

This indicates that the radiative flux is in a diffusive form that is proportional to the local gradient of temperature. This kind of approximation is called *diffusion approximation* and is applicable to the interior of stars.

14.4 Radiative-convective equilibrium

The radiative equilibrium temperature profile is generally statically unstable as seen in Section 14.3.2. The lapse rate of temperature in radiative equilibrium is larger than that of the dry adiabat in the lower layer of the atmosphere for plausible values of the optical parameters. Thus, convection will occur in the unstable layers, so that a new equilibrium state is established instead of radiative equilibrium. To obtain the resultant new equilibrium, one needs to consider the effect of convection on the temperature profile in addition to the effect of radiation.

If convective and radiative fluxes co-exist in the free atmosphere, the energy balance over a long time average is expressed from (14.1.4) as:

$$\frac{\partial}{\partial z}\left(\overline{F^{conv}} + \overline{F^{rad}}\right) = 0. \tag{14.4.1}$$

If there is no heat source within the earth, the net flux must vanish:

$$\overline{F^{conv}} + \overline{F^{rad}} = 0. \tag{14.4.2}$$

In general, in the framework of a vertically one-dimensional structure, one needs to introduce a convective model to obtain an expression for convective flux $\overline{F^{conv}}$, since convection is inherently two- or three-dimensional. There are basically two types of convective models: the first is to directly give the expression of $\overline{F^{conv}}$, and the second is to assume the resultant temperature profile. An example of the former method is the *mixing length theory*, while one for the latter is *convective adjustment*. These were originally based on convection in a dry atmosphere. One needs to model a moist circulation that includes the phase changes of water to apply to the real atmosphere. The modeling of moist convection will be considered in Chapter 15. Here, we will explain the above two methods for a dry atmosphere.

14.4.1 Mixing length theory

The *mixing length theory* is based on the assumptions that air parcels associated with convection is distinguished from the environmental field, and that convective energy flux is estimated as the vertical transport of heat by individual air parcels. The buoyancy force drives air parcels vertically up or down to a distance l at which they are mixed with the environment. In the general formulation of the mixing length theory, convective flux is written as

$$\overline{F^{conv}} = -\rho C_p K^{conv} \frac{\partial \overline{\theta}}{\partial z} \qquad (14.4.3)$$

where K^{conv} is the eddy diffusion coefficient and θ is potential temperature. From dimensional analysis, the eddy diffusion coefficient K^{conv} can be set to a product of length scale l and velocity scale V: $K^{conv} = lV$ (see (11.2.12)). The length scale l is thought to be a typical height scale of convection and corresponds to the typical scale of an air parcel. l is called the *mixing length*.

The choice of mixing length l depends on many factors. In the boundary layer near the ground, for instance, the mixing length is generally given proportional to z.† In a deep atmosphere, on the other hand, there is no typical length scale other than the pressure scale height $H = C_pT/g$, so that l is usually set to H. As for the velocity scale, we assume that the magnitude of vertical velocity is given by the buoyancy force

$$V = \left(l \frac{\theta'}{\overline{\theta}} g \right)^{\frac{1}{2}}, \qquad (14.4.4)$$

where θ' is the deviation of potential temperature from the environmental field possessed by an air parcel. Assuming that the vertical motion of the air parcel is adiabatic, we can relate the deviation of potential temperature to the gradient of potential temperature in the environment, as shown in Fig. 14.5:

$$\theta' = -l \frac{\partial \overline{\theta}}{\partial z}, \qquad (14.4.5)$$

where positive deviation $\theta' > 0$ (i.e., positive buoyancy) is realized only in the case $\frac{\partial \overline{\theta}}{\partial z} < 0$. Thus, we obtain the eddy diffusion coefficient:

$$K^{conv} = lV = l^2 \left(-\frac{g}{\overline{\theta}} \frac{\partial \overline{\theta}}{\partial z} \right)^{\frac{1}{2}}. \qquad (14.4.6)$$

---

†In the theory of the boundary layer, the mixing length is proportional to the height z as given by (11.3.6).

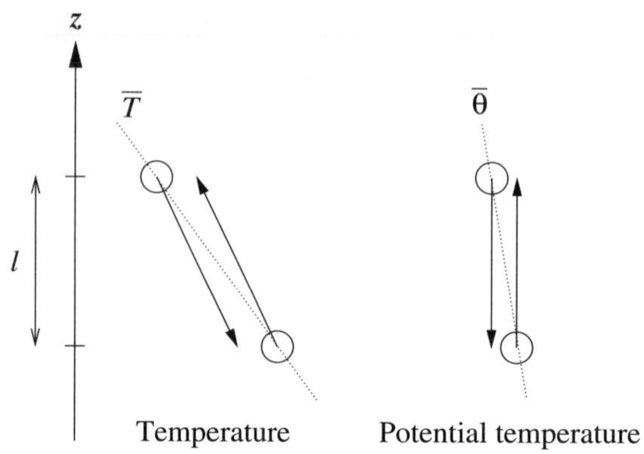

FIGURE 14.5: The thermodynamic paths of air parcels for the mixing length theory. Left: relation between temperature of the environment and that of upward and downward moving air parcels. Right: relation between potential temperature or entropy of the environment and that of air parcels. Dotted lines are environmental states, and solid arrows are the paths of air parcels. Air parcels are mixed with environmental air after vertical displacement l.

The above formula can be rewritten using temperature. Using the hydrostatic balance and the equation of state of the ideal gas, we generally have

$$\begin{aligned} \frac{1}{\overline{\theta}}\frac{\partial \overline{\theta}}{\partial z} &= \frac{1}{\overline{T}}\frac{\partial \overline{T}}{\partial z} - \frac{R_d}{C_p}\frac{1}{\overline{p}}\frac{\partial \overline{p}}{\partial z} = \frac{1}{\overline{T}}\frac{\partial \overline{T}}{\partial z} + \frac{R_d}{C_p}\frac{\rho g}{\overline{p}} \\ &= \frac{1}{\overline{T}}\left(\frac{\partial \overline{T}}{\partial z} + \Gamma_d\right). \end{aligned} \tag{14.4.7}$$

Therefore, (14.4.3) and (14.4.6) are written as

$$\overline{F^{conv}} = \rho C_p l'^2 \left(\frac{g}{\overline{T}}\right)^{\frac{1}{2}} \left[-\left(\frac{\partial \overline{T}}{\partial z} + \Gamma_d\right)\right]^{\frac{3}{2}} \tag{14.4.8}$$

$$= -\rho C_p K^{conv\prime} \left(\frac{\partial \overline{T}}{\partial z} + \Gamma_d\right), \tag{14.4.9}$$

$$K^{conv\prime} = l'^2 \left[-\frac{g}{\overline{T}}\left(\frac{\partial \overline{T}}{\partial z} + \Gamma_d\right)\right]^{\frac{1}{2}}, \tag{14.4.10}$$

where $l' = l(\overline{\theta}/\overline{T})^{1/4}$. We may think of l' as the usual mixing length by ignoring the difference in the factor.

The diffusion coefficient is positive $K^{conv} > 0$ if stratification is unstable. This means that convective flux exists only in the unstable layer if the mixing length theory is applied. We refer to this unstable layer as the convective layer. Even if convection motion occurs, stratification is still unstable; the convective flux of the mixing length theory does not completely eliminate the unstable stratification of the convective layer. In general, however, as will soon be shown, the convective layer is marginally unstable and very close to neutral. From the balance of the energy flux in the equilibrium state (14.4.2) and convective flux (14.4.8), the ratio of the lapse rate to the dry adiabat is estimated as

$$\frac{-\left(\frac{\partial \overline{T}}{\partial z}+\Gamma_d\right)}{\Gamma_d} = \left(\frac{\overline{F^{rad}}\sqrt{C_p\overline{T}}}{\rho g^2 l^2}\right)^{\frac{2}{3}}. \tag{14.4.11}$$

In the case of radiative equilibrium, $\overline{F^{rad}}$ is everywhere equal to zero. Within the convective layer, however, its value is unknown in advance. Here, we use a representative value of incident solar radiation $\overline{F^{rad}} \approx 300$ W m^{-2} to estimate the above ratio. Using $C_p \approx 10^3$ J kg^{-1} K^{-1}, $\overline{T} \approx 300$ K, $g \approx 10$ m s^{-2}, and $l \approx 8$ km, Eq. (14.4.11) is about 10^{-3}. Thus, the difference of the lapse rate from that of the dry adiabat is very small if the mixing length theory is applicable.

Now, we may relate the above ratio to the non-dimensional parameters of Golitsyn's model introduced in Section 12.4. Using $\overline{F^{rad}} \approx F_s$, $\overline{T} \approx T_e$, $l \approx C_pT_e/g$, and $\rho l \approx M_{col}$ (horizontally averaged mass of an air column per unit area),†

$$\begin{aligned}\frac{-\left(\frac{\partial \overline{T}}{\partial z}+\Gamma_d\right)}{\Gamma_d} &\approx \left(\frac{F_s}{\sqrt{C_pT_e}M_{col}g}\right)^{\frac{2}{3}} = \left(\frac{\sigma_B^{1/8}F_s^{7/8}}{C_p^{1/2}M_{col}g}\right)^{\frac{2}{3}} \\ &= (\Pi_M\Pi_g)^{\frac{2}{3}}.\end{aligned} \tag{14.4.12}$$

Since $\Pi_M \ll 1$ and $\Pi_g \ll 1$ in most of the planets, the lapse rate in the convective layer of the radiative-convective equilibrium is close to neutral. Gierasch and Goody (1968) applied the mixing length theory to the atmosphere of Mars. Matsuda and Matsuno (1978) discussed the radiative-convective equilibrium of the atmosphere of Venus using mixing length theory. They found that the lapse rate of the convective layer is close to that of the dry adiabat.

In the boundary layer, the mixing length l is thought to be proportional to the distance from the surface and is shorter than the mixing length in the free atmosphere used above. So, we can estimate the typical height at which the lapse rate is appreciably different from that of the dry adiabat. Assuming that the ratio (14.4.11) equals one at height $l = z_m$, we obtain

$$l = \left(\frac{\overline{F^{rad}}\sqrt{C_p\overline{T}}}{\rho g^2}\right)^{\frac{1}{2}} \equiv z_m. \tag{14.4.13}$$

For the same parameters used above, we have $z_m \approx 40$ m. This leads us to believe that the lapse rate of radiative-convective is different from that of the dry adiabat between the surface and the height z_m. Letting θ_m denote the potential temperature at z_m, and θ_s the surface value of potential temperature, we express convective flux as

$$\overline{F^{conv}} = -\rho C_p \frac{K^{conv}}{z_m}(\theta_m - \theta_s). \tag{14.4.14}$$

†In this section, the mass of an air column is denoted by M_{col}. In Section 12.4, it is simply denoted by M, but M denotes the mass flux here.

Introducing a typical velocity scale u_* which is driven by buoyancy in the boundary layer, we can estimate the diffusion coefficient using (14.4.4)–(14.4.6) as

$$\frac{K^{conv}}{z_m} \approx \left(-\frac{g}{\overline{\theta}}\frac{\partial\overline{\theta}}{\partial z}\right)^{\frac{1}{2}} z_m \propto u_*. \qquad (14.4.15)$$

Therefore, we can express convective flux in the boundary layer as

$$\overline{F^{conv}} = \rho C_p C_D u_*(\theta_s - \theta_m). \qquad (14.4.16)$$

In this form, the coefficient C_D is called the *bulk coefficient.* The value $C_D u_*$ is typically 0.01 m s^{-1}.†

14.4.2 Convective adjustment

We have found, based on evaluation using mixing length theory, that the lapse rate in the convective layer in radiative-convective equilibrium is very close to that of the dry adiabat Γ_d. This suggests that the temperature structure of radiative-convective equilibrium can be determined by setting the lapse rate of the convective layer to Γ_d without the explicit expression of convective flux. This method is called *convective adjustment.*

In the case of the gray radiation model, the temperature structure of radiative equilibrium is statically unstable in the layer close to the surface if absorbing constituents are more abundant in the lower layer; the layer is unstable below the level where the optical depth is equal to (14.3.28). The convective layer of radiative-convective equilibrium, therefore, exists at least between the surface and the level given by (14.3.28). Because of the energy balance and the requirement for continuity of temperature, the top level of the convective layer is located at a higher level than this. The atmospheric structure of radiative-convective equilibrium can be divided into the convective layer and the layer above the convective layer. The convective layer corresponds to the *troposphere*, the top level of the convective layer is the *tropopause*, and the layer above the tropopause is the *stratosphere.* In the stratosphere, only radiative flux exists as the energy transport; thus the stratosphere is in radiative equilibrium. The schematic relation between the temperature structure of radiative-convective equilibrium and that of radiative equilibrium is displayed in Fig. 14.6. It should be noted that the stratosphere is actually defined as a layer in which temperature increases with height as shown in Fig. 14.1. In the case of the gray radiation model, the stratosphere is simply defined as a layer in radiative equilibrium. Since absorption by ozone is not considered in this case, the temperature in the stratosphere approaches a constant value with height.

The convective adjustment method determines the temperature structure in the convective layer using the externally specified lapse rate of temperature. In the case of a dry atmosphere, mixing length theory suggests that the lapse rate in the convective layer is very close to that of the dry adiabat Γ_d. Although the lapse rate

†If this form is used in the momentum equation, C_D is called the aerodynamic drag coefficient and u_* is called the friction velocity. This corresponds to C_H of (11.3.17) in Section 11.3.3. The lateral velocity at the lowest layer $|\overline{\boldsymbol{u}}_1|$ is used in Section 11.3.3. Here we use the turbulent velocity u_* which is driven by the convection itself.

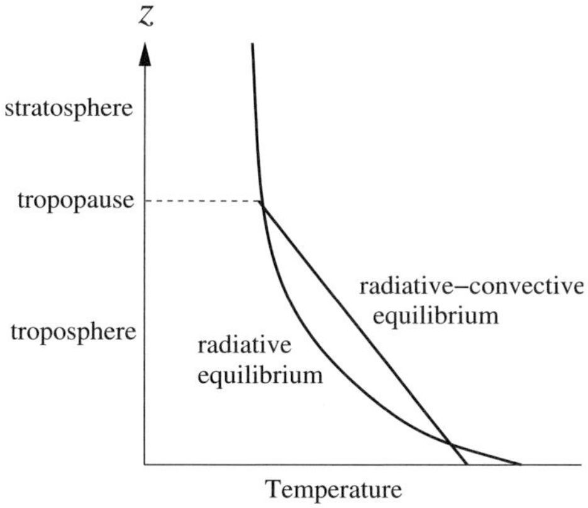

FIGURE 14.6: The schematic temperature profiles of radiative equilibrium and radiative-convective equilibrium with convective adjustment. If the gray radiation model is used with the same optical parameters, the temperature profile in the stratosphere of radiative-convective equilibrium agrees with that of radiative equilibrium.

near the surface is larger than Γ_d, the difference is appreciable only in the very thin layer below z_m, (14.4.13). Thus, in the layer below z_m, we represent the sum of thermal diffusive flux and turbulent flux due to small convective eddies as sensible heat flux from the surface to the atmosphere. We assume the formula (14.4.16) for sensible heat flux.

In the equilibrium state, the sum of convective flux, radiative flux, and sensible heat flux is constant irrespective of height as shown by (14.1.5). On the assumption described above, sensible heat flux $\overline{F^{sh}}$ exists only in the layer just above the surface. On the other hand, convective flux $\overline{F^{conv}}$ becomes zero at the surface, and radiative flux continuously varies near the surface. Thus, if we neglect the structure within the surface layer, we may set the boundary condition of convective flux at the surface as

$$\overline{F^{conv}}(0) = \overline{F^{sh}} = \overline{\rho C_p C_D u_* (T_s - T_0)}, \tag{14.4.17}$$

where (14.4.16) is used and the depth of the boundary layer is neglected: the surface temperature T_s and the temperature at the bottom of the atmosphere T_0 are used instead of potential temperature. The temperature gap between the surface and the atmosphere depends on the magnitude of the coefficient $C_D u_*$. In the limit $C_D u_* \to \infty$, for instance, the temperature gap approaches zero and $F^{\uparrow}(0) = \pi B(0) = \sigma_B T_s^4$.

The profiles of radiative flux and convective flux in the limit $C_D u_* \to \infty$ are schematically shown in Fig. 14.7. The tropopause height is denoted by z^T. Radiative flux consists of the net solar radiation flux $\overline{F^{rads}}$ and the infrared planetary radiation $\overline{F^{radl}}$. Throughout all the layer in radiative-convective equilibrium, the energy flux satisfies the relation $\overline{F^{conv}} + \overline{F^{rads}} + \overline{F^{radl}} = 0$. We assume that solar radiation is not absorbed in the atmosphere and is given by $\overline{F^{rads}} = (1 - A)F^{\odot}/4$. Convective flux $\overline{F^{conv}}$ is different from zero within the troposphere (i.e., between the surface and the tropopause, where the tropopause height is denoted by z^T).

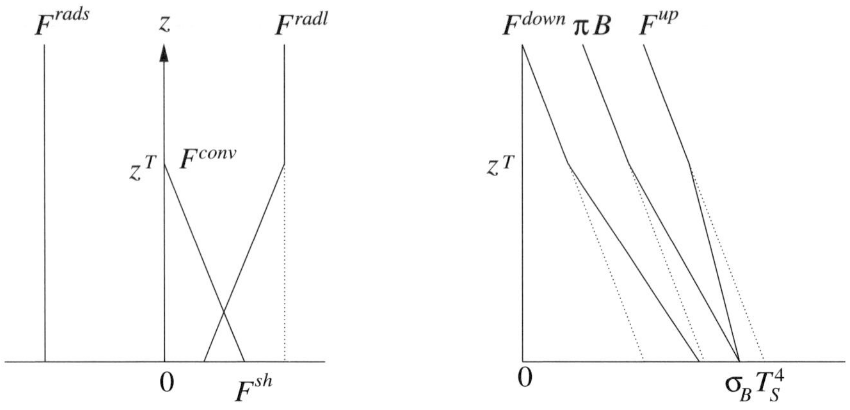

FIGURE 14.7: The schematic profiles of radiative flux and convective flux in radiative-convective equilibrium in the case of $C_D u_* \to \infty$. z^T is the tropopause. F^{rads} is solar radiative flux, F^{radl} is planetary radiative flux, F^{conv} is convective flux, and F^{sh} is sensible heat flux from the surface. F^{radl} is equal to the difference between the upward planetary radiative flux F^{up} and the downward planetary radiative flux F^{down}. πB is equal to $\sigma_B T^4$. The dotted lines show the profiles in radiative equilibrium.

At the surface, convective flux is equal to $\overline{F^{sh}}$. Long-wave radiation is given by $F^{radl} = F^{\uparrow} - F^{\downarrow}$, and is equal to $-\overline{F^{rads}} - \overline{F^{sh}}$ at the surface. F^{radl} increases with height from the surface to z^T, and is equal to $-\overline{F^{rads}}$ above z^T. In the stratosphere above z^T, the variations of the fluxes, $F^{\uparrow}$, $F^{\downarrow}$, and πB, follow those of radiative equilibrium shown in Fig. 14.3. In the troposphere below z^T, $F^{\downarrow}$ is larger than that of radiative equilibrium, and $F^{\uparrow}$ is smaller than that of radiative equilibrium. Since surface temperature and the temperature at the bottom of atmosphere is assumed to be equal, $F^{\uparrow} = \pi B$ is satisfied at the surface.

The vertical temperature profiles for the case when a more realistic radiative process is used are shown in the left panel of Fig. 14.8 (Manabe and Strickler 1964). This shows the vertical temperature profiles in radiative equilibrium, radiative-convective equilibrium with dry adiabatic adjustment, and that with a critical lapse rate of 6.5 K km^{-1}, which represents the average lapse rate in mid-latitudes. The tropopause becomes higher and the surface temperature becomes lower as the critical lapse rate becomes smaller (i.e., the upward convective energy flux is larger in the case with the dry adiabat (about 10 K km^{-1}) than in the case with the lapse rate 6.5 K km^{-1}). Different from the temperature profiles with the gray radiation model in Fig. 14.6, the temperature profiles in the stratosphere differ between the pure radiative equilibrium and the radiative-convective equilibrium. The right panel of Fig. 14.8 shows the vertical distributions of the heating rate due to radiation for each component of gases. In the troposphere, the cooling effect of water vapor on long-wave radiation is most important; net cooling amounts to about 1.5 K day^{-1} through the troposphere. In the stratosphere, solar heating due to O_3 is balanced by long-wave cooling due to CO_2.

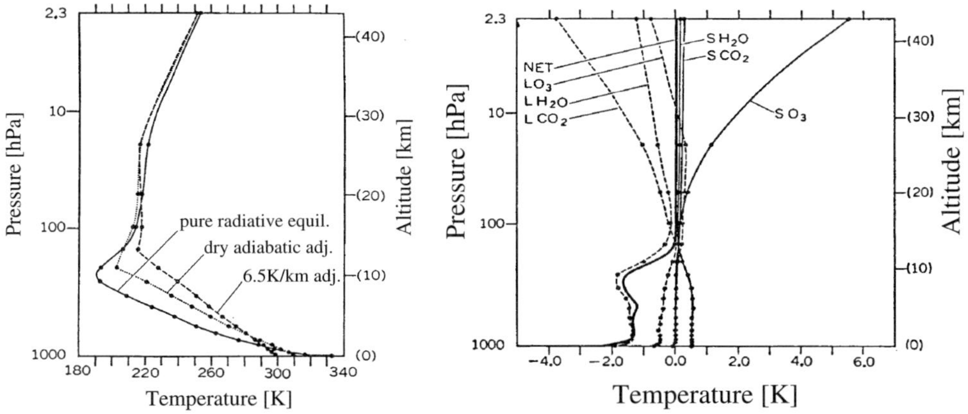

FIGURE 14.8: The vertical distributions of the temperature and heating rate in radiative equilibrium and in radiative-convective equilibrium. Left: temperature distributions of radiative equilibrium (dashed), radiative-convective equilibrium with a dry adiabatic lapse rate (dotted), and radiative-convective equilibrium with a critical lapse rate of 6.5 K km^{-1} (solid). Right: the rate of temperature change due to radiation for radiative-convective equilibrium. Heating rates due to gas components, CO_2, H_2O, and O_3, for long-wave radiation (L) and short-wave radiation (S) are shown. After Manabe and Strickler (1964) by permission of the American Meteorological Society.

14.5 Radiative-convective equilibrium in a moist atmosphere

14.5.1 Energy balance

The vertical one-dimensional energy balance of a moist atmosphere is given by the horizontal and time average of the equation of total moist energy (9.5.17):

$$\frac{\partial}{\partial z}\left(\overline{F^{conv}} + \overline{F^{rad}} + \overline{F^{sh}} + \overline{F^{lh}}\right) = 0, \tag{14.5.1}$$

where $\overline{F^{lh}}$ is the latent heat flux due to the diffusion of water vapor. In this case, the convective energy flux F^{conv} is given by (14.1.2)

$$F^{conv} = \rho w \sigma = \rho w(\sigma_d + Lq) = \rho w(C_p T + gz + Lq), \tag{14.5.2}$$

where σ is the moist static energy and σ_d is the dry static energy. If the contributions of the thermal diffusion and the diffusion of water vapor are confined in a thin layer very close to the surface, the energy balance in the free atmosphere can be written as

$$\frac{\partial}{\partial z}\left(\overline{F^{conv}} + \overline{F^{rad}}\right) = 0. \tag{14.5.3}$$

In this form, the energy balance of a moist atmosphere is apparently the same as that of a dry atmosphere; the difference is hidden in the convective flux, which has a contribution of latent heat flux in the case of a moist atmosphere.

To consider the vertical one-dimensional balance of a moist atmosphere, we need to add the equation of water vapor. The balance of water vapor is given by the horizontal and time average of (9.5.1):

$$\frac{\partial}{\partial z}\left(\overline{\rho w q}+\overline{i_z}\right) \quad = \quad -\overline{S_q}, \tag{14.5.4}$$

where S_q is the source term of water vapor and i_z is the vertical component of the diffusion of water vapor. At the surface, this flux is equal to evaporation from the surface; we assume that it is given by a bulk formula similar to that of sensible heat flux (14.4.17):

$$\overline{\rho w q}(0) \quad = \quad \overline{E_v} \quad = \quad \overline{\rho C_D u_*(q_s - q_0)}, \tag{14.5.5}$$

where q_s is specific humidity at the surface, and q_0 is that at the bottom of the atmosphere. We have assumed that coefficient C_D is the same as that of thermal diffusion (14.4.17) for simplicity.† We further assume that diffusive flux $\overline{i_z}$ is smaller than convective flux $\overline{\rho w q}$ in the free atmosphere. Neglecting the thickness of the boundary layer, we can equate the boundary value of $\overline{\rho w q}$ at the bottom of the atmosphere to (14.5.5). Thus, at the bottom of the atmosphere, the value of convective flux (14.5.2) is given by the sum of sensible heat flux and latent heat flux from the surface and is given by the following bulk formula:

$$\overline{F^{conv}}(0) \quad = \quad \overline{F^{sh}}+\overline{F^{lh}} \quad = \quad \overline{F^{sh}}+L\overline{E_v} \quad = \quad \overline{\rho C_p C_D u_*(\sigma_s - \sigma_0)}, \tag{14.5.6}$$

where $\sigma_s = C_p T_s + L q_s$ and $\sigma_0 = C_p T_0 + L q_0$.

We may have an alternative form of the energy balance of a moist atmosphere; the effect of the latent heat release of water vapor can be interpreted as the heating of the dry atmosphere. In this case, multiplying L by (14.5.4) and subtracting it from (14.5.1), we obtain

$$\frac{\partial}{\partial z}\left(\overline{F_d^{conv}}+\overline{F^{rad}}+\overline{F^{sh}}\right) \quad = \quad \overline{\rho C_p Q_m}, \tag{14.5.7}$$

where Q_m is heating due to latent heat release, given by (9.5.10). F_d^{conv} is the convective energy flux of a dry atmosphere, given by (14.1.2). If diffusive fluxes are negligible outside the boundary layer, the energy balance in the free atmosphere is written as

$$\frac{\partial}{\partial z}\left(\overline{F_d^{conv}}+\overline{F^{rad}}\right) \quad = \quad \overline{\rho C_p Q_m}. \tag{14.5.8}$$

This means the balance between latent heat release and divergence of radiative and convective fluxes.

14.5.2 Convective adjustment

One needs to take account of the transport of water vapor to obtain the convective flux of the radiative-convective equilibrium of a moist atmosphere. However,

†The two bulk coefficients are generally different. In Chapter 11, the bulk formulas of the energy and water vapor fluxes (11.3.17) have different coefficients as denoted by C_H and C_q, respectively.

if we assume that moist air parcels remain a constant moist entropy during their motion, the mixing length theory that was used for a dry atmosphere can be utilized to obtain the temperature structure of a moist atmosphere without explicitly considering the transport of water vapor. In this case, strictly speaking, we need to assume that an equal amount of water is contained in all the air parcels and, if water vapor is condensed, that the condensed water is held in the air parcels without precipitating as rain. According to the mixing length theory on these assumptions, the temperature structure in the convective layer almost becomes neutral as described in Section 14.4.1. Specifically, in this case, it is neutral to a saturated moist atmosphere and the lapse rate should be close to the moist adiabat that is given in Sections 8.1.7 or 8.2.4. An approximate equation of the moist adiabat (8.2.49) is convenient for the qualitative discussion here. The moist adiabatic lapse rate of temperature per unit height is given by

$$\Gamma_m = \Gamma_d \frac{1 + \frac{\varepsilon L}{R_d T} \frac{p^*(T)}{p}}{1 + \frac{\varepsilon^2 L^2}{C_p R_d T^2} \frac{p^*(T)}{p}}, \qquad (14.5.9)$$

which is a function of T and p, and $\Gamma_d = g/C_p$. In fact, this is the pseudo-moist adiabat. In general, we have $|\Gamma_m| < |\Gamma_d|$ (i.e., the lapse rate in the convective layer of a moist atmosphere is smaller than that of a dry atmosphere). This means that the tropopause height of a moist atmosphere is in general higher than that in a dry atmosphere if other conditions including radiative property are the same. Fig. 14.9 shows the relation between the dry adiabat and the moist adiabat. As temperature increases, the difference between the moist adiabat and the dry adiabat becomes larger, while the two lapse rates get closer as temperature gets colder or altitude gets higher.

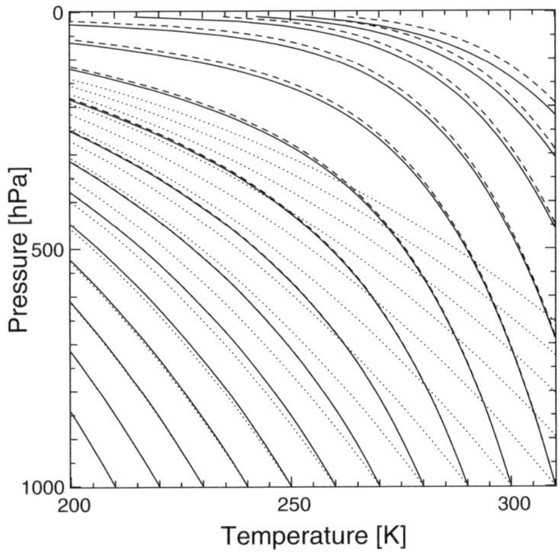

FIGURE 14.9: Temperature profiles according to the dry adiabat (dotted), the pseudo-moist adiabat (solid), and the saturated moist adiabat (dashed) whose water content is conserved as the saturation value at the surface.

In reality, precipitation occurs with moist convection and air parcels in a moist atmosphere have different water content. This means that the assumption used for the mixing length theory is not appropriate for a moist atmosphere, and that the theoretical basis for use of the moist adiabat as the lapse rate of convective adjustment is lost. In addition, moist convection has a very complicated structure if precipitation is associated. For numerical simulations of the global-scale circulation, for instance, the statistical models of the effect of cumulus convection are introduced. In the next chapter, the vertical one-dimensional structure of a moist atmosphere will be considered using a schematic model of moist convection which consists of an upward motion region and a downward motion region. According to such a simplified model, it will be found that the lapse rate of the moist convective layer is close to the moist adiabat even if precipitation is associated.

14.6 Runaway greenhouse effect

In the case of a moist atmosphere, if incident solar radiation is larger than some critical value, there is a regime in which no liquid phase is allowed at the surface. This kind of regime is caused by the *runaway greenhouse effect*, which is qualitatively explained as follows. Suppose that solar radiation is increased. In this case, the surface temperature given in the solution to radiative-convective equilibrium becomes warmer. If water vapor is just saturated at the ground, the atmosphere contains more water vapor and thus the total optical depth of the atmosphere becomes more opaque. If the atmosphere becomes more opaque, in turn, the surface temperature becomes warmer because of the greenhouse effect, and then the atmosphere contains more water vapor. This positive feedback will continue until all liquid water (i.e., the oceans) evaporates. In the very early stages of earth's history, some modeling studies show that the earth could have been in a state in which all oceans evaporated (Kasting, 1989; Abe and Matsui, 1988).

It is found that positive feedback for a runaway greenhouse effect works only if incident solar radiation is larger than some critical value. This critical point can be estimated by the following procedure. To simplify the argument, we assume that only water vapor contributes to radiation and that water vapor has a constant absorbing coefficient k which is independent of wavelength (the gray radiation). Total optical depth is given by (10.5.1):

$$\tau = \int_{\infty}^{z} k\rho q dz = \int_{0}^{p} kq\frac{dp}{g}. \tag{14.6.1}$$

We assume that specific humidity q is constant in the stratosphere irrespective of height and that it is saturated at the tropopause. Letting T_T and p_T denote temperature and pressure at the tropopause, respectively, we have specific humidity at the tropopause given by

$$q_T = \varepsilon\frac{p^*(T_T)}{p_T}. \tag{14.6.2}$$

Thus, the optical depth at the tropopause is given by (14.6.1) using (8.2.11):

$$\tau_T = kq_T \frac{p_T}{g} = k\varepsilon \frac{p^*(T_T)}{g} = \frac{k\varepsilon p_0^*}{g} \exp\left[\frac{L_0}{\varepsilon R_d}\left(\frac{1}{T_0} - \frac{1}{T}\right)\right]. \tag{14.6.3}$$

If the stratosphere is in radiative equilibrium with the gray atmosphere, we have a relation between the optical depth and temperature in the stratosphere using (14.3.16) and $\pi B = \sigma_B T^4$:

$$\tau = \frac{4}{3}\left(\frac{\sigma_B T^4}{F} - \frac{1}{2}\right), \tag{14.6.4}$$

where F is the net incident solar flux given by (14.3.8). Therefore, the water vapor at the tropopause can be in a saturated state if (14.6.3) has a solution to (14.6.4) (i.e., the runaway greenhouse effect does not occur). Fig. 14.10 shows three profiles of the gray radiative equilibrium and the tropopause temperature given by (14.6.3). For the smallest value of $F = 240$ W m^{-2}, Eq. (14.6.3) has two solutions to (14.6.4). The upper point is regarded as the tropopause. For the largest value of $F = 320$ W m^{-2}, on the other hand, no solution exists. This means that, if solar radiation is large enough, there is no solution for radiative-convective equilibrium in which the air is saturated at the tropopause. This mechanism of the runaway greenhouse effect was found by Komabayashi (1968) and Ingersoll (1969), and the critical value of solar radiation is called the *Komabayashi-Ingersoll limit* (Nakajima et al., 1992). Fig. 14.11 shows the relationship between temperature and pressure for different values of sea surface temperature (Nakajima et al., 1992). This calculation is done using a simplified gray radiation. As sea surface temperature increases, the total

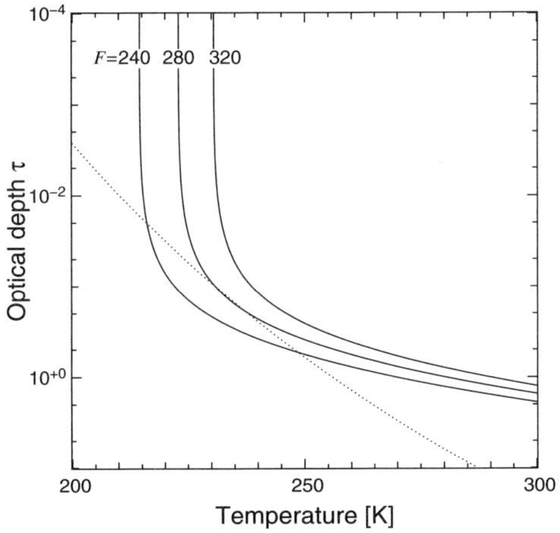

FIGURE 14.10: The relation between temperature and optical depth for the runaway greenhouse effect. The solid curves are the temperature structure of the gray radiative equilibrium for solar radiation $F = 240$, 280, and 320 W m^{-2}. The dotted curves relate the optical depth at the tropopause to the tropopause temperature under the saturation condition (14.6.3). If $F = 240$ W m^{-2}, the tropopause can be saturated, while there is no solution to the saturated tropopause if $F = 320$ W m^{-2}.

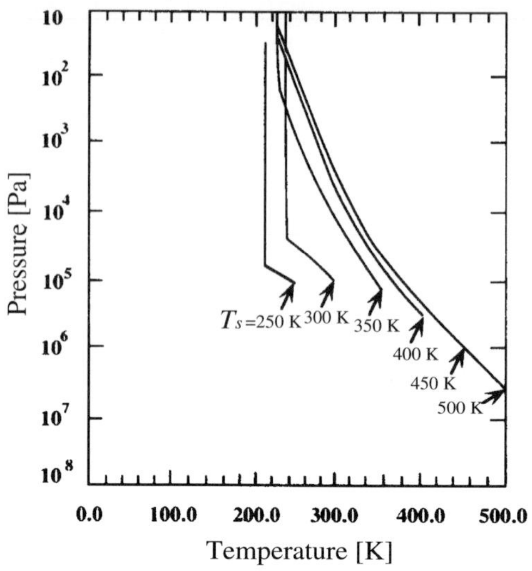

FIGURE 14.11: The relation between temperature and pressure for various sea surface temperatures. After Nakajima et al. (1992) with permission of the American Meteorological Society.

mass of the atmosphere increases due to the increased abundance of water. At a sea surface temperature of 450 K, atmospheric mass is ten times larger than the present value.

References and suggested reading

The calculation of the radiative equilibrium using the gray radiation model is shown in Chapter 9 of Goody and Yung (1989) and Chapter 2 of Houghton (2002). Radiative equilibrium and radiative-convective equilibrium using a more realistic radiation model are calculated by Manabe and Möller (1961) and Manabe and Strickler (1964). While the latter uses 6.5 K km^{-1}, the average lapse rate in mid-latitudes, for the critical lapse rate of convective adjustment, the moist adiabat that represents the averaged state in the tropics is used in radiative-convective equilibrium models (Hummel and Kuhn, 1981). Manabe et al. (1965) also use the moist adiabatic adjustment to introduce the effects of cumulus convection in their general circulation model. Ramanathan and Coakley (1978) give a good historical review of radiative-convective equilibrium, together with their relation to mixing length theory. Introduction to the runaway green house effect is found in Houghton (2002) and Abe and Matsui (1988), Kasting (1989), and Nakajima et al. (1992) give a clear description. It is also investigated by Ishiwatari et al. (2003) using a general circulation model.

Abe, Y. and Matsui, T., 1988: Evolution of an impact-generated H_2O-CO_2 atmosphere and formation of a hot proto-ocean on earth. *J. Atmos. Sci.*, **45**, 3081–3101.

Gierasch, P. J. and Goody, R., 1968: A study of the thermal and dynamical structure of the Martian lower atmosphere. *Planet. Space Sci.*, **16**, 615–646.

Goody, R. and Yung, Y., 1989: *Atmospheric Radiation*, 2nd ed. Oxford University Press, New York, 528 pp.

Houghton, J. T., 2002: *The Physics of Atmospheres*, 3rd ed., Cambridge University Press, Cambridge, UK, 320 pp.

Hummel, J. R. and Kuhn, W. R., 1981: Comparison of radiative-convective models with constant and pressure-dependent lapse rates. *Tellus*, **33**, 254–261.

Ingersoll, A. P., 1969: The runaway greenhouse: A history of water on Venus. *J. Atmos. Sci.*, **26**, 1191–1198.

Ishiwatari, M., Takehiro, S.-I., Nakajima, K., and Hayashi, Y.-Y., 2002: A numerical study on appearance of the runaway greenhouse state of a three-dimensional gray atmosphere. *J. Atmos. Sci.*, **59**, 3223–3238.

Kasting, J. F., 1989: Runaway and moist greenhouse atmospheres and the evolution of Earth and Venus. *Icarus*, **74**, 472–494.

Komabayashi, M., 1967: Discrete equilibrium temperatures of a hypothetical planet with the atmosphere and the hydrosphere of one component-two phase system under constant solar radiation. *J. Met. Soc. Japan*, **45**, 137–139.

Manabe, S. and Möller, F., 1961: On the radiative equilibrium and heat balance of the atmosphere. *Mon. Wea. Rev.*, **89**, 503–532.

Manabe, S. and Strickler, R. F., 1964: Thermal equilibrium of the atmosphere with a convective adjustment. *J. Atmos. Sci.*, **21**, 361–385.

Manabe, S., Smagorinsky, J., and Strickler, R. F., 1965: Simulated climatology of a general circulation model with a hydrologic cycle. *Mon. Wea. Rev.*, **93**, 769–798.

Matsuda, Y. and Matsuno, T., 1978: Radiative-convective equilibrium of the Venusian atmosphere. *J. Meteorol. Soc. Japan*, **60**, 245–254.

Nakajima, S., Hayashi, Y.-Y., and Abe, Y., 1992: A study on the "runaway greenhouse effect" with a one-dimensional radiative-convective equilibrium model. *J. Atmos. Sci.*, **49**, 2256–2266.

Ramanathan, V. and Coakley, J. A., Jr., 1978: Climate modeling through radiative-convective models. *Rev. Geophys. and Space Phys.*, **16**, 465–489.

U. S. Standard Atmosphere, 1976: U. S. Government Printing Office, Washington, D. C.

15

Moist convection

In the previous chapter, we examined the vertical structure of the atmosphere based on the balance between the energy transports due to radiation and convection. If convective motion is described by the mixing length theory, the effect of convection is almost equivalent to adjusting the temperature profile to an adiabatic profile. As for a moist atmosphere, convective motion has asymmetry between the upward motion and the downward motion because latent heat release is associated with upward motion. Since such an asymmetry is generally not taken into account for the mixing length theory, we must reconsider the effect of moist convection on the temperature profile in a moist atmosphere. This leads to the construction of a cumulus model that is based on the characteristic motion of moist convection in nature (i.e., cumulus convection).

In this chapter, to consider the vertical structure of a moist atmosphere, we examine the fundamental characteristics of the moist circulation that is observed as cumulus convection in the tropics. Using a simplified cumulus model, we present a schematic view of the thermodynamic structure that is associated with moist circulation. We will find that the horizontal mean temperature profile of a moist atmosphere that has asymmetry between upward motion and downward motion is also close to the moist adiabat.

In a moist atmosphere, the condensation of water vapor occurs in the layer above a particular level, called the cloud base or the lifting condensation level. Thus, the troposphere of a moist atmosphere is further divided into two layers: the upper layer is associated with water vapor condensation and is characterized by moist convection, and the lower layer is free from water vapor condensation and is characterized by dry convection. The latter is called the mixed layer and plays fundamental roles in the transport of heat and water vapor from the surface to the atmosphere. We examine the properties of the mixed layer in vertical, one-dimensional, statistical equilibrium states using the above simplified cumulus model. The concept of this chapter is related to the cumulus parameterization used for general circulation models. We will briefly describe a mass flux scheme that is a representative cumulus parameterization.

15.1 Circulation structures of convection

The vertical structure of the earth's atmosphere is maintained by a moist circulation, particularly in the tropics. We need to analyze the structure of this moist circulation to understand the vertical profiles of temperature and water vapor in the tropics. Moist circulation in the tropics is viewed as the convection associated with hydrological circulation. Moist convection is quite different from more fundamental forms of convection, such as the Bénard convection and that in a dry atmosphere. We begin by comparing moist convection with dry convection and the Bénard convection.

The energy balance of an equilibrium state of a moist atmosphere is written as the balance between convective flux and radiative flux as shown by (14.5.3); in this equation, the latent heat of water vapor is included in moist enthalpy. In a radiative-convective equilibrium state, the heating due to convection occurs in the form of supply of the latent and sensible heat fluxes from the surface and the cooling due to radiation that occurs in the free atmosphere (i.e., the troposphere). In the same way, we can consider the convection in a dry atmosphere without the hydrological cycle. If a dry atmosphere is heated from the surface and cooled in the free atmosphere, dry convection will occur in the free atmosphere and the radiative-convective equilibrium of the dry atmosphere will be established as described in Section 14.4. We furthermore consider the convection in the Boussinesq fluid. This is known as the Bénard convection, which is theoretically described in Section 5.2.1. The theory underlying the Bénard convection can be formulated as a fluid surrounded by rigid walls at the top and bottom of the fluid. The fluid is heated from the bottom boundary and cooled at the top boundary. Unlike the atmosphere, the Boussinesq fluid has a very narrow depth such that the difference between temperature and potential temperature is negligible.

We examine the temperature structures of the three types of convection by introducing a simplified convective model; we assume that the convections have a steady cycle, that upward and downward motion regions are clearly distinguished, and that the exchange between upward and downward motion regions is negligible except for the top and bottom boundary layers. These assumptions are oversimplified for quantitative discussion of the effect of convection. Nevertheless, such a conceptual model is still useful because it provides constraints for possible temperature profiles.

First, we consider the temperature structure of the Bénard convection, which is schematically shown in Fig. 15.1. The temperatures at the bottom and top boundaries are specified as T_1 and T_2, respectively. We can assume that the temperatures in upward and downward motion regions are vertically uniform and given by T_1 and T_2, respectively. In this case, the vertical profile of the horizontal mean temperature follows the dotted curve in the right panel of Fig. 15.1. In the middle of the convective layer, the horizontal mean temperature is the average between the

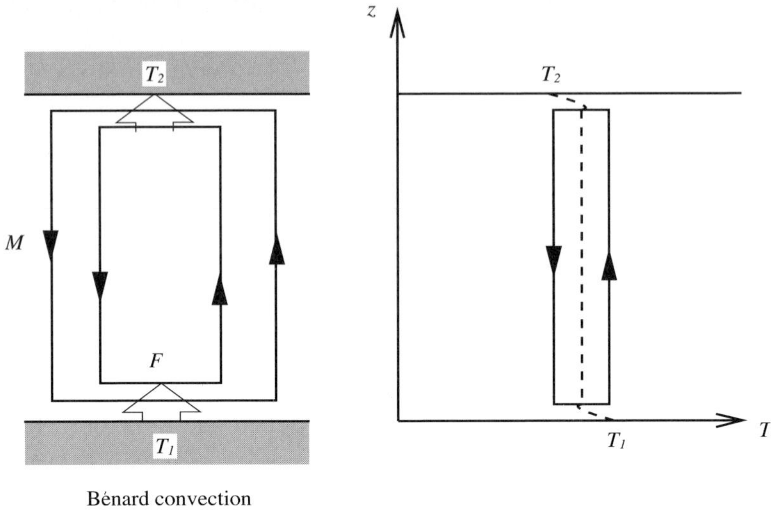

FIGURE 15.1: The schematic temperature structure and the circulation structure of the Bénard convection. T_1 and T_2 are the temperatures at the bottom and top boundaries, respectively. F is the heat flux from the surface and M is the mass flux of the circulation. The dotted line in the right panel is horizontal mean temperature.

top and bottom boundaries, $(T_1 + T_2)/2$. We also assume that there are thin boundary layers near the top and bottom boundaries, within which temperature abruptly changes from mean values to T_1 or T_2.

Horizontal mean upward heat transport F is independent of height in the equilibrium state. Convective mass flux per unit area M is defined as the mass transport in either the upward motion region or the downward motion region. In the steady state, M is equal in both regions. It is also assumed that M is independent of height except for the thin boundary layers. Using a constant density ρ_0 of the Boussinesq fluid, mass flux is related to upward velocity w^u and downward velocity w_d as

$$\frac{M}{\rho_0} = \frac{w^u}{2} = -\frac{w^d}{2}. \tag{15.1.1}$$

The factor $1/2$ comes from the assumption that the upward motion region occupies the same areal fraction as that of the downward motion region. Convective heat flux is also expressed by

$$F \equiv \overline{Tw} = \frac{T_1 w^u + T_2 w^d}{2} = \frac{M}{\rho_0}\Delta T, \tag{15.1.2}$$

where the overline denotes the horizontal average. In principle, convective motion is driven by the buoyancy force. Thus, we assume that the vertical velocity is bounded by the work done by the buoyancy force along the fluid depth H:

$$w^u = -w^d = A\sqrt{\frac{gH\Delta T}{\overline{T}}}, \tag{15.1.3}$$

where $\Delta T = T_1 - T_2$, $\overline{T} = (T_1 + T_2)/2$, and A is a coefficient which may depend on viscosity, and will be smaller than one. In this case, heat transport is expressed as

$$F = A\sqrt{\frac{gH}{\overline{T}}}\Delta T^{\frac{3}{2}}. \tag{15.1.4}$$

Coefficient A is included as an unknown factor in this equation. In Chapter 5, we estimated the heat transport associated with the Rayleigh-Taylor instability by (5.2.53). By equating the Nusselt number $Nu = FH/\kappa\Delta T$ to experimental values (see (5.2.53)), we may be able to estimate coefficient A.

Second, let us examine the radiative-convective equilibrium in a dry atmosphere. It should be noted, however, that the dry radiative-convective equilibrium that was considered in Section 14.4 does not exist in the real atmosphere of the earth in the pure sense, although dry convection itself is observed in the planetary boundary layer and in the mixed layer. The concept of a dry radiative-convective equilibrium may be, nevertheless, applicable to the atmospheres of Venus or Mars, and is a step toward understanding moist radiative-convective equilibrium. We consider an idealistic situation where a dry atmosphere exists over a horizontally uniform land mass that has constant temperature.† In this case, we assume that a troposphere has formed above the land; in the troposphere, the atmosphere is cooled by radiation and its cooling is balanced by heat transport from the surface through dry convection. The top of the layer is the tropopause. The upper part above the tropopause is in radiative equilibrium. The troposphere is occupied by dry convection, and the atmosphere is as a whole in dry radiative-convective equilibrium.

In order to clarify the argument, we assume that convective motion in the troposphere has a steady cellular circulation. Dry convection differs from the Bénard convection in that the upper rigid boundary does not exist in the case of dry convection and that cooling occurs throughout the troposphere due to radiation instead of the thermal diffusion near the top boundary in the case of the Bénard convection. In addition, the atmosphere is sufficiently deep such that adiabatic temperature decreases with height owing to the pressure effect. A schematic structure of convective circulation and temperature profiles is shown in Fig. 15.2. Since cooling is not confined in a thin layer near the upper boundary, streamlines are not concentrated near the tropopause. The atmospheric temperature near the ground is heated up close to the ground temperature T_s. Air parcels are always cooled by radiation above the lower boundary layer both in the upward and downward motion regions. Thus, the lapse rate in the upward motion region is steeper than that of the dry adiabat, while that in the downward motion region is more stable than that of the dry adiabat. If the temperature change is symmetric between the upward and downward motion regions and the deviations of the lapse rates from the dry adiabat are the same magnitude in the two regions, the horizontally averaged temperature over the upward and downward motion regions is close to that of

†This situation can be created in numerical models (e.g., Nakajima and Matsuno, 1988 and Pauluis and Held, 2002).

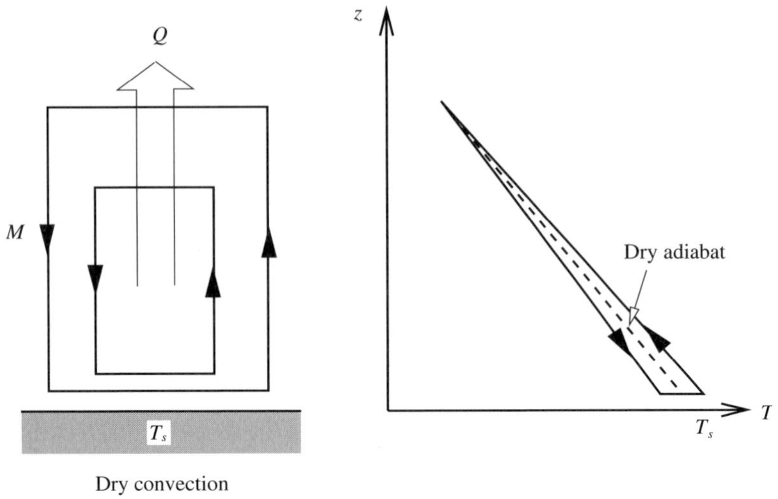

FIGURE 15.2: The same as Fig. 15.1 but for dry convection. Q is radiative cooling and M is the mass flux of the circulation. The dotted line in the right panel is the horizontal mean temperature and is close to that of the dry adiabat.

the dry adiabat. As a result, the average temperature in the convective layer will be close to the dry adiabatic temperature structure as shown in Fig. 15.2 (right).

In the case of dry convection, the deviations of the lapse rate from the dry adiabat in the upward and downward motion regions are affected by radiative cooling and the ratio of upward velocity to downward velocity. If the radiative cooling rate is denoted by Q [K s^{-1}] and the vertical velocity of the upward or downward motion is denoted by w [m s^{-1}], the difference between the realized lapse rate and the dry adiabat is given by $\Delta\Gamma = Q/w$ [K m^{-1}]. In a similar way to (15.1.4), if we assume that the energy gained by the buoyancy force is transformed to kinetic energy, we can estimate vertical velocity,

$$w = A\sqrt{\frac{gH\Delta T}{T}}, \tag{15.1.5}$$

where ΔT is the temperature difference between the upward and downward motion regions, H is the depth of the troposphere, and A is an appropriate coefficient satisfying $A < 1$. Since the convective energy flux is equal to the radiative cooling rate

$$w\Delta T = QH, \tag{15.1.6}$$

thus we obtain

$$\Delta T = \left(\frac{TQ^2H}{A^2 g}\right)^{\frac{1}{3}}, \tag{15.1.7}$$

and the difference of the lapse rate between the upward and downward motion regions

$$\Delta\Gamma = \frac{Q}{w} = A^{-1}\left(\frac{TQ}{gH^2}\right)^{\frac{1}{3}}. \quad (15.1.8)$$

For the values $T = 300$ K, $Q = 2$ K day^{-1}, $H = 8$ km, and $A = 1$, we obtain a very small difference $\Delta T = 0.05$ K and $\Delta\Gamma = 0.006$ K km^{-1}. Thus, the lapse rate of the dry convective region is close to that of the dry adiabat; this result is similar to the temperature difference estimated by the mixing length theory (14.4.11).

In the third case, we consider the characteristics of moist convection. Here, we introduce the following simplifications to clarify its differences from the Bénard convection or from dry convection. By neglecting the mixed layer, the convective region is assumed to be just above the surface up to the tropopause and is divided into upward and downward motion regions. (The structure of the mixed layer will be considered in Section 15.4.) The latent heat release of water vapor occurs only in the upward motion region, and the upward motion region is everywhere saturated; thus the upward motion region is thought to be the cloud region. There is no latent heat release or absorption in the downward motion region.

On the above assumptions, we can find that moist convection has an asymmetry between the upward and downward motion regions because of latent heat release; this characteristic is different from dry convection. If upward motion is sufficiently fast, the temperature in the upward motion region is close to the moist adiabat since the cooling due to radiation is small. If downward motion is sufficiently fast, on the other hand, the temperature in the downward motion region becomes closer to the dry adiabat. In Fig. 15.3, the dry adiabat of the downward motion region thus obtained is indicated by the dotted line, which starts from the top of the

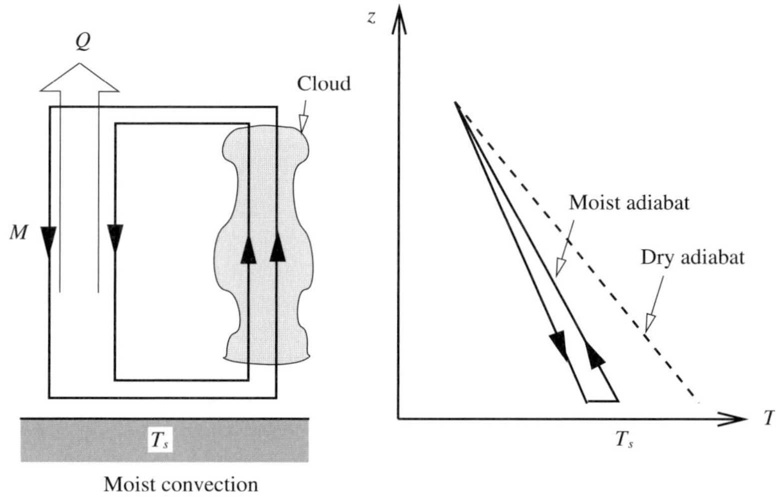

FIGURE 15.3: The same as Fig. 15.1 but for moist convection. Q is the radiative cooling rate and the hatched region is the cloud region. The dotted line is the dry adiabat and the temperature profile of the upward motion region is given by the moist adiabat.

convective layer. If the temperature profile of the downward motion region follows such a dry adiabatic profile, however, the temperature in the downward motion region is warmer than that of the upward motion region so that air parcels in the upward motion region cannot gain buoyancy. We conclude from this consideration that downward motion must be slow enough to be cooled by radiation; the temperature in the downward motion region must be colder than the moist adiabatic temperature in the upward motion region (i.e., the upward velocity must be faster than the downward velocity). This also indicates that the area of the upward motion region is smaller than that of the downward motion region.

In reality, in the tropics the temperature difference between the upward motion region inside cumulus clouds and the downward motion region in the environment of cumulus is very small, so that the horizontal mean temperature profile is almost prescribed by the moist adiabat of the upward branch of cumulus convection particularly in low latitudes. In Section 15.3, we further examine the thermodynamic structure of moist circulation in more detail.

15.2 Static stability of a moist atmosphere

15.2.1 Conditional instability

Before examination of the circulation structure of moist convection, we review the static stability of a moist atmosphere. The *parcel method* is used to consider the static stability of a moist atmosphere. As described in Section 2.3, the parcel method is based on the following assumptions; an air parcel ascends or descends without exchanging heat and moisture with its environment and keeps its entropy and water vapor component at the same values as those of the originating level. It is also assumed that the air parcel does not affect the stratification of the environment and that its pressure is always the same as that of the environment. If the displacement of the air parcel is infinitesimal, the stability criterion is given by the moist adiabat for a saturated air parcel, while it is given by the dry adiabat for an unsaturated air parcel. Here, we use the lapse rate of temperature with respect to pressure $\gamma = dT/dp$; the following argument still holds for the lapse rate with respect to height $\Gamma = -dT/dz$. The lapse rate of the dry adiabat is denoted by γ_d, and that of the moist adiabat is denoted by γ_m. Precisely, γ_d depends on water vapor, but we use an approximate value $1/\rho C_{pd}$ as given by (1.1.57), for simplicity. γ_m is a function of temperature, pressure, and water vapor, and is approximately given by (8.2.49). Let γ denote the lapse rate of the environment. If γ is larger than γ_d, the atmospheric state is *absolutely unstable*; in this case, the atmosphere is always statically unstable regardless of whether the air parcel is saturated or not. Contrary to this, if γ is smaller than γ_m, the atmospheric state is *absolutely stable*; the atmosphere is always statically stable regardless of whether the air parcel is saturated or not. In the intermediate case when γ is between γ_d and γ_m, the atmospheric state is *conditionally unstable*. That is, the static stability of the

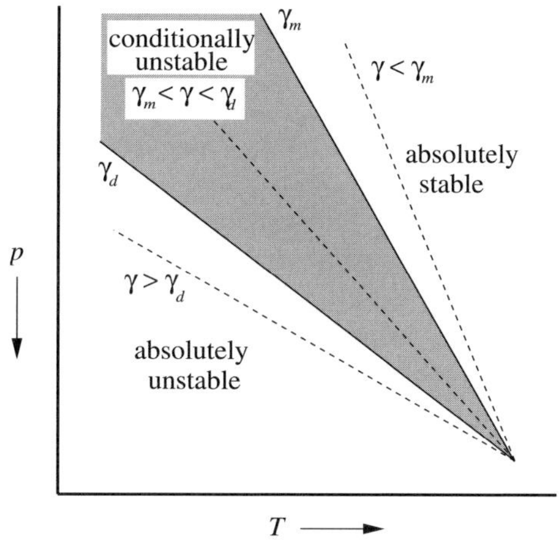

FIGURE 15.4: The stability criterion of a moist atmosphere showing the relation between temperature structure and stability in the temperature-pressure diagram. The hatched region between the dry adiabat γ_d and the moist adiabat γ_m is conditionally unstable.

atmosphere is categorized as

$$\gamma > \gamma_d \quad : \quad \text{absolutely unstable,} \tag{15.2.1}$$

$$\gamma_m < \gamma < \gamma_d \quad : \quad \text{conditionally unstable,} \tag{15.2.2}$$

$$\gamma < \gamma_m \quad : \quad \text{absolutely stable,} \tag{15.2.3}$$

as shown in Fig. 15.4. If stratification is conditionally unstable, actual stability depends on whether the air parcel is saturated or not. If the air parcel is not saturated, stratification is stable for an infinitesimal displacement. If the air parcel is saturated, on the other hand, it is unstable for an infinitesimal displacement. Even if the air parcel is not saturated initially, however, it can be unstable for a finite displacement. During ascending motion, the temperature of the air parcel follows the dry adiabat up to the saturation level and becomes colder than the environment. Above the saturation level, however, it follows the most adiabat. If the air parcel furthermore ascends, it becomes warmer than the environment; in the end, the air parcel gains buoyancy and its upward motion is accelerated. This case is schematically shown later in Fig. 15.5.

15.2.2 Convective available potential energy: CAPE

The stability of a finite displacement of an air parcel in a conditional unstable environment can be formulated by the buoyancy experienced by the air parcel as it is lifted upward. Let us consider an air parcel which is displaced upward from level z_B. The air parcel is assumed to be unsaturated initially but contains a moderate amount of water vapor. No heat or moisture exchange is allowed between the air parcel and the environment, and its pressure is always the same as that of the environment during its ascent. If the density of the air parcel is smaller than that of the environment, it gains positive buoyancy, while if its density is greater, it gains negative buoyancy. The buoyancy of the air parcel is written as

$$B = -\frac{\rho_p - \overline{\rho}}{\overline{\rho}} g, \tag{15.2.4}$$

where ρ_p and $\overline{\rho}$ are the densities of the air parcel and the environment, respectively. In the case that the environment is conditionally unstable and the air parcel is not saturated at the starting level, the buoyancy of the air parcel is negative as long as the upward displacement of the air parcel is small enough. If the air parcel is lifted upward continuously, it becomes saturated and follows the moist adiabat. The level where the air parcel begins to be saturated is called the *lifting condensation level*, which is denoted by z_{LCL}. If the air parcel is lifted further, the temperature of the air parcel becomes warmer than that of the environment, and it gains buoyancy. This height is called the *level of free convection* and is denoted by z_{LFC}. Since the lapse rate of the environment generally tends to be smaller at higher levels near the tropopause and the temperature increases with height in the stratosphere, the buoyancy of the air parcel becomes smaller and eventually reaches zero at a sufficiently high level. The level where buoyancy vanishes is denoted by z_T. Generally, an integral of buoyancy between the starting level z_B and z_T is called *convective available potential energy* (*CAPE*). The CAPE W is given by

$$W = \int_{z_B}^{z_T} B\,dz = -\int_{z_B}^{z_T} \frac{\rho_p - \overline{\rho}}{\overline{\rho}} g\,dz, \tag{15.2.5}$$

or

$$W = -\int_{p_T}^{p_B} \frac{\rho_p - \overline{\rho}}{\overline{\rho}} dp, \tag{15.2.6}$$

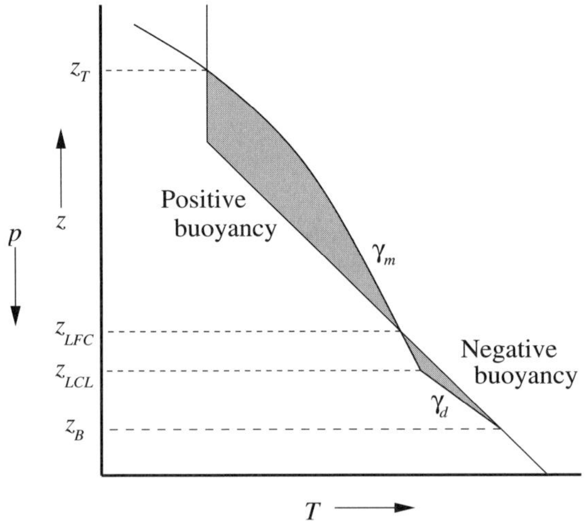

FIGURE 15.5: The finite displacement of an air parcel in a conditionally unstable environment. The thick solid curve is the temperature change of the air parcel, and the thin curve is environmental temperature. The buoyancy in the hatched region is used for the calculation of CAPE. If the effect of water vapor is neglected, warmer temperature means increased buoyancy.

where the hydrostatic balance in the environment is used and pressures at the levels z_B and z_T are denoted by p_B and p_T, respectively. The relation between the temperature change of the air parcel and buoyancy is depicted in Fig. 15.5.

If the air parcel is lifted upward and becomes buoyant in a conditionally unstable atmosphere, the buoyancy force works on the air parcel; the work applied to the air parcel per unit mass is given by CAPE. Assuming that all the work is converted to kinetic energy, we can estimate the possible vertical velocity of the air parcel as

$$w_{max} = \sqrt{2W}. \tag{15.2.7}$$

In reality, of course, all the CAPE is not convertible to kinetic energy due to dissipative processes. This estimation only gives the maximum of the upward velocity of an air parcel lifted upward in a given state of the atmosphere.

CAPE is not solely determined by the thermal structure of the environment but also depends on the originating level of an air parcel and the processes the air parcel experiences along its displacement. The buoyancy of an air parcel depends on how the water content of the air parcel changes. Two extreme cases are the moist adiabatic process in which all the water content is conserved and the pseudo-moist adiabatic process in which water falls as precipitation if it exceeds the saturation vapor pressure. For the former case, the air parcel keeps its moist entropy and mass concentration of water during its ascent. In particular, the sum of all the water substance (i.e., vapor and liquid water or ice) does not change even after the air parcel is saturated.

15.2.3 Saturation condition in the direction of vertical motion

In the earth atmosphere, an air parcel can be saturated only if it is displaced in the upward direction and cannot be saturated if it is displaced in the downward direction. As shown below, this relation may be different for other types of gas components. Here, we examine the saturation condition of an air parcel for adiabatic displacement in the vertical direction.

In general, an air parcel can be saturated either when temperature decreases or when pressure increases, as can be seen from the equation of saturation specific humidity (8.2.22). If the effect of temperature is dominant, the air parcel becomes saturated when it is displaced upward, while if the effect of pressure is dominant, it becomes saturated when it is displaced downward. The relation between the pressure and temperature of the air parcel for adiabatic motion is given by the adiabat (14.5.9). We consider the adiabatic process in the case that water vapor content is sufficiently small. Under this condition, the lapse rate is close to the dry adiabat:

$$\frac{dp}{dT} = \frac{1}{\gamma_d} = \frac{C_{pd}p}{R_d T}. \tag{15.2.8}$$

Since the partial pressure of water vapor is given from (8.2.25), it is approximated as

$$p_v = \frac{\varepsilon^{-1}q}{1+(\varepsilon^{-1}-1)q}p \approx \varepsilon^{-1}qp. \tag{15.2.9}$$

Since the specific humidity of the air parcel q is conserved until it is saturated, the change in partial pressure of water vapor is given by an equation similar to (15.2.8):

$$\frac{dp_v}{dT} = \frac{C_{pd}p_v}{R_dT}. \qquad (15.2.10)$$

At saturation point, partial pressure is equal to the saturation pressure of water vapor: $p_v = p^*$. The saturation pressure is given by the Clapeyron-Clausius equation (8.1.64) as

$$\frac{dp^*}{dT} = \frac{\varepsilon Lp^*}{R_dT^2}. \qquad (15.2.11)$$

By comparing the change in saturation condition and the adiabat near the saturation point of the air parcel, we can judge whether the air parcel is saturated during upward displacement or during downward displacement. At the saturation point $p_v = p^*$, we have

$$\frac{\frac{dp^*}{dT}}{\frac{dp_v}{dT}} = \frac{\varepsilon L}{C_{pd}T}. \qquad (15.2.12)$$

If this ratio is greater than one, the air parcel is saturated if it is displaced upward, while if this ratio is smaller than one, the air parcel is saturated if it is displaced downward. These relations are schematically shown in Fig. 15.6. In the case of the earth's atmosphere, since

$$\frac{\varepsilon L}{C_{pd}T} \approx 5.2,$$

for $T = 300$K, then the air parcel can be saturated only if it is displaced upward, in general. This ratio, however, may be smaller than one if thermodynamic variables are different.

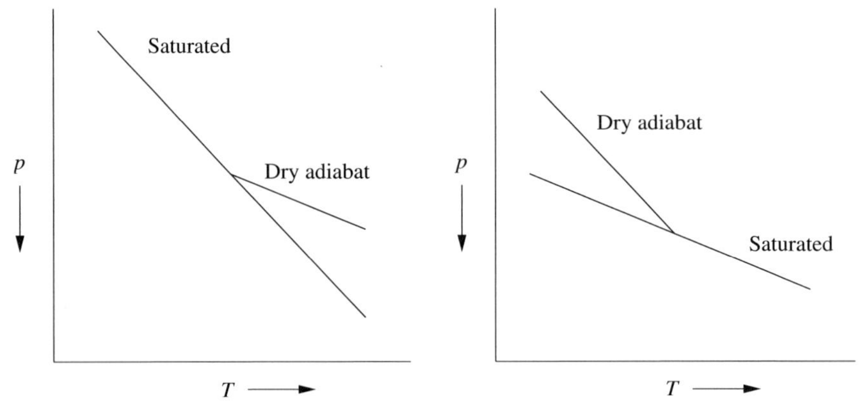

FIGURE 15.6: The relation between the saturation condition and the dry adiabat. Left: the air parcel is saturated in its upward displacement. Right: the air parcel is saturated in its downward displacement. The left panel corresponds to the case of the earth atmosphere.

15.3 Circulation structure of moist convection

15.3.1 Thermal structure

Let us now consider the circulation structure of moist convection in more detail, following the schematic model shown in Fig. 15.3. Here, we assume that moist circulation is modeled like Fig. 15.7 (Satoh and Hayashi, 1992), in which circulation is steady and the convective layer, or the troposphere, is divided into two regions: the upward motion region corresponding to the interior of cumulus clouds and the downward motion region in the environment of cumulus. At a first consideration, mass exchange between the upward and downward motion regions is allowed only in the top and bottom boundary layers. Although boundary layers are required at the top and bottom of the convective layer, these inner structures are not explicitly treated. We assume that the fractional area of the upward motion region f is smaller than that of the downward motion region:

$$f \ll 1. \tag{15.3.1}$$

The following simplifications are introduced to the equations of mass, energy (static energy), and water vapor within the respective regions. The mass flux M_c of the circulation associated with cumulus convection is defined by

$$M_c \equiv f\rho^u(z)w^u(z), \tag{15.3.2}$$

where w^u and ρ^u are the vertical velocity and the density in the upward motion region. In this simple model, no mass exchange is allowed between the upward and downward motion regions, and the mass flux M_c is constant irrespective of height:

$$M_c = \text{const.} \tag{15.3.3}$$

Mass exchange occurs only at the top and bottom boundary layers as shown in Fig. 15.7. In the steady state, since the horizontally averaged mass flux $\overline{\rho w}$ is zero, the magnitude of mass flux in the downward motion region is equal to (15.3.2):

$$M_c = -(1-f)\rho^d(z)w^d(z). \tag{15.3.4}$$

Using these relations for M_c, we can express the vertical flux of ϕ, which is a physical quantity per unit mass, as

$$\begin{aligned}\overline{\rho w\phi}(z) &\equiv f\rho^u(z)w^u(z)\phi^u(z) + (1-f)\rho^d(z)w^d(z)\phi^d(z) \\ &= M_c[\phi^u(z) - \phi^d(z)].\end{aligned} \tag{15.3.5}$$

The upward motion region is assumed to be the pseudo-moist adiabat; water vapor is assumed to be saturated everywhere in the upward motion region and condensed water to immediately fall to the ground. In contrast, water vapor is conserved during downward motion

$$q^u(z) = q^*(\overline{p}(z), T^u(z)), \tag{15.3.6}$$

$$q^d = \text{const.}, \tag{15.3.7}$$

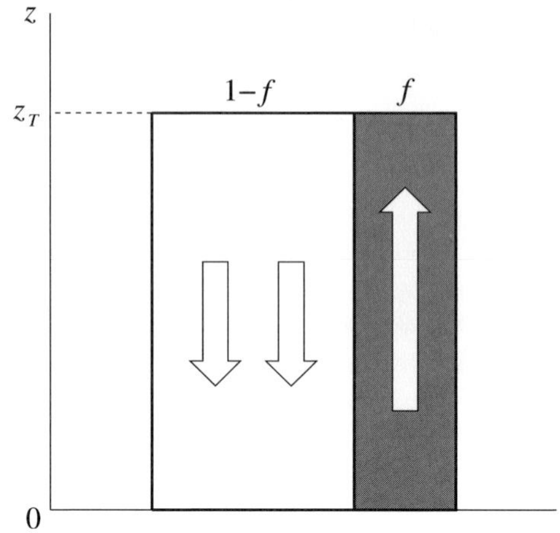

FIGURE 15.7: Schematic figure of a simple model of moist convection. f is the fractional area of the upward motion region, and z_T is tropopause height.

where q is specific humidity, $q^*(p, T)$ is saturation specific humidity at pressure p and temperature T, given by (8.2.22). At the tropopause, both specific humidities in the upward and downward motion regions are equal and, thus, q^d is given by the tropopause value in the upward motion region:

$$q^d = q^u(z_T). \tag{15.3.8}$$

In the upward motion region, the effect of radiative cooling is neglected and motion is assumed to be adiabatic. Then, from (8.2.17), moist static energy σ is constant irrespective of height in the upward motion region:

$$\sigma^u = C_pT^u(z) + Lq^u(z) + gz = \text{const.} \tag{15.3.9}$$

In the downward motion region, however, moist static energy changes due to radiative cooling. Thus, it can be written as

$$\sigma^d(z) = C_pT^d(z) + Lq^d + gz. \tag{15.3.10}$$

The vertical energy flux associated with moist convection $F^{conv} \equiv \overline{\rho w \sigma}$ is given by (15.3.5):

$$F^{conv}(z) \equiv \overline{\rho w \sigma}(z) = M_c[\sigma^u - \sigma^d(z)]. \tag{15.3.11}$$

The solution to the radiative-convective equilibrium in a moist atmosphere is given by the balance between radiative flux and convective flux as shown by (14.5.3). In the equilibrium state, we assume that there is no net energy flux at the surface. In this case, vertical energy flux is everywhere zero irrespective of height: in the troposphere ($0 < z < z_T$),

$$F^{rad}(z) + F^{conv}(z) = 0, \tag{15.3.12}$$

and in the stratosphere ($z > z_T$),

$$F^{rad}(z) = 0 \tag{15.3.13}$$

(i.e., the stratosphere is in radiative equilibrium). At the tropopause, in particular, convective energy flux vanishes:

$$F^{conv}(z_T) = 0. \tag{15.3.14}$$

Using this and (15.3.11), or (15.3.8)–(15.3.10), we obtain

$$\sigma^u = \sigma^d(z_T), \tag{15.3.15}$$
$$T^u(z_T) = T^d(z_T). \tag{15.3.16}$$

The temperatures in the upward and downward motion regions take the same value at the tropopause. As shown below, however, at the bottom of the two regions ($z = 0$), the two temperatures are generally different.

The balance in the radiative-equilibrium state is rewritten using radiative cooling Q^{rad} and convective heating Q^{conv}, which are respectively defined as

$$Q^{rad}(z) = \frac{d}{dz}F^{rad}(z), \tag{15.3.17}$$
$$Q^{conv}(z) = -\frac{d}{dz}F^{conv}(z) = C_p M_c \left[\frac{d}{dz}T^d(z) + \frac{g}{C_p}\right], \tag{15.3.18}$$

where (15.3.11) and (15.3.10) are used. The equation of balance (14.5.3) is rewritten as

$$Q^{rad}(z) = C_p M_c \left[\frac{d}{dz}T^d(z) + \frac{g}{C_p}\right]. \tag{15.3.19}$$

This means that the adiabatic warming in the environment of the cumulus region is balanced by the radiative cooling Q^{rad} in the radiative-convective equilibrium.

In the above simple model, one needs to externally specify the value of mass flux M_c to obtain an equilibrium solution. An additional equation such as the balance of kinetic energy is required to determine a unique solution, which will be explained in the next subsection. Here, we show possible solutions by giving parametric values to mass flux. We use the gray radiation model with no scattering with $F_s = 350$ W/m^2, $\tau_s^* = 2.0$, and $\alpha = 1.0$ in (14.3.22), and surface temperature is simply given by $T_s = T^u(0)$. Fig. 15.8 shows the temperatures and the lapse rates in the radiative-convective equilibrium solutions for $M_c = 1.7$, 2.6, 4.0, and 6.5 $\times 10^{-3}$ kg m^{-2} s^{-1}. Values in the upward motion region are indicated by dashed curves and those in the downward motion region by solid curves. As M_c becomes larger, the temperature in the downward motion region becomes warmer. In particular, the temperature in the downward motion region is everywhere warmer than that in the upward motion region (i.e., $T^u < T^d$); at the largest value $M_c = 6.5 \times 10^{-3}$ kg m^{-2} s^{-1}. This implies that the buoyancy in the upward motion region is everywhere

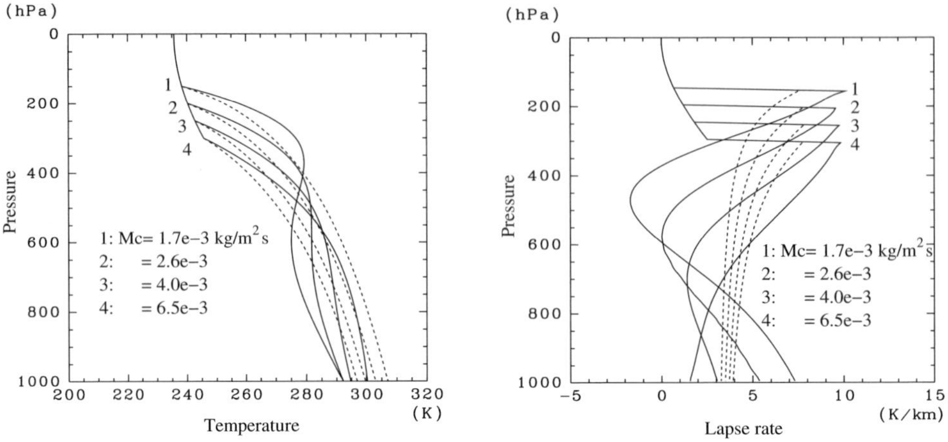

FIGURE 15.8: The vertical profiles of temperatures (left) and the lapse rates $-\frac{dT}{dz}$ (right) of the radiative-convective equilibrium using the simple cumulus model. Solid curves are those of the horizontal average (the downward motion region in the case of the troposphere), and dashed curves are those of the upward motion regions. The values of mass flux are M_c = 1.7, 2.6, 4.0, and 6.5 $\times 10^{-3}$ kg m^{-2} s^{-1} (from Satoh and Hayashi, 1992).

negative if the contribution of water vapor to buoyancy is negligible. This means that solutions for such large values of M_c are inappropriate for radiative-convective equilibrium states. As for smaller values of M_c, the temperature in the upward motion region becomes warmer and is thought to be realizable. Just below the tropopause, however, the temperature in the upward motion region is in general colder compared with the temperature in the downward motion region. This characteristic is also unfavorable for the buoyancy of air parcels in the upward motion region. The constraint that mass flux be constant with height and time is the origin for this unrealistic characteristic, which will be described after consideration of the determination of the value of M_c in the next subsection.

The vertical profiles of temperature of the radiative-convective equilibrium using the cumulus model shown in Fig. 15.8 are further examined using the change in the thermodynamic quantities of an air parcel associated with moist circulation. Fig. 15.9 shows the vertical profiles of temperature T, specific humidity q, and moist static energy σ in the case of $M_c = 2.9 \times 10^{-3}$ kg m^{-2} s^{-1}. In Fig. 15.9, the direction of circulation is designated by arrows.

The circulation of an air parcel along the path of convective motion is described by the simple model of moist convection shown in Fig. 15.7. During upward motion (Fig. 15.9, A $\rightarrow$ B), the air parcel is always saturated by releasing latent heat. Its upward velocity is so fast that cooling due to radiation is negligible. Thus, moist static energy σ^u is conserved; temperature T^u changes according to the moist adiabat; and humidity q^u is always at the saturated value. During downward motion (Fig. 15.9, B $\rightarrow$ C), on the other hand, there is no release or absorption of latent heat. Downward velocity is slow enough that it is cooled by radiation. Thus, static energy σ^d decreases as the air parcel goes down; temperature T^d changes according

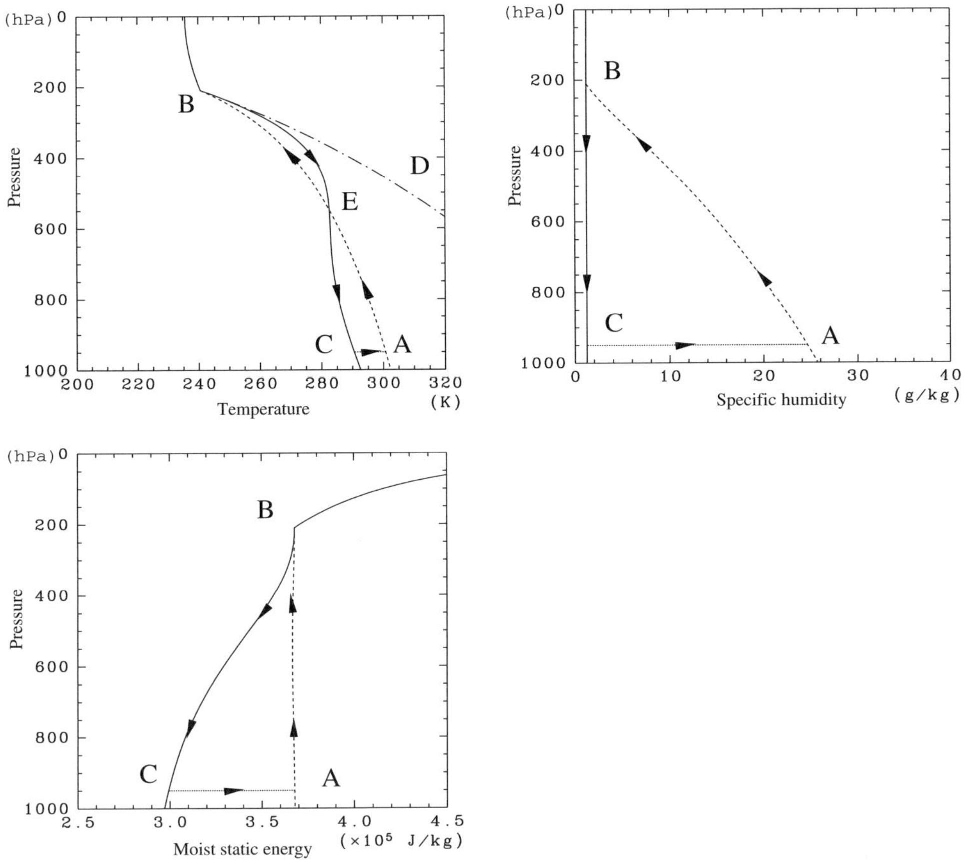

FIGURE 15.9: Changes in temperature T (top left), specific humidity q (top right), and moist static energy h (bottom) along the circulation path of moist convection in the case of $M_c = 2.9 \times 10^{-3}$ kg m^{-2} s^{-1}. Solid curves are the horizontal average (the downward motion region in the troposphere), and dashed curves are profiles of the upward motion regions. Dotted curves are the change in boundary layer. Symbols A and C are at the lowest level of the upward motion region and the downward motion region, respectively, and B is the tropopause. The arrows show the direction of circulation. Dashed-dotted curves (B $\rightarrow$ D) represents the dry adiabat, and E is the height at which the air parcel loses its buoyancy in the upward motion region (from Satoh and Hayashi, 1992).

to the lapse rate which is smaller than that of the dry adiabat (Fig. 15.9, B $\rightarrow$ D), while humidity q^d is conserved. When the air parcel enters the mixed layer, sensible and latent heat fluxes are supplied from the surface (Fig. 15.9, C $\rightarrow$ A). Thus, in the mixed layer, temperature changes from $T^d(0)$ to $T^u(0)$, specific humidity changes from q^d to $q^u(0)$, and moist static energy changes from $\sigma^d(0)$ to σ^u. Then, the state of the air parcel returns to its initial state.

The lapse rate in the upward motion region is equal to the moist adiabat, while the lapse rate in the downward motion region depends on the profiles of radiative cooling, and on the magnitude of the mass flux M_c. As M_c is larger and thus vertical velocity w^d is larger (see (15.3.4)), the time required for the air parcel to travel from the tropopause to the surface (the advective time) becomes shorter, and the temperature change due to radiative cooling becomes smaller. In this case, therefore, the lapse rate in the downward motion region becomes closer to that of the dry adiabat. As M_c becomes smaller, advective time becomes larger and the temperature change due to radiation becomes larger. Thus, the lapse rate becomes larger when compared with that of the dry adiabat.

Near the tropopause, the lapse rate in the downward motion region is about 10 K km^{-1} irrespective of M_c. Since radiative cooling is small near the tropopause, the lapse rate is close to that of the dry adiabat as shown by (15.3.19). Since the lapse rate in the upward motion region matched that of the moist adiabat, the temperature in the upward motion region is generally lower than the temperature in the downward motion region just below the tropopause.

The profiles of specific humidity are given by (15.3.6) and (15.3.7) for the upward and downward motion regions, respectively. The specific humidity in the downward motion region thus obtained has the observed characteristic that specific humidity rapidly decreases just above the mixed layer, but its value is much smaller than observed. Over a statistically long time average, in general, specific humidity gradually increases with height, and the sharp decrease at the top of the mixed layer becomes obscure. The actual increase of the specific humidity in the downward motion region is due to water supply by shallow convection.

15.3.2 Mass flux

Mass flux M_c is related to various balance equations. First, we derive relations to satisfy the requirement of energy and water vapor fluxes at the ground. Let sensible heat flux at the surface be denoted by F^{sh}, and evaporation at the surface by E_v. The relation of energy flux (14.5.6) is expressed as

$$M_c[\sigma^u - \sigma^d(0)] = F^{sh} + LE_v, \tag{15.3.20}$$

and the relation of water vapor flux (14.5.5) is expressed as

$$M_c[q^u(0) - q^d] = E_v. \tag{15.3.21}$$

Using (15.3.20) and (15.3.21), we obtain the following three relations for M_c:

$$M_c = \frac{F^{sh} + LE_v}{\sigma^u - \sigma^d(0)} \tag{15.3.22}$$

$$= \frac{F^{sh}}{C_p[T^u(0) - T^d(0)]} \tag{15.3.23}$$

$$= \frac{E_v}{q^u(0) - q^d}. \tag{15.3.24}$$

Up to this point, only the balances of static energy and water vapor are used; we have not introduced any assumptions additional to those in the previous subsection.

To obtain an appropriate value for M_c, we need an additional requirement for buoyancy; for this purpose, we can use the balance of kinetic energy which contains the condition of buoyancy. In this simple model where M_c is independent of height, the balance of the kinetic energy of the air column per unit area is given as

$$M_c W - D = 0, \tag{15.3.25}$$

where $M_c W$ is the production rate of kinetic energy and D is the dissipation rate per unit column. W is the total work done by buoyancy, or CAPE defined by (15.2.5), and is given in this case by

$$W \equiv -\int_0^{z_T} \frac{\rho^u(z) - \rho^d(z)}{\rho^d(z)} g dz \approx \int_0^{z_T} \frac{T^u(z) - T^d(z)}{T^d(z)} g dz. \tag{15.3.26}$$

Here, just for simplicity, we have neglected the contribution of water vapor to buoyancy, and thus the difference of densities between the upward and downward motion regions, ρ^u and ρ^d, are approximated by the temperature difference between the two regions. Since the dissipation rate D is always positive, the necessary condition is $W > 0$ in order for the solution to be appropriate (i.e., the work done by buoyancy must be positive). If D is expressed by the quantities used in this model based on appropriate closure assumptions, (15.3.25) gives an additional relation between M_c and W. It is difficult, however, to quantify all the dissipation process within the troposphere; a precise expression for D is unknown in such a simple framework of the cumulus model. Observationally, the mass flux associated with the environment of cumulus is about 3×10^{-3} kg m^{-2} s^{-1}, and the temperature difference between the cloud region and the environment is about $\Delta T \approx 3$ K, so we may estimate as $W \approx (\Delta T/T) g z_T \approx 10^3$ J kg^{-1}. Thus, from the product of the two, we may have $D \approx 30$ W m^{-2}. According to the above model for $M_c = 3 \times 10^{-3}$ kg m^{-2} s^{-1}, we also have $W = 500$ W m^{-2}, which is close to the value estimated from observation.

The work W given by (15.2.5) is also a simplified form of the so-called *cloud work function*. Arakawa and Schubert (1974), whose parameterization is used in general circulation models, employ the assumption of quasi-equilibrium

$$\frac{dW}{dt} = 0, \tag{15.3.27}$$

to obtain the vertical distribution of M_c. In radiative-convective equilibrium problems, however, since the cloud work function is statistically time-independent, the assumption of quasi-equilibrium does not give any constraint for determination of M_c.

In any case, if we obtain the value of the dissipation rate D, a unique equilibrium solution to the radiative-convective equilibrium can be determined using (15.3.25). Here, we give a crude estimation of D from scaling analysis. Using a representative velocity scale of turbulence v' and its length scale l, the dissipation rate of kinetic energy in turbulence per unit mass is estimated as v'^3/l. Thus, the dissipation rate per unit column may be written as

$$D = \int_0^{z_T} \rho \frac{\overline{v'^3}}{l} dz \tag{15.3.28}$$

If we assume that the dissipation of kinetic energy occurs mainly in the upward motion region and approximate $v' \approx w^u$, D is expressed by M_c using (15.3.2):

$$D \approx \int_0^{z_T} f\rho^u(z)\frac{w^u(z)^3}{l^u}dz = M_c^3 \int_0^{z_T} \frac{dz}{f^2\rho^u(z)^2 l^u}. \tag{15.3.29}$$

where l^u is the length scale of turbulence in clouds. We define a coefficient C whose dimension is kg m^{-3} by

$$C \equiv \left(\int_0^{z_T} \frac{dz}{f^2\rho^u(z)^2 l^u}\right)^{-1/2} \approx f\langle\rho\rangle\sqrt{\frac{l^u}{z_T}}, \tag{15.3.30}$$

where $\langle\rho\rangle$ is the vertically averaged density. From (15.3.25) and (15.3.29), we have

$$M_c = C\sqrt{W}. \tag{15.3.31}$$

Using the values $f \approx |w^d/w^u| \approx 10^{-3}$, $z^T \approx 10^4$ m, $l^u \approx 10^3$ m and $\langle\rho\rangle \approx 0.5$ kg/m^3, we may have a typical value $C \approx 10^{-4}$ kg/m^3. Eq. (15.3.31) can be used as an additional equation for determination of a unique solution.

15.3.3 Implications for improved cumulus models

The simple cumulus model introduced in Fig. 15.7 is a highly idealistic model of the cumulus convection observed in reality. It is intended for the steady state for consideration of radiative-convective equilibrium. The results from the simple cumulus model have the following problems: First, the buoyancy near the tropopause is negative. Second, near the top of the mixed layer, the temperature in the downward motion region is colder than that of the upward motion region; this is contradicted by the view that the top of the mixed layer is capped by an inversion layer. Third, the temperature difference between the upward and downward motion regions is as much as 10 K, which is much larger than observation (less than about 3 K). These problems come about because of defects in the simple cumulus model. However, this information contains insights into what should be added to the simple model to get more realistic results.

One cause of the above problems is the assumption that M_c is set constant irrespective of height. In reality, M_c has vertical dependency since individual clouds entrain the environmental air. Clouds have time dependency and horizontal structure, which are also attributable to the vertical profile of mass flux. In reality, the environmental air of cumulus is cooled by radiation and stratification has both temporal and horizontal variation. The top level of cumulus is determined by the condition that the buoyancy of air parcels in cumulus vanishes with instantaneous stratification. Buoyancy drives the convective mass flux in the cumulus and then the downward motion is associated in the environment. Such a downdraft is generated below the top level of cumulus, and brings adiabatic warming in the environment. In general, the convective mass flux increases with height due to entrainment of environmental air. Even if convective mass flux is assumed to be constant in each cumulus cloud, the vertical profile of M_c is generated if all the convective mass flux associated with each cumulus cloud is averaged in time and domain. These collective effects will explain the vertical profile of M_c at the equilibrium state. In such a view (i.e., with time and horizontal variation), the apparent contradiction that buoyancy is negative just below the tropopause will be resolved in the time and domain-averaged equilibrium state and the temperature difference between the upward and downward motion regions will be much reduced. As a result, the temperature in the downward motion region will be much closer to that of the moist adiabat of the upward motion region. It is expected that the distribution of M_c is sensitive to the cooling profile of radiation.

The lack of an inversion layer just above the mixed layer, or the boundary layer in the simple model, will not be resolved in the framework of a one-dimensional horizontal uniform perspective, even though time dependency is taken into consideration. This is because air parcels in the mixed layer cannot have buoyancy if it is capped by an inversion layer. The co-existence of the mixed layer capped by the inversion layer and tall cumulonimbus clouds rooted in the mixed layer is hardly realized under conditions of a horizontally uniform boundary. Thus, it is naturally thought that the inversion layer occurs in the horizontally inhomogeneous condition. For instance, if the surface temperature has large-scale variation, cumulus activity is enhanced at the warm surface temperature region, while only shallow cumulus clouds are generated at the cold region, which is capped by downward motion in the free atmosphere.

The cumulus model considered in this section has been deliberately simplified to ease investigation of the relation between atmospheric vertical structure and cumulus convection. Many more elements must be introduced to achieve realistic circulation. In particular, an appropriate evaluation of the statistical effects of cumulus convection is required for general circulation models, in which the effects of many cumulus clouds must be considered within a region whose domain size is approximately 100 km. For example, the cumulus scheme proposed by Arakawa and Schubert (1974) is nowadays frequently used in many general circulation models. In Arakawa and Schubert, instead of a single type of cumulus convection, a spectrum of cumulus clouds with various size distributions is assumed. Each cumulus cloud entrains the environmental air during its ascent; the entrainment determines the mixing of the cloud parcel and then the top level of the clouds. It is thought that the size distribution of a cumulus cloud is determined when buoyancy W (i.e., the cloud work function) is in quasi-equilibrium. A similar condition is given by Eq. (15.3.19) for radiative-convective equilibrium experiments. Further improvements have been added for cumulus parameterization, such as the introduction of the effects of downdraft due to rain and the effects of ice. A prognostic method for mass flux is also sought.

15.4 Mixed layer

In the simple cumulus model depicted in Fig. 15.7, we assumed that the entire troposphere is occupied by the upward motion region and the downward motion region, and that only a thin boundary layer is allowed between the surface and the atmosphere. In the real tropics, the cloud base is located at about a 1 km height in general, and water vapor condensation does not occur below the cloud base. Such a layer between the cloud base and the surface boundary layer is called the *mixed layer*. The boundary layer has a typical thickness of a few tens of meters and has a large gradient of temperature and water vapor. In the mixed layer, entropy and specific humidity are well homogenized in the vertical direction due to convective mixing. Above the cloud base, a shallow cumulus layer is sometimes observed, called the *trade wind cumulus*. The trade wind cumulus can be viewed as a visible part of the mixed layer where condensation and evaporation of water vapor occurs. In general, however, the upper boundary of the mixed layer is defined as the cloud base either of shallow cumulus clouds or tall cumulonimbus clouds. The height at the cloud base corresponds to the *lifting condensation level* (LCL), where air parcels lifted upward in the mixed layer begin to condensate.

In order to consider the energy and water budgets in the mixed layer, we modify the simple cumulus model in the previous section by adding a mixed layer below the cumulus upward and downward motion regions as shown in Fig. 15.10. The cloud base agrees with the top of the mixed layer and its height is denoted by z_M. In the mixed layer, static energy and specific humidity are assumed to be constant irrespective of height; they are denoted by σ_M and q_M, respectively. In this case, the temperature profile in the mixed layer follows the dry adiabat. We also assume that physical quantities in the mixed layer at z_M are continuously connected to those of the upward motion region:

$$\sigma_M = \sigma^u(z_M), \qquad q_M = q^u(z_M). \tag{15.4.1}$$

The energy flux and the water vapor flux from the surface, F^{sh} and E_v, are given by the bulk formula (14.4.17), (14.5.6):

$$F^{sh} = \rho_0 C_D v_* C_p(T_s - T_0) = M_b C_p(T_s - T_0), \tag{15.4.2}$$

$$E_v = \rho_0 C_D v_*(q_s - q_M) = M_b(q_s - q_M). \tag{15.4.3}$$

Here, we simply assume that the bulk coefficients for heat and water vapor take the same value $C_D u^*$, and define

$$M_b \equiv \rho_0 C_D v_*. \tag{15.4.4}$$

Typically, $M_b \approx 0.01$ kg m^{-2} s^{-1}. T_0 and ρ_0 are temperature and density at the bottom of the atmosphere (i.e., at the top of the boundary layer), respectively. At the surface, we assume that water vapor is saturated:

$$q_s = q^*(p_s, T_s). \tag{15.4.5}$$

From (15.4.2) and (15.4.3), we have

$$F^{sh} + LE_v = M_b(\sigma_s - \sigma_M). \tag{15.4.6}$$

On the other hand, water vapor flux and convective energy flux at the top of the mixed layer z_M are written as

$$\overline{\rho w q}(z_M) = M_c[q^u(z_M) - q^d] = M_c(q_M - q^d), \tag{15.4.7}$$

$$\overline{F^{conv}}(z_M) = M_c[\sigma^u(z_M) - \sigma^d(z_M)] = M_c[\sigma_M - \sigma^d(z_M)]. \tag{15.4.8}$$

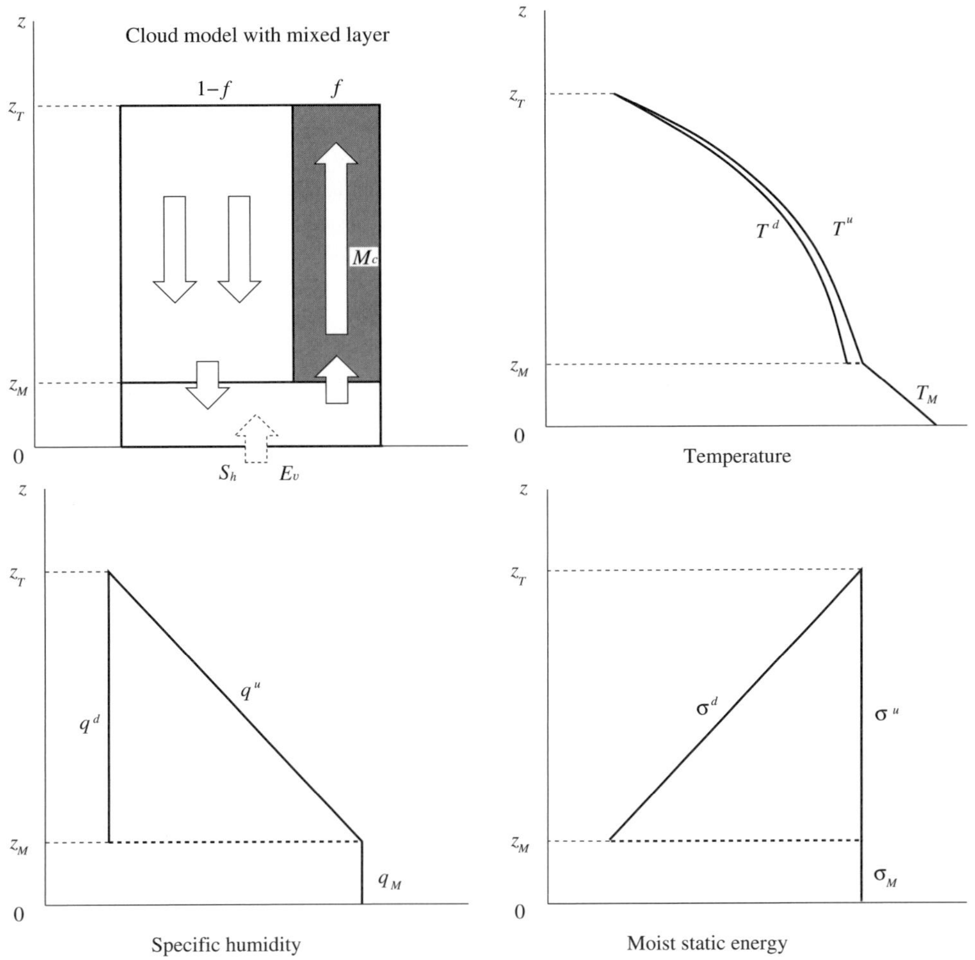

FIGURE 15.10: A simple cumulus model with the mixed layer (upper left), and the structures of temperature (upper right), specific humidity (lower left), and moist static energy (lower right). z_T is the tropopause height and z_M is the height of the mixed layer. The convective layer $z_M < z < z_T$ is composed of upward and downward motions, in which mass flux is denoted by M_c. In the mixed layer $0 < z < z_M$, sensible heat flux S_h and evaporation E_v are supplied from the surface, and energy and water are exchanged with the convective layer through M_c.

The supply of water vapor from the surface to the mixed layer is equal to the upward transport of water vapor through the cloud base height z_M. Note that we neglect the lateral transport of water vapor or the evaporation of rain within the mixed layer in this simple model. Thus, the budget of water vapor in the mixed layer is written as

$$E_v = \overline{\rho w q}(z_M). \tag{15.4.9}$$

From (15.4.3) and (15.4.7), the specific humidity in the mixed layer is expressed as

$$q_M = \frac{M_b q_s + M_c q_d}{M_b + M_c} \approx \frac{M_b}{M_b + M_c} q_s, \tag{15.4.10}$$

where $q_d \ll q_M$ and $M_c < M_b$ is used in this approximation. We have $M_c/M_b \approx 0.3$ for typical values $M_c \approx 0.003$ kg m^{-2} s^{-1} and $M_b \approx 0.01$ kg m^{-2} s^{-1}. If the temperature at the bottom of atmosphere is given by the surface temperature, $q_s = q^*(p_s, T_0)$ represents the specific humidity at the bottom of the atmosphere. Therefore, the ratio q_M/q_s corresponds to relative humidity at the bottom of the atmosphere, and is written as

$$r = \frac{q_M}{q_s} \approx \frac{M_b}{M_b + M_c}. \tag{15.4.11}$$

For $M_c/M_b = 0.3$, we have the estimation of relative humidity at the bottom of the atmosphere as $r = 80\%$. Since specific humidity q is constant irrespective of height in the mixed layer, relative humidity increases with height up to 100% at the top of the mixed layer (i.e., the lifting condensation level).

In a similar way, we consider the energy budget in the mixed layer. The total energy fluxes at the surface, the top of the mixed layer, and the tropopause are written as

$$\overline{F^{rad}}(0) + F^{sh} + LE_v = 0, \tag{15.4.12}$$
$$\overline{F^{rad}}(z_M) + \overline{F^{conv}}(z_M) = 0, \tag{15.4.13}$$
$$\overline{F^{rad}}(z_T) = 0. \tag{15.4.14}$$

Using (15.4.8), the energy balance in the mixed layer and that in the convective layer are written respectively as

$$F^{sh} + LE_v - M_c[\sigma_M - \sigma^d(z_M)] = \overline{F^{rad}}(z_M) - \overline{F^{rad}}(0) \equiv R_M, \tag{15.4.15}$$

$$M_c[\sigma_M - \sigma^d(z_M)] = \overline{F^{rad}}(z_T) - \overline{F^{rad}}(z_M) \equiv R_C, \tag{15.4.16}$$

where R_M and R_C are divergences of the energy flux due to radiation in the mixed layer and the convective layer, respectively. They correspond to the vertical integrals of radiative cooling in respective layers. Using (15.4.9) and (15.4.7) to eliminate the contribution of water vapor, we obtain the equations of the energy balance as

$$F^{sh} - M_c C_p[T_M(z_M) - T^d(z_M)] = R_M, \tag{15.4.17}$$
$$LE_v + M_c C_p[T_M(z_M) - T^d(z_M)] = R_C. \tag{15.4.18}$$

In reality, the temperature difference between the upward and downward motion regions is very small. If we equate the temperature in the upward motion region to that in the downward motion region at the cloud base, and set $T_M(z_M) = T^d(z_M)$, we obtain $F^{sh} = R_M$ and $LE_v = R_C$. On this assumption, the *Bowen ratio*, which is defined as the ratio of sensible heat flux to latent heat flux, is written as

$$b \equiv \frac{F^{sh}}{LE_v} = \frac{R_M}{R_C}, \quad \text{if } T_M(z_M) = T^d(z_M). \tag{15.4.19}$$

This relation is derived by Sarachik (1978) and implies that the portion of sensible heat supply from the surface is exactly cooled by radiation. However, the temperature difference between the upward and downward motion regions becomes relatively large if the simple steady cumulus model is used, as shown in Fig. 15.10. The estimation of (15.4.19) does not generally hold.

References and suggested reading

The vertical structure of a moist atmosphere is considered by Sarachik (1978) and Lindzen et al. (1982) by taking account of the circulation structure of moist convection. Their arguments are reconsidered by Satoh and Hayashi (1992), who place a constraint on moist circulation in the radiative-convective equilibrium. Section 15.3 follows Satoh and Hayashi (1992). The theory of the mixed layer in Section 15.4 was originally given by Sarachik (1978). The moist circulation structure used in this chapter is a so-called two-column model, which is an extreme simplification of moist convection in the real atmosphere. The more elaborate cumulus model follows that given by Arakawa and Schubert (1974), whose model is the representative cumulus parameterization used for general circulation models. There have been many reviews on cumulus parameterization published recently. Among them, Emanuel and Raymond (1993) and Smith (1997) give up-to-date information. See also Arakawa (2004). The stability of a moist atmosphere discussed in Section 15.2 can be analyzed by using the various quantities or graphical diagrams proposed by many authors. They are closely related to definitions of the moist thermodynamic variables given in Chapter 8. See chapter 6 of Emanuel (1994), for instance.

Arakawa, A., 2004: The cumulus parameterization problem: Past, present, and future. *J. Climate*, printing.

Arakawa, A. and Schubert, W. H., 1974: Interactions of cumulus cloud ensemble with the large-scale environment, Part I. *J. Atmos. Sci.*, **31**, 674–701.

Emanuel, K. A. and Raymond, D. J., 1993: *The Representation of Cumulus Convection in Numerical Modeling of the Atmosphere*, Meteorological Monograph No. 24. American Meteorological Society, Boston, 246 pp.

Emanuel, K., 1994: *Atmospheric Convection.* Oxford University Press, New York, 580 pp.

Lindzen, R. S., Hou, A. Y., and Farrell, B. F., 1982: The role of convective model choice in calculating the climate impact of doubling CO_2. *J. Atmos. Sci.*, **39**, 1189–1205.

Nakajima, K. and Matsuno, T., 1988: Numerical experiments concerning the origin of cloud clusters in the tropical atmosphere. *J. Meteorol. Soc. Japan*, **66**, 309–329.

Pauluis, O. and Held, I., 2002: Entropy budget of an atmosphere in radiative-convective equilibrium. Part I: Maximum work and frictional dissipation. *J. Atmos. Sci.*, **59**, 125–139.

Sarachik, E. S., 1978: Tropical sea surface temperature: An interactive one-dimensional atmosphere-ocean model. *Dyn. Atmos. Oceans*, **2**, 455–469.

Satoh, M. and Hayashi, Y.-Y., 1992: Simple cumulus models in one-dimensional radiative convective equilibrium problems. *J. Atmos. Sci.*, **49**, 1202–1220.

Smith, R. K., 1997: *The Physics and Parameterization of Moist Atmospheric Convection*, NATO Advanced Study Institute on the Physics and Parameterization. Kluwer Academic Publishers, Dordrecht, 498 pp.

16

Low-latitude circulations

The low-latitude circulations of the atmosphere are associated with hydrological circulations, and the large-scale circulations in low latitudes can be characterized by an organized moist convection. In low latitudes, convection occurs in the form of moist convection associated with latent heat release. If the surface boundary has a laterally inhomogeneous condition with a horizontal scale of the order of several 1000 km, moist convection is organized to have a large-scale circulation structure that has the specified horizontal scale. Examples of large-scale circulations in low latitudes are the *Walker circulation* and the *Hadley circulation*. The Walker circulation is viewed as a large-scale circulation forced by a longitudinal forcing along the equator, while the Hadley circulation is viewed as that forced by a latitudinal forcing between the equator and higher latitudes. These are in principle two-dimensional structures in a horizontal-vertical section, where the horizontal direction can be either latitudinal or longitudinal. In reality, however, the observed circulation is not at all two-dimensional. Fig. 16.1 shows the horizontal distribution of vertical velocity at the middle level 500 hPa. The data source is the same as shown in Appendix A3. Upward motions are seen in low latitudes in every season, though they are not horizontally uniform. The actual distribution of upward and downward motions has such a three-dimensional structure.

In this chapter, bearing in mind that circulation has a complicated structure in the real atmosphere, we consider the general properties of the two-dimensional cellular structures of steady large-scale circulations as representative circulations in low latitudes. We begin with large-scale circulation in a nonrotating frame, which is thought to correspond to the Walker circulation. Next, in order to investigate the dynamics of the Hadley circulation, we move on to the large-scale circulation in a uniformly rotating frame (f-plane), and, finally, the zonally symmetric large-scale circulation of a sphere.

The dynamics of the Hadley circulation are directly related to understanding the zonally averaged meridional structure of the atmosphere. The Hadley circulation prevails from the tropics to the mid-latitudes, and has a zonally uniform meridional

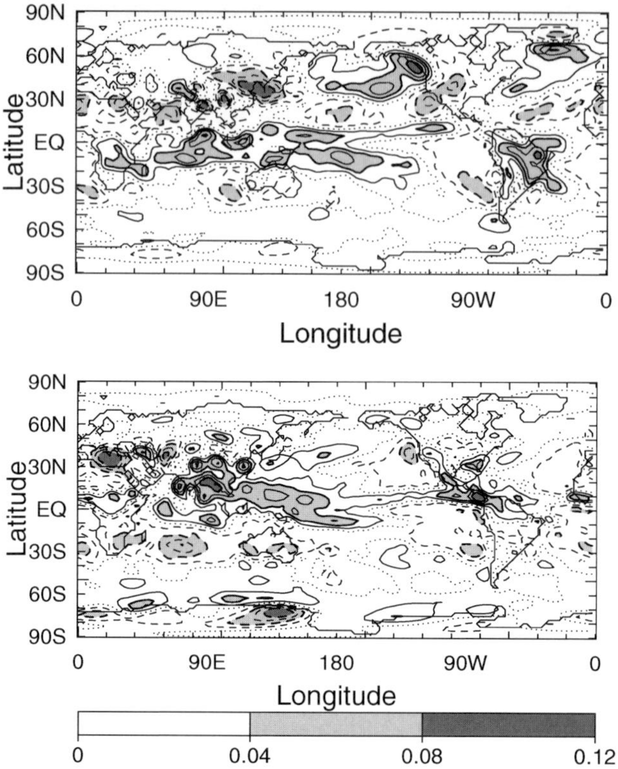

FIGURE 16.1: The horizontal distribution of the vertical velocity ω at height 500 hPa. Monthly mean of January (top) and July (bottom). The contour interval is 0.02 Pa s^{-1}, and the contour interval with thick curves is 0.04 Pa s^{-1}. Solid: negative values (upward motion), dashed: positive values (downward motion), and dotted: zero.

cellular structure with the upward branch in the tropics and the downward branch in the subtropics. The Hadley circulation is not only a representative structure of the meridional field in lower latitudes but also plays a key role in the interaction with mid-latitude circulation. Meridional circulation primarily consists of the Hadley cell and the Ferrel cell, which will be considered in Chapter 17.

16.1 Dynamics of Walker circulation

16.1.1 Large-scale circulation as an organized moist convection

Fig. 16.2 shows the longitudinal-vertical section of the vertical velocity along the equator, which is analyzed from observed data and corresponds to Fig. 16.1. In the equatorial Pacific, upward motions are located around the longitudes 90–120E, while downward motions are located around the longitudes 90–120W. The upward motion region resides in the warmer sea surface temperature region. Such an east-west circulation on the equator is referred to as the Walker circulation. An aspect of the dynamics of the Walker circulation is described by the thermally forced problem

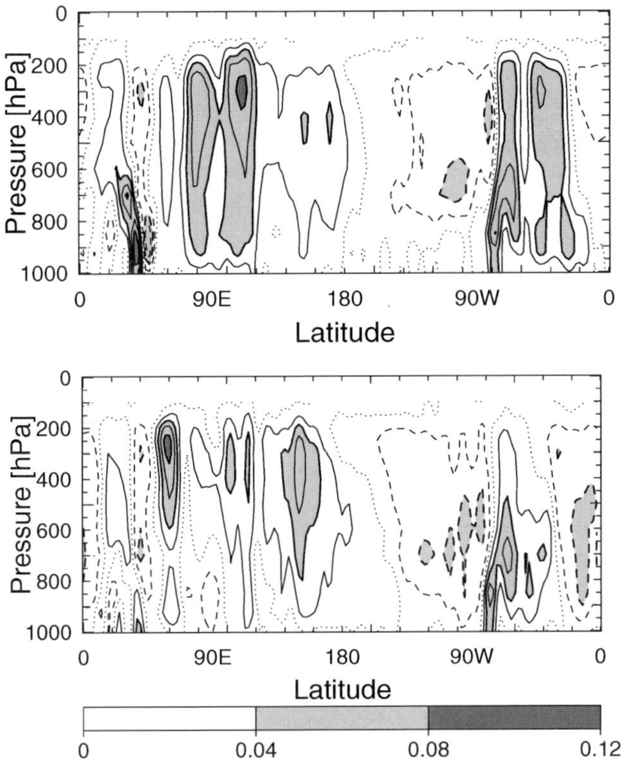

FIGURE 16.2: The same as Fig. 16.1 but for the longitudinal-vertical cross section of the vertical velocity ω along the equator.

on the equator given in Section 6.5. Since diabatic heating due to latent heat release is associated with the upward motion regions, the Kelvin wave response prevails along the equator eastward of the heating region. Although the equatorial Kelvin wave response in the zonal-height section is regarded as the Walker circulation, the argument of Section 6.5 is based on linear theory and simplified damping effects are introduced. In this section, by introducing a more realistic hydrological process, the dynamics of the Walker circulation are considered by viewing it as a circulation driven by a horizontal temperature gradient.

If a vertically one-dimensional radiative equilibrium state is calculated at each latitude using the local condition of solar radiation and distributions of absorbing quantities including water vapor, the temperature structure of the radiative equilibrium is generally statically unstable at any latitude. This implies that the radiative effect always has a tendency to generate convectively unstable states in the atmosphere. In reality, however, local convection does not necessarily occur in all places, since active and suppressed convective regions are distributed due to lateral advection associated with large-scale circulations. Here, local convection in the case of a moist atmosphere is in the form of cumulus convection whose lateral scale is of the order a few kilometers. In general, a large-scale circulation appears as a result of horizontal filtering with an appropriate length scale which covers both active and suppressed convective regions. Although the averaging scale is not always clearly defined, it is much larger than individual cumulus scales. More active cumulus convection corresponds to the region of the upward branch of large-scale circulation, whereas the suppressed convective region corresponds to the region of the downward branch.

If external forcing, such as surface temperature, has inhomogeneity, a large-scale circulation will be organized in the horizontal scale of the applied forcing. If the external forcing is horizontally uniform, on the other hand, the typical horizontal scale cannot be defined in advance. Nevertheless, it may be expected that interactions of individual cumulus clouds will drive the organized circulation of cumulus convection to form a large-scale circulation.

As the simplest framework, let us first consider moist convection on a uniform surface temperature in a nonrotating frame. Such moist convection can be compared with the Bénard convection in a Boussinesq fluid or with the dry convection described in Section 15.1. In reality, cumulus convection predominates in the low-latitudes where the surface temperature is relatively uniform and the effect of rotation is small. In these regions, however, cumulus convection is very variable and does not have a cellular structure like the Bénard convection. It is difficult to make a theory on the moist convection on a uniform surface temperature in a nonrotating frame. Fig. 16.3 shows a numerical example of moist convection in such a situation. Because of interaction in the precipitating process, each convective plume has a finite lifetime and is very variable.

We may think of moist convection on a uniform surface temperature in a rotating frame in a similar way. Although we can undertake theoretical analysis in the case of the Boussinesq fluid as described in Section 5.2.2, moist convection in the real atmosphere does not have such an organized structure. Instead, organization of the cloud clusters associated with easterly waves or tropical cyclones might be considered as large-scale structures of moist convection in a rotating frame.

Second, let us consider moist convection on a differentially distributed surface temperature. We may say that Walker circulation and Hadley circulation are large-

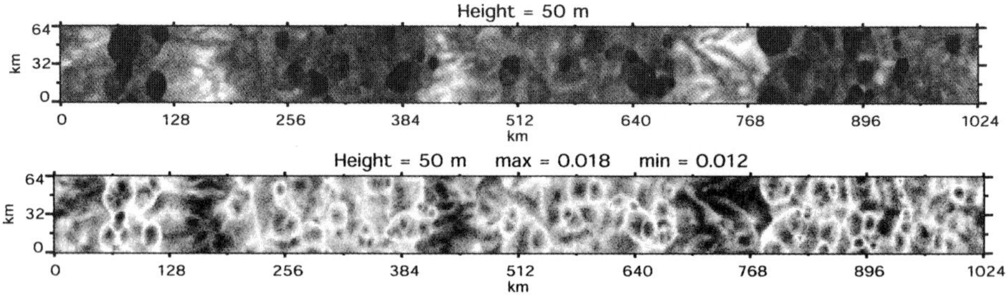

FIGURE 16.3: Distributions of potential temperature deviation and specific humidity for a three-dimensional numerical simulation of the radiative-convective equilibrium. For height at 50 m. Gray scales range from -1.0 K to 1.0 K for potential temperature and from 12 to 18 g kg^{-1} for specific humidity. After Tompkins (2001) by permission of the American Meteorological Society.

scale circulations forced by the variation in surface temperature on a scale of a few thousand kilometers. For instance, a horizontally forced circulation in a nonrotating frame has a structure similar to the Walker circulation, which is thought to be forced by a longitudinal surface temperature variation near the equator. In contrast, the latitudinal difference in surface temperature can force a meridional cellular circulation, which corresponds to the Hadley circulation. The Hadley circulation is affected by the rotation of the earth, and is viewed as a horizontally forced circulation in the rotating frame, as will be studied in Section 16.2.

16.1.2 Large-scale circulation in a nonrotating frame

If the boundary conditions are horizontally uniform, there is no reason for the upward motion of moist convection to be located at a specific position. If the surface temperature has large-scale variation, on the other hand, the upward motion region tends to concentrate in the warmer surface temperature region and moist convection is suppressed in the colder region. In this case, large-scale circulation will be driven so as to have a horizontal scale comparable to that of the specified surface temperature variation. Such a large-scale circulation in a nonrotating frame can be viewed as a model of Walker circulation.

The organization of large-scale circulation is demonstrated using a horizontal-height two-dimensional numerical model. We consider a domain with horizontal length L and assume that the surface temperature difference in L is given by ΔT_s. We take 10,000 km as a typical length of L in the case of Walker circulation. We show below numerical examples of large-scale circulations in the domain $L = 10{,}000$ km with $\Delta T_s = 0$, 0.5, and 1 K. The numerical model is a primitive equation model in hydrostatic balance. A uniform cooling rate -2 K day^{-1} is given below the level $\sigma = 0.1$, where σ denotes a vertical coordinate defined as pressure divided by surface pressure (the σ-coordinates; see Section 3.3.2). For comparison, experiments are carried out both for a moist atmosphere and a dry atmosphere. We use a relatively coarse resolution with grid interval 100 km, so that individual clouds cannot be resolved in the model. Thus, two methods of convective scheme are used: First, the effect of local moist convection is parameterized by moist adiabatic adjustment at each grid point. Second, no parameterization except for an explicit large-scale circulation is used. Fig. 16.4 shows the time sequences of the distribution of precipitation for a moist atmosphere with moist convective adjustment. The precipitating area corresponds to the upward motion region. It is shown that precipitation is more concentrated on the warmer side as ΔT_s becomes larger.

Although the above model is very simple and idealized, we can argue some of the properties of tropical circulations. Despite the concentration of precipitation, the temperature in the free atmosphere is horizontally homogenized except for the lower boundary layer. The difference of temperature in the free atmosphere is rapidly smoothed out due to the propagation of gravity waves. As a result, a stable layer is established near the surface as ΔT_s becomes larger since the surface temperature is colder than the free atmosphere above. At the top of the boundary layer, water vapor is saturated and stratiform clouds prevail in the uppermost layer of the stable layer. This saturated layer is thought to correspond to the trade cumulus.

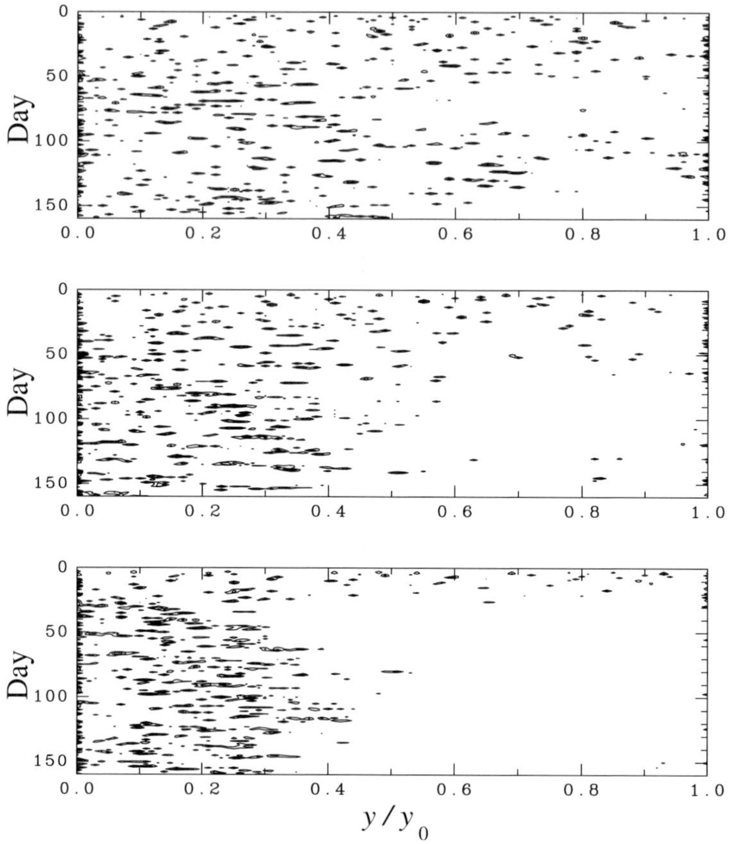

FIGURE 16.4: The time sequences of the distribution of precipitation for the horizontal surface temperature difference $\Delta T_s = 0$, 0.5, and 1 K in the two-dimensional model, showing the concentration of precipitation in the warmer region of the surface temperature. The surface temperature is warmer in the left-hand side of each panel.

In the two-dimensional time mean field, the following streamfunction can be defined:

$$\Psi = \int_z^\infty \rho v \, dz = \int_0^p v \frac{dp}{g} = \frac{p_s}{g} \int_\sigma^1 v \, d\sigma, \tag{16.1.1}$$

where hydrostatic balance is used. The dimension of the streamfunction is in kg s^{-1} m^{-1}. Except for the lower boundary layer, the vertical temperature structure is almost the same as that of the upward branch of large-scale circulation, and its lapse rate is described by the moist adiabatic lapse rate Γ_m. Using the difference between the lapse rates of the dry adiabat and the moist adiabat $\Gamma_d - \Gamma_m \approx$ 2 K km^{-1}, and the radiative cooling rate $Q^{rad}/C_p\rho = 2$ K day^{-1} in (15.3.19),

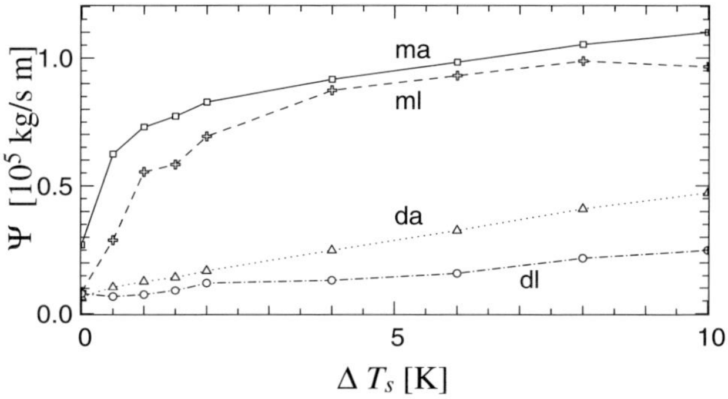

FIGURE 16.5: The dependency of the maximum values of the streamfunction [kg s^{-1} m^{-1}] (ordinate) on the surface temperature difference ΔT_s [K] (abscissa). A comparison between a moist atmosphere and a dry atmosphere both with the convective adjustment and without any convective scheme. ma: moist atmosphere with convective adjustment, ml: moist atmosphere without any parameterization, da: dry atmosphere with convective adjustment, and dl: dry atmosphere without any parameterization.

downward mass flux in the clear region on the colder side of temperature is given by

$$M = \frac{Q^{rad}}{C_p(\Gamma_d - \Gamma_m)}. \tag{16.1.2}$$

The strength of the streamfunction is maximized if the upward motion of large-scale circulation is concentrated in a small area at the warmest surface temperature:

$$\Psi_{\max} = ML = \frac{QL}{C_p(\Gamma_d - \Gamma_m)} \approx 1 \times 10^5 \text{ kg s}^{-1} \text{ m}^{-1}. \tag{16.1.3}$$

Fig. 16.5 shows the dependency of the maximum values of the streamfunction on horizontal surface temperature difference. As the temperature difference increases, the streamfunction consists of a single cell covering the whole domain. In the case of a moist atmosphere, the maximum value of the streamfunction is very close to this uppermost value at a temperature difference of 2 K. In the case of a dry atmosphere, however, the maximum value of the streamfunction is much smaller than that of the moist case, though it becomes stronger as the temperature difference becomes larger. Note that even when the temperature difference is close to zero, convective motions result in smaller scale cells; this is why the streamfunction has a finite value at $\Delta T_s = 0$.

16.2 Dynamics of Hadley circulation

The circulations of the atmosphere over low and mid-latitudes are characterized by a relatively zonally symmetric meridional circulation called the *Hadley circulation.* In practice, the Hadley circulation is defined as the zonal average of meridional flows. The Hadley circulation can be viewed as a large-scale meridional circulation driven by the underlying meridional surface temperature difference. Because of the rotation of the earth, angular momentum transport is associated with meridional circulation. The seasonal change in the distribution of the zonal-mean mass streamfunction is shown in Fig. A3.2.

In this section, we consider the dynamics of Hadley circulation by regarding Hadley circulation as a large-scale convective cell in a rotating frame. As a slightly different model of Hadley circulation, we also examine the dynamical balances of Hadley circulation as a radiatively driven circulation.

16.2.1 Cyclostrophic balance

In a rotating frame, lateral temperature difference does not necessarily drive meridional circulation as shown in Fig. 16.5, since there may be a state in the thermal wind balance with the prescribed surface temperature gradient. If the surface temperature has a meridional gradient, we can assume there exists a basic state that has a thermal wind balanced by the surface temperature gradient and has no meridional flow. The thermal structure in this case is given by the local radiative-convective equilibrium at each latitude. If the latitudinal distribution of solar radiation is specified using a latitudinal distribution function $f(\varphi)$ as in (13.2.3), for instance, a local radiative-convective equilibrium state can be determined by the vertically one-dimensional model at each latitude. The thermal structure of this atmospheric state is completely determined by the latitudinal distribution of the surface temperature and the vertical profiles of temperature. This thermal structure without a meridional flow can be thought of as the basic state for Hadley circulation. In this case, the energy balance at the surface can be arbitrarily constrained. One may assume a *swamp boundary condition*, under which no heat capacity is allowed at the ground. Instead of this, one may externally specify a distribution of the surface temperature. Although the energy budget in the atmosphere will not be closed if the surface temperature is prescribed, residual energy at the surface is regarded as energy transport in the ocean.

Here, we simply specify the distribution of the surface temperature to find the distribution of zonal winds that are in thermal wind balance with surface temperature. Since rotation becomes zero near the equator, we need to consider the variation of the Coriolis parameter (i.e., we assume cyclostrophic balance instead of thermal wind balance in a uniform rotation). As a basic state, both profiles of the troposphere and stratosphere can be determined by the radiative-convective equilibrium at each latitude. In the case of a moist atmosphere, temperature profiles in the radiative-convective equilibrium are approximately given by the moist adiabat, starting with surface temperature. Since the moist adiabat depends on temperature and pressure, the lapse rate changes in the vertical and horizontal directions. For simplicity, however, we neglect the variation of the lapse rate and assume that it is given by a constant Γ. Surface temperature is denoted by T_s, and the temperature gap between the surface and the atmosphere is neglected. We further assume that tropopause height H_T is externally specified as a constant value irrespective of latitude. The exact distribution of tropopause height could be determined if a realistic radiation model is used to calculate the radiative-convective equilibrium; but we do not take this approach.

First, we consider the case that surface temperature distribution is given by

$$T_s(\varphi) \quad = \quad T_E - \Delta T_s \sin^2 \varphi. \tag{16.2.1}$$

Using the Boussinesq approximation, the meridional structure of potential temperature is given by

$$\theta(\varphi, z) \quad = \quad T_s(\varphi) + (\Gamma - \Gamma_d)z = T_E - \Delta T_s \sin^2 \varphi + \gamma z, \tag{16.2.2}$$

where $\Gamma_d = g/C_p$, and $\gamma = \Gamma - \Gamma_d$ is the vertical lapse rate for potential temperature. Cyclostrophic balance with this thermal structure is written in spherical coordinates as†

$$\frac{\partial}{\partial z}\left(2\Omega \sin \varphi u + \frac{\tan \varphi}{R} u^2\right) \quad = \quad -\frac{g}{\theta_0}\frac{1}{R}\frac{\partial \theta}{\partial \varphi}. \tag{16.2.3}$$

We can solve it for zonal velocity u by substituting (16.2.2) into this equation and using the boundary condition $u = 0$ at $z = 0$:

$$u \quad = \quad \Omega R \cos \varphi \left(\sqrt{1 + \frac{2g\Delta T_s}{\Omega^2 a^2 \theta_0} z} - 1\right). \tag{16.2.4}$$

This atmospheric structure in the cyclostrophic balance cannot be realized in fact. Angular momentum corresponding to (16.2.4) is given by

$$l \quad = \quad uR\cos\varphi + \Omega R^2 \cos^2 \varphi = \Omega R^2 \cos^2 \varphi \sqrt{1 + \frac{2g\Delta T_s}{\Omega^2 a^2 \theta_0} z}. \tag{16.2.5}$$

From this, l takes the maximum value at the uppermost layer of the equator $\varphi = 0$. This value is larger than the values at the surface; l takes the maximum value ΩR^2 at the equator and the minimum value zero at the poles (at $z = 0$). If a small diffusion of the angular momentum is introduced to this cyclostrophic balance state, all the values of angular momentum in the atmosphere must lie within the range between the maximum and minimum values at the surface. This means that angular momentum distribution (16.2.5) is physically impossible (i.e., a state with cyclostrophic balance cannot be realized if diffusive transport is introduced and, therefore, meridional circulation must occur). This statement is referred to as *Hide's theorem* (Hide, 1969).

16.2.2 Large-scale circulation in a rotating system

From the above consideration, we conclude that meridional flow must occur in the low latitudes of the atmosphere (i.e., Hadley circulation inevitably exists). The Hadley circulation can be viewed as a large-scale circulation driven by a surface temperature gradient in a rotating system. Thus, we first consider a simple case in a uniformly rotating frame in order to clarify the dynamics of Hadley circulation.

†The cyclostrophic balance is introduced as (2.4.24). Eq. (16.2.3) is the corresponding thermal wind balance in a rotating frame.

Let us consider a two-dimensional meridional circulation on the f-plane in a limited domain. We assume that surface temperature has a difference ΔT_s in the domain length L and that moist convection occurs mainly in the small concentrated region in the warmer side and is suppressed in the rest of the domain. In this case, it is expected that large-scale circulation occurs with a horizontal length L, which is the same length scale as the surface temperature gradient, if no rotation is applied. The moist convective region corresponds to the upward motion region of large-scale circulation, while the suppressed region corresponds to the downward motion region. Large-scale circulation has a two-dimensional steady flow in a statistical sense and is characterized as an *overturning circulation* in a vertical-horizontal section.

Since the overturning circulation is affected by rotation, zonal flow, which has a component normal to the meridional section, has a distribution affected by the Coriolis force. This can be described by angular momentum conservation. If friction is negligible in the free atmosphere, angular momentum is conserved along the lateral flow in the upper layer. We define the origin of the y-axis at the most concentrated region of the upward motion, and take the y-axis in the direction of the flow in the upper layer. The velocity component in this direction is denoted by v. The x-axis is perpendicular to circulation, and the velocity component in the x-direction is denoted by u. Angular momentum conservation reads

$$u - fy = 0. \tag{16.2.6}$$

In the lower layer, angular momentum is not conserved due to the frictional force in the boundary layer. We assume that u is zero at $z = 0$ as a first approximation. In the lateral flow region in the upper layer, temperature is rapidly homogenized. In contrast to a nonrotating system in which temperature is horizontally homogenized, the thermal wind balance gives a constraint on temperature in the case of a rotating system; the horizontal temperature gradient is related to the vertical shear of velocity:

$$f\frac{\partial u}{\partial z} = -\frac{g}{\theta_0}\frac{\partial \theta}{\partial y}, \tag{16.2.7}$$

where θ is potential temperature and θ_0 is a typical value of potential temperature. The thermal wind balance at the middle level is approximately written as

$$f\frac{u}{H} = -\frac{g}{\theta_0}\frac{\partial \hat{\theta}}{\partial y}, \tag{16.2.8}$$

where H is the vertical depth of circulation and $\hat{\theta}$ is the vertical average of potential temperature. Using (16.2.6) and (16.2.8), we can estimate the difference of potential temperature

$$\Delta\hat{\theta} = \frac{f^2\theta_0}{2gH}y^2, \tag{16.2.9}$$

which is the difference from the vertical average of potential temperature in the upward motion region $\hat{\theta}(0)$. If the distance from the origin of the y-axis is large enough such that

$$\Delta\hat{\theta} \quad > \quad \frac{y}{L}\Delta T_s, \tag{16.2.10}$$

then the vertical average of potential temperature becomes lower than surface temperature and it becomes convectively unstable. This implies that the horizontal extent of large-scale circulation cannot reach such a region. Therefore, the horizontal length of large-scale circulation is given by

$$y_{max} \quad = \quad \frac{2gH}{f^2\theta_0}\frac{\Delta T_s}{L}, \tag{16.2.11}$$

which is given from (16.2.9) and (16.2.10). This is a typical length scale of large-scale convection in the f-plane. Fig. 16.6 shows a schematic diagram of the relation between the surface temperature and the potential temperature in the middle layer of the atmosphere. This relation determines the horizontal length of large-scale circulation in the f-plane. We have $y_{max} \approx 130$ km for the parameters of mid-latitude values: $H = 10$ km, $f = \Omega$ (Ω is the angular velocity of the earth), $\theta_0 = 300$ K, and $\Delta T_s/L = 10$ K/10,000 km. It does not seem, however, that such a two-dimensional steady circulation with this horizontal length exists in the real atmosphere. Although squall lines or spiral bands of cyclones have a two-dimensional structure, their structures are more complicated than the steady circulation considered here. The above estimation should be regarded as no more than an introduction to the Hadley circulation, which now follows.

As the above overturning flow gets closer to the equator and the Coriolis parameter f becomes smaller, horizontal length increases. Horizontal length is infinite at

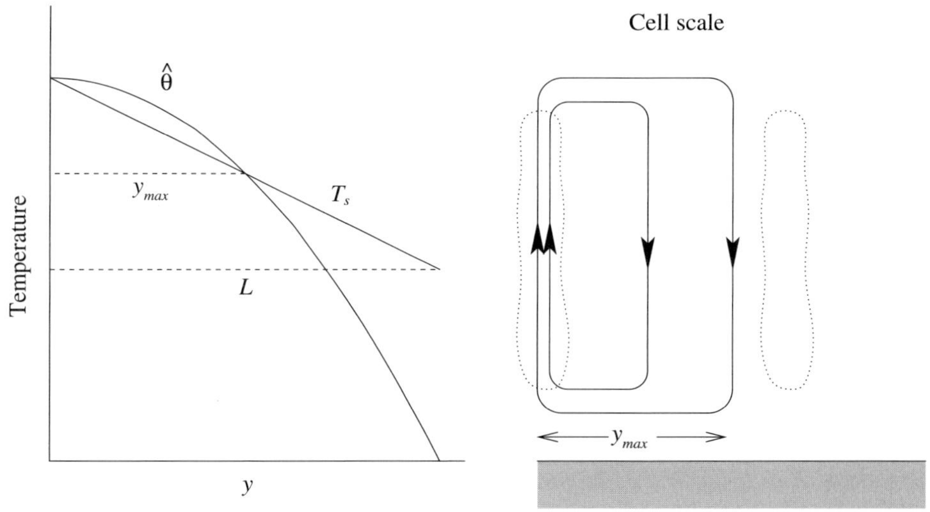

FIGURE 16.6: The horizontal scale of large-scale circulation in the f-plane, and the relation between surface temperature T_s and potential temperature in the middle of the atmosphere $\hat{\theta}$. L is the horizontal scale of the variation of surface temperature and y_{max} is the maximum lateral length of the overturning circulation.

the equator. This means that we must take into account the latitudinal variation of f to obtain a finite horizontal scale. The horizontal scale of the Hadley circulation is given in this situation. We also need to take into account spherical geometry using the cyclostrophic balancc instead of the thermal wind balance.

Let us consider the Hadley circulation symmetric about the equator on the same latitudinal distribution of surface temperature as (16.2.1). In this case, since surface temperature has a maximum at the equator, we assume that the upward motion of Hadley circulation is concentrated at the equator. Angular momentum conservation along the upper layer flow is written as

$$uR\cos\varphi + \Omega R^2\cos^2\varphi = \Omega R^2, \tag{16.2.12}$$

where R is the radius of the earth. From this, the zonal wind distribution in the upper layer is given by

$$u = u_M \equiv \frac{\Omega R\sin^2\varphi}{\cos\varphi}. \tag{16.2.13}$$

The vertical average of potential temperature is assumed to be in cyclostrophic balance with this zonal wind profile:

$$2\Omega\sin\varphi\frac{\partial u}{\partial z} + \frac{\tan\varphi}{R}\frac{\partial u^2}{\partial z} = -\frac{g}{\theta_0}\frac{1}{R}\frac{\partial\hat{\theta}}{\partial\varphi}. \tag{16.2.14}$$

We can estimate the difference between the zonal velocities at the levels $z = 0$ and H from the above equation. By neglecting the zonal velocity near the surface, we obtain the cyclostrophic relation with respect to the zonal velocity at H as

$$\frac{1}{H}\left(2\Omega\sin\varphi u + \frac{\tan\varphi}{R}u^2\right) = -\frac{g}{\theta_0}\frac{1}{R}\frac{\partial\hat{\theta}}{\partial\varphi}. \tag{16.2.15}$$

Substituting (16.2.13) into this, and integrating with respect to latitude, we obtain the difference between the vertically averaged potential temperature $\hat{\theta}$ and the equatorial value as

$$\frac{\Delta\hat{\theta}}{\theta_0} = \frac{\Omega^2R^2}{gH}\frac{\sin^4\varphi}{2\cos^2\varphi}. \tag{16.2.16}$$

Large-scale circulation extends as long as the vertically averaged potential temperature is larger than surface temperature. The latitude where the two temperatures agree is the horizontal extent of Hadley circulation (Fig. 16.7). Thus, the latitudinal width of Hadley circulation φ_H is given by

$$\tan\varphi_H = \left(\frac{2gH}{\Omega^2R^2}\frac{\Delta T_s}{\theta_0}\right)^{\frac{1}{2}}. \tag{16.2.17}$$

We have $\varphi_H \approx 17°$ in the case $\Delta T_s/\theta_0 = 0.1$. This width is relatively smaller than the width of Hadley circulation in the real atmosphere $\sim 30°$, but is a first approximation to reality. The difference is partly derived from the strong constraint of angular momentum conservation (16.2.12).

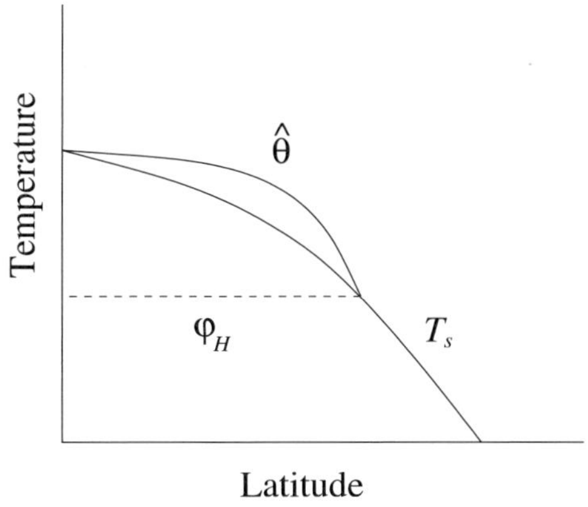

FIGURE 16.7: The latitudinal extent of Hadley circulation as a relation between surface temperature T_s and potential temperature in the middle level of the atmosphere $\hat{\theta}$. The latitudinal width of Hadley circulation is denoted by φ_H.

The strength of Hadley circulation can be estimated by assuming that upward motion is concentrated in a small region near the equator and downward motion is uniform in the rest of Hadley circulation region. The meridional mass streamfunction of Hadley circulation is defined by

$$\Psi = 2\pi R\cos\varphi \int_z^\infty \overline{\rho v}\, dz = 2\pi R\cos\varphi \int_0^p \overline{v}\, \frac{dp}{g}, \tag{16.2.18}$$

where $(\bar{\ })$ denotes the zonal average; in this case it is not important since we are considering an axisymmetric circulation. In the Hadley circulation region, meridional flow is poleward $v > 0$ in the upper half-layer of the troposphere and is equatorward $v < 0$ in the lower half-layer (the sign is for the northern hemisphere). The altitude of the boundary of the two layers is denoted by z_H, which is assumed to be independent of latitude. At each latitude, the streamfunction takes a maximum value $\Psi_M(\varphi)$ at height z_H. Using mass continuity in a steady state

$$\frac{1}{R\cos\varphi}\frac{\partial}{\partial\varphi}(\cos\varphi\overline{\rho v}) + \frac{\partial}{\partial z}\overline{\rho w} = 0, \tag{16.2.19}$$

we have an expression

$$\Psi_M(\varphi) = 2\pi R\cos\varphi \int_{z_H}^\infty \overline{\rho v}\, dz = 2\pi R^2 \int_\varphi^{\varphi_H} \overline{\rho w}\cos\varphi\, d\varphi. \tag{16.2.20}$$

Assuming that the downward mass flux of Hadley circulation $M = \overline{\rho w}$ is given as a constant value by (16.1.3), and the outermost contour of the mass streamfunction is closed at the latitude φ_H, we have

$$\Psi_M(\varphi) = 2\pi R^2 M(\sin\varphi_H - \sin\varphi). \tag{16.2.21}$$

From this, we see that the mass streamfunction takes the largest value at the equator, $2\pi R^2 M\sin\varphi_H$. We may have the upward mass flux of Hadley circulation by dividing total mass flux by an area of the upward motion region. However, we have assumed that the upward motion region is concentrated in a small equatorial band. In reality, the upward motion region is broader by as much as 10 degrees in latitude. The upward motion region of Hadley circulation corresponds to the *Intertropical Convergence Zone* (*ITCZ*). It is a fundamental and unresolved problem of large-scale circulation to theoretically determine the width of the ITCZ.

Fig. 16.8 compares the meridional streamfunctions obtained by a set of idealistic

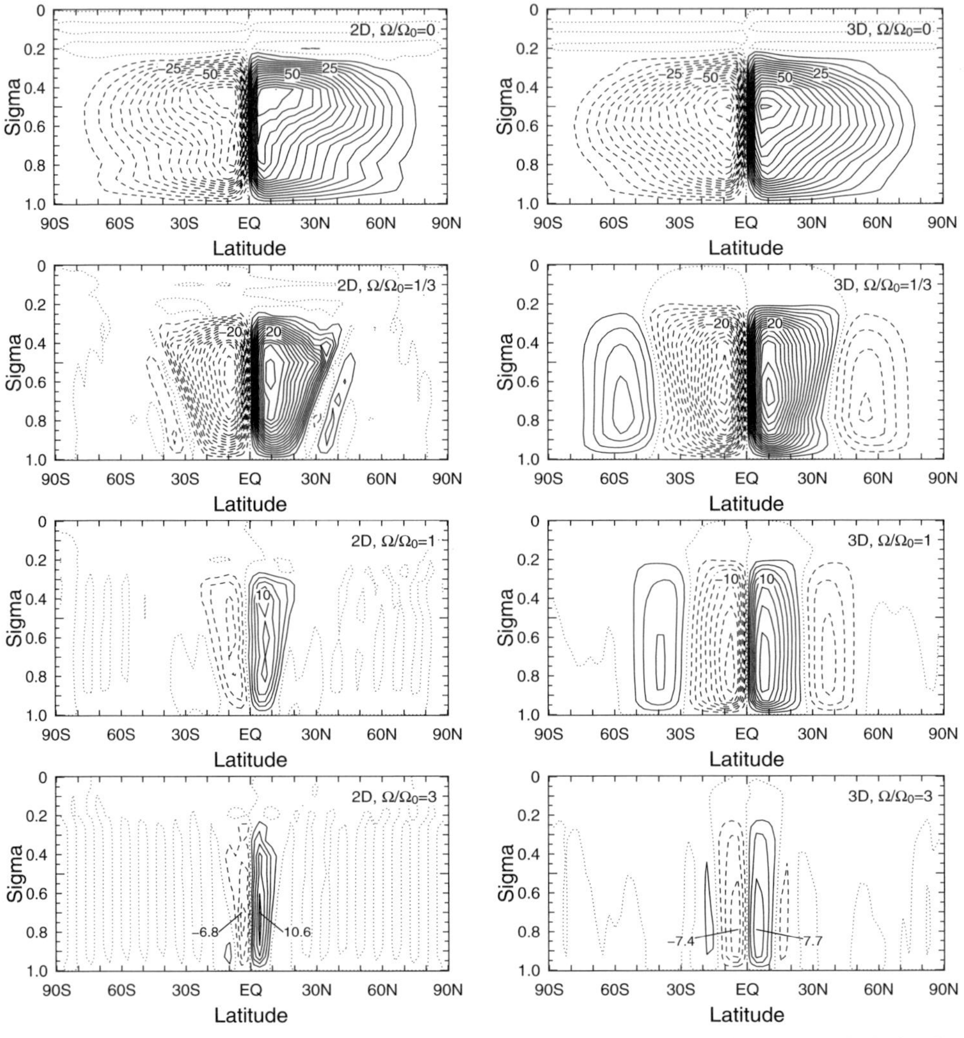

FIGURE 16.8: Comparison of the streamfunctions given by a general circulation model. Left: the two-dimensional axisymmetric model; right: the three-dimensional model. From the top to the bottom, the figures correspond to the case with the rotation rate $\Omega/\Omega_0 = 0$, $1/3$, 1, and 3, respectively. The unit is 10^{10} kg m^{-1} and the contour interval is 5×10^{10} kg m^{-1} for $\Omega/\Omega_0 = 0$, and 2×10^{10} kg m^{-1} otherwise. After Satoh et al. (1995).

numerical experiments. These show the dependency on rotation rates for the two-dimensional axisymmetric model and for the three-dimensional model (Satoh et al., 1995). Both cases are calculated using a general circulation model that includes a hydrological cycle over an ocean surface with a latitudinal sea surface temperature distribution (16.2.1). (These are the so-called *aqua-planet* experiment; see Section 24.3.) Rotation rates are changed as $\Omega/\Omega_0 = 0$, 1/3, 1, and 3, where Ω_0 is the terrestrial rotation rate. In the case of no rotation ($\Omega/\Omega_0 = 0$), streamfunctions are almost the same between the two-dimensional model and the three-dimensional model and they cover the entire hemisphere. In the case of slow rotation with $\Omega/\Omega_0 = 1/3$, the shape of the streamfunctions of Hadley circulation is very similar between the two models, and they reach about latitude 45° from the equator. As the rotation rate is faster than the terrestrial case with $\Omega/\Omega_0 = 1$, the width of the Hadley circulation of the two-dimensional model becomes smaller than that of the three-dimensional model. This result implies that the above theory is more applicable as the rotation rate is slower than the terrestrial case.

16.2.3 The Held and Hou model

In the previous section, we obtained the width and intensity of Hadley circulation by assuming that Hadley circulation is an axisymmetric large-scale circulation in a moist atmosphere. Held and Hou (1980), in contrast, consider Hadley circulation as a circulation driven by the diabatic forcing due to Newtonian radiation. Although the model of Held and Hou is formulated without explicitly considering the effect of latent heat release, it is useful for understanding the basic balance of Hadley circulation.

The diabatic heating of Newtonian radiation is given proportional to the difference between temperature and its reference value. In general, the profile of the reference temperature is arbitrarily specified. Using potential temperature, the reference profile of the Held and Hou model is given by

$$\begin{aligned}\theta_e(\varphi, z) &= \theta_0 \left[1 - \frac{2}{3}\Delta_H P_2(\varphi) + \Delta_V \left(\frac{z}{H} - \frac{1}{2}\right)\right] \\ &= \theta_0 \left[1 - \Delta_H \left(\sin^2\varphi - \frac{1}{3}\right) + \Delta_V \left(\frac{z}{H} - \frac{1}{2}\right)\right], \end{aligned} \tag{16.2.22}$$

where Δ_H and Δ_V represent the ratios of the deviations of potential temperature in latitudinal and vertical directions from the basic value θ_0, respectively. The latitudinal profile is based on that of solar heating (13.2.6). In the vertical direction, we set $\Delta_V > 0$, so that Newtonian radiation tends to stabilize stratification. At each latitude, this profile is regarded as a radiative-convective equilibrium state without a large-scale circulation (i.e., a balanced state between convection and radiation at each local latitude). The interpretation of this reference state is not straightforward, however, since Hadley circulation suppresses local convection in the downward motion region. A local radiative-convective equilibrium has no counterpart in the real atmosphere. We should proceed to the following consideration with this reservation. Note that *radiative equilibrium* generally has unstable stratification in the lower layers as seen in Chapter 14. Thus, the reference state with $\Delta_V > 0$

cannot be regarded as a radiative equilibrium state instead of a radiative-convective one.

Using Newtonian-type diabatic heating, the equation of potential temperature in a zonally symmetric two-dimensional system is expressed as

$$\frac{\partial \theta}{\partial t} + \frac{1}{R\cos\varphi}\frac{\partial}{\partial \varphi}(v\theta\cos\varphi) + \frac{\partial}{\partial z}(w\theta) = -\frac{\theta - \theta_e}{\tau_R}, \tag{16.2.23}$$

where τ_R is the damping time of radiation. We integrate this equation over the whole domain of Hadley circulation between $z = 0$ and $z = H$ in the vertical direction and $\varphi = 0$ and $\varphi = \varphi_H$ in the latitudinal direction. Integration on the left-hand side becomes zero at a steady state, since no normal flux is allowed at the boundary of Hadley circulation. Therefore, the domain-averaged heat balance is expressed as

$$\int_0^{\varphi_H} \hat{\theta}\cos\varphi\, d\varphi = \int_0^{\varphi_H} \hat{\theta}_e \cos\varphi\, d\varphi, \tag{16.2.24}$$

where $\hat{\theta}$ is the vertical average of potential temperature, and $\hat{\theta}_e$ is that of the reference value. If the reference profile of potential temperature is given by (16.2.22), we obtain

$$\hat{\theta}_e = \frac{1}{H}\int_0^H \theta_e\, dz = \theta_0\left[1 - \Delta_H\left(\sin^2\varphi - \frac{1}{3}\right)\right]. \tag{16.2.25}$$

We assume that potential temperature is equal to the reference value at the boundary of Hadley circulation: $\hat{\theta}(\varphi_H) = \hat{\theta}_e(\varphi_H)$. The zonal wind in the upper layer has the same profile as (16.2.13) which is derived from conservation of angular momentum. Temperature distribution is constrained by cyclostrophic balance (16.2.14). Thus, the latitudinal difference of the vertically averaged potential temperature is the same as (16.2.16):

$$\frac{\hat{\theta}(0) - \hat{\theta}(\varphi)}{\theta_0} = \frac{\Omega^2 R^2}{gH}\frac{\sin^4\varphi}{2\cos^2\varphi}. \tag{16.2.26}$$

Eqs. (16.2.24), (16.2.25), and (16.2.26) determine the latitudinal distribution of potential temperature, which is depicted in Fig. 16.9. The width of Hadley circulation is determined by the constraint that the two areas enclosed by the curves $\hat{\theta}$ and $\hat{\theta}_e$ must be equal due to the energy balance in the Newtonian radiation model. This method to ascertain the width of Hadley circulation is called the *equal area method*. A similar method for the balance of energy was considered with the energy budget model (EBM) in Chapter 13 (see Fig. 13.4). In the case of the EBM, the balance between solar and planetary radiations was considered. In the present case, however, such a relation for radiation is not directly introduced. Substituting (16.2.25) and (16.2.26) into (16.2.24) and integrating over the domain of Hadley circulation, we obtain the equation for the width of Hadley circulation. Introducing $\mu_H = \sin\varphi_H$, we have a relation for the Hadley width:

$$\frac{1}{3}(4R_H - 1)\mu_H^3 - \frac{\mu_H^5}{1-\mu_H^2} - \mu_H + \frac{1}{2}\ln\left(\frac{1+\mu_H}{1-\mu_H}\right) = 0, \tag{16.2.27}$$

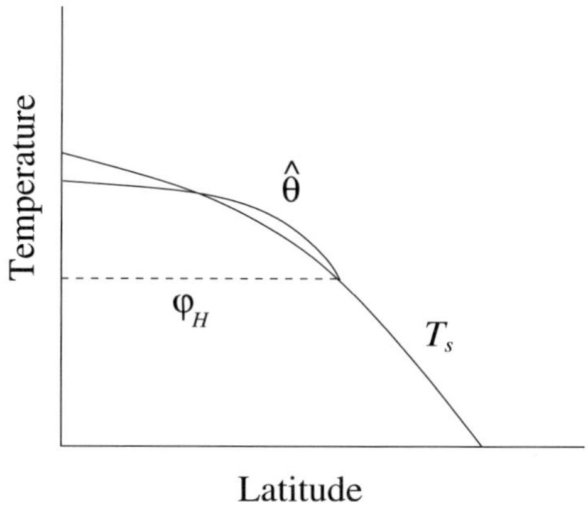

FIGURE 16.9: The latitudinal extent of Hadley circulation in the Newtonian radiation model, and the relation between the reference potential temperature θ_e and the vertically averaged potential temperature $\hat{\theta}$. The areas enclosed by the two curves are equal. The width of Hadley circulation is given by the latitude φ_H.

where

$$R_H \equiv \frac{gH}{\Omega^2 R^2}\Delta_H, \tag{16.2.28}$$

is a parameter related to latitudinal temperature difference.

If the width φ_H is small enough that $\sin\varphi \approx \varphi$ and $\cos\varphi \approx 1$, Eqs. (16.2.25) and (16.2.26) are written as

$$\frac{\hat{\theta}(\varphi)}{\theta_0} = \frac{\hat{\theta}(0)}{\theta_0} - \frac{\Omega^2 R^2}{2gH}\varphi^4, \qquad \frac{\hat{\theta}_e(\varphi)}{\theta_0} = \frac{\hat{\theta}_e(0)}{\theta_0} - \Delta_H \varphi^2. \tag{16.2.29}$$

Substituting these equations into (16.2.24) with $\cos\varphi \approx 1$, the width of Hadley circulation is given by

$$\varphi_H = \left(\frac{5}{3}\frac{gH}{\Omega^2 R^2}\Delta_H\right)^{\frac{1}{2}} = \left(\frac{5}{3}R_H\right)^{\frac{1}{2}}. \tag{16.2.30}$$

This equation corresponds to (16.2.17) in the case of a moist atmosphere.

In this model, latitudinal energy transport exists only in the region between the equator and the latitude φ_H where meridional circulation exists. Energy transport is calculated by the vertical average of the potential temperature equation (16.2.23), which is written as

$$\frac{\partial \hat{\theta}}{\partial t} + \frac{1}{R\cos\varphi}\frac{\partial}{\partial\varphi}\left(\cos\varphi \frac{1}{H}\int_0^H v\theta\, dz\right) = -\frac{\hat{\theta} - \hat{\theta}_e}{\tau_R}. \tag{16.2.31}$$

This corresponds to (13.3.20) of the EBM. Assuming a steady balanced state and integrating with respect to latitude, we obtain latitudinal energy transport as

$$\frac{1}{H}\int_0^H v\theta\, dz = -\frac{R_H}{\cos\varphi}\int_0^\varphi \frac{\hat{\theta} - \hat{\theta}_e}{\tau_R}\cos\varphi\, d\varphi. \tag{16.2.32}$$

In the case of $R_H \ll 1$, in particular, substitution of (16.2.29) yields

$$\frac{1}{\theta_0}\int_0^H v\theta dz = \frac{5}{18}\left(\frac{5}{3}\right)^{\frac{1}{2}} \frac{HR\Delta_H}{\tau_R} R_H^{\frac{3}{2}} \left[\frac{\varphi}{\varphi_H} - 2\left(\frac{\varphi}{\varphi_H}\right)^3 + \left(\frac{\varphi}{\varphi_H}\right)^5\right]. \tag{16.2.33}$$

Latitudinal transport becomes zero at $\varphi = 0$ and φ_H, and takes its maximum value at $\varphi = \varphi_H/\sqrt{5}$. From comparison of (16.2.31) and (13.3.20) of the EBM, $\hat{\theta}_e$ corresponds to solar flux $\tilde{T}_R$, while $\hat{\theta}$ corresponds to planetary flux $\tilde{T}$. In reality, solar flux is almost equal to planetary flux at about 45° (Fig. 13.1). As for the Held and Hou model, however, since no heat transport is allowed on the poleward side of Hadley circulation, the curve of $\hat{\theta}_e$ crosses that of $\hat{\theta}$ within the region of Hadley circulation (Fig. 16.9). In this case, energy transport is maximized within the Hadley circulation; this characteristic is unrealistic. One should be careful if the Held and Hou model is used for the interpretation of energy transport.

In the case of the Held and Hou model, the distribution of mass flux can be determined from constraint of the energy balance. It is assumed that the latitudinal flow is concentrated within the thin bottom and upper boundary layers, and that basic stratification does not change from that of the reference state. In this case, energy transport is written as

$$\frac{1}{\theta_0}\int_0^H v\theta\, dz = V\Delta_V, \tag{16.2.34}$$

where V is latitudinal mass flux, and is given from the continuity equation as

$$V = -\int_0^{H/2} v\, dz = \int_{H/2}^H v\, dz. \tag{16.2.35}$$

Thus, from (16.2.33), we have

$$V(\varphi) = \frac{5}{18}\left(\frac{5}{3}\right)^{\frac{1}{2}} \frac{HR\Delta_H}{\tau_R\Delta_V} R_H^{\frac{3}{2}} \left[\frac{\varphi}{\varphi_H} - 2\left(\frac{\varphi}{\varphi_H}\right)^3 + \left(\frac{\varphi}{\varphi_H}\right)^5\right]. \tag{16.2.36}$$

Angular momentum transport is similarly calculated. The zonally averaged momentum equation is written as

$$\frac{\partial u}{\partial t} + \frac{1}{R\cos^2\varphi}\frac{\partial}{\partial\varphi}(uv\cos^2\varphi) + \frac{\partial}{\partial z}(wu) + fv = -\frac{\partial \tau_x}{\partial z}, \tag{16.2.37}$$

where τ_x is the vertical momentum flux due to friction, and lateral diffusive flux is neglected. The vertical average of the above equation in the steady state is written as

$$\frac{1}{R\cos^2\varphi}\frac{\partial}{\partial\varphi}\left(\frac{1}{H}\int_0^H uv\, dz\cos^2\varphi\right) = \frac{\tau_x(0)}{H}, \tag{16.2.38}$$

where $\tau_x(0)$ is diffusion flux at the surface. Note that the contribution of the Coriolis parameter vanishes because of (16.2.35). Diffusion flux is proportional to the magnitude of the zonal wind near the surface with an opposite sign:

$$\tau_x(0) = -Cu(0). \tag{16.2.39}$$

If the sign of $\tau_x(0)$ is known, the direction of the surface wind can be determined. A characteristic behavior of $\tau_x(0)$ can be seen by assuming that the zonal wind is given by the angular momentum conserving flow (16.2.13) in the upper layer and that the zonal wind near the surface is almost negligible compared with that in the upper layer. Consistent with the previous assumption, we assume that meridional flow is also concentrated in the layers near both the surface and tropopause. In the case of $R_H \ll 1$, relative angular momentum transport is written as

$$\begin{aligned}\frac{1}{H}\int_0^H uv\,dz &= \frac{u_M V}{H}\\ &= \frac{25}{54}\frac{\Omega R^2 \Delta_H}{\tau_R \Delta_V} R_H^2 \frac{\varphi^2}{\varphi_H}\left[\frac{\varphi}{\varphi_H} - 2\left(\frac{\varphi}{\varphi_H}\right)^3 + \left(\frac{\varphi}{\varphi_H}\right)^5\right].\end{aligned}$$

From (16.2.38) and (16.2.39), diffusive flux is given as

$$\begin{aligned}-\tau_x(0) &= Cu(z=0,\varphi)\\ &= -\frac{25}{18}\frac{\Omega R H \Delta_H}{\tau_R \Delta_V} R_H^2\left[\frac{\varphi}{\varphi_H} - \frac{10}{3}\left(\frac{\varphi}{\varphi_H}\right)^3 + \frac{7}{3}\left(\frac{\varphi}{\varphi_H}\right)^5\right],\end{aligned}$$

where $\cos\varphi \approx 1$ is used. This flux changes sign at $\varphi = \sqrt{3/7}\varphi_H$. Thus, flow in the lowest layer is easterly on the equatorial side of Hadley circulation, and westerly on the polar side. This distribution is not similar to that which is characteristic of the real atmosphere; in the real atmosphere, the lower flow is everywhere easterly in the Hadley circulation region.

The Held and Hou model has the following difficulties if it is applied to the real atmosphere. First, the model is based on the assumption that the profile of the upper zonal wind is determined by conservation of angular momentum. In reality, the zonal wind is smaller than that determined by angular momentum conservation. This fact suggests the necessity for consideration of momentum transport due to asymmetric components.

Second, it is unclear what constitutes the counterpart of the reference potential temperature of Newtonian radiation in the real atmosphere. As already described, the reference potential temperature is defined as the radiative-convective equilibrium state in the case when Hadley circulation does not exist. If Hadley circulation exists, the radiative-convective equilibrium state does not have any meaning, and radiation cannot be modeled by Newtonian radiation with such a reference temperature. It may be true that the width of Hadley circulation is determined if the latitudinal profile of the vertically averaged reference potential temperature is related to that of solar radiation. However, it is difficult to assume that radiative-convective equilibrium has a stably stratified vertical profile. For a more realistic application, it is simpler to use the condition that surface temperature is prescribed rather than that the profile of solar radiation is given. In this case, the width of Hadley circulation is determined by the diagram depicted in Fig. 16.7. It is not necessary, however, to assume that the latitudinal profile of surface temperature is $\sin^2 \varphi$ (i.e., similar to that of solar radiation). A more realistic profile of surface temperature can be given to consider that which is characteristic of Hadley circulation.

In the Held and Hou model, the width of Hadley circulation is determined by the constraint that the energy budget is closed within the Hadley circulation region. The equal area method as shown in Fig. 16.9 determines the width; the integral of the difference between potential temperature and its reference temperature should be zero. In the energy budget of the real atmosphere, however, solar radiation is larger than planetary radiation equatorward of 45°, and planetary radiation is larger than solar radiation poleward of 45° (i.e., energy transport extends from the equator to the pole) so that the energy budget is not closed if the global atmosphere is not considered. In the Hadley circulation region, energy inflow is positive, and its inflow is balanced by energy outflow from the Hadley circulation region to mid and high latitudes. Energy transport to the mid-latitudes is needed for constraint on the energy budget.

References and suggested reading

The view that large-scale circulations in low latitudes can be interpreted as the organization of moist convection is argued by Emanuel et al. (1994). There are many publications giving observational facts about Walker circulation and its relation to El Niño; the textbook by Philander (1990) is a good reference for this topic. Although there are a few theories that quantify the Walker circulation, Bretherton and Sobel (2002) among others try to interpret the dynamics of the Walker circulation using simplified models, in which cloud-radiation feedback plays a major role. Recently, the statistical behaviors of cumulus convection over a large domain have been explicitly calculated by Tompkins (2001) and Grabowski and Moncrieff (2002).

The theory for an axisymmetric Hadley circulation was established by Schneider (1977) and Held and Hou (1980), and was extended by Lindzen and Hou (1988) and Hou and Lindzen (1992), and is summarized in chapter 7 of Lindzen (1990). Satoh (1994) interpreted and reconsidered the Held and Hou model as a role in radiative-convective equilibrium.

Bretherton, C. S. and Sobel, A. H., 2002: A simple model of a convectively coupled Walker circulation using the weak temperature gradient approximation. *J. Climate*, **15**, 2907–2920.

Emanuel, K. A., Neelin, J. D., and Bretherton, C. S., 1994: On large-scale circulations in convecting atmospheres. *Q. J. Roy. Meteorol. Soc.*, **120**, 1111–1143.

Grabowski, W. W. and Moncrieff, M. W., 2002: Large-scale organization of tropical convection in two-dimensional explicit numerical simulations: Effects of interactive radiation. *Q. J. Roy. Meteorol. Soc.*, **128**, 2349–2376.

Held, I. M. and Hou, A. Y., 1980: Nonlinear axially symmetric circulations in a nearly inviscid atmosphere. *J. Atmos. Sci.*, **37**, 515–533.

Hide, R., 1969: Dynamics of the atmospheres of the major planets with an appendix on the viscous boundary layer at the rigid boundary surface of an electrically conducting rotating fluid in the presence of a magnetic field. *J. Atmos. Sci.*, **26**, 841–853.

Hou, A. Y. and Lindzen, R. S., 1992: The influence of concentrated heating on the Hadley circulation. *J. Atmos. Sci.*, **49**, 1233–1241.

Lindzen, R. S., 1990: *Dynamics in Atmospheric Physics.* Cambridge University Press, Cambridge, UK, 310 pp.

Lindzen, R. S. and Hou, A. Y., 1988: Hadley circulations for zonally averaged heating centered off the equator. *J. Atmos. Sci.*, **45**, 2416–2427.

Philander, S. G., 1990: *El Niño, La Niña, and the Southern Oscillation.* Academic Press, San Diego, 289 pp.

Satoh, M., 1994: Hadley circulations in radiative-convective equilibrium in an axially symmetric atmosphere. *J. Atmos. Sci.*, **51**, 1947–1968.

Satoh, M., Shiobara, M., and Takahashi, M., 1995: Hadley circulations and their roles in the global angular momentum budget in two- and three-dimensional models. *Tellus*, **47A**, 548–560.

Schneider, E. K., 1977: Axially symmetric steady-state models of the basic state for instability and climate studies. II: Nonlinear calculations. *J. Atmos. Sci.*, **34**, 280–296.

Tompkins, A. M., 2001: Organization of tropical convection in low vertical wind shears: The role of water vapor. *J. Atmos. Sci.*, **58**, 529–545.

17

Circulations on a sphere

In this chapter, some characteristics of circulations on a sphere, particularly the latitudinal distribution of zonal-mean angular momentum, are considered using a one-layer model on a sphere. Mid-latitude circulations on the scale of extratropical cyclones are almost geostrophic and nondivergent and their motions are almost isentropic. As shown in Section 3.4, shallow water equations can be used for description of the motion along isentropic surfaces. Thus, isentropic motions in the mid-latitudes can be studied using shallow water equations or barotropic equations.

Low-latitude circulations can also be examined by using shallow water equations, where they are characterized by divergent motions associated with latent heat release. Roughly speaking, the direction of meridional flow in the upper layer is opposite to that in the lower layer in low latitudes (i.e., the flow is baroclinic). If such divergent or convergent motions exist, each layer is relatable to the shallow water system.

In the first section, we formulate divergent shallow water equations on a sphere in general form. It will be shown that the same method for studying the characteristics of the Hadley circulation described in Section 16.2 is applicable to the shallow water system. Then, Hadley circulation is formulated in a one-layer model and its role in angular momentum transport is examined. Next, the propagation of Rossby waves is considered using the nondivergent barotropic system, based on the assumption that Rossby waves are excited in the region of baroclinic instability. This is an aspect of the interaction between low and mid-latitude circulations through the angular momentum budget. We will not explicitly consider the source of Rossby waves in this chapter; it will be studied in the next chapter. In the last section, the energy spectrum and meridional scale of jet streams are examined in terms of two-dimensional turbulence.

The one-layer model on a sphere is a basis for general circulation modeling. A method for numerical discretization on a sphere will be described in Chapter 21. Characteristic flows of shallow water or nondivergent system on a sphere can be used for testing validation of general circulation models (see Chapter 24).

17.1 Shallow water equations on a sphere

First, we present various forms of shallow water equations on a sphere. Let u denote the longitudinal velocity component, v the latitudinal velocity component, and η the surface height of shallow water. The shallow water equations on a sphere are generally given as

$$\frac{\partial u}{\partial t} + \frac{u}{R\cos\varphi}\frac{\partial u}{\partial \lambda} + \frac{v}{R}\frac{\partial u}{\partial \varphi} - \frac{uv\tan\varphi}{R} - 2\Omega v\sin\varphi = -\frac{g}{R\cos\varphi}\frac{\partial \eta}{\partial \lambda} + F_\lambda, \tag{17.1.1}$$

$$\frac{\partial v}{\partial t} + \frac{u}{R\cos\varphi}\frac{\partial v}{\partial \lambda} + \frac{v}{R}\frac{\partial v}{\partial \varphi} + \frac{u^2\tan\varphi}{R} + 2\Omega u\sin\varphi = -\frac{g}{R}\frac{\partial \eta}{\partial \varphi} + F_\varphi, \tag{17.1.2}$$

$$\frac{\partial \eta}{\partial t} + \frac{1}{R\cos\varphi}\frac{\partial}{\partial \lambda}(u\eta) + \frac{1}{R\cos\varphi}\frac{\partial}{\partial \varphi}(v\eta\cos\varphi) = Q, \tag{17.1.3}$$

where R is the radius of the earth, (F_λ, F_φ) is an external force, and Q is a source term of mass. These terms will be given with appropriate assumptions (see Section 17.2). Here, we assume conservation of mass: $Q = 0$.

The vector-invariant form of shallow water equations is given from (17.1.1) and (17.1.2) using vorticity ζ as

$$\frac{\partial u}{\partial t} = (\zeta + 2\Omega\sin\varphi)v - \frac{1}{R\cos\varphi}\frac{\partial}{\partial \lambda}\left(g\eta + \frac{u^2+v^2}{2}\right) + F_\lambda, \tag{17.1.4}$$

$$\frac{\partial v}{\partial t} = -(\zeta + 2\Omega\sin\varphi)u - \frac{1}{R}\frac{\partial}{\partial \varphi}\left(g\eta + \frac{u^2+v^2}{2}\right) + F_\varphi. \tag{17.1.5}$$

Vorticity ζ and divergence δ are given by

$$\zeta = \frac{1}{R\cos\varphi}\frac{\partial v}{\partial \lambda} - \frac{1}{R\cos\varphi}\frac{\partial(u\cos\varphi)}{\partial \varphi} = \nabla_H^2\psi, \tag{17.1.6}$$

$$\delta = \frac{1}{R\cos\varphi}\frac{\partial u}{\partial \lambda} + \frac{1}{R\cos\varphi}\frac{\partial(v\cos\varphi)}{\partial \varphi} = \nabla_H^2\chi, \tag{17.1.7}$$

where ψ is the streamfunction, χ is the velocity potential, and ∇_H^2 is the Laplacian on a sphere, expressed as

$$\nabla_H^2 = \left[\frac{1}{R^2\cos^2\varphi}\frac{\partial^2}{\partial \lambda^2} + \frac{1}{R^2\cos\varphi}\frac{\partial}{\partial \varphi}\left(\cos\varphi\frac{\partial}{\partial \varphi}\right)\right]. \tag{17.1.8}$$

ψ and χ are related to the horizontal components of velocity as

$$u = -\frac{1}{R}\frac{\partial \psi}{\partial \varphi} + \frac{1}{R\cos\varphi}\frac{\partial \chi}{\partial \lambda}, \tag{17.1.9}$$

$$v = \frac{1}{R\cos\varphi}\frac{\partial \psi}{\partial \lambda} + \frac{1}{R}\frac{\partial \chi}{\partial \varphi}. \tag{17.1.10}$$

The vorticity equation and the divergence equation can be directly derived from the equations of motion (17.1.4) and (17.1.5):

$$\begin{aligned}\frac{\partial\zeta}{\partial t} &= -\frac{1}{R\cos\varphi}\frac{\partial}{\partial\lambda}\left[(\zeta+2\Omega\sin\varphi)u+F_\varphi\right]\\ &\quad-\frac{1}{R\cos\varphi}\frac{\partial}{\partial\varphi}\left\{\left[(\zeta+2\Omega\sin\varphi)v+F_\lambda\right]\cos\varphi\right\},\end{aligned}\tag{17.1.11}$$

$$\begin{aligned}\frac{\partial\delta}{\partial t} &= \frac{1}{R\cos\varphi}\frac{\partial}{\partial\lambda}\left[(\zeta+2\Omega\sin\varphi)v+F_\lambda\right]\\ &\quad-\frac{1}{R\cos\varphi}\frac{\partial}{\partial\varphi}\left\{\left[(\zeta+2\Omega\sin\varphi)u+F_\varphi\right]\cos\varphi\right\}\\ &\quad-\nabla_H^2\left(g\eta+\frac{u^2+v^2}{2}\right).\end{aligned}\tag{17.1.12}$$

These are flux-form equations. Advective-form equations are given by

$$\frac{\partial\zeta}{\partial t}+\frac{u}{R\cos\varphi}\frac{\partial\zeta}{\partial\lambda}+\frac{v}{R}\frac{\partial\zeta}{\partial\varphi}+\frac{2\Omega\cos\varphi}{R}v+(\zeta+2\Omega\sin\varphi)\delta = F_\zeta,\tag{17.1.13}$$

$$\begin{aligned}\frac{\partial\delta}{\partial t}-\frac{v}{R\cos\varphi}\frac{\partial}{\partial\lambda}(\zeta+2\Omega\sin\varphi)+\frac{u}{R}\frac{\partial}{\partial\varphi}(\zeta+2\Omega\sin\varphi)&\\ -(\zeta+2\Omega\sin\varphi)\zeta+\nabla_H^2\left(g\eta+\frac{u^2+v^2}{2}\right) &= F_\delta,\end{aligned}\tag{17.1.14}$$

where F_ζ and F_δ are the dissipation terms of vorticity and divergence, respectively, given by

$$F_\zeta = \frac{1}{R\cos\varphi}\frac{\partial F_\varphi}{\partial\lambda}-\frac{1}{R\cos\varphi}\frac{\partial(F_\lambda\cos\varphi)}{\partial\varphi},\tag{17.1.15}$$

$$F_\delta = \frac{1}{R\cos\varphi}\frac{\partial F_\lambda}{\partial\lambda}+\frac{1}{R\cos\varphi}\frac{\partial(F_\varphi\cos\varphi)}{\partial\varphi}.\tag{17.1.16}$$

If the frictional force is expressed as convergence of the viscous stress tensor on a sphere with a constant viscous coefficient ν, it can be written in the following form (see the appendix to this chapter, Section 17.5):

$$F_\lambda = \nu\left[\left(\nabla_H^2+\frac{2}{R^2}\right)u-\frac{2\sin\varphi}{R^2\cos^2\varphi}\frac{\partial v}{\partial\lambda}-\frac{u}{R^2\cos^2\varphi}\right],\tag{17.1.17}$$

$$F_\varphi = \nu\left[\left(\nabla_H^2+\frac{2}{R^2}\right)v+\frac{2\sin\varphi}{R^2\cos^2\varphi}\frac{\partial u}{\partial\lambda}-\frac{v}{R^2\cos^2\varphi}\right].\tag{17.1.18}$$

In this case, the following relations can be derived:

$$F_\zeta = \nu\left(\nabla_H^2+\frac{2}{R^2}\right)\zeta,\tag{17.1.19}$$

$$F_\delta = \nu\left(\nabla_H^2+\frac{2}{R^2}\right)\delta.\tag{17.1.20}$$

It is convenient to use absolute vorticity ζ_a and potential vorticity q, which are defined by

$$\zeta_a = \zeta + 2\Omega \sin\varphi, \qquad q = \frac{\zeta_a}{\eta}. \tag{17.1.21}$$

From (17.1.13), the equation of absolute vorticity is written as

$$\frac{d\zeta_a}{dt} + \zeta_a \delta = F_\zeta, \tag{17.1.22}$$

where

$$\frac{d}{dt} = \frac{\partial}{\partial t} + \frac{u}{R\cos\varphi}\frac{\partial}{\partial\lambda} + \frac{v}{R}\frac{\partial}{\partial\varphi}. \tag{17.1.23}$$

Using $Q = 0$, the equation of mass (17.1.3) is rewritten as

$$\frac{d\eta}{dt} + \eta\delta = 0. \tag{17.1.24}$$

Thus, the equation of potential vorticity is given by

$$\frac{dq}{dt} = \frac{F_\zeta}{\eta}. \tag{17.1.25}$$

Multiplying this equation by η and using (17.1.15), we obtain the flux-form equation of potential vorticity as

$$\frac{\partial}{\partial t}(\eta q) + \frac{1}{R\cos\varphi}\frac{\partial}{\partial\lambda}(\eta q u - F_\varphi) + \frac{1}{R\cos\varphi}\frac{\partial}{\partial\varphi}[(\eta q v + F_\lambda)\cos\varphi] = 0. \tag{17.1.26}$$

This is the conservative form of potential vorticity. Since $\eta q = \zeta_a$, this equation can be viewed as the flux-form equation of absolute vorticity.

We next introduce relative angular momentum l and absolute angular momentum l_a by

$$l = uR\cos\varphi, \qquad l_a = uR\cos\varphi + \Omega R^2\cos^2\varphi. \tag{17.1.27}$$

Multiplying the zonal component of the equation of motion (17.1.1) by $\cos\varphi$, we obtain the conservation of angular momentum:

$$\frac{dl}{dt} = \frac{d}{dt}(u\cos\varphi) - 2\Omega v\sin\varphi\cos\varphi = -\frac{g}{R}\frac{\partial\eta}{\partial\lambda} + \cos\varphi\, F_\lambda. \tag{17.1.28}$$

Multiplying this equation by η and using (17.1.3), we obtain the flux-form equation of angular momentum:

$$\begin{aligned}\frac{\partial}{\partial t}(\eta u\cos\varphi) &+ \frac{1}{R\cos\varphi}\frac{\partial}{\partial\lambda}[\eta u(u + \Omega R\cos\varphi)\cos\varphi] \\ &+ \frac{1}{R\cos\varphi}\frac{\partial}{\partial\varphi}[\eta v(u + \Omega\cos\varphi)\cos^2\varphi] \\ = &-\frac{g}{R}\frac{\partial}{\partial\lambda}\frac{\eta^2}{2} + \cos\varphi\,\eta F_\lambda.\end{aligned} \tag{17.1.29}$$

With the aid of (17.1.4), this equation is further rewritten using vorticity as

$$\begin{aligned}\frac{\partial}{\partial t}(u\cos\varphi) &= (\zeta + 2\Omega\sin\varphi)v\cos\varphi \\ &\quad -\frac{1}{R}\frac{\partial}{\partial\lambda}\left(g\eta + \frac{u^2+v^2}{2}\right) + \cos\varphi\, F_\lambda. \end{aligned} \tag{17.1.30}$$

From the equations of angular momentum (17.1.29) and (17.1.30), the zonally averaged budget of angular momentum is written, respectively, as

$$\frac{\partial}{\partial t}(\overline{\eta u}\cos\varphi) + \frac{1}{R\cos\varphi}\frac{\partial}{\partial\varphi}(\overline{\eta uv}\cos^2\varphi + \overline{\eta v}\Omega\cos^3\varphi) = \cos\varphi\,\overline{\eta F_\lambda} \tag{17.1.31}$$

$$\frac{\partial}{\partial t}(\overline{u}\cos\varphi) - \overline{\zeta_a v}\cos\varphi = \cos\varphi\,\overline{F_\lambda} \tag{17.1.32}$$

where the zonal average of a quantity A is defined as

$$\overline{A} = \frac{1}{2\pi}\int_0^{2\pi} A\,d\lambda. \tag{17.1.33}$$

If the dissipation term is written as (17.1.17), we can express the frictional term as

$$\cos\varphi\overline{F_\lambda} = \nu\frac{1}{R^2\cos\varphi}\frac{\partial}{\partial\varphi}\left[\cos^3\varphi\frac{\partial}{\partial\varphi}\left(\frac{\overline{u}}{\cos\varphi}\right)\right], \tag{17.1.34}$$

where (17.5.1) from the appendix is used. In this form, $\overline{u}/(R\cos\varphi)$ is relative angular velocity. If angular velocity is constant over a sphere, there is no loss of angular momentum due to the frictional force. This is consistent with the fact that a fluid is in steady state if circulation is a rigid body rotation.

17.2 The Hadley cell model

Motion in the upper layer of Hadley circulation can be modeled by shallow water equations on a sphere. We consider the zonally symmetric flow of the shallow water model. Neglecting longitudinal dependency in (17.1.4), (17.1.5), and (17.1.3), we have the shallow water equations for zonal symmetric flow:

$$\frac{\partial u}{\partial t} = (\zeta + 2\Omega\sin\varphi)v + F_\lambda, \tag{17.2.1}$$

$$\frac{\partial v}{\partial t} = -(\zeta + 2\Omega\sin\varphi)u - \frac{1}{R}\frac{\partial}{\partial\varphi}\left(g\eta + \frac{u^2+v^2}{2}\right) + F_\varphi, \tag{17.2.2}$$

$$\frac{\partial \eta}{\partial t} = -\frac{1}{R\cos\varphi}\frac{\partial}{\partial\varphi}(v\,\eta\cos\varphi) + Q. \tag{17.2.3}$$

From (17.1.6), vorticity in zonal symmetric flow is reduced to

$$\zeta = -\frac{1}{R\cos\varphi}\frac{\partial(u\cos\varphi)}{\partial\varphi}. \tag{17.2.4}$$

If we assume that the source or sink terms of momentum are given by diffusion-type equations (17.1.17) and (17.1.18), the frictional forces in zonal symmetric flow become (see (17.5.1))

$$F_\lambda = \nu \frac{1}{R^2 \cos^2\varphi} \frac{\partial}{\partial\varphi}\left[\cos^3\varphi \frac{\partial}{\partial\varphi}\left(\frac{u}{\cos\varphi}\right)\right], \tag{17.2.5}$$

$$F_\varphi = \nu \frac{1}{R^2 \cos^2\varphi} \frac{\partial}{\partial\varphi}\left[\cos^3\varphi \frac{\partial}{\partial\varphi}\left(\frac{v}{\cos\varphi}\right)\right]. \tag{17.2.6}$$

From (17.2.5), momentum transport due to the frictional force occurs when angular velocity $\omega = u/(R\cos\varphi)$ has latitudinal variation. Only for flow with rigid body rotation, does the frictional force vanish. We may use another type of the source or sink of the momentum given by the Rayleigh friction:

$$F_\lambda = -\kappa_M u, \tag{17.2.7}$$

$$F_\varphi = -\kappa_M v, \tag{17.2.8}$$

where κ_M is constant.

For zonal symmetric flow, the conservation of angular momentum is given by neglecting the longitudinal dependency in (17.1.28):

$$\frac{d}{dt}(u\cos\varphi + \Omega R\cos^2\varphi) = \cos\varphi\, F_\lambda. \tag{17.2.9}$$

The total angular momentum integrated over the whole domain on a sphere is conserved if the frictional force is given by the diffusion-type equation (17.2.5), since the latitudinal integral $\int \cos\varphi d\varphi \cdot F_\lambda \cos\varphi$ vanishes. In the case of Rayleigh friction (17.2.7), however, total angular momentum is not conserved since the integral of $F_\lambda \cos\varphi$ does not generally vanish. In this case, since the integral of $F_\lambda \cos\varphi$ is proportional to that of relative angular momentum $u\cos\varphi$, the frictional force has the effect of reducing zonal motion relative to the earth rotation. It can be thought that this type of frictional force expresses the bulk effect of momentum transport between the ground and the atmosphere.

In order to consider Hadley circulation using the shallow water model, we assume that the mass sink term has a distribution similar to that of the potential temperature (16.2.22) used in the Held and Hou model described in Section 16.2.3; surface height η is relaxed to the following latitudinal reference profile η_e:

$$Q = -\kappa_T(\eta - \eta_e), \tag{17.2.10}$$

$$\eta_e = \eta_0\left(1 - \frac{2}{3}\Delta_H P_2(\varphi)\right) = \eta_0\left[1 - \Delta_H\left(\sin^2\varphi - \frac{1}{3}\right)\right], \tag{17.2.11}$$

where κ_T is the inverse of relaxation time, η_0 is reference surface height, and Δ_H is the typical scale of latitudinal variation of η_e.

Once the relaxation-type mass source is given as (17.2.10), the balance of the steady state is described as

$$0 = (\zeta + 2\Omega\sin\varphi)v + F_\lambda, \tag{17.2.12}$$

$$0 = -(\zeta + 2\Omega\sin\varphi)u - \frac{1}{R}\frac{\partial}{\partial\varphi}\left(g\eta + \frac{u^2+v^2}{2}\right) + F_\varphi, \tag{17.2.13}$$

$$0 = -\frac{1}{R\cos\varphi}\frac{\partial}{\partial\varphi}(v\,\eta\cos\varphi) - \kappa_T(\eta - \eta_e). \tag{17.2.14}$$

From (17.2.14), if there is no meridional flow $v = 0$, the water depth is equal to the relaxation profile: $\eta = \eta_e$. In this case, since $F_\varphi = 0$ according to (17.2.6) or (17.2.8), (17.2.13) is rewritten as

$$\frac{1}{R\tan\varphi}u^2 + 2\Omega\sin\varphi\,u = -\frac{g}{R}\frac{\partial\eta_e}{\partial\varphi}. \tag{17.2.15}$$

Substituting (17.2.11) into η_e, we can solve it for u as

$$u = \Omega R\cos\varphi\left(\sqrt{1+\frac{2g\Delta_H}{\Omega^2R^2}} - 1\right) \approx \frac{g\Delta_H}{\Omega R}\cos\varphi, \tag{17.2.16}$$

where $\frac{g\Delta_H}{\Omega^2 R^2} \ll 1$ is used. This profile of u has a maximum at the equator $\varphi = 0$.

Next, we assume that $v \neq 0$ from the equator to the latitude φ_H when meridional flow exists. If the friction term is negligible $F_\lambda = 0$, (17.2.12) is reduced to

$$\zeta + 2\Omega\sin\varphi = 0 \tag{17.2.17}$$

(i.e., absolute vorticity is zero). It is thought that this zero-vorticity region corresponds to the latitudinal extent of the Hadley cell in the shallow water system. The similar argument given in Section 16.2.3 is applicable to various quantities in the region of Hadley circulation. Using (17.2.4), Eq. (17.2.17) can be rewritten as

$$-\frac{1}{R\cos\varphi}\frac{\partial}{\partial\varphi}(u\cos\varphi + \Omega R\cos^2\varphi) = 0. \tag{17.2.18}$$

Thus, the absolute angular momentum l_a given by (17.1.27) is constant. If we assume that $u = 0$ at the equator $\varphi = 0$, the zonal wind profile is given by

$$u = \Omega R\frac{\sin^2\varphi}{\cos\varphi}. \tag{17.2.19}$$

If we neglect v in (17.2.13), the profile of η is given by

$$\eta = \eta(0) - \frac{u^2}{2g} = \eta(0) - \frac{\Omega^2R^2}{2g}\frac{\sin^4\varphi}{\cos^2\varphi}, \tag{17.2.20}$$

where $\eta(0)$ is water depth at the equator. Since water depth η must be continuous, $\eta = \eta_e$ is required at the polar boundary of the Hadley cell $\varphi = \varphi_H$. The width of the Hadley cell φ_H can be given by the mass balance (17.2.14). Multiplying $\cos\varphi$ by (17.2.14), and integrating from $\varphi = 0$ to φ_H, we have

$$\int_0^{\varphi_H}\eta\cos\varphi\,d\varphi = \int_0^{\varphi_H}\eta_e\cos\varphi\,d\varphi. \tag{17.2.21}$$

This equation is equivalent to (16.2.27) if one notes the definitions $\mu_H = \sin\varphi_H$ and $R_H = g\eta_0\Delta_H/\Omega^2R^2$. In particular, for the case $\varphi_H \ll 1$, the Hadley width is approximated as $\varphi_H = (5R_H/3)^{1/2}$, which is the same as (16.2.30).

In the case of Rayleigh friction, (17.2.7) is used for the friction term. We use the following set of equations, which are approximations of (17.2.12)–(17.2.14):

$$0 = (\zeta + 2\Omega\sin\varphi)v - \kappa_M u, \tag{17.2.22}$$

$$\begin{aligned} 0 &= -(\zeta + 2\Omega\sin\varphi)u - \frac{1}{R}\frac{\partial}{\partial\varphi}\left(g\eta + \frac{u^2}{2}\right) - \kappa_M v \\ &= -2\Omega\sin\varphi - \frac{\tan\varphi}{R^2}u^2 - \frac{g}{R}\frac{\partial\eta}{\partial\varphi} - \kappa_M v, \end{aligned} \tag{17.2.23}$$

$$0 = -\frac{\eta_0}{R\cos\varphi}\frac{\partial}{\partial\varphi}(v\cos\varphi) - \kappa_T(\eta - \eta_e). \tag{17.2.24}$$

Following Held and Phillips (1990), we solve the above equations using the following parameters: $g\eta_0 = 10^3$ m^2 s^{-2}, $\Delta_H = 20$, $\kappa_T^{-1} = 10$ days, $\Omega = 2\pi$ day^{-1}, and $R = 6.37 \times 10^4$ m. The solutions to the above equation set for various values of the coefficient κ_M are shown in Fig. 17.1. The coefficients of Rayleigh friction are given as $\kappa_M^{-1} = 40, 20, 10, 5$ days, and $\kappa_M = 0$. At the limit of no friction $\kappa_M \to 0$, which is denoted by NF, the Hadley cell is extended to the latitude $\varphi_H = 21.4°$. As shown by (17.2.18) and (17.2.17), within the Hadley cell region $|\varphi| < \varphi_H$, absolute angular momentum is constant and absolute vorticity is zero. As κ_M becomes larger (stronger friction), the meridional wind becomes stronger and the Hadley cell width becomes wider. In this case, however, the polar boundary of the Hadley cell becomes less clear.

Let us introduce diffusion-type friction in addition to Rayleigh friction. Let us add (17.2.5) and (17.2.6) to the right-hand sides of (17.2.22) and (17.2.23), respectively. The distribution of zonal winds is shown in Fig. 17.2 for the diffusion coefficients $\nu = 0$, 1×10^5, 2×10^5, and 4×10^5 m^2 s^{-1}. It can be seen that zonal wind is westerly at the equator; the role of the latitudinal diffusion which brings the equatorial westerly is called the *Gierasch effect*, which is a possible mechanism to explain the super-rotation of Venus's atmosphere. The balance of zonal wind at the equator is written as

$$0 = -\kappa_M u + \nu\frac{1}{R^2\cos^2\varphi}\frac{\partial}{\partial\varphi}\left[\cos^3\varphi\frac{\partial}{\partial\varphi}\left(\frac{u}{\cos\varphi}\right)\right]. \tag{17.2.25}$$

From this, we can see that u becomes stronger as κ_M becomes smaller. If shallow water equations are used as a model for two layers with opposite meridional flows, the supply of angular momentum from the lower layer is parameterized as Rayleigh friction. The larger κ_M corresponds to the larger momentum transport from the lower layer to the upper layer.

Observed zonal wind profiles in the upper layer (200 hPa) are shown in Fig. 17.3. Actually, equatorial winds are almost always easterly in every season, particularly in boreal summer. The profile of zonal winds almost follows that of the angular momentum conserving flow $R\Omega\sin^2\varphi$ in low latitudes.

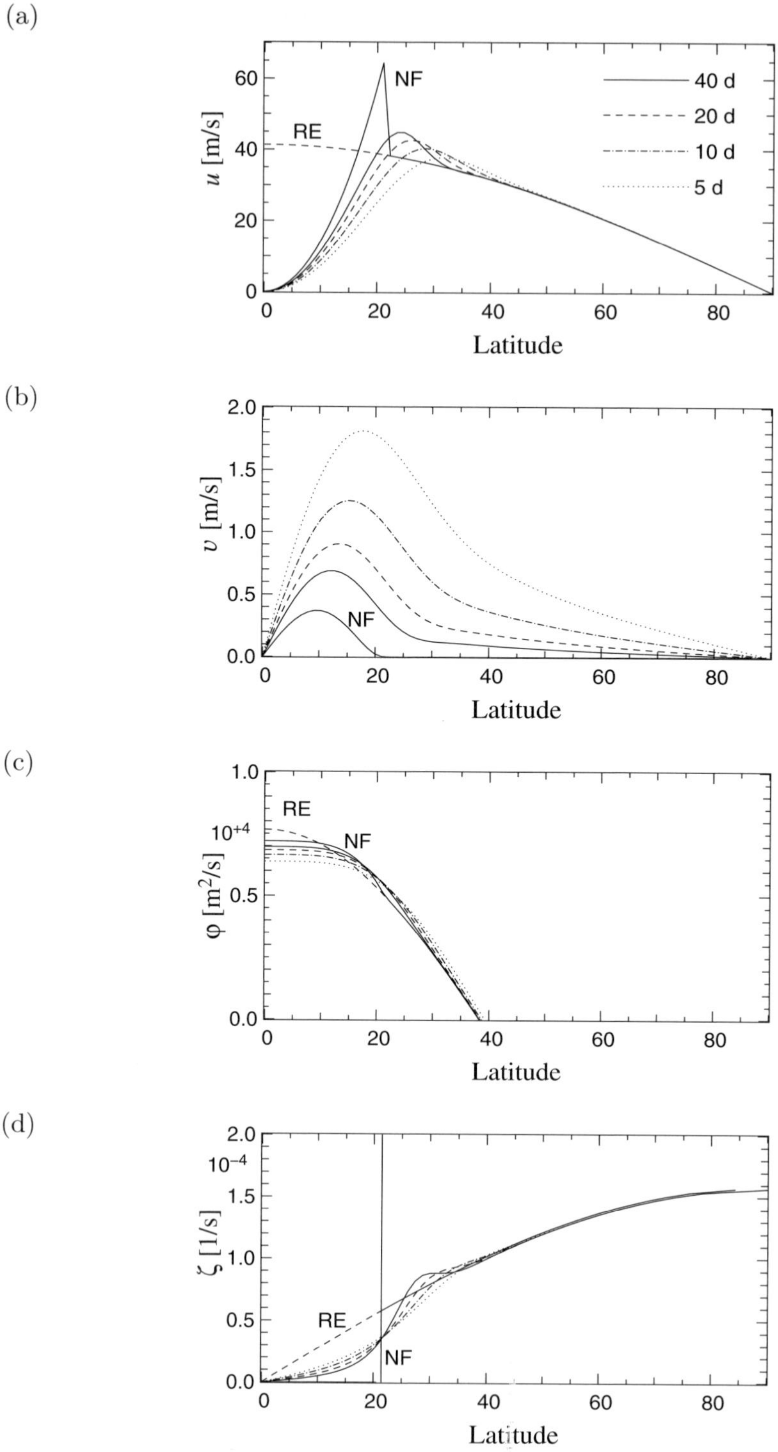

FIGURE 17.1: From top to bottom: latitudinal distributions of (a) zonal wind u [m], (b) meridional wind v [m], (c) surface height $\phi = g\eta$ [m^2 s^{-1}], and (d) absolute vorticity ζ [s^{-1}] given by the axisymmetric shallow water model using Rayleigh friction: $\kappa_M^{-1} = 40$ days (solid), 20 days (dash), 10 days (dash-dot), and 5 days (dot). NF is the profile for $\kappa_M = 0$ and RE is the relaxation profile.

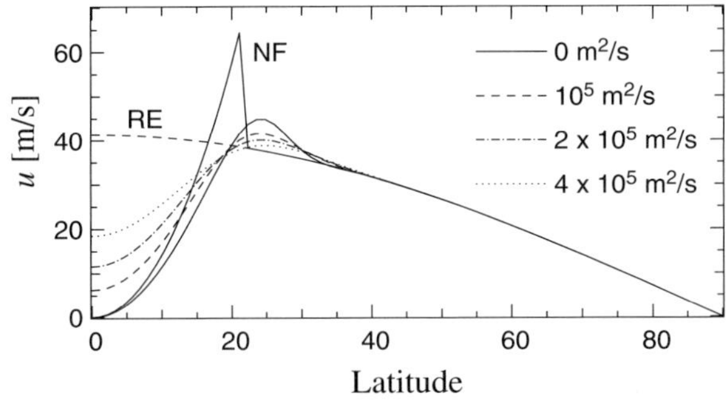

FIGURE 17.2: The dependency of zonal wind on the diffusion coefficient ν given by the axisymmetric shallow water model with diffusion-type friction: $\nu = 0$ (solid), 1×10^5 (dash), 2×10^5 (dash-dot), and 4×10^5 (dot). The coefficient of Rayleigh friction is $\kappa_M^{-1} = 40$ days. NF is the profile for $\kappa_M = 0$ and RE is the relaxation profile.

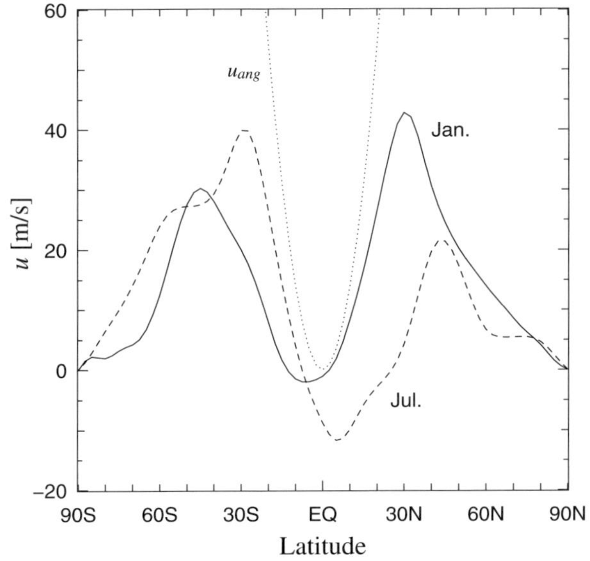

FIGURE 17.3: The observed latitudinal profiles of zonal winds at 200 hPa. Solid curve: January; dashed curve: July. The dotted curve is the zonal wind for the angular momentum conserving flow $u_{ang} = R\Omega \sin^2 \varphi$. See Appendix A3 for data source.

17.3 Mid-latitude circulations

Since geostrophic motions prevail in mid-latitudes, some characteristics of mid-latitude circulations are described by the non-divergent one-layer model or the barotropic model. In this section, we begin by summarizing barotropic equations including several forms of angular momentum equations. In the following subsections, the propagation of Rossby waves is considered in our discussion of zonal wind distribution in terms of wave and mean flow interactions.

17.3.1 Momentum balance of barotropic flow

In the nondivergent system, divergence defined by (17.1.7) equals zero:

$$\delta = \frac{1}{R\cos\varphi}\frac{\partial u}{\partial\lambda} + \frac{1}{R\cos\varphi}\frac{\partial(v\cos\varphi)}{\partial\varphi} = 0. \tag{17.3.1}$$

In this case, the velocity potential χ can be set as a constant from (17.1.7). Thus, from (17.1.9) and (17.1.10), velocity components are expressed using the streamfunction:

$$u = -\frac{1}{R}\frac{\partial\psi}{\partial\varphi}, \qquad v = \frac{1}{R\cos\varphi}\frac{\partial\psi}{\partial\lambda}. \tag{17.3.2}$$

The relation between vorticity and the streamfunction is given by (17.1.6):

$$\zeta = \frac{1}{R\cos\varphi}\frac{\partial v}{\partial\lambda} - \frac{1}{R\cos\varphi}\frac{\partial(u\cos\varphi)}{\partial\varphi} = \nabla_H^2\psi, \tag{17.3.3}$$

and the vorticity equation (17.1.13) is written as

$$\frac{\partial\zeta}{\partial t} + \frac{u}{R\cos\varphi}\frac{\partial\zeta}{\partial\lambda} + \frac{v}{R}\frac{\partial\zeta}{\partial\varphi} + \frac{2\Omega\cos\varphi}{R}v = F_\zeta. \tag{17.3.4}$$

This equation can be rewritten using the streamfunction as

$$\frac{\partial}{\partial t}\nabla_H^2\psi - \frac{1}{R^2\cos\varphi}\frac{\partial\psi}{\partial\varphi}\frac{\partial\nabla_H^2\psi}{\partial\lambda} + \frac{1}{R^2\cos\varphi}\frac{\partial\psi}{\partial\lambda}\cdot\frac{\partial\nabla_H^2\psi}{\partial\varphi} + \frac{2\Omega}{R^2}\frac{\partial\psi}{\partial\lambda} = F_\zeta. \tag{17.3.5}$$

The dissipation term F_ζ is expressed as (17.1.19), for instance.

Under nondivergent conditions, the flux form of the angular momentum equation (17.1.29) becomes†

$$\begin{aligned}\frac{\partial}{\partial t}(u\cos\varphi) &+ \frac{1}{R\cos\varphi}\frac{\partial}{\partial\lambda}[u(u+\Omega R\cos\varphi)\cos\varphi] \\ &+\frac{1}{R\cos\varphi}\frac{\partial}{\partial\varphi}[v(u+\Omega\cos\varphi)\cos^2\varphi] = -\frac{g}{R}\frac{\partial\eta}{\partial\lambda} + \cos\varphi\,F_\lambda.\end{aligned} \tag{17.3.6}$$

Since $\overline{v} = 0$, the zonal average of the angular momentum equation is reduced to

$$\frac{\partial}{\partial t}(\overline{u}\cos\varphi) + \frac{1}{R\cos\varphi}\frac{\partial}{\partial\varphi}(\overline{uv}\cos^2\varphi) = \cos\varphi\,\overline{F_\lambda}. \tag{17.3.7}$$

An alternative form of conservation of angular momentum is given using vorticity as (17.1.32), which becomes

$$\frac{\partial}{\partial t}(\overline{u}\cos\varphi) - \overline{\zeta v}\cos\varphi = \cos\varphi\,\overline{F_\lambda}. \tag{17.3.8}$$

† η plays the role of pressure in barotropic equations. Even if divergence δ equals zero, η is not constant.

At this point, it should be noted that, from comparison of (17.3.7) and (17.3.8), we obtain

$$\frac{1}{R\cos\varphi}\frac{\partial}{\partial\varphi}(\overline{uv}\cos^2\varphi) = -\overline{\zeta v}\cos\varphi. \tag{17.3.9}$$

This relation is called *Taylor's identity*, and corresponds to Eq. (5.5.176) of the quasi-geostrophic equations.

From (17.3.4), the equation of absolute vorticity $\zeta_a = \zeta + 2\Omega\sin\varphi$ is written as

$$\frac{d\zeta_a}{dt} = F_\zeta. \tag{17.3.10}$$

In particular, in the case that $F_\zeta = 0$, ζ_a is conservative in the Lagrangian sense. Multiplying the above equation by ζ_a, we obtain the conservation of *absolute enstrophy* $\zeta_a^2/2$:

$$\frac{d}{dt}\frac{\zeta_a^2}{2} = \zeta_a F_\zeta. \tag{17.3.11}$$

From this, the equation of enstrophy $\zeta^2/2$ (i.e., the square of *relative* vorticity) is written as

$$\frac{\partial}{\partial t}\frac{\zeta^2}{2} + \frac{u}{R\cos\varphi}\frac{\partial}{\partial\lambda}\frac{\zeta^2}{2} + \frac{v}{R}\frac{\partial}{\partial\varphi}\frac{\zeta^2}{2} + \frac{2\Omega\cos\varphi}{R}v\zeta = \zeta_a F_\zeta. \tag{17.3.12}$$

From the corresponding flux-form equation, the zonal average of the enstrophy equation is given by

$$\frac{\partial}{\partial t}\frac{\overline{\zeta^2}}{2} + \frac{1}{R\cos\varphi}\frac{\partial}{\partial\varphi}\left(\frac{\overline{v\zeta^2}}{2}\cos\varphi\right) + \frac{2\Omega\cos\varphi}{R}\overline{v\zeta} = \overline{\zeta_a F_\zeta}. \tag{17.3.13}$$

Using this equation and angular momentum conservation (17.3.8) to eliminate the term $\overline{v\zeta}$, we obtain

$$\frac{\partial}{\partial t}\left(\overline{u}\cos\varphi + \frac{R}{2\Omega}\frac{\overline{\zeta^2}}{2}\right) + \frac{1}{R\cos\varphi}\frac{\partial}{\partial\varphi}\left(\frac{R}{2\Omega}\frac{\overline{v\zeta^2}}{2}\cos\varphi\right)$$
$$= \cos\varphi\,\overline{F_\lambda} + \frac{R}{2\Omega}\overline{\zeta_a F_\zeta}. \tag{17.3.14}$$

This is the equation of *pseudo-momentum* of the barotropic system. It is found that the domain integral of the pseudo-momentum,

$$A = \overline{u}\cos\varphi + \frac{R}{2\Omega}\frac{\overline{\zeta^2}}{2}, \tag{17.3.15}$$

conserves if the dissipation terms F_λ and F_ζ are negligible.

17.3.2 Weak nonlinear theory of Rossby waves

Propagation of mid-latitude disturbances in the nondivergent system on a sphere can be described by the theory of Rossby waves. Let us consider the characteristics of propagation of Rossby waves and the roles of Rossby waves in the momentum balance of mid-latitude circulations in the following subsections.

Let us divide the flow field into a zonally uniform basic field $u_B(\varphi)$ and the deviation from it. Expanding the deviation field into a series of the amplitude of the disturbance, we obtain

$$u = u_B + u' + u^{(2)} + \cdots , \tag{17.3.16}$$

$$v = v' + v^{(2)} + \cdots , \tag{17.3.17}$$

where ($'$) denotes deviation from the zonal mean ($\overline{\ }$). We assume that the amplitude of the disturbance is much smaller than that of the basic field, such that, for instance, $u'/u_B = O(\varepsilon)$ and $u^{(2)}/u_B = O(\varepsilon^2)$ where $\varepsilon \ll 1$. Note that, since u is time-dependent, $\overline{u(t)}$ is not equal to the basic field u_B, but its difference is second order. Similarly, vorticity and divergence are expanded as

$$\zeta = \zeta_B + \zeta' + \zeta^{(2)} + \cdots , \tag{17.3.18}$$

$$\psi = \psi_B + \psi' + \psi^{(2)} + \cdots . \tag{17.3.19}$$

The first-order terms of the vorticity equation (17.3.4) are collected as

$$\frac{\partial \zeta'}{\partial t} + \frac{\overline{u}}{R\cos\varphi}\frac{\partial \zeta'}{\partial \lambda} + \hat{\beta} v' = F'_\zeta, \tag{17.3.20}$$

where $\hat{\beta}$ is a generalized β effect expressed by

$$\hat{\beta} = \frac{2\Omega\cos\varphi}{R} + \frac{1}{R}\frac{\partial \zeta_B}{\partial \varphi} = \frac{2\Omega\cos\varphi}{R} - \frac{1}{R^2}\frac{\partial}{\partial\varphi}\left[\frac{1}{\cos\varphi}\frac{\partial}{\partial\varphi}(u_B\cos\varphi)\right]. \tag{17.3.21}$$

Multiplying (17.3.20) by ζ' gives

$$\frac{\partial}{\partial t}\frac{\zeta'^2}{2} + \frac{\overline{u}}{R\cos\varphi}\frac{\partial}{\partial\lambda}\frac{\zeta'^2}{2} + \hat{\beta} v'\zeta' = \zeta' F'_\zeta. \tag{17.3.22}$$

Here, the third term on the left-hand side is written by using the nondivergent condition (17.3.1)

$$v'\zeta' = \frac{1}{R\cos\varphi}\frac{\partial}{\partial\lambda}\left(\frac{v'^2 - u'^2}{2}\right) - \frac{1}{R\cos^2\varphi}\frac{\partial}{\partial\varphi}(u'v'\cos^2\varphi). \tag{17.3.23}$$

Therefore, we have

$$\frac{\partial}{\partial t}\frac{\zeta'^2}{2} + \frac{1}{R\cos\varphi}\frac{\partial}{\partial\lambda}\left(\frac{\zeta'^2}{2}\overline{u} + \hat{\beta}\frac{v'^2 - u'^2}{2}\right) - \frac{\hat{\beta}}{R\cos^2\varphi}\frac{\partial}{\partial\varphi}(u'v'\cos^2\varphi)$$
$$= \zeta' F'_\zeta. \tag{17.3.24}$$

In the case $\hat{\beta} \neq 0$, this becomes

$$\begin{aligned}
\frac{\partial}{\partial t}\left(\frac{\cos\varphi}{\hat{\beta}}\frac{\zeta'^2}{2}\right) + \frac{1}{R\cos\varphi}\frac{\partial}{\partial\lambda}\left(\frac{\cos\varphi}{\hat{\beta}}\frac{\zeta'^2}{2}\overline{u} + \cos\varphi\frac{v'^2 - u'^2}{2}\right) & \\
- \frac{1}{R\cos\varphi}\frac{\partial}{\partial\varphi}(u'v'\cos^2\varphi) &= \frac{\cos\varphi}{\hat{\beta}}\zeta' F'_\zeta, \qquad (17.3.25)
\end{aligned}$$

or

$$\frac{\partial A}{\partial t} + \nabla_H \cdot \boldsymbol{F} = \frac{\cos\varphi}{\hat{\beta}}\zeta' F'_\zeta, \qquad (17.3.26)$$

where

$$A = \frac{\cos\varphi}{\hat{\beta}}\frac{\zeta'^2}{2}, \qquad (17.3.27)$$

$$\boldsymbol{F} = \left(A\overline{u} + \frac{1}{2}\cos\varphi\,(v'^2 - u'^2),\ \cos\varphi\, u'v'\right). \qquad (17.3.28)$$

In the case $F'_\zeta = 0$ in (17.3.26), A is conservative in flux-form sense.

Next, let us consider the second-order balance of the zonal-mean equation of angular momentum conservation. From (17.3.8), we have

$$\frac{\partial}{\partial t}(\overline{u^{(2)}}\cos\varphi) - \overline{\zeta' v'}\cos\varphi = \cos\varphi\,\overline{F_\lambda^{(2)}}. \qquad (17.3.29)$$

The zonal average of the equation of enstrophy (17.3.22) is given by

$$\frac{\partial}{\partial t}\frac{\overline{\zeta'^2}}{2} + \hat{\beta}\overline{v'\zeta'} = \overline{\zeta' F'_\zeta}. \qquad (17.3.30)$$

Using (17.3.29) and (17.3.30) to eliminate $\overline{v'\zeta'}$, we obtain

$$\frac{\partial}{\partial t}\left(\overline{u^{(2)}}\cos\varphi + \frac{\cos\varphi}{\hat{\beta}}\frac{\overline{\zeta'^2}}{2}\right) = \cos\varphi\,\overline{F_\lambda^{(2)}} + \frac{\cos\varphi}{\hat{\beta}}\overline{\zeta' F'_\zeta}. \qquad (17.3.31)$$

Here, we define

$$\overline{A} \equiv \frac{\cos\varphi}{\hat{\beta}}\frac{\overline{\zeta'^2}}{2}. \qquad (17.3.32)$$

The quantity $-\overline{A}$ is called *pseudo-angular momentum* and (17.3.31) is called the conservation of pseudo-angular momentum. This equation is similar to the exact equation of the non-linear system (17.3.14). If there is no basic zonal flow $u_B = 0$, since $\hat{\beta} = 2\Omega\cos\varphi/R$ from (17.3.21), this equation agrees with the exact solution up to second-order terms.

In the case that frictional forces vanish $F_\lambda = F_\zeta = 0$, (17.3.31) can be solved as

$$\overline{u^{(2)}}\cos\varphi + \overline{A} = \text{const.} \qquad (17.3.33)$$

(i.e., the quantity on the left-hand side is constant irrespective of time). As an example, we consider the case when a disturbance of vorticity is given around some latitude initially under the constraint $\overline{u^{(2)}} = 0$. The pseudo-angular momentum at the initial state is denoted by $\overline{A_0}$ (> 0). After sufficient time has passed, the disturbance will propagate to remote latitudes, and pseudo-angular momentum $\overline{A}$ will tend to zero around the latitude where the initial disturbance is given. Thus, from conservation of pseudo-angular momentum (17.3.33), we have

$$\overline{A_0} = \overline{u^{(2)}} \cos\varphi. \tag{17.3.34}$$

This means that the westerly $\overline{u^{(2)}} > 0$ is induced at the initially disturbed latitudes.

17.3.3 Propagation of Rossby waves: WKBJ theory

Let us formulate the propagation of Rossby waves on a sphere using the WKBJ theory (see Chapter 4). Using (17.3.2) and (17.3.3), the linearized vorticity equation (17.3.20) is expressed with the streamfunction ψ:

$$\frac{\partial}{\partial t}\nabla_H^2\psi + \frac{\overline{u}}{R\cos\varphi}\frac{\partial}{\partial\lambda}\nabla_H^2\psi + \frac{\hat{\beta}}{R\cos\varphi}\frac{\partial\psi}{\partial\lambda} = 0. \tag{17.3.35}$$

Here, we assumed no dissipation $F'_\zeta = 0$. Introducing a phase function Θ and a small parameter ε, we expand the streamfunction ψ as

$$\psi = \sum_{n=0}^{\infty} \varepsilon^n \psi_n(\varphi,\lambda,t) e^{i\frac{\Theta(\varphi,\lambda,t)}{\varepsilon}}, \tag{17.3.36}$$

where the wave numbers (k, l) and the frequency ω are defined by

$$\omega \equiv -\frac{1}{\varepsilon}\frac{\partial\Theta}{\partial t}, \quad k \equiv \frac{1}{\varepsilon}\frac{1}{R\cos\varphi}\frac{\partial\Theta}{\partial\lambda}, \quad l \equiv \frac{1}{\varepsilon}\frac{1}{R}\frac{\partial\Theta}{\partial\varphi}. \tag{17.3.37}$$

Substituting (17.3.36) into the vorticity equation (17.3.35), we write down the equations for each order of ε. From the $O(\varepsilon^0)$ equation, we obtain the dispersion relation†

$$P \equiv (\omega - \overline{u}k)\left(k^2 + l^2\right) + k\hat{\beta} = 0. \tag{17.3.39}$$

This corresponds to the dispersion relation of Rossby waves on the assumption of the β-plane approximation. It can be solved for ω:

$$\omega = \overline{u}k - \frac{\hat{\beta}k}{k^2 + l^2} \equiv \Omega(k, l; \varphi). \tag{17.3.40}$$

---

†Actually, the $O(\varepsilon^0)$ equation is slightly different from (17.3.39), and given by

$$(\omega - \overline{u}k)\left(k^2 + l^2 - i\frac{\tan\varphi}{R}l\right) + k\hat{\beta} = 0. \tag{17.3.38}$$

To derive the precise dispersion relation, we need to use Mercator coordinates (x, y):

$$dx = Rd\lambda, \qquad dy = \frac{Rd\varphi}{\cos\varphi}.$$

From this, group velocity is expressed as

$$c_{g\lambda} = \frac{\partial \Omega}{\partial k} = \overline{u} + \frac{\hat{\beta}(k^2 - l^2)}{(k^2 + l^2)^2} = \overline{u} + (k^2 - l^2)\frac{\hat{\omega}^2}{\hat{\beta}k^2}, \tag{17.3.41}$$

$$c_{g\varphi} = \frac{\partial \Omega}{\partial l} = \frac{2\hat{\beta}kl}{(k^2 + l^2)^2} = \frac{2l}{k}\frac{\hat{\omega}^2}{\hat{\beta}}, \tag{17.3.42}$$

where

$$\hat{\omega} = -\frac{\hat{\beta}k}{k^2 + l^2} \tag{17.3.43}$$

is called *intrinsic frequency*, or *Doppler-shifted frequency*. Using (17.3.37), the conservations of wave numbers are given as

$$\frac{\partial (k \cos \varphi)}{\partial t} = -\frac{1}{R}\frac{\partial \omega}{\partial \lambda}, \qquad \frac{\partial l}{\partial t} = -\frac{1}{R}\frac{\partial \omega}{\partial \varphi}, \tag{17.3.44}$$

$$\frac{\partial (k \cos \varphi)}{\partial \varphi} = \frac{\partial l}{\partial \lambda}. \tag{17.3.45}$$

If we regard the dispersion relation as a function $\omega = \Omega(k \cos \varphi, l; \lambda, \varphi, t)$, its derivatives with respect to λ, φ, and t are given, respectively, as

$$\begin{aligned}\frac{\partial \omega}{\partial \lambda} &= \frac{\partial (k \cos \varphi)}{\partial \lambda}\frac{\partial \Omega}{\partial (k \cos \varphi)} + \frac{\partial l}{\partial \lambda}\frac{\partial \Omega}{\partial l} + \frac{\partial \Omega}{\partial \lambda} \\ &= \frac{c_{g\lambda}}{\cos \varphi}\frac{\partial (k \cos \varphi)}{\partial \lambda} + c_{g\varphi}\frac{\partial (k \cos \varphi)}{\partial \varphi} + \frac{\partial \Omega}{\partial \lambda},\end{aligned} \tag{17.3.46}$$

$$\begin{aligned}\frac{\partial \omega}{\partial \varphi} &= \frac{\partial (k \cos \varphi)}{\partial \varphi}\frac{\partial \Omega}{\partial (k \cos \varphi)} + \frac{\partial l}{\partial \varphi}\frac{\partial \Omega}{\partial l} + \frac{\partial \Omega}{\partial \varphi} \\ &= \frac{c_{g\lambda}}{\cos \varphi}\frac{\partial l}{\partial \lambda} + c_{g\varphi}\frac{\partial l}{\partial \varphi} + \frac{\partial \Omega}{\partial \varphi},\end{aligned} \tag{17.3.47}$$

$$\begin{aligned}\frac{\partial \omega}{\partial t} &= \frac{\partial (k \cos \varphi)}{\partial t}\frac{\partial \Omega}{\partial (k \cos \varphi)} + \frac{\partial l}{\partial t}\frac{\partial \Omega}{\partial l} + \frac{\partial \Omega}{\partial t} \\ &= -\frac{c_{g\lambda}}{R \cos \varphi}\frac{\partial \omega}{\partial \lambda} - \frac{c_{g\varphi}}{R}\frac{\partial \omega}{\partial \varphi} + \frac{\partial \Omega}{\partial t}.\end{aligned} \tag{17.3.48}$$

Since Ω is independent of λ and t due to (17.3.40). From (17.3.44), the following equations of wave numbers are obtained:

$$\frac{\partial (k \cos \varphi)}{\partial t} + \frac{c_{g\lambda}}{R \cos \varphi}\frac{\partial (k \cos \varphi)}{\partial \lambda} + \frac{c_{g\varphi}}{R}\frac{\partial (k \cos \varphi)}{\partial \varphi} = 0, \tag{17.3.49}$$

$$\frac{\partial l}{\partial t} + \frac{c_{g\lambda}}{R \cos \varphi}\frac{\partial l}{\partial \lambda} + \frac{c_{g\varphi}}{R}\frac{\partial l}{\partial \varphi} = -\frac{1}{R}\frac{\partial \Omega}{\partial \varphi}, \tag{17.3.50}$$

$$\frac{\partial \omega}{\partial t} + \frac{c_{g\lambda}}{R \cos \varphi}\frac{\partial \omega}{\partial \lambda} + \frac{c_{g\varphi}}{R}\frac{\partial \omega}{\partial \varphi} = 0. \tag{17.3.51}$$

The path of a wave packet moving with group velocity $(c_{g\lambda}, c_{g\varphi})$ is called the *ray*. Eqs. (17.3.49) and (17.3.51) state that $k \cos \varphi$ and ω are conserved along rays. In contrast, the meridional wave number l is not conserved along rays. From (17.3.40), l is given by

$$l^2 = -\frac{\hat{\beta}k}{\omega - \overline{u}k} - k^2, \tag{17.3.52}$$

from which we find that l changes with latitude φ since $\overline{u}$ is a function of φ.

Next, the $O(\varepsilon^1)$ terms of the vorticity equation (17.3.35) give

$$PA_1 - \frac{\partial P}{\partial \omega}\frac{\partial \psi_0}{\partial t} + \frac{\partial P}{\partial k}\frac{1}{R\cos\varphi}\frac{\partial \psi_0}{\partial \lambda} + \frac{\partial P}{\partial l}\frac{1}{R}\frac{\partial \psi_0}{\partial \varphi} + D\psi_0 = 0, \tag{17.3.53}$$

where

$$\begin{aligned} D &= \frac{1}{2}\left[-2\frac{\partial^2 P}{\partial\omega\partial k}\frac{\partial k}{\partial t} - 2\frac{\partial^2 P}{\partial\omega\partial l}\frac{\partial l}{\partial t} + \frac{\partial^2 P}{\partial k^2}\frac{1}{R\cos\varphi}\frac{\partial k}{\partial\lambda}\right. \\ &\quad \left. + 2\frac{\partial^2 P}{\partial k\partial l}\frac{1}{R\cos\varphi}\frac{\partial l}{\partial\lambda} + \frac{\partial^2 P}{\partial l^2}\frac{1}{R\cos\varphi}\frac{\partial(l\cos\varphi)}{\partial\varphi}\right] \\ &= -2k\frac{\partial k}{\partial t} - 2l\frac{\partial l}{\partial t} + \frac{\omega - 3\overline{u}k}{R\cos\varphi}\frac{\partial k}{\partial\lambda} + \frac{2\overline{u}l}{R\cos\varphi}\frac{\partial l}{\partial\lambda} + \frac{\omega - \overline{u}k}{R\cos\varphi}\frac{\partial(l\cos\varphi)}{\partial\varphi}. \end{aligned} \tag{17.3.54}$$

Using $P = 0$ and (17.3.45) to rewrite (17.3.53), and multiplying the result by $-(k^2 + l^2)/\hat{\beta}\psi_0\cos\varphi$, we obtain the equation of amplitude:

$$\frac{\partial \mathcal{A}}{\partial t} + \frac{1}{R\cos\varphi}\frac{\partial}{\partial\lambda}(c_{g\lambda}\mathcal{A}) + \frac{1}{R\cos\varphi}\frac{\partial}{\partial\varphi}(\cos\varphi\, c_{g\varphi}\mathcal{A}) = 0, \tag{17.3.55}$$

where

$$\mathcal{A} = \frac{(k^2 + l^2)^2\cos\varphi\,\psi_0^2}{2\hat{\beta}} \tag{17.3.56}$$

is pseudo-angular momentum, (17.3.32). Eq. (17.3.55) states that $\mathcal{A}$ is conserved along rays.

The kinetic energy of the disturbance E is expressed by

$$E = \frac{u'^2 + v'^2}{2} = \frac{|\nabla_H\psi|^2}{2}. \tag{17.3.57}$$

Using the WKBJ approximation, averaging over the phase gives

$$\begin{aligned} \langle E\rangle &= \frac{1}{2\pi}\int_0^{2\pi} E\,d\theta \\ &= -\frac{1}{2\pi}\int_0^{2\pi}(k^2 + l^2)\psi_0^2\cos^2\theta\,d\theta = -\frac{(k^2 + l^2)\psi_0^2}{2}, \end{aligned} \tag{17.3.58}$$

where $\theta = \Theta/\varepsilon$. Thus, from (17.3.43) and (17.3.56), pseudo-angular momentum is expressed as

$$\mathcal{A} = \frac{k\langle E\rangle}{\hat{\omega}}. \tag{17.3.59}$$

Substituting (17.3.59) into (17.3.55), and using the conservation of the wave number $k\cos\varphi$, (17.3.49), we obtain the conservation of the wave action of Rossby waves on the sphere:

$$\frac{\partial}{\partial t}\frac{\langle E\rangle}{\hat{\omega}} + \frac{1}{R\cos\varphi}\frac{\partial}{\partial\lambda}\left(c_{g\lambda}\frac{\langle E\rangle}{\hat{\omega}}\right) + \frac{1}{R\cos\varphi}\frac{\partial}{\partial\varphi}\left(\cos\varphi\, c_{g\varphi}\frac{\langle E\rangle}{\hat{\omega}}\right) = 0. \tag{17.3.60}$$

17.3.4 Latitudinal propagation of Rossby waves

We next consider the latitudinal propagation of Rossby waves based on the equation of the latitudinal wave number; (17.3.52) is rewritten as

$$\tilde{l}^2 = \tilde{k}_s^2 - \tilde{k}^2, \tag{17.3.61}$$

where $\tilde{k} = k\cos\varphi$, $\tilde{l} = l\cos\varphi$, and

$$\tilde{k}_s^2 = -\frac{\tilde{\beta}\tilde{k}}{\omega - \tilde{u}\tilde{k}} = \frac{\cos^2\varphi}{\overline{u} - c}\left(\frac{2\Omega\cos\varphi}{R} - \frac{1}{R^2}\frac{\partial}{\partial\varphi}\left[\frac{1}{\cos\varphi}\frac{\partial}{\partial\varphi}(\overline{u}\cos\varphi)\right]\right), \tag{17.3.62}$$

in which $\tilde{u} = \frac{\overline{u}}{\cos\varphi}$, $\tilde{\beta} = \hat{\beta}\cos\varphi$, and $c = \omega/k$ is the phase speed in the λ-direction. From (17.3.61), we have

$$\tilde{l} = \pm\sqrt{\tilde{k}_s^2 - \tilde{k}^2}. \tag{17.3.63}$$

Since $\tilde{k}$ is conserved along rays due to (17.3.49), we can fix the value of $\tilde{k}$ to consider the propagation of a wave packet. Thus, if the meridional distribution of the zonal winds $\overline{u}$ is given, the characteristics of wave propagation are described by the distribution of $\tilde{k}_s$. The regions where the wave packet can propagate are the latitudinal belts satisfying $\tilde{l}^2 > 0$. This means $\tilde{k}_s > \tilde{k}$ from (17.3.61). The latitude where $\tilde{l}^2 = 0$ is satisfied is called the *turning latitude*, and the latitude where $\tilde{l}^2 = \infty$ is satisfied is called the *critical latitude*. In general, we have $\tilde{k}_s \to 0$ in the limit $\varphi \to \pm\frac{\pi}{2}$, so that $\tilde{l}^2 < 0$ near the poles. Thus, the turning latitude $\tilde{l} = 0$ always exists and the wave packet does not reach the poles.

For stationary waves with $c = 0$, we have $\overline{u} = \hat{\beta}/k_s^2$. Thus, the two components of group velocity (17.3.41) and (17.3.42) are written as

$$c_{g\lambda} = \cos\varphi\frac{2\tilde{\beta}\tilde{k}^2}{\tilde{k}_s^4}, \qquad c_{g\varphi} = \cos\varphi\frac{2\tilde{\beta}\tilde{k}\tilde{l}}{\tilde{k}_s^4}. \tag{17.3.64}$$

This means that $(c_{g\lambda}, c_{g\varphi})$ is parallel to the wave number vector $(\tilde{k}, \tilde{l})$. If we define the angle between the wave number vector and the λ-direction by α, we have

$$\tilde{k}_s\cos\alpha = \tilde{k} = \text{const.} \tag{17.3.65}$$

Thus, $\cos\alpha$ becomes smaller in the region of larger $\tilde{k}_s$, whereas $\cos\alpha$ becomes larger in the region of smaller $\tilde{k}_s$. In the stationary case $c = 0$, α is equal to the direction of group velocity. In this case, $\tilde{k}_s$ plays the role of reflectivity.

As an illustrative example of the notions introduced above, Fig. 17.4 shows reflectivity $\tilde{k}_s R$ when there exists a distribution of zonal winds with two jet streams around 30° and 60°:

$$\overline{u} = 35\sin^3(\pi\mu) + 10\sin^2(\pi\mu^5) - 5(1-\mu^2)^2, \tag{17.3.66}$$

where $\mu = \sin\varphi$. Reflectivity $\tilde{k}_s$ becomes larger if the latitude becomes closer to the maximum of the jet steams, and also becomes larger near the equator. For a specified value of the wave number $\tilde{k}R = 1, 2, \cdots$, the wave can propagate only in the region $\tilde{k}_s > \tilde{k}$. It can be seen that $\tilde{k}_s^2$ is negative on the polar side of latitude 73.3° and the equatorial side of latitude 9.9°; wave packets with any wave number $\tilde{k}$ cannot enter these latitude belts. The wave number $\tilde{k}R = 3$ is shown by the dotted line in Fig. 17.4 (b). A wave with $\tilde{k}R = 3$ can propagate only in the region where the curve of $\tilde{k}_s R$ is larger than the dotted line. There are two wave propagation regions; one is latitudes lower than about 45° and the other is around 60°. Since there are two turning latitudes at both the equatorial and the polar side of the wave propagation region around 60°, Rossby waves are trapped in this latitude belt if a Rossby wave packet exists around 60°. This kind of wave propagation region is called the *wave duct.*

Wave packets of stationary waves $c = 0$ do not propagate into the region of an easterly $\overline{u} < 0$. The critical latitude is the latitude where the zonal wind vanishes $\overline{u} = 0$. In the wave propagation region between the critical latitude and the turning latitude, all the wave packets eventually propagate toward the critical latitude since those wave packets propagating toward the turning latitude turn back to the direction to the critical latitude. As shown below, it takes an infinite amount of time for wave packets to reach the critical latitude. Thus, we may say that all wave packets are absorbed into the background field near the critical latitude.

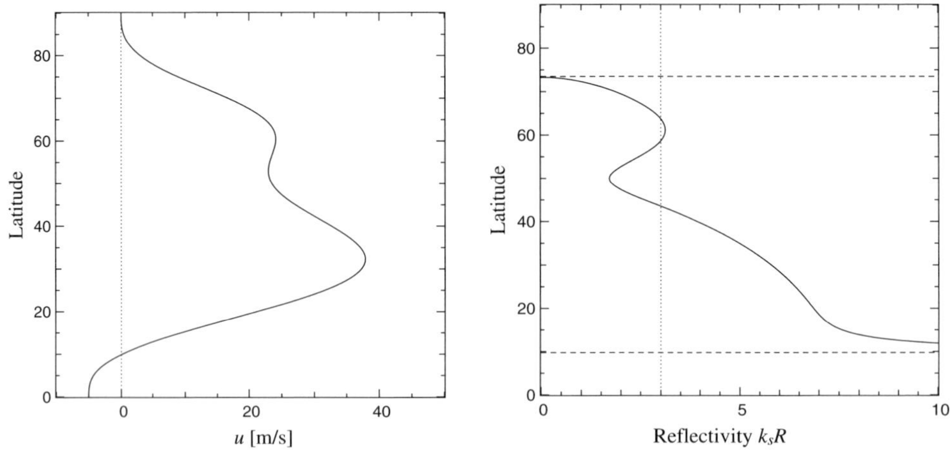

FIGURE 17.4: The distributions of zonal winds $\overline{u}$ (left), and reflectivity $\tilde{k}_s R$ (right). The dotted line is the wave number $\tilde{k}_s R = 3$. The dashed lines are boundaries between the region where $\tilde{k}_s$ is positive and that where $\tilde{k}_s$ is negative.

Let us consider a wave packet that is stationary and has zonally uniform properties in the equation of amplitude (17.3.55); thus we have

$$\cos\varphi\, c_{g\varphi}\mathcal{A} \quad = \quad \text{const.} \tag{17.3.67}$$

Using (17.3.42), (17.3.56), and $\tilde{k} = \text{const.}$, we obtain

$$\psi_0 \quad \propto \quad \tilde{l}^{-\frac{1}{2}}. \tag{17.3.68}$$

Therefore, amplitude becomes smaller near the critical latitude where $\tilde{l}$ approaches infinity, while amplitude tends to infinity near the turning latitude where $\tilde{l}$ approaches zero. Letting $y = y_c$ denote the critical latitude, we generally have

$$\overline{u} - c \quad \propto \quad y - y_c. \tag{17.3.69}$$

In this case, near the critical latitude, from (17.3.61) and (17.3.42),

$$l \propto (y - y_c)^{-\frac{1}{2}}, \qquad c_{g\varphi} \propto l^{-3} \propto (y - y_c)^{\frac{3}{2}}. \tag{17.3.70}$$

Thus, we have

$$\int \frac{dy}{c_{g\varphi}} \quad \propto \quad (y - y_c)^{-\frac{1}{2}} \to \infty \tag{17.3.71}$$

(i.e., it takes an infinite amount of time for a wave packet to reach the critical latitude).

It can be shown, however, that the WKBJ approximation breaks down near the critical latitude. The change in amplitude ψ_0 in the latitudinal direction is, from (17.3.68) and (17.3.70),

$$\psi_0 \quad \propto \quad (y - y_c)^{\frac{1}{4}}. \tag{17.3.72}$$

From this, the ratio between the scale of change in amplitude and wavelength is given by

$$\frac{1}{l\psi_0}\frac{d\psi_0}{dy} \quad \propto \quad (y - y_c)^{-\frac{1}{2}}. \tag{17.3.73}$$

This ratio becomes infinite as $y \to y_c$. Since this means that the amplitude oscillates very rapidly in one wavelength, the WKBJ approximation breaks down near the critical latitude.

17.3.5 Angular momentum change

Let us evaluate the angular momentum change when Rossby waves propagate in the meridional direction using the linearized vorticity equation (17.3.20). Following Held and Hoskins (1985), we take an example in which the basic state of zonal winds $\overline{u}$ and disturbance of initial vorticity are given as

$$\overline{u} \quad = \quad A\sin\left[\frac{3\pi}{2}(1 + \sin\varphi)\right] + B\cos^2\varphi, \tag{17.3.74}$$

$$\zeta'(\lambda, \varphi) \quad = \quad C\cos\varphi\,\cos(M\lambda)\,\exp\left[-\left(\frac{\varphi - \varphi_0}{\Delta\varphi}\right)^2\right], \tag{17.3.75}$$

where $A = 18$ m s^{-1}, $B = 14$ m s^{-1}, $C = 5.0 \times 10^{-5}$ s^{-1}, $\varphi = 45°$, $\Delta\varphi = 10°$, and $M = 6$. The vorticity equation is given from (17.3.20) in the form:

$$\frac{\partial \zeta'}{\partial t} + \frac{\overline{u}}{R\cos\varphi}\frac{\partial \zeta'}{\partial \lambda} + \hat{\beta}v' \quad = \quad \nu\nabla_H^2\zeta'. \tag{17.3.76}$$

Here, $\hat{\beta}$ is defined by (17.3.21). The diffusion-type dissipation term is assumed; however, following Held and Hoskins (1985), the second term in (17.1.19) is omitted.

The change in zonal winds can be given by time integration of (17.3.31). It can be thought that the pseudo-angular momentum associated with the initial vorticity disturbance $\overline{A}$ will be dissipated in a sufficiently large t, where $\overline{A}$ is given by (17.3.32). From (17.3.31), we have the change in zonal winds as

$$\left[\overline{u^{(2)}}(t) - \overline{u^{(2)}}(0)\right]\cos\varphi \quad = \quad \overline{A}(t=0) + \frac{\nu\cos\varphi}{\hat{\beta}}\int_0^t \overline{\zeta'\nabla_H^2\zeta'}\,dt, \tag{17.3.77}$$

where the term $F_\lambda^{(2)}$ is neglected. In general, we have $\overline{A}(t=0) > 0$ at the region where the initial disturbance is given. If $\overline{A}$ becomes smaller as the disturbance evolves and propagates, the change in zonal winds is given by (17.3.34). Therefore, westerlies are induced as the initial disturbance decays, if the contribution of the dissipation term is negligible. An example of such a change in zonal winds is given by Fig. 17.5. This figure shows the change in pseudo-angular momentum $\overline{A}$ and total change in zonal winds $\lim_{t\to\infty}\overline{u^{(2)}}(t)\cos\varphi$ for $\nu = 10^4$ m^2 s^{-1}.

17.3.6 Barotropic instability

The possible distributions of zonal winds can be constrained under the condition that zonal winds are stable in terms of barotropic instability. Let the latitudinal distribution of zonal winds be $\overline{u}(\varphi)$. From (17.1.21) and (17.1.6), the absolute angular momentum is written as

$$\zeta_a \quad = \quad -\frac{1}{R\cos\varphi}\frac{\partial(\overline{u}\cos\varphi)}{\partial\varphi} + 2\Omega\sin\varphi. \tag{17.3.78}$$

If the gradient of absolute vorticity does not change sign, zonal flow is barotropically stable. In addition, inertial instability does not occur if $\zeta_a > 0$ for $\varphi > 0$ and $\zeta_a < 0$ for $\varphi < 0$. Thus, in order for the basic field to be stable, the condition,

$$\frac{\partial\zeta_a}{\partial\varphi} \quad = \quad -\frac{\partial}{\partial\varphi}\left[\frac{1}{R\cos\varphi}\frac{\partial(\overline{u}\cos\varphi)}{\partial\varphi}\right] + 2\Omega\cos\varphi \quad > \quad 0, \tag{17.3.79}$$

must be satisfied. If $\overline{u}$ does not satisfy this condition (i.e., if absolute vorticity decreases poleward in some region), the flow is unstable and disturbances will develop; then, the arrangement of zonal winds will occur so as to satisfy the condition (17.3.79). As a result, the resultant distribution of zonal winds is marginally stable or neutral for barotropic instability. In this case, the neutral state is given by a constant profile of absolute vorticity in the latitudinal direction. This type of neutralization can be called *barotropic adjustment*, similar to convective adjustment, or baroclinic adjustment, which will be described in Section 18.2.2:

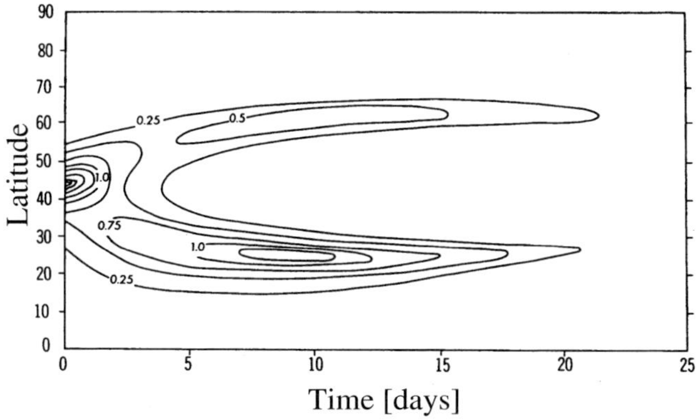

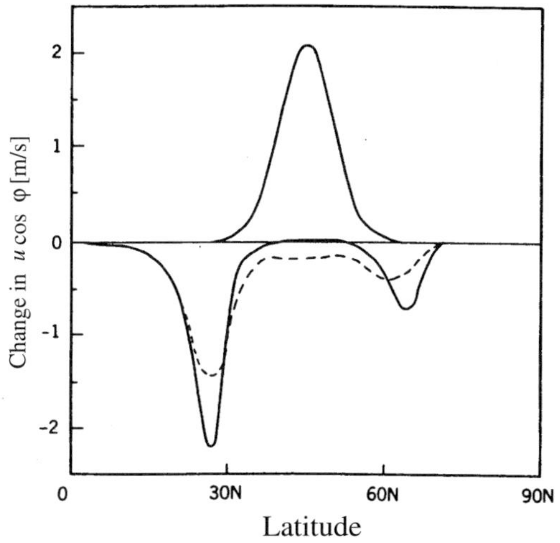

FIGURE 17.5: Left: the change in pseudo-angular momentum $\overline{A}$. The contour interval is 0.25 m s^{-1}. Right: the total change in zonal winds. Solid: $\nu = 10^4$ m^2 s^{-1}, and dashed: $\nu = 10^5$ m^2 s^{-1}. Reprinted from Held and Hoskins (1985) by permission of Elsevier (copyright, 2003).

As an example, we consider the case in which barotropic adjustment occurs in the latitude band $\varphi_1 < \varphi < \varphi_2$. We refer to zonal wind distribution after the adjustment as u_{adj}, and that of absolute vorticity as ζ_{adj}. For a barotropic adjustment state, we have

$$\frac{\partial \zeta_{adj}}{\partial \varphi} = 0. \qquad (17.3.80)$$

It is required that zonal winds be continuous at the two latitudes φ_1 and φ_2, $\overline{u}_{adj} = \overline{u}$, while vorticity does not need to be continuous. From (17.3.78) and $\zeta_{adj} =$ const., zonal wind distribution is expressed as

$$\overline{u}_{adj} \cos \varphi = -\zeta_{adj} R \sin \varphi + \Omega R \sin^2 \varphi + C, \qquad (17.3.81)$$

where C is constant. Using zonal winds at latitudes φ_1 and φ_2, $\overline{u}_1 = \overline{u}(\varphi_1)$, $\overline{u}_2 = \overline{u}(\varphi_2)$, the adjusted vorticity and zonal winds are given, respectively, by

$$\begin{aligned}\zeta_{adj} &= -\frac{\overline{u}_1 \cos\varphi_1 - \overline{u}_2 \cos\varphi_2}{R(\sin\varphi_1 - \sin\varphi_2)} + \Omega(\sin\varphi_1 + \sin\varphi_2), \\ \overline{u}_{adj} \cos\varphi &= -\frac{\overline{u}_1 \cos\varphi_1(\sin\varphi - \sin\varphi_2) - \overline{u}_2 \cos\varphi_2(\sin\varphi_1 - \sin\varphi)}{\sin\varphi_1 - \sin\varphi_2} \\ &\quad + \Omega R(\sin\varphi - \sin\varphi_2)(\sin\varphi - \sin\varphi_2), \end{aligned} \tag{17.3.82}$$

and

$$C = -\frac{\overline{u}_1 \cos\varphi_1 \sin\varphi_2 - \overline{u}_2 \cos\varphi_2 \sin\varphi_1}{\sin\varphi_1 - \sin\varphi_2} + \Omega R \sin\varphi_1 \sin\varphi_2.$$

Since angular momentum is conserved during the adjustment process, we have

$$\int_{\varphi_1}^{\varphi_2} \overline{u} \cos\varphi \, d\varphi = \int_{\varphi_1}^{\varphi_2} \overline{u}_{adj} \cos\varphi \, d\varphi. \tag{17.3.83}$$

From (17.3.82) and (17.3.83), we obtain the relation between φ_2 and φ_1. The set of latitudes φ_1 and φ_2 cannot uniquely be determined from this procedure alone.

Fig. 17.6 shows an example of zonal wind distribution which is barotropically unstable. The basic distribution of zonal winds is specified as a jet stream type:

$$\overline{u} = u_0 \sin^3(\pi \sin^2\varphi), \tag{17.3.84}$$

where $u_0 = 50$ m s^{-1}. Zonal winds have a maximum at $\varphi = 45°$. At $\varphi = 57.8°$, the absolute vorticity ζ_a becomes maximum, and the gradient of angular momentum changes its sign. Thus, the region near the jet maximum is unstable in terms of barotropic instability. After the disturbance has fully developed, zonal winds will be close to (17.3.81) near the adjusted region. The latitudinal width of the disturbances due to barotropic instability is extended until any local unstable profiles are dissolved. The figure shows the distributions of zonal wind and absolute vorticity after adjustment by the dashed-dotted curves where the adjusted region is between 47.4° and 82.8°.

17.4 Turbulence on a sphere

17.4.1 Turbulence on a β-plane

The mid-latitude jet stream of the atmosphere sometimes emerges as a multiple jet structure. Such a multiple jet structure is familiar in the Jovian atmosphere The theory of two-dimensional turbulence on a sphere gives an explanation for a multiple jet structure. Before describing the two-dimensional turbulence on a sphere, we overview the two-dimensional turbulence on a β-plane. We have already considered the two-dimensional turbulence without the effect of rotation in Section 11.1.2; the energy cascade and energy spectrum on two-dimensional turbulence are derived from similarity theory. The turbulence on a β-plane can be considered as an extension of two-dimensional turbulence.

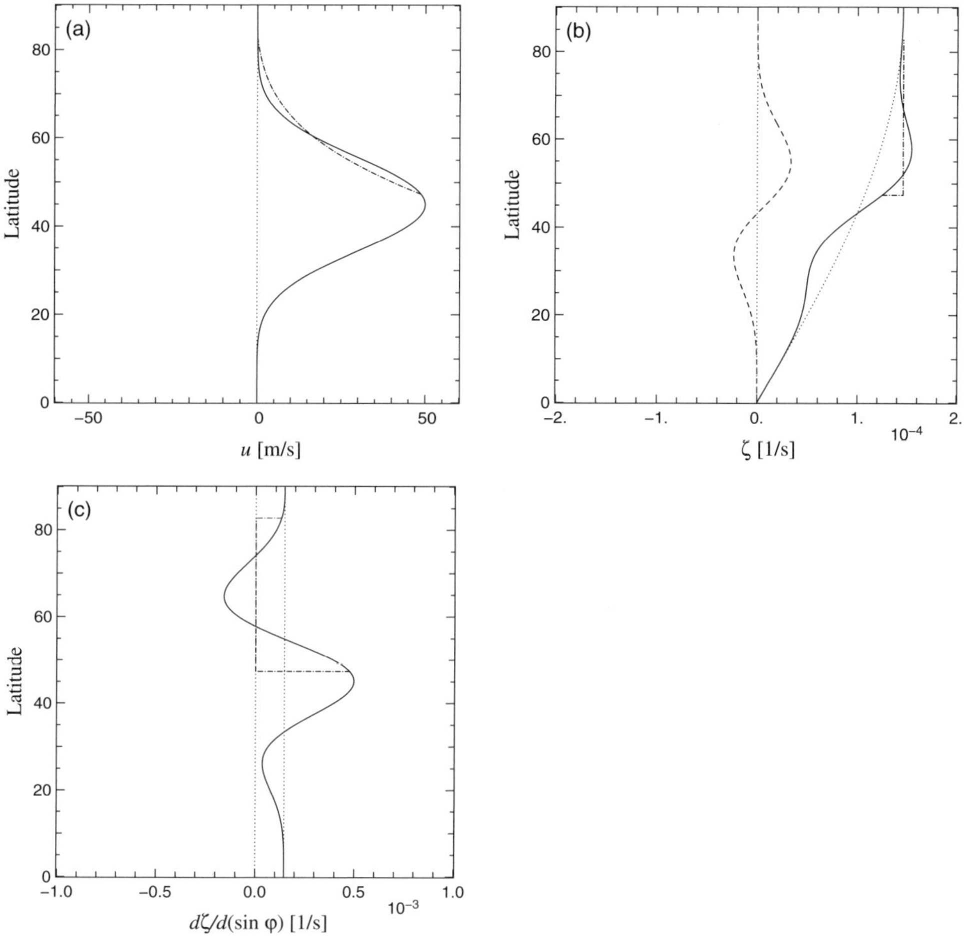

FIGURE 17.6: Latitudinal distributions of (a) zonal winds $\overline{u}$, (b) absolute vorticity ζ_a, and (c) gradient of absolute vorticity $d\zeta_a/d(\sin\varphi)$. In (b), the dashed curve is relative vorticity and the dotted curve is the Coriolis component. In (c), the dashed curve is relative vorticity, and the dotted line is the β term. In (a) and (b), dashed-dotted curves express the profiles in the adjustment region.

The vorticity equation on a β-plane is given by

$$\frac{\partial}{\partial t}\nabla^2\psi + J(\psi, \nabla^2\psi) + \beta\frac{\partial \psi}{\partial x} = 0, \qquad (17.4.1)$$

where ψ is the streamfunction and J is the Jacobian defined by

$$J(A, B) = \frac{\partial A}{\partial x}\frac{\partial B}{\partial y} - \frac{\partial A}{\partial y}\frac{\partial B}{\partial x}. \qquad (17.4.2)$$

The second term on the left-hand side of (17.4.1) is the nonlinear term, and the third term is the β effect. In order to examine the possible statistical equilibrium states of the system described by (17.4.1), we start from the following two extreme regimes.

(I) If the nonlinear term is larger than the β effect, we can omit the β term in (17.4.1) to obtain

$$\frac{\partial}{\partial t}\nabla^2\psi + J(\psi, \nabla^2\psi) = 0. \tag{17.4.3}$$

Thus, the argument of two-dimensional turbulence in Section 11.1.2 is applicable to the statistical equilibrium of this system. In this case, energy is transferred to smaller wave number regions and the upward energy cascade occurs. Note that (17.4.3) is satisfied in the case on an f-plane (i.e., turbulence in the rotating frame). On an f-plane, it is thought that the upward cascade occurs in scales that are smaller than the Rossby radius of deformation.

(II) In contrast, if the β term is larger than the nonlinear term, neglecting the nonlinear term in (17.4.1) gives

$$\frac{\partial}{\partial t}\nabla^2\psi + \beta\frac{\partial\psi}{\partial x} = 0. \tag{17.4.4}$$

This is a linear equation and can be described by superpositions of linear Rossby waves. In this case, if a basic solution is written as

$$\psi = \psi_k \exp[i(\boldsymbol{k}\cdot\boldsymbol{x} - \omega t)], \tag{17.4.5}$$

the dispersion relation is given by

$$\omega = -\frac{\beta k_x}{|\boldsymbol{k}|^2}, \tag{17.4.6}$$

where $\boldsymbol{k} = (k_x, k_y)$. If motions are described by such a linear equation, no energy transfer occurs between different wave numbers and energy cascade does not occur.

As shown above, it can be expected that the cascades of energy and enstrophy are different between regimes (I) and (II). This means that it is important to know whether the flow field of a given system described by (17.4.1) is closer to (I) or (II). To characterize the flow field, we introduce a parameter that represents the ratio of the nonlinear term to the β term:

$$\varepsilon = \frac{\text{nonlinear term}}{\beta \text{ term}} = \frac{U}{\beta L^2}, \tag{17.4.7}$$

where U is a characteristic scale of mean velocity and L is a characteristic horizontal scale of motion. U can be defined by the square of mean kinematic energy, and L is the inverse of the average wave number weighted by the energy spectrum $E(k)$:

$$U^2 = \frac{1}{S}\int \frac{|\nabla\psi|^2}{2}\, dS, \tag{17.4.8}$$

$$\frac{1}{L} = \langle k\rangle = \frac{\int kE(k)\, dk}{\int E(k)\, dk}, \tag{17.4.9}$$

where S is the area of the system. In the case of $\varepsilon \gg 1$, the flow field is closer to regime (I), whereas in the case of $\varepsilon \ll 1$, the flow field is essentially linear and closer to regime (II). If total energy is conserved, U is independent of time and is determined by the initial value. Therefore, the parameter ε changes with the value of L.

Based on the fact that the turbulence on a β-plane has two extreme characteristic regimes described by (I) and (II), the following picture of time evolution can be shown:

1. First, initial energy is given as $\varepsilon \gg 1$.

2. Based on the two-dimensional turbulence of regime (I), upward cascade of energy occurs.

3. The typical length scale L becomes larger and $\langle k \rangle$ becomes smaller.

4. Then, ε also becomes smaller.

5. If ε is small enough such that $\varepsilon \approx 1$, Rossby waves emerge as described by (II).

According to the above scenario, the two-dimensional turbulence on a β-plane experiences a transition from (I) to (II).

As the flow field evolves from (I) to (II), it is expected that the timescale of the upward cascade of energy becomes slower. The typical timescale of regime (I) is given by

$$T \approx \frac{L}{U}. \tag{17.4.10}$$

This indicates that as L becomes larger due to the energy cascade, timescale T becomes larger. Although the timescale of regime (II) is constrained by the dispersion relation of Rossby waves (17.4.6), its value cannot be determined solely by the dispersion relation. For instance, if the upward cascade of energy continues and keeps isotropy at $k_x = k_y$, ω would get larger because of (17.4.6). It is known, however, that ω becomes even smaller in regime (II) according to the weakly nonlinear theory whose details we do not describe here (Rhines, 1975). Because of the dispersion relation, if ω becomes smaller, the zonal wave number k_x becomes smaller compared with k_y, which brings the anisotropy of turbulence. In regime (II), therefore, turbulence is no longer isotropic and a zonal flow emerges. The characteristic length of the width of zonal winds is given by setting $\varepsilon = 1$ in (17.4.7); thus

$$L = \sqrt{\frac{U}{\beta}}, \tag{17.4.11}$$

This length is called the *Rhines scale*. Based on numerical calculations of dissipative turbulence on a β-plane, Rhines (1975) demonstrates the above scenario.

The Rhines scale is different from the Rossby radius of deformation, which is the natural horizontal length of a rotating fluid. The Rossby radius of deformation is given by

$$L_R = \frac{c}{f}, \tag{17.4.12}$$

where c is the speed of a gravity wave and f is the Coriolis parameter. In a stratified fluid, the speed of a gravity wave is given by $c = NH$ where H is the depth of atmosphere and N is the buoyancy frequency. In a shallow water system, it is given by $c = \sqrt{gH}$. The typical horizontal scale of turbulence on a β-plane corresponds to the latitudinal length scale of zonal winds (i.e., the width of the jet), when zonal winds have a band structure. The width of the jet can be closer to either the Rhines scale or the Rossby length in extreme cases.

17.4.2 Two-dimensional turbulence on a sphere

Arguments similar to turbulence on a β-plane can be applicable to fluid motions on a sphere. From (17.3.4), the vorticity equation on a sphere is written as

$$\frac{\partial \zeta}{\partial t} + \frac{1}{R^2} J(\psi, \zeta) + \frac{2\Omega}{R^2}\frac{\partial \psi}{\partial \lambda} = \nu \left(\nabla_H^2 + \frac{2}{R^2} \right)^N \zeta, \tag{17.4.13}$$

where R is the radius of the sphere and the dissipation term on the right-hand side is assumed to be the N-th power of the Laplacian of vorticity instead of (17.1.19). Such a high power of N with $N \geq 2$ is required for numerical calculations. The Jacobian on a sphere is written as

$$J(A, B) = \frac{\partial A}{\partial \lambda}\frac{\partial B}{\partial \varphi} - \frac{\partial A}{\partial \varphi}\frac{\partial B}{\partial \lambda}. \tag{17.4.14}$$

The equation system described by (17.4.13) has conservative quantities; total energy, total enstrophy, and angular momentum are respectively defined by

$$E = \frac{1}{2}\int |\boldsymbol{u}|^2 \, dS = \frac{1}{2}\int |\nabla \psi|^2 \, dS, \tag{17.4.15}$$

$$Q = \frac{1}{2}\int |\zeta|^2 \, dS = \frac{1}{2}\int |\nabla^2 \psi|^2 \, dS, \tag{17.4.16}$$

$$A = \int u \cos\varphi \, dS, \tag{17.4.17}$$

where only those parts that are relative to rigid body rotation Ω are considered.

Expanding the streamfunction using spherical harmonics gives

$$\psi(\lambda, \varphi, t) = \sum_{n=0}^{\infty} \sum_{m=-n}^{n} \psi_n^m(t) P_n^m(\sin\varphi) e^{im\lambda}, \tag{17.4.18}$$

where ψ_n^m are amplitudes of the streamfunction with zonal wave number m and latitudinal wave number $n - m$, and P_n^m are the associated Legendre functions.

Integrating the energy (17.4.15) by part, we obtain

$$E(t) \quad = \quad \frac{1}{2}\int \psi\nabla^2\psi\, dS \quad = \quad \frac{1}{2R^2}\sum_{n=0}^{\infty}\sum_{m=-n}^{n} n(n+1)|\psi_n^m(t)|^2. \qquad (17.4.19)$$

From this, the energy spectrum defined as a function of total wave number n is given by

$$E(n,t) \quad = \quad \frac{n(n+1)}{2R^2}\sum_{m=-n}^{n}|\psi_n^m(t)|^2. \qquad (17.4.20)$$

The observed power spectra obtained from aircraft data are shown in Fig. 17.7. This shows the relations between power spectra density and horizontal wavelength for zonal wind, meridional wind, and potential temperature. Between about 1000 km and 3000 km wavelength, all spectra have a slope near -3. This implies an enstrophy cascade from longer to shorter scales of the two-dimensional turbulence described in Section 11.1.2. The spectra have a different slope regime ($-5/3$) at shorter wavelengths below about 400 km.

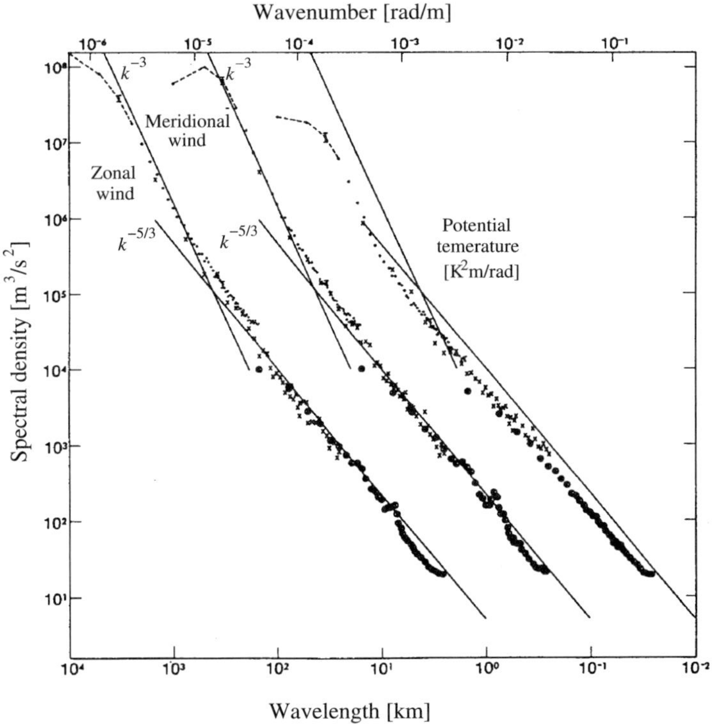

FIGURE 17.7: The variance power spectra of wind and potential temperature near the tropopause from aircraft data. The spectra for meridional wind and potential temperature are shifted to the 10^1 and 10^2 shorter scales on the right. After Nastrom and Gage (1985) by permission of the American Meteorological Society.

Yoden and Yamada (1993) calculated dissipative turbulence on a sphere. Fig. 17.8 shows the results of their calculations on the energy spectrum. Time is non-dimensionalized by $T = R/U$ where U is the average velocity defined by the square of kinetic energy. Numerical dissipation is given by (17.4.13) with $N = 2$ and the dissipation coefficient is set to $\nu = 10^{-6}$. The figure shows the ensemble average of 48 cases. When the rotation velocity Ω is equal to zero, the energy spectrum has a dependency n^{-6} for $n > 10$ and n^{-3} for $n < 10$. As Ω becomes larger, the wave

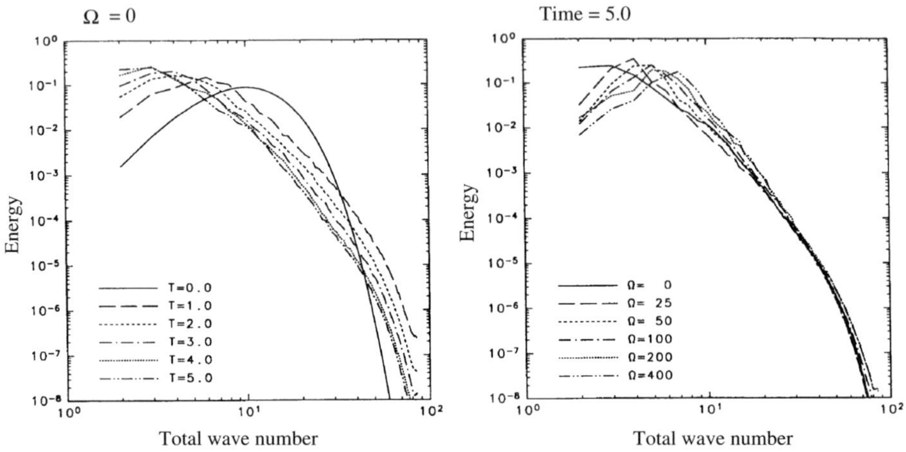

FIGURE 17.8: The energy spectrum of two-dimensional turbulence on a sphere. Left: without rotation $\Omega = 0$. Right: dependency on the rotation velocity Ω. The ensemble average of 48 cases is shown. After Yoden and Yamada (1993) by permission of the American Meteorological Society.

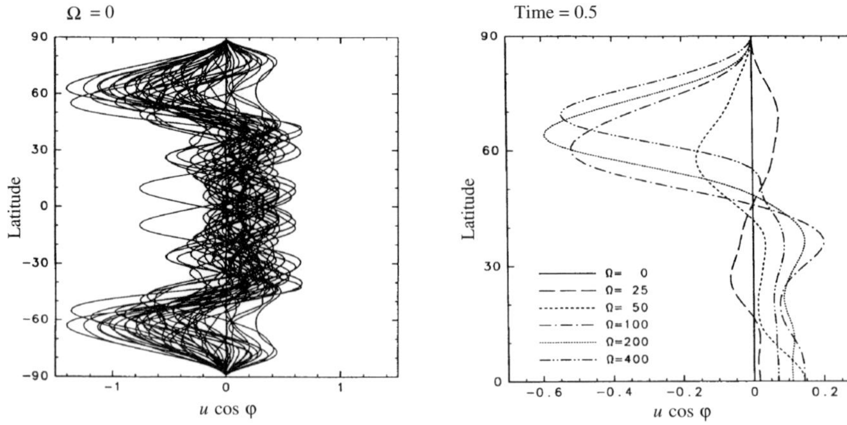

FIGURE 17.9: The zonal-mean angular momentum $u \cos \varphi$ of two-dimensional turbulence on a sphere. Left: 48 cases for $\Omega = 100$ at $t = 5$. Right: dependency on Ω at $t = 5$. The ensemble average is defined as a result of averaging 48 cases. After Yoden and Yamada (1993) by permission of the American Meteorological Society.

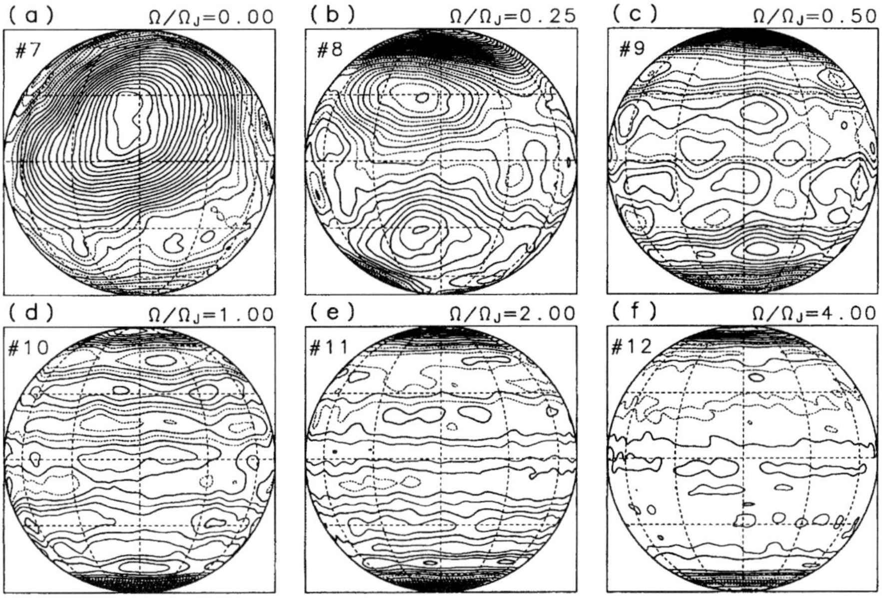

FIGURE 17.10: The streamfunctions for the two-dimensional turbulence experiment on a sphere at 1000 Jovian days. The contour interval is 2.5×10^8 m^2 s^{-1}. Negative values are dotted. After Nozawa and Yoden (1997) by permission of the American Institute of Physics.

number of the maximum energy spectrum n_{max} becomes larger. This suggests that upward energy cascade is suppressed in the smaller scale. Wave numbers larger than n_{max} have a dependency as n^{-6}. Fig. 17.9 shows snapshots of the latitudinal distributions of the zonal-mean angular momentum for different cases and their ensemble averages for different rotation rates. The multiple jet structure appears in these snapshot figures. It is interesting that zonal winds are easterly in high latitudes for the cases of large rotation rates.

It is argued that the above theory for two-dimensional turbulence on a sphere can be used to explain the jet structure of the Jovian atmosphere. Such numerical experiments were conducted by Williams (1978) and were revisited by Nozawa and Yoden (1997). Fig. 17.10 shows the streamfunctions on a sphere with dependency on the rotation rate. The Jovian case is $\Omega/\Omega_J = 1$. As the rotation rate increases, the zonal band structure becomes dominant.

17.5 Appendix: Expressions of friction terms on a sphere

The Laplacian of a vector quantity $\boldsymbol{v} = (v_\lambda, v_\varphi, v_r)$ is expressed as (A1.5.5) in Appendix A1. Let us substitute $\boldsymbol{v} = (u, v, 0)$ and assume that the horizontal components u and v are proportional to the radius r to help us consider the Laplacian on a sphere. We have, for instance, for u,

$$\begin{aligned}\nabla^2 u &= \left[\frac{1}{r^2\cos^2\varphi}\frac{\partial^2}{\partial\lambda^2} + \frac{1}{r^2\cos\varphi}\frac{\partial}{\partial\varphi}\left(\cos\varphi\frac{\partial}{\partial\varphi}\right) + \frac{1}{r^2}\frac{\partial}{\partial r}\left(r^2\frac{\partial}{\partial r}\right)\right]u \\ &= \left[\frac{1}{r^2\cos^2\varphi}\frac{\partial^2}{\partial\lambda^2} + \frac{1}{r^2\cos\varphi}\frac{\partial}{\partial\varphi}\left(\cos\varphi\frac{\partial}{\partial\varphi}\right)\right]u + \frac{2}{r^2}u.\end{aligned}$$

By replacing r by R, we obtain

$$\nabla^2 u = \left(\nabla_H^2 + \frac{2}{R^2}\right)u.$$

A similar expression also holds for v. Thus, from (A1.5.5), we have

$$\begin{aligned}\nabla^2\boldsymbol{v} = \Bigg(&\left(\nabla_H^2 + \frac{2}{R^2}\right)u - \frac{2\sin\varphi}{R^2\cos^2\varphi}\frac{\partial v}{\partial\lambda} - \frac{u}{R^2\cos^2\varphi}, \\ &\left(\nabla_H^2 + \frac{2}{R^2}\right)v - \frac{v}{R^2\cos^2\varphi} + \frac{2\sin\varphi}{R^2\cos^2\varphi}\frac{\partial u}{\partial\lambda}, \\ &-\frac{2}{R^2\cos\varphi}\frac{\partial u}{\partial\lambda} - \frac{2}{R^2\cos\varphi}\frac{\partial(v\cos\varphi)}{\partial\varphi}\Bigg).\end{aligned}$$

In particular, using the expression for divergence δ in (17.1.7), we can find $(\nabla^2\boldsymbol{v})_r = -2\delta$. The following relation also holds if u is independent of λ:

$$\left(\nabla_H^2 + \frac{2}{R^2}\right)u - \frac{u}{R^2\cos^2\varphi} = \frac{1}{R^2\cos^2\varphi}\frac{\partial}{\partial\varphi}\left[\cos^3\varphi\frac{\partial}{\partial\varphi}\left(\frac{u}{\cos\varphi}\right)\right].$$

This transformation corresponds to that of the frictional forces in the spherical coordinates given by (A1.5.9). If we assume that the velocity components are uniform in the λ-direction and $\sigma'_{\lambda r}$ is negligible in these equations, we obtain

$$f_\lambda = \frac{1}{\rho}\frac{1}{R\cos^2\varphi}\frac{\partial}{\partial\varphi}(\cos^2\varphi\,\sigma'_{\lambda\varphi}) = \frac{\nu}{\rho}\left(\nabla_H^2 u - \frac{u}{R^2\cos^2\varphi}\right).$$

Here, the symbol of the dissipation coefficient is written as ν instead of η and the zonal winds are represented by u instead of v_λ. If the dependency of velocity components on λ is negligible, the stress tensor is given from (A1.5.8) as

$$\sigma'_{\lambda\varphi} = \nu\frac{\cos\varphi}{R}\frac{\partial}{\partial\varphi}\left(\frac{u}{\cos\varphi}\right).$$

If we express the dissipation term using vorticity and divergence, we obtain (17.1.19) and (17.1.20). These relations can be easily derived if the three-dimensional vector form is used. That is, since we generally have

$$\nabla^2\boldsymbol{A} = \nabla(\nabla\cdot\boldsymbol{A}) - \nabla\times\nabla\times\boldsymbol{A},$$

then

$$\nabla\times\nabla^2\boldsymbol{v} = -\nabla\times\nabla\times\nabla\times\boldsymbol{v} = -\nabla\times\nabla\times\boldsymbol{\omega} = \nabla^2\boldsymbol{\omega}, \quad (17.5.1)$$

$$\nabla\cdot\nabla^2\boldsymbol{v} = \nabla\cdot\nabla(\nabla\cdot\boldsymbol{v}) = \nabla^2\delta. \quad (17.5.2)$$

Note here that u and v are proportional to the radius r whereas $\boldsymbol{\omega}$ and δ are independent of r, when mapping onto a two-dimensional spherical surface. As for vorticity, since from (A1.5.3),

$$\begin{aligned}\boldsymbol{\omega} &= (\omega_\lambda, \omega_\varphi, \omega_r) = \left(-\frac{2v}{R}, \frac{2u}{R}, \frac{1}{R\cos\varphi}\frac{\partial v}{\partial\lambda} - \frac{1}{R\cos\varphi}\frac{\partial(u\cos\varphi)}{\partial\varphi}\right) \\ &= \left(-\frac{2v}{R}, \frac{2u}{R}, \zeta\right),\end{aligned}$$

the radial component of the Laplacian of the vorticity vector gives (17.1.19) using (17.5.1). For divergence, since

$$\nabla\cdot\nabla^2\boldsymbol{v} = \nabla_H\cdot(\nabla^2\boldsymbol{v})_H + \frac{1}{r^2}\frac{\partial}{\partial r}\left[r^2(\nabla^2\boldsymbol{v})_r\right] = \nabla_H\cdot(\nabla^2\boldsymbol{v})_H - \frac{2\delta}{r^2},$$

we obtain using (17.5.2) and $\nabla^2\delta = \nabla_H^2\delta$

$$\nabla_H\cdot(\nabla^2\boldsymbol{v})_H = \left(\nabla_H^2 + \frac{2}{R^2}\right)\delta, \qquad (17.5.3)$$

from which we have (17.1.20).

References and suggested reading

The Hadley circulation model of the shallow water equations in Section 17.2 follows Held and Phillips (1990), who discuss the applicability of the shallow water model to the real atmosphere. The propagation of Rossby waves on a sphere and their angular momentum transport is reviewed by Held and Hoskins (1985). Wave-mean flow interaction in the middle atmosphere and the properties of wave propagation are summarized by Andrews et al. (1987). The basic theory of wave propagation called the ray theory is referred to in the references cited in Chapter 4. The two-dimensional turbulence on a β-plane described in Section 17.4 is numerically studied and theoretically interpreted by Rhines (1975) and McWilliams (1984). Two-dimensional turbulence on a sphere is examined by Williams (1978) who was the first to give an interpretation of the jets in the Jovian atmosphere, and was followed by Yoden and Yamada (1993) and Nozawa and Yoden (1997).

Andrews, D. G., Holton, J. R., and Leovy, C. B., 1987: *Middle Atmosphere Dynamics*. Academic Press, San Diego, 489 pp.

Held, I. M. and Hoskins, B. J., 1985: Large-scale eddies and the general circulation of the troposphere. *Advances in Geophys.*, **28**, 3–31.

Held, I. M. and Phillips, P. J., 1990: A barotropic model of the interaction between the Hadley cell and a Rossby wave. *J. Atmos. Sci.*, **47**, 856–869.

McWilliams, J., 1984: The emergence of isolated coherent vortices in turbulent flow. *J. Fluid Mech.*, **146**, 21–43.

Nastrom, G. D. and Gage, K. S., 1985: A climatology of atmospheric wavenumber spectra of wind and temperature observed by commercial aircraft. *J. Atmos. Sci.*, **42**, 950–960.

Nozawa, T. and Yoden, S., 1997: Formation of zonal band structure in forced two-dimensional turbulence on a rotating sphere. *Phys. Fluids*, **9**, 2081–2093.

Rhines, P. B., 1975: Waves and turbulence on a beta-plane. *J. Fluid Mech.*, **69**, 417–443.

Williams, G. P., 1978: Planetary circulations. I: Barotropic representation of Jovian and terrestrial turbulence. *J. Atmos. Sci.*, **35**, 1399–1426.

Yoden, S. and Yamada, M., 1993: A numerical experiment on two-dimensional decaying turbulence on a rotating sphere. *J. Atmos. Sci.*, **38**, 631–643.

18

Mid-latitude circulations

Some properties of mid-latitude circulation are described in this chapter by comparing zonally averaged meridional circulations using various averaging methods to study angular momentum and energy budgets. In particular, baroclinic waves and their associated meridional circulations are examined.

In general, Eulerian mean circulation associated with baroclinic waves is indirect where its direction is opposite to that of Hadley circulation. Such an indirect circulation in mid-latitudes is called the *Ferrel cell*. The Ferrel cell plays an important role in the global angular momentum budget. On the other hand, if circulation is zonally averaged on isentropic surfaces, a hemispheric one-cellular direct circulation emerges with no indirect circulation. The hemispheric circulation in isentropic coordinates is relevant to heat transport. It is also related to material transport, which will be described in Chapter 19.

The statistical equilibrium states of baroclinic waves are also discussed. The dynamics of tropopause height and static stability in mid-latitudes are closely related to the statistics of baroclinic waves. The chapter concludes with the description of the typical life cycle of extratropical cyclones caused by baroclinic instability.

18.1 Meridional circulation

18.1.1 Eulerian mean circulation

We start with the zonally averaged circulation using pressure coordinates. In order to describe mid-latitude circulation in particular, we use the quasi-geostrophic equations on a sphere. The zonal-mean quasi-geostrophic equations in spherical and pressure coordinates are given by†

---

†The equation set in pressure coordinates (3.3.42)–(3.3.44) is rewritten as the corresponding set in quasi-geostrophic approximations (7.4.25)–(7.4.29) using spherical coordinate equations (3.3.15)–(3.3.18).

$$\frac{\partial \overline{u}}{\partial t} + \frac{1}{R\cos^2\varphi}\frac{\partial}{\partial \varphi}(\cos^2\varphi\overline{u'v'}) - f\overline{v} = -\overline{G_x}, \tag{18.1.1}$$

$$f\frac{\partial \overline{u}}{\partial p} - \frac{\mathcal{R}_p}{R}\frac{\partial \overline{\theta}}{\partial \varphi} = 0, \tag{18.1.2}$$

$$\frac{1}{R\cos\varphi}\frac{\partial}{\partial \varphi}(\overline{v}\cos\varphi) + \frac{\partial \overline{\omega}}{\partial p} = 0, \tag{18.1.3}$$

$$\frac{\partial \overline{\theta}}{\partial t} + \frac{1}{R\cos\varphi}\frac{\partial}{\partial \varphi}(\cos\varphi\overline{v'\theta'}) + \overline{\omega}\frac{\partial \overline{\theta}}{\partial p} = \overline{\mathcal{Q}}. \tag{18.1.4}$$

Eq. (18.1.1) is the momentum equation in the longitudinal direction, (18.1.2) is thermal wind balance, (18.1.3) is the continuity equation, and (18.1.4) is the equation of potential temperature. $(\overline{\ })$ denotes the zonal average on isobaric surfaces and $(')$ denotes the deviation from zonal average. The diabatic heating $\mathcal{Q}$ and the coefficient $\mathcal{R}_p$ are defined by

$$\mathcal{Q} = Q\left(\frac{p_0}{p}\right)^{\kappa}, \quad \mathcal{R}_p = \frac{R_d}{p}\left(\frac{p}{p_0}\right)^{\kappa} = \frac{R_d}{p_0}\left(\frac{p_0}{p}\right)^{\frac{1}{\gamma}}, \tag{18.1.5}$$

where Q is the original diabatic term, $\kappa = R_d/C_p$, and $\gamma = C_p/C_v$. The thermal wind balance (18.1.2) is given from the hydrostatic balance (3.3.43) using the specific volume $\alpha = \mathcal{R}_p\theta$. The frictional forcing G_x is given by

$$G_x = \frac{1}{\rho}\frac{\partial \tau_x}{\partial z} = \frac{\partial \tau_{px}}{\partial p}, \tag{18.1.6}$$

where τ_x is the stress tensor and $\tau_{px} = -g\tau_x$. We assume that the contributions from topographic torque are included in G_x (Peixoto and Oort, 1992).

From the continuity equation (18.1.3), the meridional streamfunction Ψ can be defined as

$$\overline{v} = \frac{g}{2\pi R\cos\varphi}\frac{\partial \Psi}{\partial p}, \quad \overline{\omega} = -\frac{g}{2\pi R^2\cos\varphi}\frac{\partial \Psi}{\partial \varphi}. \tag{18.1.7}$$

Assuming $\Psi = 0$ at the top of the atmosphere $p = 0$, we integrate the first equation of (18.1.7) vertically to obtain

$$\Psi = 2\pi R\cos\varphi\int_0^p \overline{v}\frac{dp}{g}. \tag{18.1.8}$$

Over a long time average, in particular, vertical velocity is $\overline{\omega} = 0$ at the surface $p = p_s$. Thus, the total vertical integral gives $\Psi(p_s) = 0$.

We consider a long time-averaged state such that tendency terms can be neglected (i.e., a statistical equilibrium state). In this case, the angular momentum balance is given from (18.1.1) as

$$\frac{1}{R\cos^2\varphi}\frac{\partial}{\partial \varphi}(\cos^2\varphi\overline{u'v'}) - f\overline{v} = -\overline{G_x}, \tag{18.1.9}$$

in which the latitudinal transport of relative angular momentum is expressed as $\overline{u'v'}$. Here, we define a streamfunction of the transport of zonal-mean relative angular momentum by

$$\Psi_{ang} = 2\pi R^2 \cos^2\varphi \int_0^p \overline{u'v'} \frac{dp}{g}. \tag{18.1.10}$$

Integrating (18.1.9) from the top of the atmosphere to a pressure level p yields

$$\frac{\partial}{\partial\varphi}\Psi_{ang} - R^2\cos\varphi f\Psi = -\frac{2\pi R^3}{g}\cos^2\varphi\overline{\tau_{px}}. \tag{18.1.11}$$

Generally, the frictional force in the atmosphere has an appreciable value only in the lower boundary layer. Thus, the angular momentum balance in the free atmosphere is expressed by neglecting the right-hand side of (18.1.11) as

$$\frac{\partial}{\partial\varphi}\Psi_{ang}(p) - R^2\cos\varphi f\Psi(p) = 0. \tag{18.1.12}$$

In the upper layer of the troposphere, in general, angular momentum is transported from lower latitudes to mid-latitudes, so that angular momentum transport $\overline{u'v'}$ is convergent in mid-latitudes while it is divergent in lower latitudes. In this situation, we have $f\Psi(p) < 0$ in mid-latitudes from (18.1.12). This means that equatorward mass transport exists in the upper layer of mid-latitudes. This kind of zonal-mean circulation is called *indirect circulation*, or the *Ferrel cell*, and is a general characteristic of mid-latitude circulation.

If we extend the integration of (18.1.11) to the surface $p = p_s$, we obtain

$$\frac{\partial}{\partial\varphi}\Psi_{ang}(p_s) = -2\pi R^3\cos^2\varphi C\overline{u}(p_s), \tag{18.1.13}$$

where $\Psi(p_s) = 0$ and the bulk method

$$g\tau_{px} = C\overline{u}(p_s), \tag{18.1.14}$$

is used following (16.2.39). C is a constant. In mid-latitudes, since the left-hand side of (18.1.13) is positive in general, surface winds are westerly (i.e., the surface winds in the latitudes where indirect circulation resides must be westerly). This situation is schematically depicted in the left panel of Fig. 18.1.

From (18.1.4), thermal balance over a long time average is represented by

$$\frac{1}{R\cos\varphi}\frac{\partial}{\partial\varphi}(\cos\varphi\overline{v'\theta'}) + \overline{\omega}\frac{\partial\overline{\theta}}{\partial p} = \overline{Q}. \tag{18.1.15}$$

The second term on the left-hand side is the adiabatic heating or cooling due to vertical winds. Stratification is stable in mid-latitudes, $\frac{\partial\overline{\theta}}{\partial p} < 0$, in general. Since the equator side of the indirect circulation region has downward motions $\overline{\omega} > 0$ and the polar side of the indirect circulation region has upward motions $\overline{\omega} < 0$, adiabatic warming occurs in the equator side and adiabatic cooling occurs in the

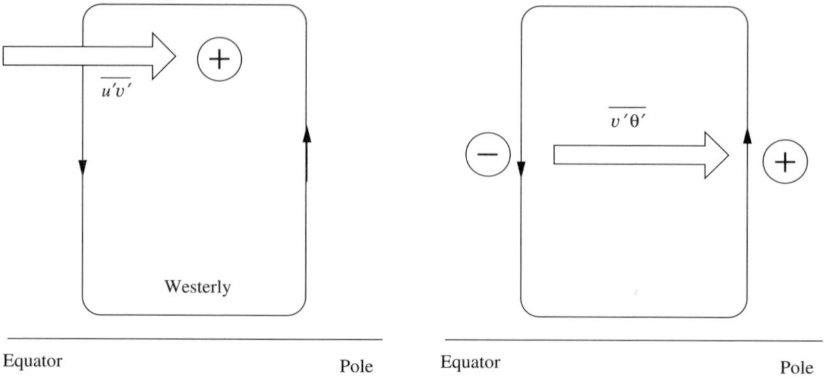

FIGURE 18.1: The schematic relations between the indirect cell and angular momentum transport or heat transport. Left: the direction of the Ferrel cell and angular momentum transport. Right: heat flux due to the Ferrel cell. The symbol + represents the convergence of heat flux or angular momentum flux, and the symbol − represents divergence.

polar side of indirect circulation. If the diabatic term on the right-hand side is negligible, the eddy heat flux $\overline{v'\theta'}$ must be poleward in mid-latitudes to balance the adiabatic heating term; thus heat flux is divergent in the equator side and convergent in the polar side. This type of heat flux is characteristic of baroclinic waves. Although the diabatic effects due to radiative cooling and latent heat release are important, these effects are not so strong as to overcome the direction of the eddy heat flux. This relation between heat flux and indirect circulation is schematically shown in the right panel of Fig. 18.1.

Fig. 18.2 shows the observational meridional distributions of the mass streamfunction Ψ, the eddy momentum transport $\overline{u'v'}$ and the eddy heat transport $\overline{v'T'}$. The schematic relations shown in Fig. 18.1 can be confirmed from the three panels of this figure. In general, eddy momentum transport becomes greater in upper layers, while eddy heat transport becomes greater in the lower layers.

18.1.2 Angular momentum balance between the Hadley cell and the Ferrel cell

We next consider the angular momentum transport between the Hadley cell in low latitudes and the Ferrel cell in mid-latitudes. Meridional circulations over low and mid-latitudes are related through the angular momentum budget. We use primitive equations on a sphere to consider Hadley circulation, instead of the quasi-geostrophic equations used in the previous section. The equation for zonal-mean zonal winds in pressure coordinates on a sphere is written as

$$\frac{\partial \overline{u}}{\partial t} + \frac{1}{R\cos^2\varphi}\frac{\partial}{\partial \varphi}\left(\cos^2\varphi\,\overline{uv}\right) + \frac{\partial}{\partial p}\left(\overline{u\omega}\right) - 2\Omega\sin\varphi\,\overline{v} = -\overline{G_x}. \quad (18.1.16)$$

From this, using angular momentum $l = uR\cos\varphi + \Omega R^2\cos^2\varphi$, we obtain the

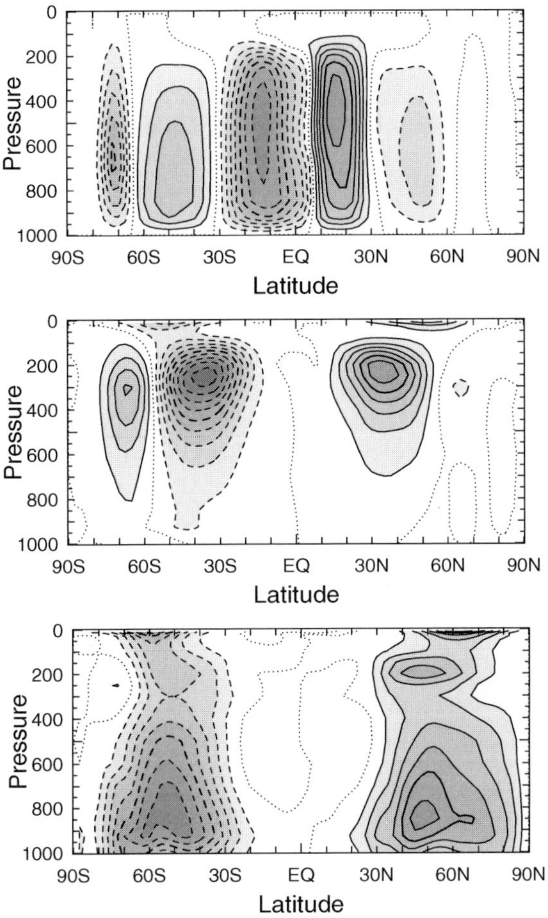

FIGURE 18.2: The meridional distributions of the mass streamfunction Ψ (top), eddy momentum transport $\overline{u'v'}$ (middle), and eddy heat transport $\overline{v'T'}$ (bottom). The contour interval of each figure is 10^{10} kg s^{-1}, 5 m^2 s^{-2}, and 2 K m s^{-1}, respectively: solid curves represent positive values, dashed curves negative values, and dotted curves zero. The data are from annual averages between 1985 and 1994 based on the same source as Appendix A3.

equation of angular momentum in flux form as

$$\frac{\partial \bar{l}}{\partial t} + \frac{1}{R\cos\varphi}\frac{\partial}{\partial\varphi}\left(\cos\varphi\,\overline{lv}\right) + \frac{\partial}{\partial p}\left(\overline{l\omega} + R\cos\varphi\,\overline{\tau_{px}}\right) \quad = \quad 0, \tag{18.1.17}$$

where (18.1.6) and (18.1.3) are used. Thus, angular momentum transport across a latitudinal belt φ is given by

$$F_{ang}(\varphi) \quad = \quad 2\pi R\cos\varphi \int_0^{p_s} \overline{lv}\,\frac{dp}{g} \quad \approx \quad 2\pi R^2\cos^2\varphi \int_0^{p_s} \overline{uv}\,\frac{dp}{g}, \tag{18.1.18}$$

whose approximation is based on the assumption that net mass latitudinal transport is zero over a long time average. This means that net angular momentum transport is expressed by the transport of relative angular momentum, defined by $uR\cos\varphi$. In this case, the streamfunction of zonal-mean angular momentum flux can be defined by

$$\Psi_{ang}(\varphi, p) \quad = \quad 2\pi a^2 \cos^2\varphi \int_0^p \overline{uv}\,\frac{dp}{g}. \tag{18.1.19}$$

Thus, F_{ang} and Ψ_{ang} are related as

$$F_{ang}(\varphi) \quad = \quad \Psi_{ang}(\varphi, p_s). \tag{18.1.20}$$

This indicates that the surface value of Ψ_{ang} is equal to total angular momentum transport across the latitude belt φ.

Although the above relation is a general form of angular momentum transport irrespective of latitude, the corresponding formula in mid-latitudes becomes a simpler relation if quasi-geostrophic approximation is used. The equation of zonal winds in quasi-geostrophic approximation is described by (18.1.1), which is given by neglecting vertical advection in the primitive equation system (18.1.16) and all momentum transport except for the eddy component. Using (18.1.3), we rewrite (18.1.1) in the following form for angular momentum conservation:

$$\begin{aligned} \frac{\partial \bar{l}}{\partial t} + \frac{1}{R\cos\varphi}\frac{\partial}{\partial\varphi}\left(R\cos^2\varphi\,\overline{u'v'} + \Omega R^2\cos^3\varphi\,\bar{v}\right) & \\ + \frac{\partial}{\partial p}\left(\Omega R^2\cos^2\varphi\,\bar{\varpi} + R\cos\varphi\,\overline{\tau_{px}}\right) &= 0. \end{aligned} \tag{18.1.21}$$

According to this approximation, the vertical transport of angular momentum is mainly explained by zonal-mean advection $\bar{\varpi}$ of the rigid body component $l_0 \equiv \Omega R^2\cos^2\varphi$ in the free atmosphere. In this case, the lateral transport of angular momentum is expressed by (18.1.10).

Fig. 18.3 shows the mass streamfunction and the zonal-mean zonal winds, and Fig. 18.4 shows the streamfunction of the relative component of angular momentum flux. As shown in Fig. 18.3, the boundary between the Hadley and Ferrel cells is located at the latitude where the zonal-mean zonal winds are zero at the surface. Since the low-level winds of the Hadley circulation are easterly, angular momentum is supplied from the surface to the atmosphere in the Hadley circulation region. Angular momentum is transported upward within the Hadley cell, and transported to the Ferrel cell by the eddy component $\overline{u'v'}$. The fact that the contour of the mass streamfunction is almost vertical at the boundary between the two cells indicates that the zonal-mean component of lateral transport $\overline{u}\,\overline{v}$ is small at the boundary. The exchange of angular momentum between the two cells occurs mainly in the upper layer of the troposphere between 200 hPa and 500 hPa (Fig. 18.4). Within the Ferrel cell, eddy components have a net downward angular momentum transport. This has a tendency to reduce westerly shear and maintains surface westerlies. At the surface in mid-latitudes, angular momentum is transferred from the atmosphere to the surface.

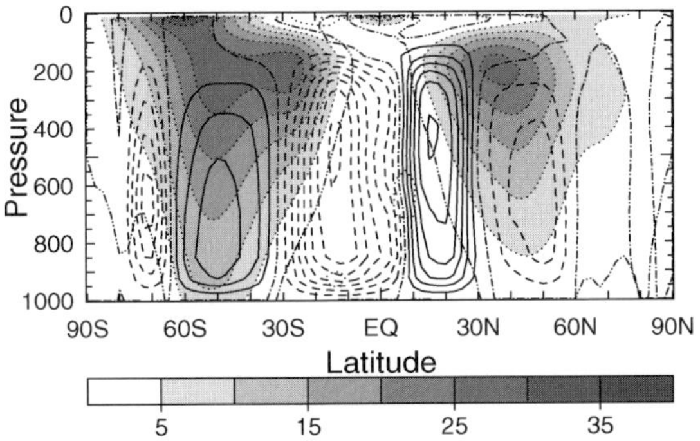

FIGURE 18.3: The meridional structure of the mass streamfunction and zonal-mean zonal winds. The contour interval of the streamfunction is 10^{10} kg s^{-1}: solid curves represent positive values, dashed curves negative values, and dashed-dotted curves zero. The contour interval of zonal winds is 5 m s^{-1}. The dashed-dotted-dotted curves are contours of 0 m s^{-1}. The region of positive zonal winds is shown by shading. The source of data is the same as Fig. 18.2.

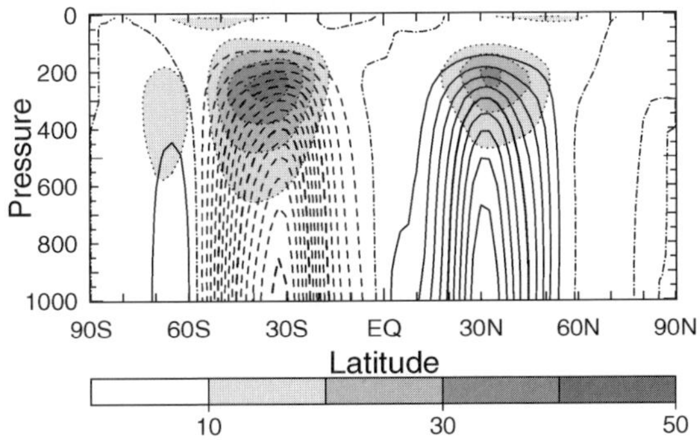

FIGURE 18.4: The meridional structure of the momentum transport due to eddies $\overline{u'v'}$ and the streamfunction of the relative component of zonal-mean angular momentum transport. The contour interval of the streamfunction is 2×10^{18} kg m^2 s^{-2}: solid curves represent positive values, dashed curves negative values, and dashed-dotted curves zero. The contour interval of momentum transport is 10 m^2 s^{-2}. The region where the absolute value of $\overline{u'v'}$ is greater than 10 m^2 s^{-2} is shown by shading. The contours of the zero value of momentum transport are omitted. The source of data is the same as Fig. 18.2.

The statistically balanced state of angular momentum transport shown in Fig. 18.4 is modeled as Fig. 18.5, in which only angular momentum transport between four regions in the meridional section are considered. The regions A and B are in the Hadley cell, while the regions C and D are in the Ferrel cell. The regions A and D are in the boundary layers where the vertical turbulent diffusion of momentum is dominant. The regions B and C are in the free atmosphere, where vertical diffusion is negligible. The latitude of the boundary between AB and CD is denoted by φ_H, and the polar boundary of CD is denoted by φ_F. The pressure level between the layers AD and BC is denoted by p_B, and that of the top level of BC is denoted by p_T. The level p_M corresponds to the top of the mixing layer, and the level p_T

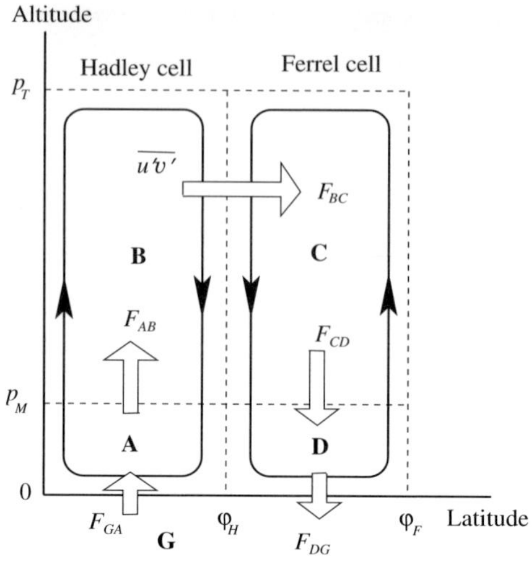

FIGURE 18.5: A schematic model of angular momentum transport between the Hadley and Ferrel cells. See text for symbols.

corresponds to the tropopause. Although there is a distinct difference between the tropopause in low latitudes and that in mid-latitudes in reality, we neglect the difference in this model by assuming that the latitudinal variation of the tropopause does not play a significant role in the angular momentum budget. We also assume that the lateral transport of angular momentum across the latitude φ_H occurs only in the free atmosphere at the boundary between B and C and that angular momentum transport between A and D is negligible. According to Fig. 18.4, this can be satisfied if p_M is taken as about 800 hPa. We further assume that there is no lateral transport at the polar boundary of CD nor at the equatorial boundary of AB, and neglect transport due to seasonal change or exchange between the stratosphere and the troposphere. The surface is denoted by G, and angular momentum transport from G to A is denoted by F_{GA}. In a similar way, transport from A to B, B to C, C to D, and D to G is denoted by F_{AB}, F_{BC}, F_{CD}, and F_{DG}, respectively. The angular momentum balance can be written as

$$F_M \equiv F_{GA} = F_{AB} = F_{BC} = F_{CD} = F_{DG}. \tag{18.1.22}$$

In the respective regions, the main contribution from transport is the stress flux $\overline{\tau_{px}}$ in F_{GA} and F_{DG}, vertical momentum advection $\overline{l\omega}$ in F_{AB} and F_{CD}, and lateral transport $\overline{lv}$ in F_{BC}.

18.1.3 Transformed Eulerian mean circulation

In the heat balance of Eulerian mean formulation (18.1.15), the eddy component $\overline{v'\theta'}$ and vertical transport by zonal-mean circulation $\overline{\omega}\frac{\partial\overline{\theta}}{\partial p}$ are the main terms. In this formulation, however, the relation between these two main terms and the diabatic term $\overline{Q}$ is not clear. If the Transformed Eulerian Mean (TEM) equations are used, the relation between the diabatic term and meridional circulation is more direct.

The TEM equations in quasi-geostrophic approximation are derived in Section 7.4.2. Here, we use the TEM equations in pressure coordinates on a sphere in a similar way to Section 18.1.1. Using (18.1.1)–(18.1.4), the TEM equations in quasi-geostrophic approximation on a sphere are written as

$$\frac{\partial \overline{u}}{\partial t} - f\overline{v}^* = -\overline{G_x} + \frac{1}{R\cos\varphi}\nabla\cdot\boldsymbol{F}, \tag{18.1.23}$$

$$f\frac{\partial \overline{u}}{\partial p} - \frac{\mathcal{R}_p}{R}\frac{\partial\overline{\theta}}{\partial\varphi} = 0, \tag{18.1.24}$$

$$\frac{1}{R\cos\varphi}\frac{\partial}{\partial\varphi}(\overline{v}^*\cos\varphi) + \frac{\partial\overline{\omega}^*}{\partial p} = 0, \tag{18.1.25}$$

$$\frac{\partial\overline{\theta}}{\partial t} + \overline{\omega}^*\frac{\partial\overline{\theta}}{\partial p} = \overline{\mathcal{Q}}. \tag{18.1.26}$$

The vector $\boldsymbol{F}$ is the *Eliassen-Palm flux* (*EP-flux*), which satisfies

$$\nabla\cdot\boldsymbol{F} = \frac{1}{R\cos\varphi}\frac{\partial}{\partial\varphi}(F_\varphi\cos\varphi) + \frac{\partial F_p}{\partial p}, \tag{18.1.27}$$

$$F_\varphi = -R\cos\varphi\overline{u'v'}, \tag{18.1.28}$$

$$F_p = fR\cos\varphi\frac{\overline{v'\theta'}}{\frac{\partial\overline{\theta}}{\partial p}}. \tag{18.1.29}$$

$\overline{v}^*$ and $\overline{\omega}^*$ are called the *residual circulation* defined by

$$\overline{v}^* = \overline{v} - \frac{\partial}{\partial p}\left(\frac{\overline{v'\theta'}}{\frac{\partial\overline{\theta}}{\partial p}}\right), \quad \overline{\omega}^* = \overline{\omega} + \frac{1}{R\cos\varphi}\frac{\partial}{\partial\varphi}\left(\frac{\overline{v'\theta'}\cos\varphi}{\frac{\partial\overline{\theta}}{\partial p}}\right). \tag{18.1.30}$$

As can be seen from (18.1.25), residual circulation is nondivergent, so that the streamfunction Ψ^* can be introduced as

$$\overline{v}^* = \frac{g}{2\pi R\cos\varphi}\frac{\partial\Psi^*}{\partial p}, \quad \overline{\omega}^* = -\frac{g}{2\pi R^2\cos\varphi}\frac{\partial\Psi^*}{\partial\varphi}. \tag{18.1.31}$$

The value of Ψ^* can be calculated from the vertical integral of $\overline{v}^*$ by setting Ψ^* to zero at the top of the atmosphere. Using (18.1.30) and (18.1.8), the streamfunction of the residual circulation Ψ^* is related to the Eulerian mean streamfunction Ψ as

$$\Psi^* = 2\pi R\cos\varphi\int_0^p \overline{v}^*\frac{dp}{g} = \Psi - \frac{2\pi R\cos\varphi}{g}\frac{\overline{v'\theta'}}{\frac{\partial\overline{\theta}}{\partial p}}. \tag{18.1.32}$$

Fig. 18.6 shows the meridional distribution of the annual-mean residual circulation calculated using (18.1.32). This figure also shows the meridional distribution of potential temperature. In a statistical steady state, as seen from (18.1.26), diabatic heating $\mathcal{Q}$ is balanced by the vertical advection of potential temperature due to residual circulation. In this sense, residual circulation can be viewed as thermal

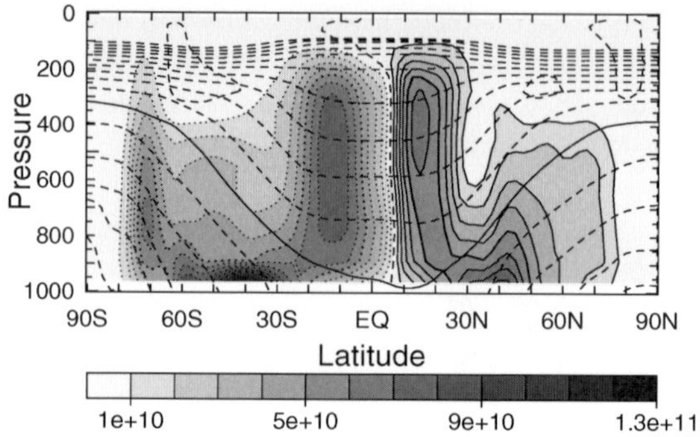

FIGURE 18.6: The meridional distributions of the mass streamfunction of residual circulation Ψ^*. The contour interval of the streamfunction is 10^{10} kg s^{-1}: solid curves represent positive values, dashed curves negative values, and dashed-dotted curves zero. The zonal-mean potential temperature is also shown by dashed curves with a contour interval of 20 K. (The 300 K-contour is solid.) Data are from the annual average of 1993 based on the same source as Appendix A3.

circulation. Since the troposphere is stably stratified in general with $\frac{\partial\overline{\theta}}{\partial p} < 0$, the direction of $\overline{\omega}$ can be determined if the distribution of $\mathcal{Q}$ is specified in the meridional section. In the troposphere, the diabatic term is negative $\mathcal{Q} < 0$ due to radiative cooling except for the tropics and the lower layers of the baroclinic zone in mid-latitudes. In the tropics, we have $\mathcal{Q} > 0$ because of latent heat release and have $\mathcal{Q} > 0$ in mid-latitudes because of the latent heat release due to precipitation associated with extratropical baroclinic waves. Except for these regions in the tropics and mid-latitudes, the diabatic term is negative so that residual circulation is downward $\overline{\omega}^* > 0$ over wide areas.

18.1.4 Isentropic mean circulation

Equations averaged along isentropic surfaces are analogous to the TEM equations, as described in Section 7.5. Isentropic mean meridional circulation can also be viewed as thermal circulation, similarly to TEM circulation. From (3.3.63)–(3.3.66), the hydrostatic equations in isentropic coordinates on a sphere are written as

$$\frac{du}{dt} - \frac{uv}{R}\tan\varphi - fv = -\frac{1}{R\cos\varphi}\frac{\partial M}{\partial\lambda} + G_\lambda, \tag{18.1.33}$$

$$\frac{dv}{dt} + \frac{u^2}{R}\tan\varphi + fu = -\frac{1}{R}\frac{\partial M}{\partial\varphi} + G_\varphi, \tag{18.1.34}$$

$$0 = \frac{\partial M}{\partial\theta} - \frac{C_pT}{\theta}, \tag{18.1.35}$$

$$\frac{\partial\rho_\theta}{\partial t} + \frac{1}{R\cos\varphi}\frac{\partial}{\partial\lambda}(\rho_\theta u) + \frac{1}{R\cos\varphi}\frac{\partial}{\partial\varphi}(\rho_\theta v\cos\varphi) + \frac{\partial}{\partial\theta}(\rho_\theta\dot{\theta}) = 0, \tag{18.1.36}$$

$$\dot{\theta} = \mathcal{Q}. \tag{18.1.37}$$

In this section, M represents the Montgomery function, or static energy. Partial derivatives with respect to time, latitude, and longitude are taken along isentropic surfaces. To consider the zonal average of the above equations, we define the zonal mean along isentropic surfaces by $\overline{A}$, and introduce the following notations:

$$A' = A - \overline{A}, \qquad \overline{A}^* = \frac{\overline{\rho_\theta A}}{\overline{\rho_\theta}}, \tag{18.1.38}$$

from which we have

$$\overline{\rho_\theta A} = \overline{\rho_\theta}\overline{A} + \overline{\rho'_\theta A'} = \overline{\rho_\theta}\overline{A}^*. \tag{18.1.39}$$

First, using (18.1.39), the zonal average of the continuity equation (18.1.36) becomes

$$\frac{\partial \overline{\rho_\theta}}{\partial t} + \frac{1}{R\cos\varphi}\frac{\partial}{\partial\varphi}(\overline{\rho_\theta}\,\overline{v}^*\cos\varphi) + \frac{\partial}{\partial\theta}(\overline{\rho_\theta}\,\overline{\dot{\theta}}^*) = 0. \tag{18.1.40}$$

We assume that $\rho_\theta = 0$ below the surface of the ground if isentropic surfaces intersect with the ground.

Second, by using the vector invariant form (3.3.74), the zonal average of the momentum equation (18.1.33) becomes

$$\frac{\partial \overline{u}}{\partial t} - \overline{v\omega_{a\theta}} = -\overline{\dot{\theta}\frac{\partial u}{\partial \theta}} + \overline{G_\lambda}. \tag{18.1.41}$$

Here, absolute vorticity $\omega_{a\theta}$ is written as

$$\omega_{a\theta} = \frac{1}{R\cos\varphi}\frac{\partial v}{\partial\lambda} - \frac{1}{R}\frac{\partial(u\cos\varphi)}{\partial\varphi} + f = \rho_\theta P, \tag{18.1.42}$$

where P is potential vorticity. Thus, (18.1.41) is rewritten as

$$\frac{\partial \overline{u}}{\partial t} = \overline{\rho_\theta v P} - \overline{\dot{\theta}\frac{\partial u}{\partial \theta}} + \overline{G_\lambda}. \tag{18.1.43}$$

We should note that the right-hand side is equal to the zonal average of the latitudinal flux of PV that appeared in (3.3.78), the zonal average of which is written as

$$\frac{\partial}{\partial t}(\overline{\rho_\theta}\overline{P}^*) + \frac{1}{R\cos\varphi}\frac{\partial}{\partial\varphi}\left(\overline{\rho_\theta v P} - \overline{\dot{\theta}\frac{\partial u}{\partial \theta}} + \overline{G_\lambda}\right) = 0. \tag{18.1.44}$$

Making use of expression (3.3.69), we can write the equation of momentum (18.1.33) in flux form by multiplying by ρ_θ:

$$\begin{aligned}\frac{\partial}{\partial t}(\rho_\theta u) &+ \frac{1}{R\cos^2\varphi}\frac{\partial}{\partial\lambda}(\rho_\theta u^2) + \frac{1}{R\cos^2\varphi}\frac{\partial}{\partial\varphi}(\rho_\theta uv\cos^2\varphi) \\ &+ \frac{\partial}{\partial\theta}(\rho_\theta u\dot{\theta}) - f\rho_\theta v \\ = &- \frac{1}{gR\cos\varphi}\frac{\partial}{\partial\lambda}\left(\frac{\kappa}{\kappa+1}p\pi\right) + \frac{1}{gR\cos\varphi}\frac{\partial}{\partial\theta}\left(p\frac{\partial M}{\partial\lambda}\right) + \rho_\theta G_\lambda.\end{aligned} \tag{18.1.45}$$

The zonal average of this becomes

$$\frac{\partial}{\partial t}(\overline{\rho_\theta u}) + \frac{1}{R\cos^2\varphi}\frac{\partial}{\partial\varphi}(\overline{\rho_\theta uv}\cos^2\varphi) + \frac{\partial}{\partial\theta}(\overline{\rho_\theta u\dot{\theta}}) - f\overline{\rho_\theta v}$$
$$= \frac{1}{gR\cos\varphi}\frac{\partial}{\partial\theta}\left(\overline{p\frac{\partial M}{\partial\lambda}}\right) + \overline{\rho_\theta G_\lambda}. \quad (18.1.46)$$

This equation can be rewritten in a form analogous to the TEM equation (18.1.23):

$$\frac{\partial\overline{u}}{\partial t} + \overline{v}^*\left[\frac{1}{R\cos\varphi}\frac{\partial}{\partial\varphi}(\overline{u}\cos\varphi) - f\right] + \overline{\dot{\theta}}^*\frac{\partial\overline{u}}{\partial\theta} - \overline{G_\lambda}$$
$$= -\frac{1}{\overline{\rho_\theta}}\frac{\partial}{\partial t}(\overline{\rho'_\theta u'}) + \frac{1}{\overline{\rho_\theta}R\cos\varphi}\nabla_\theta\cdot\boldsymbol{F}_\theta, \quad (18.1.47)$$

where

$$\nabla_\theta\cdot\boldsymbol{F}_\theta = \frac{1}{R\cos\varphi}\left(\frac{\partial}{\partial\varphi}F_{\theta,\varphi}\cos\varphi\right) + \frac{\partial}{\partial\theta}F_{\theta,\theta}, \quad (18.1.48)$$

$$F_{\theta,\varphi} = -R\cos\varphi\overline{(\rho_\theta v)'u'}, \quad (18.1.49)$$

$$F_{\theta,\theta} = \frac{1}{g}\overline{p'\frac{\partial M'}{\partial\lambda}} - R\cos\varphi\overline{(\rho_\theta\dot{\theta})'u'}. \quad (18.1.50)$$

In the case of statistically steady states, (18.1.40) becomes nondivergent, such that the meridional mass streamfunction Ψ_θ can be introduced to satisfy

$$\overline{\rho_\theta}\,\overline{v}^* = -\frac{1}{2\pi R\cos\varphi}\frac{\partial\Psi_\theta}{\partial\theta}, \quad (18.1.51)$$

$$\overline{\rho_\theta}\,\overline{\dot{\theta}}^* = \frac{1}{2\pi R^2\cos\varphi}\frac{\partial\Psi_\theta}{\partial\varphi}. \quad (18.1.52)$$

From (18.1.51), the streamfunction is given by

$$\Psi_\theta = 2\pi R\cos\varphi\int_\theta^\infty \overline{\rho_\theta}\,\overline{v}^*\,d\theta, \quad (18.1.53)$$

where we set the streamfunction at the top of the atmosphere $\theta = \infty$ to zero. In the case that the isentropic surface intersects with the ground, we assume $\rho_\theta = 0$ in the region where θ is lower than the potential temperature at the ground. A typical distribution of the mass streamfunction in isentropic coordinates is depicted in Fig. 18.7. The dashed curve is the zonal-mean potential temperature at the surface. Above this curve, the streamlines are down-gradient to the isentropes except for the equatorial region, while, below the curve, the streamlines are equatorward and up-gradient to the isentropes. This indicates that diabatic heating affects the equatorial region and the surface boundary layer.

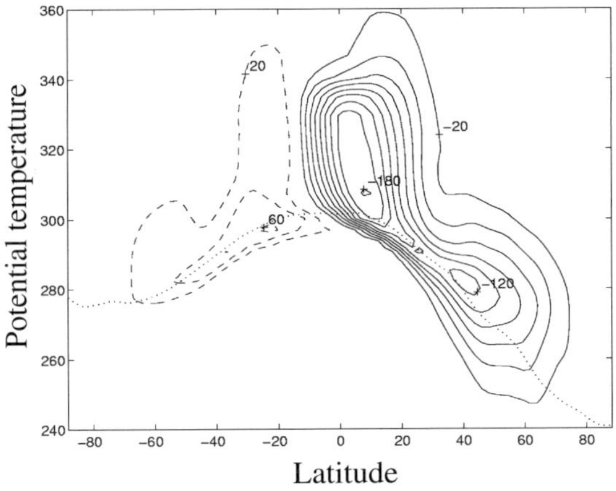

FIGURE 18.7: The meridional distribution of the streamfunction of zonal-mean mass flux in isentropic coordinates. The unit is 10^9 kg s^{-1}. The data are based on results from a general circulation model. After Held and Schneider (1999) by permission of the American Meteorological Society.

In isentropic coordinates, zonal-mean mass transport just above the ground is related to heat flux. To show this, we estimate mass transport below a specified isentropic surface near the ground. We take θ_I as the value of potential temperature, the surface of which lies just above the ground and does not intersect the ground. The potential temperature at the ground is denoted by θ_s. The corresponding pressure on the isentropic surface θ_I and that on the ground are denoted by p_I and p_s, respectively. Mass transport between the ground and the surface θ_I is given as

$$\begin{aligned}\int_{\theta_B}^{\theta_I} \rho_\theta v\, d\theta &= -\int_{\theta_B}^{\theta_I} \frac{1}{g}\frac{\partial p}{\partial \theta} v\, d\theta = \int_{p_s}^{p_I} v\,\frac{dp}{g} \equiv \tilde{v}\frac{p_s - p_I}{g} \\ &\approx \rho_{\theta s}\tilde{v}(\theta_I - \theta_s),\end{aligned} \tag{18.1.54}$$

where $\tilde{v}$ is the average of the meridional wind in the boundary layer between θ_I and θ_s, and ρ_θ is assumed to be constant in the layer. If $\tilde{v}$ is approximated by the geostrophic component $\tilde{v}_g$, we have $\overline{\tilde{v}_g} = 0$. By neglecting the longitudinal dependency of $\rho_{\theta s}$, we obtain the zonal average of the above equation as

$$\int_{\theta_B}^{\theta_I} \overline{\rho_\theta}\,\overline{v}^*\, d\theta \approx \overline{\rho_{\theta s}\tilde{v}(\theta_I - \theta_s)} \approx -\rho_{\theta s}\overline{\tilde{v}_g'\theta_s'}. \tag{18.1.55}$$

This indicates that, if heat flux is poleward $\overline{\tilde{v}_g'\theta_s'} > 0$ near the ground, mass flux in the boundary layer is equatorward.† This status is schematically shown in Fig. 18.8.

A similar relation holds near the tropopause. Here, we define the tropopause as a surface with a constant potential vorticity.‡ We assume that the tropopause

†The signs are for the northern hemisphere throughout this section.

‡The tropopause is defined in Chapter 14. We will investigate various perspectives of the tropopause in Section 18.2.3.

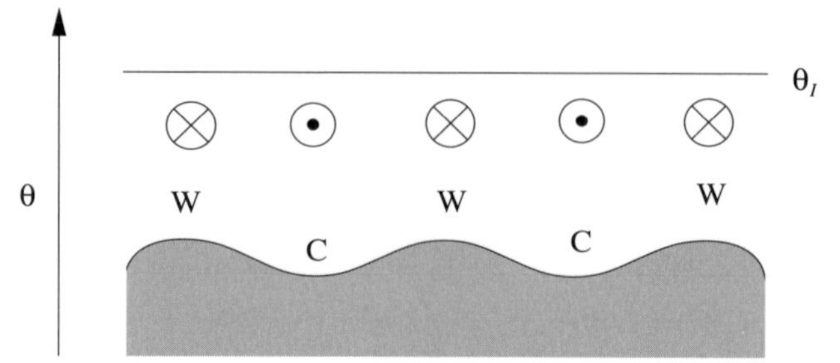

FIGURE 18.8: The relation between mass transport and heat transport near the ground. The vertical axis is potential temperature and the shaded region is the ground. The air is warmer where the topography of the ground is convex (W), and colder where concave (C). The direction shown by ⊙ is poleward, and the opposite direction shown by ⊗ is equatorward. In this case, zonal-mean heat transport is poleward, while zonal-mean mass transport is equatorward near the ground.

is confined between two isentropic surfaces θ_- and θ_+ ($\theta_- < \theta_+$) and denote the potential temperature at the tropopause by θ_T. The density above the tropopause (in the stratosphere) is denoted by $\rho_{\theta+}$, and that below the tropopause (in the troposphere) by $\rho_{\theta-}$. Since stratification in the stratosphere is more stable than that in the troposphere, we generally have $\rho_{\theta+} < \rho_{\theta-}$. Mass transport in the layer between the isentropic surfaces θ_- and θ_+ is approximately given by

$$\begin{aligned}\int_{\theta_-}^{\theta_+} \rho_\theta v \, d\theta &= \int_{\theta_T}^{\theta_+} \rho_\theta v \, d\theta + \int_{\theta_-}^{\theta_T} \rho_\theta v \, d\theta \\ &\approx \rho_{\theta+}\tilde{v}(\theta_+ - \theta_T) + \rho_{\theta-}\tilde{v}(\theta_T - \theta_-). \end{aligned} \tag{18.1.56}$$

Thus, the zonally averaged mass transport near the tropopause is written as

$$\int_{\theta_-}^{\theta_+} \overline{\rho_\theta}\,\overline{v}^* \, d\theta \approx \overline{\rho_{\theta+}\tilde{v}(\theta_+ - \theta_T)} + \overline{\rho_{\theta-}\tilde{v}(\theta_T - \theta_-)} \approx (\rho_{\theta-} - \rho_{\theta+})\overline{\tilde{v}_g\theta'_T}. \tag{18.1.57}$$

In general, heat flux is poleward near the tropopause in the mid-latitude baroclinic zone as shown in Fig. 18.2. In this case, since $\overline{\tilde{v}_g\theta'_T} > 0$ and $\rho_{\theta-} - \rho_{\theta+} > 0$, mass transport near the tropopause is poleward. Fig. 18.9 shows this kind of situation.

The mass transport in the middle level of the troposphere is relatable to the angular momentum balance (18.1.43). In statistical steady states, we have

$$\overline{\rho_\theta v P} = \overline{\dot{\theta}\frac{\partial u}{\partial \theta}} - \overline{G_\lambda}. \tag{18.1.58}$$

Since G_λ is negligible in the free atmosphere and vertical advection is also negligible under geostrophic approximation, the right-hand side generally vanishes in this balance; thus we have

$$\overline{\rho_\theta v P} = 0. \tag{18.1.59}$$

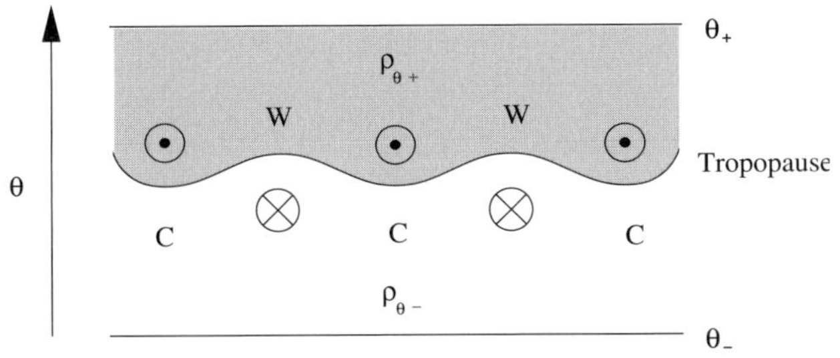

FIGURE 18.9: The same as Fig. 18.8 but for the relation between mass transport and heat transport near the tropopause. Since density in the stratosphere is smaller than in the troposphere in general, mass transport is poleward if heat transport is poleward near the tropopause.

Using the definition

$$\hat{A} = A - \overline{A}^*, \tag{18.1.60}$$

we obtain

$$\begin{aligned}\overline{\rho_\theta v P} &= \overline{\rho_\theta v P}^* = \overline{\rho_\theta}\ \overline{v}^* \overline{P}^* + \overline{\rho_\theta \hat{v} \hat{P}}^* \\ &= \overline{v}^* \overline{\omega_{a\theta}} + \overline{\rho_\theta \hat{v}\hat{P}}^* \quad \approx \quad f\overline{v}^* + \overline{\rho_\theta \hat{v} \hat{P}}^* .\end{aligned} \tag{18.1.61}$$

Thus, (18.1.59) is reduced to

$$\overline{v}^* = -\frac{\rho_\theta}{f}\overline{\hat{v}\hat{P}}^* . \tag{18.1.62}$$

In general, we can assume that the direction of eddy transport of PV along isentropic surfaces in the free atmosphere is opposite to the gradient of zonal-mean PV contours. This implies $\overline{\hat{v}\hat{P}}^* < 0$ since PV increases with latitude. Thus, we have $\overline{v}^* > 0$ (i.e., meridional circulation is poleward in the free atmosphere).

The above description is only for zonal averaged circulation. The motions of individual air parcels are different from mean circulation. We will discuss the Lagrangian perspective in Chapter 19.

18.2 Meridional thermal structure

18.2.1 Stability of a zonal symmetric state

In Chapter 16, we examined the conditions needed for Hadley circulation to exist by giving a simplified thermal structure (16.2.2). The discussion was based on axisymmetric circulation where eddies in mid and high latitudes are prohibited. In this section, we investigate the stability of the thermal structure in the extratropics under the same thermal conditions.

We consider the quasi-Boussinesq system in which the meridional potential temperature profile is given by (16.2.2). It is a hypothetical state where radiative-convective equilibrium is satisfied at each latitude and no meridional circulation exists. The zonal wind profile balanced by this potential temperature field comprises the cyclostrophic winds given by (16.2.3). Here, for simplicity, we assume that this balanced state comprises geostrophic winds instead of cyclostrophic winds,

$$2\Omega \sin\varphi \frac{\partial u}{\partial z} = -\frac{g}{\theta_0}\frac{1}{R}\frac{\partial \theta}{\partial \varphi}. \tag{18.2.1}$$

Substituting the potential temperature profile (16.2.2) into θ on the right-hand side of (18.2.1), we obtain

$$\frac{\partial u}{\partial z} = \frac{g\Delta T_s}{\Omega R \theta_0}\cos\varphi. \tag{18.2.2}$$

This zonal wind profile has constant shear. We consider the stability of the above state based on the necessary conditions for baroclinic instability, which are given by (5.5.163).

Using (5.5.148), the gradient of quasi-geostrophic potential vorticity is written as

$$\begin{aligned} \Pi_y &= \beta - \frac{\partial^2 u}{\partial y^2} - \frac{1}{\rho_s}\frac{\partial}{\partial z}\left(\rho_s \frac{f^2}{N^2}\frac{\partial u}{\partial z}\right) \\ &= \beta - \frac{\partial^2 u}{\partial y^2} + \frac{f}{\rho_s}\frac{\partial}{\partial z}\left(\rho_s S\right), \end{aligned} \tag{18.2.3}$$

where

$$S \equiv -\frac{f}{N^2}\frac{\partial u}{\partial z} = \left(\frac{\partial \theta}{\partial y}\right) \Big/ \left(\frac{\partial \theta}{\partial z}\right) \tag{18.2.4}$$

represents inclination of an isentropic surface, $f = 2\Omega \sin\varphi$ is the Coriolis parameter with $y = R\varphi$, and $\beta = \Omega\cos\varphi / R$. We assume that density is given by

$$\rho_s = \rho_0 \exp\left(-\frac{z}{H}\right), \tag{18.2.5}$$

where H is the scale height of density. Using the potential temperature profile (16.2.2) and the thermal wind profile (18.2.1), the potential vorticity gradient is given by

$$\Pi_y = \beta + \frac{1}{H}\frac{f^2}{N^2}\frac{\partial u}{\partial z} = \beta\left(1 + \frac{4\Delta T_s}{\gamma H}\sin^2\varphi\right), \tag{18.2.6}$$

where (18.2.2) is used. γ represents the vertical gradient of potential temperature that is used in (16.2.2); γH is the difference in potential temperature between the ground and the tropopause at height H. In general, the inequality $4\Delta T_s > \gamma H$ holds, such that we have $\Pi_y > 0$ in the inner region of the atmosphere. On the other hand, we generally have $\frac{\partial u}{\partial z} > 0$ near the ground $z = 0$. Under these conditions, (5.5.163) can be satisfied for a suitable choice of disturbance P, such that the necessary condition for instability is satisfied. Thus, this basic state is unstable with respect to baroclinic instability.

In this case, the Richardson number of the potential temperature field (16.2.2) is given by

$$Ri = \frac{N^2}{\left(\frac{\partial u}{\partial z}\right)^2} = \frac{\Omega^2 R^2 \theta_0 \gamma}{g\Delta T_s^2}\frac{1}{\cos^2\varphi} = \frac{1}{\cos^2\varphi}\frac{\Omega^2 R^2}{gH}\frac{\gamma H \theta_0}{\Delta T_s^2}. \tag{18.2.7}$$

Using representative values of the atmosphere, $\Omega^2 R^2/gH \approx 2.2$ and $\gamma H\theta_0/\Delta T_s^2 \approx 12$, we generally have $Ri \gg 1$.

18.2.2 Baroclinic adjustment

As shown above, the local radiative-convective equilibrium state, represented by the thermal profile (16.2.2) and the thermal wind balance (18.2.2), is unstable for baroclinic instability, and thus does not exist as a statistical equilibrium state. In the extratropics of the real atmosphere, therefore, eddies due to baroclinic instability persistently evolve such that the meridional thermal structure becomes different from that assumed in the basic state. The question is then what is the meridional structure of temperature and other quantities in the new, statistical equilibrium state.

There are a few but no satisfactory theories that explain the meridional structure of the statistical equilibrium state in which nonlinear baroclinic waves are fully developed. Among them, the theory called *baroclinic adjustment* is frequently invoked, though it is not well understood how much it describes the meridional structure of the atmosphere. Since the basic field given by local radiative-convective equilibrium satisfies the necessary conditions for baroclinic instability, nonaxisymmetric baroclinic eddies evolve in the real atmosphere. The statistical equilibrium state as a result of baroclinic instability is different from the basic field. Baroclinic adjustment hypothesizes that the statistical equilibrium state is maintained in a marginally unstable state for the necessary conditions of baroclinic instability. This idea is similar to convective adjustment in Section 14.4.2 or barotropic adjustment in Section 17.3.6. The convective adjustment method assumes that convective instability occurs in a statically unstable layer and that the resulting stratification of the layer becomes vertically isentropic. The barotropic adjustment method in turn assumes that the resultant absolute vorticity field is horizontally uniform. In the case of baroclinic adjustment, baroclinic instability occurs when both surface temperature and potential vorticity in the free atmosphere have latitudinal gradients. The adjusted state is given as the one in which these gradients are eliminated by eddies.

Baroclinic adjustment assumes homogeneity of potential vorticity in the free atmosphere. Baroclinic waves are basically adiabatic, such that air parcels move along isentropic surfaces. Thus, it is thought that baroclinic waves homogenize the potential vorticity on each isentropic surface. From this view, the resultant field of baroclinic equilibration is described by a uniform potential vorticity on isentropic surfaces. This speculation implies that baroclinic adjustment corresponds to the barotropic adjustment on each isentropic surface. Sun and Lindzen (1994) introduced such a homogenization mechanism to consider the meridional structure of the atmosphere. Mixing on isentropic surfaces and its relation to the tropopause will be further discussed in Chapter 19.

Here, we consider an example of baroclinic adjustment using the quasi-geostrophic approximation from the previous subsection. We consider the case when the latitudinal dependency of u is negligible in the equation of Π_y, (18.2.3). The necessary conditions for instability no longer hold if we specify a state in which $S(0) = 0$ and

Π_y does not change its sign in the free atmosphere. We assume that the equilibrium state of baroclinic adjustment is given by the state where these two conditions hold. It will be found, however, that the equilibrium state is not uniquely determined. In order to obtain the thermal structure of the equilibrium state, we introduce a parameter z_A and assume that the stability parameter $S(z)$ changes between the ground $z = 0$ and z_A. The equilibrium state is given by the state that satisfies

$$S(0) \quad = \quad 0; \quad \text{and} \quad \Pi_y \quad = \quad 0, \quad \text{for} \quad 0 < z < z_A. \tag{18.2.8}$$

In the upper layer, $z > z_A$, the equilibrium state remains the same as the basic state. Setting (18.2.3) to zero and neglecting the latitudinal gradient of u, we obtain the solution for S which satisfies the above constraints as

$$S_A(z) \quad = \quad S_A(0)e^{\frac{z}{H}} + \frac{\beta H}{f}\left(1 - e^{\frac{z}{H}}\right), \tag{18.2.9}$$

where S_A is the stability parameter S after baroclinic adjustment. Letting S_0 denote the stability parameter of the basic state, since $S_A(z_A) = S_0(z_A)$, we can express the height as

$$z_A \quad = \quad H \ln \left[\frac{1 - \frac{f S_0(z_A)}{\beta H}}{1 - \frac{f S_A(0)}{\beta H}} \right]. \tag{18.2.10}$$

Substituting $S_A(0) = 0$ and assuming that S_0 is constant in the basic state, we obtain

$$z_A \quad = \quad H \ln\left(1 - \frac{f S_0}{\beta H}\right) \quad = \quad H \ln\left(1 + \frac{h}{H}\right), \tag{18.2.11}$$

where

$$h \quad = \quad -\frac{f S_0}{\beta} \tag{18.2.12}$$

is a parameter that characterizes the baroclinic instability of the Charney problem (5.5.113). This parameter is rewritten as

$$h \quad = \quad -\frac{f}{\beta\gamma}\frac{\partial T_s}{\partial y}, \tag{18.2.13}$$

where (18.2.4) and the potential temperature distribution of the basic field (16.2.2) are used.

Although the distribution of S_A is given by (18.2.9), the distribution of potential temperature is not uniquely determined since S_A contains both the vertical gradient and the latitudinal gradient of potential temperature. The latter is related to the vertical shear of zonal winds through the thermal wind balance. If just the vertical shear is variable and the stability parameter N^2 is constant, the zonal wind profile at the baroclinically adjusted state in the layer $z < z_A$ is given by

$$u(z) \quad = \quad -\frac{\beta H N^2}{f^2}\left[z - z_A - H\left(e^{\frac{z}{H}} - 1\right)\right] + \left.\frac{\partial u}{\partial z}\right|_0 (z_A - H), \tag{18.2.14}$$

where $\frac{\partial u}{\partial z}\big|_0$ represents constant vertical shear in the basic state. On the other hand, if just the vertical gradient of potential temperature is variable and the zonal wind is prescribed, the Brunt-Väisälä frequency of the baroclinically adjusted state is given by

$$N^2(z) = -\frac{f^2}{\beta H}\frac{\partial u}{\partial z}\frac{1}{1-e^{\frac{z}{H}}}. \tag{18.2.15}$$

This has a singularity since $N^2 \to \infty$ as $z \to 0$. In reality, both vertical shear and stratification will be changed by baroclinic instability, such that the profiles of the zonal wind and potential temperature become different from those of the initial state.

The zonal velocity profiles for the baroclinic adjustment model with constant N^2 and the profiles of N^2 for the baroclinic adjustment model with constant shear are shown in Fig. 18.10. The dotted line is the profile in the basic state, which is assumed to have linear shear and constant static stability. The solid curve is the vertical profile after baroclinic adjustment. At the adjusted state with the N^2 constant model, vertical shear vanishes at the ground, while it approaches that of the basic state at z_A. For the constant shear model, N^2 diverges toward infinity near the ground. In this calculation, the typical values at latitude 30° are used: $h = 8.0$ km and $z_A = 5.6$ km for $f = 7.27 \times 10^{-5}$ s^{-1}, $\beta = 1.98 \times 10^{-11}$ s^{-1} m^{-1}, $N = 0.01$ s^{-1}, and $\frac{\partial u}{\partial z}\big|_0 = 3.0 \times 10^{-3}$ s^{-1}.

18.2.3 Tropopause height

Tropopause height is a representative index of the thermal structure of the atmosphere. The tropopause is clearly defined by the vertically one-dimensional radiative-convective equilibrium model, where the boundary between the stratosphere and the troposphere is called the tropopause. In this model, the stratosphere is the layer where only the radiative process contributes, while the tropopause is the layer where radiative and convective processes determine the temperature structure. Thus, the two layers are clearly defined in this model. In reality, however, it is difficult to identify the region where the convective process is active, so that this definition of the tropopause is not usable in practice.

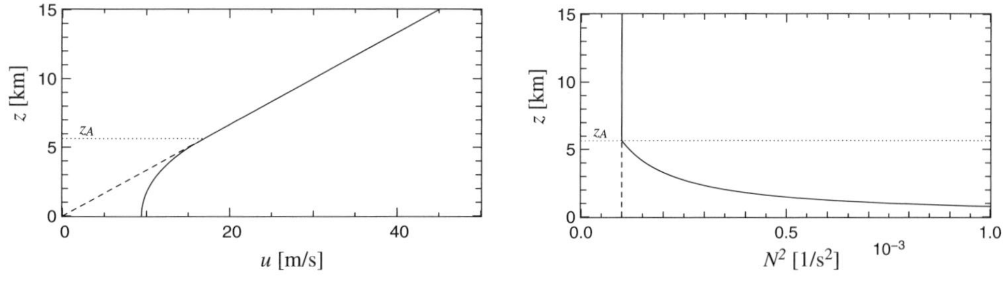

FIGURE 18.10: The vertical profiles of zonal winds (left) and the square of the Brunt-Väisälä frequency (right) before/after baroclinic adjustment. Dotted: basic state; solid: a state after baroclinic adjustment.

In general, two kinds of definitions based on thermal and dynamical properties are used to identify the tropopause. The thermal definition uses the lapse rate of temperature; the thermal tropopause is defined as the lowest level at which the lapse rate becomes less than 2 K km^{-1}, provided that the average lapse rate between this level and all higher levels within 2 km does not exceed 2 K km^{-1}. Although the thermal tropopause is simply defined by the temperature structure, it is not a material surface. In contrast, the surface of a particular value of potential vorticity is used to define the dynamical tropopause. For instance, the surface of potential vorticity at 2 PVU (PVU $= 10^{-6}$ m^2 K s^{-1} kg^{-1}) is used as the tropopause. Values around 2–4 PVU are used depending on the purpose. Since potential vorticity is conserved along a fluid motion if motions are adiabatic and frictionless, air in the dynamical tropopause stays at the same surface in such conditions; thus, the dynamical tropopause is thought of as a material surface. The dynamical tropopause is not used in lower latitudes, however, since potential vorticity approaches zero near the equator.

Fig. 18.11 shows the meridional distribution of the lapse rate of temperature. The meridional distribution of potential vorticity will be shown in Fig. 19.1. For both the thermal and dynamical tropopause, the tropopause is located at distinctly different altitudes between the tropics and the extratropics. From this, it is thought that tropopause height is constrained by different mechanism in the two latitudinal regions. In the framework of one-dimensional radiative-convective equilibrium with convective adjustment, the temperature structure is determined if the lapse rate in the troposphere is specified. In this case, the tropopause is defined as the top of the adjusted layer. Making use of this idea and noticing the relationship between tropopause height and lapse rate, we will have a constraint on the latitudinal variation of tropopause height.

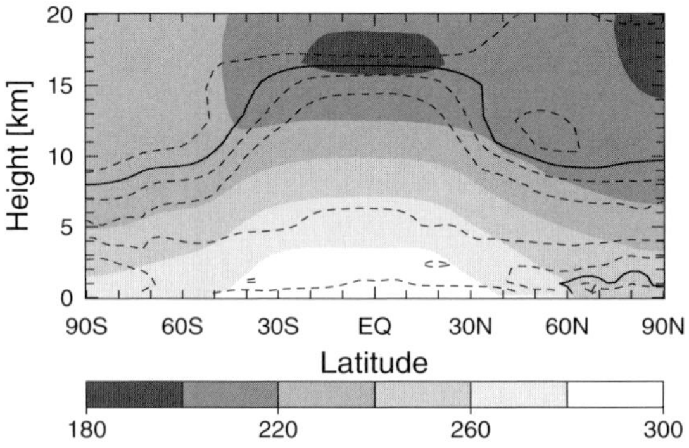

FIGURE 18.11: The meridional distribution of the lapse rate of temperature. The solid and dashed curves are the lapse rate with contour interval with 2 K km^{-1}. In particular, the solid curve is a contour of -2 K km^{-1} and corresponds to the tropopause except for that in lower layers near the Arctic where a stable layer prevails. The shaded scale represents temperature distribution. The lapse rate is calculated from the monthly mean temperature distribution in January 1993.

In low latitudes, vertical stratification is maintained by the moist process (i.e., cumulus convection in the Intertropical Convergence Zone (ITCZ) and Hadley circulation region). This means that the lapse rate Γ is determined by the moist adiabat, $\Gamma = \Gamma_m$, in the whole low-latitudinal region covered by Hadley circulation. As latitude becomes higher and surface temperature becomes lower, tropopause height becomes gradually lower until the extratropics where tropopause height is abruptly changed. In mid and high latitudes, the moist process is no longer dominant, so that the lapse rate is different from the moist adiabat.

In order to depict how tropopause height is determined in the extratropics, we introduce the simple model presented by Held (1982). As investigated in the previous subsections, we assume that baroclinic adjustment determines a statistical equilibrium state as a result of baroclinic instability in mid latitudes. The uppermost level z_A of baroclinic adjustment is given by (18.2.10) or (18.2.11). We assume that this height corresponds to tropopause height H_T in the extratropics. Substituting (18.2.13) into (18.2.11) gives

$$H_T = H \ln\left(1 + \frac{h}{H}\right) = H \ln\left[1 - \frac{f}{\beta(\Gamma_d - \Gamma)}\frac{\partial T_s}{\partial y}\right], \tag{18.2.16}$$

in which it is assumed that the lapse rate Γ remains unchanged from that of the basic state. If surface temperature T_s is specified, this equation gives relation between lapse rate Γ and tropopause height H_T.

One additional constraint is required to determine tropopause height. The radiative condition can be used for this purpose (i.e., radiative equilibrium holds in the stratosphere above the tropopause). To see how the tropopause is determined, we use the gray radiation model and assume that the atmosphere consists of the stratosphere in radiative equilibrium and the troposphere whose lapse rate of temperature is constant Γ. What we need to know is the dependency of tropopause height H_T on surface temperature T_s and lapse rate Γ. We specify the vertical distribution of the optical depth of the gray atmosphere by a function $\tau(z)$. In the stratosphere, from (14.3.2) and (14.3.3), upward radiation flux $F^{\uparrow}$ and downward radiation flux $F^{\downarrow}$ satisfy

$$\frac{2}{3}\frac{dF^{\uparrow}}{d\tau} = F^{\uparrow} - \pi B, \qquad -\frac{2}{3}\frac{dF^{\downarrow}}{d\tau} = F^{\downarrow} - \pi B, \tag{18.2.17}$$

where $\pi B = \sigma_B T^4$. When net incident solar radiation flux F is given, we can use (14.3.15) and (14.3.16) to solve the functions of radiative fluxes on optical depth τ:

$$F^{\uparrow} = \frac{F}{2}\left(\frac{3}{2}\tau + 2\right), \quad F^{\downarrow} = \frac{F}{2}\cdot\frac{3}{2}\tau, \quad \pi B = \frac{F}{2}\left(\frac{3}{2}\tau + 1\right). \tag{18.2.18}$$

Thus, the temperature profile in the stratosphere is

$$T(z) = \left[\frac{F}{2\sigma_B}\left(\frac{3}{2}\tau(z) + 1\right)\right]^{\frac{1}{2}}, \qquad \text{for } z \geq H_T. \tag{18.2.19}$$

In particular, the temperature at the tropopause is given by substituting $z = H_T$.

Therefore, the temperature profile in the troposphere is

$$T(z) = \left[\frac{F}{2\sigma_B}\left(\frac{3}{2}\tau(H_T)+1\right)\right]^{\frac{1}{2}} + \Gamma(H_T - z), \qquad \text{for } z \geq H_T. \tag{18.2.20}$$

For simplicity, the ground temperature is assumed to be equal to that of the lowest level of the atmosphere: $T_s = T(0)$. If we integrate $F^{\uparrow}$ in (18.2.17) from the bottom to top of the atmosphere using the temperature profiles (18.2.19) and (18.2.20), we will have the outward long-wave radiative flux at the top of the atmosphere F_T^{radl}. At the equilibrium state, this flux must be equal to net incident solar radiation F. This constraint is what we call the radiation condition. The equilibrium solution can be found by changing H_T if Γ is given as a known parameter. Thus, the relation between tropopause height and lapse rate is written in the form

$$H_T = \mathcal{H}(\Gamma) = \tilde{\mathcal{H}}(\Gamma; F, \{\tau\}), \tag{18.2.21}$$

where $\tilde{\mathcal{H}}$ on the right-hand side indicates that tropopause height depends on solar radiation F and the vertical distribution of τ together with the lapse rate Γ. The surface temperature is given by (18.2.20) as

$$T_s = T(0) = \left[\frac{F}{2\sigma_B}\left(\frac{3}{2}\tau(H_T)+1\right)\right]^{\frac{1}{2}} + \Gamma H_T, \tag{18.2.22}$$

which is a function of Γ using (18.2.21).

One may consider the condition that T_s is externally specified instead of depending on solar flux F. Under such a condition, F must be determined so as to satisfy (18.2.21) and (18.2.22). In general, this solar flux is different from actual solar flux. In such a case, one should regard this difference as contributing to heat transport and the tendency toward heat capacity in the ocean.

Fig. 18.12 shows an example of the dependency of tropopause height on lapse rate for a fixed solar radiation condition (Held 1982). The radiation calculation is based on the gray radiation model. The radiation constraint given by (18.2.21) is denoted by $R(\Gamma)$. The figure shows another constraint on tropopause height given by baroclinic adjustment. Tropopause height is the depth at which baroclinic adjustment occurs and is given by (18.2.16) with $D_{MID} = H_T$. The latitudinal temperature gradient $\frac{\partial T_s}{\partial y}$ is fixed at those values which yield $h/H = 1$ and 2 when $\Gamma = 6.5\ \text{K km}^{-1}$. The scale height is set equal to $H = 7.5$ km. This result indicates that tropopause height is expected to be in the range around $H_T = 9$–10 km, though the results depend on many parameters.

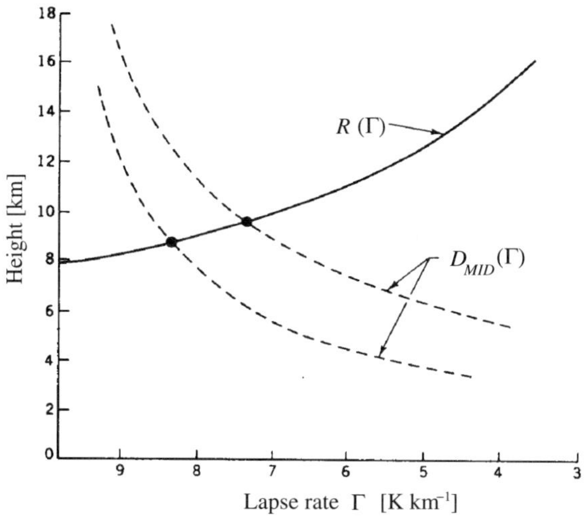

FIGURE 18.12: The tropopause height given by the radiation constraint is shown by a solid curve with $R(\Gamma)$. The abscissa is the lapse rate Γ [K km^{-1}]. The dotted curves D_{MID} represent tropopause height given by the dynamical constraint in mid-latitudes. The upper curve is for $h/H = 2$ and the lower curve is for $h/H = 1$. After Held (1982) with permission of the American Meteorological Society.

18.3 Life cycle experiments of extratropical cyclones

Mid-latitude circulation involves the evolution of extratropical cyclones. The development of extratropical cyclones is explained by the linear baroclinic instability theory described in Section 5.5. The whole process of this development, involving nonlinear processes and successive decay of extratropical cyclones, is known as the life cycle of extratropical cyclones. In this section, the life cycle of extratropical cyclones is described based on experiments using a general circulation model.

Let us imagine how atmospheric circulation would evolve if it started in a state of local radiative-convective equilibrium at each latitude. It is thought that both symmetric meridional circulation and asymmetric disturbance would grow; the former is Hadley circulation in low latitudes, and the latter is baroclinic instability in mid and high latitudes. When these circulations are fully developed, the atmosphere will achieve another statistical equilibrium state of general circulation. Circulations in the mid and high latitudes of such an equilibrium state cannot be described only by the linear growth of baroclinic instability. The statistical equilibrium state is a long time average of a balanced state between external forcing and the dissipation processes after baroclinic instability has developed. The baroclinic adjustment explained in Section 18.2.2 is a possible approach to understanding statistical equilibrium in the extratropics. However, the theory of baroclinic adjustment is not definitive, since the equilibrium state is not uniquely determined. As another approach, one can consider the time evolution of baroclinic instability from its initial development to the fully matured stage. Some of the characteristics of the extratropical atmospheric structure can be gained by following the whole life cycle of extratropical cyclones (i.e., linear growth of baroclinic instability, nonlinear saturation of cyclones, and their decay). This kind of approach is called the problem with the *life cycle of extratropical cyclones.* It is an initial value problem that concerns how disturbances evolve for a given specific initial state. In general, however, it is not clear how the statistical equilibrium state of the atmosphere is related to the evolution of the atmospheric field as an initial value problem. The life cycle problem of extratropical cyclones is nevertheless very informative, since the dynamics of extratropical cyclones can be described in terms of the atmospheric mean field.

Various initial states can be chosen for the life cycle problem. As mentioned above, one can use a locally radiative-convective equilibrium state with zonal winds in the thermal wind balance. Alternatively, the zonally uniform state with a symmetric Hadley circulation can be used as an initial state. Generally, to investigate the effects of baroclinic instability on the atmospheric mean field, one uses initial states with no meridional flow, giving a concentrated jet in the upper layer of the subtropics instead of the Hadley circulation. The top panels of Fig. 18.13 are two examples of initial states; these show the meridional distributions of the zonal winds and potential temperature. In Fig. 18.13 (a), surface wind is set to zero and surface pressure is uniform, initially. In Fig. 18.13 (b), in contrast, it has a surface wind variation (i.e., a barotropic component). If we add random disturbance to these initial states, the most unstable baroclinic wave will emerge. The bottom figures of Fig. 18.13 are the meridional distributions of zonal-mean zonal winds and potential temperature at 7 days after the initial states. As shown below, disturbances with zonal wave number seven are predominant at this time.

The above two examples are frequently referred to as the two most characteristic life cycles of extratropical cyclones. Thorncroft et al. (1993) symbolically referred to the life cycle experiment in Fig. 18.13 (a), (c) as *LC1* and that of Fig. 18.13 (b), (d) as *LC2*. Fig. 18.14 shows time evolution of surface pressure of LC1, while Fig. 18.15 shows that of LC2. In both cases, disturbances with zonal wave number seven are predominant. As extratropical cyclones develop, the centers of low pressure move poleward, and thus a low pressure belt is formed in higher latitudes at the final stage. On the other hand, when the barotropic component is added as in LC2, vortices remain for a long time after the development of extratropical cyclones,

The life cycle of the LC1-type extratropical cyclone experiences the following stages: At the linearly developing stage, the disturbances in extratropical cyclones concentrate in the lower layer. Then, this is followed by the upward propagation of these disturbances. After that, the cyclones enter into the decaying stage associated with Rossby wave radiation from upper vortices and the convergence of angular momentum in the upper layer. At lower latitudes, in contrast, Rossby waves that propagate from mid-latitudes are absorbed.

The above mentioned stages of the life cycle are schematically shown by Eliassen-Palm flux (EP-flux) analysis. We introduced the expressions of EP-flux in the spherical coordinates as (18.1.27)–(18.1.29). Here, for simplicity, we use Cartesian coordinates without any metric terms. The linearized equation of quasi-geostrophic

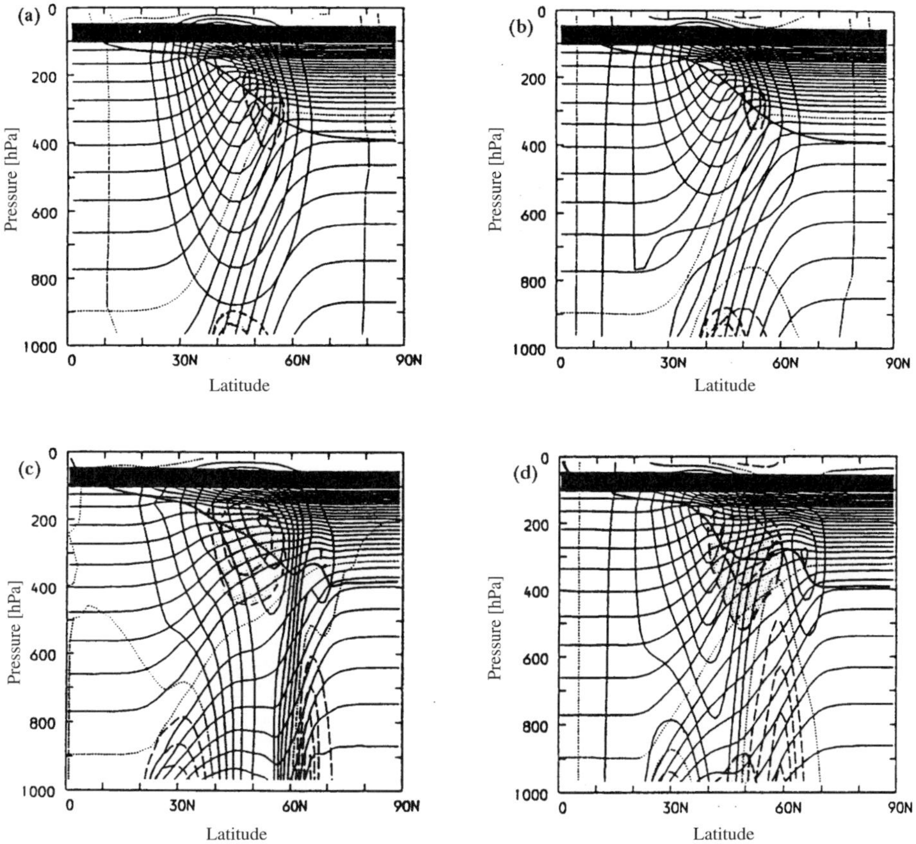

FIGURE 18.13: The meridional structures of zonal winds and potential temperature of the initial state for the life cycle experiment (a), (b) and that at day 7 (c), (d). (c) corresponds to the initial state with no surface winds (a), and (d) corresponds to the initial state with the barotropic component (d). The contour intervals are 5 m s^{-1} and 5 K for zonal winds and potential temperature, respectively. The zonal wind surface with 0 m s^{-1} is shown as a dotted curve. The potential vorticity surface with PV = 2 PVU is given as solid curves. After Thorncroft et al. (1993) by permission of the *Q. J. Roy. Meteorol. Soc.*.

potential vorticity is written as

$$\frac{\partial q'}{\partial t} = -\overline{u}\frac{\partial q'}{\partial x} - v'\frac{\partial \overline{q}}{\partial y}. \tag{18.3.1}$$

Multiplying by q', averaging over the longitude, and dividing the result by $\partial\overline{q}/\partial y$, we obtain

$$\frac{\partial A}{\partial t} = -\overline{v'q'} = -\nabla \cdot \boldsymbol{F}, \tag{18.3.2}$$

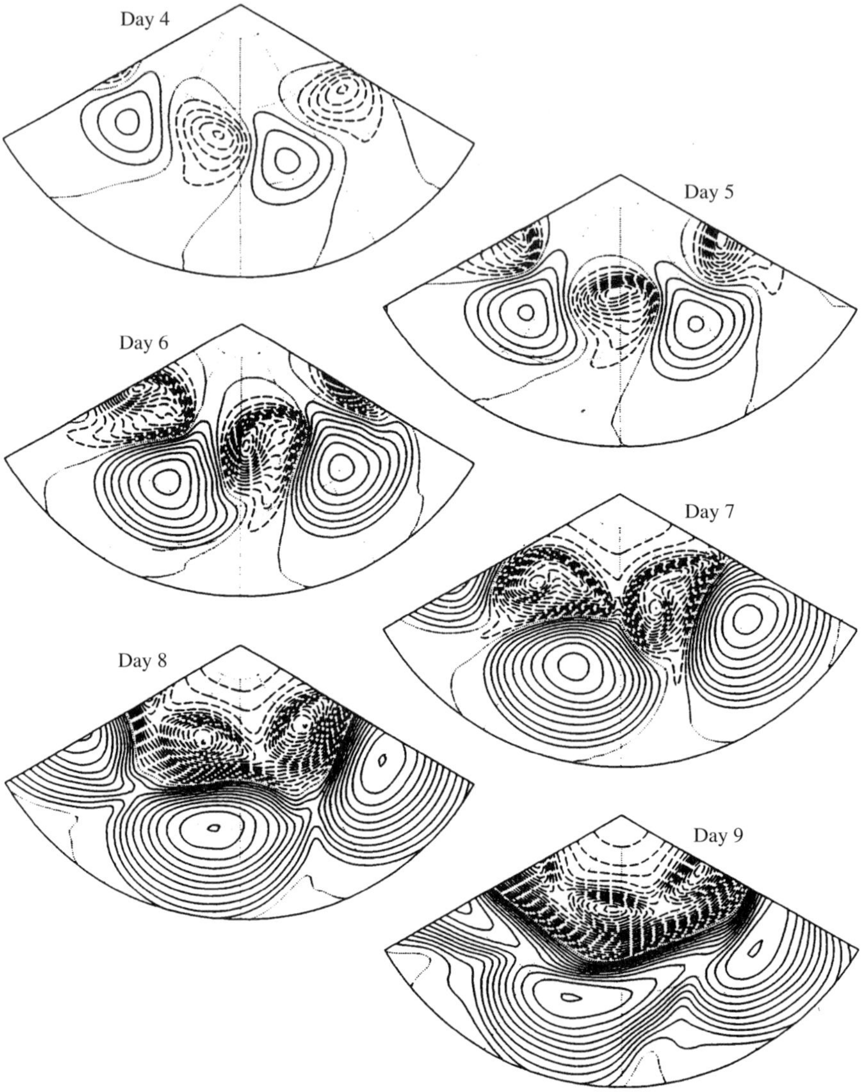

FIGURE 18.14: The time evolution of surface pressure for the life cycle experiment LC1 between days 4 and 9. The contour interval is 4 hPa with the 1,000 hPa contour dotted. The outermost circle is the latitudinal circle at 20°, and lines of constant latitude and longitude are drawn every 20 and 30 degrees, respectively. After Thorncroft et al. (1993) by permission of the *Q. J. Roy. Meteorol. Soc.*.

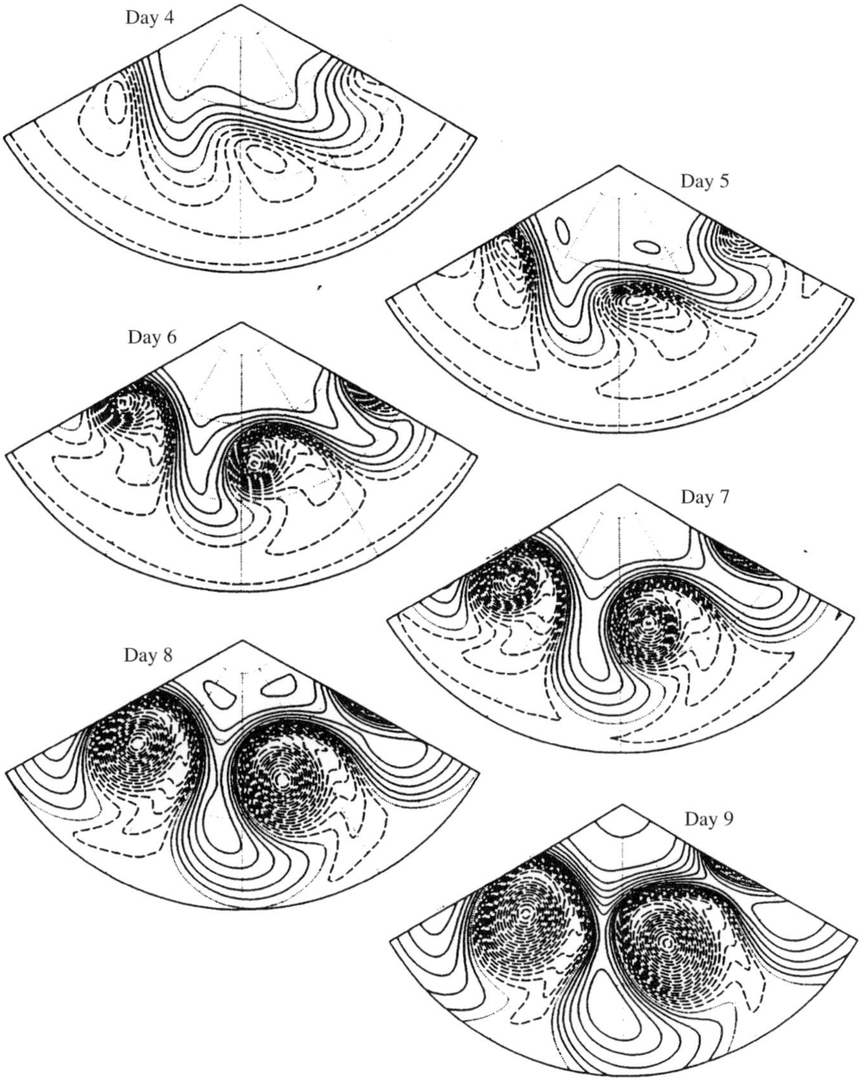

FIGURE 18.15: The same as Fig. 18.14 but for the life cycle experiment of LC2. After Thorncroft et al. (1993) by permission of the *Q. J. Roy. Meteorol. Soc.*.

where

$$A = \frac{1}{2}\frac{\overline{q'^2}}{\frac{\partial \overline{q}}{\partial y}} = \frac{1}{2}\overline{\eta'^2}\frac{\partial \overline{q}}{\partial y}, \qquad \boldsymbol{F} = \left(-\overline{u'v'}, \frac{f_0}{N^2}\overline{v'\theta'}\right). \tag{18.3.3}$$

A is *wave activity*, $\eta' = -q'/(\partial \overline{q}/\partial y)$ is displacement in the latitudinal direction, and $\boldsymbol{F}$ is the *EP flux vector* (or *EP-flux*). If the notion of group velocity is applicable and only a single wave packet with group velocity $\boldsymbol{C}_g$ exists, EP-flux is related as $\boldsymbol{F} = \boldsymbol{C}_g A$. When the sign of $\partial \overline{q}/\partial y$ is positive and wave activity propagates into the region where the EP-flux is convergent, the potential vorticity flux $\overline{v'q'}$ is negative according to (18.3.2) (i.e., the potential vorticity flux is in the opposite direction to the gradient of potential vorticity). Variance in potential vorticity $\overline{q'^2}$ increases and displacement of air parcels also increases.† When there is no dissipation and all wave activities vanish from the region considered, the signs of the above quantities are reversed. The equation of zonal-mean potential vorticity is given by

$$\frac{\partial \overline{q}}{\partial t} = -\frac{\partial}{\partial y}\overline{v'q'}. \tag{18.3.4}$$

The equation of potential temperature is also given by (18.3.1) and (18.3.4) by replacing $\overline{q}$ and q' by $\overline{\theta}$ and θ', respectively.‡ The above equations give a complete set of equations for the evolution of disturbances and the zonal-mean field.

Evolution of the EP-flux $\boldsymbol{F}$ at three different stages of the life cycle LC1 is shown in Fig. 18.16. In the first stage, $\boldsymbol{F}$ is almost vertical and confined to lower layers. As growth in the lower layer is saturated, flux in the middle layer of the troposphere becomes larger and a strong convergent area exists at the upper troposphere. This indicates that wave activity is propagating upward. This second stage is completed by day 8, and is followed by the quasi-horizontal direction of the EP-flux appearing just below the tropopause; the convergent region in the second stage turns to be divergent around (50°N, 350 hPa) and strong convergence exists around (30°N, 150 hPa). In the third stage, wave activity propagates quasi-horizontally from mid-latitudes to the subtropics. The schematic pattern of the three stages of EP-flux is summarized in Fig. 18.17. As investigated in Section 17.3, Rossby wave propagation and its absorption are associated with angular momentum transport, and result in a change of zonal wind distribution on a sphere.

†If $\boldsymbol{F} > 0$, we have $\partial A/\partial t > 0$ and $\overline{v'q'} < 0$ from (18.3.2). In the case $\partial \overline{q}/\partial y > 0$, $\overline{q'^2}$ and $\overline{\eta'^2}$ increase with time due to (18.3.3). We also have $\overline{v'q'} \propto -\partial \overline{q}/\partial y$, where $\overline{v'q'} = -\overline{\eta' v'}\partial \overline{q}/\partial y$.

‡At the lower boundary, the equation of potential temperature is given by

$$\frac{\partial \theta'}{\partial t} = -\overline{u}\frac{\partial \theta'}{\partial x} - v'\frac{\partial \overline{\theta}}{\partial y}, \qquad \frac{\partial \overline{\theta}}{\partial t} = -\frac{\partial}{\partial y}\overline{v'\theta'}.$$

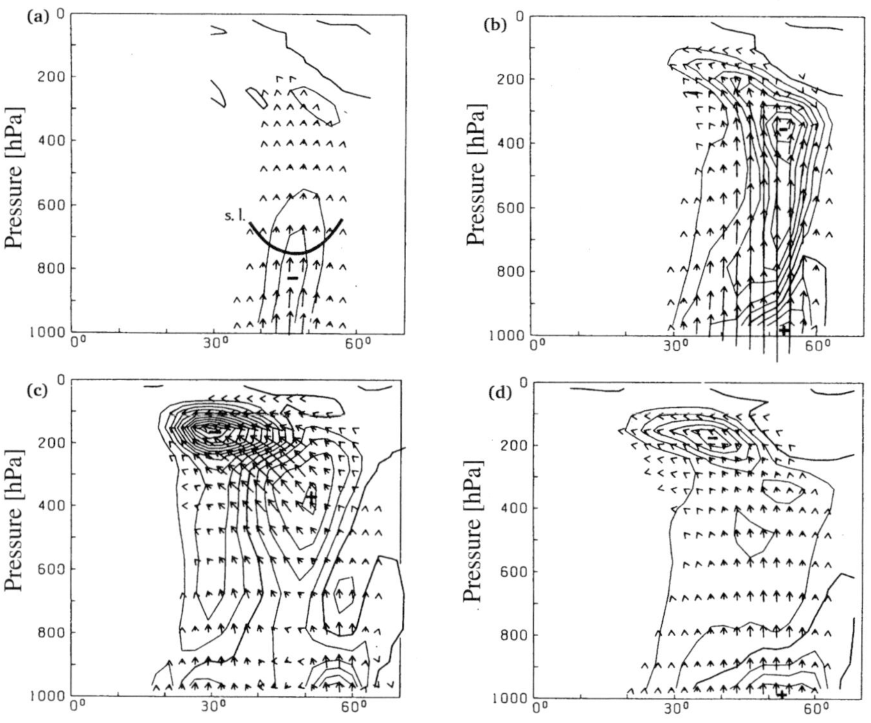

FIGURE 18.16: The meridional sections of EP-flux (arrows) and the convergent/divergent region (+ and − enclosed by contours) for the three different stages of the life cycle of extratropical cyclones LC1 at (a) day 0, (b) day 5, and (c) day 8. (d) is the time-averaged field. After Edmon et al. (1980) by permission of the American Meteorological Society.

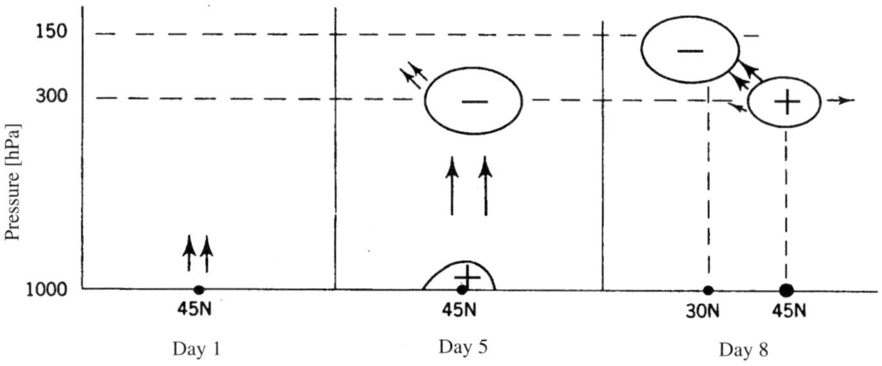

FIGURE 18.17: Schematic figure of the evolution of EP-flux (arrows) and the convergent/divergent region (+ and −) for the three different stages of the life cycle of extratropical cyclones at days 1, 5, and 8. After Held and Hoskins (1985) by permission of Elsevier (copyright, 2003).

References and suggested reading

Mid-latitude circulation is characterized by the dynamics of synoptic-scale disturbances represented by extratropical circulations. Palmén and Newton (1969) is a classical textbook about the various characteristics of extratropical circulations and their roles in general circulation. Lorenz (1967) gives a historical review of the zonal-mean circulation of the atmosphere. The observational facts about Eulerian mean circulation described in Section 18.1.1 are fully discussed by Peixoto and Oort (1992). The theoretical basis behind angular momentum and energy budgets of description of the Eulerian mean and its relation to the transformed Eulerian mean circulation in Section 18.1.3 are summarized in Chapter 10 of Holton (1992). Isentropic mean meridional circulation (Section 18.1.4) is described by Townsend and Johnson (1985), and an interesting interpretation is given by Held and Schneider (1999). The barotropic adjustment theory was first proposed by Stone (1978) using the two-layer model, and was extended to continuously stratified fluid by Lindzen and Farrell (1980) and Gutowski (1985) among others. The simple theory underlying the mid-latitude tropopause in Section 18.2.3 is based on Held (1982). The life cycle of extratropical cyclones described in Section 18.3 is investigated by Simmons and Hoskins (1978), and further examined by Thorncroft et al. (1993) by introducing two paradigms called LC1 and LC2. The relation between the general circulation and the life cycle of extratropical cyclones is clearly reviewed by Held and Hoskins (1985). For various perspectives on EP-flux and its observational facts, see Edmon et al. (1980).

Andrews, D. G. and McIntyre, M. E., 1976: Planetary waves in horizontal and vertical shear: The generalized Eliassen-Palm relation and the mean zonal acceleration. *J. Atmos. Sci.*, **33**, 2031–2048.

Edmon, Jr., H. J., Hoskins, B. J., and McIntyre, M. E., 1980: Eliassen-Palm cross sections for the troposphere. *J. Atmos. Sci.*, **37**, 2600–2612.

Gutowski, Jr., W. J., 1985: Baroclinic adjustment and midlatitude temperature profiles. *J. Atmos. Sci.*, **42**, 1733–1745.

Held, I. M., 1982: On the height of the tropopause and the static stability of the troposphere. *J. Atmos. Sci.*, **39**, 412–417.

Held, I. M. and Hoskins, B. J., 1985: Large-scale eddies and the general circulation of the troposphere. *Advances in Geophys.*, **28**, 3–31.

Held, I. M. and Schneider, T., 1999: The surface branch of the zonally averaged mass transport circulation in the troposphere. *J. Atmos. Sci.*, **56**, 1688–1697.

Holton, J. R., 1992: *An Introduction to Dynamic Meteorology*, 3rd ed. Academic Press, San Diego, 507 pp.

Lindzen, R. S. and Farrell, B., 1980: The role of polar regions in global climate, and a new parameterization of global heat transport. *Mon. Wea. Rev.*, **108**, 2064–2078.

Lorenz, E. N., 1967: *The Nature and Theory of the General Circulation of the Atmosphere.* World Meteorological Organization, Geneva, 161 pp.

Palmén, E. and Newton, C. W., 1969: *Atmospheric Circulation Systems.* Academic Press, San Diego, 603 pp.

Peixoto, P. J. and Oort, A. H., 1992: *Physics of Climate.* American Institute of Physics, New York, 520 pp.

Simmons, A. J. and Hoskins, B. J., 1978: The life cycles of some nonlinear baroclinic waves. *J. Atmos. Sci.*, **35**, 414–432.

Stone, P. H., 1978: Baroclinic adjustment. *J. Atmos. Sci.*, **35**, 561–571.

Sun, D.-Z. and Lindzen, R. S., 1994: A PV view of the zonal mean distribution of temperature and wind in the extratropical troposphere. *J. Atmos. Sci.*, **51**, 757–772.

Thorncroft, C. D., Hoskins, B. J., and McIntyre, M. E., 1993: Two paradigms of baroclinic-wave life-cycle behavior. *Q. J. Roy. Meteorol. Soc.*, **119**, 17–55.

Townsend, R. D. and Johnson, D. R., 1985: A diagnostic study of the isentropic zonally averaged mass circulation during the first GARP global experiment. *J. Atmos. Sci.*, **42**, 1565–1579.

19

Global mixing

We have examined the meridional circulation of the atmosphere in terms of the Hadley circulation in low latitudes and the circulations associated with baroclinic waves in mid-latitudes. Both circulations have a hemispheric one-cellular structure if viewed as transformed Eulerian mean or isentropic mean circulations. One-cellular circulation is directly described as a thermally driven circulation, and is also related to material transport in the Lagrangian sense. It must be remarked, however, that one-cellular circulation is not a static overturning flow. For instance, associated with baroclinic waves, latitudinally meandering flows along isentropic surfaces predominate at mid-latitudes. The one-cellular circulation of the atmosphere is characterized by these isentropic flows and the cross isentropic circulations in the meridional section. Isentropic flows are by definition adiabatic, while cross isentropic flows are related to diabatic heating or cooling. The former results in *isentropic mixing* through north-south meandering air motions.

In this chapter, we describe the global-scale mixing of the atmosphere that is characterized by thermally driven meridional circulations and isentropic flows. First, the meridional structure of potential vorticity and potential temperature is reviewed. The meridional section of the atmosphere can be divided into three regions that have different characteristics depending on the relation between isentropic surfaces, ground surface, and tropopause. Second, the Lagrangian circulation of the atmospheric general circulation is examined using trajectory analysis. In the last section, the mass exchange between the troposphere and stratosphere is overviewed.

19.1 Potential vorticity and potential temperature

Under adiabatic and frictionless conditions, potential vorticity P and potential temperature θ are conserved in the Lagrangian sense. If potential vorticity is tracked on isentropic surfaces, therefore, it gives an approximate Lagrangian view of air motions. In contrast, by tracking the potential temperature on constant potential vorticity surfaces, we will obtain similar Lagrangian motions.

In the hydrostatic balance, potential vorticity is expressed by (3.3.70) as

$$P = \frac{\zeta_{a\theta}}{\rho_\theta} = \frac{\zeta_\theta + f}{\rho_\theta} = -g(\zeta_\theta + f)\frac{\partial \theta}{\partial p}, \qquad (19.1.1)$$

where $\zeta_{a\theta}$ and ζ_θ are absolute and relative vorticities in isentropic coordinates, respectively, ρ_θ is density in isentropic coordinates given by (3.3.62), and $f = 2\Omega \sin\varphi$ is the Coriolis parameter. For large-scale motions at mid-latitudes, since $\zeta_\theta \ll f$ in general, we may have an approximation:

$$P \approx -gf\frac{\partial \theta}{\partial p}. \qquad (19.1.2)$$

In the troposphere, typical values are $\frac{\partial \theta}{\partial p} \approx -\frac{1}{\rho g}\frac{\partial \theta}{\partial z} \approx 10^{-4}$ K Pa^{-1} and $P \approx 10^{-7}$ K m^2 kg^{-1} s^{-1}. The unit 10^{-6} K m^2 kg^{-1} s^{-1} is called one Potential Vorticity Unit, or 1 PVU.

The observed meridional distributions of zonal-mean potential vorticity and zonal-mean potential temperature are shown in Fig. 19.1. Potential vorticity is close to zero in the Hadley circulation region in the tropics, and its gradient is relatively small along isentropic surfaces in mid-latitudes 30–60°. The homogenization of potential vorticity along isentropic surfaces is caused by the mixing of air parcels along isentropic surfaces. If air parcels are mixed by advective flow, no change in entropy occurs. Thus, it is called *isentropic mixing.*

The isentropic motions of air parcels in the troposphere are interrupted by two boundaries: the ground surface and the tropopause. In general, isentropic surfaces incline and intersect with the ground. Since air motions cannot pass through the ground, the ground plays the role of a *mixing barrier.* The tropopause is a boundary

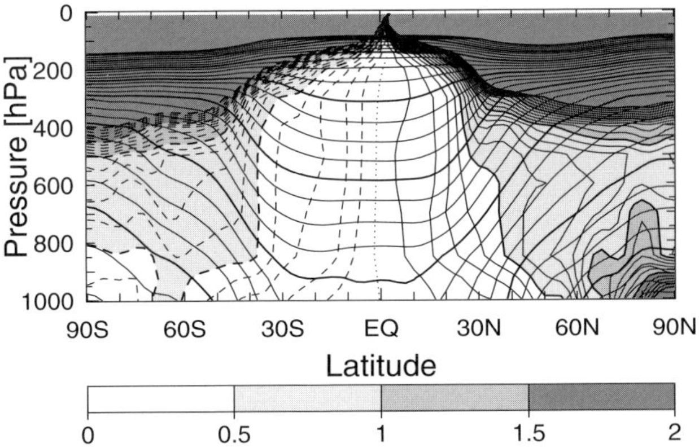

FIGURE 19.1: The meridional distributions of zonal-mean potential vorticity and zonal-mean potential temperature. The contour interval of potential vorticity is 0.1 PVU and that of potential temperature is 5 K with the thick curves being 20 K. Potential vorticity is calculated from the monthly mean data in January of 1993. Data are based on the same source as Appendix A3.

between the stratosphere and the troposphere and again plays the role of a mixing barrier. Although various definitions of the tropopause exist as described in Section 18.2.3, a constant PV surface is frequently used as the dynamic tropopause in mid-latitudes but not for that of lower latitudes. Since PV is conserved along air parcels for adiabatic and frictionless motion, air parcels in the troposphere do not easily pass through the tropopause.

As schematically shown in Fig. 19.2 (a), the atmosphere can be divided into three regions according to the relation between the tropopause and isentropic surfaces; these regions are called the *Overworld*, the *Middleworld*, and the *Underworld*. The Overworld is the region whose isentropic surfaces are above the tropopause, the Middleworld is the region whose isentropic surfaces intersect with the tropopause, and the Underworld is the region whose isentropic surfaces are below the tropopause. In the real atmosphere, almost all isentropic surfaces in the Underworld also intersect with the ground. This means that the highest isentropic surface in the Underworld is in contact with the ground near the equator and touches the tropopause at the poles.

Fig. 19.2 (b) displays the schematic motions of air parcels in the meridional section. In the mid-latitudes, in particular, since the potential temperature of air parcels is well conserved, air parcels move along isentropic surfaces over a short-range timescale, such as ten days. As air parcels move along isentropic surfaces, they also undulate with an amplitude about 1,000 km in the latitudinal direction. Thus, air parcels have an oscillatory motion in the meridional section as shown in Fig. 19.2 (b). If air parcels are tracked for an even longer time, thermally driven meridional circulation will emerge and the cross isentropic motion occurs as described in Section 18.1.4.

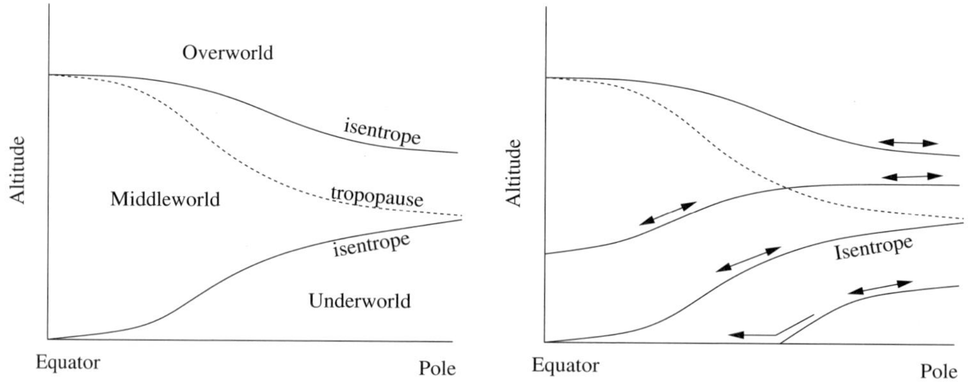

FIGURE 19.2: (a) Definitions of the three regions in the meridional section according to the position of isentropic surfaces. The dashed curves are the tropopause and solid curves are isentropic surfaces, (b) The relation between the motions of air parcels and isentropic surfaces. The meridional motions of air parcels are shown by arrows.

19.2 Lagrangian circulation

If the velocity field $\boldsymbol{v}(\boldsymbol{x}, t)$ is known, the trajectories of air parcels can be calculated. The location of an air parcel at any time t is given by integrating the relation between position and velocity,

$$\frac{d\boldsymbol{x}}{dt} = \boldsymbol{v}(\boldsymbol{x}, t), \tag{19.2.1}$$

and given the initial conditions in which the air parcel is located at $\boldsymbol{x}_0$ at time t_0. Here, we use spherical coordinates $(\lambda, \varphi, \zeta)$ where λ is longitude, φ is latitude, and ζ is an appropriate vertical coordinate. The position of an air parcel is calculated from zonal wind u, latitudinal wind v, and vertical velocity $\dot{\zeta}$ by

$$\frac{d\lambda}{dt} = \frac{u(\lambda, \varphi, \zeta, t)}{R\cos\varphi}, \quad \frac{d\varphi}{dt} = \frac{v(\lambda, \varphi, \zeta, t)}{R}, \quad \frac{d\zeta}{dt} = \dot{\zeta}. \tag{19.2.2}$$

Pressure p is used as the vertical coordinate ζ if observed data are used, while in modeling studies the vertical coordinate of the model such as $\sigma = p/p_s$ is used where p_s is the surface pressure. The potential temperature θ can be the vertical coordinate in isentropic coordinates. Latitude-longitude coordinates have singular points at the poles. Thus, other coordinates such as stereographic coordinates are used near the poles, and the velocity field and the positions on the trajectory are transformed into the new coordinates.

When the trajectories of a large number of air parcels are calculated using the global data of the velocity field, the set of trajectories is usually referred to as Lagrangian circulation. In this case, Lagrangian circulation is differently expressed depending on the initial positions of air parcels or the duration of integration time for the trajectories.

In practice, it is hard to determine the precise trajectories of air parcels, since vertical velocities are generally unknown from observational data. In the observed global data or the output data from an atmospheric general circulation model, vertical motions associated with cumulus convection are not explicitly resolved, since cumulus convection is expressed in some parameterized forms. This means that vertical velocity from global data is different from that of the real vertical wind. If an air parcel approaches an active cumulus region in the tropics, for instance, the precise position of the air parcel cannot be traced any longer since vertical velocity is not properly represented. Remember that the calculation of trajectory always involves large uncertainty of the vertical position in the tropical region.

As an example of Lagrangian circulation, we present a trajectory calculation given by a dynamical core experiment for a general circulation model, which is a simple experimental set for understanding the dynamical properties of the atmosphere (see Chapter 24). The dynamical core experiment used here is slightly different from that in Chapter 24 because diabatic heating is directly given near the equator. The radiative process is represented by Newtonian cooling, and explicit vertical motions associated with cumulus convection are not included (Satoh, 1999). Although the physical process is simplified, the mid-latitude circulation of the model

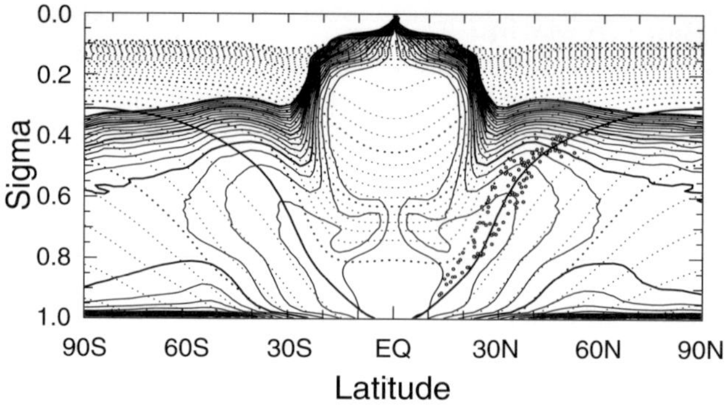

FIGURE 19.3: The zonal-mean potential temperature and zonal-mean potential vorticity for the dynamical core experiment. The contour interval of potential temperature is 5 K, and that of potential vorticity is 0.1 PVU. Backward trajectories are calculated for air parcels that are released from the latitude 32.1° and potential temperature 296 K. The positions of air parcels at 10 days before the arrival at this point are marked by dots. The vertical ordinate is altitude in the sigma-coordinate.

has characteristics similar to the general circulation of the real atmosphere; this is because the dynamics at mid-latitudes are well described by dry dynamics. Fig. 19.3 shows the meridional distributions of zonal-mean potential temperature and potential vorticity (PV). While PV is close to zero in a wide range near the equator, it has an intrusion of relatively higher absolute values of PV in the lower layers of low latitudes around the subtropical jet near the latitude 30°. At mid-latitudes, PV is relatively homogenized along isentropic surfaces.

Air parcels that started at mid-latitudes in the troposphere behave relatively adiabatically. They meander on isentropic surfaces and oscillate between the lower layer in the subtropics and the tropopause region at high latitudes. In Fig. 19.3, the distribution of air parcels that started from latitude 32.1° on the isentropic surface at 296 K is added in the meridional section. Air parcels spread throughout the depth of the troposphere on the isentropic surface in about ten days. At mid-latitudes, the values of potential temperature possessed by the air parcels gradually decrease due to radiative cooling. Since deviation from the original isentropic surface is small in the 10-day excursion, the motions of air parcels are almost on the same isentropic surface. On the other hand, although PV is also conserved if the motions are adiabatic and frictionless, the contours of PV in the region in which the air parcels spread are relatively broad; this reflects the homogenization of PV by isentropic mixing. Near the tropopause and the lower layer in the subtropics, however, air parcels enter the steeper gradient regions of PV; this indicates that the PV values of air parcels are changing in these regions. Near the tropopause, for instance, when an air parcel that has a relatively smaller value of PV in the troposphere reaches the tropopause region, it comes into contact with an air parcel that has larger values of PV in the stratosphere. In such a case, smaller scale turbulence occurs and irreversible mixing processes will change the value of PV. Fig. 19.4 shows the relation between the PV distributions and the locations of air parcels on isentropic surfaces and meridional sections.

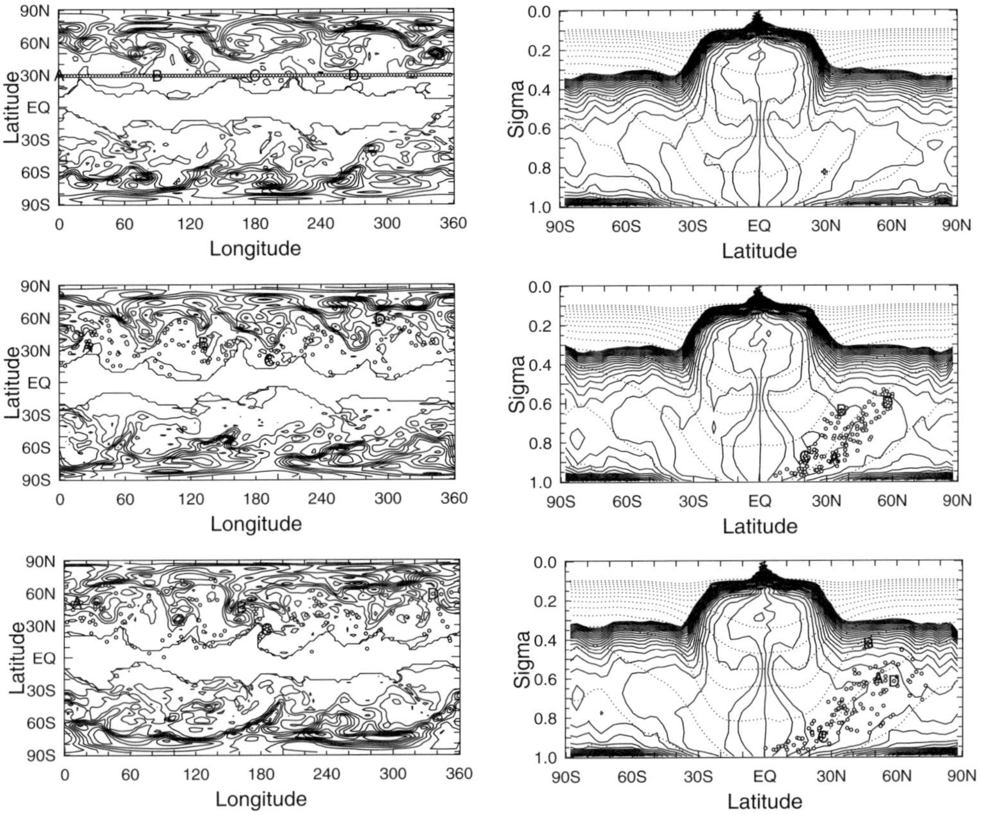

FIGURE 19.4: The relation between the locations of air parcels and the distributions of PV: horizontal distribution (left) and meridional distribution (right). In the horizontal distribution, the PV contours on the isentropic surface 291.5 K are shown. In the meridional distribution, both the contours of potential temperature and potential vorticity are shown. The top two panels are the distributions when the air parcels are released (day 0), the middle and bottom panels are those at day 5 and 10 from the release of the air parcels, respectively. After Satoh (1999).

Fig. 19.5 shows the meridional distribution of the Lagrangian circulation which is calculated as the zonal average of Lagrangian motions during the 10 days based on the above experiment. It should be noted that the zonal-mean trajectories shown by the arrows do not represent the actual motion of air parcels. Since the motions of air parcels are almost isentropic in about 10 days, the Lagrangian mean trajectories of the parcels that started from lower latitudes are directed poleward, while those of the parcels that started from higher latitudes are directed equatorward, as can be seen from trajectories on the 300-K isentropic surface, for example. For a longer duration than 10 days, however, air parcels depart from the original isentropic surface such that the Lagrangian mean trajectories cross isentropic surfaces. At mid-latitudes, the trajectories are downward. The trajectories in the lower layers of the Underworld are interesting in that isentropic surfaces intersect with the ground. Air parcels in the Underworld rapidly move toward the lower latitudes

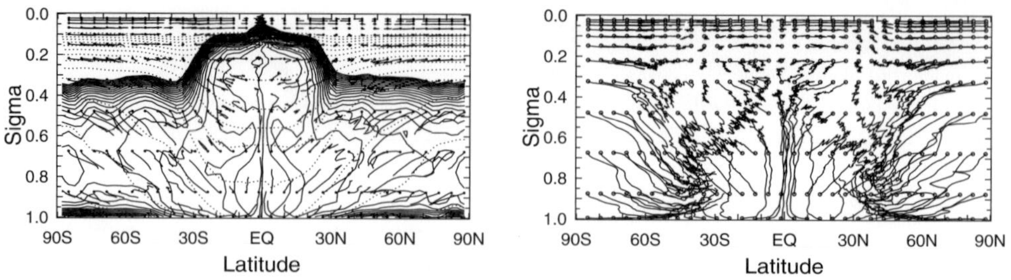

FIGURE 19.5: The meridional distribution of Lagrangian mean circulation for 10 (left) and 60 (right) days. The contours of potential temperature and potential vorticity are shown in the figure for day 10. The contour intervals are 5 K and 0.1 PVU, respectively. After Satoh (1999).

along the surface, such that trajectories are equatorward in the lower boundary layer.

An interesting phenomenon of air motions along isentropic surfaces is the transport of water vapor in the middle level of the troposphere. If water vapor is transported from lower latitudes to mid and high latitudes along an isentropic surface, it is condensed into rain at some latitude in general. The region of water vapor condensation is characterized by latent heat release associated with extratropical cyclones. Actually, the meridional distribution of diabatic heating shows that diabatic heating has two maxima: the deep tropics and the middle layer in mid-latitudes. The latter region reflects the precipitation zone associated with the warm front of extratropical cyclones. In contrast to this, when air is transported from high latitudes to low latitudes, since water vapor content in high latitudes is small, dry air intrudes into low latitudes. This type of transport of dry air occurs in a narrow region of the low latitudes about 100 km in horizontal scale and 100 m in vertical, and is called the *dry intrusion*. As a result, the vertical structure of water vapor has a very fine layer structure, and has a strong impact on radiative transfer (Pierrehumbert, 1999).

19.3 Stratosphere Troposphere Exchange (STE)

Mass exchange between the stratosphere and the troposphere (STE) can be evaluated by tracing the Lagrangian motions of air parcels. To estimate the exchange, we need to define the tropopause as the boundary between the stratosphere and the troposphere using some clear criterion. Mass flux across the tropopause is given by calculating the motions of air parcels relative to the tropopause, by setting air parcels on the tropopause initially. In mid and high latitudes, the surface of a particular value of potential vorticity can be used as the tropopause. If potential vorticity is conserved along the air trajectories under adiabatic and frictionless flows, the constant potential vorticity surface is a material surface and no STE occurs. For synoptic-scale disturbances in mid-latitudes, potential vorticity is well conserved in a short range timescale (say, a few days). In general, however, potential vorticity is not conserved for a longer time and mass exchange exists between the stratosphere and the troposphere.

STE may be categorized into the following two aspects. The first is that caused by isentropic mixing, and the second is by diabatic motions that intersect isentropic surfaces. In fact, these two occur simultaneously, so that they cannot be considered separately in a strict sense. Since the potential vorticity surface and the isentropic surface are almost parallel near the poles, however, it is formally thought that only the second effect contributes to STE. Generally, the stirring on isentropic surfaces is so rapid that isentropic mixing of potential vorticity easily occurs. When stratospheric air intrudes into the troposphere in a filamentary shape, for instance, air parcels with a larger value of potential vorticity in the stratosphere descend equatorward, and then become closer to air with lower values of potential vorticity in the troposphere. Thus, mixing occurs between air parcels with a large gradient in potential vorticity, and the potential vorticity of air parcels from the stratosphere become smaller. As a result, the potential vorticity of the intruded air parcels becomes smaller than the potential vorticity at the tropopause, and net mass transport from the stratosphere to the troposphere occurs. Fig. 19.6 schematically shows this process of STE on the isentropic surface. The tropopause is defined as a constant potential vorticity surface which is indicated by a solid curve on the isentropic surface. Air parcels that start from the tropopause at time $t-\Delta t$ denoted by a dashed curve are transformed to the solid curve or dotted curve connected to the solid one at time t. The narrow filamentary region enveloped by the dotted curve is well mixed with the environment in the troposphere, and the value of potential vorticity changes. Then, the tropopause at time t becomes the solid curve. In this case, the mass between the material surface at t of the air parcels that start from the tropopause at $t - \Delta t$ and the tropopause defined by a

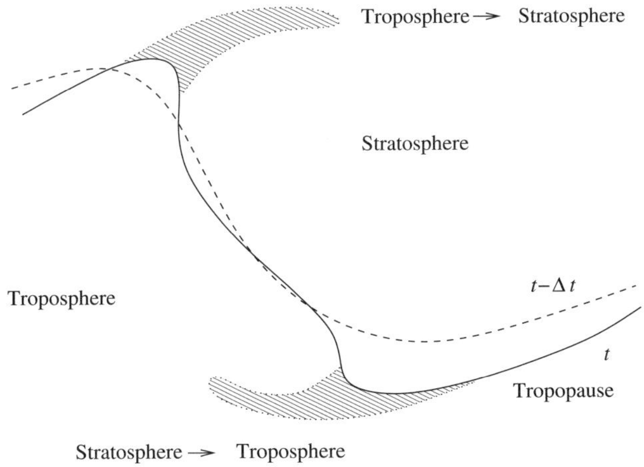

FIGURE 19.6: Schematic of mass exchange between the stratosphere and the troposphere (STE) on an isentropic surface. The dashed curve is the tropopause at time $t - \Delta t$ and the solid curve is that at time t. The dotted curve is a part of the material surface of air parcels that started from the tropopause at $t - \Delta t$. The air on the dotted curve experiences a change in potential vorticity and deviates from the tropopause. The hatched region corresponds to STE.

particular value of the potential vorticity is STE generated within the time interval Δt. The upper hatched region in the figure is mass inflow from the troposphere to the stratosphere, while the lower hatched region is mass inflow from the stratosphere to the troposphere. Fig. 19.7 shows an observational example of STE. Evolution of the PV contours on the 320 K isentropic surface is depicted with 24-hour intervals. The 2-PVU isoline shown by the bold curve is elongated and becomes isolated. This indicates that stratospheric air composed of air possessing higher PV values than 2 PVU is mixed with tropospheric air.

The second effect of STE, diabatic heating, is not strictly distinguished from the first effect (i.e., isentropic mixing), if we look closely at individual STE processes. It is a useful concept, however, since it can be used to estimate the bulk effect of STE. For instance, let us consider the case where the rate of change in potential temperature due to radiation at the lower stratosphere in high latitudes is given uniformly by Q [K s^{-1}]. This kind of heating actually occurs in the lower stratosphere of the northern hemispheric polar region in winter where subsidence motion exists (Fig. 19.8). In this case, air parcels in the lower stratosphere are cooled and their potential temperature gradually decreases, so that these air parcels are taken

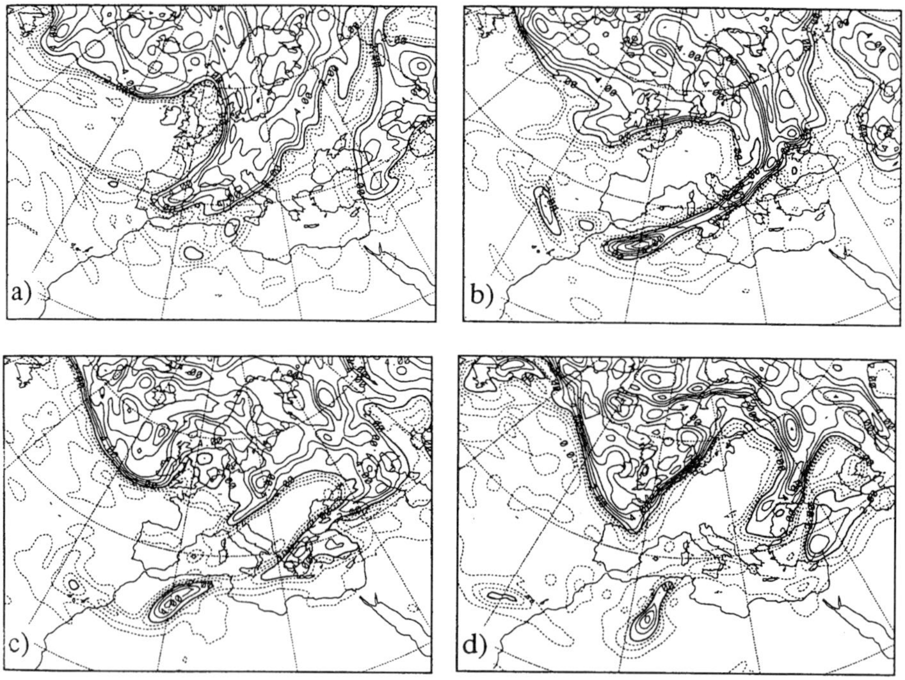

FIGURE 19.7: Evolution of the PV distribution on the $\theta = 320$ K surface for 1200UT on November (a) 10, (b) 11, (c) 12, and (d) 13, 1991. The bold curve is PV = 2 PVU, lower contour values are dashed at 0.5-PVU intervals, and higher contour values are solid at 1-PVU intervals. After Appenzeller et al. (1996), reproduced by permission of the American Geophysical Union.

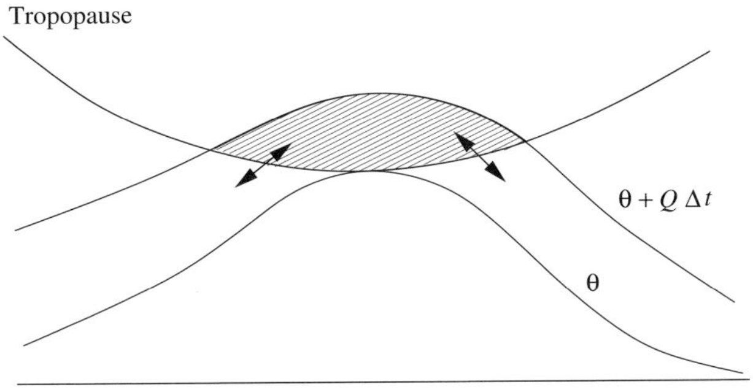

FIGURE 19.8: An example of STE when diabatic heating exists. Air parcels in the region between the isentropic surface $\theta+Q\Delta t$ and the tropopause in the polar region are taken into the troposphere during the time interval Δt. Precisely, isentropic mixing occurs in the directions indicated by the arrows and mass inflow from the troposphere to the stratosphere is associated with it. The net STE is, however, the air inflow into the troposphere designated by the hatched region of the stratosphere.

into the troposphere in the end. Let θ denote the minimum value of potential temperature at the tropopause. During the time interval Δt, air parcels in the region between the isentropic surface $\theta + Q\Delta t$ and the tropopause are taken into the troposphere.

Typical indicators of air exchange from the stratosphere to the troposphere are *tropopause folds*. Associated with the development of extratropical cyclones, the upper air decreases along a cold front behind extratropical cyclones. In this case, upper air parcels with high values of potential vorticity descend and the troposphere is lowered. This kind of deformation of the tropopause occurs locally so that the tropopause will have a folded structure.

References and suggested reading

The isentropic views of the meridional section described in Section 19.1 are given by Hoskins (1991). The Lagrangian circulations through trajectory calculations were initially shown by Kida (1977) using a general circulation model. The concept of dry intrusion mentioned in Section 19.2 is frequently invoked in various contexts. Browning (1997) discusses the dry intrusion associated with extratropical cyclones. The effect of dry intrusion on the water vapor field in the subtropics is reviewed by Pierrehumbert (1999). A more comprehensive review of STE that was described in Section 19.3 is given by Holton et al. (1995).

Appenzeller, C., Davies, H. C., and Norton, W. A., 1996: Fragmentation of stratospheric intrusions. *J. Geophys. Res.*, **101**, 1435–1456.

Browning, K. A., 1997: The dry intrusion perspective of extra-tropical cyclone development. *Meteorol. Appl.*, **4**, 317–324.

Holton, J. R., Haynes, P. H., McIntyre, M. E., Douglass, A. R., Rood, R. B., and Pfister, L., 1995: Stratosphere-troposphere exchange. *Reviews of Geophysics*, **33**, 403–439.

Hoskins, B. J., 1991: Towards a PV-θ view of the general circulation. *Tellus*, **43A/B**, 27–35.

Kida, H., 1977: A numerical investigation of the atmospheric general circulation and stratospheric-tropospheric mass exchange. II: Lagrangian motion of the atmosphere. *J. Meteorol. Soc. Japan*, **55**, 71–88.

Pierrehumbert, R. T., 1999: Subtropical water vapor as a mediator of rapid global climate change. in: *Mechanisms of Global Change at Millennial Time Scales*, (eds.) P. U. Clark, R. S. Webb, and L. D. Keigwin, American Geophysical Union, Washington, D. C. Geophysical Monograph Series **112**, 394 pp.

Satoh, M., 1999: Relation between the meridional distribution of potential vorticity and the Lagrangian mean circulation in the troposphere. *Tellus*, **51A**, 833–853.

Part III

General Circulation Modeling

In Part III, fundamental numerical techniques needed to construct an atmospheric general circulation model are described. Since atmospheric flows are nonlinear, numerical experiments are powerful and inevitable approaches to the study of the characteristics of the atmosphere. Atmospheric general circulation models (AGCMs) are numerical models that simulate three-dimensional global atmospheric flows. If an AGCM is combined with an oceanic general circulation model and a land surface model using various physical processes, they form the coupled model that is used for climate prediction of global warming.

Atmospheric circulation modeling mainly consists of two parts: the dynamical process and the physical process. The dynamical part describes the fluid motions of the atmosphere based on appropriate discretized forms of dynamical equations. This is the central part of AGCMs and is called the dynamical core. As described at the beginning of Chapter 10, the physical part includes representations of the hydrological process, radiation, and turbulence. There is a wide range of techniques for the numerical implementation of these processes. In this part, we concentrate on the dynamical core of AGCMs, and do not touch on the numerical procedure of the physical processes, whose basis has already been introduced in Part I.

This part is intended as a technical guide to making a practical model. Descriptions are restricted to the spectrum model that is widely used and has a clear model structure. More precisely, the spectrum method referred to in this part is the spectral transform method using spherical harmonics.

Chapter 20 summarizes the basic equations used for AGCMs. These equations were introduced in Chapter 3 in Part I. Other types of models such as grid models are briefly mentioned in Chapter 20. In Chapter 21, the spherical spectrum method, which is the discretization in horizontal directions on a sphere, is described, and vertical discretizations are explained in Chapter 22. The description of the dynamical framework of AGCMs is concluded with the time integration method in Chapter 23. In order to validate AGCMs and study of dynamical characteristics of the atmosphere, some standard experiments are listed in Chapter 24.

20

Basic equations of general circulation models

In this first chapter of Part 3, we present the basic equations for an atmospheric general circulation model. We consider the model governed by primitive equations on a sphere in σ coordinates. Primitive equations have already been described in Chapter 3. Here, we summarize these equations in spherical coordinates for convenience of numerical discretization in the following chapters.

The typical resolution of the currently used general circulation model is about 100 km in the horizontal direction and about a few km in the vertical direction. This means that the aspect ratio of the typical resolution is about 1/100, which represents a quasi-two-dimensional atmospheric motion. It can be thought that a meteorologically important phenomenon at such a small aspect ratio is approximately in hydrostatic balance in the vertical direction. Thus, primitive equations in hydrostatic balance are generally used as the governing equations of general circulation models.

Recently, however, much higher computer performance than ever before can be used, such as massively parallel computers. In such circumstances, a very high resolution simulation with less than 10 km in the horizontal scale is becoming achievable for global models. In these models, the above assumption is no longer acceptable, and nonhydrostatic equations will take the place of hydrostatic primitive equations for the framework of general circulation models. These models may well be called the next generation atmospheric general circulation model, and will be briefly described in Section 20.1.

20.1 Overview

The general circulation model is a numerical model that can calculate three-dimensional atmospheric motions on a globe. Primitive equations based on hydrostatic balance in the vertical direction are most commonly used as the governing equations of general circulation models. Although it is expected that nonhydrostatic equations will be adopted for next generation general circulation models in the near future, we concentrate on hydrostatic primitive equations. We summarize the primitive equations in this chapter, and describe the numerical methods based on the primitive equations in the following chapters.

The spatial discretization of general circulation models is categorized into two groups: the spectral method and the grid method. In the case of the spectral method, variables are expanded in spectral space only in the horizontal dimensions (i.e., the spherical surfaces) and are based on the grid method in the vertical direction using the finite discretized form. The spectral method used in meteorology is normally based on the transform method in which nonlinear terms are evaluated in the grid space. One of the advantages of the spectral method is that the conservation of physical quantities is automatically guaranteed. In contrast, if the grid method is used, the discretization forms should be carefully devised in order to keep the conservation. Even for spectral models, since the finite discretization method is used in the vertical direction, we need to take care of the discretized expressions of energy transform terms to conserve the total energy. Spectral expressions on a sphere will be described in Chapter 21, while vertical discretization will be discussed in Chapter 22.

The time integration method of numerical models is in general constrained by the fastest motions contained in the governing equations, such as sound waves and gravity waves. If the implicit scheme is used, the time step of time integration becomes free from the constraint of such waves. For general circulation models, the semi-implicit method is used for the treatment of gravity waves. The time integration scheme with the semi-implicit method will be described in Chapter 23.

One can further extend the time step using the *semi-Lagrangian scheme*, where advection is no longer a constraint. In the field of numerical weather prediction, one major branch is the semi-Lagrangian method (Staniforth and Côté, 1991). There is no restriction on the time step for advection and waves if both the semi-Lagrangian method and the semi-implicit method are used. Normally, the time step is optimized based on the condition that the accuracy of the prognostic fields is kept within a satisfactory range. A defect of the semi-Lagrangian method is that the global integral of physical quantities is not generally conserved, so that it is thought that semi-Lagrange models is not suitable for long time integration like climate study. We do not describe the semi-Lagrangian method in this book; the dynamical framework using the semi-Lagrangian method is very different from Eulerian spectral models.

Along with the recent development of computer facilities, the resolution of general circulation models can be much increased. The horizontal resolution of currently used general circulation models is about 100 km. The recent development of computer technology will allow us to increase the horizontal resolution to just below 10 km in the near future. New types of general circulation models are being developed to maximize the performance of computer power.

One of the most uncertain factors in the reliability of currently used general circulation models is the use of cumulus parameterization. Since the horizontal extent of cumulus convection is about 1 km, the effects of cumulus convection must be statistically treated in general circulation models with horizontal resolutions of about 100 km. However, it is very difficult to appropriately parameterize all the statistical effects of cumulus convection, though many kinds of cumulus parameterizations are being used in current models. As the horizontal resolution of numerical models approaches 1 km, individual clouds can be directly resolved in the models, so that it is expected that we will no longer need to use such cumulus parameterization based on statistical hypothesis. Thus, the likely horizontal resolution of next generation general circulation models is a few kilometers. We expect the use of models with 10-km resolution or less will come within the range of our computer facilities. With such finer resolution models, the assumption of hydrostatic balance is no longer acceptable. We must switch governing equation of the general circulation models from hydrostatic primitive equations to non-hydrostatic equations. As for vertical resolution, we do not have a suitable measure of its appropriateness.

The change in designs of computers imposes a restriction on the algorithm of numerical models. In general, we tend to use massively parallel computers with distributed memories. It is considered that, as the number of computers becomes larger, such an algorithm requiring data transformations between all the computers becomes less scalable. We have mainly focused on spectrum models with the transform method for the numerical framework of general circulation models in this book. The transform method requires Legendre transformation at each time step between the grid space and the spectral space. It is thought in general that the Legendre transformation is not an appropriate choice for the algorithm on massive parallel computers with distributed memories. if the model resolution becomes very fine. Hence, we need to choose different numerical frameworks for higher resolution general circulation models in the future. A possible choice is the use of grid models on a globe. Although latitudinal-longitudinal grid models have been used as the other type of general circulation model, a new framework of the conservative flux-form semi-Lagrangian method has also been used for general circulation models recently (Lin and Rood, 1997). However, since the simple latitudinal-longitudinal grid model has a very inhomogeneous grid spacing between the equator and the poles, this type of grid model is not suitable for higher resolution global models. We instead need to use quasi-homogeneous resolution grid models. As examples of such quasi-homogeneous grids, we can use the icosahedral grid or the cubic grid (Fig. 20.1: Heikes and Randall, 1995; Stuhne and Peltier, 1996; McGregor, 1996).

20.2 Basic equations

20.2.1 Primitive equations

We use primitive equations with hydrostatic balance in the vertical direction as the basic equations for a general circulation model. Primitive equations are given by (3.3.15)–(3.3.18) in Chapter 3, and are expressed in height z coordinates. For the general circulation model, the σ coordinates system is mainly adopted since it is easy to incorporate the topography at the lowest boundary. σ represents

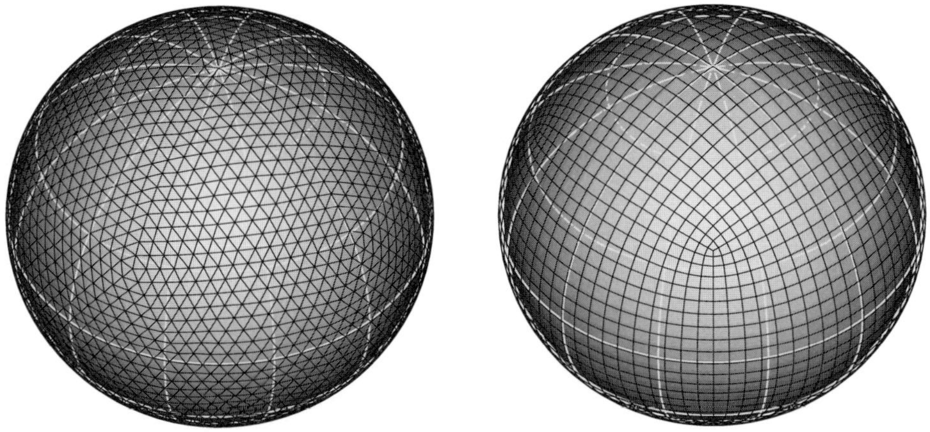

FIGURE 20.1: Left: the icosahedral grid, right: the conformal cubic grid.

the pressure normalized by surface pressure and can be a monotonic function of pressure in generic form. The transformation to σ coordinates is summarized in Section 3.3.4. Here, we only consider the simplest form

$$\sigma = \frac{p}{p_s}, \tag{20.2.1}$$

where p_s is the surface pressure. In this case, top and bottom boundary conditions are written as

$$\begin{cases} \sigma = 1, & \text{at} \quad z = z_s, \quad \text{i.e.,} \quad p = p_s, \\ \sigma = 0, & \text{at} \quad z = \infty, \quad \text{i.e.,} \quad p = 0. \end{cases} \tag{20.2.2}$$

where z_s is the surface height.

Here, we present the primitive equations in σ coordinates on a sphere. We use longitude λ and latitude φ on spherical surfaces. The equations of horizontal motion, hydrostatic balance, continuity, and thermal energy in σ coordinates are respectively given by (3.3.54), (3.3.55), (3.3.56), and (3.3.61):

$$\frac{d\boldsymbol{v}_H}{dt} = -f\boldsymbol{k} \times \boldsymbol{v_H} - \nabla_\sigma \Phi - R_d T \nabla_\sigma \pi + \boldsymbol{F_H}, \tag{20.2.3}$$

$$0 = \frac{\partial \Phi}{\partial \sigma} + \frac{R_d T}{\sigma}, \tag{20.2.4}$$

$$\frac{d\pi}{dt} = -\nabla_\sigma \cdot \boldsymbol{v_H} - \frac{\partial \dot{\sigma}}{\partial \sigma}, \tag{20.2.5}$$

$$\frac{dT}{dt} = \kappa T \left(\frac{\dot{\sigma}}{\sigma} - \frac{\partial \dot{\sigma}}{\partial \sigma} - \nabla_\sigma \cdot \boldsymbol{v_H} \right) + Q, \tag{20.2.6}$$

where $\pi = \ln p_s$, $\kappa = R_d/C_p$, and $\boldsymbol{k}$ is a unit vector in the vertical direction. $\boldsymbol{v}_H = (u, v)$ is the longitudinal and latitudinal components of velocity. $\boldsymbol{F}_H = (F_\lambda, F_\varphi)$ is the frictional force and Q is diabatic heating. ∇_σ is the two-dimensional gradient operator along a σ surface. The time derivative is written as

$$\begin{aligned}\frac{dA}{dt} &= \frac{\partial A}{\partial t} + \boldsymbol{v}_H \cdot \nabla_\sigma A + \dot{\sigma}\frac{\partial A}{\partial \sigma} \\ &= \frac{\partial A}{\partial t} + \frac{u}{R\cos\varphi}\frac{\partial A}{\partial \lambda} + \frac{v}{R}\frac{\partial A}{\partial \varphi} + \dot{\sigma}\frac{\partial A}{\partial \sigma}\end{aligned} \tag{20.2.7}$$

where the derivatives with respect to t, λ, and φ are taken along a constant σ surface. The Lagrangian derivative of the velocity $\boldsymbol{v}_H$ is given by

$$\frac{d\boldsymbol{v}_H}{dt} = \left(\frac{du}{dt} + \frac{uv}{R}\tan\varphi,\ \frac{dv}{dt} - \frac{u^2}{R}\tan\varphi\right). \tag{20.2.8}$$

The horizontal divergence along a σ surface is written as

$$D = \nabla_\sigma \cdot \boldsymbol{v}_H = \frac{1}{R\cos\varphi}\frac{\partial u}{\partial \lambda} + \frac{1}{R\cos\varphi}\frac{\partial(v\cos\varphi)}{\partial \varphi}. \tag{20.2.9}$$

Thus, (20.2.3)–(20.2.6) are written explicitly in the following form:

$$\begin{aligned}\frac{\partial u}{\partial t} &= -\frac{u}{R\cos\varphi}\frac{\partial u}{\partial \lambda} - \frac{v}{R}\frac{\partial u}{\partial \varphi} - \dot{\sigma}\frac{\partial u}{\partial \sigma} + \frac{uv}{R}\tan\varphi + 2\Omega\sin\varphi\cdot v \\ &\quad -\frac{1}{R\cos\varphi}\frac{\partial \Phi}{\partial \lambda} - \frac{R_d T}{R\cos\varphi}\frac{\partial \pi}{\partial \lambda} + F_\lambda,\end{aligned} \tag{20.2.10}$$

$$\begin{aligned}\frac{\partial v}{\partial t} &= -\frac{u}{R\cos\varphi}\frac{\partial v}{\partial \lambda} - \frac{v}{R}\frac{\partial v}{\partial \varphi} - \dot{\sigma}\frac{\partial v}{\partial \sigma} - \frac{u^2}{R}\tan\varphi - 2\Omega\sin\varphi\cdot u \\ &\quad -\frac{1}{R}\frac{\partial \Phi}{\partial \varphi} - \frac{R_d T}{R}\frac{\partial \pi}{\partial \varphi} + F_\varphi,\end{aligned} \tag{20.2.11}$$

$$0 = \frac{\partial \Phi}{\partial \sigma} + \frac{R_d T}{\sigma}, \tag{20.2.12}$$

$$\frac{\partial \pi}{\partial t} = -\frac{u}{R\cos\varphi}\frac{\partial \pi}{\partial \lambda} - \frac{v}{R}\frac{\partial \pi}{\partial \varphi} - D - \frac{\partial \dot{\sigma}}{\partial \sigma}, \tag{20.2.13}$$

$$\frac{\partial T}{\partial t} = -\frac{u}{R\cos\varphi}\frac{\partial T}{\partial \lambda} - \frac{v}{R}\frac{\partial T}{\partial \varphi} - \dot{\sigma}\frac{\partial T}{\partial \sigma} + \kappa T\left(\frac{\dot{\sigma}}{\sigma} - \frac{\partial \dot{\sigma}}{\partial \sigma} - D\right) + Q. \tag{20.2.14}$$

We also need the equation of state, which is written for dry air as

$$p = \rho R_d T. \tag{20.2.15}$$

The case where the hydrological cycle and phase change are included will be summarized in Section 20.2.6.

20.2.2 Alternative forms of the equations of motion

The velocity components in the longitude-latitude coordinates $\boldsymbol{v}_H = (u, v)$ are not continuous at the poles. Hence, it is convenient to define the following nonsingular form of velocity components by multiplying by $\cos\varphi$:

$$U = u\cos\varphi, \qquad V = v\cos\varphi. \tag{20.2.16}$$

We define $\boldsymbol{V_H} = (U, V)$. Since U and V become zero at the poles, singularity at the poles is avoided.

We replace the latitudinal coordinate with $\mu = \sin\varphi$. In this case, the gradient on a σ surface is defined by

$$\nabla_\sigma^\mu \equiv \frac{1}{\cos\varphi}\nabla_\sigma = \left(\frac{1}{R(1-\mu^2)}\frac{\partial}{\partial\lambda}, \frac{1}{R}\frac{\partial}{\partial\mu}\right). \tag{20.2.17}$$

Using this, we have the following forms of derivatives for a scalar quantity A:

$$\boldsymbol{v_H}\cdot\nabla_\sigma A = \boldsymbol{V_H}\cdot\nabla_\sigma^\mu A, \tag{20.2.18}$$

$$\begin{aligned}\nabla_\sigma\cdot(\boldsymbol{v_H}A) &= \frac{1}{a\cos\varphi}\frac{\partial}{\partial\lambda}(uA) + \frac{1}{a\cos\varphi}\frac{\partial}{\partial\varphi}(vA\cos\varphi)\\ &= \frac{1}{a(1-\mu^2)}\frac{\partial}{\partial\lambda}(UA) + \frac{1}{a}\frac{\partial}{\partial\mu}(VA)\\ &= \nabla_\sigma^\mu\cdot(\boldsymbol{V_H}A).\end{aligned} \tag{20.2.19}$$

The Laplacian operator on a sphere is written as

$$\begin{aligned}\nabla_\sigma^2 &= \frac{1}{R^2}\left\{\frac{1}{\cos^2\varphi}\frac{\partial^2}{\partial\lambda^2} + \frac{1}{\cos\varphi}\frac{\partial}{\partial\varphi}\left(\cos\varphi\frac{\partial}{\partial\varphi}\right)\right\}\\ &= \frac{1}{R^2}\left\{\frac{1}{1-\mu^2}\frac{\partial^2}{\partial\lambda^2} + \frac{\partial}{\partial\mu}\left[(1-\mu^2)\frac{\partial}{\partial\mu}\right]\right\}.\end{aligned} \tag{20.2.20}$$

Multiplying (20.2.10) and (20.2.11) by $\cos\varphi$ and changing variables u, v to U, V, we can rewrite the equations of motion as

$$\begin{aligned}\frac{\partial U}{\partial t} &= -\frac{1}{R(1-\mu^2)}\left[-U\frac{\partial U}{\partial\lambda} - (1-\mu^2)V\frac{\partial U}{\partial\mu}\right] - \dot\sigma\frac{\partial U}{\partial\sigma}\\ &\quad +2\Omega\mu V - \frac{1}{R}\frac{\partial\Phi}{\partial\lambda} - \frac{R_dT}{R}\frac{\partial\pi}{\partial\lambda} + F_\lambda\cos\varphi,\end{aligned} \tag{20.2.21}$$

$$\begin{aligned}\frac{\partial V}{\partial t} &= \frac{1}{R(1-\mu^2)}\left[-U\frac{\partial V}{\partial\lambda} - (1-\mu^2)V\frac{\partial V}{\partial\mu} - \mu(U^2+V^2)\right] - \dot\sigma\frac{\partial V}{\partial\sigma}\\ &\quad -2\Omega\mu U - \frac{1-\mu^2}{R}\frac{\partial\Phi}{\partial\mu} - \frac{R_dT(1-\mu^2)}{R}\frac{\partial\pi}{\partial\mu} + F_\varphi\cos\varphi.\end{aligned} \tag{20.2.22}$$

20.2.3 The equation for sigma velocity

Since hydrostatic balance is assumed for the vertical momentum equation, vertical velocity is not a predictable variable in the governing system of primitive equations (20.2.10)–(20.2.14). The vertical velocity in σ coordinates, $\dot\sigma$ (or *sigma velocity*), can be calculated using boundary conditions and the continuity equation (20.2.13).

The boundary condition for $\dot\sigma$ at $\sigma = 0, 1$ is given by

$$\dot\sigma = 0. \tag{20.2.23}$$

The continuity equation (20.2.13) is rewritten as

$$\frac{\partial \pi}{\partial t} = -\boldsymbol{V_H} \cdot \nabla^{\mu}_{\sigma} \pi - D - \frac{\partial \dot{\sigma}}{\partial \sigma}, \tag{20.2.24}$$

Integrating this from $\sigma = 0$ to 1 gives

$$\frac{\partial \pi}{\partial t} = -\int_0^1 \boldsymbol{V_H} \cdot \nabla^{\mu}_{\sigma} \pi \, d\sigma - \int_0^1 D d\sigma. \tag{20.2.25}$$

Integration (20.2.24) from arbitrary σ to $\sigma = 1$, we also have

$$(1-\sigma)\frac{\partial \pi}{\partial t} = -\int_\sigma^1 \boldsymbol{V_H} \cdot \nabla^{\mu}_{\sigma} \pi \, d\sigma - \int_\sigma^1 D d\sigma + \dot{\sigma}, \tag{20.2.26}$$

or, alternatively, integrating from $\sigma = 0$ to σ gives

$$\sigma \frac{\partial \pi}{\partial t} = -\int_0^\sigma \boldsymbol{V_H} \cdot \nabla^{\mu}_{\sigma} \pi d\sigma - \int_0^\sigma D d\sigma - \dot{\sigma}. \tag{20.2.27}$$

From (20.2.25), (20.2.26), or (20.2.27), we have the following different forms of the equation for $\dot{\sigma}$:

$$\begin{aligned}
\dot{\sigma} &= (1-\sigma)\frac{\partial \pi}{\partial t} + \int_\sigma^1 \boldsymbol{V_H} \cdot \nabla^{\mu}_{\sigma} \pi d\sigma + \int_\sigma^1 D d\sigma \\
&= -(1-\sigma)\int_0^1 D d\sigma + \int_\sigma^1 D d\sigma - (1-\sigma)\int_0^1 \boldsymbol{V_H} \cdot \nabla^{\mu}_{\sigma} \pi d\sigma \\
&\quad + \int_\sigma^1 \boldsymbol{V_H} \cdot \nabla^{\mu}_{\sigma} \pi d\sigma \\
&= -\sigma \frac{\partial \pi}{\partial t} - \int_0^\sigma \boldsymbol{V_H} \cdot \nabla^{\mu}_{\sigma} \pi d\sigma - \int_0^\sigma D d\sigma \\
&= \sigma \int_0^1 \boldsymbol{V_H} \cdot \nabla^{\mu}_{\sigma} \pi d\sigma - \int_0^\sigma \boldsymbol{V_H} \cdot \nabla^{\mu}_{\sigma} \pi d\sigma + \sigma \int_0^1 D d\sigma - \int_0^\sigma D d\sigma.
\end{aligned} \tag{20.2.28}$$

20.2.4 The thermodynamic equation

The thermodynamic equation (20.2.14) can be rewritten by using T' and Φ', which are the deviations of temperature and geopotential from horizontally uniform values $\overline{T}$ and $\overline{\Phi}$, respectively: $T = \overline{T} + T'$ and $\Phi = \overline{\Phi} + \Phi'$. Here, $\overline{T}$ and $\overline{\Phi}$ are not necessarily the horizontally mean values, but can be appropriately chosen. Thus, (20.2.14) can be written as

$$\begin{aligned}
\frac{\partial T}{\partial t} &= -\boldsymbol{V_H} \cdot \nabla^{\mu}_{\sigma} T - \dot{\sigma}\frac{\partial T}{\partial \sigma} + \kappa T \left(\frac{\dot{\sigma}}{\sigma} - \frac{\partial \dot{\sigma}}{\partial \sigma} - D \right) + Q \\
&= -\nabla^{\mu}_{\sigma} \cdot (\boldsymbol{V_H} T') + T' D + \dot{\sigma}\gamma - \kappa T \left(D + \frac{\dot{\sigma}}{\sigma} \right) + Q,
\end{aligned} \tag{20.2.29}$$

where γ is static stability defined by

$$\gamma \equiv \kappa \frac{T}{\sigma} - \frac{\partial T}{\partial \sigma} = -\sigma^{\kappa} \frac{\partial (T\sigma^{-\kappa})}{\partial \sigma}. \tag{20.2.30}$$

Since the potential temperature is written as

$$\theta = T\left(\frac{p_0}{p}\right)^{\kappa} = T\sigma^{-\kappa}\left(\frac{p_0}{p_s}\right)^{\kappa}, \tag{20.2.31}$$

the Brunt-Väisälä frequency N is related to static stability γ as

$$N^2 \equiv g\frac{\partial \ln \theta}{\partial z} = g\frac{\partial \sigma}{\partial z}\frac{\partial \ln \theta}{\partial \sigma} = \frac{g^2 \sigma}{R_d T^2}\gamma, \tag{20.2.32}$$

where hydrostatic balance (20.2.12) is used.

Using (20.2.24), the thermodynamic equation (20.2.29) can be rewritten as

$$\frac{\partial T}{\partial t} = -\nabla_{\sigma}^{\mu} \cdot (\boldsymbol{V_H} T') + T'D + \dot{\sigma}\gamma + \kappa T\left(\frac{\partial \pi}{\partial t} + \boldsymbol{V_H} \cdot \nabla_{\sigma}^{\mu}\pi\right) + Q. \tag{20.2.33}$$

20.2.5 Vorticity and divergence equations

Vorticity and divergence equations can be derived from the equations of motion (20.2.21) and (20.2.22). Vorticity ζ and divergence D are related to the streamfunction ψ and the velocity potential χ, respectively, as

$$\boldsymbol{v_H} = \boldsymbol{k} \times \nabla_{\sigma}\psi + \nabla_{\sigma}\chi, \tag{20.2.34}$$

$$\zeta = \boldsymbol{k} \cdot \nabla_{\sigma} \times \boldsymbol{v_H} = \nabla_{\sigma}^2 \psi, \tag{20.2.35}$$

$$D = \nabla_{\sigma} \cdot \boldsymbol{v_H} = \nabla_{\sigma}^2 \chi. \tag{20.2.36}$$

These are written in spherical coordinates using U, V, and μ as

$$U = -\frac{1-\mu^2}{R}\frac{\partial \psi}{\partial \mu} + \frac{1}{R}\frac{\partial \chi}{\partial \lambda}, \tag{20.2.37}$$

$$V = \frac{1}{R}\frac{\partial \psi}{\partial \lambda} + \frac{1-\mu^2}{R}\frac{\partial \chi}{\partial \mu}, \tag{20.2.38}$$

$$\zeta = \frac{1}{R(1-\mu^2)}\frac{\partial V}{\partial \lambda} - \frac{1}{R}\frac{\partial U}{\partial \mu}, \tag{20.2.39}$$

$$D = \frac{1}{R(1-\mu^2)}\frac{\partial U}{\partial \lambda} + \frac{1}{R}\frac{\partial V}{\partial \mu}. \tag{20.2.40}$$

Noting (20.2.39), we obtain the vorticity equation by

$$\text{Vorticity equation} = \boldsymbol{k} \cdot \nabla_{\sigma}^{\mu} \times \begin{pmatrix} \text{Eq. (20.2.21)} \\ \text{Eq. (20.2.22)} \end{pmatrix}.$$

Using the relation

$$\begin{aligned}
&\frac{1}{R(1-\mu^2)}\frac{\partial}{\partial\lambda}\left\{\frac{1}{R(1-\mu^2)}\left[-U\frac{\partial V}{\partial\lambda}-(1-\mu^2)V\frac{\partial V}{\partial\mu}-\mu(U^2+V^2)\right]\right\}\\
&\qquad -\frac{1}{R}\frac{\partial}{\partial\mu}\left\{\frac{1}{R(1-\mu^2)}\left[-U\frac{\partial U}{\partial\lambda}-(1-\mu^2)V\frac{\partial U}{\partial\mu}\right]\right\}\\
&\quad = -\left[\frac{1}{R(1-\mu^2)}\frac{\partial}{\partial\lambda}(U\zeta)+\frac{1}{R}\frac{\partial}{\partial\mu}(V\zeta)\right] = -\nabla_\sigma^\mu\cdot(\zeta\boldsymbol{V_H}),
\end{aligned}$$

we find that the vorticity equation is written as

$$\frac{\partial\zeta}{\partial t} = -\left[\frac{1}{R(1-\mu^2)}\frac{\partial}{\partial\lambda}A+\frac{1}{R}\frac{\partial}{\partial\mu}B\right], \tag{20.2.41}$$

where

$$A = (\zeta+2\Omega\mu)U+\dot\sigma\frac{\partial V}{\partial\sigma}+\frac{R_dT'}{R}(1-\mu^2)\frac{\partial\pi}{\partial\mu}-F_\varphi\cos\varphi, \tag{20.2.42}$$

$$B = (\zeta+2\Omega\mu)V-\dot\sigma\frac{\partial U}{\partial\sigma}-\frac{R_dT'}{R}\frac{\partial\pi}{\partial\lambda}+F_\lambda\cos\varphi. \tag{20.2.43}$$

If we rewrite the Coriolis term as

$$\begin{aligned}
-\nabla_\sigma^\mu\cdot(2\Omega\mu\boldsymbol{V_H}) &= -2\Omega\left(\mu\nabla_\sigma^\mu\cdot\boldsymbol{V_H}+\boldsymbol{V_H}\cdot\nabla_\sigma^\mu\mu\right)\\
&= -2\Omega\left(\mu D+\frac{V}{R}\right),
\end{aligned} \tag{20.2.44}$$

we obtain the following form of the vorticity equation from (20.2.41):

$$\frac{\partial\zeta}{\partial t} = -\left[\frac{1}{R(1-\mu^2)}\frac{\partial}{\partial\lambda}A'+\frac{1}{R}\frac{\partial}{\partial\mu}B'\right]-2\Omega\left(\mu D+\frac{V}{R}\right), \tag{20.2.45}$$

$$A' = \zeta U+\dot\sigma\frac{\partial V}{\partial\sigma}+\frac{R_dT'}{R}(1-\mu^2)\frac{\partial\pi}{\partial\mu}-F_\varphi\cos\varphi, \tag{20.2.46}$$

$$B' = \zeta V-\dot\sigma\frac{\partial U}{\partial\sigma}-\frac{R_dT'}{R}\frac{\partial\pi}{\partial\lambda}+F_\lambda\cos\varphi. \tag{20.2.47}$$

In a similar way, since (20.2.40), the divergence equation can be calculated by

$$\text{Divergence equation} = \nabla_\sigma^\mu\cdot\begin{pmatrix}\text{Eq. (20.2.21)}\\ \text{Eq. (20.2.22)}\end{pmatrix}.$$

We can show

$$\begin{aligned}
&\frac{1}{R(1-\mu^2)}\frac{\partial}{\partial\lambda}\left\{\frac{1}{R(1-\mu^2)}\left[-U\frac{\partial U}{\partial\lambda}-(1-\mu^2)V\frac{\partial U}{\partial\mu}\right]\right\}\\
&\qquad +\frac{1}{R}\frac{\partial}{\partial\mu}\left\{\frac{1}{R(1-\mu^2)}\left[-U\frac{\partial V}{\partial\lambda}-(1-\mu^2)V\frac{\partial V}{\partial\mu}-\mu(U^2+V^2)\right]\right\}
\end{aligned}$$

$$
\begin{aligned}
&= \frac{1}{R(1-\mu^2)}\frac{\partial}{\partial\lambda}(V\zeta) - \frac{1}{R}\frac{\partial}{\partial\mu}(U\zeta) \\
&\quad -\left\{\frac{1}{a^2(1-\mu^2)}\frac{\partial^2}{\partial\lambda^2} + \frac{1}{a^2}\frac{\partial}{\partial\mu}\left[(1-\mu^2)\frac{\partial}{\partial\mu}\right]\right\}\frac{U^2+V^2}{2(1-\mu^2)} \\
&= \boldsymbol{k}\cdot\nabla_\sigma^\mu \times (\zeta \boldsymbol{V_H}) - \nabla_\sigma^2 \frac{U^2+V^2}{2(1-\mu^2)}.
\end{aligned}
$$

Thus the divergence equation can be written as

$$
\frac{\partial D}{\partial t} = \frac{1}{R(1-\mu^2)}\frac{\partial}{\partial\lambda}B - \frac{1}{R}\frac{\partial}{\partial\mu}A - \nabla_\sigma^2(\Phi' + R_d\overline{T}\pi + E), \tag{20.2.48}
$$

where

$$
E = \frac{U^2+V^2}{2(1-\mu^2)} \tag{20.2.49}
$$

is the kinetic energy. Using the expression of the Coriolis term

$$
\begin{aligned}
\boldsymbol{k}\cdot\nabla_\sigma^\mu \times (2\Omega\mu \boldsymbol{V_H}) &= 2\Omega(\mu\nabla_\sigma^\mu \times \boldsymbol{V_H} + \nabla_\sigma^\mu \times \boldsymbol{V_H}) \\
&= 2\Omega\left(\mu\zeta - \frac{U}{R}\right),
\end{aligned} \tag{20.2.50}
$$

we can rewrite (20.2.48) as

$$
\begin{aligned}
\frac{\partial D}{\partial t} &= \frac{1}{R(1-\mu^2)}\frac{\partial}{\partial\lambda}B' - \frac{1}{R}\frac{\partial}{\partial\mu}A' + 2\Omega\left(\mu\zeta - \frac{U}{R}\right) \\
&\quad -\nabla_\sigma^2(\Phi' + R_d\overline{T}\pi + E).
\end{aligned} \tag{20.2.51}
$$

20.2.6 The moisture equation

We also have an equation for a scalar quantity in addition to the above basic equations. Specific humidity q is an example of a scalar quantity. The general form of the equation of q is written as

$$
\frac{dq}{dt} = S_q, \tag{20.2.52}
$$

where S_q is a source term of specific humidity; it includes convergence of moisture diffusion or change due to condensation. This equation can be rewritten as

$$
\begin{aligned}
\frac{\partial q}{\partial t} &= -\boldsymbol{v_H}\cdot\nabla_\sigma q - \dot{\sigma}\frac{\partial q}{\partial\sigma} + S_q \\
&= -\nabla_\sigma^\mu\cdot(\boldsymbol{V_H} q) + qD - \dot{\sigma}\frac{\partial q}{\partial\sigma} + S_q.
\end{aligned} \tag{20.2.53}
$$

If water vapor is contained, the equation of state (20.2.15) is modified as

$$
p = \rho R_d T_v, \tag{20.2.54}
$$

where $T_v = (1 + 0.608q)T$ is virtual temperature given by (8.2.38).

20.2.7 Summary of governing equations

Let us summarize the above equation set for a general circulation model; (20.2.41), (20.2.48), (20.2.12), (20.2.33), (20.2.25), (20.2.28), and (20.2.53) are

$$\frac{\partial \zeta}{\partial t} = -\left[\frac{1}{R(1-\mu^2)}\frac{\partial}{\partial \lambda}A + \frac{1}{R}\frac{\partial}{\partial \mu}B\right], \tag{20.2.55}$$

$$\frac{\partial D}{\partial t} = \frac{1}{R(1-\mu^2)}\frac{\partial}{\partial \lambda}B - \frac{1}{R}\frac{\partial}{\partial \mu}A - \nabla_\sigma^2(\Phi' + R_d\overline{T}\pi + E), \tag{20.2.56}$$

$$0 = \frac{\partial \Phi}{\partial \sigma} + \frac{R_d T}{\sigma}, \tag{20.2.57}$$

$$\begin{aligned}\frac{\partial T}{\partial t} &= -\left[\frac{1}{R(1-\mu^2)}\frac{\partial}{\partial \lambda}(UT') + \frac{1}{R}\frac{\partial}{\partial \mu}(VT')\right] \\ &\quad + T'D + \dot{\sigma}\gamma + \kappa T\left(\frac{\partial \pi}{\partial t} + \boldsymbol{V_H}\cdot\nabla_\sigma^\mu \pi\right) + Q,\end{aligned} \tag{20.2.58}$$

$$\frac{\partial \pi}{\partial t} = -\int_0^1 \boldsymbol{V_H}\cdot\nabla_\sigma^\mu \pi d\sigma - \int_0^1 D d\sigma. \tag{20.2.59}$$

$$\dot{\sigma} = -\sigma\frac{\partial \pi}{\partial t} - \int_0^\sigma \boldsymbol{V_H}\cdot\nabla_\sigma^\mu \pi d\sigma - \int_0^\sigma D d\sigma, \tag{20.2.60}$$

$$\frac{\partial q}{\partial t} = -\left[\frac{1}{R(1-\mu^2)}\frac{\partial}{\partial \lambda}(Uq) + \frac{1}{R}\frac{\partial}{\partial \mu}(Vq)\right] + qD - \dot{\sigma}\frac{\partial q}{\partial \sigma} + S_q, \tag{20.2.61}$$

where $\mu = \sin\varphi$ and the following symbols are defined:

$$\boldsymbol{V_H} = (U, V) = (u\cos\varphi, v\cos\varphi) = \boldsymbol{v_H}\cos\varphi, \tag{20.2.62}$$

$$\zeta = \frac{1}{R(1-\mu^2}\frac{\partial V}{\partial \lambda} - \frac{1}{R}\frac{\partial U}{\partial \mu}, \tag{20.2.63}$$

$$D = \frac{1}{R(1-\mu^2)}\frac{\partial U}{\partial \lambda} + \frac{1}{R}\frac{\partial V}{\partial \mu}, \tag{20.2.64}$$

$$A = (\zeta + 2\Omega\mu)U + \dot{\sigma}\frac{\partial V}{\partial \sigma} + \frac{R_d T'}{R}(1-\mu^2)\frac{\partial \pi}{\partial \mu} - F_\varphi\cos\varphi, \tag{20.2.65}$$

$$B = (\zeta + 2\Omega\mu)V - \dot{\sigma}\frac{\partial U}{\partial \sigma} - \frac{R_d T'}{R}\frac{\partial \pi}{\partial \lambda} + F_\lambda\cos\varphi, \tag{20.2.66}$$

$$E = \frac{U^2 + V^2}{2(1-\mu^2)}, \tag{20.2.67}$$

$$\gamma = -\sigma^\kappa\frac{\partial(T\sigma^{-\kappa})}{\partial \sigma}. \tag{20.2.68}$$

In (20.2.55)–(20.2.61), the predictable variables are ζ, D, T, π, and q, while $\dot{\sigma}$ and Φ are the diagnostic variables. The velocity components U and V are solved from ζ and D through the streamfunction ψ and the velocity potential χ. That is, from (20.2.35) and (20.2.36), we have

$$\zeta = \nabla_\sigma^2\psi, \qquad D = \nabla_\sigma^2\chi, \tag{20.2.69}$$

then the Poisson equation must be solved for ψ and χ. The velocity components can be calculated from (20.2.37), (20.2.38):

$$U = -\frac{1-\mu^2}{R}\frac{\partial\psi}{\partial\mu} + \frac{1}{R}\frac{\partial\chi}{\partial\lambda}, \quad (20.2.70)$$

$$V = \frac{1}{R}\frac{\partial\psi}{\partial\lambda} + \frac{1-\mu^2}{R}\frac{\partial\chi}{\partial\mu}. \quad (20.2.71)$$

References and suggested reading

As standard textbooks of atmospheric numerical models, we first refer to Haltiner and Williams (1980) in which basic equations and various techniques of numerical models are described. Krishnamurti et al. (1998) is more specific to the numerical methods of a global spectral model, and is close to the approach taken in part III of this book. Durran (1998) is a comprehensive book on the discretization methods that appear in geophysical fluid dynamics.

Durran, D. R., 1998: *Numerical Methods for Wave Equations in Geophysical Fluid Dynamics.* Springer-Verlag, New York, 465 pp.

Haltiner, G. J. and Williams, R. T., 1980: *Numerical Prediction and Dynamic Meteorology*, 2nd ed. John Wiley & Sons, New York, 477 pp.

Heikes, R. and Randall, D. A., 1995: Numerical integration of the shallow-water equations on a twisted icosahedral grid. Part I: Basic design and results of tests. *Mon. Wea. Rev.*, **123**, 1862–1880.

Krishnamurti, T. N., Bedi, H. S., and Hardiker, V. M., 1998: *An Introduction to Global Spectral Modeling.* Oxford University Press, New York, 253 pp.

Lin, S.-J. and Rood, R. B., 1997: An explicit flux-form semi-Lagrangian shallow-water model on the sphere. *Q. J. Roy. Meteorol. Soc.*, **123**, 2477–2498.

McGregor, J. L., 1996: Semi-Lagrangian advection on conformal-cubic grids. *Mon. Wea. Rev.*, **124**, 1311–1322.

Staniforth, A. and Côté, J., 1991: Semi-Lagrangian integration schemes for atmospheric models – A review. *Mon. Wea. Rev.*, **119**, 1847–1859.

Stuhne, G. R. and Peltier, W. R., 1996: Vortex erosion and amalgamation in a new model of large scale flow on the sphere. *J. Comp. Phys.*, **128**, 58–81.

21

Spectral method on a sphere

One of the dynamical frameworks of currently used atmospheric general circulation models is the spectral method in which all the variables are expanded using spherical harmonics. The spectral method used for atmospheric general circulation models is different from a pure spectrum method in which only the coefficients in spectrum space are integrated as used in fluid dynamics. Instead, the values at the grid points and the coefficients in spectral space are used for calculating nonlinear terms and these values in different spaces are transformed at each time step; this method is called the *transform method.*

In this chapter, after a brief introduction of the spectrum method, we first summarize the mathematical formulation of the spherical harmonics and necessary integral formulas for later use. Then, we describe the spectrum method using the nondivergent barotropic system on a sphere. This clarifies the treatment of nonlinear terms, and the difference between the interaction coefficients method and the transform method are explained. We also describe conservation in the spectral form of the barotropic equation. Subsequently, we discuss the spectrum method of shallow water equations and the primitive equations on a sphere.

Spherical spectral expansion plays a fundamental role not only in numerical techniques but also in global atmospheric dynamics. The normal modes of waves on a sphere are analyzed using spherical harmonics in Section 4.7.2. Nonlinear interaction between different spherical modes is essential for two-dimensional turbulence on a sphere as described in Section 17.4.2.

21.1 The spectrum method

To introduce the spectrum method, we use the following differential equation

$$\frac{\partial}{\partial t}\psi(x,t) \quad = \quad F(\psi), \tag{21.1.1}$$

where F is an operator involving spatial derivatives of ψ. This equation can be solved for ψ by integrating in time on appropriate boundary and initial conditions.

The function $\psi(x)$ can be expanded by a series of appropriate orthogonal functions ϕ_k such that

$$\psi(x,t) = \sum_{k=1}^{\infty} \psi_k(t)\phi_k(x), \tag{21.1.2}$$

where ψ_k's are called *expansion coefficients*. In the case of the spectrum method, coefficients ψ_k are integrated in time instead of integrating ψ in physical space. From orthogonality, expansion functions satisfy

$$\int_S \phi_k(x)\phi_l(x)\,dx = a_k\delta_{kl}, \tag{21.1.3}$$

where a_k is the norm of ϕ_k.

If the above equations are to be solved numerically, the number of expansion functions must be truncated with a finite number N. In this case, let a numerical solution for ψ be denoted by ϕ:

$$\phi(x) = \sum_{k=1}^{N} \psi_k(t)\phi_k(x). \tag{21.1.4}$$

Since the summation is over a finite number, (21.1.4) is not an exact solution to (21.1.1) unless $\phi_k(x)$ are eigenfunctions of F. To seek an approximate solution, we define the *residual* by

$$\mathcal{R}(\phi) \equiv \frac{\partial}{\partial t}\phi - F(\phi). \tag{21.1.5}$$

We can obtain equations for the expansion coefficients by minimizing the residual $\mathcal{R}(\phi)$ using an appropriate method. There are several methods for minimizing the residual. In particular, the method called the *Galerkin approximation* requires that the residual $\mathcal{R}$ should be orthogonal to all the expansion functions ϕ_k:

$$\int_S \mathcal{R}(\phi(x))\phi_k(x)\,dx = 0, \quad \text{for} \quad k = 1,\cdots,N. \tag{21.1.6}$$

Substituting (21.1.4) into this equation, we obtain

$$\frac{d\psi_k}{dt} = \frac{1}{a_k}\int_S F(\phi(x))\phi_k(x)\,dx. \tag{21.1.7}$$

This is a basic equation for integrating ψ_k in time. The remaining problem is how to express the right-hand side.

To solve the primitive equations on a sphere summarized in the previous chapter, we use spherical harmonics as expansion functions. The properties of spherical harmonics are described in Section 21.2. The tendency F involves nonlinear terms such as advection terms. To calculate the nonlinear tendency term in the spectrum method, one may use the interaction coefficient method or the transform method. These will be explained in Section 21.4.

21.2 Spectral expansion on a sphere

In this section, we summarize the spectral expansion method on a sphere.

21.2.1 Spherical harmonics

A function f that satisfies the Laplace equation

$$\nabla^2 f = \left(\frac{\partial^2}{\partial x^2} + \frac{\partial^2}{\partial y^2} + \frac{\partial^2}{\partial z^2} \right) f = 0, \tag{21.2.1}$$

is called the *harmonic function.* In spherical coordinates (λ, φ, r), the Laplacian ∇^2 takes the form:

$$\nabla^2 f = \frac{1}{r^2}\Lambda f + \frac{1}{r}\frac{\partial^2}{\partial r^2}(rf) = \frac{1}{r^2}\left[\Lambda f + \frac{\partial}{\partial r}\left(r^2 \frac{\partial}{\partial r} f\right)\right], \tag{21.2.2}$$

in which the operator Λ is written as

$$\begin{aligned} \Lambda &= \frac{1}{\cos^2\varphi}\frac{\partial^2}{\partial \lambda^2} + \frac{1}{\cos\varphi}\frac{\partial}{\partial \varphi}\left(\cos\varphi \frac{\partial}{\partial \varphi}\right) \\ &= \frac{1}{1-\mu^2}\frac{\partial^2}{\partial \lambda^2} + \frac{\partial}{\partial \mu}\left[(1-\mu^2)\frac{\partial}{\partial \mu}\right], \end{aligned} \tag{21.2.3}$$

where $\mu = \sin\varphi$ is defined. As a class of harmonic functions,

$$f = r^n Y_n(\lambda, \mu) \tag{21.2.4}$$

is called a *solid harmonic*, where $Y_n(\lambda, \mu)$ is an n-th order *spherical harmonic.* Substituting (21.2.4) into (21.2.2), we obtain the differential equation for Y_n:

$$[\Lambda + n(n+1)]\, Y_n(\lambda, \mu) = 0. \tag{21.2.5}$$

We define the two-dimensional Laplacian operator on a sphere by

$$\nabla_H^2 \equiv \frac{1}{R^2}\Lambda = \frac{1}{R^2}\left\{\frac{1}{1-\mu^2}\frac{\partial^2}{\partial \lambda^2} + \frac{\partial}{\partial \mu}\left[(1-\mu^2)\frac{\partial}{\partial \mu}\right]\right\}, \tag{21.2.6}$$

where R is the radius of the sphere. Using this, (21.2.5) is written as

$$\left[\nabla_H^2 + \frac{n(n+1)}{R^2}\right] Y_n = 0. \tag{21.2.7}$$

Since ∇^2 is different from Λ in dimension by $1/r^2$, we introduce the factor $1/R^2$ in ∇_H^2. We also define the gradient and divergence operators on a sphere:

$$\nabla_H A = \left(\frac{1}{R\sqrt{1-\mu^2}}\frac{\partial A}{\partial \lambda}, \frac{\sqrt{1-\mu^2}}{R}\frac{\partial A}{\partial \mu} \right), \tag{21.2.8}$$

$$\nabla_H \cdot \boldsymbol{A} = \frac{1}{R}\left[\frac{1}{\sqrt{1-\mu^2}}\frac{\partial A_\lambda}{\partial \lambda} + \frac{\partial[\sqrt{1-\mu^2}A_\mu]}{\partial \mu} \right], \tag{21.2.9}$$

where A is a scalar and $\boldsymbol{A} = (A_\lambda, A_\mu, A_r)$ is a vector.

The expression of spherical harmonic Y_n can be determined by the separation method of variables as

$$Y_n(\lambda,\mu) = \Theta(\lambda)\Phi(\mu). \tag{21.2.10}$$

Substituting this into (21.2.5) and dividing by $\Theta\Phi/(1-\mu^2)$, we obtain

$$(1-\mu^2)\left\{\frac{1}{\Phi}\frac{d}{d\mu}\left[(1-\mu^2)\frac{d}{d\mu}\Phi\right]+n(n+1)\right\} = -\frac{1}{\Theta}\frac{d^2}{d\lambda^2}\Theta. \tag{21.2.11}$$

This is the separation form: the left-hand side is a function of μ and the right-hand side is a function of λ. Equating the above equation to a constant C, we have

$$\frac{d^2\Theta}{d\lambda^2}+C\Theta = 0. \tag{21.2.12}$$

The solution to this is written as

$$\Theta = C_1e^{i\sqrt{C}\lambda}+C_2e^{-i\sqrt{C}\lambda}, \tag{21.2.13}$$

where C_1 and C_2 are constant. From periodicity $\Theta(\lambda+2\pi)=\Theta(\lambda)$, we have $C=m^2$ for $m=0,1,2,\cdots$. Therefore, (21.2.13) can be written as

$$\Theta = Ae^{im\lambda}+Be^{-im\lambda} = A'\cos m\lambda+B'\sin m\lambda, \tag{21.2.14}$$

where A, B and A', B' are constant. Substituting (21.2.14) into (21.2.11), we obtain the equation for Φ as

$$\frac{d}{d\mu}\left[(1-\mu^2)\frac{d}{d\mu}\Phi\right]+\left[n(n+1)-\frac{m^2}{1-\mu^2}\right]\Phi = 0. \tag{21.2.15}$$

The solution to this equation is an *associated Legendre function*, whose normalized form is given as (see Section 21.8)

$$\tilde{P}_n^m(\mu) = \sqrt{(2n+1)\frac{(n-m)!}{(n+m)!}}\frac{(1-\mu^2)^{\frac{m}{2}}}{2^nn!}\frac{d^{n+m}}{d\mu^{n+m}}(\mu^2-1)^n, \tag{21.2.16}$$

where m is an integer with $0\le m\le n$. $\tilde{P}_n^m$ is a normalized associated Legendre function, and satisfies the orthogonality:†

$$\int_{-1}^{1}\tilde{P}_n^m(\mu)\tilde{P}_l^m(\mu)d\mu = 2\delta_{nl}. \tag{21.2.17}$$

The n index indicates the degree and the m index indicates the order of the associated Legendre function.

† We define normalization such that the integral in the range $-1\le\mu\le1$ is 2. In some literature, normalization is defined such that the integral is 1.

Substituting (21.2.14), (21.2.16) into (21.2.10), we find that spherical harmonic Y_n can be expanded as

$$Y_n(\lambda, \mu) = \sum_{m=0}^{n} \tilde{P}_n^m(\mu)(A_n^m e^{im\lambda} + B_n^m e^{-im\lambda}), \tag{21.2.18}$$

where A_n^m and B_n^m are expansion coefficients. If we define $\tilde{P}_n^m$ for the indices $-n \leq m \leq 0$ by†

$$\tilde{P}_n^m(\mu) = (-1)^m \tilde{P}_n^{-m}(\mu), \tag{21.2.19}$$

we may expand spherical harmonic in the form

$$Y_n(\lambda, \mu) = \sum_{m=-n}^{n} a_n^m \tilde{P}_n^m(\mu) e^{im\lambda}. \tag{21.2.20}$$

If we define a function for all the indices $-n \leq m \leq n$,

$$Y_n^m(\lambda, \mu) = \tilde{P}_n^m(\mu) e^{im\lambda}, \tag{21.2.21}$$

then the following orthogonality relation holds:

$$\int_0^{2\pi} \int_{-1}^{1} Y_n^m(\lambda, \mu) Y_n^{m*}(\lambda, \mu) d\mu d\lambda = 4\pi \delta_{mm'} \delta_{nn'}, \tag{21.2.22}$$

where Y_n^{m*} is the complex conjugate of Y_n^m and can be written from (21.2.19) as

$$Y_n^{m*}(\lambda, \mu) = \tilde{P}_n^m(\mu) e^{-im\lambda} = (-1)^m Y_n^{-m}(\lambda, \mu). \tag{21.2.23}$$

The associated Legendre function $\tilde{P}_n^m(\mu)$ has $n - |m|$ zero points in the range $-1 < \mu < 1$. Since the trigonometric functions $\cos m\lambda$ and $\sin m\lambda$ have $2m$ zero points in the longitudinal direction, a spherical harmonic function $Y_n^m(\lambda, \mu)$ divides the sphere into $2m(n - |m| + 1)$ pieces if $m \neq 0$. Fig. 21.1 shows examples of profiles of Y_n^m.

Spherical harmonics have the following property for rotational transformation of the coordinates. If new coordinates (λ', μ') are generated by the operation of rotation $\mathcal{R}$, spherical harmonics of n-th degree in the new coordinates are expressed by a linear combination of the same spherical harmonics of n-th degree in the original coordinates:

$$Y_n^m(\lambda, \mu) = \sum_{m'=-M}^{M} D_{mm'}^n(\mathcal{R}) Y_n^{m'}(\lambda', \mu'), \tag{21.2.24}$$

This can be confirmed by (21.2.5) where the spherical harmonic $Y_n^{m'}$ is the angular part of solid harmonics of the n-th degree. The operation of rotation does not change the degree n.

†Another convention $\tilde{P}_n^m(\mu) = \tilde{P}_n^{-m}(\mu)$ is used in some literature.

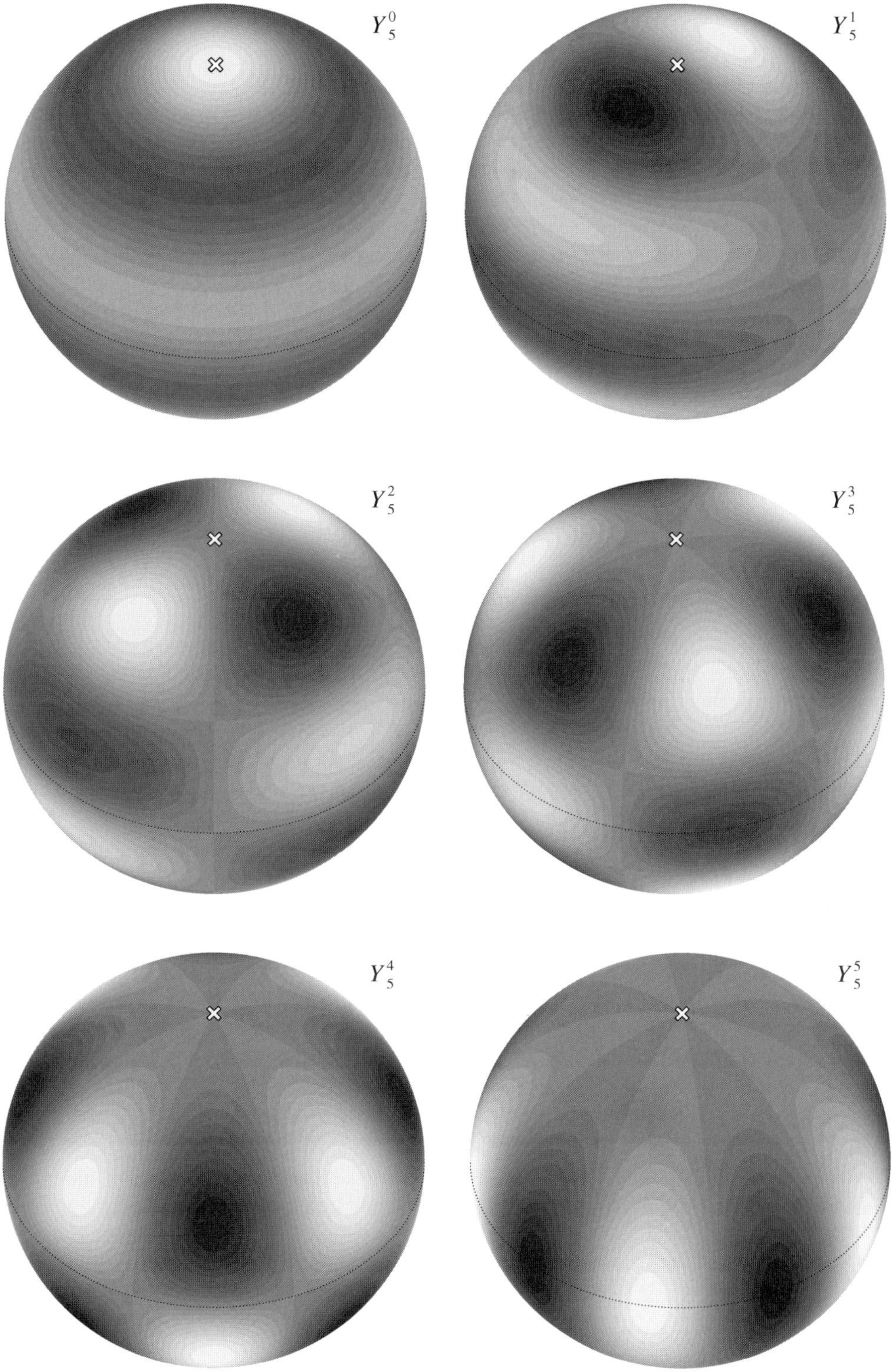

FIGURE 21.1: Examples of spherical harmonics. The distributions of Y_5^m $(0 \leq m \leq 5)$ on the sphere. The cross is the pole and the dotted curve is the equator. The gray scale represents the amplitude of Y_n^m.

21.2.2 Spectral expansion by spherical harmonics

In general, a smooth function on a sphere Ψ can be written as a superposition of spherical harmonics Y_n^m:

$$\Psi(\lambda,\mu) = \sum_{n=0}^{\infty}\sum_{m=-n}^{n} \Psi_n^m Y_n^m(\lambda,\mu), \tag{21.2.25}$$

where Ψ_n^m are the expansion coefficients. From the orthogonality relation (21.2.22), the expansion coefficients are given by

$$\Psi_n^m = \frac{1}{4\pi}\int_0^{2\pi}\int_{-1}^{1} \Psi(\lambda,\mu) Y_n^{m*}(\lambda,\mu) d\mu d\lambda. \tag{21.2.26}$$

In general, Ψ_n^m are complex. Since (21.2.23), we have

$$\Psi^* = \sum_{n=0}^{\infty}\sum_{m=-n}^{n} \Psi_n^{m*} Y_n^{m*} = \sum_{n=0}^{\infty}\sum_{m=-n}^{n} (-1)^m \Psi_n^{-m*} Y_n^m. \tag{21.2.27}$$

Thus, in the case when Ψ is a real function,

$$\Psi_n^m = (-1)^m \Psi_n^{-m*} \tag{21.2.28}$$

must be satisfied, so that a pair of terms with indices m and $-m$ in the sum of (21.2.27) is written as

$$\begin{aligned}
\Psi_n^m Y_n^m + \Psi_n^{-m} Y_n^{-m} &= \Psi_n^m \tilde{P}_n^m e^{im\lambda} + (-1)^m \Psi_n^{m*} \tilde{P}_n^{-m} e^{-im\lambda} \\
&= \Psi_n^m \tilde{P}_n^m e^{im\lambda} + \Psi_n^{m*} \tilde{P}_n^m e^{-im\lambda} \\
&= 2Re(\Psi_n^m e^{im\lambda}) \tilde{P}_n^m \\
&= (A_n'^m \cos m\lambda + B_n'^m \sin m\lambda) \tilde{P}_n^m,
\end{aligned} \tag{21.2.29}$$

where $A_n'^m$ and $B_n'^m$ are real coefficients: for $m \geq 1$

$$A_n'^m = 2Re(\Psi_n^m), \qquad B_n'^m = -2Im(\Psi_n^m), \tag{21.2.30}$$

and for $m = 0$

$$A_n'^0 = Re(\Psi_n^0) = \Psi_n^0, \qquad B_n'^0 = Im(\Psi_n^0) = 0. \tag{21.2.31}$$

These coefficients can be calculated from the real and imaginary parts of (21.2.26):

$$Re(\Psi_n^m) = \frac{1}{4\pi}\int_0^{2\pi}\int_{-1}^{1} \Psi(\lambda,\mu) \cos m\lambda \tilde{P}_n^m(\lambda,\mu) d\mu d\lambda, \tag{21.2.32}$$

$$Im(\Psi_n^m) = \frac{1}{4\pi}\int_0^{2\pi}\int_{-1}^{1} \Psi(\lambda,\mu) \sin m\lambda \tilde{P}_n^m(\lambda,\mu) d\mu d\lambda. \tag{21.2.33}$$

Using the coefficients $A_n'^m$ and $B_n'^m$, (21.2.25) can be rewritten as

$$\Psi(\lambda,\mu) = \sum_{n=0}^{\infty}\sum_{m=-n}^{n} (A_n'^m \cos m\lambda + B_n'^m \sin m\lambda) \tilde{P}_n^m(\mu). \tag{21.2.34}$$

21.2.3 Truncation of spectral expansion

Let us consider a numerical method of the expansion by spherical harmonics. The function Ψ in (21.2.25) is expanded as an infinite number of terms with indices $-n < m < n$ and $n < \infty$. For numerical calculation, n and m in (21.2.25) must be truncated at a finite number. In general, the truncations for m and n can be arbitrarily chosen, and their choices depend on the problem considered. A general form of truncation is shown in Fig. 21.2(a); this is called the *pentagonal truncation* characterized by M, L, and N; M is the maximum zonal wave number, L is the maximum number of n at $m = 0$, and N is the maximum number of n at $m = M$. As special cases of truncation, *triangular truncation* is the case when $M = N = L$, and *parallelogramic truncation* is the case when $N = M + L$. *Rhomboidal truncation* is a special case of parallelogramic truncation when $N = 2L$ and $M = L$. A slightly generalized case with $L = N > M$ is called *trapezoidal truncation.* Triangular truncation and rhomboidal truncation are the two most frequently used truncations. These truncations are shown in Fig. 21.2.

For triangular truncation, (21.2.25) is truncated as

$$\Psi(\lambda, \mu) = \sum_{n=0}^{M} \sum_{m=-n}^{n} \Psi_n^m Y_n^m(\lambda, \mu) = \sum_{m=-M}^{M} \sum_{n=|m|}^{M} \Psi_n^m Y_n^m(\lambda, \mu), \qquad (21.2.35)$$

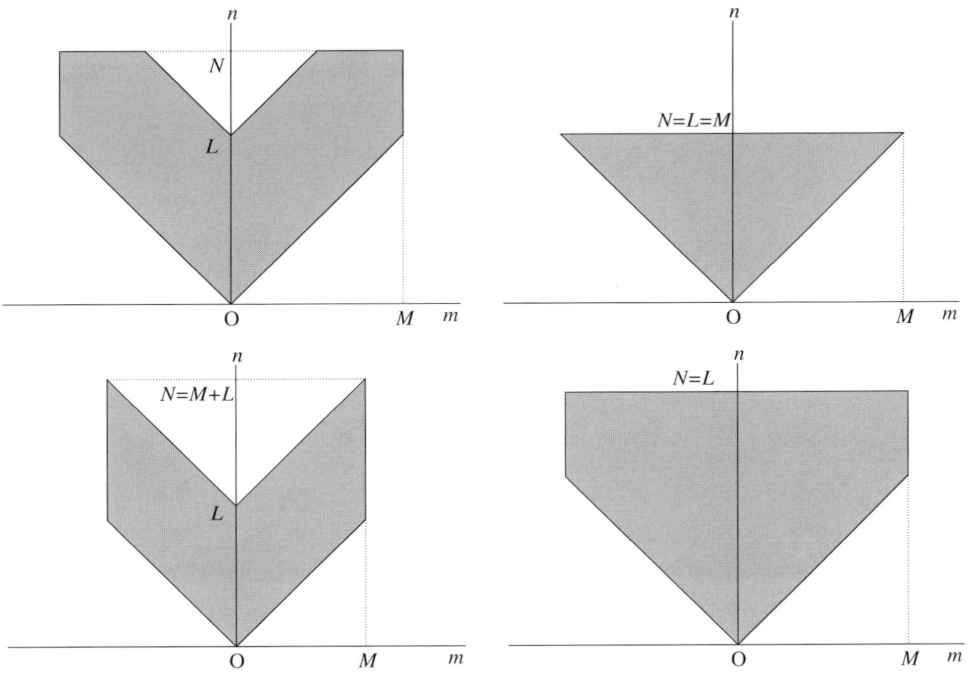

FIGURE 21.2: Top left: pentagonal truncation; top right: triangular truncation; bottom left: parallelogramic truncation; bottom right: trapezoidal truncation. Rhomboidal truncation is a special case of parallelogramic truncation with $M = L$.

while, for parallelogramic truncation, it is truncated as

$$\Psi(\lambda,\mu) = \sum_{m=-M}^{M} \sum_{n=|m|}^{|m|+L} \Psi_n^m Y_n^m(\lambda,\mu). \tag{21.2.36}$$

Usually, triangular truncation and rhomboidal truncation are denoted by symbols T and R, respectively. A specific truncation is indicated by a combination of the symbol T or R and the truncation value of the zonal wave number M. For example, the triangular truncation with $M = 21$ is denoted by T21. The number of waves, or the number of independent spherical harmonics, of T21 is $M^2/2$. This number almost corresponds to that of R15 since the maximum zonal wave number 15 is approximately equal to $\sqrt{M^2/2} = 14.8$.

Eq. (21.2.24) indicates that, if the coordinates are rotated, spherical harmonics in the rotated coordinates are expressed by a linear combination of the spherical harmonics of the same degree n as the original coordinates. In the case of triangular truncation, all spherical harmonics of the degree n are included in the expansion. From (21.2.24), the expansion coefficients are related as

$$\Psi'^{m'}_n = \sum_{m=-M}^{M} D^n_{mm'} \Psi^m_n. \tag{21.2.37}$$

That is, the expansion coefficients in rotated coordinates are expressed by those of the original coordinates. This means that any function Ψ expanded with triangular truncation is invariant by rotation. Thus, it can be said that resolution is uniform on a sphere if triangular truncation is used.

Any truncations other than triangular truncation do not have such an invariant property under rotation. If the direction of the poles is changed, representation of the truncated field becomes different. One may think, however, that this characteristic is not a requirement for atmospheric models, since a strong latitudinal contrast exists in the atmosphere such that zonal symmetry is more important than spherical uniformity. Rhomboidal truncation might be more appropriate since the number of latitudinal modes is L for any zonal wave number m $(0 \leq m \leq M)$. Others may think, however, that triangular truncation is better as the resolution becomes finer. As smaller scale waves are resolved, the direction of propagation of smaller waves becomes isotropic; thus the horizontally uniform resolution of triangular truncation is a more suitable choice than rhomboidal truncation, which has a relatively higher resolution in the latitudinal direction.

Fig. 21.3 compares triangular and rhomboidal truncations with the same total wave numbers. In the case of triangular truncation with T21, finer zonal waves are resolved around the equator. In the case of rhomboidal truncation R15, relatively higher resolution in the latitudinal direction is represented. Rhomboidal truncation possesses higher resolution at higher latitudes.

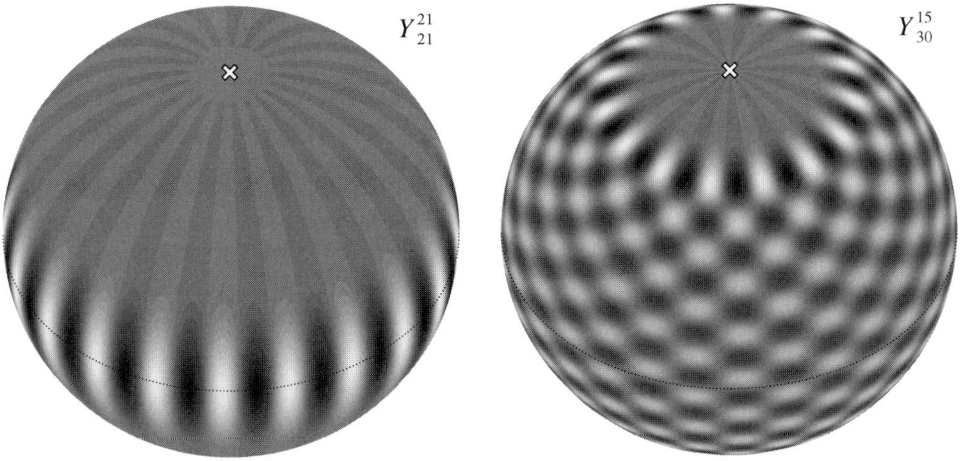

FIGURE 21.3: Spherical harmonics Y_{21}^{21} and Y_{30}^{15} are the finest waves represented by triangular truncation T21 and rhomboidal truncation R15, respectively.

21.3 Quadrature on a sphere

We need integrations of grid point values with respect to latitude and longitude to perform the spectrum method, particularly for the transform method described in the next section. These are the following two forms of integrals:

$$\int_{-1}^{1} f(\mu)d\mu, \tag{21.3.1}$$

$$\int_{0}^{2\pi} g(\lambda)d\lambda, \tag{21.3.2}$$

where $f(\mu)$ is a continuous function in the range $-1 \leq \mu \leq 1$ and $g(\lambda)$ is a continuous and periodic function with a period 2π. Let us explain the numerical integral methods called the *Gauss-Legendre method* and the *Gauss trapezoidal quadrature.* If f is a polynomial of μ and if g is a finite Fourier series, these integrals are exact provided that the positions and number of grid points are appropriately chosen.

21.3.1 Gauss-Legendre method

We assume the following form of numerical quadrature for the integral (21.3.1):

$$\int_{-1}^{1} f(\mu)d\mu \approx \sum_{n=1}^{N} w_n f(\mu_n). \tag{21.3.3}$$

Here, μ_n $(n = 1, \cdots, N)$ are points located in the order $-1 \leq \mu_1 < \cdots < \mu_N \leq 1$, and each of w_n $(n = 1, \cdots, N)$ denotes a weight at the point μ_n. The *Gauss-Legendre method* provides a procedure to find the most accurate integration of (21.3.3) by finding the best choice of the points μ_n and the weights w_n.

Let us assume that $f(\mu)$ is a polynomial of degree $(2N-1)$, given by

$$f(\mu) = \sum_{i=0}^{2N-1} a_i \mu^i. \tag{21.3.4}$$

In this case, we have

$$\int_{-1}^{1} f(\mu)d\mu = \sum_{i=0}^{2N-1} \frac{a_i}{i+1} 2\delta_{i,odd} = \sum_{j=1}^{N} \frac{a_{2j-1}}{j}, \tag{21.3.5}$$

where $\delta_{i,odd} = 1$ if i is an odd number and is zero otherwise. Since there are N values of the coefficients a_{2j-1} in this summation, they are expressed by the N point values of f at μ_n $(n = 1, \cdots, N)$:

$$a_{2j-1} = \sum_{n=1}^{N} b_{jn} f(\mu_n). \tag{21.3.6}$$

The coefficients b_{jk} form an $N \times N$ matrix. Substituting this into (21.3.5) gives

$$\int_{-1}^{1} f(\mu)d\mu = \sum_{n=0}^{N} w_n f(\mu_n), \tag{21.3.7}$$

where

$$w_n = \sum_{j=1}^{N} \frac{b_{jn}}{j}. \tag{21.3.8}$$

This means that the integral (21.3.3) is exact if N point values at μ_n are given as above.

We then find values of weights w_n if a polynomial $f(\mu)$ is given. Let us introduce the Lagrange interpolation polynomial $L(\mu)$:

$$L(\mu) = \Phi_N(\mu) \sum_{n=1}^{N} \frac{f(\mu_n)}{(\mu - \mu_n)\Phi'_N(\mu_n)}, \tag{21.3.9}$$

where

$$\Phi_N(\mu_n) = \prod_{l=1}^{N} (\mu - \mu_l), \tag{21.3.10}$$

$$\Phi'_N(\mu_n) = \left.\frac{d\Phi_N(\mu)}{d\mu}\right|_{\mu=\mu_n} = \prod_{l=1, l\neq n}^{N} (\mu_n - \mu_l). \tag{21.3.11}$$

Substituting Φ_N and Φ'_N into (21.3.9), we may write

$$L(\mu) = \sum_{n=1}^{N} f(\mu_n) \prod_{l=1, l\neq n}^{N} \frac{\mu - \mu_l}{\mu_n - \mu_l}, \tag{21.3.12}$$

which is a polynomial of degree $N-1$. Since

$$L(\mu_n) = f(\mu_n), \quad \text{for} \quad n = 1, \cdots, N, \tag{21.3.13}$$

we can find that $f(\mu) - L(\mu)$ has zero points at μ_n $(n = 1, \cdots, N)$. Since $f(\mu) - L(\mu)$ is a polynomial of degree $(2N-1)$, it can be factorized as

$$f(\mu) - L(\mu) = \Phi_N(\mu) S(\mu), \tag{21.3.14}$$

where $S(\mu)$ is a polynomial of degree $(N-1)$. Thus, we have

$$\begin{aligned} \int_{-1}^{1} f(\mu) d\mu &= \int_{-1}^{1} L(\mu) d\mu + \int_{-1}^{1} \Phi_N(\mu) S(\mu) d\mu \\ &= \sum_{n=1}^{N} w_n f(\mu_n) + R_N, \end{aligned} \tag{21.3.15}$$

where

$$w_n = \int_{-1}^{1} \frac{\Phi_N(\mu)}{(\mu - \mu_n)\Phi_N'(\mu_n)} d\mu, \tag{21.3.16}$$

$$R_N = \int_{-1}^{1} \Phi_N(\mu) S(\mu) d\mu. \tag{21.3.17}$$

If $R_N = 0$ holds in (21.3.15), the integral is equivalent to (21.3.7). In order to have $R_N = 0$ as an identity, we must require that Φ_N be orthogonal to S. Since $S(\mu)$ is a polynomial of degree $(N-1)$, it can be written as a summation of the Legendre polynomials of degree $(N-1)$ given by (21.8.12). We always have $R_N = 0$, therefore, by setting Φ_N to the Legendre polynomial of degree N,

$$\Phi_N(\mu) = \tilde{P}_N(\mu). \tag{21.3.18}$$

Here, we use a normalized Legendre function $\tilde{P}_n$ which is reduced from the normalized associated Legendre function $\tilde{P}_n^m$ (21.2.16) with $m = 0$. It is different from the Legendre function P_n defined by (21.8.12). As shown above, μ_n $(n = 1, \cdots, N)$ should be located at the zero points of the Legendre polynomial $\tilde{P}_N(\mu)$ and the weights w_n be given by

$$w_n = \int_{-1}^{1} \frac{\tilde{P}_N(\mu)}{(\mu - \mu_n)\tilde{P}_N'(\mu_n)} d\mu. \tag{21.3.19}$$

μ_n are called *Gaussian latitudes*, and w_n are called *Gaussian weights*. As shown in Section 21.9, Gaussian weights can be expressed as

$$w_n = \frac{2(2N-1)(1-\mu_n^2)}{[N\tilde{P}_{N-1}(\mu_n)]^2}. \tag{21.3.20}$$

This form is suitable for numerical calculation.

21.3.2 Gauss trapezoidal quadrature

The numerical integration of a periodic function $g(\lambda)$, (21.3.2), can be given by the *Gauss trapezoidal quadrature*. To show this, we first try to express the integral in the following form:

$$\frac{1}{2\pi}\int_0^{2\pi} g(\lambda)d\lambda \approx \frac{1}{N}\sum_{n=1}^{N} g(\lambda_n), \tag{21.3.21}$$

where λ_n are longitudinal points located with an equal interval, given by

$$\lambda_n = \frac{2\pi}{N}(n-1), \quad \text{for} \quad n = 1, \cdots, N. \tag{21.3.22}$$

We assume that $g(\lambda)$ is given by a Fourier series with a maximum zonal wave number M:

$$g(\lambda) = \sum_{m=-M}^{M} g_m e^{im\lambda}. \tag{21.3.23}$$

In this case, we show that the identity (21.3.21) holds if

$$N \geq M+1 \tag{21.3.24}$$

is satisfied.

The left-hand side of (21.3.21) is given from integration of (21.3.23) as

$$\frac{1}{2\pi}\int_0^{2\pi} g(\lambda)d\lambda = \sum_{m=-M}^{M} g_m \int_0^{2\pi} e^{im\lambda}d\lambda = g_0. \tag{21.3.25}$$

On the other hand, the right-hand side of (21.3.21) is given by substituting $\lambda = \lambda_n$ into (21.3.23):

$$\begin{aligned}\frac{1}{N}\sum_{n=1}^{N} g(\lambda_n) &= \frac{1}{N}\sum_{m=-M}^{M} g_m \sum_{n=1}^{N} e^{im\lambda_n} \\ &= g_0 + \sum_{m=-M, m\neq 0}^{M} \left(g_m \frac{1}{N}\sum_{n=1}^{N} \alpha_m^n \right),\end{aligned} \tag{21.3.26}$$

where

$$\alpha_m = e^{\frac{2\pi i m}{N}}. \tag{21.3.27}$$

The second term on the right-hand side of (21.3.26) should be zero in order that (21.3.21) is exact. If $\alpha_m \neq 1$,

$$\frac{1}{N}\sum_{n=1}^{N} \alpha_m^n = \frac{1}{N}\frac{1-\alpha_m^N}{1-\alpha_m} = 0, \tag{21.3.28}$$

and if $\alpha_m = 1$,

$$\frac{1}{N}\sum_{n=1}^{N}\alpha_m^n = 1. \qquad (21.3.29)$$

If the maximum zonal wave number M of a Fourier series is greater than or equal to the number of grid points in the longitudinal direction N (i.e., $N \leq M$), there is a wave number $m = N$ in the range $-M \leq m \leq M$. However, since $\alpha_m = 1$ for this wave number m, (21.3.28) breaks down. If $N \geq M+1$, on the other hand, since $\alpha_m \neq 1$ for m in the range $-M \leq m \leq M$ or $m \neq 0$, (21.3.28) always holds. That is, we have an exact quadrature

$$\frac{1}{2\pi}\int_0^{2\pi} g(\lambda)d\lambda = \frac{1}{N}\sum_{n=1}^{N} g(\lambda_n), \qquad (21.3.30)$$

if the inequality (21.3.24) is satisfied.

21.4 Spectral integration method

A difficulty of the spectral method for solving a partial differential equation lies in the treatment of nonlinear terms. There are two approaches: the interaction coefficients method and the transform method. The latter enables us to use the spectral method for practical purposes. We mainly describe the transform method after briefly reviewing the interaction coefficients method, and summarize the characteristics of the two methods. In this section, in order to describe how to solve with the spectral method, we use a nondivergent barotropic equation.

21.4.1 Nondivergent barotropic equation

The nondivergent barotropic equation is the basis of primitive equations; it can be obtained from the vorticity equation of primitive equations. In (20.2.45), by setting divergence D, vertical velocity $\dot{\sigma}$, deviation of temperature T', and frictional terms F_λ and F_φ to zero, we obtain the following form of nondivergent barotropic equation:

$$\frac{\partial}{\partial t}\nabla_H^2\psi = F(\psi) - \frac{2\Omega}{R^2}\frac{\partial\psi}{\partial\lambda}, \qquad (21.4.1)$$

where ψ is the streamfunction, $\mu = \sin\varphi$, and

$$\begin{aligned} F(\psi) &= -\frac{1}{R}\left[\frac{1}{1-\mu^2}\frac{\partial}{\partial\lambda}(U\zeta) + \frac{\partial}{\partial\mu}(V\zeta)\right] \\ &= \frac{1}{R^2}\left(\frac{\partial\nabla_H^2\psi}{\partial\lambda}\frac{\partial\psi}{\partial\mu} - \frac{\partial\psi}{\partial\lambda}\frac{\partial\nabla_H^2\psi}{\partial\mu}\right) \qquad (21.4.2) \end{aligned}$$

is the nonlinear term. The vorticity and velocity components are related to the streamfunction as

$$\zeta = \nabla_H^2\psi, \qquad (21.4.3)$$

$$U = -\frac{1-\mu^2}{R}\frac{\partial \psi}{\partial \mu}, \tag{21.4.4}$$

$$V = \frac{1}{R}\frac{\partial \psi}{\partial \lambda}. \tag{21.4.5}$$

It can be seen from (21.4.1) and (21.4.2) that the nondivergent barotropic equation is described by a single variable ψ.

To numerically solve the nondivergent barotropic equation with the spectral method, we expand the streamfunction ψ by spherical harmonics. We only consider triangular truncation here. Then the streamfunction ψ is written as

$$\psi(\lambda, \mu, t) = R^2 \sum_{m=-M}^{M} \sum_{n=|m|}^{M} \psi_n^m(t) Y_n^m(\lambda, \mu). \tag{21.4.6}$$

It should be noted at this point that it is unknown whether the truncated streamfunction ψ satisfies (21.4.1). To distinguish the truncated streamfunction from the exact solution, we designate (21.4.6) by ϕ and define the residual $\mathcal{R}(\phi)$ as

$$\frac{\partial}{\partial t}\nabla_H^2 \phi - F(\phi) + \frac{2\Omega}{R^2}\frac{\partial \phi}{\partial \lambda} = \mathcal{R}(\phi). \tag{21.4.7}$$

We need to introduce a constraint on the residual $\mathcal{R}(\phi)$ to obtain equations for the expansion coefficients ψ_n^m. To do this, we use the *Galerkin approximation* (i.e., the residual $\mathcal{R}$ should be orthogonal to every spherical harmonic Y_n^m):

$$\frac{1}{4\pi}\int_{-1}^{1} d\mu \int_0^{2\pi} d\lambda\, \mathcal{R}(\phi) Y_n^{m*} = 0. \tag{21.4.8}$$

Under this condition, we obtain

$$\begin{aligned} 0 &= \frac{1}{4\pi}\int_{-1}^{1} d\mu \int_0^{2\pi} d\lambda \left[\frac{\partial}{\partial t}\nabla_H^2 \phi - F(\phi) + \frac{2\Omega}{R^2}\frac{\partial \phi}{\partial \lambda}\right] Y_n^{m*} \\ &= -n(n+1)\frac{d\psi_n^m}{dt} - F_n^m + 2i\Omega m \psi_n^m. \end{aligned} \tag{21.4.9}$$

From this, the equations for the expansion coefficients of the spectral form of the nondivergent barotropic equation are given by

$$\frac{d\psi_n^m}{dt} = -\frac{1}{n(n+1)}F_n^m + \frac{2i\Omega m}{n(n+1)}\psi_n^m. \tag{21.4.10}$$

F_n^m are coefficients of the nonlinear term and the following relations hold:

$$F_n^m = \frac{1}{4\pi}\int_{-1}^{1} d\mu \int_0^{2\pi} d\lambda\, F(\lambda, \mu, t) Y_n^{m*}(\lambda, \mu), \tag{21.4.11}$$

$$F(\lambda, \mu, t) = \sum_{m=-M}^{M} \sum_{n=|m|}^{M} F_n^m(t) Y_n^m(\lambda, \mu). \tag{21.4.12}$$

21.4.2 Interaction coefficients method

In principle, we may directly calculate the expansion coefficients of the nonlinear term by substituting (21.4.2) into (21.4.11). The result is written as

$$\begin{aligned} F_n^m &= \frac{i}{2}\int_{-1}^{1} d\mu \cdot \tilde{P}_n^m \sum_{m_1}\sum_{n_1}\sum_{m_2}\sum_{n_2} \delta_{m,m_1+m_2} \\ &\times [n_2(n_2+1) - n_1(n_1+1)]\, m_1 \tilde{P}_{n_1}^{m_1} \frac{d\tilde{P}_{n_2}^{m_2}}{d\mu} \psi_{n_1}^{m_1} \psi_{n_2}^{m_2}. \end{aligned}$$

By averaging the above equation and the one with replaced indices 1 and 2, we obtain

$$F_n^m = \frac{i}{4}\sum_{m_1}\sum_{n_1}\sum_{m_2}\sum_{n_2} L(n_1, n_2, n; m_1, m_2, m)\psi_{n_1}^{m_1}\psi_{n_2}^{m_2}, \tag{21.4.13}$$

where

$$\begin{aligned} L(n_1, n_2, n; m_1, m_2, m) &= \delta_{m,m_1+m_2}[n_2(n_2+1) - n_1(n_1+1)] \\ \times \int_{-1}^{1} d\mu \cdot \tilde{P}_n^m &\left(m_1 \tilde{P}_{n_1}^{m_1} \frac{d\tilde{P}_{n_2}^{m_2}}{d\mu} - m_2 \tilde{P}_{n_2}^{m_2} \frac{d\tilde{P}_{n_1}^{m_1}}{d\mu} \right). \end{aligned} \tag{21.4.14}$$

These are called *interaction coefficients*, which represent how a pair of two modes with wave numbers (m_1, n_1) and (m_2, n_2) interact to generate a third mode with the wave number (m, n).

It is known that there are selection rules under which interaction coefficients are different from zero:

$$m = m_1 + m_2, \quad n_1 \neq n_2, \quad |n_1 - n_2| \leq n \leq n_1 + n_2, \tag{21.4.15}$$

and they must satisfy the symmetry condition

$$L(n_2, n_1, n; m_2, m_1, m) = L(n_1, n_2, n; m_1, m_2, m). \tag{21.4.16}$$

It can be shown that the number of the interaction coefficients is $O(M^5)$. This means that the interaction coefficients method requires a huge number of numerical calculations as wave numbers M increase.

21.4.3 Transform method

As an alternative to the interaction coefficients method, the *transform method* needs less computations to calculate the nonlinear term for multidimensional flows. The transform method requires both transformations from the spectral space to the grid space and from the grid space to the spectral space at each time step. The expansion coefficients of the nonlinear term can be given by transformation from the grid point values of the nonlinear term, which are calculated as the product of grid point values. Grid point values are transformed from the spectral space at the previous time step.

To describe the transform method, we use the nondivergent barotropic equation (21.4.1) and the expansion of the streamfunction ψ in the form (21.4.6). We also represent vorticity ζ and velocity components U and V by a series of spherical harmonics as

$$\psi(\lambda,\mu,t) = R^2 \sum_{m=-M}^{M} \sum_{n=|m|}^{M} \psi_n^m(t) Y_n^m(\lambda,\mu), \tag{21.4.17}$$

$$\zeta(\lambda,\mu,t) = \sum_{m=-M}^{M} \sum_{n=|m|}^{M} \zeta_n^m(t) Y_n^m(\lambda,\mu), \tag{21.4.18}$$

$$U(\lambda,\mu,t) = R \sum_{m=-M}^{M} \sum_{n=|m|}^{M+1} U_n^m(t) Y_n^m(\lambda,\mu), \tag{21.4.19}$$

$$V(\lambda,\mu,t) = R \sum_{m=-M}^{M} \sum_{n=|m|}^{M} V_n^m(t) Y_n^m(\lambda,\mu). \tag{21.4.20}$$

First, we derive the relations between ζ_n^m, U_n^m, V_n^m, and ψ_n^m.† From (21.4.3), we have

$$\zeta = \nabla_H^2 \psi = -\sum_m \sum_n n(n+1)\psi_n^m Y_n^m, \tag{21.4.21}$$

thus

$$\zeta_n^m = -n(n+1)\psi_n^m. \tag{21.4.22}$$

From (21.4.4), using the formula (21.8.21) in the appendix, we have

$$\begin{aligned} U &= -\frac{1-\mu^2}{R}\frac{\partial\psi}{\partial\mu} = R\sum_m\sum_n \psi_n^m[n\varepsilon_{n+1}^m Y_{n+1}^m - (n+1)\varepsilon_n^m Y_{n-1}^m] \\ &= R\sum_m \sum_{n=|m|}^{M+1} \left[(n-1)\varepsilon_n^m\psi_{n-1}^m - (n+2)\varepsilon_{n+1}^m\psi_{n+1}^m\right] Y_n^m, \end{aligned} \tag{21.4.23}$$

where $Y_n^m = 0$ for $n < m$. Thus, we obtain

$$U_n^m = (n-1)\varepsilon_n^m\psi_{n-1}^m - (n+2)\varepsilon_{n+1}^m\psi_{n+1}^m. \tag{21.4.24}$$

From (21.4.5), we promptly have

$$V_n^n = im\psi_n^m. \tag{21.4.25}$$

†In the case of the *divergent* barotropic equation, the range of summation indices should be $\sum_{n=|m|}^{M+1}$. This is because the term $\frac{\mu^2-1}{R}\frac{\partial\xi}{\partial\mu}$ is added to the right-hand side of (21.4.5) and a new term of the order $M+1$ is required from the recurrence relation. See (21.6.21).

Since the maximum zonal wave number of U, V, and ζ is M, the products $U\zeta$ and $V\zeta$ can be expanded by the Fourier series truncated up to the zonal wave number $2M$. Then we write these nonlinear terms as

$$U\zeta = R \sum_{m=-2M}^{2M} A_m(\mu) e^{im\lambda}, \tag{21.4.26}$$

$$V\zeta = R \sum_{m=-2M}^{2M} B_m(\mu) e^{im\lambda}. \tag{21.4.27}$$

From this, we obtain the values of A_m and B_m at the latitude μ as

$$A_m(\mu) = \frac{1}{2\pi R} \int_0^{2\pi} U(\lambda,\mu)\zeta(\lambda,\mu) e^{-im\lambda} d\lambda, \tag{21.4.28}$$

$$B_m(\mu) = \frac{1}{2\pi R} \int_0^{2\pi} V(\lambda,\mu)\zeta(\lambda,\mu) e^{-im\lambda} d\lambda. \tag{21.4.29}$$

From (21.4.11), therefore, the expansion coefficients of F are written as

$$\begin{aligned} F_n^m &= \frac{1}{4\pi} \int_{-1}^{1} d\mu \int_0^{2\pi} d\lambda \cdot e^{-im\lambda} \tilde{P}_n^m \\ &\quad \times \left\{ -\frac{1}{R(1-\mu^2)} \left[\frac{\partial}{\partial \lambda}(U\zeta) + (1-\mu^2) \frac{\partial}{\partial \mu}(V\zeta) \right] \right\} \\ &= -\frac{1}{2} \int_{-1}^{1} \frac{d\mu}{1-\mu^2} \left[imA_m + (1-\mu^2) \frac{dB_m}{d\mu} \right] \tilde{P}_n^m \\ &= -\frac{1}{2} \int_{-1}^{1} \frac{d\mu}{1-\mu^2} \left[imA_m \tilde{P}_n^m - B_m (1-\mu^2) \frac{d\tilde{P}_n^m}{d\mu} \right], \end{aligned} \tag{21.4.30}$$

where we have used $B_m(\pm 1) = 0$ since $V = 0$ at $\mu = \pm 1$ in (21.4.27). To solve the spectral expression of the nondivergent barotropic equation (21.4.10), we use the expansion coefficients of $U\zeta$ and $V\zeta$ to calculate the nonlinear term F_n^m. This procedure is called the transform method.

The calculation procedure of the transform method is summarized as follows. First, if U, V, and ζ are given in real space on a sphere, A_m and B_m are calculated from (21.4.28) and (21.4.29), respectively. Next, F_n^m is given from (21.4.30). The value of the next time step ψ_n^m is given by integrating (21.4.10) in time. If, in turn, ψ_n^m is updated, ζ_n^m, U_n^m, and V_n^m are calculated from (21.4.22), (21.4.24), and (21.4.25). Finally, ζ, U, and V in real space are given by using (21.4.18), (21.4.19), and (21.4.20).

We need to know the degree of the integrand in (21.4.30) to accurately integrate it. Since (21.4.30) is equivalent to expression of the interaction coefficients method (21.4.13), the degree of the integrand of (21.4.30) is given by the maximum degree that appears in the interaction coefficients L, (21.4.14). That is, we need to know the maximum degree with respect to μ of

$$\mathcal{P}(\mu) = \tilde{P}_n^m \left(m_1 \tilde{P}_{n_1}^{m_1} \frac{d\tilde{P}_{n_2}^{m_2}}{d\mu} - m_2 \tilde{P}_{n_2}^{m_2} \frac{d\tilde{P}_{n_1}^{m_1}}{d\mu} \right). \tag{21.4.31}$$

Let p_l denote a polynomial of degree l. From the definition (21.2.16), the associated Legendre functions are expressed as

$$\tilde{P}_n^m = (1-\mu^2)^{\frac{m}{2}} p_{n-m}.$$

Using the recurrence relation (21.8.21) and the condition $m = m_1 + m_2$ in (21.4.15), we obtain

$$\begin{aligned} \mathcal{P}(\mu) &= (1-\mu^2)^{\frac{m}{2}} p_{n-m} \cdot (1-\mu^2)^{\frac{m_1}{2}-1} p_{n_1+1-m_1} \cdot (1-\mu^2)^{\frac{m_2}{2}} p_{n_2-m_2} \\ &= (1-\mu^2)^{m-1} p_{n+n_1+n_2-2m+1} \quad = \quad p_{n+n_1+n_2-1}. \end{aligned}$$

Since the integers n, n_1, and n_2 are smaller or equal to M, the maximum degree of (21.4.31) is $(3M-1)$.

Note that in the divergent barotropic equation or the primitive equation, the meridional wind is expanded as

$$V = R \sum_{m=-M}^{M} \sum_{n=|m|}^{M+1} V_n^m Y_n^m, \tag{21.4.32}$$

instead of (21.4.20). Thus, the maximum degree of the integrand of F is $3M$.

As shown above, in the case of the nondivergent barotropic equation, the integrand of (21.4.30) is a polynomial of degree $(3M-1)$ and the integrands of (21.4.28) and (21.4.29) are the Fourier series with maximum zonal wave number $3M$. In this case, as shown in the next subsection, in order to exactly calculate an integral, we need to have $3M/2$ grid numbers in the latitudinal direction and $(3M+1)$ grid numbers in the longitudinal direction using the Gauss-Legendre integral formula and the Gauss trapezoidal quadrature. This means that the total calculation is $O(M^3)$ in the case of the transform method, and is much faster than that of the interaction coefficient method, which is $O(M^5)$.

21.4.4 Procedure of the transform method

The transform method requires integrations of grid point values with respect to latitude and longitude for transformation to spectral coefficients at every time step. Eqs. (21.4.28) and (21.4.29) contain the integral with respect to longitude λ, and (21.4.30) contains the integral with respect to latitude μ. According to the Gauss-Legendre method and the Gauss trapezoidal quadrature, if $f(\mu)$ is a polynomial of degree $(2N-1)$, and if $g(\lambda)$ is a Fourier series truncated at a zonal wave number

M, these integrals are exactly calculated by the following schemes:

$$\int_{-1}^{1} f(\mu)d\mu = \sum_{n=1}^{N} w_n f(\mu_n), \tag{21.4.33}$$

$$\frac{1}{2\pi}\int_{0}^{2\pi} g(\lambda)d\lambda = \frac{1}{N}\sum_{n=1}^{N} g(\lambda_n), \tag{21.4.34}$$

where μ_n are Gaussian latitudes, w_n Gaussian weights, and λ_n N-equidistance points in longitude.

Now, we can integrate the nonlinear term F_n^m using the above quadratures. In the case of the nondivergent barotropic equation, first, F_n^m in (21.4.30) can be accurately calculated using the values at Gaussian latitudes μ_j ($j = 1, \cdots, J$ where $J = 3M/2$) and the formula (21.3.7), such that

$$F_n^m = -\frac{1}{2}\sum_{j=1}^{J} \frac{w_j}{1-\mu_j^2}\left[imA_m\tilde{P}_n^m(\mu_j) - B_m(1-\mu_j^2)\frac{d\tilde{P}_n^m(\mu_j)}{d\mu}\right]. \tag{21.4.35}$$

Second, $A_m(\mu_j)$ and $B_m(\mu_j)$ are given by the integrals, (21.4.28) and (21.4.29). Since U, V, and ζ are expanded by the Fourier series truncated at the zonal wave number M, these integrals may contain a zonal wave number $3M$ at most. If we take I points in the longitudinal direction with equal intervals in the range $0 \le \lambda \le 2\pi$ and $I \ge 3M+1$, the integral (21.3.21) becomes exact according to the Gaussian trapezoidal quadrature. Thus, the integrals (21.4.28) and (21.4.29) can be accurately calculated by

$$A_m(\mu_j) = \frac{1}{RI}\sum_{i=1}^{I} U(\lambda_l,\mu_j)\zeta(\lambda_l,\mu_j)e^{-im\lambda_i}, \tag{21.4.36}$$

$$B_m(\mu_j) = \frac{1}{RI}\sum_{i=1}^{I} V(\lambda_l,\mu_j)\zeta(\lambda_l,\mu_j)e^{-im\lambda_i}. \tag{21.4.37}$$

On the other hand, the grid point values of ζ, U, and V can be given from spectral coefficients ζ_n^m, U_n^m, and V_n^m, respectively, using (21.4.18)–(21.4.20):

$$(\zeta_m(\mu_j), U_m(\mu_j), V_m(\mu_j)) = \sum_{n=|m|}^{M} (\zeta_n^m, U_n^m, V_n^m)\tilde{P}_n^m(\mu_j), \tag{21.4.38}$$

$$(\zeta(\lambda_l,\mu_j), U(\lambda_l,\mu_j), V(\lambda_l,\mu_j)) = \sum_{m=-M}^{M} (\zeta_m(\mu_j), U_m(\mu_j), V_m(\mu_j))e^{im\lambda_l}. \tag{21.4.39}$$

This means that the grid point values are calculated via two steps: the Fourier transform (21.4.39) and the Legendre transform (21.4.38).

The numerical procedure of the transform method for the nondivergent barotropic equation can be summarized as follows:

(1) First, we assume that $\psi_n^m(t)$ is given.

(2) Calculate ζ_n^m, U_n^m, and V_n^m using (21.4.22), (21.4.24), and (21.4.25).

(3) Calculate $\zeta_m(\mu_j)$, $U_m(\mu_j)$, and $V_m(\mu_j)$ using the Fourier transform (21.4.39).

(4) Calculate $\zeta(\lambda_l, \mu_j)$, $U(\lambda_l, \mu_j)$, and $V(\lambda_l, \mu_j)$ using the Legendre transform (21.4.38).

(5) Calculate the products $\zeta(\lambda_l, \mu_j)U(\lambda_l, \mu_j)$ and $\zeta(\lambda_l, \mu_j)V(\lambda_l, \mu_j)$ at the grid points.

(6) Calculate $A_m(\mu_j)$ and $B_m(\mu_j)$ using (21.4.36) and (21.4.37).

(7) The nonlinear term F_n^m can be given by (21.4.35).

(8) Add linear terms to F_n^m so that the tendency of ψ_n^m can be obtained.

(9) The next step value $\psi_n^m(t + \Delta t)$ can be temporally integrated using (21.4.10), and return to (1).

The operation counts of each time step are $O(M^2)$ in (2) and (5), and $O(M^3)$ in the Legendre transform (3) and (7). If the fast Fourier Transform (FFT) is used, the operation time of the Fourier transform in (2), (4), and (6) is $O(M^2 \log M)$. Therefore, the total operation time of the transform method is much smaller than that of the integration coefficients method, $O(M^5)$.

21.4.5 Grid number of pentagonal truncation

In subsections 21.4.2 through 21.4.4, we have concentrated on the nondivergent barotropic equation with triangular truncation. Let us now generalize the procedure to the pentagonal truncation introduced in Section 21.2.3. To do this, we need to know the maximum degree of the polynomials of the integrand of the Legendre transform under the prescribed truncation. If vorticity and zonal winds are expanded as†

$$\zeta = \sum_{m_1}\sum_{n_1} \zeta_{n_1}^{m_1} Y_{n_1}^{m_1}, \tag{21.4.40}$$

$$U = \sum_{m_2}\sum_{n_2} U_{n_2}^{m_2} Y_{n_2}^{m_2}. \tag{21.4.41}$$

The second-order nonlinear term is given by

$$\zeta U = \sum_{m_1}\sum_{n_1}\sum_{m_2}\sum_{n_2} \zeta_{n_1}^{m_1} U_{n_2}^{m_2} Y_{n_1}^{m_1} Y_{n_2}^{m_2}. \tag{21.4.42}$$

The product $Y_{n_1}^{m_1} Y_{n_2}^{m_2}$ can be expanded by spherical harmonics as

$$Y_{n_1}^{m_1} Y_{n_2}^{m_2} = \sum_{m}\sum_{n} C_{m_1 n_1 m_2 n_2 m n} Y_n^m, \tag{21.4.43}$$

†The upper limit of n_2 is larger by one than that of n_1.

where the C's are interaction coefficients. Fig. 21.4 shows the indices of pentagonal truncation. Let (m_1, n_1) and (m_2, n_2) be independent pairs of indices of pentagonal truncation and Y_n^m be a product of $Y_{n_1}^{m_1}$ and $Y_{n_2}^{m_2}$. The indices (m, n) fall in the region of the outer closed lines in the figure. If the maximum value of n for a given m is denoted by $n(m)$, the indices (n, m) are in the ranges of $-2M \le m \le 2M$ and $|m| \le n \le n(m)$.

The expansion coefficients of the nonlinear term ζU can be given by

$$(\zeta U)_n^m = \frac{1}{4\pi}\int_{-1}^{1} d\mu \int_0^{2\pi} d\lambda\, (\zeta U) Y_n^{m*}. \tag{21.4.44}$$

Let us examine the degree of this integrand. Since ζU can be expanded as

$$\zeta U = \sum_{m=-2M}^{2M} \sum_{n'=|m|}^{n(m)} C_{n'}^m Y_{n'}^m, \tag{21.4.45}$$

(21.4.44) is written as

$$(\zeta U)_n^m = \frac{1}{2} \sum_{n'=|m|}^{n(m)} C_{n'}^m \int_{-1}^{1} d\mu \tilde{P}_{n'}^m \tilde{P}_n^m, \tag{21.4.46}$$

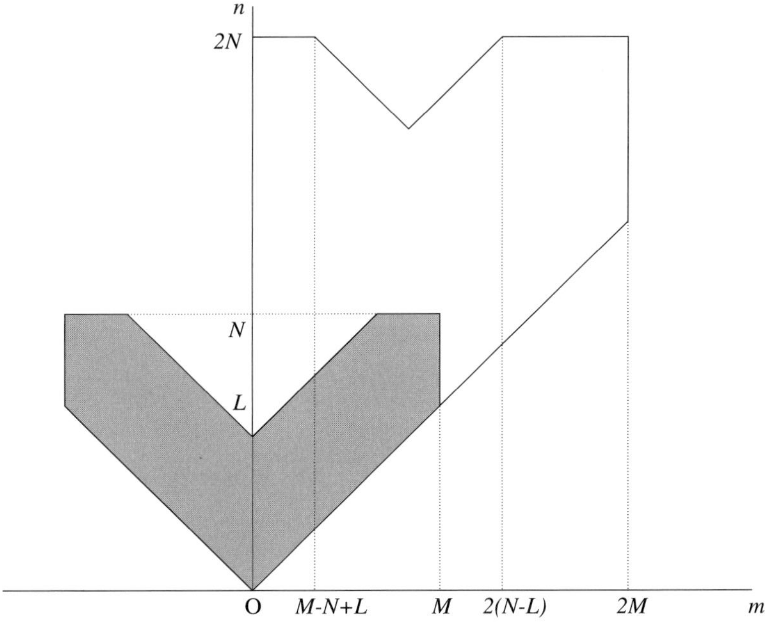

FIGURE 21.4: The range of the indices of the interaction in the case of pentagonal truncation. If the indices (m_1, n_1) and (m_2, n_2) are in the hatched region of the pentagonal truncation, the indices of Y_n^m as a product of $Y_{n_1}^{m_1}$ and $Y_{n_2}^{m_2}$ are in the range of the outer closed boundary. Only the region $m \ge 0$ is shown.

where $\tilde{P}_{n'}^m \tilde{P}_n^m$ is a polynomial of degree $(n + n')$. Thus, it can be shown that the maximum degree satisfies $(n+n')_{max} \leq 2N+L$. The maximum zonal wave number of the integrand of (21.4.44) is $\lambda = 3M$. Therefore, the numbers of grid points in the longitudinal direction I and the latitudinal direction J must be chosen as

$$I \geq 3M + 1, \tag{21.4.47}$$

$$J \geq \frac{2N + L + 1}{2}. \tag{21.4.48}$$

If grid numbers satisfy these conditions, numerical integrations are always exact.

21.4.6 Aliasing

In numerical calculations, waves with a smaller scale than the resolvable scale cannot be represented. If such waves are generated by nonlinear interaction, these waves are falsely represented as waves with a larger scale than the resolution. This phenomena is called *aliasing*. The spectrum model is free from aliasing, in particular if the interaction coefficients method is used. As for the transform method, if a sufficient number of grid points are used, all the integrals are accurately calculated, so that the transform method is equivalent to the interaction coefficients method and is also free from aliasing. The sufficient condition for aliasing-free grid points is to take I grid points with $I \geq 3M + 1$ in the longitude and J grid points with $J \geq (2N + L + 1)/2$ in the latitude; under these conditions, second-order nonlinear terms can be accurately integrated without aliasing. If the number of grid points is insufficient, on the other hand, the transform method generates the aliasing error.

The above discussion cannot be applicable to primitive equations where third-order nonlinear terms like $\dot{\sigma}\frac{\partial U}{\partial \sigma}$ are contained ($\dot{\sigma}$ itself is the second order). There is no theoretical answer to the sufficient number of grid points in primitive equations. Experience says that if there are sufficient grid points for the representation of second-order nonlinear terms, primitive equations can be integrated without numerical instability.

21.5 Conservation laws of the nondivergent barotropic equation

The nondivergent barotropic equation has the following conservative quantities if no friction is included: vorticity, enstrophy, kinetic energy, and angular momentum. Here, conservation means that global mean quantities are constant regardless of time. As will be shown below, these quantities are also conserved in the spectral model with truncation.

The nondivergent barotropic equation is given by (21.4.1). The spectral expression is given by (21.4.10), where nonlinear terms are expressed as (21.4.13) and (21.4.14). Here, we distinguish the solution to the truncated spectral model ϕ from the exact solution ψ. In general, the equation for ϕ has a residual $\mathcal{R}(\phi)$ as shown in (21.4.7). The residual $\mathcal{R}$ satisfies orthogonality (21.4.8).

21.5.1 Vorticity

First, the conservation of vorticity is a trivial property. Global mean vorticity Z always satisfies

$$Z = \frac{1}{4\pi}\int_{-1}^{1} d\mu \int_{0}^{2\pi} d\lambda \cdot \nabla_H^2 \psi = 0. \tag{21.5.1}$$

If it is truncated in the wave number domain, truncated vorticity also satisfies

$$\begin{aligned} Z &= \frac{1}{4\pi}\int_{-1}^{1} d\mu \int_{0}^{2\pi} d\lambda \cdot \nabla_H^2 \phi \\ &= \frac{1}{4\pi}\int_{-1}^{1} d\mu \int_{0}^{2\pi} d\lambda \cdot \sum_m \sum_n [-n(n+1)\psi_n^m Y_n^m] = 0, \end{aligned} \tag{21.5.2}$$

which is of course conserved. Note that the global mean of spherical harmonic Y_n^m is zero except for Y_0^0.

21.5.2 Angular momentum

Global mean angular momentum AM is expressed by using (21.4.4) as

$$AM = \frac{1}{4\pi}\int_{-1}^{1} d\mu \int_{0}^{2\pi} d\lambda \cdot R\cos\varphi(u + R\cos\varphi \cdot \Omega) \tag{21.5.3}$$

$$= \frac{1}{4\pi}\int_{-1}^{1} d\mu \int_{0}^{2\pi} d\lambda \cdot (\mu^2 - 1)\frac{\partial \psi}{\partial \mu} + \frac{2}{3}R^2\Omega. \tag{21.5.4}$$

Thus, the tendency of angular momentum is written as

$$\frac{dAM}{dt} = \frac{1}{4\pi}\int_{-1}^{1} d\mu \int_{0}^{2\pi} d\lambda \cdot \frac{\partial}{\partial t}\left[(\mu^2 - 1)\frac{\partial \psi}{\partial \mu}\right]. \tag{21.5.5}$$

In order to calculate the right-hand side, we integrate the vorticity equation (21.4.1) in the longitudinal direction:

$$\int_0^{2\pi} d\lambda \cdot \frac{\partial}{\partial t}\nabla_H^2\psi = -\int_0^{2\pi} d\lambda \cdot \frac{1}{R}\frac{\partial}{\partial \mu}(V\nabla_H^2\psi),$$

and then integrate it in the latitudinal direction from $\mu = -1$ to μ using the expression of the Laplacian (21.2.6):

$$\begin{aligned} \int_{-1}^{\mu} d\mu \int_0^{2\pi} d\lambda \cdot \frac{\partial}{\partial t}\nabla_H^2\psi &= \frac{1}{R^2}\int_0^{2\pi} d\lambda \cdot \frac{\partial}{\partial t}\left[(\mu^2 - 1)\frac{\partial \psi}{\partial \mu}\right] \\ &= -\int_0^{2\pi} d\lambda \cdot \frac{1}{R} V\nabla_H^2\psi. \end{aligned}$$

Furthermore, by integrating this once more from $\mu = -1$ to 1, we have

$$\int_{-1}^{1} d\mu \int_{0}^{2\pi} d\lambda \cdot \frac{\partial}{\partial t}\left[(\mu^2-1)\frac{\partial \psi}{\partial \mu}\right] = -R\int_{-1}^{1} d\mu \int_{0}^{2\pi} d\lambda \cdot V\nabla_H^2\psi$$
$$= -\int_{-1}^{1} d\mu \int_{0}^{2\pi} d\lambda \cdot \frac{\partial \psi}{\partial \lambda}\nabla_H^2\psi$$
$$= -\int_{-1}^{1} d\mu \int_{0}^{2\pi} d\lambda \cdot \left[\nabla_H \cdot \left(\frac{\partial \psi}{\partial \lambda}\nabla_H\psi\right) - \frac{\partial}{\partial \lambda}\frac{|\nabla_H\psi|^2}{2}\right] = 0.$$

Thus, from (21.5.5), we have shown the conservation of angular momentum:

$$\frac{d}{dt}AM = 0. \tag{21.5.6}$$

Next, we show the above conservation in truncated spectral space. Substituting (21.4.41) into (21.5.3) gives

$$\begin{aligned} AM &= \frac{1}{4\pi}\int_{-1}^{1} d\mu \int_{0}^{2\pi} d\lambda \cdot R^2 \sum_m \sum_n U_n^m Y_n^m + \frac{2}{3}R^2\Omega \\ &= R^2 U_0^0 + \frac{2}{3}R^2\Omega = -\frac{2}{\sqrt{3}}R^2\psi_1^0 + \frac{2}{3}R^2\Omega, \end{aligned} \tag{21.5.7}$$

where

$$U_0^0 = -2\varepsilon_1^0\psi_1^0 = -\frac{2}{\sqrt{3}}\psi_1^0$$

from (21.4.24). Therefore, we have

$$\frac{d}{dt}AM = -\frac{2}{\sqrt{3}}R^2\frac{d\psi_1^0}{dt}. \tag{21.5.8}$$

Using (21.4.10), we have

$$\frac{d\psi_1^0}{dt} = -\frac{1}{2}F_1^0 = -\frac{i}{8}\sum_{n_1}\sum_{n_2}\sum_m L(n_1,n_2,1;m,-m,0)\psi_{n_1}^m\psi_{n_2}^{-m}. \tag{21.5.9}$$

It can be shown that $L(n_1, n_2, 1; m, -m, 0) = 0$ by using (21.4.14). Thus, we have shown from (21.5.8) that

$$\frac{d}{dt}AM = 0. \tag{21.5.10}$$

If a horizontal hyper-diffusion,

$$-\nu\left[\nabla_H^k - \left(\frac{-2}{R^2}\right)^k\right]\nabla_H^2\psi$$

with $k > 1$, is added to the nondivergent barotropic equation (21.4.1), this becomes in spectral form

$$-\nu\left\{[-n(n+1)]^k - (-2)^k\right\}[-n(n+1)\psi_n^m].$$

Since the contribution of this term to ψ_1^0 is zero, angular momentum is also conserved (see (17.1.19)).

21.5.3 Kinetic energy

The global mean kinetic energy KE is given by

$$KE = \frac{1}{4\pi}\int_{-1}^{1} d\mu \int_{0}^{2\pi} d\lambda \cdot \frac{1}{2}\left|\nabla_H \psi\right|^2 . \tag{21.5.11}$$

Multiplying ψ by (21.4.1) and integrating on a sphere gives

$$\frac{1}{4\pi}\int_{-1}^{1} d\mu \int_{0}^{2\pi} d\lambda\ \psi \left[\frac{\partial}{\partial t}\nabla_H^2 \psi - F(\psi) + \frac{2\Omega}{R^2}\frac{\partial \psi}{\partial \lambda}\right] = 0. \tag{21.5.12}$$

The first term is the tendency of kinetic energy:

$$\frac{d}{dt}KE = -\frac{1}{4\pi}\int_{-1}^{1} d\mu \int_{0}^{2\pi} d\lambda\ \psi \frac{\partial}{\partial t}\nabla_H^2 \psi. \tag{21.5.13}$$

It can be shown that the second and third terms vanish. Thus, the kinetic energy is conserved in this analytic system:

$$\frac{d}{dt}KE = 0. \tag{21.5.14}$$

The conservation of kinetic energy in the truncated spectral model can be similarly shown by multiplying (21.4.7) by ϕ and integrating on a sphere:

$$\frac{1}{4\pi}\int_{-1}^{1} d\mu \int_{0}^{2\pi} d\lambda\ \phi \left[\frac{\partial}{\partial t}\nabla_H^2 \phi - F(\phi) + \frac{2\Omega}{R^2}\frac{\partial \phi}{\partial \lambda} - \mathcal{R}(\phi)\right] = 0. \tag{21.5.15}$$

Here, (21.5.13) also holds for ϕ, and the second and third terms vanish. Since ϕ is a superposition of Y_n^m, the residual term is expressed using (21.4.8) as

$$\frac{1}{4\pi}\int_{-1}^{1} d\mu \int_{0}^{2\pi} d\lambda\ \phi \mathcal{R}(\phi) = 0. \tag{21.5.16}$$

Thus, we also have

$$\frac{d}{dt}KE = 0 \tag{21.5.17}$$

(i.e., the kinetic energy is conserved in the truncated spectral model). From (21.5.11), kinetic energy is written as

$$\begin{aligned} KE &= -\frac{1}{8\pi}\int_{-1}^{1} d\mu \int_{0}^{2\pi} d\lambda \cdot \phi \nabla_H^2 \phi \\ &= \frac{1}{2}\sum_m \sum_n (-1)^m n(n+1)\psi_n^m \psi_n^{-m}, \end{aligned} \tag{21.5.18}$$

where the relation $\psi_n^{m*} = (-1)^m \psi_n^{-m}$ is used.

21.5.4 Enstrophy

Enstrophy is the square of vorticity. The global average of enstrophy EN is given by

$$EN = \frac{1}{4\pi}\int_{-1}^{1} d\mu \int_{0}^{2\pi} d\lambda \cdot \left|\nabla_H^2\psi\right|^2 . \tag{21.5.19}$$

Multiplying (21.4.1) by $\nabla_H^2\psi$ and integrating on a sphere, we have

$$\frac{1}{4\pi}\int_{-1}^{1} d\mu \int_{0}^{2\pi} d\lambda\, \nabla_H^2\psi \left[\frac{\partial}{\partial t}\nabla_H^2\psi - F(\psi) + \frac{2\Omega}{R^2}\frac{\partial\psi}{\partial\lambda}\right] = 0. \tag{21.5.20}$$

The first term is the tendency of enstrophy, which is written as

$$\frac{1}{2}\frac{d}{dt}EN = \frac{1}{4\pi}\int_{-1}^{1} d\mu \int_{0}^{2\pi} d\lambda\, \nabla_H^2\psi \frac{\partial}{\partial t}\nabla_H^2\psi, \tag{21.5.21}$$

and the second and third terms vanish. Thus, enstrophy is conserved:

$$\frac{d}{dt}EN = 0. \tag{21.5.22}$$

The conservation of enstrophy in the truncated spectral model can be shown in the same way as the conservation of kinetic energy. Multiplying (21.4.7) by $\nabla_H^2\phi$ gives

$$\frac{1}{4\pi}\int_{-1}^{1} d\mu \int_{0}^{2\pi} d\lambda\, \nabla_H^2\phi \left[\frac{\partial}{\partial t}\nabla_H^2\phi - F(\phi) + \frac{2\Omega}{R^2}\frac{\partial\phi}{\partial\lambda} - \mathcal{R}(\phi)\right] = 0. \tag{21.5.23}$$

The second and third terms are identical to zero. Since $\nabla_H^2\phi$ is a superposition of Y_n^m, the term including the residual becomes

$$\frac{1}{4\pi}\int_{-1}^{1} d\mu \int_{0}^{2\pi} d\lambda\, \nabla_H^2\phi\mathcal{R}(\phi) = 0, \tag{21.5.24}$$

where (21.4.8) is used. Thus, we have

$$\frac{d}{dt}EN = 0 \tag{21.5.25}$$

(i.e., enstrophy is conserved in the truncated spectral model). From (21.5.19), enstrophy is written in spectral form as

$$\begin{aligned} EN &= \sum_m\sum_n n^2(n+1)^2\psi_n^m\psi_n^{m*} \\ &= \sum_m\sum_n (-1)^m n^2(n+1)^2\psi_n^m\psi_n^{-m}. \end{aligned} \tag{21.5.26}$$

The square of absolute vorticity $Z_a = \nabla_H^2\psi + 2\Omega\mu$ is also conserved on a sphere. This can easily be shown by writing (21.4.1) as

$$\frac{\partial}{\partial t}Z_a = J(\psi, Z_a). \tag{21.5.27}$$

The integral of the Jacobian is identically zero.

21.6 Shallow water spectral model

We next step to the spectrum model with shallow water equations on a sphere. Different from the nondivergent barotropic model, the shallow water model has divergence motion. The shallow water model on a sphere is expressed in vorticity divergence form as (17.1.11) and (17.1.12), together with (17.1.3). If there is no dissipation or mass source, these are written as

$$\frac{\partial \zeta}{\partial t} = -\frac{1}{R(1-\mu^2)}\frac{\partial}{\partial \lambda}A - \frac{1}{R}\frac{\partial}{\partial \mu}B, \tag{21.6.1}$$

$$\frac{\partial D}{\partial t} = \frac{1}{R(1-\mu^2)}\frac{\partial}{\partial \lambda}B - \frac{1}{R}\frac{\partial}{\partial \mu}A - \nabla_H^2\left[g\eta + \frac{U^2+V^2}{2(1-\mu^2)}\right], \tag{21.6.2}$$

$$\frac{\partial \eta}{\partial t} = -\frac{1}{R(1-\mu^2)}\frac{\partial}{\partial \lambda}(U\eta) - \frac{1}{R(1-\mu^2)}\frac{\partial}{\partial \varphi}(V\eta), \tag{21.6.3}$$

where

$$A = (\zeta + 2\Omega\mu)U, \qquad B = (\zeta + 2\Omega\mu)V. \tag{21.6.4}$$

Let us define $\phi \equiv g\eta$ and its deviation $\phi' = \phi - \overline{\phi}$, where $\overline{\phi}$ is a constant value. Then, (21.6.3) is rewritten as

$$\frac{\partial \phi'}{\partial t} = -\frac{1}{R(1-\mu^2)}\frac{\partial}{\partial \lambda}(U\phi') - \frac{1}{R(1-\mu^2)}\frac{\partial}{\partial \varphi}(V\phi') - \overline{\phi}D. \tag{21.6.5}$$

Shallow water equations (21.6.1)–(21.6.3) corresponds to (20.2.55), (20.2.56), and (20.2.59) of primitive equations.

Vorticity and divergence are expressed by the streamfunction ψ and the velocity potential χ, respectively. In spectral form, it is easy to obtain ψ and χ from vorticity and divergence. The velocity components U and V are also calculated from ψ and χ. These relations are given by (20.2.69) together with (20.2.70) and (20.2.71):

$$\zeta = \nabla_H^2\psi, \tag{21.6.6}$$

$$D = \nabla_H^2\chi, \tag{21.6.7}$$

$$U = -\frac{1-\mu^2}{R}\frac{\partial \psi}{\partial \mu} + \frac{1}{R}\frac{\partial \chi}{\partial \lambda}, \tag{21.6.8}$$

$$V = \frac{1}{R}\frac{\partial \psi}{\partial \lambda} + \frac{1-\mu^2}{R}\frac{\partial \chi}{\partial \mu}. \tag{21.6.9}$$

We use triangular truncation as the spectral expansion in this section. Let the maximum zonal wave number be denoted by M, the number of grid points in the longitudinal direction by I, and that in the latitudinal direction by J. These must satisfy $J \geq 3M/2$ and $I \geq 3M+1$ in order to be free from alias. The locations of the latitudinal grid points are denoted by μ_j $(j = 1, \cdots, J)$, and those of the longitudinal grid points are denoted by λ_i $(i = 1, \cdots, I)$. The spectral expansion of a variable X is given as

$$X(\lambda, \mu, t) = \sum_{m=-M}^{M}\sum_{n=|m|}^{M} X_n^m(t)Y_n^m(\lambda, \mu), \tag{21.6.10}$$

and its inverse form is written as

$$\begin{aligned} X_n^m(t) &= \frac{1}{4\pi}\int_{-1}^{1} d\mu \int_0^{2\pi} d\lambda \cdot e^{-im\lambda}\tilde{P}_n^m(\mu)X(\lambda,\mu,t) \\ &= \frac{1}{2}\int_{-1}^{1} d\mu \tilde{P}_n^m(\mu)X_m(\mu,t) \\ &= \frac{1}{2}\sum_{j=i}^{J} w_j \tilde{P}_n^m(\mu_j)X_m(\mu_j,t), \end{aligned} \tag{21.6.11}$$

where w_j are the Gaussian weights and X_m is a Fourier component given by

$$\begin{aligned} X_m(\mu_j,t) &= \frac{1}{2\pi}\int_0^{2\pi} d\lambda \cdot e^{-im\lambda}X(\lambda,\mu_j,t) \\ &= \frac{1}{I}\sum_{i=i}^{I} e^{-im\lambda_i}X(\lambda_j,\mu_i,t). \end{aligned} \tag{21.6.12}$$

In the shallow water spectral model, the streamfunction, the velocity component, and surface displacement are prognostic variables; these are expanded by spherical harmonics as

$$\psi(\lambda,\mu,t) = R^2\sum_{m=-M}^{M}\sum_{n=|m|}^{M}\psi_n^m(t)Y_n^m(\lambda,\mu), \tag{21.6.13}$$

$$\chi(\lambda,\mu,t) = R^2\sum_{m=-M}^{M}\sum_{n=|m|}^{M}\chi_n^m(t)Y_n^m(\lambda,\mu), \tag{21.6.14}$$

$$\phi'(\lambda,\mu,t) = \sum_{m=-M}^{M}\sum_{n=|m|}^{M}\phi_n'^m(t)Y_n^m(\lambda,\mu). \tag{21.6.15}$$

The spectral coefficients $\psi_n^m(t)$, $\chi_n^m(t)$, and $\phi_n'^m(t)$ are temporally integrated. Vorticity ζ and divergence D are also expanded as

$$\zeta(\lambda,\mu,t) = \sum_{m=-M}^{M}\sum_{n=|m|}^{M}\zeta_n^m(t)Y_n^m(\lambda,\mu), \tag{21.6.16}$$

$$D(\lambda,\mu,t) = \sum_{m=-M}^{M}\sum_{n=|m|}^{M}D_n^m(t)Y_n^m(\lambda,\mu). \tag{21.6.17}$$

These coefficients are related as

$$\zeta_n^m = -n(n+1)\psi_n^m, \tag{21.6.18}$$

$$D_n^m = -n(n+1)\chi_n^m. \tag{21.6.19}$$

The velocity components U and V are expanded in a form similar to (21.4.19) and (21.4.20):

$$U(\lambda,\mu,t) = R\sum_{m=-M}^{M}\sum_{n=|m|}^{M+1} U_n^m(t)Y_n^m(\lambda,\mu), \tag{21.6.20}$$

$$V(\lambda,\mu,t) = R\sum_{m=-M}^{M}\sum_{n=|m|}^{M+1} V_n^m(t)Y_n^m(\lambda,\mu), \tag{21.6.21}$$

As in (21.4.24), these coefficients are related as

$$U_n^m = (n-1)\varepsilon_n^m\psi_{n-1}^m - (n+2)\varepsilon_{n+1}^m\psi_{n+1}^m + im\chi_n^m, \tag{21.6.22}$$

$$V_n^m = im\psi_n^m - (n-1)\varepsilon_n^m\chi_{n-1}^m + (n+2)\varepsilon_{n+1}^m\chi_{n+1}^m. \tag{21.6.23}$$

For the nonlinear terms (21.6.4), we designate their Fourier component by

$$A_m(\mu_j) = \frac{1}{IR}\sum_{i=1}^{I} e^{-im\lambda_i}[\zeta(\lambda_i,\mu_j)+2\Omega\mu_j]\,U(\lambda_i,\mu_j), \tag{21.6.24}$$

$$B_m(\mu_j) = \frac{1}{IR}\sum_{i=1}^{I} e^{-im\lambda_i}[\zeta(\lambda_i,\mu_j)+2\Omega\mu_j]\,V(\lambda_i,\mu_j), \tag{21.6.25}$$

which corresponds to (21.4.36) and (21.4.37) of the nondivergent equations. Let the nonlinear terms of the vorticity equation and the divergence equation be denoted by F_ζ and F_D. Noting (21.4.35), we can expand these nonlinear terms as

$$F_{\zeta n}^{\ m} = -\frac{1}{2}\sum_{j=1}^{J}\frac{w_j}{1-\mu_j^2} \times\left[imA_m(\mu_j)\tilde{P}_n^m(\mu_j) - B_m(\mu_j)(1-\mu_j^2)\frac{d\tilde{P}_n^m(\mu_j)}{d\mu}\right], \tag{21.6.26}$$

$$F_{D n}^{\ m} = \frac{1}{2}\sum_{j=1}^{J}\frac{w_j}{1-\mu_j^2} \times\left[imB_m(\mu_j)\tilde{P}_n^m(\mu_j) + A_m(\mu_j)(1-\mu_j^2)\frac{d\tilde{P}_n^m(\mu_j)}{d\mu}\right]. \tag{21.6.27}$$

The expansion coefficients of kinetic energy in the divergence equation (21.6.2) are also given as

$$\begin{aligned} E_n^m &= \frac{1}{4\pi}\int_{-1}^{1}d\mu\int_0^{2\pi}d\lambda\cdot e^{-im\lambda}\tilde{P}_n^m\frac{U^2+V^2}{2(1-\mu^2)} \\ &= \frac{1}{2I}\sum_{i=1}^{I}\sum_{j=1}^{J}w_je^{-im\lambda_i}\tilde{P}_n^m(\mu_j)\frac{U(\lambda_i,\mu_j)^2+V(\lambda_i,\mu_j)^2}{2(1-\mu_j^2)}. \end{aligned} \tag{21.6.28}$$

For any variable α, the Fourier components of the mass fluxes $U\alpha$ and $V\alpha$ are given by

$$A_{\alpha,m}(\mu_j) = \frac{1}{I}\sum_{i=1}^{I} e^{im\lambda_i} U(\lambda_i,\mu_j)\alpha(\lambda_i,\mu_j), \tag{21.6.29}$$

$$B_{\alpha,m}(\mu_j) = \frac{1}{I}\sum_{i=1}^{I} e^{im\lambda_i} V(\lambda_i,\mu_j)\alpha(\lambda_i,\mu_j), \tag{21.6.30}$$

thus the divergence of mass transport of α is written as

$$\begin{aligned}\mathcal{A}_{\alpha}{}_n^m &= \frac{1}{4\pi}\int_{-1}^{1} d\mu \int_0^{2\pi} d\lambda \cdot e^{-im\lambda}\tilde{P}_n^m \\ &\quad \times\left\{-\frac{1}{R(1-\mu^2)}\left[\frac{\partial}{\partial\lambda}(U\alpha) + (1-\mu^2)\frac{\partial}{\partial\mu}(V\alpha)\right]\right\} \\ &= -\frac{1}{2}\int_{-1}^{1}\frac{d\mu}{1-\mu^2}\left[imA_{\alpha,m}\tilde{P}_n^m - B_{\alpha,m}(1-\mu^2)\frac{d\tilde{P}_n^m}{d\mu}\right] \\ &= -\frac{1}{2}\sum_{j=1}^{J}\frac{w_j}{1-\mu_j^2} \\ &\quad \times\left[imA_{\alpha,m}(\mu_j)\tilde{P}_n^m(\mu_j) - B_{\alpha,m}(\mu_j)(1-\mu_j^2)\frac{d\tilde{P}_n^m(\mu_j)}{d\mu}\right].\end{aligned} \tag{21.6.31}$$

The mass transport of ϕ' in (21.6.5) can be given by $\mathcal{A}_{\phi'}{}_n^m$.

Thus, we arrive at the spectral expression of shallow water equations (21.6.1), (21.6.2), and (21.6.5); these are summarized as

$$\frac{\partial\psi_n^m}{\partial t} = -\frac{1}{n(n+1)}F_{\zeta}{}_n^m, \tag{21.6.32}$$

$$\frac{\partial\chi_n^m}{\partial t} = -\frac{1}{n(n+1)}F_{D}{}_n^m - \frac{1}{R^2}\left(\phi_n'^m + E_n^m\right), \tag{21.6.33}$$

$$\frac{\partial\phi_n'^m}{\partial t} = \mathcal{A}_{\phi'}{}_n^m - \overline{\phi}D_n^m. \tag{21.6.34}$$

21.7 Primitive equation spectral model

21.7.1 Equations of spectral models

Finally, we conclude this chapter with the spectral expression of primitive equations on a sphere. The equation set is summarized in Chapter 20. Since the basic relations are analogous to the shallow water spectral model, we only show the results of the equations.

We start from (20.2.55)–(20.2.61). Using triangular truncation, we expand the variables as

$$(\psi, \chi, T', \pi, \Phi, q) = \sum_{m=-M}^{M} \sum_{n=|m|}^{M} (R^2\psi_n^m, R^2\chi_n^m, T_n'^m, \pi_n^m, q_n^m) Y_n^m(\lambda, \mu). \tag{21.7.1}$$

The expansion of vorticity ζ, divergence χ, and velocity components U and V is expressed as

$$(\zeta, D) = \sum_{m=-M}^{M} \sum_{n=|m|}^{M} (\zeta_n^m, D_n^m) Y_n^m(\lambda, \mu), \tag{21.7.2}$$

$$(U, V) = R \sum_{m=-M}^{M} \sum_{n=|m|}^{M+1} (U_n^m, V_n^m) Y_n^m(\lambda, \mu). \tag{21.7.3}$$

The expansion coefficients ζ_n^m, χ_n^m, U_n^m, V_n^m, ψ_n^m, and χ_n^m are functions of σ and t. Relations between these expansion coefficients are the same as those of the shallow water model: (21.6.18), (21.6.19), (21.6.22), and (21.6.23).

The corresponding spectral equations are given as follows:

$$\frac{\partial \zeta_n^m}{\partial t} = -\frac{1}{n(n+1)} F_{\zeta n}^{\ m}, \tag{21.7.4}$$

$$\frac{\partial D_n^m}{\partial t} = -\frac{1}{n(n+1)} F_{D n}^{\ m} - \frac{1}{R^2}(\Phi_n^m + R_d \overline{T} \pi_n^m + E_n^m), \tag{21.7.5}$$

$$0 = \frac{\partial \Phi_n^m}{\partial \sigma} + \frac{R_d T_n^m}{\sigma}, \tag{21.7.6}$$

$$\frac{\partial T_n^m}{\partial t} = A_{T'}{}_n^m + H_n^m + \frac{Q_n^m}{C_p}, \tag{21.7.7}$$

$$\frac{\partial \pi_n^m}{\partial t} = Z_n^m + n(n+1) \int_0^1 \chi_n^m \, d\sigma, \tag{21.7.8}$$

$$\frac{\partial q_n^m}{\partial t} = A_{q}{}_n^m + R_n^m + S_n^m. \tag{21.7.9}$$

$F_{\zeta n}^{\ m}$ and $F_{D n}^{\ m}$ are written in a form similar to (21.6.26) and (21.6.27):

$$F_{\zeta n}^{\ m} = -\frac{1}{2} \sum_{j=1}^{J} \frac{w_j}{1-\mu_j^2} \times \left[im A_m(\mu_j) \tilde{P}_n^m(\mu_j) - B_m(\mu_j)(1-\mu_j^2) \frac{d\tilde{P}_n^m(\mu_j)}{d\mu} \right], \tag{21.7.10}$$

$$F_{D n}^{\ m} = \frac{1}{2} \sum_{j=1}^{J} \frac{w_j}{1-\mu_j^2} \times \left[im B_m(\mu_j) \tilde{P}_n^m(\mu_j) + A_m(\mu_j)(1-\mu_j^2) \frac{d\tilde{P}_n^m(\mu_j)}{d\mu} \right]. \tag{21.7.11}$$

A_m and B_m are the Fourier components A and B, (20.2.65) and (20.2.67). The advection terms, $\mathcal{A}_{T'}$ and $\mathcal{A}_q$, are given in the same form as (21.6.31). For instance, we have

$$\begin{aligned}\mathcal{A}_{T'}{}_n^m &= -\frac{1}{2}\sum_{j=1}^{J}\frac{w_j}{1-\mu_j^2}\\ &\quad\times\left[imA_{T',m}(\mu_j)\tilde{P}_n^m(\mu_j) - B_{T',m}(\mu_j)(1-\mu_j^2)\frac{d\tilde{P}_n^m(\mu_j)}{d\mu}\right].\end{aligned} \tag{21.7.12}$$

We also write

$$(H_n^m, Z_n^m, R_n^m) = \frac{1}{2}\sum_{j=1}^{J} w_j(H_m(\mu_j), Z_m(\mu_j), R_m(\mu_j))\tilde{P}_n^m(\mu_j), \tag{21.7.13}$$

where

$$H_m(\mu_j) = \frac{1}{I}\sum_{i=1}^{I} e^{-im\lambda_i}\left\{T'D + \dot{\sigma}\gamma + \kappa T\left(\frac{\partial\pi}{\partial t} + \boldsymbol{V_H}\cdot\nabla_\sigma^\mu\pi\right)\right\}(\lambda_i,\mu_j), \tag{21.7.14}$$

$$Z_m(\mu_j) = \frac{1}{I}\sum_{i=1}^{I} e^{-im\lambda_i}\left(-\int_0^1 \boldsymbol{V_H}d\sigma\cdot\nabla_\sigma^\mu\pi\right)(\lambda_i,\mu_j), \tag{21.7.15}$$

$$R_m = \frac{1}{I}\sum_{i=1}^{I} e^{-im\lambda_i}\left(qD - \dot{\sigma}\frac{\partial q}{\partial\sigma}\right)(\lambda_i,\mu_j). \tag{21.7.16}$$

Q_n^m and S_n^m are the spectral coefficients of Q and S, respectively.

21.7.2 Horizontal diffusion and conservation

In two-dimensional turbulence, as described in Section 11.1.2, energy is transferred toward the lower wave number region and enstrophy is transferred toward the higher wave number region. In the spectral model, since spectral expansion is truncated at a finite wave number, it is expected that enstrophy is accumulated in the higher wave number region. Hence, the spectral model uses such a dissipation form that is effective in the higher wave number region. In general, the following forms of dissipation are added to the vorticity and divergence equations, (20.2.55) and (20.2.56):

$$-\mathcal{D}_M\zeta = K_{MH}\left[(-1)^{N_D+1}\nabla_H^2{}^{N_D} + \left(\frac{2}{R^2}\right)^{N_D}\right]\zeta, \tag{21.7.17}$$

$$-\mathcal{D}_M D = K_{MH}\left[(-1)^{N_D+1}\nabla_H^2{}^{N_D} + \left(\frac{2}{R^2}\right)^{N_D}\right]D. \tag{21.7.18}$$

With this form, the rigid body rotation expressed by the component $n = 1$ is not damped by dissipation. Note that the case with $N_D = 1$ is equivalent to the case using (17.1.19) and (17.1.20) and that (17.4.13) is different from the above form if $N_D \neq 1$. Similarly, the dissipation term of temperature and that of moisture are given by

$$-\mathcal{D}_H T = K_{HH}(-1)^{N_D+1}\nabla_H^{2\ N_D} T, \tag{21.7.19}$$
$$-\mathcal{D}_E q = K_{EH}(-1)^{N_D+1}\nabla_H^{2\ N_D} q, \tag{21.7.20}$$

which are added to (20.2.58) and (20.2.61), respectively. In the spectral expression, the above dissipation terms are written as

$$-\mathcal{D}_M \zeta_n^m = -K_{MH}\frac{[n(n+1)]^{N_D} - 2^{N_D}}{R^{2N_D}}\zeta_n^m, \tag{21.7.21}$$
$$-\mathcal{D}_M D_n^m = -K_{MH}\frac{[n(n+1)]^{N_D} - 2^{N_D}}{R^{2N_D}}D_n^m, \tag{21.7.22}$$
$$-\mathcal{D}_H T_n^m = -K_{HH}\frac{[n(n+1)]^{N_D}}{R^{2N_D}}T_n^m, \tag{21.7.23}$$
$$-\mathcal{D}_E q_n^m = -K_{EH}\frac{[n(n+1)]^{N_D}}{R^{2N_D}}q_n^m. \tag{21.7.24}$$

The diffusion coefficients K_{MH}, K_{HH}, and K_{EH} are expressed by a typical damping time τ_D on the highest wave number N, such as

$$K_{MH} = \frac{R^{2N_D}}{[N(N+1)]^{N_D}}\frac{1}{\tau_D}. \tag{21.7.25}$$

Typically, τ_D is set to one day for T42, and τ_D is set smaller as horizontal resolution becomes finer (see Chapter 24).

As described in Sections 21.5.1–21.5.4, the nondivergent barotropic spectral model has exact conservations of many variables. The primitive equation spectral model does not logically have such conservations if the wave number is truncated. These conservations may hold as an approximation, the extent of which is given by numerical experiments (Bourke, 1972).

21.8 Appendix A: Associated Legendre functions

The latitudinal distribution of the solution to the Laplace equation must satisfy (21.2.15), and can be rewritten as

$$(1-\mu^2)\frac{d^2}{d\mu^2}\Phi - 2\mu\frac{d}{d\mu}\Phi + \left[n(n+1) - \frac{m^2}{1-\mu^2}\right]\Phi = 0. \tag{21.8.1}$$

To solve this equation, we find a solution for Φ in the form

$$\Phi = (1-\mu^2)^{\frac{m}{2}}\Psi. \tag{21.8.2}$$

Substituting this into (21.8.1), we obtain

$$(1-\mu^2)\frac{d^2\Psi}{d\mu^2} - 2(m+1)\mu\frac{d\Psi}{d\mu} + [n(n+1) - m(m+1)]\Psi = 0. \qquad (21.8.3)$$

The solution to (21.8.3) can be related to the *Legendre functions*, which are solutions to the following differential equation, called the *Legendre equation*:

$$\frac{d}{d\mu}\left[(1-\mu^2)\frac{d}{d\mu}y\right] + n(n+1)y = 0, \qquad (21.8.4)$$

or

$$(1-\mu^2)\frac{d^2y}{d\mu^2} - 2\mu\frac{dy}{d\mu} + n(n+1)y = 0. \qquad (21.8.5)$$

Differentiating (21.8.5) with respect to μ by m times, we obtain

$$\begin{aligned}(1-\mu^2)\frac{d^{m+2}y}{d\mu^{m+2}} &- 2(m+1)\mu\frac{d^{m+1}y}{d\mu^{m+1}} \\ &+[n(n+1) - m(m+1)]\frac{d^my}{d\mu^m} = 0. \end{aligned} \qquad (21.8.6)$$

From comparison between (21.8.3) and (21.8.6), we find a solution

$$\Phi = \frac{d^my}{d\mu^m}. \qquad (21.8.7)$$

The solution to the Legendre equation (21.8.4) or (21.8.5) is constructed by the following procedure. Since $\mu = 0$ is not a singular point of (21.8.4), the solution y can be expanded around $\mu = 0$ as

$$y = A_0 + A_1\mu + A_2\mu^2 + \cdots. \qquad (21.8.8)$$

Substituting this series into (21.8.5) and comparing the coefficients of μ in this equation, we have

$$(k+1)(k+2)A_{k+2} + (n-k)(n+k+1)A_k = 0, \qquad (21.8.9)$$

where $(n-k)(n+k+1) = n(n+1) - k(k+1)$ is used. In order to obtain two independent solutions, we first set $A_0 = 1$ and $A_1 = 1$. Then we have

$$y_1 = 1 - \frac{n(n+1)}{2!}\mu^2 + \frac{n(n+1)(n-2)(n+3)}{4!}\mu^4 - \cdots. \qquad (21.8.10)$$

Next, setting $A_0 = 0$ and $A_1 = 1$, we have

$$y_2 = \mu - \frac{(n-1)(n+2)}{3!}\mu^3 + \frac{(n-1)(n+2)(n-3)(n+4)}{5!}\mu^5 - \cdots. \qquad (21.8.11)$$

Since $\frac{A_{2k+2}}{A_{2k}} \to 1$ and $\frac{A_{2k+1}}{A_{2k-1}} \to 1$ as $k \to \infty$, the above series converge if $|\mu| < 1$. If n is an even integer, y_1 is a finite series up to the term μ^n and the coefficient of μ^n is given by

$$\frac{n(n+1)(n-2)\cdots(2n-1)}{n!}.$$

If n is an odd integer, y_2 is a finite series up to μ^n and the coefficient of μ^n is given by

$$\frac{(n-1)(n+2)\cdots(2n-1)}{n!}.$$

Multiplying $1 \cdot 3 \cdot 5 \cdots (n-2)$ to y_1 and $1 \cdot 3 \cdot 5 \cdots (n-1)$ to y_2, we obtain the *Legendre polynomials* as the solutions to (21.8.4):

$$\begin{aligned} P_n(\mu) &= \frac{1 \cdot 3 \cdot 5 \cdots (2n-1)}{n!} \\ &\quad \times \left[\mu^n - \frac{n(n-1)}{2(2n-1)}\mu^{n-2} + \frac{n(n-1)(n-2)(n-3)}{2 \cdot 4 \cdot (2n-1) \cdot (2n-3)}\mu^{n-4} - \cdots\right] \\ &= \frac{1}{2^n} \sum_{l=0}^{[n/2]} (-1)^l \frac{(2n-2l)!}{l!(n-l)!(n-2l)!}\mu^{n-2l}, \end{aligned} \tag{21.8.12}$$

where $[n/2]$ is an integer just below or equal to $n/2$: $[n/2] = n/2$ if n is even and $[n/2] = (n-1)/2$ if n is odd. Using the above coefficients, we obtain the *Rodrigues formula*:

$$P_n(\mu) = \frac{1}{2^n n!} \frac{d^n}{d\mu^n}(\mu^2 - 1)^n. \tag{21.8.13}$$

From (21.8.2) and (21.8.7), therefore, we find that solutions to (21.2.15) or (21.8.1) are written as

$$\Phi = P_n^m(\mu) = \frac{(1-\mu^2)^{\frac{m}{2}}}{2^n n!} \frac{d^{n+m}}{d\mu^{n+m}}(\mu^2 - 1)^n. \tag{21.8.14}$$

These are called *associated Legendre functions.* Since P_n is a polynomial of degree n, P_n^m is different from zero only if $0 \leq m \leq n$. Fig. 21.5 shows some examples of Legendre polynomials.

The associated Legendre functions (21.8.14) satisfy the following orthogonality:

$$\int_{-1}^{1} P_n^m(\mu) P_l^m(\mu) d\mu = \frac{2}{2n+1} \frac{(n+m)!}{(n-m)!} \delta_{nl}. \tag{21.8.15}$$

From this, we may define normalized associated Legendre functions by

$$\begin{aligned} \tilde{P}_n^m(\mu) &= \sqrt{(2n+1)\frac{(n-m)!}{(n+m)!}} P_n^m(\mu) \\ &= \sqrt{(2n+1)\frac{(n-m)!}{(n+m)!}} \frac{(1-\mu^2)^{\frac{m}{2}}}{2^n n!} \frac{d^{n+m}}{d\mu^{n+m}}(\mu^2 - 1)^n. \end{aligned} \tag{21.8.16}$$

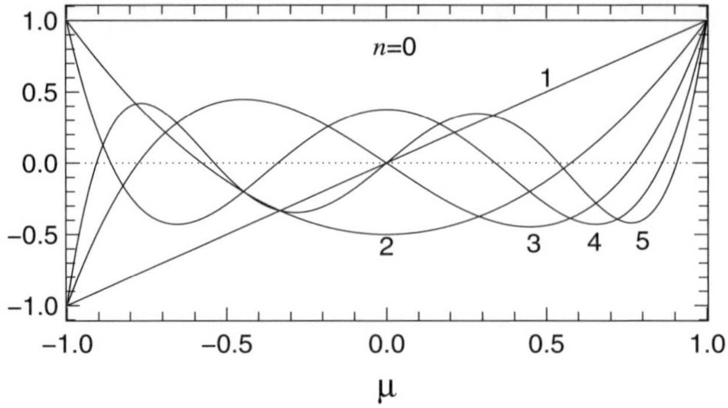

FIGURE 21.5: The profiles of Legendre polynomials $P_n(\mu)$ for $n = 0$ to 5. The abscissa is μ.

For these functions, orthogonality is written as

$$\int_{-1}^{1} \tilde{P}_n^m(\mu)\tilde{P}_l^m(\mu)d\mu = 2\delta_{nl}. \tag{21.8.17}$$

$\tilde{P}_n^m(\mu)$ has $n - |m|$ zero points in the region $-1 < \mu < 1$, and has a symmetry

$$\tilde{P}_n^m(\mu) = (-1)^{n-m}\tilde{P}_n^m(-\mu). \tag{21.8.18}$$

Although $\tilde{P}_n^m(\mu)$ are defined for $0 \le m \le n$, we can generalize functions to the indices $-n \le m \le 0$ by

$$\tilde{P}_n^m(\mu) = (-1)^m\tilde{P}_n^{-m}(\mu). \tag{21.8.19}$$

Fig. 21.6 shows some examples of associated Legendre functions.

The following recurrence equations hold between normalized associated Legendre functions:

$$\mu\tilde{P}_n^m(\mu) = \varepsilon_{n+1}^m\tilde{P}_{n+1}^m(\mu) + \varepsilon_n^m\tilde{P}_{n-1}^m(\mu), \tag{21.8.20}$$

$$(\mu^2 - 1)\frac{d\tilde{P}_n^m}{d\mu}(\mu) = n\varepsilon_{n+1}^m\tilde{P}_{n+1}^m(\mu) - (n+1)\varepsilon_n^m\tilde{P}_{n-1}^m(\mu), \tag{21.8.21}$$

where

$$\varepsilon_n^m = \sqrt{\frac{n^2 - m^2}{4n^2 - 1}}. \tag{21.8.22}$$

In order to numerically calculate $\tilde{P}_n^m(\mu)$, we first obtain from (21.8.16)

$$\tilde{P}_m^m(\mu) = \sqrt{\frac{2m+1}{(2m)!}\frac{(2m)!}{2^m m!}}(1-\mu^2)^{\frac{m}{2}} = \sqrt{\frac{(2m+1)!!}{(2m)!!}}(1-\mu^2)^{\frac{m}{2}}, \tag{21.8.23}$$

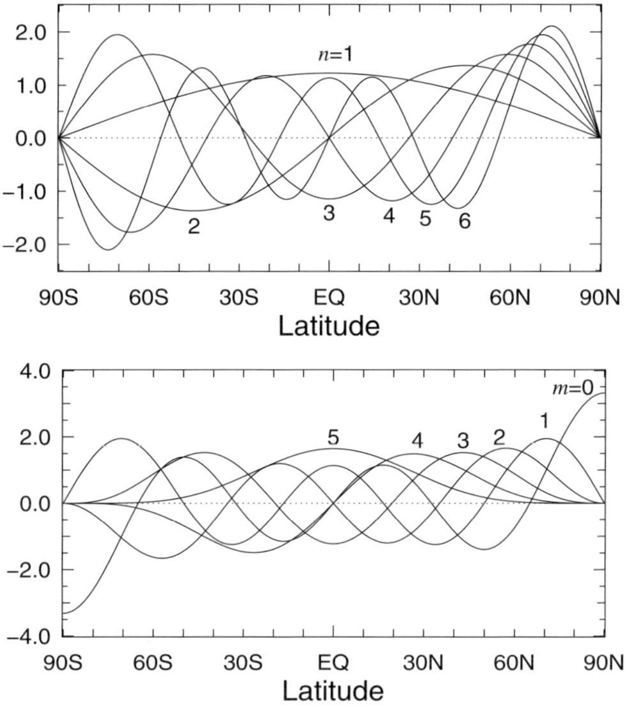

FIGURE 21.6: Latitudinal profiles of normalized associated Legendre functions $\tilde{P}_n^m(\sin\varphi)$. Top: $\tilde{P}_n^1$ $(1 \leq n \leq 6)$, and bottom: $\tilde{P}_5^m$ $(0 \leq m \leq 5)$. The abscissa is latitude φ.

$$\begin{aligned}\tilde{P}_{m+1}^m(\mu) &= \sqrt{\frac{2m+3}{(2m+1)!}}\frac{(2m+2)!}{2^{m+1}(m+1)!}\mu(1-\mu^2)^{\frac{m}{2}} \\ &= \sqrt{\frac{(2m+3)!!}{(2m)!!}}\mu(1-\mu^2)^{\frac{m}{2}}, \end{aligned} \qquad (21.8.24)$$

where $k!! = k \cdot (k-2) \cdot (k-4) \cdots$. Using the recurrence equation (21.8.20), we subsequently obtain $\tilde{P}_n^m(\mu)$ for $n \geq m+2$. We also calculate $\frac{d\tilde{P}_n^m}{d\mu}$ from (21.8.21).

21.9 Appendix B: Gaussian weights

Gaussian weights can be expressed as a convenient form suitable for numerical calculation. From (21.2.16), the coefficient of the term of the highest degree in $\tilde{P}_k(\mu)$ is given by

$$A_k = \sqrt{2k+1}\frac{(2k)!}{2^k(k!)^2}. \qquad (21.9.1)$$

Thus,

$$\tilde{P}_{k+1}(\mu) - \frac{A_{k+1}}{A_k}\mu\tilde{P}_k(\mu) \qquad (21.9.2)$$

is a polynomial of degree k, so that it may be expanded as

$$\tilde{P}_{k+1}(\mu) - a_k\mu\tilde{P}_k(\mu) \;=\; b_k\tilde{P}_k(\mu) + b_{k-1}\tilde{P}_{k-1}(\mu) + \cdots, \qquad (21.9.3)$$

where $a_k = A_{k+1}/A_k$. Using the recurrence relation (21.8.20), the left-hand side is a summation of the Legendre polynomials of degree $\geq k-1$. Thus, we have $b_r = 0$ for $r = 1, \cdots, k-2$, so that (21.9.3) can be written as

$$\tilde{P}_{k+1}(\mu) \;=\; (a_k\mu + b_k)\tilde{P}_k(\mu) + b_{k-1}\tilde{P}_{k-1}(\mu). \qquad (21.9.4)$$

Multiplying $\tilde{P}_{k+1}(\mu)$ or $\tilde{P}_{k-1}(\mu)$ by the above equation, and integrating by using the orthogonal relation (21.8.17), we have

$$2 \;=\; a_k\int_{-1}^{1}\tilde{P}_{k+1}(\mu)\mu\tilde{P}_k(\mu)\mu d\mu, \qquad (21.9.5)$$

$$0 \;=\; a_k\int_{-1}^{1}\tilde{P}_k(\mu)\mu\tilde{P}_{k-1}(\mu)\mu d\mu + 2b_{k-1}. \qquad (21.9.6)$$

Replacing k with $k-1$ in (21.9.5) and using (21.9.6), we obtain

$$b_{k-1} \;=\; -\frac{a_k}{a_{k-1}}. \qquad (21.9.7)$$

Substituting this into (21.9.4), we have

$$\mu\tilde{P}_k(\mu) \;=\; \frac{\tilde{P}_{k+1}(\mu)}{a_k} + \frac{\tilde{P}_{k-1}(\mu)}{a_{k-1}} - \frac{b_k}{a_k}\tilde{P}_k(\mu). \qquad (21.9.8)$$

Multiplying this by $\tilde{P}_k(x)$, and substituting it with that when μ and x are exchanged, we obtain

$$(\mu - x)\tilde{P}_k(\mu)\tilde{P}_k(x) \;=\; \frac{\tilde{P}_k(x)\tilde{P}_{k+1}(\mu) - \tilde{P}_k(\mu)\tilde{P}_{k+1}(x)}{a_k} - \frac{\tilde{P}_{k-1}(x)\tilde{P}_k(\mu) - \tilde{P}_{k-1}(\mu)\tilde{P}_k(x)}{a_{k-1}}. \qquad (21.9.9)$$

In the case of $k = 0$, the contribution from the second term on the right-hand side vanishes. Thus, summing up from $k = 0$ to $N-1$, we obtain

$$\sum_{k=0}^{N-1}\tilde{P}_k(\mu)\tilde{P}_k(x) \;=\; \frac{\tilde{P}_{N-1}(x)\tilde{P}_N(\mu) - \tilde{P}_{N-1}(\mu)\tilde{P}_N(x)}{a_{N-1}(\mu - x)}. \qquad (21.9.10)$$

Substituting the zero points of $\tilde{P}_N(\mu)$, μ_n $(n = 1, \cdots, N)$, into x gives

$$\frac{\tilde{P}_N(\mu)\tilde{P}_{N-1}(\mu_n)}{a_{N-1}(\mu - \mu_n)} = \sum_{k=0}^{N-1} \tilde{P}_k(\mu)\tilde{P}_k(\mu_n). \tag{21.9.11}$$

Substituting this further into (21.3.19), we finally obtain

$$\begin{aligned} w_n &= \frac{a_{N-1}}{\tilde{P}'_N(\mu_n)\tilde{P}_{N-1}(\mu_n)} \sum_{k=0}^{N-1} \tilde{P}_k(\mu_n) \int_{-1}^{1} \tilde{P}_k(\mu) d\mu \\ &= \frac{2a_{N-1}}{\tilde{P}'_N(\mu_n)\tilde{P}_{N-1}(\mu_n)}, \end{aligned} \tag{21.9.12}$$

where only the term $k = 0$ remains after integration and $\tilde{P}_0(\mu) = 1$ is used. We have

$$a_{N-1} = \frac{A_N}{A_{N-1}} = \frac{\sqrt{4N^2 - 1}}{N}. \tag{21.9.13}$$

Since from the recurrence relations (21.8.20) and (21.8.21),

$$(\mu^2 - 1)\frac{d\tilde{P}_N(\mu)}{d\mu} = N\mu\tilde{P}_N(\mu) - (2N+1)\varepsilon_N^0 \tilde{P}_{N-1}(\mu), \tag{21.9.14}$$

substituting $\mu = \mu_n$ into this gives

$$\tilde{P}'_N(\mu_n) = \frac{1}{1 - \mu_n^2} \frac{N(2N+1)}{\sqrt{4N^2 - 1}} \tilde{P}_{N-1}(\mu_n). \tag{21.9.15}$$

Substituting (21.9.13) and (21.9.15) into (21.9.12), we obtain the final form of Gaussian weights:

$$w_n = \frac{2(2N-1)(1 - \mu_n^2)}{[N\tilde{P}_{N-1}(\mu_n)]^2}, \tag{21.9.16}$$

which is a more useful expression for Gaussian weights.† Fig. 21.7 shows examples of relations of Gaussian latitudes and Gaussian weights for $N = 16$, 32, 64, and 128.

---

†The approach described in this subsection is based on the Christoffel-Darboux identity. The factor in Gaussian weights depends on the normalization. In some literature, the factor 2 does not appear in the expression of Gaussian weights.

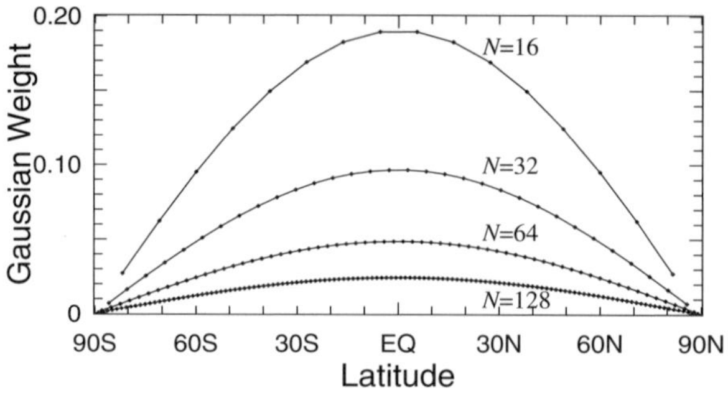

FIGURE 21.7: Relations of Gaussian latitudes and Gaussian weights for N = 16, 32, 64, and 128.

References and suggested reading

The basics of spectral methods are fully discussed in Gottlieb and Orszag (1977) and Canuto et al. (1988). The transform method was developed by Eliassen et al. (1970) and Orszag (1970), and was extended to the barotropic model by Bourke (1972). These are reviewed in Bourke et al. (1977). The spherical method on a sphere is described in detail in a textbook by Krishnamurti et al. (1998).

Bourke, W., 1972: An efficient, one-level primitive-equation spectral model. *Mon. Wea. Rev.*, **100**, 683–689.

Bourke, W., MacAveney, K., Puri, K., and Thuring, R., 1977: Global modeling of atmospheric flow by spectral methods. *Methods in Computational Physics*, Vol. 17, Academic Press, New York, pp. 267–324.

Canuto, C., Hussaini, M. Y., Quarteroni, A., and Zang, T. A., 1988: *Spectral Methods in Fluid Dynamics.* Springer-Verlag, New York, 567 pp.

Eliassen, E., Machenhauer, B., and Rasmussen, E., 1970: *On a numerical method for integration of the hydrodynamical equations with a spectral representation of the horizontal fields*, Report No. 2. Department of Meteorology, Copenhagen University, Denmark, 35 pp.

Gottlieb, D. and Orszag, S. A., 1977: *Numerical Analysis of Spectral Methods: Theory and Application*, CBMS No. 26. SIAM, Philadelphia, 170 pp.

Krishnamurti, T. N., Bedi, H. S., and Hardiker, V. M., 1998: *An Introduction to Global Spectral Modeling.* Oxford University Press, New York, 253 pp.

Orszag, S. A., 1970: Transform method for the calculation of vector-coupled sums: Application to the spectral form of the vorticity equation. *J. Atmos. Sci.*, **27**, 890–895.

22

Vertical discretization

Although the spectral method can be used on the spherical surface of general circulation models, it is difficult to extend to the vertical direction since various boundary conditions are applied at the top and bottom of the atmosphere. Thus, the grid method is normally used for vertical discretization of general circulation models. We have seen that the spectral method of shallow water equations on a sphere naturally assures conservation of various physical quantities. It will be shown that conservation in the primitive equation model is achieved by a proper vertical discretization. Here we use the flux-form finite volume method for vertical discretization. We require the domain integral conservation of energy and that of potential temperature. We will also find a discretization that minimizes an error in pressure gradient force when the atmosphere is everywhere isentropic.

22.1 Conservation of primitive equations

In this section, we use equations with Cartesian coordinates for simplicity in order to clarify the conservation of primitive equations. We start with momentum equations, the hydrostatic equation, the continuity equation, and the thermodynamic equation given by (20.2.3), (20.2.4), (20.2.5), and (20.2.6), respectively:

$$\frac{d\boldsymbol{v}_H}{dt} = -f\boldsymbol{k}\times\boldsymbol{v}_H - \nabla_\sigma\Phi - R_dT\nabla_\sigma \ln p_s + \boldsymbol{F_H}, \qquad (22.1.1)$$

$$0 = \frac{\partial\Phi}{\partial\sigma} + \frac{R_dT}{\sigma}, \qquad (22.1.2)$$

$$\frac{d\pi}{dt} = -\nabla_\sigma\cdot\boldsymbol{v}_H - \frac{\partial\dot{\sigma}}{\partial\sigma}, \qquad (22.1.3)$$

$$\frac{ds}{dt} = \frac{C_pQ}{T}, \qquad (22.1.4)$$

where the equation of entropy is used instead of the thermodynamic equation (20.2.6): s is the specific entropy. We use σ-coordinates when the vertical coordinate is defined as $\sigma = p/p_s$, where p_s is surface pressure. We also define $\pi = \ln p_s$.

The boundary conditions are $\dot{\sigma} = 0$ at $\sigma = 0$ (the top of the atmosphere) and $\dot{\sigma} = 0$ at $\sigma = 1$ (the surface). The tendency of surface pressure $\frac{\partial \pi}{\partial t}$ and vertical velocity $\dot{\sigma}$ can be given by integrating (22.1.3). Noting (20.2.25), we have[†]

$$\frac{\partial \pi}{\partial t} = -\int_0^1 \boldsymbol{v}_H \cdot \nabla_\sigma \pi \, d\sigma - \int_0^1 \nabla_\sigma \cdot \boldsymbol{v}_H d\sigma. \tag{22.1.5}$$

From (20.2.28), we have

$$\dot{\sigma} = -\sigma \frac{\partial \pi}{\partial t} - \int_0^\sigma (\nabla_\sigma \cdot \boldsymbol{v}_H + \boldsymbol{v}_H \cdot \nabla_\sigma \pi) d\sigma. \tag{22.1.6}$$

We also define the tendency of pressure by

$$\omega \equiv \frac{dp}{dt} = p_s \dot{\sigma} + \sigma \dot{p_s} = p_s \dot{\sigma} + \sigma \left(\frac{\partial p_s}{\partial t} + \boldsymbol{v}_H \cdot \nabla_\sigma p_s \right), \tag{22.1.7}$$

which is vertical velocity in the σ-coordinate.

22.1.1 Conservation of mass

If hydrostatic balance is applicable, the vertical integral of a quantity per unit mass A is written as

$$\int_{z_0}^\infty \rho A dz = \frac{1}{g} \int_0^{p_s} A dp = \frac{p_s}{g} \int_0^1 A d\sigma, \tag{22.1.8}$$

where z_0 is surface height. In the case of $A = 1$, this gives the mass per unit area p_s/g. Thus, if the area integral of the surface pressure p_s is conserved, the total mass is also conserved.

The continuity equation (22.1.3) is rewritten in flux form as

$$\frac{\partial p_s}{\partial t} + \nabla_\sigma \cdot (p_s \boldsymbol{v}_H) + \frac{\partial}{\partial \sigma}(p_s \dot{\sigma}) = 0. \tag{22.1.9}$$

The vertical integral of this equation is given by

$$\frac{\partial p_s}{\partial t} = -\int_0^1 \nabla_\sigma \cdot (p_s \boldsymbol{v}_H) d\sigma, \tag{22.1.10}$$

and the vertical and spherical integral of the continuity equation is reduced to

$$\frac{d}{dt} \int_S p_s dS = 0, \tag{22.1.11}$$

where dS denotes an area element on a sphere. This also holds under the periodic condition or if there is no flux at the boundaries. Eq. (22.1.11) means conservation of the total mass of the atmosphere.

[†]Since we use Cartesian coordinates for simplicity, the original vector $\boldsymbol{V_H}$ is replaced by $\boldsymbol{v_H}$ and ∇_σ^μ by ∇_σ throughout this section.

22.1.2 Constraint on the vertical integral of pressure gradient

Using hydrostatic balance (22.1.2), the pressure gradient (22.1.1) is rewritten as

$$-\frac{1}{\rho}\nabla_z p \quad = \quad -\nabla_\sigma \Phi - R_d T \nabla_\sigma \ln p_s \quad = \quad -\nabla_\sigma \Phi + \frac{\sigma}{p_s}\frac{\partial \Phi}{\partial \sigma}\nabla_\sigma p_s. \quad (22.1.12)$$

Integrating this in the vertical direction using (22.1.8) yields

$$\begin{aligned} -\int_{z_0}^{\infty} \nabla_z p\, dz &= -\frac{p_s}{g}\int_0^1 \left(\nabla_\sigma \Phi - \frac{\sigma}{p_s}\frac{\partial \Phi}{\partial \sigma}\nabla_\sigma p_s\right) d\sigma \\ &= -\frac{1}{g}\int_0^1 \left[\nabla_\sigma (p_s \Phi) - \frac{\partial}{\partial \sigma}(\Phi\sigma)\nabla_\sigma p_s\right] d\sigma \\ &= -\frac{1}{g}\left(\nabla_\sigma \int_0^1 p_s \Phi\, d\sigma - \Phi_s \nabla_\sigma p_s\right) \\ &= -\frac{1}{g}\left[\nabla_\sigma \int_0^1 p_s(\Phi - \Phi_s) d\sigma + p_s \nabla_\sigma \Phi_s\right], \end{aligned} \quad (22.1.13)$$

where Φ_s is the geopotential height at the surface and is equal to gz_0. On a flat surface where Φ_s is constant, a line integral along a closed curve C is written as

$$-\oint_C \left(\nabla_z \int_{z_0}^{\infty} p\, dz\right) ds \quad = \quad -\frac{1}{g}\oint_C \left(\nabla_\sigma \int_0^1 p_s \Phi\, d\sigma\right) ds \quad = \quad 0 \quad (22.1.14)$$

(i.e., if the surface has no topography, any line integral of the vertical integral of pressure gradient along a closed curve must be zero). This condition reduces to conservation of angular momentum if the closed curve is taken as a latitudinal circle.

22.1.3 Conservation of total energy

The sum of the domain integrals of kinetic energy and enthalpy is equal to domain-integral total energy under hydrostatic balance (Section 12.1.3). The conservation of total energy in pressure coordinates is given by (3.3.51). Here, we derive the corresponding conservation law in σ-coordinates, and consider vertical discretization based on conservation.

The inner product of the momentum equation (22.1.1) and $p_s \boldsymbol{v}_H$ gives the equation of kinetic energy:

$$p_s \frac{d}{dt}\frac{\boldsymbol{v_H}^2}{2} \quad = \quad -p_s \boldsymbol{v}_H \cdot (\nabla_\sigma \Phi + \sigma\alpha\nabla_\sigma p_s) + p_s \boldsymbol{v}_H \cdot \boldsymbol{F_H}, \quad (22.1.15)$$

where $\alpha = R_d T/p$ is the specific volume. Using (22.1.9), this is written in flux form as

$$\begin{aligned} &\frac{\partial}{\partial t}\left(p_s \frac{\boldsymbol{v_H}^2}{2}\right) + \nabla_\sigma \cdot \left(p_s \boldsymbol{v}_H \frac{\boldsymbol{v_H}^2}{2}\right) + \frac{\partial}{\partial \sigma}\left(p_s \dot{\sigma}\frac{\boldsymbol{v_H}^2}{2}\right) \\ &= \quad -p_s \boldsymbol{v}_H \cdot (\nabla_\sigma \Phi + \sigma\alpha\nabla_\sigma p_s) + p_s \boldsymbol{v}_H \cdot \boldsymbol{F_H}. \end{aligned} \quad (22.1.16)$$

The first term on the right-hand side of (22.1.15) or (22.1.16) is the work done by the pressure gradient force and can be rewritten as

$$\begin{aligned}
&-p_s\boldsymbol{v}_H\cdot(\nabla_\sigma\Phi+\sigma\alpha\nabla_\sigma p_s)\\
&= -\nabla_\sigma\cdot(p_s\boldsymbol{v}_H\Phi)+\Phi\nabla_\sigma\cdot(p_s\boldsymbol{v}_H)-p_s\sigma\alpha\boldsymbol{v}_H\cdot\nabla_\sigma p_s\\
&= -\nabla_\sigma\cdot(p_s\boldsymbol{v}_H\Phi)-\Phi\left[\frac{\partial}{\partial\sigma}(p_s\dot{\sigma})+\frac{\partial p}{\partial t}\right]-p_s\sigma\alpha\boldsymbol{v}_H\cdot\nabla_\sigma p_s\\
&= -\nabla_\sigma\cdot(p_s\boldsymbol{v}_H\Phi)-\frac{\partial}{\partial\sigma}(p_s\dot{\sigma}\Phi)+p_s\dot{\sigma}\frac{\partial\Phi}{\partial\sigma}-\Phi\frac{\partial p_s}{\partial t}-p_s\sigma\alpha\boldsymbol{v}_H\cdot\nabla_\sigma p_s\\
&= -\nabla_\sigma\cdot(p_s\boldsymbol{v}_H\Phi)-\frac{\partial}{\partial\sigma}(p_s\dot{\sigma}\Phi)+(\sigma p_s\alpha-\Phi)\frac{\partial p_s}{\partial t}\\
&\quad -p_s\left[\sigma\left(\frac{\partial p_s}{\partial t}+\boldsymbol{v}_H\cdot\nabla_\sigma p_s\right)+p_s\dot{\sigma}\right]\alpha\\
&= -\nabla_\sigma\cdot(p_s\boldsymbol{v}_H\Phi)-\frac{\partial}{\partial\sigma}\left(p_s\dot{\sigma}\Phi+\Phi\sigma\frac{\partial p_s}{\partial t}\right)-p_s\omega\alpha, \qquad (22.1.17)
\end{aligned}$$

where the continuity equation (22.1.9), hydrostatic balance $\frac{\partial\Phi}{\partial\sigma}=-p_s\alpha$, and the definition of ω (22.1.7) are used. Using (22.1.17), the equation of kinetic energy (22.1.16) is written as

$$\begin{aligned}
&\frac{\partial}{\partial t}\left(p_s\frac{\boldsymbol{v_H}^2}{2}\right)+\nabla_\sigma\cdot\left[p_s\boldsymbol{v}_H\left(\frac{\boldsymbol{v_H}^2}{2}+\Phi\right)\right]\\
&\quad+\frac{\partial}{\partial\sigma}\left[p_s\dot{\sigma}\left(\frac{\boldsymbol{v_H}^2}{2}+\Phi\right)+\Phi\sigma\frac{\partial p_s}{\partial t}\right] = -p_s\omega\alpha+p_s\boldsymbol{v}_H\cdot\boldsymbol{F_H}. \qquad (22.1.18)
\end{aligned}$$

The equation of enthalpy $h=C_pT$ can be derived from the equation of entropy (22.1.4) using the thermodynamic relation (see (3.3.48)):

$$C_p\frac{dT}{dt} = \alpha\omega+C_pQ. \qquad (22.1.19)$$

Note that the thermodynamic equation (20.2.6) can be given from this using the relation between ω and $\dot{\sigma}$, (22.1.7). Multiplying this by p_s, we obtain the flux-form enthalpy equation as

$$\frac{\partial}{\partial t}(p_sC_pT)+\nabla_\sigma\cdot(p_s\boldsymbol{v}_HC_pT)+\frac{\partial}{\partial\sigma}(p_s\dot{\sigma}C_pT) = p_s\omega\alpha+p_sC_pQ. \qquad (22.1.20)$$

This can be further rewritten as

$$\begin{aligned}
\frac{\partial T}{\partial t} &= -\nabla_\sigma\cdot(\boldsymbol{v}_HT')+T'D-\dot{\sigma}\sigma^\kappa\frac{\partial(T\sigma^{-\kappa})}{\partial\sigma}\\
&\quad+\kappa T\left(\frac{\partial\pi}{\partial t}+\boldsymbol{v}_H\cdot\nabla_\sigma\pi\right)+Q, \qquad (22.1.21)
\end{aligned}$$

which corresponds to (20.2.33).

The equation of total energy is given by the sum of (22.1.18) and (22.1.20):

$$\frac{\partial}{\partial t}\left(p_s\frac{\boldsymbol{v_H}^2}{2}+C_pT\right)+\nabla_\sigma\cdot\left[p_s\boldsymbol{v}_H\left(\frac{\boldsymbol{v_H}^2}{2}+C_pT+\Phi\right)\right]$$
$$+\frac{\partial}{\partial\sigma}\left[p_s\dot{\sigma}\left(\frac{\boldsymbol{v_H}^2}{2}+C_pT+\Phi\right)+\Phi\sigma\frac{\partial p_s}{\partial t}\right] = p_s(\boldsymbol{v}_H\cdot\boldsymbol{F_H}+C_pQ). \tag{22.1.22}$$

The vertical integral of (22.1.22) gives the column-integrated total energy:

$$\frac{\partial}{\partial t}\left[p_s\Phi_s+\int_0^1 p_s\left(\frac{\boldsymbol{v_H}^2}{2}+C_pT\right)d\sigma\right]$$
$$+\nabla_\sigma\cdot\int_0^1 p_s\boldsymbol{v}_H\left(\frac{\boldsymbol{v_H}^2}{2}+C_pT+\Phi\right)d\sigma = \int_0^1 p_s(\boldsymbol{v}_H\cdot\boldsymbol{F_H}+C_pQ)d\sigma. \tag{22.1.23}$$

Thus, if $\boldsymbol{F_H}=0$ and $Q=0$, the domain integral of total energy is conserved:

$$\frac{d}{dt}\int_S dS\left[p_s\Phi_s+\int_0^1 p_s\left(\frac{\boldsymbol{v_H}^2}{2}+C_pT\right)d\sigma\right] = 0. \tag{22.1.24}$$

22.1.4 Conservation of potential temperature

Eq. (22.1.4) states that entropy s is conserved in the Lagrangian sense if $Q=0$. Since $s=C_p\ln\theta$, potential temperature θ is also conserved if $Q=0$:

$$\frac{d\theta}{dt} = 0. \tag{22.1.25}$$

In this case, an arbitrary function of potential temperature $f(\theta)$ is also conserved:

$$\frac{d}{dt}f(\theta) = 0. \tag{22.1.26}$$

This can be written in flux form as

$$\frac{\partial}{\partial t}[p_sf(\theta)]+\nabla_\sigma\cdot[p_s\boldsymbol{v}_Hf(\theta)]+\frac{\partial}{\partial\sigma}[p_s\dot{\sigma}f(\theta)] = 0. \tag{22.1.27}$$

The domain integral of this becomes

$$\frac{d}{dt}\int_S dS\int_0^1 p_sf(\theta)d\sigma = 0. \tag{22.1.28}$$

As shown above, the domain integral of any function of potential temperature $f(\theta)$ is analytically conserved. In numerical models, however, a finite number of functions of θ are conserved if the equation is numerically discretized. Normally, the following set of functions is chosen as the conservative variables for numerical models:

$$f(\theta) = \theta, \qquad g(\theta) = \theta^2. \tag{22.1.29}$$

Any other functions can be chosen and a corresponding numerical scheme can be constructed. One might choose entropy as a conserved quantity; in this case

$$f(\theta) = \theta, \qquad g(\theta) = \ln\theta, \tag{22.1.30}$$

are two conserved quantities.

22.2 Conservation of vertically discretized equations

We require that the quantities described in Section 22.1 should be conserved in vertically discretized form. We concentrate on vertical discretization in this chapter, and do not touch on horizontal and time discretizations.

We define the vertical levels as shown in Fig. 22.1; ζ, D, T, and q are defined at *integer levels*, while $\dot{\sigma}$ is defined at the *half-integer levels*. This type of definition of the vertical level is called the *Lorenz grid*.† The total number of integer levels is denoted by K. The vertical coordinate σ is defined at both the integer and half-integer levels in the order:

$$\hat{\sigma}_{\frac{1}{2}} = 1 > \sigma_1 > \cdots > \sigma_K > \hat{\sigma}_{K+\frac{1}{2}} = 0. \tag{22.2.1}$$

To define the positions of these levels, the half-integer levels $\hat{\sigma}_{k+\frac{1}{2}}$ ($k = 0, 1, \cdots, K$) are specified first and then the integer levels σ_k ($k = 1, \cdots, K$) are interpolated from the half-integer levels. As shown below, this interpolation is related to the requirement for conservation of energy. The thickness of the layer is denoted by

$$\Delta\sigma_k = \hat{\sigma}_{k-\frac{1}{2}} - \hat{\sigma}_{k+\frac{1}{2}}, \qquad \text{for} \quad k = 1, \cdots, K, \tag{22.2.2}$$

which satisfy

$$\sum_{k=1}^{K} \Delta\sigma_k = 1. \tag{22.2.3}$$

22.2.1 Conservation of mass

The vertically discretized form of the conservation of mass (22.1.9) is given by

$$\frac{\partial p_s}{\partial t} + \nabla_\sigma \cdot (p_s \boldsymbol{v}_k) + \frac{1}{\Delta\sigma_k}\left(p_s\dot{\sigma}_{k-\frac{1}{2}} - p_s\dot{\sigma}_{k+\frac{1}{2}}\right) = 0. \tag{22.2.4}$$

This equation is the starting point for the following discretizations. The vertical integral $\int_0^1 d\sigma$ becomes $\sum_{k=1}^{K} \Delta\sigma_k$ in discretized form. Thus, the sum of the above equation in the vertical direction gives

$$\frac{\partial p_s}{\partial t} = -\sum_{k=1}^{K} \nabla_\sigma \cdot (p_s \boldsymbol{v}_k)\Delta\sigma_k, \tag{22.2.5}$$

† Recent numerical models are based on the *Charney-Phillips grid*, in which ζ, D, $\dot{\sigma}$, and q are defined at the integer levels, while T is defined at the half-integer levels. It is argued that the Charney-Phillips grid has the advantage of numerical stability in numerical models with hydrostatic approximation compared with the Lorenz grid.

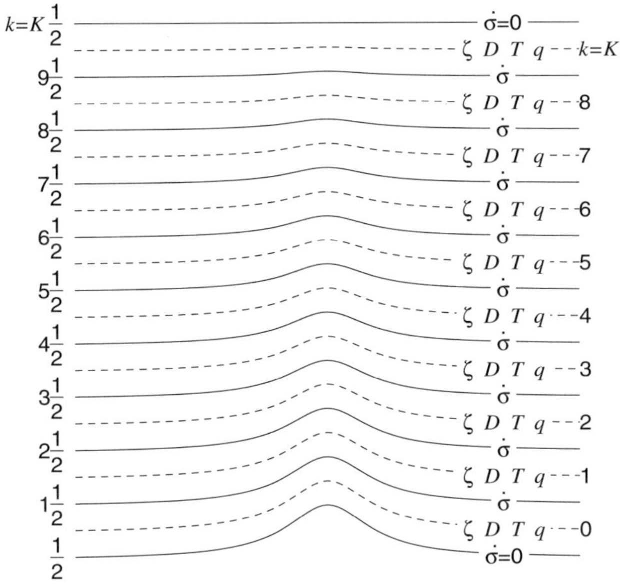

FIGURE 22.1: Vertical levels for the Lorenz grid and arrangement of variables. Solid curves represent half-integer levels, and dotted curves the integer levels.

where $\dot{\sigma}_{\frac{1}{2}} = \dot{\sigma}_{K+\frac{1}{2}} = 0$ is used. Eq. (22.2.5) corresponds to (22.1.10). The domain integral of (22.2.5) gives the conservation of total mass, which corresponds to (22.1.11). Similarly, (22.1.5) can be written as

$$\frac{\partial \pi}{\partial t} = -\sum_{k=1}^{K} (\nabla_\sigma \cdot \boldsymbol{v}_k + \boldsymbol{v}_k \cdot \nabla_\sigma \pi) \Delta\sigma_k, \tag{22.2.6}$$

where $\pi = \ln p_s$. If p_s is chosen as a prognostic variable, the domain integral of p_s is conserved, but if π is chosen as a prognostic variable, the conservation of mass is no longer satisfied since the domain integral of the right-hand side of (22.2.6) does not vanish in general.

The discretized form of the vertical velocity of σ (22.1.6) is given by

$$\dot{\sigma}_{k+\frac{1}{2}} = -\hat{\sigma}_{k+\frac{1}{2}} \frac{\partial \pi}{\partial t} - \sum_{l=k+1}^{K} (\nabla_\sigma \cdot \boldsymbol{v}_l + \boldsymbol{v}_l \cdot \nabla_\sigma \pi) \Delta\sigma_l, \tag{22.2.7}$$

$$\dot{\sigma}_{k-\frac{1}{2}} = -\hat{\sigma}_{k-\frac{1}{2}} \frac{\partial \pi}{\partial t} - \sum_{l=k}^{K} (\nabla_\sigma \cdot \boldsymbol{v}_l + \boldsymbol{v}_l \cdot \nabla_\sigma \pi) \Delta\sigma_l, \tag{22.2.8}$$

in which $\frac{\partial \pi}{\partial t}$ is given by (22.2.6). Either of the two equations can be used for calculation of vertical velocity $\dot{\sigma}$.

22.2.2 Transformation to the flux form

For any variable A, the relation between the advective form and the flux form is given by

$$p_s \frac{dA}{dt} = \frac{\partial}{\partial t}(p_s A) + \nabla_\sigma \cdot (p_s \boldsymbol{v}_H A) + \frac{\partial}{\partial \sigma}(p_s \dot{\sigma} A). \tag{22.2.9}$$

The corresponding discretized form for the vertical divergence term is written as

$$\begin{aligned} p_s \left(\frac{dA}{dt}\right)_k &= \frac{\partial}{\partial t}(p_s A_k) + \nabla_\sigma \cdot (p_s \boldsymbol{v}_H A_k) \\ &+ \frac{1}{\Delta\sigma_k}\left(p_s \dot{\sigma}_{k-\frac{1}{2}} \hat{A}_{k-\frac{1}{2}} - p_s \dot{\sigma}_{k+\frac{1}{2}} \hat{A}_{k+\frac{1}{2}}\right), \end{aligned} \tag{22.2.10}$$

where A_k is the value of A at the integer level k, and $\hat{A}_{k-\frac{1}{2}}$ is that at the half-integer level $k - \frac{1}{2}$. Suppose that the values of A are given at the integer levels first, and the values at the half-integer levels $\hat{A}_{k-\frac{1}{2}}$ are appropriately interpolated from the values at the integer levels A_k. Summing up (22.2.10) in the vertical direction yields

$$p_s \sum_{k=1}^{K} \left(\frac{dA}{dt}\right)_k \Delta\sigma_k = \frac{\partial}{\partial t} \sum_{k=1}^{K} p_s A_k \Delta\sigma_k + \sum_{k=1}^{K} \nabla_\sigma \cdot (p_s \boldsymbol{v}_H A_k) \Delta\sigma_k. \tag{22.2.11}$$

The domain integral of the second term on the right vanishes. Thus, a discretized form such as (22.2.10) guarantees the conservation of A regardless of the form of $\hat{A}_{k-\frac{1}{2}}$. We can make use of this arbitrariness to introduce other constraints.

From (22.2.10), we can derive an advective-form equation that has the same conservative property as the flux-form equation. Since

$$\begin{aligned} &\frac{\partial}{\partial t}(p_s A_k) + \nabla_\sigma \cdot (p_s \boldsymbol{v}_H A_k) + \frac{1}{\Delta\sigma_k}\left(p_s \dot{\sigma}_{k-\frac{1}{2}} \hat{A}_{k-\frac{1}{2}} - p_s \dot{\sigma}_{k+\frac{1}{2}} \hat{A}_{k+\frac{1}{2}}\right) \\ &= p_s \left(\frac{\partial}{\partial t} + \boldsymbol{v}_H \cdot \nabla_\sigma\right) A_k + A_k \left[\frac{\partial}{\partial t} p_s + \nabla_\sigma \cdot (\boldsymbol{v}_H p_s)\right] \\ &\quad + \frac{1}{\Delta\sigma_k}\left(p_s \dot{\sigma}_{k-\frac{1}{2}} \hat{A}_{k-\frac{1}{2}} - p_s \dot{\sigma}_{k+\frac{1}{2}} \hat{A}_{k+\frac{1}{2}}\right) \\ &= p_s \left(\frac{\partial}{\partial t} + \boldsymbol{v}_H \cdot \nabla_\sigma\right) A_k \\ &\quad + \frac{1}{\Delta\sigma_k}\left[p_s \dot{\sigma}_{k-\frac{1}{2}} \left(\hat{A}_{k-\frac{1}{2}} - A_k\right) + p_s \dot{\sigma}_{k+\frac{1}{2}} \left(A_k - \hat{A}_{k+\frac{1}{2}}\right)\right], \end{aligned} \tag{22.2.12}$$

where (22.2.4) is used, thus we obtain

$$\begin{aligned} \left(\frac{dA}{dt}\right)_k &= \frac{\partial}{\partial t} A_k + \boldsymbol{v}_k \cdot \nabla_\sigma A_k \\ &+ \frac{1}{\Delta\sigma_k}\left[\dot{\sigma}_{k-\frac{1}{2}} \left(\hat{A}_{k-\frac{1}{2}} - A_k\right) + \dot{\sigma}_{k+\frac{1}{2}} \left(A_k - \hat{A}_{k+\frac{1}{2}}\right)\right]. \end{aligned} \tag{22.2.13}$$

Up to now, the discretized form of $\hat{A}_{k+\frac{1}{2}}$ has not been specified. To determine the form, we introduce a constraint in which a function $F(A)$ is conserved. We define

$$F_k \equiv F(A_k), \qquad F'_k \equiv \frac{dF(A_k)}{dA}. \tag{22.2.14}$$

Multiplying (22.2.12) by F'_k, and using the conservation of mass, (22.2.4), we obtain

$$\begin{aligned}
F'_k p_s \left(\frac{dA}{dt}\right)_k &= p_s \left(\frac{\partial}{\partial t} + \boldsymbol{v}_H \cdot \nabla_\sigma\right) F_k \\
&\quad + \frac{1}{\Delta\sigma_k}\left[p_s \dot{\sigma}_{k-\frac{1}{2}} F'_k \left(\hat{A}_{k-\frac{1}{2}} - A_k\right) + p_s \dot{\sigma}_{k+\frac{1}{2}} F'_k \left(A_k - \hat{A}_{k+\frac{1}{2}}\right)\right] \\
&= \frac{\partial}{\partial t}(p_s F_k) + \nabla_\sigma \cdot (p_s \boldsymbol{v}_k F_k) \\
&\quad + \frac{1}{\Delta\sigma_k}\left\{p_s \dot{\sigma}_{k-\frac{1}{2}}\left[F'_k\left(\hat{A}_{k-\frac{1}{2}} - A_k\right) + F_k\right]\right. \\
&\quad \left. - p_s \dot{\sigma}_{k+\frac{1}{2}}\left[-F'_k\left(A_k - \hat{A}_{k+\frac{1}{2}}\right) + F_k\right]\right\}.
\end{aligned} \tag{22.2.15}$$

In order for the flux in (22.2.15) to be canceled out when it is vertically integrated, the values of F at the half-integer levels must satisfy

$$\hat{F}_{k-\frac{1}{2}} = F'_k\left(\hat{A}_{k-\frac{1}{2}} - A_k\right) + F_k, \tag{22.2.16}$$

$$\hat{F}_{k+\frac{1}{2}} = -F'_k\left(A_k - \hat{A}_{k+\frac{1}{2}}\right) + F_k, \tag{22.2.17}$$

from which, by replacing k with $k-1$ in (22.2.17), we obtain

$$F'_k\left(\hat{A}_{k-\frac{1}{2}} - A_k\right) + F_k = -F'_{k-1}\left(A_{k-1} - \hat{A}_{k-\frac{1}{2}}\right) + F_{k-1},$$

Thus, we have

$$\hat{A}_{k-\frac{1}{2}} = \frac{(F'_{k-1}A_{k-1} - F_{k-1}) - (F'_k A_k - F_k)}{F'_{k-1} - F'_k}. \tag{22.2.18}$$

This corresponds to the discretized form of

$$A \equiv \frac{d(F'A - F)}{dF'}. \tag{22.2.19}$$

If we choose $F(A) = A^2$ as a function of F, (22.2.18) is reduced to

$$\hat{A}_{k-\frac{1}{2}} = \frac{1}{2}(A_{k-1} + A_k). \tag{22.2.20}$$

In this case, the domain integrals of A and A^2 are conserved. If we choose $F(A) = \ln A$, on the other hand, we have

$$\hat{A}_{k-\frac{1}{2}} = \frac{-\ln A_{k-1} + \ln A_k}{\frac{1}{A_{k-1}} - \frac{1}{A_k}}. \tag{22.2.21}$$

This form may be used if one wants to have the conservation of entropy $C_p \ln\theta$.

22.2.3 Constraint on the vertical integral of the pressure gradient

The pressure gradient force in the σ-coordinate is written by multiplying (22.1.12) by p_s/g:

$$
\begin{aligned}
-\frac{p_s}{g\rho}\nabla_z p &= \frac{p_s}{g}\left(-\nabla_\sigma \Phi + \frac{\sigma}{p_s}\frac{\partial \Phi}{\partial \sigma}\nabla_\sigma p_s\right) \\
&= -\frac{1}{g}\left[\nabla_\sigma (p_s \Phi) - \frac{\partial (\Phi\sigma)}{\partial \sigma}\nabla_\sigma p_s\right].
\end{aligned} \tag{22.2.22}
$$

From this relation, we can choose the discretized form of the pressure gradient as

$$
\begin{aligned}
-\frac{p_s}{g}\left(\frac{1}{\rho}\nabla_z p\right)_k &= -\frac{1}{g}\Big[\nabla_\sigma (p_s \Phi_k) \\
&\quad - \frac{1}{\Delta\sigma_k}\left(\hat{\Phi}_{k-\frac{1}{2}}\hat{\sigma}_{k-\frac{1}{2}} - \hat{\Phi}_{k+\frac{1}{2}}\hat{\sigma}_{k+\frac{1}{2}}\right)\nabla_\sigma p_s\Big].
\end{aligned} \tag{22.2.23}
$$

Thus, the corresponding form of the vertical integral of the pressure gradient, (22.1.13), is given by

$$
\begin{aligned}
&\sum_{k=1}^{K}\left\{-\frac{1}{g}\left[\nabla_\sigma (p_s \Phi_k) - \frac{1}{\Delta\sigma_k}\left(\hat{\Phi}_{k-\frac{1}{2}}\hat{\sigma}_{k-\frac{1}{2}} - \hat{\Phi}_{k+\frac{1}{2}}\hat{\sigma}_{k+\frac{1}{2}}\right)\nabla_\sigma p_s\right]\right\}\Delta\sigma_k \\
&= -\frac{1}{g}\left[\sum_{k=1}^{K}\nabla_\sigma (p_s \Phi_k) - \hat{\Phi}_{\frac{1}{2}}\nabla_\sigma p_s\right] \\
&= -\frac{1}{g}\left\{\nabla_\sigma\left[\sum_{k=1}^{K} p_s(\Phi_k - \hat{\Phi}_{\frac{1}{2}})\Delta\sigma_k\right] + p_s \nabla_\sigma \hat{\Phi}_{\frac{1}{2}}\right\}.
\end{aligned} \tag{22.2.24}
$$

If the ground surface has no topography, the line integral of (22.2.24) along a closed curve vanishes, so that the constraint on the vertical integral of the pressure gradient is satisfied. At this point, we are free to choose the discretized form of $\hat{\Phi}$. In particular, $\hat{\Phi}_{\frac{1}{2}}$ does not necessarily agree with the geopotential height at the surface Φ_s.

We may rewrite the pressure gradient in a convenient form. Eq. (22.2.23) is rewritten as

$$
\begin{aligned}
-\frac{p_s}{g}\left(\frac{1}{\rho}\nabla_z p\right)_k &= -\frac{p_s}{g}\Big\{\nabla_\sigma \Phi_k \\
&\quad + \left[\Phi_k - \frac{1}{\Delta\sigma_k}\left(\hat{\Phi}_{k-\frac{1}{2}}\hat{\sigma}_{k-\frac{1}{2}} - \hat{\Phi}_{k+\frac{1}{2}}\hat{\sigma}_{k+\frac{1}{2}}\right)\right]\nabla_\sigma \pi\Big\},
\end{aligned} \tag{22.2.25}
$$

The second term in the brace on the right-hand side is thought of as a discretized form of the following hydrostatic balance:

$$
p_s \sigma \alpha = \Phi - \frac{\partial (\Phi\sigma)}{\partial \sigma}. \tag{22.2.26}
$$

Thus, it is natural to define

$$(\sigma\alpha)_k \equiv \frac{1}{p_s}\left[\Phi_k - \frac{1}{\Delta\sigma_k}\left(\hat{\Phi}_{k-\frac{1}{2}}\hat{\sigma}_{k-\frac{1}{2}} - \hat{\Phi}_{k+\frac{1}{2}}\hat{\sigma}_{k+\frac{1}{2}}\right)\right]. \tag{22.2.27}$$

Using this symbol, (22.2.25) can be written as

$$-\frac{p_s}{g}\left(\frac{1}{\rho}\nabla_z p\right)_k = -\frac{p_s}{g}\left[\nabla_\sigma \Phi_k + (\sigma\alpha)_k \nabla_\sigma p_s\right]. \tag{22.2.28}$$

The acceleration term in the momentum equation is discretized using the relation (22.2.12) such as

$$\begin{aligned} p_s\left(\frac{d\boldsymbol{v}}{dt}\right)_k &= p_s\left(\frac{\partial}{\partial t} + \boldsymbol{v}_k \cdot \nabla_\sigma\right)\boldsymbol{v}_k \\ &+\frac{1}{\Delta\sigma_k}\left[p_s\dot{\sigma}_{k-\frac{1}{2}}\left(\hat{\boldsymbol{v}}_{k-\frac{1}{2}} - \boldsymbol{v}_k\right) + p_s\dot{\sigma}_{k+\frac{1}{2}}\left(\boldsymbol{v}_k - \hat{\boldsymbol{v}}_{k+\frac{1}{2}}\right)\right]. \end{aligned} \tag{22.2.29}$$

The inner product of this with $\boldsymbol{v}_k$ gives the equation of kinetic energy corresponding to (22.1.16). In order that $\boldsymbol{v}_k^2$ be conservative with respect to vertical advection, the half-level value of velocity is interpolated as (22.2.20); that is

$$\hat{\boldsymbol{v}}_{k-\frac{1}{2}} = \frac{1}{2}(\boldsymbol{v}_{k-1} + \boldsymbol{v}_k). \tag{22.2.30}$$

Thus, the discretized form of the equation of kinetic energy (22.1.16) becomes

$$\begin{aligned} &\frac{\partial}{\partial t}\left(p_s\frac{\boldsymbol{v}_k^2}{2}\right) + \nabla_\sigma \cdot \left(p_s\boldsymbol{v}_k\frac{\boldsymbol{v}_k^2}{2}\right) + \frac{1}{\Delta\sigma_k}\left(p_s\dot{\sigma}_{k-\frac{1}{2}}\frac{\hat{\boldsymbol{v}}_{k-\frac{1}{2}}^2}{2} - p_s\dot{\sigma}_{k+\frac{1}{2}}\frac{\hat{\boldsymbol{v}}_{k+\frac{1}{2}}^2}{2}\right) \\ &= -p_s\boldsymbol{v}_k \cdot \left[\nabla_\sigma\Phi_k + (\sigma\alpha)_k\nabla_\sigma p_s\right] + p_s\boldsymbol{v}_k \cdot \boldsymbol{F}_k. \end{aligned} \tag{22.2.31}$$

Now, let us investigate the vertical integral of this equation. The left-hand side is

$$\frac{\partial}{\partial t}\left[\frac{1}{g}\sum_{k=1}^{K}\left(\frac{1}{2}p_s\boldsymbol{v}_k^2\right)\Delta\sigma_k\right] + \nabla_\sigma \cdot \left[\frac{1}{g}\sum_{k=1}^{K}\left(p_s\boldsymbol{v}_k\frac{\boldsymbol{v}_k^2}{2}\right)\Delta\sigma_k\right]. \tag{22.2.32}$$

The work done by the pressure gradient on the right-hand side of (22.2.31) should be rewritten in the form (22.1.17). Thus, the corresponding discretized form can be rewritten as follows:

$$\begin{aligned} &-p_s\boldsymbol{v}_k \cdot \left[\nabla_\sigma\Phi_k + (\sigma\alpha)_k\nabla_\sigma p_s\right] \\ &= -\nabla_\sigma \cdot (p_s\boldsymbol{v}_k\Phi_k) + \Phi_k\nabla_\sigma \cdot (p_s\boldsymbol{v}_k) - p_s(\sigma\alpha)_k\boldsymbol{v}_k \cdot \nabla_\sigma p_s \\ &= -\nabla_\sigma \cdot (p_s\boldsymbol{v}_k\Phi_k) - \Phi_k\left[\frac{1}{\Delta\sigma_k}(p_s\dot{\sigma}_{k-\frac{1}{2}} - p_s\dot{\sigma}_{k+\frac{1}{2}}) + \frac{\partial p_s}{\partial t}\right] \\ &\quad -p_s(\sigma\alpha)_k\boldsymbol{v}_k \cdot \nabla_\sigma p_s \\ &= -\nabla_\sigma \cdot (p_s\boldsymbol{v}_k\Phi_k) - \frac{1}{\Delta\sigma_k}\left(p_s\dot{\sigma}_{k-\frac{1}{2}}\hat{\Phi}_{k-\frac{1}{2}} - p_s\dot{\sigma}_{k+\frac{1}{2}}\hat{\Phi}_{k+\frac{1}{2}}\right) \\ &\quad +\frac{1}{\Delta\sigma_k}\left[p_s\dot{\sigma}_{k-\frac{1}{2}}\left(\hat{\Phi}_{k-\frac{1}{2}} - \Phi_{k-\frac{1}{2}}\right) + p_s\dot{\sigma}_{k+\frac{1}{2}}\left(\Phi_{k+\frac{1}{2}} - \hat{\Phi}_{k+\frac{1}{2}}\right)\right] \\ &\quad -\Phi_k\frac{\partial p_s}{\partial t} - p_s(\sigma\alpha)_k\boldsymbol{v}_k \cdot \nabla_\sigma p_s, \end{aligned}$$

where the continuity equation (22.2.4) is used. Using the definition (22.2.27), this is further rewritten as

$$\begin{aligned}
&= -\nabla_\sigma \cdot (p_s \boldsymbol{v}_k \Phi_k) - \frac{1}{\Delta\sigma_k}\left(p_s \dot\sigma_{k-\frac{1}{2}} \hat\Phi_{k-\frac{1}{2}} - p_s \dot\sigma_{k+\frac{1}{2}} \hat\Phi_{k+\frac{1}{2}}\right) \\
&\quad + [p_s(\sigma\alpha)_k - \Phi_k] \frac{\partial p_s}{\partial t} - p_s \left\{ (\sigma\alpha)_k \left(\frac{\partial}{\partial t} + \boldsymbol{v}_k \cdot \nabla_\sigma\right) p_s \right. \\
&\quad \left. - \frac{1}{\Delta\sigma_k}\left[\dot\sigma_{k-\frac{1}{2}}\left(\hat\Phi_{k-\frac{1}{2}} - \Phi_{k-\frac{1}{2}}\right) + \dot\sigma_{k+\frac{1}{2}}\left(\Phi_{k+\frac{1}{2}} - \hat\Phi_{k+\frac{1}{2}}\right)\right]\right\} \\
&= -\nabla_\sigma \cdot (p_s \boldsymbol{v}_k \Phi_k) \\
&\quad - \frac{1}{\Delta\sigma_k}\left[\left(p_s \dot\sigma_{k-\frac{1}{2}} + \hat\sigma_{k-\frac{1}{2}} \frac{\partial p_s}{\partial t}\right) \hat\Phi_{k-\frac{1}{2}} \right. \\
&\quad \left. - \left(p_s \dot\sigma_{k+\frac{1}{2}} + \hat\sigma_{k+\frac{1}{2}} \frac{\partial p_s}{\partial t}\right) \hat\Phi_{k+\frac{1}{2}}\right] - p_s(\omega\alpha)_k,
\end{aligned} \tag{22.2.33}$$

where we have defined

$$\begin{aligned}
(\omega\alpha)_k &\equiv (\sigma\alpha)_k \left(\frac{\partial}{\partial t} + \boldsymbol{v}_k \cdot \nabla_\sigma\right) p_s \\
&\quad - \frac{1}{\Delta\sigma_k}\left[\dot\sigma_{k-\frac{1}{2}}\left(\hat\Phi_{k-\frac{1}{2}} - \Phi_k\right) + \dot\sigma_{k+\frac{1}{2}}\left(\Phi_k - \hat\Phi_{k+\frac{1}{2}}\right)\right].
\end{aligned} \tag{22.2.34}$$

It can be seen that (22.2.33) corresponds to the discretized form of (22.1.17).

22.2.4 Hydrostatic balance

We need to obtain a specific discretization for hydrostatic balance in order to have the conservation of energy. If we note the analytic form of hydrostatic balance

$$\frac{\partial \Phi}{\partial \sigma} = -p_s \alpha, \tag{22.2.35}$$

it is natural to express the difference of Φ between two layers by the values of $p_s\alpha$ at these layers. We introduce coefficients $A_{k-\frac{1}{2}}$ and $B_{k-\frac{1}{2}}$ for $k = 2, \cdots, K$, such that

$$\Phi_k - \Phi_{k-1} = A_{k-\frac{1}{2}} p_s \alpha_k + B_{k-\frac{1}{2}} p_s \alpha_{k-1}. \tag{22.2.36}$$

Considering the discrete form of hydrostatic balance in the equation of momentum (22.2.26), we may set

$$[\Phi\delta\sigma - \delta(\Phi\sigma)]_k \equiv \Phi_k \Delta\sigma_k - \hat\sigma_{k-\frac{1}{2}} \hat\Phi_{k-\frac{1}{2}} + \hat\sigma_{k+\frac{1}{2}} \hat\Phi_{k+\frac{1}{2}} = C_k p_s \alpha_k, \tag{22.2.37}$$

where C_k is constant. In particular, at the lowest level, we have

$$\begin{aligned}\Phi_1 - \hat{\Phi}_{\frac{1}{2}} &= \sum_{k=1}^{K}[\Phi\delta\sigma - \delta(\Phi\sigma)]_k - \sum_{k=2}^{K}\hat{\sigma}_{k-\frac{1}{2}}(\Phi_k - \Phi_{k-1}) \\ &= p_s\alpha_1(C_1 - \hat{\sigma}_{\frac{3}{2}}B_{\frac{3}{2}}) \\ &\quad + \sum_{k=2}^{K} p_s\alpha_k(C_k - \hat{\sigma}_{k-\frac{1}{2}}A_{k-\frac{1}{2}} - \hat{\sigma}_{k+\frac{1}{2}}B_{k+\frac{1}{2}}). \qquad (22.2.38)\end{aligned}$$

Thus, if we choose

$$C_k = \hat{\sigma}_{k-\frac{1}{2}}A_{k-\frac{1}{2}} + \hat{\sigma}_{k+\frac{1}{2}}B_{k+\frac{1}{2}}, \qquad (22.2.39)$$

the values of α in upper layers do not affect $\Phi_1 - \hat{\Phi}_{\frac{1}{2}}$. That is, the difference of geopotential in the lowest level is expressed by local quantities as

$$\Phi_1 - \hat{\Phi}_{\frac{1}{2}} = p_s\alpha_1 A_{\frac{1}{2}}. \qquad (22.2.40)$$

Using (22.2.36), (22.2.37), and (22.2.39), we have

$$\Phi_k - \hat{\Phi}_{k-\frac{1}{2}} = A_{k-\frac{1}{2}}p_s\alpha_k, \qquad \hat{\Phi}_{k+\frac{1}{2}} - \Phi_k = B_{k+\frac{1}{2}}p_s\alpha_k. \qquad (22.2.41)$$

From this, the interpolation of Φ_k is given by

$$\Phi_k = \frac{A_{k-\frac{1}{2}}\hat{\Phi}_{k+\frac{1}{2}} + B_{k+\frac{1}{2}}\hat{\Phi}_{k-\frac{1}{2}}}{A_{k-\frac{1}{2}} + B_{k+\frac{1}{2}}}. \qquad (22.2.42)$$

We have not yet specified the coefficients $A_{k-\frac{1}{2}}$ and $B_{k-\frac{1}{2}}$. We use the arbitrariness of these coefficients to minimize the error in the pressure gradient. From (22.2.27), (22.2.37), and (22.2.39), we have

$$(\sigma\alpha)_k = \frac{1}{\Delta\sigma_k}\left(\hat{\sigma}_{k-\frac{1}{2}}A_{k-\frac{1}{2}} + \hat{\sigma}_{k+\frac{1}{2}}B_{k+\frac{1}{2}}\right)\alpha_k. \qquad (22.2.43)$$

Substituting this into (22.2.28), we obtain the expression of the pressure gradient as

$$-\left(\frac{1}{\rho}\nabla_z p\right)_k = -\nabla_\sigma\Phi_k - \frac{1}{\Delta\sigma_k}\left(\hat{\sigma}_{k-\frac{1}{2}}A_{k-\frac{1}{2}} + \hat{\sigma}_{k+\frac{1}{2}}B_{k+\frac{1}{2}}\right)\alpha_k\nabla_\sigma p_s. \qquad (22.2.44)$$

If the atmosphere is at rest, the pressure gradient must be zero. This must hold even when there is a topography (i.e., $\nabla_\sigma p_s \neq 0$). In this case, the geopotential is a function of pressure, $\Phi = \Phi(p)$ and

$$\nabla_\sigma\Phi_k = \left(\frac{d\Phi}{dp}\right)_k\left(\frac{\partial p}{\partial p_s}\right)_\sigma\nabla_\sigma p_s. \qquad (22.2.45)$$

If we consider a special field that satisfies

$$\left(\frac{d\Phi}{dp}\right)_k = -\alpha_k, \qquad (22.2.46)$$

we have, by setting (22.2.44) to zero,

$$\left(\frac{\partial p}{\partial p_s}\right)_\sigma = -\frac{1}{\Delta\sigma_k}\left(\hat{\sigma}_{k-\frac{1}{2}}A_{k-\frac{1}{2}} + \hat{\sigma}_{k+\frac{1}{2}}B_{k+\frac{1}{2}}\right). \tag{22.2.47}$$

Here, in order to obtain expressions of the coefficients $A_{k-\frac{1}{2}}$ and $B_{k-\frac{1}{2}}$, we introduce a function of pressure, $f(p)$. Multiplying (22.2.47) by $f'(p)$, we have

$$\left(\frac{\partial f(p)}{\partial p_s}\right)_\sigma = -\frac{f'(p)}{\Delta\sigma_k}\left(\hat{\sigma}_{k-\frac{1}{2}}A_{k-\frac{1}{2}} + \hat{\sigma}_{k+\frac{1}{2}}B_{k+\frac{1}{2}}\right), \tag{22.2.48}$$

or

$$\begin{aligned}\left[\frac{\partial(p_s f(p))}{\partial p_s}\right]_\sigma &= -\frac{1}{\Delta\sigma_k}\Big[\hat{\sigma}_{k-\frac{1}{2}}(f_k + p_s A_{k-\frac{1}{2}} f'_k) \\ &\quad - \hat{\sigma}_{k+\frac{1}{2}}(f_k - p_s B_{k+\frac{1}{2}} f'_k)\Big].\end{aligned} \tag{22.2.49}$$

If we note

$$\left[\frac{\partial(p_s f(p))}{\partial p_s}\right]_\sigma = \left[\frac{\partial(\sigma f(p))}{\partial\sigma}\right]_{p_s}, \tag{22.2.50}$$

the discretized form of $\hat{f}$ should be chosen as

$$\hat{f}_{k-\frac{1}{2}} = f_k + p_s A_{k-\frac{1}{2}} f'_k, \qquad \hat{f}_{k+\frac{1}{2}} = f_k - p_s B_{k+\frac{1}{2}} f'_k. \tag{22.2.51}$$

From this, if we choose the coefficients as

$$A_{k-\frac{1}{2}} = \frac{\hat{f}_{k-\frac{1}{2}} - f_k}{p_s f'_k}, \qquad B_{k+\frac{1}{2}} = \frac{f_k - \hat{f}_{k+\frac{1}{2}}}{p_s f'_k}, \tag{22.2.52}$$

the discretization (22.2.49) becomes equivalent to

$$\left[\frac{\partial(p_s f_k)}{\partial p_s}\right]_\sigma = \left(\frac{\Delta(\sigma f)}{\Delta\sigma}\right)_k. \tag{22.2.53}$$

This can easily be integrated with respect to p_s. The values of f at the integer levels can be given by

$$f_k = \frac{1}{p_s}\left[\frac{\Delta(\sigma\int f dp_s)}{\Delta\sigma}\right]_k = \left[\frac{\Delta(\int f dp)}{\Delta p}\right]_k. \tag{22.2.54}$$

This in turn is used to obtain the values of pressure at the integer levels using the relation $f_k = f(p_k)$.

Substituting (22.2.52) into (22.2.43), we have

$$\begin{aligned}(\sigma\alpha)_k &= \frac{1}{p_s\Delta\sigma_k}\left[\hat{\sigma}_{k-\frac{1}{2}}(\hat{f}_{k-\frac{1}{2}} - f_k) + \hat{\sigma}_{k+\frac{1}{2}}(f_k - \hat{f}_{k-\frac{1}{2}})\right]\frac{\alpha_k}{f'_k} \\ &= \frac{1}{p_s}\left\{\left[\frac{\Delta(\sigma f(p))}{\Delta\sigma}\right]_k - f_k\right\}\frac{\alpha_k}{f'_k} = \left(\frac{\partial f}{\partial p_s}\right)_k \frac{\alpha_k}{f'_k}.\end{aligned} \tag{22.2.55}$$

Thus, the pressure gradient (22.2.44) can be written as

$$-\left(\frac{1}{\rho}\nabla_z p\right)_k = -\nabla_\sigma \Phi_k - \frac{\alpha_k}{f'_k}\nabla_\sigma f_k. \tag{22.2.56}$$

Making use of (22.2.36), we express the vertical difference of the pressure gradient as

$$\begin{aligned} &-\left[\left(\frac{1}{\rho}\nabla_z p\right)_k - \left(\frac{1}{\rho}\nabla_z p\right)_{k-1}\right] \\ &= -\nabla_\sigma(\Phi_k - \Phi_{k-1}) - \left[\frac{\alpha_k}{f'_k}\nabla_\sigma f_k - \frac{\alpha_{k-1}}{f'_{k-1}}\nabla_\sigma f_{k-1}\right] \\ &= -\left[(\hat{f}_{k-\frac{1}{2}} - f_k)\nabla_\sigma \frac{\alpha_k}{f'_k} + (f_{k-1} - \hat{f}_{k-\frac{1}{2}})\nabla_\sigma \frac{\alpha_{k-1}}{f'_{k-1}} \right. \\ &\quad \left. + \left(\frac{\alpha_k}{f'_k} - \frac{\alpha_{k-1}}{f'_{k-1}}\right)\nabla_\sigma f_{k-1}\right]. \end{aligned} \tag{22.2.57}$$

This means that the contribution of the pressure gradient vanishes if α_k/f'_k is constant everywhere in a static state.

If we choose the function $f(p) = \Pi(p) = (p/p_{00})^\kappa$ where Π is the Exner function, $\kappa = R_d/C_p$, and $p_{00} = 1{,}000$ hPa, we have

$$\frac{\alpha}{f'(p)} = \frac{\alpha}{d\Pi/dp} = \frac{\alpha p}{\kappa \Pi} = \frac{C_p T}{\Pi} = C_p\theta, \tag{22.2.58}$$

thus the right-hand side of (22.2.57) vanishes for an isentropic atmosphere (i.e., there is no error in the pressure gradient if the atmosphere is in an isentropic static state).

For the rest of this section, we summarize expressions in the case $f = \Pi$. The interpolation coefficients A and B are given by (22.2.52) with $\Pi'(p) = \kappa\Pi/p$:

$$A_{k-\frac{1}{2}} = \frac{\hat{\Pi}_{k-\frac{1}{2}} - \Pi_k}{\kappa p_s \Pi_k/p_k} = \frac{\sigma_k}{\kappa}\left[\left(\frac{\hat{\sigma}_{k-\frac{1}{2}}}{\sigma_k}\right)^\kappa - 1\right] \equiv \frac{\sigma_k}{\kappa}a_k, \tag{22.2.59}$$

$$B_{k+\frac{1}{2}} = \frac{\Pi_k - \hat{\Pi}_{k+\frac{1}{2}}}{\kappa p_s \Pi_k/p_k} = \frac{\sigma_k}{\kappa}\left[1 - \left(\frac{\hat{\sigma}_{k+\frac{1}{2}}}{\sigma_k}\right)^\kappa\right] \equiv \frac{\sigma_k}{\kappa}b_k, \tag{22.2.60}$$

where

$$a_k = \left(\frac{\hat{\sigma}_{k-\frac{1}{2}}}{\sigma_k}\right)^\kappa - 1, \qquad b_k = 1 - \left(\frac{\hat{\sigma}_{k+\frac{1}{2}}}{\sigma_k}\right)^\kappa. \tag{22.2.61}$$

Using (22.2.36) and (22.2.40), hydrostatic balance is given by

$$\begin{aligned} \Phi_k - \Phi_{k-1} &= \frac{\sigma_k}{\kappa}\left[\left(\frac{\hat{\sigma}_{k-\frac{1}{2}}}{\sigma_k}\right)^\kappa - 1\right]p_s\alpha_k + \frac{\sigma_k}{\kappa}\left[1 - \left(\frac{\hat{\sigma}_{k-\frac{1}{2}}}{\sigma_k}\right)^\kappa\right]p_s\alpha_{k-1} \\ &= C_p a_k T_k + C_p b_{k-1} T_{k-1}, \end{aligned} \tag{22.2.62}$$

$$\Phi_1 - \hat{\Phi}_{\frac{1}{2}} = \frac{\sigma_1}{\kappa}\left[\left(\frac{\hat{\sigma}_{\frac{1}{2}}}{\sigma_1}\right)^\kappa - 1\right]p_s\alpha_1 = C_p a_1 T_1. \tag{22.2.63}$$

The values of σ at the integer levels can be determined by (22.2.54):

$$\begin{aligned} \Pi_k &= \left[\frac{\Delta(\int \Pi dp)}{\Delta p}\right]_k = \frac{1}{\kappa+1}\left[\frac{\Delta(p\Pi)}{\Delta p}\right]_k \\ &= \frac{1}{\kappa+1}\frac{\hat{\sigma}_{k-\frac{1}{2}}\Pi_{k-\frac{1}{2}} - \hat{\sigma}_{k+\frac{1}{2}}\Pi_{k+\frac{1}{2}}}{\Delta\sigma_k}. \end{aligned} \tag{22.2.64}$$

From this, we have

$$\sigma_k = \left[\frac{1}{\kappa+1}\left(\frac{\hat{\sigma}^{\kappa+1}_{k-\frac{1}{2}} - \hat{\sigma}^{\kappa+1}_{k+\frac{1}{2}}}{\hat{\sigma}_{k-\frac{1}{2}} - \hat{\sigma}_{k+\frac{1}{2}}}\right)\right]^{\frac{1}{\kappa}}. \tag{22.2.65}$$

From (22.2.43), the coefficients in the pressure gradient $(\sigma\alpha)_k$ are given by

$$\begin{aligned} (\sigma\alpha)_k &= \frac{1}{\Delta\sigma_k}\left\{\hat{\sigma}_{k-\frac{1}{2}}\frac{\sigma_k}{\kappa}\left[\left(\frac{\hat{\sigma}_{k-\frac{1}{2}}}{\sigma_k}\right)^{\kappa} - 1\right]\right. \\ &\quad \left. + \hat{\sigma}_{k+\frac{1}{2}}\frac{\sigma_k}{\kappa}\left[1 - \left(\frac{\hat{\sigma}_{k+\frac{1}{2}}}{\sigma_k}\right)^{\kappa}\right]\right\}\alpha_k \\ &= \frac{C_p}{p_s\Delta\sigma_k}\left(a_k\hat{\sigma}_{k-\frac{1}{2}} + b_k\hat{\sigma}_{k+\frac{1}{2}}\right)T_k = \frac{C_p}{p_s}\kappa_k T_k, \end{aligned} \tag{22.2.66}$$

where

$$\kappa_k = \frac{1}{\Delta\sigma_k}\left(a_k\hat{\sigma}_{k-\frac{1}{2}} + b_k\hat{\sigma}_{k+\frac{1}{2}}\right). \tag{22.2.67}$$

From comparison between (22.2.66) and the equation of state $p\alpha = \kappa C_p T$, κ_k can be regarded as the value of κ at each level calculated from the equation of state.

In the thermodynamic equation (22.1.20), the transformation term $p_s\omega\alpha$ should be of the same form as the discretization of the equation of kinetic energy (22.2.34). Thus, we have

$$\begin{aligned} (\omega\alpha)_k &\equiv (\sigma\alpha)_k\left(\frac{\partial}{\partial t} + \boldsymbol{v}_k\cdot\nabla_\sigma\right)p_s \\ &\quad -\frac{1}{\Delta\sigma_k}\left[\dot{\sigma}_{k-\frac{1}{2}}\left(\hat{\Phi}_{k-\frac{1}{2}} - \Phi_k\right) + \dot{\sigma}_{k+\frac{1}{2}}\left(\Phi_k - \hat{\Phi}_{k+\frac{1}{2}}\right)\right] \\ &= \frac{\alpha_k}{\Delta\sigma_k}\left\{A_{k-\frac{1}{2}}\left[\hat{\sigma}_{k-\frac{1}{2}}\left(\frac{\partial}{\partial t} + \boldsymbol{v}_k\cdot\nabla_\sigma\right)p_s + p_s\dot{\sigma}_{k-\frac{1}{2}}\right]\right. \\ &\quad \left. + B_{k+\frac{1}{2}}\left[\hat{\sigma}_{k+\frac{1}{2}}\left(\frac{\partial}{\partial t} + \boldsymbol{v}_k\cdot\nabla_\sigma\right)p_s + p_s\dot{\sigma}_{k+\frac{1}{2}}\right]\right\} \\ &= C_p\kappa_k T_k\left(\frac{\partial}{\partial t} + \boldsymbol{v}_k\cdot\nabla_\sigma\right)\pi + \frac{C_p}{\Delta\sigma_k}\left(a_k\dot{\sigma}_{k-\frac{1}{2}} + b_k\dot{\sigma}_{k+\frac{1}{2}}\right)T_k. \end{aligned} \tag{22.2.68}$$

22.2.5 The thermodynamic equation

The final step is to seek a discretized form of the thermodynamic equation. The central point of this section is to find the interpolation of $\hat{T}_{k+\frac{1}{2}}$, temperature at the half-integer levels. To obtain this form, we use a constraint that the domain average of potential temperature is conserved under the adiabatic condition, $Q = 0$.

From (22.2.68), the equation of enthalpy (22.1.20) is discretized as

$$\begin{aligned} \frac{\partial}{\partial t}(p_s C_p T_k) + \nabla_\sigma \cdot (p_s \boldsymbol{v}_k C_p T_k) + \frac{C_p}{\Delta\sigma_k}\left(p_s \dot{\sigma}_{k-\frac{1}{2}} \hat{T}_{k-\frac{1}{2}} - p_s \dot{\sigma}_{k+\frac{1}{2}} \hat{T}_{k+\frac{1}{2}}\right) \\ = p_s C_p \kappa_k T_k \left(\frac{\partial}{\partial t} + \boldsymbol{v}_k \cdot \nabla_\sigma\right)\pi + \frac{p_s C_p}{\Delta\sigma_k}\left(a_k \dot{\sigma}_{k-\frac{1}{2}} + b_k \dot{\sigma}_{k+\frac{1}{2}}\right) T_k. \end{aligned} \tag{22.2.69}$$

The discretization of $\hat{T}_{k+\frac{1}{2}}$ is not yet specified at this point. Using (22.2.13), the above equation is rewritten in advective form as

$$\begin{aligned} &\left(\frac{\partial}{\partial t} + \boldsymbol{v}_k \cdot \nabla_\sigma\right) T_k \\ &= -\frac{1}{\Delta\sigma_k}\left[\dot{\sigma}_{k-\frac{1}{2}}(\hat{T}_{k-\frac{1}{2}} - T_k) + \dot{\sigma}_{k+\frac{1}{2}}(T_k - \hat{T}_{k+\frac{1}{2}})\right] \\ &\quad + \kappa_k T_k \left(\frac{\partial}{\partial t} + \boldsymbol{v}_k \cdot \nabla_\sigma\right)\pi + \frac{1}{\Delta\sigma_k}\left(a_k \dot{\sigma}_{k-\frac{1}{2}} + b_k \dot{\sigma}_{k+\frac{1}{2}}\right) T_k. \end{aligned} \tag{22.2.70}$$

We define the potential temperature at the integer level by $\theta_k = T_k/\Pi_k$. Since

$$\begin{aligned} &\left(\frac{\partial}{\partial t} + \boldsymbol{v}_k \cdot \nabla_\sigma\right) T_k \\ &= \Pi_k \left(\frac{\partial}{\partial t} + \boldsymbol{v}_k \cdot \nabla_\sigma\right)\theta_k + \theta_k \left(\frac{\partial}{\partial t} + \boldsymbol{v}_k \cdot \nabla_\sigma\right)\Pi_k \\ &= \Pi_k \left(\frac{\partial}{\partial t} + \boldsymbol{v}_k \cdot \nabla_\sigma\right)\theta_k + \kappa T_k \left(\frac{\partial}{\partial t} + \boldsymbol{v}_k \cdot \nabla_\sigma\right)\pi, \end{aligned} \tag{22.2.71}$$

the equation of potential temperature can be written using (22.2.70) as

$$\begin{aligned} \left(\frac{\partial}{\partial t} + \boldsymbol{v}_k \cdot \nabla_\sigma\right)\theta_k = & -\frac{1}{\Pi_k}\frac{1}{\Delta\sigma_k}\left\{\hat{\sigma}_{k-\frac{1}{2}}\left[\hat{T}_{k-\frac{1}{2}} - (1 + a_k)T_k\right]\right. \\ & \left. + \hat{\sigma}_{k+\frac{1}{2}}\left[(1 - b_k)T_k - \hat{T}_{k+\frac{1}{2}}\right]\right\}. \end{aligned} \tag{22.2.72}$$

On the other hand, the conservation of potential temperature $\frac{d\theta}{dt} = 0$ is rewritten as the advective form using (22.2.13):

$$\left(\frac{\partial}{\partial t} + \boldsymbol{v}_k \cdot \nabla_\sigma\right)\theta_k = -\frac{1}{\Delta\sigma_k}\left[\dot{\sigma}_{k-\frac{1}{2}}\left(\hat{\theta}_{k-\frac{1}{2}} - \theta_k\right) + \dot{\sigma}_{k+\frac{1}{2}}\left(\theta_k - \hat{\theta}_{k+\frac{1}{2}}\right)\right], \tag{22.2.73}$$

where the discretization of $\hat{\theta}_{k+\frac{1}{2}}$ has yet to be determined. If (22.2.72) is equivalent to (22.2.73), the domain average of potential temperature is conserved under the adiabatic condition. Thus, we have

$$\hat{T}_{k-\frac{1}{2}} - (1+a_k)T_k = \Pi_k(\hat{\theta}_{k-\frac{1}{2}} - \theta_k), \tag{22.2.74}$$

$$(1-b_k)T_k - \hat{T}_{k+\frac{1}{2}} = \Pi_k(\theta_k - \hat{\theta}_{k+\frac{1}{2}}). \tag{22.2.75}$$

From this, we obtain the discretizations of T and θ at the half-integer levels by

$$\begin{aligned}\hat{T}_{k-\frac{1}{2}} &= \frac{a_k\Pi_{k-1}T_k + b_{k-1}\Pi_k T_{k-1}}{\Pi_{k-1} - \Pi_k} \\ &= a_k\left[1 - \left(\frac{\sigma_k}{\sigma_{k-1}}\right)^{\kappa}\right]^{-1} T_k + b_{k-1}\left[\left(\frac{\sigma_{k-1}}{\sigma_k}\right)^{\kappa} - 1\right]^{-1} T_{k-1},\end{aligned} \tag{22.2.76}$$

$$\begin{aligned}\hat{\theta}_{k-\frac{1}{2}} &= \frac{a_k\Pi_k\theta_k + b_{k-1}\Pi_{k-1}\theta_{k-1}}{\Pi_{k-1} - \Pi_k} \\ &= a_k\left[\left(\frac{\sigma_{k-1}}{\sigma_k}\right)^{\kappa} - 1\right]^{-1} \theta_k + b_{k-1}\left[1 - \left(\frac{\sigma_k}{\sigma_{k-1}}\right)^{\kappa}\right]^{-1} \theta_{k-1}.\end{aligned} \tag{22.2.77}$$

From comparison between the thermodynamic equation (22.2.70) and (22.1.21), we find that the term in (22.1.21)

$$-\dot{\sigma}\sigma^{\kappa}\frac{\partial(T\sigma^{-\kappa})}{\partial\sigma} \tag{22.2.78}$$

has the following discretized form:

$$\begin{aligned}&-\frac{1}{\Delta\sigma_k}\left[\dot{\sigma}_{k-\frac{1}{2}}(\hat{T}_{k-\frac{1}{2}} - T_k) + \dot{\sigma}_{k+\frac{1}{2}}(T_k - \hat{T}_{k+\frac{1}{2}})\right] \\ &\quad + \frac{1}{\Delta\sigma_k}\left(a_k\dot{\sigma}_{k-\frac{1}{2}} + b_k\dot{\sigma}_{k+\frac{1}{2}}\right)T_k \\ &= -\frac{1}{\Delta\sigma_k}\left\{\dot{\sigma}_{k-\frac{1}{2}}\left[\hat{T}_{k-\frac{1}{2}} - (1+a_k)T_k\right] + \dot{\sigma}_{k+\frac{1}{2}}\left[(1-b_k)T_k - \hat{T}_{k+\frac{1}{2}}\right]\right\} \\ &= \frac{1}{2}\left(\hat{\sigma}^{\kappa}_{k-\frac{1}{2}}\dot{\sigma}_{k-\frac{1}{2}}\frac{T_k\sigma_k^{-\kappa} - \hat{T}_{k-\frac{1}{2}}\hat{\sigma}^{-\kappa}_{k-\frac{1}{2}}}{\Delta\sigma_k/2}\right. \\ &\qquad \left. + \hat{\sigma}^{\kappa}_{k+\frac{1}{2}}\dot{\sigma}_{k+\frac{1}{2}}\frac{\hat{T}_{k+\frac{1}{2}}\hat{\sigma}^{-\kappa}_{k+\frac{1}{2}} - T_k\sigma_k^{-\kappa}}{\Delta\sigma_k/2}\right).\end{aligned} \tag{22.2.79}$$

We have used (22.2.61) for this derivation.

22.3 Summary

To conclude this chapter, we summarize the vertically discretized form of primitive equations. Primitive equations with vorticity and divergence equations are given by (20.2.45), (20.2.51), (20.2.57), (20.2.58), (20.2.59) in Chapter 20 and (22.1.6) in Chapter 22. Vertical discretizations of these equations are written as

$$\frac{\partial \zeta_k}{\partial t} = -\left[\frac{1}{R(1-\mu^2)}\frac{\partial}{\partial \lambda}A_k + \frac{1}{R}\frac{\partial}{\partial \mu}B_k\right], \tag{22.3.1}$$

$$\begin{aligned}\frac{\partial D_k}{\partial t} &= \frac{1}{R(1-\mu^2)}\frac{\partial}{\partial \lambda}B_k - \frac{1}{R}\frac{\partial}{\partial \mu}A_k \\ &\quad -\nabla_\sigma^2(\Phi_k' + C_p\kappa_k\overline{T}_k\pi + E_{kin\,k}),\end{aligned} \tag{22.3.2}$$

$$\begin{aligned}\frac{\partial T_k}{\partial t} &= -\left[\frac{1}{R(1-\mu^2)}\frac{\partial}{\partial \lambda}(U_kT_k') + \frac{1}{R}\frac{\partial}{\partial \mu}(V_kT_k')\right] + T_k'D_k \\ &\quad -\frac{1}{\Delta\sigma_k}\left\{\dot{\sigma}_{k-\frac{1}{2}}\left[\hat{T}_{k-\frac{1}{2}} - (1+a_k)T_k\right]\right. \\ &\quad \left. + \dot{\sigma}_{k+\frac{1}{2}}\left[(1-b_k)T_k - \hat{T}_{k+\frac{1}{2}}\right]\right\} \\ &\quad +\kappa_kT_k\left(\frac{\partial \pi}{\partial t} + \boldsymbol{V}_{Hk}\cdot\nabla_\sigma^\mu\pi\right) + Q_k,\end{aligned} \tag{22.3.3}$$

$$\frac{\partial \pi}{\partial t} = -\sum_{k=1}^{K}(\boldsymbol{V}_{Hk}\cdot\nabla_\sigma^\mu\pi)\,\Delta\sigma_k - \sum_{k=1}^{K}D_k\Delta\sigma_k, \tag{22.3.4}$$

$$\dot{\sigma}_{k-\frac{1}{2}} = -\hat{\sigma}_{k-\frac{1}{2}}\frac{\partial \pi}{\partial t} - \sum_{l=k}^{K}(D_l + \boldsymbol{V}_{Hl}\cdot\nabla_\sigma^\mu\pi)\Delta\sigma_l, \tag{22.3.5}$$

$$\Phi_k - \Phi_{k-1} = C_pa_kT_k + C_pb_{k-1}T_{k-1}, \tag{22.3.6}$$

$$\Phi_1 - \hat{\Phi}_{\frac{1}{2}} = C_pa_1T_1, \tag{22.3.7}$$

where

$$\sigma_k = \left[\frac{1}{\kappa+1}\left(\frac{\hat{\sigma}_{k-\frac{1}{2}}^{\kappa+1} - \hat{\sigma}_{k+\frac{1}{2}}^{\kappa+1}}{\hat{\sigma}_{k-\frac{1}{2}} - \hat{\sigma}_{k+\frac{1}{2}}}\right)\right]^{\frac{1}{\kappa}}, \tag{22.3.8}$$

$$a_k = \left(\frac{\hat{\sigma}_{k-\frac{1}{2}}}{\sigma_k}\right)^\kappa - 1, \tag{22.3.9}$$

$$b_k = 1 - \left(\frac{\hat{\sigma}_{k+\frac{1}{2}}}{\sigma_k}\right)^\kappa, \tag{22.3.10}$$

$$\kappa_k = \frac{1}{\Delta\sigma_k}\left(a_k\hat{\sigma}_{k-\frac{1}{2}} + b_k\hat{\sigma}_{k+\frac{1}{2}}\right), \tag{22.3.11}$$

$$\hat{T}_{k-\frac{1}{2}} = a_k\left[1 - \left(\frac{\sigma_k}{\sigma_{k-1}}\right)^\kappa\right]^{-1}T_k + b_{k-1}\left[\left(\frac{\sigma_{k-1}}{\sigma_k}\right)^\kappa - 1\right]^{-1}T_{k-1}. \tag{22.3.12}$$

References and suggested reading

This chapter closely follows Arakawa and Suarez (1983). Besides the Lorenz grid described in this chapter, the *Charney-Phillips grid* is also used for numerical models. A similar vertical discretization that guarantees conservation has been proposed (Arakawa and Moorthi, 1988; Arakawa and Konor, 1996).

Arakawa, A. and Suarez, M. J., 1983: Vertical differencing of the primitive equations in sigma coordinates. *Mon. Wea. Rev.*, **111**, 34–45.

Arakawa, A. and Moorthi, S., 1988: Baroclinic instability in vertically discrete system. *J. Atmos. Sci.*, **45**, 1688–1707.

Arakawa, A. and Konor, C., 1996: Vertical differencing of the primitive equations based on the Charney-Phillips grid in hybrid sigma-p vertical coordinates. *Mon. Wea. Rev.*, **124**, 511–528.

23

Time integration

The time integration method of spectral general circulation models is described in this chapter. We start with a general concept of time integration schemes in Section 23.1. In particular, we concentrate mainly on the *leapfrog scheme* for temporal integration and its stability analysis.

To temporally integrate a numerical model, one must be careful about the fast motions or waves contained in the numerical system. The fast waves contained in primitive equations are gravity waves and Lamb waves. If one uses an explicit scheme, the time increment for integration is constrained by such fast waves. Instead, if one uses an implicit scheme, the constraint is relaxed. General circulation models normally adopt the semi-implicit scheme in which only the terms related to such fast waves are implicitly treated. In Section 23.2, we outline the concept of the semi-implicit scheme.

In general, one needs to solve a Poisson equation in order to use the semi-implicit scheme. The Poisson equation can be easily solved under the spectral method, while it requires a matrix inversion for the finite discrete method. In our approach, physical fields are expanded with the spectral method on a sphere and discretized in the vertical direction. Thus, the semi-implicit method requires a one-dimensional matrix inversion in the vertical. To construct a matrix, we need to make use of the discretized expressions derived in the previous chapter. In Section 23.3, the time integration procedure with the semi-implicit method for primitive equation models is explicitly described.

23.1 Time integration scheme and stability

Time discretization schemes can be classified from various perspectives: such as the explicit scheme versus the implicit scheme, the recursive scheme versus the nonrecursive scheme, and the two-level scheme versus the three-level scheme. In general, what is required for the time discretization method of atmospheric general circulation models is efficiency rather than accuracy. Here, the scheme is called efficient if the number of computational counts required for a given integration time is smaller. The choice of time discretization scheme depends on the nature of the time variability of atmospheric circulation. Since various physical processes, particularly the latent heat release associated with cumulus convection, are highly nonlinear, atmospheric general circulation inherently has discontinuity as its time derivative. This suggests that higher order accuracy in the direction of time is not an important requirement in some circumstances. This implies that recursive schemes, with which time integrations are repeated many times in one step, are not normally used for general circulation models. For instance, although the 4-th order Runge-Kutta scheme is a frequently used integration scheme for computational fluid dynamics, it is not usually adopted for general circulation models.

The three-level leapfrog scheme is one of the most generally used time integration schemes in general circulation models. Such basic dynamics of the governing equations as advection or oscillation are discretized in time based on the leapfrog scheme. In this case, we need a numerical filter to suppress a spurious oscillation called the numerical mode. In contrast, dissipation or frictional terms are treated by using the forward scheme. In this section we briefly describe the stability of these time discretization schemes using simple examples.

23.1.1 Wave equation

We first consider an ordinary differential equation for oscillatory motion of the form

$$\frac{dF(t)}{dt} = i\omega F(t). \qquad (23.1.1)$$

To discretize this in time, we use the values of F at three time levels $n-1$, n, and $n+1$, which are denoted by

$$F^{n-1} = F(t-\Delta t), \quad F^{n} = F(t), \quad \text{and} \quad F^{n+1} = F(t+\Delta t), \qquad (23.1.2)$$

respectively, where Δt is the time increment. A general form of the three-level discretization scheme of (23.1.1) is written as

$$\frac{F^{n+1}-F^{n-1}}{2\Delta t} = i\omega(aF^{n+1}+bF^{n}+cF^{n-1}), \qquad (23.1.3)$$

where $a+b+c=1$ has to be satisfied for the scheme to be appropriate. In the case of $a=0$, (23.1.3) can be directly integrated for F^{n+1}: this is called the *explicit scheme*. In the case of $a \neq 0$, on the other hand, F^{n+1} appears both on the left and right-hand sides. This equation cannot be solved in the above form, so that some manipulations are required to solve for F^{n+1}. This is called the *implicit scheme*.

The stability of the scheme can be investigated using the *amplification factor*:

$$\lambda = \frac{F^{n+1}}{F^{n}}. \qquad (23.1.4)$$

Generally, λ takes a complex value. In order for the scheme to be stable, $|\lambda| \leq 1$ must be satisfied. In the following arguments, we study the stability of some special cases by setting

$$\nu = \omega\Delta t, \tag{23.1.5}$$

which corresponds to the *Courant number* of partial differential equations.

The first case is $a = b = 0$ and $c = 1$. This is equivalent to the two-level forward scheme with the time step $2\Delta t$. Substituting (23.1.4) into (23.1.3) gives

$$\lambda^2 = 2i\nu + 1 \tag{23.1.6}$$

(i.e., $|\lambda| > 1$). The forward scheme is always unstable for the oscillation equation.

The next case is $a = 1$ and $b = c = 0$. This scheme is backward and implicit. The amplification factor satisfies

$$\lambda^2 = \frac{1}{1 - 2i\nu}, \tag{23.1.7}$$

which implies $|\lambda| < 1$; thus the scheme is always stable (i.e., the time step $2\Delta t$ can be arbitrarily chosen regardless of ω). The fact that there is no constraint on the time step is a general characteristic of the implicit scheme.

Another special case of the implicit scheme with $b = 0$ and $a = c = 1/2$ is called the *trapezoidal scheme*. In this case, since

$$\frac{F^{n+1} - F^{n-1}}{2\Delta t} = i\omega\frac{F^{n+1} + F^{n-1}}{2}, \tag{23.1.8}$$

we obtain using (23.1.4)

$$\lambda^2 - 1 = i\nu(\lambda^2 + 1),$$

that is,

$$\lambda^2 = \frac{1 + i\nu}{1 - i\nu} = \frac{(1 - \nu^2) + 2i\nu}{1 + \nu^2}. \tag{23.1.9}$$

Therefore, the amplification factor satisfies

$$|\lambda^2| = 1. \tag{23.1.10}$$

The trapezoidal scheme is also always stable for all values of ν, similar to the backward scheme.

Finally, the *leapfrog scheme* is the case when $a = c = 0$ and $b = 1$. In this case, we have

$$\frac{F^{n+1} - F^{n-1}}{2\Delta t} = i\omega F^n. \tag{23.1.11}$$

Substituting (23.1.4) into this yields

$$\lambda^2 - 1 = 2i\nu\lambda.$$

Thus,

$$\lambda \;=\; i\nu \pm \sqrt{1-\nu^2}. \tag{23.1.12}$$

From this, we find that the scheme is stable only if $|\nu| \leq 1$.

Among the above schemes, only the leapfrog scheme is the three-level scheme that connects the values of the three time steps F^{n+1}, F^n, and F^{n-1}. From (23.1.12), we have $\lambda = \pm 1$ as $\nu \to 0$. The discretized solution for $\lambda = -1$ oscillates with successive steps in time, and is clearly different from the true solution; this solution is called the *numerical mode.* It does not approach the true solution even if $\Delta t \to 0$. This is a general characteristic of three-level schemes. Even for the general case of three-level schemes, the right-hand side of $\Delta t\times$ (23.1.3) approaches zero as $\nu \to 0$, and we have $\lambda = \pm 1$ in this limit; thus it contains the numerical mode that corresponds to $\lambda = -1$. The numerical mode is derived from the fact that even step values and odd step values cannot be related. In the case of the two-time level scheme, on the other hand, both the forward and backward schemes do not have the numerical mode.

23.1.2 Damping equation

The ordinary differential equation for decaying motion is given by

$$\frac{dF(t)}{dt} \;=\; -\kappa F(t), \tag{23.1.13}$$

where $\kappa > 0$. In the same way as (23.1.3), this equation is discretized with the three-level scheme as

$$\frac{F^{n+1} - F^{n-1}}{2\Delta t} \;=\; -\kappa(aF^{n+1} + bF^n + cF^{n-1}), \tag{23.1.14}$$

where $a + b + c = 1$. We will examine the stability of the schemes by setting

$$\alpha \;=\; \kappa\Delta t. \tag{23.1.15}$$

The above scheme is reduced to the two-time level forward scheme if $a = b = 0$ and $c = 1$. In this case, the amplification factor is given by

$$\lambda^2 \;=\; 1 - 2\alpha. \tag{23.1.16}$$

This scheme is stable if $0 \leq \alpha \leq 1$ is satisfied. The scheme becomes the backward implicit one when $a = 1$ and $b = c = 0$. In this case, the amplification factor is given by

$$\lambda^2 \;=\; \frac{1}{1+2\alpha}. \tag{23.1.17}$$

This scheme is always stable regardless of α. The leapfrog scheme is given by setting $a = c = 0$ and $b = 1$. In this case, we have

$$\lambda^2 + 2\alpha\lambda - 1 \;=\; 0, \tag{23.1.18}$$

from which we obtain

$$\lambda = \alpha \pm \sqrt{\alpha^2 + 1}. \tag{23.1.19}$$

Thus, since a solution $\lambda > 1$ always exists, the leapfrog scheme is unstable for the damping equation. In general, therefore, the forward scheme or the backward scheme is used for the decaying equation.

23.1.3 Time filter

A three-level scheme such as the leapfrog scheme contains the numerical mode. One way of preventing the numerical mode is by using a time filter (Asselin, 1972). The filtered value at the n-th step $\overline{F^n}$ is given by

$$\overline{F^n} = F^n + \beta\left(\overline{F^{n-1}} - 2F^n + F^{n+1}\right), \tag{23.1.20}$$

where β is a smoothing parameter (a value around $\beta = 0.05$ is used in general). If the time filter is applied to the oscillation equation that is discretized with the leapfrog scheme, (23.1.11) becomes

$$F^{n+1} = \overline{F^{n-1}} + 2i\Delta t\omega F^n. \tag{23.1.21}$$

The effect of the time filter is analyzed by assuming that the ratio of filtered value to nonfiltered value is constant irrespective of time steps:

$$\frac{\overline{F^n}}{F^n} = \frac{\overline{F^{n-1}}}{F^{n-1}} = X. \tag{23.1.22}$$

Substituting (23.1.4) into (23.1.20) and (23.1.21), we obtain

$$X = 1 + \beta\left(\lambda^{-1}X - 2 + \lambda\right), \tag{23.1.23}$$

$$\lambda^2 = X + 2i\nu\lambda. \tag{23.1.24}$$

Eliminating X from these equations, we have the amplification factor λ by

$$\lambda = \beta + i\nu \pm \sqrt{(1-\beta)^2 - \nu^2}. \tag{23.1.25}$$

If we take $|\beta| < 1$, the amplification factor approaches $\lambda = 1$ or $-1 + 2\beta$ as $\nu \to 0$. Thus, the amplification factor for the numerical mode becomes $|\lambda| < 1$, which implies the scheme is stable.

23.2 The semi-implicit scheme

If an explicit time integration scheme is used, the time step of numerical models is constrained by the wave speed of the fastest mode in the system. The fastest mode contained in the primitive equation model is the external gravity wave mode, or Lamb waves, whose phase speed is about 300 m s^{-1}. The time step Δt must be short enough to satisfy the *Courant-Fredrichs-Lewy condition* (*CFL condition*) for such a fast wave in order that the model runs stably with the explicit scheme. In general, the CFL condition is given by

$$\frac{c\Delta t}{\Delta x} \leq 1, \tag{23.2.1}$$

where Δx is a grid interval and the left-hand side is called the *Courant number*. It is thought, however, that external gravity waves are not important for large-scale circulations of the atmosphere. It is not worth spending computational resources on using such a small time step for unimportant waves. If implicit time integration schemes are used, the time step can be extended without the CFL constraint. However, the fully implicit scheme, in which all tendency terms are treated implicitly, requires complicated manipulations and a lot of computational time in general. To diminish complexity, one can integrate implicitly only those terms related to fast modes. Other terms relating to slow motions can be integrated explicitly by using such a scheme as the leapfrog scheme. This time integration scheme is called the *semi-implicit scheme.*

23.2.1 Stability analysis

The concept of the semi-implicit scheme is clearly described using a simple ordinary differential equation:

$$\frac{dF(t)}{dt} = i\omega_s F(t) + i\omega_f F(t), \tag{23.2.2}$$

where ω_s and ω_f are real frequencies satisfying $\omega_s \ll \omega_f$. The frequencies ω_s and ω_f represent the slow and fast modes, respectively. If (23.2.2) is solved with the explicit scheme, the time step must satisfy $\omega_f \Delta t < 1$. Instead, let us use the implicit scheme only for the fast mode term $i\omega_f F$. If the trapezoidal implicit scheme is used, (23.2.2) is discretized as

$$\frac{F^{n+1} - F^{n-1}}{2\Delta t} = i\omega_s F^n + i\omega_f \frac{F^{n+1} + F^{n-1}}{2}. \tag{23.2.3}$$

This is a simple example of the semi-implicit scheme; only the second term of the tendency terms is treated implicitly.

Setting $\lambda = F^{n+1}/F^n$ in the same way as (23.1.4) and substituting it into (23.2.3), we obtain the amplification factor

$$\lambda = \frac{i\omega_s \Delta t \pm \sqrt{1 + \omega_f^2 \Delta t^2 - \omega_s^2 \Delta t^2}}{1 - i\omega_f \Delta t}. \tag{23.2.4}$$

From this, if the inside of the root is positive, which is satisfied as long as $\omega_s < \omega_f$, the amplitude is constant $|\lambda|^2 = 1$ (i.e., the scheme is neutral). For a general case including the case with $\omega_s \leq \omega_f$, the stability condition is given by

$$\omega_s^2 \Delta t^2 \leq 1 + \omega_f^2 \Delta t^2. \tag{23.2.5}$$

Thus, the time step Δt is constrained by the slow mode ω_s, while it is not constrained by the fast mode ω_f. Note that in (23.2.4) we have $\lambda = \pm 1$ as $\Delta t \to 0$. The solution that corresponds to $\lambda = -1$ is the numerical mode and will be predominant as time integration proceeds. This mode should be filtered out by using the time filter described in Section 23.1.3 or any other filtering method.

23.2.2 The shallow water model

Let us apply the semi-implicit scheme to the shallow water model, which is a reduced system of the primitive equation model. Through this procedure, it makes it easier to introduce the semi-implicit scheme to the primitive equation model. Shallow water equations are given by

$$\frac{\partial u}{\partial t} + u\frac{\partial u}{\partial x} + v\frac{\partial u}{\partial x} - fv + \frac{\partial \phi}{\partial x} = 0, \tag{23.2.6}$$

$$\frac{\partial v}{\partial t} + u\frac{\partial v}{\partial x} + v\frac{\partial v}{\partial x} + fu + \frac{\partial \phi}{\partial y} = 0, \tag{23.2.7}$$

$$\frac{\partial \phi}{\partial t} + u\frac{\partial \phi}{\partial x} + v\frac{\partial \phi}{\partial x} + \phi\left(\frac{\partial u}{\partial x} + \frac{\partial v}{\partial y}\right) = 0. \tag{23.2.8}$$

Here, we define $\phi = gh$ and assume that the Coriolis parameter f is constant. Only inertio-gravity waves exist as waves in this system. As the effect of rotation becomes smaller, the characteristics of inertio-gravity waves get closer to those of pure gravity waves, and the phase speed becomes faster and approaches $c = \sqrt{\phi}$. Assuming that the flow speed is much slower than c, we introduce the *semi-implicit scheme* by discretizing the terms related to the pure gravity waves implicitly (i.e., the pressure gradient force term and the divergence term of mass). If the trapezoidal scheme is used for these two terms, the discretized equations become

$$\delta_t u + \left(u\frac{\partial u}{\partial x} + v\frac{\partial u}{\partial x} - fv\right)^n + \frac{\partial}{\partial x}\overline{\phi}^t = 0, \tag{23.2.9}$$

$$\delta_t v + \left(u\frac{\partial v}{\partial x} + v\frac{\partial v}{\partial x} + fu\right)^n + \frac{\partial}{\partial y}\overline{\phi}^t = 0, \tag{23.2.10}$$

$$\delta_t \phi + \left(u\frac{\partial \phi}{\partial x} + v\frac{\partial \phi}{\partial x}\right)^n + \phi\left(\frac{\partial}{\partial x}\overline{u}^t + \frac{\partial}{\partial y}\overline{v}^t\right) = 0, \tag{23.2.11}$$

where

$$\delta_t F = \frac{F^{n+1} - F^{n-1}}{2\Delta t}, \tag{23.2.12}$$

$$\overline{F}^t = \frac{F^{n+1} + F^{n-1}}{2}. \tag{23.2.13}$$

Let us consider the stability of linearized equations (23.2.9)–(23.2.11) in the case of no rotation, $f = 0$. Letting the velocity components of the basic field be denoted by U and V, the depth by H, and $c_0^2 \equiv gH$, we rewrite the equation set as

$$\delta_t u + \left(U\frac{\partial u}{\partial x} + V\frac{\partial u}{\partial x}\right)^n + \frac{\partial}{\partial x}\overline{\phi}^t = 0, \qquad (23.2.14)$$

$$\delta_t v + \left(U\frac{\partial v}{\partial x} + V\frac{\partial v}{\partial x}\right)^n + \frac{\partial}{\partial y}\overline{\phi}^t = 0, \qquad (23.2.15)$$

$$\delta_t \phi + \left(U\frac{\partial \phi}{\partial x} + V\frac{\partial \phi}{\partial x}\right)^n + c_0^2\left(\frac{\partial}{\partial x}\overline{u}^t + \frac{\partial}{\partial y}\overline{v}^t\right) = 0. \qquad (23.2.16)$$

By setting

$$\begin{pmatrix} u \\ v \\ \phi \end{pmatrix} = \begin{pmatrix} \hat{u}(t) \\ \hat{v}(t) \\ \hat{\phi}(t) \end{pmatrix} \exp(ikx + ily) \equiv \boldsymbol{X}(t)\exp(ikx + ily) \qquad (23.2.17)$$

and substituting this into (23.2.14)–(23.2.16), we have a vector-form equation

$$\delta_t \boldsymbol{X} = -i(kU + lV)\boldsymbol{X}^n - iA\overline{\boldsymbol{X}}^n, \qquad (23.2.18)$$

where A is a matrix given by

$$A = \begin{pmatrix} 0 & 0 & k \\ 0 & 0 & l \\ kc_0^2 & lc_0^2 & 0 \end{pmatrix}.$$

Let λ be an eigenvalue of A, which satisfies

$$det|\lambda I - A| = \begin{vmatrix} \lambda & 0 & k \\ 0 & \lambda & l \\ kc_0^2 & lc_0^2 & \lambda \end{vmatrix} = \lambda^3 - (k^2 + l^2)c_0^2\lambda = 0,$$

where I is the identity matrix. From this, we have three eigenvalues λ_j $(j = 1, 2, 3)$ by

$$\lambda = 0, \ \pm\sqrt{c_0^2(k^2 + l^2)}. \qquad (23.2.19)$$

The matrix that consists of the corresponding eigenvectors P satisfies

$$P^{-1}AP = \begin{pmatrix} \lambda_1 & 0 & 0 \\ 0 & \lambda_2 & 0 \\ 0 & 0 & \lambda_3 \end{pmatrix}. \qquad (23.2.20)$$

Thus, if we define

$$\boldsymbol{Y} = P^{-1}\boldsymbol{X}, \qquad (23.2.21)$$

and $\boldsymbol{Y} = (Y_1, Y_2, Y_3)^T$, (23.2.18) is rewritten as

$$\delta_t Y_j^n = -i(kU + lV)Y_j^n - i\lambda_j\overline{Y_j}^t, \quad \text{for} \quad j = 1, 2, 3. \qquad (23.2.22)$$

This equation is in the same form as (23.2.3). Therefore, from (23.2.5), the stability condition of (23.2.22) is given by

$$(kU + lV)^2 \Delta t^2 \quad \leq \quad 1 + \min[c_0^2(k^2 + l^2)\Delta t^2, 0],$$

that is,

$$(kU + lV)^2 \Delta t^2 \quad \leq \quad 1. \tag{23.2.23}$$

This means that the time step Δt is constrained only by the magnitude of the flow speed, but not by the phase speed of gravity waves.

23.3 The semi-implicit method for the primitive equation model

23.3.1 Introduction of the semi-implicit scheme

To introduce the semi-implicit scheme to the primitive equation model, we begin by summarizing primitive equations without external forcing. The equations are given by (20.2.45), (20.2.51), (20.2.57), (20.2.58), and (20.2.59) in Chapter 20, and (22.1.6) in Chapter 22:

$$\frac{\partial \zeta}{\partial t} = -\left[\frac{1}{R(1-\mu^2)}\frac{\partial}{\partial \lambda}A' + \frac{1}{R}\frac{\partial}{\partial \mu}B'\right] - 2\Omega\left(\mu D + \frac{V}{R}\right), \tag{23.3.1}$$

$$\begin{aligned}\frac{\partial D}{\partial t} &= \frac{1}{R(1-\mu^2)}\frac{\partial}{\partial \lambda}B' - \frac{1}{R}\frac{\partial}{\partial \mu}A' + 2\Omega\left(\mu\zeta - \frac{U}{R}\right)\\ &\quad -\nabla_\sigma^2(\Phi' + R_d\overline{T}\pi + E_{kin}),\end{aligned} \tag{23.3.2}$$

$$\begin{aligned}\frac{\partial T}{\partial t} &= -\left[\frac{1}{R(1-\mu^2)}\frac{\partial}{\partial \lambda}(UT') + \frac{1}{R}\frac{\partial}{\partial \mu}(VT')\right]\\ &\quad +T'D - \dot{\sigma}\sigma^{\kappa}\frac{\partial}{\partial \sigma}(T\sigma^{-\kappa}) + \kappa T\left(\frac{\partial \pi}{\partial t} + \boldsymbol{V}_H \cdot \nabla_\sigma^\mu \pi\right),\end{aligned} \tag{23.3.3}$$

$$\frac{\partial \pi}{\partial t} = -\int_0^1 (\boldsymbol{V}_H \cdot \nabla_\sigma^\mu \pi)\, d\sigma - \int_0^1 D d\sigma, \tag{23.3.4}$$

$$\dot{\sigma} = -\sigma\frac{\partial \pi}{\partial t} - \int_0^\sigma (D + \boldsymbol{V}_H \cdot \nabla_\sigma^\mu \pi) d\sigma, \tag{23.3.5}$$

$$0 = \frac{\partial \Phi}{\partial \sigma} + \frac{R_d T}{\sigma}, \tag{23.3.6}$$

where

$$A' = \zeta U + \dot{\sigma}\frac{\partial V}{\partial \sigma} + \frac{R_d T'}{R}(1-\mu^2)\frac{\partial \pi}{\partial \mu}, \tag{23.3.7}$$

$$B' = \zeta V - \dot{\sigma}\frac{\partial U}{\partial \sigma} - \frac{R_d T'}{R}\frac{\partial \pi}{\partial \lambda}, \tag{23.3.8}$$

$$E_{kin} = \frac{U^2 + V^2}{2(1-\mu^2)}, \tag{23.3.9}$$

$$U = -\frac{(1-\mu^2)}{R}\frac{\partial \psi}{\partial \mu} + \frac{1}{R}\frac{\partial \xi}{\partial \lambda}, \tag{23.3.10}$$

$$V = \frac{1}{R}\frac{\partial \psi}{\partial \lambda} + \frac{(1-\mu^2)}{R}\frac{\partial \xi}{\partial \mu}, \tag{23.3.11}$$

and $\zeta = \nabla_\sigma^2 \psi$, $D = \nabla_\sigma^2 \xi$. The overbar values $\overline{\Phi}$ and $\overline{T}$ are functions of σ and constant on each of σ surfaces. One may use average temperatures on σ surfaces to represent $\overline{T}$ (or the spectral coefficient T_0^0). However, it is more appropriate in terms of stability to use a constant value, say $\overline{T} = 300$ K, irrespective of height; otherwise, it causes a numerical instability known as the SHB (Simmons-Hoskins-Burridge) instability (Simmons et al. 1978). Linearizing (23.3.1)–(23.3.5) about a static atmosphere as a basic state, we rewrite the equation set as

$$\frac{\partial \zeta}{\partial t} = -2\Omega\mu D - \frac{2\Omega}{R}\left[\frac{\partial \psi}{\partial \lambda} + (1-\mu^2)\frac{\partial \xi}{\partial \mu}\right], \tag{23.3.12}$$

$$\frac{\partial D}{\partial t} = 2\Omega\mu\zeta - \frac{2\Omega}{R}\left[-(1-\mu^2)\frac{\partial \psi}{\partial \mu} + \frac{\partial \xi}{\partial \lambda}\right] \underline{- \nabla_\sigma^2(\Phi' + R_d\overline{T}\pi)}, \tag{23.3.13}$$

$$\frac{\partial T}{\partial t} = \underline{-\dot{\sigma}\sigma^\kappa \frac{\partial}{\partial \sigma}(\overline{T}\sigma^{-\kappa})} + \underline{\kappa\overline{T}\frac{\partial \pi}{\partial t}}, \tag{23.3.14}$$

$$\frac{\partial \pi}{\partial t} = \underline{-\int_0^1 D d\sigma}, \tag{23.3.15}$$

$$\dot{\sigma} = \underline{-\sigma\frac{\partial \pi}{\partial t}} - \underline{\int_0^\sigma D d\sigma}, \tag{23.3.16}$$

$$0 = \underline{\frac{\partial \Phi}{\partial \sigma}} + \underline{\frac{R_d T}{\sigma}}, \tag{23.3.17}$$

where the terms related to the gravity waves are denoted by underlines. Considering this linearization, we divide the terms of the original equations into those related to gravity waves and other terms. In the following equations, the terms related to the gravity waves are explicitly written and other terms are abbreviated by capital symbols:

$$\frac{\partial \zeta}{\partial t} = E - \mathcal{D}_M\zeta, \tag{23.3.18}$$

$$\frac{\partial D}{\partial t} = -\nabla_\sigma^2(\Phi + R_d\overline{T}\pi) + F - \mathcal{D}_M D, \tag{23.3.19}$$

$$\frac{\partial T}{\partial t} = \left[-\dot{\sigma_D}\sigma^\kappa\frac{\partial}{\partial \sigma}(\overline{T}\sigma^{-\kappa}) + \kappa\overline{T}\frac{\partial \pi_D}{\partial t}\right] + H - \mathcal{D}_H T, \tag{23.3.20}$$

$$\frac{\partial \pi}{\partial t} = \frac{\partial \pi_A}{\partial t} + \frac{\partial \pi_D}{\partial t} = Z + \frac{\partial \pi_D}{\partial t}, \tag{23.3.21}$$

$$Z \equiv \frac{\partial \pi_A}{\partial t} = -\int_0^1 (\boldsymbol{V}_H \cdot \nabla_\sigma^\mu \pi)\, d\sigma, \tag{23.3.22}$$

$$\frac{\partial \pi_D}{\partial t} = -\int_0^1 D d\sigma, \tag{23.3.23}$$

$$\dot{\sigma} = \dot{\sigma}_A + \dot{\sigma}_B, \tag{23.3.24}$$

$$\dot{\sigma}_A = -\sigma\frac{\partial \pi_A}{\partial t} - \int_0^\sigma (\boldsymbol{V}_H \cdot \nabla_\sigma^\mu \pi) d\sigma, \tag{23.3.25}$$

$$\dot{\sigma}_D = -\sigma\frac{\partial \pi_D}{\partial t} - \int_0^\sigma D d\sigma, \tag{23.3.26}$$

$$0 = \frac{\partial \Phi}{\partial \sigma} + \frac{R_d T}{\sigma}, \tag{23.3.27}$$

where the nongravity wave terms are defined as

$$E = -\left[\frac{1}{R(1-\mu^2)}\frac{\partial}{\partial \lambda}A' + \frac{1}{R}\frac{\partial}{\partial \mu}B'\right] - 2\Omega\left(\mu D + \frac{V}{R}\right), \tag{23.3.28}$$

$$F = \left[\frac{1}{R(1-\mu^2)}\frac{\partial}{\partial \lambda}B' - \frac{1}{R}\frac{\partial}{\partial \mu}A'\right] + 2\Omega\left(\mu\zeta - \frac{U}{R}\right) - \nabla_\sigma^2 E_{kin}, \tag{23.3.29}$$

$$\begin{aligned} H = &-\left[\frac{1}{R(1-\mu^2)}\frac{\partial}{\partial \lambda}(UT') + \frac{1}{R}\frac{\partial}{\partial \mu}(VT')\right] + T'D \\ &-\dot{\sigma}_A\sigma^\kappa\frac{\partial}{\partial \sigma}(\overline{T}\sigma^{-\kappa}) - \dot{\sigma}\sigma^\kappa\frac{\partial}{\partial \sigma} + (T'\sigma^{-\kappa}) \\ &+\kappa T'\frac{\partial \pi_D}{\partial t} + \kappa T\left(\frac{\partial \pi_A}{\partial t} + \boldsymbol{V}_H \cdot \nabla_\sigma^\mu \pi\right). \end{aligned} \tag{23.3.30}$$

In (23.3.18)–(23.3.20) we added the horizontal diffusion terms $\mathcal{D}_M$ and $\mathcal{D}_H$, which are diffusion on σ surfaces of momentum and temperature, respectively. These terms are given by (21.7.17), (21.7.18), and (21.7.19).

23.3.2 Matrix expression of vertical discretization

The vertically discretized equations of the semi-implicit method can be expressed in matrix form (Hoskins and Simmons, 1975). Hereafter, a column vector consisting of K-components in the vertical levels is denoted by a bold letter. We use the following notations for a quantity defined at the integer levels A_k, and for a quantity defined at the half-integer levels $B_{k+\frac{1}{2}}$:

$$\boldsymbol{A} = (A_1, A_2, \cdots, A_K)^T, \tag{23.3.31}$$

$$\boldsymbol{B} = (B_{\frac{1}{2}}, B_{1+\frac{1}{2}}, \cdots, B_{K-\frac{1}{2}})^T. \tag{23.3.32}$$

The top level value $B_{K+\frac{1}{2}}$ is given by additional conditions. A square matrix with $K \times K$-components is denoted by an underline (_). Using these notations, (23.3.23), (23.3.26), and (23.3.27) are expressed using a coefficient vector $\boldsymbol{C}$ and coefficient matrices $\underline{S}$ and $\underline{W}$ as

$$\frac{\partial \pi_D}{\partial t} = -\boldsymbol{C} \cdot \boldsymbol{D}, \tag{23.3.33}$$

$$\dot{\boldsymbol{\sigma}}_D = \underline{S}\,\boldsymbol{D}, \tag{23.3.34}$$

$$\boldsymbol{\Phi} - \boldsymbol{\Phi}_s = \underline{W}\,\boldsymbol{T}. \tag{23.3.35}$$

The first term on the right-hand side of (23.3.20) can be written as

$$\left[\dot{\sigma}_D \sigma^\kappa \frac{\partial}{\partial \sigma}(\overline{T}\sigma^{-\kappa})\right]_k = \left(\underline{Q}\,\dot{\boldsymbol{\sigma}}_D\right)_k = \left(\underline{Q}\,\underline{S}\,\boldsymbol{D}\right)_k, \tag{23.3.36}$$

where $\underline{Q}$ is a coefficient matrix. $\boldsymbol{D}$, $\dot{\boldsymbol{\sigma}}$, $\boldsymbol{\Phi}$, and $\boldsymbol{T}$ are vectors whose components are D_k, $\dot{\sigma}_{k+\frac{1}{2}}$, Φ_k, and T_k, respectively. The vector $\boldsymbol{\Phi}_s$ consists of the geopotential height at the surface $\Phi_{s,k} = \hat{\Phi}_{\frac{1}{2}}$. The coefficient vector $\boldsymbol{C}$ and the coefficient matrices $\underline{S}$ and $\underline{W}$ are independent of time. Although the matrix $\underline{Q}$ contains $\overline{T}$, it is also independent of time if $\overline{T}$ is chosen as a constant value irrespective of time to prevent the SHB instability.

Expressions of $\boldsymbol{C}$, $\underline{S}$, $\underline{W}$, and $\underline{Q}$ can be given by using the vertical discretization described in Chapter 22. First, since (23.3.23) is written as (22.2.6):

$$\frac{\partial}{\partial t}\pi_D = -\sum_{k=1}^{K} D_k \Delta\sigma_k, \tag{23.3.37}$$

from which we obtain (23.3.33) by setting

$$C_k = \Delta\sigma_k. \tag{23.3.38}$$

Second, from (22.2.7), Eq. (23.3.26) is written as

$$\dot{\sigma}_{D\,k-\frac{1}{2}} = -\sigma_{k-\frac{1}{2}}\frac{\partial \pi_D}{\partial t} - \sum_{l=k}^{K} D_l \Delta\sigma_l. \tag{23.3.39}$$

Substituting (23.3.33) into (23.3.39) yields

$$\begin{aligned}\dot{\sigma}_{D\,k-\frac{1}{2}} &= -\sigma_{k-\frac{1}{2}}(-\boldsymbol{C}\cdot\boldsymbol{D}) - \sum_{l=k}^{K} D_l\Delta\sigma_l \\ &= \Big(\sigma_{k-\frac{1}{2}}\Delta\sigma_1, \cdots, \sigma_{k-\frac{1}{2}}\Delta\sigma_{k-1}, \sigma_{k-\frac{1}{2}}\Delta\sigma_k - \Delta\sigma_k, \\ &\qquad \cdots, \sigma_{k-\frac{1}{2}}\Delta\sigma_K - \Delta\sigma_K\Big)\,\boldsymbol{D}.\end{aligned} \tag{23.3.40}$$

Thus, we obtain (23.3.34) by setting the matrix $\underline{S}$ to

$$S_{kl} = \sigma_{k-\frac{1}{2}}\Delta\sigma_l - \Delta\sigma_l\delta_{k\le l} = \begin{cases} \sigma_{k-\frac{1}{2}}\Delta\sigma_l & ;\quad k < l, \\ \sigma_{k-\frac{1}{2}}\Delta\sigma_l - \Delta\sigma_l & ;\quad k \ge l, \end{cases} \tag{23.3.41}$$

where $\delta_{k\le l}$ is 1 if $k \le l$ and is 0 otherwise. The above equation is for the levels $k = 0, \cdots, K-1$. We also have $\dot{\sigma}_{D,K+\frac{1}{2}} = 0$ at the top boundary $k = K$.

Third, we find the discretized form of the hydrostatic equation (23.3.27) using (22.2.61):

$$\Phi_1 - \hat{\Phi}_{\frac{1}{2}} = C_p a_1 T_1, \tag{23.3.42}$$

$$\Phi_k - \Phi_{k-1} = C_p a_k T_k + C_p b_{k-1} T_{k-1}, \tag{23.3.43}$$

where

$$a_k = \left(\frac{\hat{\sigma}_{k-\frac{1}{2}}}{\sigma_k}\right)^{\kappa} - 1, \qquad b_k = 1 - \left(\frac{\hat{\sigma}_{k+\frac{1}{2}}}{\sigma_k}\right)^{\kappa}. \tag{23.3.44}$$

Eqs. (23.3.42) and (23.3.43) can be combined into

$$\begin{aligned}\Phi_k - \hat{\Phi}_{\frac{1}{2}} &= C_p \sum_{l=1}^{k} a_l T_l + C_p \sum_{l=1}^{k-1} b_l T_l = C_p \sum_{l=1}^{k-1} (a_l + b_l) T_l + C_p a_k T_k \\ &= C_p (a_1 + b_1, a_2 + b_2, \cdots, a_{k-1} + b_{k-1}, a_k, 0, \cdots, 0) \boldsymbol{T}.\end{aligned} \tag{23.3.45}$$

Thus, we obtain (23.3.35) by setting the matrix $\underline{W}$ to

$$W_{kl} = C_p a_l \delta_{k \geq l} + C_p b_l \delta_{k-1 \geq l} = \begin{cases} C_p(a_k + b_k) & ; \quad k > l, \\ C_p a_k & ; \quad k = l, \\ 0 & ; \quad k < l. \end{cases} \tag{23.3.46}$$

Finally, since (22.2.79) corresponds to (22.2.78), the first term on the right-hand side of the thermodynamic equation (23.3.20) can be written as

$$\begin{aligned}\left[\dot{\sigma}_D \sigma^{\kappa} \frac{\partial(\overline{T}\sigma^{-\kappa})}{\partial \sigma}\right]_k &= \frac{1}{\Delta \sigma_k} \left\{ \dot{\sigma}_{D,k-\frac{1}{2}} \left[\hat{\overline{T}}_{k-\frac{1}{2}} - (1 + a_k)\overline{T}_k \right] \right. \\ &\quad \left. + \dot{\sigma}_{D,k+\frac{1}{2}} \left[(1 - b_k)\overline{T}_k - \hat{\overline{T}}_{k+\frac{1}{2}} \right] \right\}.\end{aligned} \tag{23.3.47}$$

Thus, we obtain the discretized form of (23.3.36) as

$$\begin{aligned}Q_{kl} &= \frac{1}{\Delta \sigma_k} \left\{ \left[\hat{\overline{T}}_{k-\frac{1}{2}} - (1 + a_k)\overline{T}_k \right] \delta_{kl} + \left[(1 - b_k)\overline{T}_k - \hat{\overline{T}}_{k+\frac{1}{2}} \right] \delta_{k,l+1} \right\} \\ &= \begin{cases} \frac{1}{\Delta \sigma_k} \left[\hat{\overline{T}}_{k-\frac{1}{2}} - (1 + a_k)\overline{T}_k \right] & ; \quad k = l, \\ \frac{1}{\Delta \sigma_k} \left[(1 - b_k)\overline{T}_k - \hat{\overline{T}}_{k+\frac{1}{2}} \right] & ; \quad k = l + 1, \\ 0 & ; \quad \text{otherwise.} \end{cases}\end{aligned} \tag{23.3.48}$$

23.3.3 The method for solving the semi-implicit scheme

The matrix form of the vertically discretized equations (23.3.18)–(23.3.27) is summarized as

$$\frac{\partial}{\partial t}\boldsymbol{\zeta} = \boldsymbol{E} - \mathcal{D}_M \boldsymbol{\zeta}, \tag{23.3.49}$$

$$\frac{\partial}{\partial t}\boldsymbol{D} = -\nabla_\sigma^2(\boldsymbol{\Phi} + \boldsymbol{G}\pi) + \boldsymbol{F} - \mathcal{D}_M \boldsymbol{D}, \tag{23.3.50}$$

$$\frac{\partial}{\partial t}\boldsymbol{T} = -\underline{h}\,\boldsymbol{D} + \boldsymbol{H} - \mathcal{D}_H \boldsymbol{T}, \tag{23.3.51}$$

$$\frac{\partial}{\partial t}\pi = Z - \boldsymbol{C} \cdot \boldsymbol{D}, \tag{23.3.52}$$

$$\boldsymbol{\Phi} - \boldsymbol{\Phi}_s = \underline{W}\boldsymbol{T}, \tag{23.3.53}$$

where

$$G_k = C_p \kappa_k \overline{T}_k, \qquad \underline{h} = \underline{Q}\,\underline{S} - \frac{1}{C_p} \boldsymbol{G}\boldsymbol{C}^T. \tag{23.3.54}$$

Eqs. (23.3.50), (23.3.51), and (23.3.52) involve three different terms related to gravity waves, diffusion, and other processes. We discretize these equations in the time direction using the semi-implicit method:

$$\delta_t \boldsymbol{D} = -\nabla_\sigma^2 \left(\overline{\boldsymbol{\Phi}}^t + \boldsymbol{G}\overline{\pi}^t \right) + \boldsymbol{F}^t - \mathcal{D}_M \boldsymbol{D}^{t+\Delta t}, \tag{23.3.55}$$

$$\delta_t \boldsymbol{T} = -\underline{h}\,\overline{\boldsymbol{D}}^t + \boldsymbol{H}^t - \mathcal{D}_H \boldsymbol{T}^{t+\Delta t}, \tag{23.3.56}$$

$$\delta_t \pi = Z^t - \boldsymbol{C} \cdot \overline{\boldsymbol{D}}^t, \tag{23.3.57}$$

where the implicit scheme is used for gravity wave terms and the backward scheme is used for diffusion terms. We have defined the following notations:

$$\delta_t A = \frac{A^{t+\Delta t} - A^{t-\Delta t}}{2\Delta t}, \qquad \overline{A}^t = \frac{A^{t+\Delta t} + A^{t-\Delta t}}{2}. \tag{23.3.58}$$

From this, we also have

$$\overline{A}^t = A^{t-\Delta t} + \Delta t \delta_t A, \tag{23.3.59}$$

$$\delta_t A = \frac{\overline{A}^t - A^{t-\Delta t}}{\Delta t}, \tag{23.3.60}$$

$$A^{t+\Delta t} = 2\overline{A}^t - A^{t-\Delta t}. \tag{23.3.61}$$

The diffusion terms in the above equations are also implicit. Substituting the relation

$$A^{t+\Delta t} = A^{t-\Delta t} + 2\Delta t \delta_t A \tag{23.3.62}$$

into the terms proportional to $\mathcal{D}_M$, $\mathcal{D}_H$ in (23.3.55) and (23.3.56) to yield

$$(1 + 2\Delta t \mathcal{D}_M)\delta_t \boldsymbol{D} = -\nabla_\sigma^2 \left(\boldsymbol{\Phi}_s + \underline{W}\,\overline{\boldsymbol{T}}^t + \boldsymbol{G}\overline{\pi}^t \right) + \boldsymbol{F}^t - \mathcal{D}_M \boldsymbol{D}^{t-\Delta t}, \tag{23.3.63}$$

$$(1 + 2\Delta t \mathcal{D}_H)\delta_t \boldsymbol{T} = -\underline{h}\,\overline{\boldsymbol{D}}^t + \boldsymbol{H}^t - \mathcal{D}_H \boldsymbol{T}^{t-\Delta t}, \tag{23.3.64}$$

where (23.3.35) is used. By eliminating $\overline{\boldsymbol{T}}^t$ and $\overline{\pi}^t$ from (23.3.63) using (23.3.59) and substituting (23.3.64), we have

$$\begin{aligned}
&(1 + 2\Delta t \mathcal{D}_H)\,(1 + 2\Delta t \mathcal{D}_M)\,\delta_t \boldsymbol{D} \\
&= -(1 + 2\Delta t \mathcal{D}_H)\,\nabla_\sigma^2 \Big[\boldsymbol{\Phi}_s + \underline{W} \left(\boldsymbol{T}^{t-\Delta t} + \Delta t \delta_t \boldsymbol{T} \right) \\
&\qquad + \boldsymbol{G} \left(\pi^{t-\Delta t} + \Delta t \delta_t \pi \right) \Big] + \boldsymbol{F}^t - \mathcal{D}_M \boldsymbol{D}^{t-\Delta t}
\end{aligned}$$

$$
\begin{aligned}
= & -\nabla_\sigma^2 \Big\{ (1+2\Delta t \mathcal{D}_H)\, \boldsymbol{\Phi}_s + \underline{W} \Big[(1+2\Delta t \mathcal{D}_H)\, \boldsymbol{T}^{t-\Delta t} \\
& + \Delta t \left(-\underline{h}\, \overline{\boldsymbol{D}}^t + \boldsymbol{H}^t - \mathcal{D}_H \boldsymbol{T}^{t-\Delta t} \right) \Big] \\
& + (1+2\Delta t \mathcal{D}_H)\, \boldsymbol{G} \left[\pi^{t-\Delta t} + \Delta t \left(Z^t - \boldsymbol{C} \cdot \overline{\boldsymbol{D}}^t \right) \right] \Big\} \\
& + (1+2\Delta t \mathcal{D}_H) \left(\boldsymbol{F}^t - \mathcal{D}_M \boldsymbol{D}^{t-\Delta t} \right). \qquad (23.3.65)
\end{aligned}
$$

Using (23.3.60), therefore, we have the equation for $\overline{\boldsymbol{D}}^t$:

$$
\begin{aligned}
& \Big\{ (1+2\Delta t \mathcal{D}_H)(1+2\Delta t \mathcal{D}_M)\, \underline{I} - \left[\underline{W}\, \underline{h} + (1+2\Delta t \mathcal{D}_H)\, \boldsymbol{G}\, \boldsymbol{C}^T \right] \Delta t^2 \nabla_\sigma^2 \Big\} \overline{\boldsymbol{D}}^t \\
= & \ (1+2\Delta t \mathcal{D}_H)(1+\Delta t \mathcal{D}_M)\, \boldsymbol{D}^{t-\Delta t} + \Delta t\, (1+2\Delta t \mathcal{D}_H)\, \boldsymbol{F}^t \\
& -\Delta t \nabla_\sigma^2 \Big\{ (1+2\Delta t \mathcal{D}_H)\, \boldsymbol{\Phi}_s + \Delta t \underline{W} \left[(1+\Delta t \mathcal{D}_H)\, \boldsymbol{T}^{t-\Delta t} + \boldsymbol{H}^t \right] \\
& + (1+2\Delta t \mathcal{D}_H)\, \boldsymbol{G} \left(\pi^{t-\Delta t} + \Delta t Z^t \right) \Big\}, \qquad (23.3.66)
\end{aligned}
$$

where $\underline{I}$ is the identity matrix. If we expand (23.3.66) into the spectral form on the sphere, we can replace ∇_σ^2 by a factor $-\frac{n(n+1)}{R^2}$. Thus, all the expansion coefficients of $\overline{\boldsymbol{D}}^t$ can be solved as

$$
\begin{aligned}
\overline{(\boldsymbol{D}_n^m)}^t = & \ \underline{L}^{-1} (1+2\Delta t \mathcal{D}_H)(1+\Delta t \mathcal{D}_M)\, (\boldsymbol{D}_n^m)^{t-\Delta t} \\
& + \Delta t\, (1+2\Delta t \mathcal{D}_H)\, (\boldsymbol{F}_n^m)^t - \Delta t \nabla_\sigma^2 \Big\{ (1+2\Delta t \mathcal{D}_H)\, \boldsymbol{\Phi}_{s\,n}^{\ m} \\
& + \Delta t \underline{W} \left[(1+\Delta t \mathcal{D}_H)\, (\boldsymbol{T}_n^m)^{t-\Delta t} + (\boldsymbol{H}_n^m)^t \right] \\
& + (1+2\Delta t \mathcal{D}_H)\, \boldsymbol{G} \left((\pi_n^m)^{t-\Delta t} + \Delta t (Z_n^m)^t \right) \Big\}, \qquad (23.3.67)
\end{aligned}
$$

where

$$
\begin{aligned}
\underline{L} = & \Big\{ (1+2\Delta t \mathcal{D}_H)(1+2\Delta t \mathcal{D}_M)\, \underline{I} \\
& + \left[\underline{W}\, \underline{h} + (1+2\Delta t \mathcal{D}_H)\, \boldsymbol{G}\, \boldsymbol{C}^T \right] \Delta t^2 \frac{n(n+1)}{R^2} \Big\}. \qquad (23.3.68)
\end{aligned}
$$

The diffusion coefficients $\mathcal{D}_M$ and $\mathcal{D}_H$ are constant in the spectral domain.

Since we have obtained $\overline{(\boldsymbol{D}_n^m)}^t$ from the above procedure, we can calculate the values at $t+\Delta t$ using (23.3.61):

$$
(\boldsymbol{D}_n^m)^{t+\Delta t} = 2\overline{(\boldsymbol{D}_n^m)}^t - (\boldsymbol{D}_n^m)^{t-\Delta t}. \qquad (23.3.69)
$$

We also obtain the expansion coefficients of temperature and surface pressure using

(23.3.64) and (23.3.57) as

$$
\begin{aligned}
(\boldsymbol{T}_n^m)^{t+\Delta t} &= (1+2\Delta t \mathcal{D}_H)^{-1}\Big\{(\boldsymbol{T}_n^m)^{t-\Delta t} \\
&\quad +2\Delta t\left[-\underline{h}\,\overline{(\boldsymbol{D}_n^m)}^{t} + (\boldsymbol{H}_n^m)^t - \mathcal{D}_H(\boldsymbol{T}_n^m)^{t-\Delta t}\right]\Big\}, \qquad (23.3.70) \\
(\pi_n^m)^{t+\Delta t} &= (\pi_n^m)^{t-\Delta t} + 2\Delta t\left[(Z_n^m)^t - \boldsymbol{C}\cdot\overline{(\boldsymbol{D}_n^m)}^{t}\right]. \qquad (23.3.71)
\end{aligned}
$$

These constitute the time integration scheme with the semi-implicit method for expansion coefficients.

References and suggested reading

Mesinger and Arakawa (1976) give a fundamental argument on numerical schemes used in atmospheric models, including temporal schemes and grid arrangement. A detailed discussion on time integration schemes including the semi-implicit method is given in Durran (1998). The semi-implicit method for multilevel models is described in Hoskins and Simmons (1975). The details of matrix expression of this chapter follow Kanamitsu et al. (1983).

Asselin, R., 1972: Frequency filter for time integrations. *Mon. Wea. Rev.*, **100**, 487–490.

Durran, D. R., 1998: *Numerical Methods for Wave Equations in Geophysical Fluid Dynamics.* Springer-Verlag, New York, 465 pp.

Hoskins, B. J. and Simmons, A. J., 1975: A multi-layer spectral model and the semi-implicit method. *Q. J. Roy. Meteorol. Soc.*, **101**, 637–655.

Kanamitsu, M., Tada, K., Kudo, T., Sato, N., and Isa, S., 1983: Description of the JMA operational spectral model. *J. Meteorol. Soc. Japan*, **61**, 812–828.

Mesinger, F. and Arakawa, A., 1976: *Numerical Methods Used in Atmospheric Models*, GARP No. 170. WMO/ICSU Joint Organizing Committee, Geneva.

Simmons, A. J., Hoskins, B. J., and Burridge, D., 1978: Stability of the semi-implicit method of time integration. *Mon. Wea. Rev.*, **106**, 405–412.

24

Standard experiments of atmospheric general circulation models

Atmospheric general circulation models were originally developed for such practical applications as weather forecasts and climate prediction. They can also be used to investigate the basic properties of atmospheric circulations by performing numerical experiments in suitable conditions. This kind of usage is similar to certain kinds of laboratory experiments: baroclinic disturbances can be studied using a rotating annulus by changing the temperature difference between inner and outer annuli or rotation speed; and various aspects of atmospheric general circulation can also be examined using atmospheric general circulation models by artificially changing their governing parameters.

In this chapter, we describe some examples of atmospheric general circulation models by changing their numerical setups. Historically, the experiments shown below were designed for specific research into atmospheric dynamics. Once the behaviors of dynamics were well understood, these experiments became good case studies for nonlinear problems and could be used to study the appropriateness of numerical models. For example, the life cycle experiment of extratropical cyclones described in Section 18.3 was originally considered for the study of the nonlinear development of baroclinic instability (Simmons and Hoskins, 1978). As simulations of the development of cyclones can be reproduced by many other numerical models, this numerical simulation can be used as a bench mark test of the atmospheric general circulation models. The spectral method described in this book is only one type of numerical scheme of atmospheric general circulation models; such other types of dynamical framework as the grid method and the semi-Lagrangian method are also used for studying dynamic meteorology. The performances of different dynamical frameworks of atmospheric general circulation models and the effects of newly incorporated numerical schemes can be evaluated using such kinds of experiments if the behavior of the solutions is well understood. In this sense, these experiments are viewed as a standard set suitable for numerical models.

Of course, the meaning of "standard" depends on the purpose of the numerical experiment. Each experiment has its own perspective of atmospheric circulation. Even if one type of experiment is successfully simulated in a model, it is not proof of model appropriateness. Numerical models cannot be thought reliable until they are tested under various circumstances. If one runs a model with many kinds of experiments, they will reveal both the advantages and deficiencies of the model.

We have explained the dynamical part of the primitive equation model on a sphere in Chapters 20–23. In order to simulate realistic atmospheric circulation, we need to incorporate such physical processes as cumulus convection, the radiation process, and the turbulent process. Since global models do not have sufficient resolution for accurate representation of physical processes, many uncertainties are introduced to take account of effects of physical processes in a coarse grid scale. Cumulus parameterization is one of the main factors of model uncertainty. In order to discuss the performance of numerical models, therefore, it is better to evaluate different parts of the models. First, we need to discuss the performance of the dynamical part of the models by omitting all the physical processes. We also need to examine the performance of individual physical processes separately. The dynamical part of atmospheric general circulation models without physical processes is called the *dynamical core*. Standard experiments should include both types of experiments: for models with full physics and for the dynamical core model. In the first two sections, we describe examples of a *dynamical core experiment* proposed by Polvani et al. (2003) and Held and Suarez (1994). Then, we describe an *aqua-planet experiment* as an example of the model with full physics. The aqua-planet is an idealistic earth-like planet covered completely by ocean. These three experiments are highly idealized. Finally, we briefly describe the realistic experiments for model comparison called AMIP (Atmospheric Model Intercomparison Project).

24.1 Life cycle of extratropical cyclones experiment

We described the relation between the evolution of extratropical cyclones and meridional circulation in Section 18.3. If a disturbance is added to a zonally symmetric thermally balanced state with a jet stream, baroclinic waves grow: first, there is a linear developing stage and a nonlinear occluding stage then follows. Such an experiment can be used for testing the dynamical core of general circulation models. As an initial disturbance, Simmons and Hoskins (1978) and Thorncroft et al. (1993) applied the fastest growing linear solution to the initial zonal field. Such a mode is numerically calculated. In contrast, Polvani et. al. (2003) proposed an analytic function localized in longitude as an initial disturbance. Let us describe the simpler experimental setup proposed by Polvani et al. (2003).

The initial condition of Polvani et al. (2003) is a thermally balanced state between zonal winds and temperature with no meridional velocity $v = 0$ and constant surface pressure p_0. The zonal winds are given as

$$u(\varphi, p) = \begin{cases} u_0 \sin^3(\pi\mu^2) F(z), & \text{for} \quad \varphi > 0 \\ 0, & \text{for} \quad \varphi < 0, \end{cases} \tag{24.1.1}$$

where $\mu = \sin\varphi$, $z = -H\log(p/p_0)$, and

$$F(z) = \frac{1}{2}\left[1 - \tanh^3\left(\frac{z - z_0}{\Delta z}\right)\right]\sin\frac{\pi z}{z_1}. \quad (24.1.2)$$

Here, the constants used in the above formula are $u_0 = 50$ m s^{-1}, $z_0 = 22$ km, $\Delta z = 5$ km, $z_1 = 30$ km, $H = 7$ km, and $p_0 = 1{,}000$ hPa. The temperature field satisfies cyclostrophic balance and hydrostatic balance. From

$$(Rf + u\tan\varphi)u = -\frac{\partial\Phi}{\partial\varphi}, \qquad \frac{\partial\Phi}{\partial z} = \frac{R_d}{H}T, \quad (24.1.3)$$

cyclostrophic balance reads

$$\frac{\partial T}{\partial\varphi} = -\frac{H}{R_d}(Rf + 2u\tan\varphi)\frac{\partial u}{\partial z}, \quad (24.1.4)$$

in which $R = 6376$ km, $R_d = 287.04$ J kg^{-1} K^{-1}, $f = 2\Omega\sin\varphi$, and $\Omega = 7.292 \times 10^{-5}$ s^{-1}. Integrating (24.1.4) in latitude, we obtain

$$T(\varphi, p) = \int_0^{\varphi}\frac{\partial T(\varphi', z)}{\partial\varphi'}d\varphi' + T_0(z). \quad (24.1.5)$$

The temperature profile $T_0(z)$ is specified as the U.S. standard atmosphere (1976) at each p (see Section 14.1). Fig. 24.1 shows the initial zonal winds and potential temperature field in the meridional section. Initial perturbation is given to the temperature field in the following analytic form:

$$T'(\lambda, \varphi) = \hat{T}\mathrm{sech}^2\left(\frac{\lambda - \lambda_0}{\alpha}\right)\mathrm{sech}^2\left(\frac{\varphi - \varphi_0}{\beta}\right), \quad (24.1.6)$$

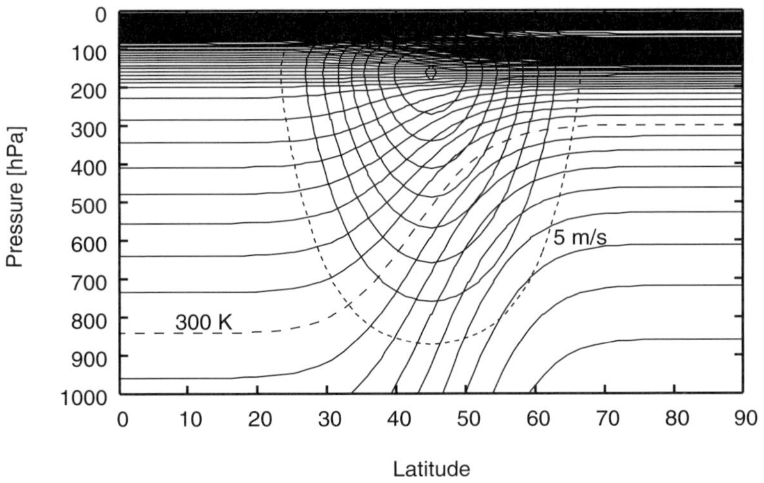

FIGURE 24.1: The initial potential temperature field and zonal wind field for extratropical cyclone experiments. The contour intervals are 5 K for potential temperature and 5 m s^{-1} for zonal winds.

where $\hat{T} = 1$ K, $\lambda_0 = 0$, $\varphi_0 = \pi/4$, $\alpha = 1/3$, and $\beta = 1/6$. The vertical profile of temperature perturbation is uniform.

Numerical diffusion is required for both the zonal wind and temperature fields. The diffusion terms are given as

$$\left(\frac{d\boldsymbol{v}}{dt}\right)_{dif} = (-1)^{n+1}\nu\nabla^{2n}\boldsymbol{v}, \qquad \left(\frac{dT}{dt}\right)_{dif} = (-1)^{n+1}\nu\nabla^{2n}T. \quad (24.1.7)$$

The order of hyperdiffusion n and the diffusion coefficient ν are specified according to the problems considered. In Polvani et al. (2003), $n = 2$ and $\nu = 2.5\times 10^{16}$ m^4 s^{-1} are used as a standard case. To show the convergence of numerical solutions, the value of the diffusion coefficient ν should remain constant regardless of the resolution of numerical models. In such a case, even though resolution increases, a finer structure will not be resolved in the model. In contrast, the diffusion coefficient should be reduced in order to represent a finer structure with higher resolution models. The timescale of diffusion is expressed as

$$\tau_{dif} = \frac{\lambda_{min}^4}{\nu}, \quad (24.1.8)$$

where λ is the smallest resolvable scale. Although there remains ambiguity for the interpretation of the resolvable scale in different model architectures, it is thought that the resolvable scale of the spectrum model with the largest wave number N is given by $\lambda_{min} = 2\pi R/N$, and that of the grid model with the grid increment Δx by $\lambda_{min} = 2\Delta x$. In general, as λ_{min} becomes half, τ_{dif} is set to be halved. As an example of numerical experiments, Fig. 24.2 shows the surface temperature distribution at day 8. This is a result from a very high resolution run of the icosahedral grid model with horizontal grid interval about $\Delta x = 3.5$ km (Tomita et al., 2003). Different from the original setup, the diffusion coefficient is set to $\nu = 9.54 \times 10^{10}$ m^4 s^{-1} which corresponds to the diffusion time 7.0 h, to represent finer structures. An occluding cyclone with spiral bands and a developing cyclone with an elongated cold front can be seen.

24.2 Held and Suarez experiment

We can reproduce a realistic general circulation by giving the realistic distributions of diabatic heating and frictional or dissipative forcing in general circulation models. Although the distributions of diabatic heating and forcing are determined by appropriate physical processes, one may obtain distributions similar to the real atmosphere by artificially prescribing their forms. The general circulation thus obtained is free from complicated physical processes and is thought to reflect the characteristics of the dynamical part of the models. In this section, we introduce a standard experimental set proposed by Held and Suarez (1994), with which the model can produce a climate state similar to that observed in some respects. In this experiment, diabatic heating and the dissipation process are given by simple arithmetic expressions. While the life cycle experiment described in the previous

FIGURE 24.2: The temperature distributions at the lowest level of the model for the extratropical cyclone experiment at day 8. The contour interval is 1 K. This result is obtained with a non-hydrostatic icosahedral grid atmospheric model (NICAM) with an average horizontal grid interval $\Delta x = 3.5$ km using the Earth Simulator. Courtesy of H. Tomita and K. Goto.

section is a test case for studying a short-range deterministic stage in the nonlinear development of extratropical cyclones of about 10 days, this Held and Suarez experiment is used for studying the statistical behavior of the atmosphere based on a log time simulation of about 1,000 days.

In the experimental setup of Held and Suarez, diabatic heating is expressed by Newtonian cooling and frictional forcing is expressed by Rayleigh friction. That is, diabatic heating and frictional forcing are respectively given by

$$Q = \left(\frac{\partial T}{\partial t}\right)_{diab} = -k_T(\varphi,\sigma)\left[T - T_{eq}(\varphi,p)\right], \tag{24.2.1}$$

$$\boldsymbol{F} = \left(\frac{\partial \boldsymbol{v}}{\partial t}\right)_{fric} = -k_v(\sigma)\boldsymbol{v}, \tag{24.2.2}$$

where φ is latitude, and altitude is denoted by the σ-coordinate. T_{eq} is a reference temperature of Newtonian cooling, given by

$$T_{eq}(\varphi,p) = \max\left\{200\text{ K}, \left[315\text{ K} - (\Delta T)_y \sin^2\varphi - (\Delta\theta)_z \ln\left(\frac{p}{p_0}\right)\cos^2\varphi\right]\left(\frac{p}{p_0}\right)^{\kappa}\right\}. \tag{24.2.3}$$

The relaxation coefficients k_T and k_v are expressed as

$$k_T(\varphi, \sigma) = k_a + (k_s - k_a) \max\left(0, \frac{\sigma - \sigma_b}{1 - \sigma_b}\right) \cos^4 \varphi, \tag{24.2.4}$$

$$k_v(\sigma) = k_f \max\left(0, \frac{\sigma - \sigma_b}{1 - \sigma_b}\right). \tag{24.2.5}$$

The following parameters are used: $\sigma_b = 0.7$, $k_f^{-1} = 1$ day, $k_a^{-1} = 40$ day, $k_s^{-1} = 4$ day, $(\Delta T)_y = 60$ K, and $(\Delta\theta)_z = 10$ K; and physical constants: surface pressure $p_0 = 1{,}000$ hPa, $\kappa = R_d/C_p = 2/7$, specific heat at constant pressure $C_p = 1004$ J kg^{-1} K^{-1}, rotational velocity $\Omega = 7.292 \times 10^{-5}$ s^{-1}, acceleration due to gravity $g = 9.8$ m s^{-2}, and radius of the earth $R = 6.371 \times 10^6$ m. At the surface boundary, no heat or momentum fluxes are allowed between the atmosphere and the surface.

Fig. 24.3 (a), (b) shows the reference temperature profile of Newtonian cooling T_{eq} and the corresponding profile of potential temperature. It is thought that T_{eq} is the temperature of radiative-convective equilibrium when no circulation exists. Thus, the reference temperature has stable stratification in the tropics, and becomes neutral as the latitude becomes higher. If meridional circulation is driven, the temperature at lower latitudes becomes colder than the reference temperature, while the temperature at mid and high latitudes becomes warmer than the reference temperature. As a result, diabatic heating is $Q > 0$ in the lower latitudes, which corresponds to latent heat release due to cumulus convection. In mid-latitudes, on the other hand, diabatic heating is $Q < 0$, which represents radiative cooling.

Since this experiment does not allow any sensible heat flux between the atmosphere and the surface, Newtonian heating (24.2.1) contains the effects of diabatic heating near the surface. As shown by (24.2.4), the relaxation time of heating gets smaller as the surface gets closer. If the relaxation time is constant irrespective of height, the temperature in the lower layer becomes colder due to advection of colder air from higher latitudes near the surface, such that stratification would become much more stable. Rayleigh friction is applied to the lower layer below the level σ_b. This effect represents surface friction. In the free atmosphere above σ_b, no explicit friction is given. It is left to individual models to use any other artificial friction or numerical damping to suppress numerical oscillations.

Under the conditions described above, a statistical equilibrium state is obtained after a sufficient amount of time integration, starting with an arbitrary initial condition. Fig. 24.3 (c), (d) shows the distributions of zonal-mean temperature and potential temperature for a 1,000-day time average. This shows that stratification is neutral in mid and high latitudes after a statistically equilibrium circulation is established. In particular, stratification is a little too strong in the lower layers. Fig. 24.4 also shows the meridional distribution of zonal-mean zonal winds. The subtropical jet has its maximum at around 40° in the upper layer. Zonal winds are easterly in the lower layer of lower latitudes as well as in that of higher latitudes. We do not assume realistic climatology in the stratosphere, such that zonal wind becomes weaker as the altitude is higher above the maximum of the subtropical jet.

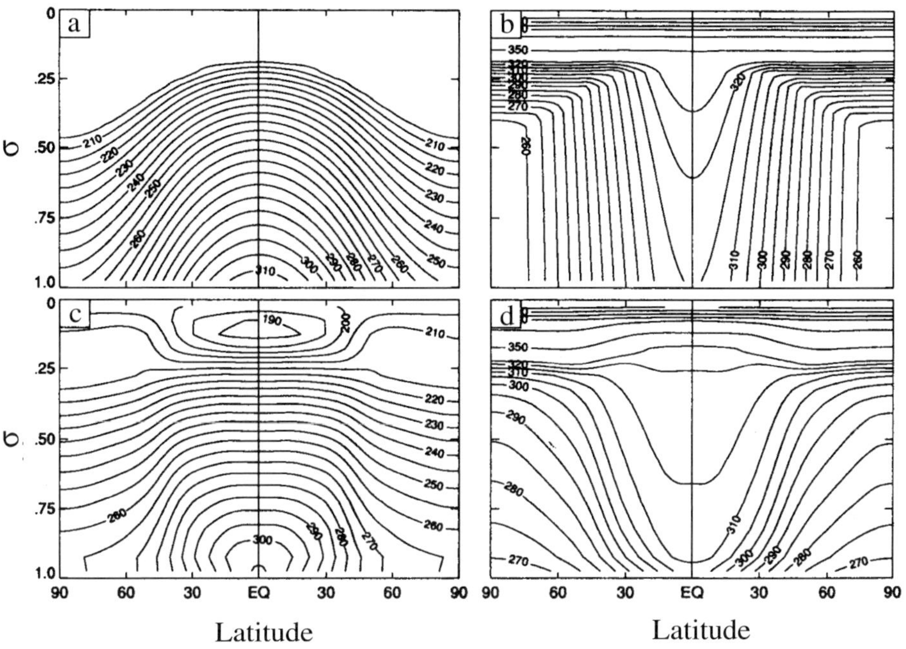

FIGURE 24.3: The meridional distributions of the reference temperature (a) and the corresponding potential temperature (b) used for Newtonian cooling in the dynamical core experiment of Held and Suarez. The lower panels are distributions of temperature (c) and potential temperature (d) of a statistically equilibrium state (1,000-day average). This is the 1,000-day average produced by the T63 spectral model. After Held and Suarez (1994) by permission of the American Meteorological Society.

Fig. 24.5 shows the kinetic energy spectrum with respect to the total spherical wave number n. The left panel is the dependency on horizontal truncation for triangular truncation n_{max} = 79, 159, and 319 (T79, T159, and T319). At the middle range of the wave number around $n \approx 20$, the energy spectrum follows a dependency with n^{-3}. This characteristic of the kinetic energy spectrum is similar to that of two-dimensional turbulence described in Sections 11.1.2 and 17.4.2. It should be noted, however, that these results are sensitive to numerical diffusions particularly in the higher wave number range. Fourth-order Laplacian diffusion is used for these calculations; the diffusion coefficients are $\nu = 1.0 \times 10^{16}$, 1.25×10^{15}, and 1.56×10^{14} m^4 s^{-1}, respectively. Decay rates for the shortest length scale R/n_{max} are 1.2, 0.57, and 0.28 hour, respectively; these are inversely proportional to horizontal resolution n_{max}. The right panel of Fig. 24.5 shows dependency on the coefficient of numerical diffusion for T79 experiments. The diffusion coefficients are set to $\nu = 1.0 \times 10^{16}$, 1.25×10^{15}, and 1.0×10^{15} m^4 s^{-1}. As diffusion gets smaller, the spectrum range close to the n^{-3} line becomes broader. At the smallest scale, however, the monotonicity of the energy spectrum breaks down and the spectrum has a small jump. It is thought that such a small numerical coefficient which results in nonmonotonic behavior of the energy spectrum is inappropriate.

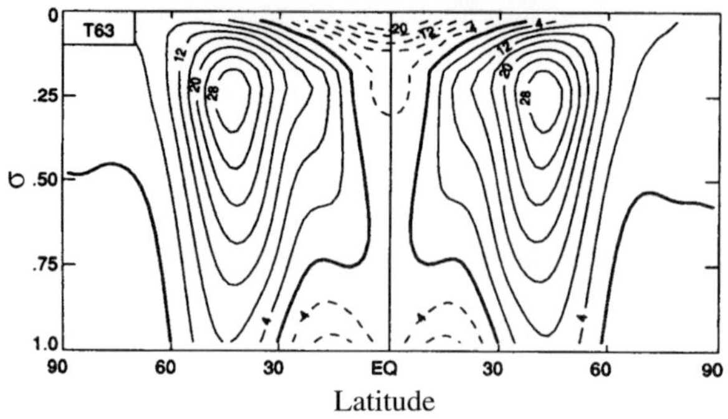

FIGURE 24.4: As in Fig. 24.3 (c) and (d) but for zonal winds. After Held and Suarez (1994) by permission of the American Meteorological Society.

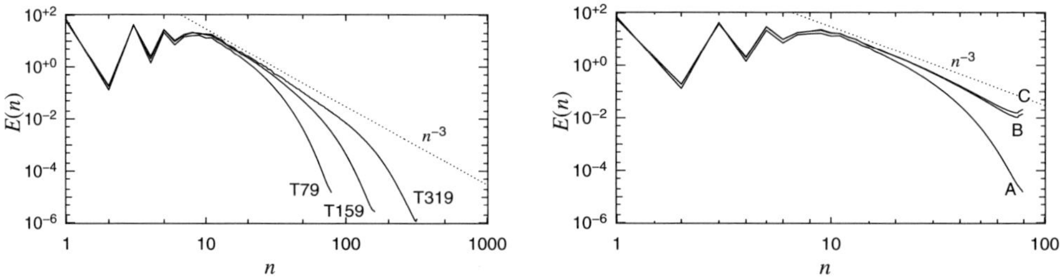

FIGURE 24.5: The kinetic energy spectrum at the 10 km high field for the Held and Suarez experiment. The unit is m^2 s^{-2} and the abscissa is the total wave number of spherical harmonics n. Left: The three solid curves are for horizontal resolutions with T79, T159, and T319 of the atmospheric general circulation model (AGCM). The dotted line indicates the dependency of n^{-3}. Right: Dependency on the coefficient of numerical diffusion. A: 10^{16} m^4 s^{-1}, B: 1.25 $\times$ 10^{15} m^4 s^{-1}, C: 10^{15} m^4 s^{-1}. This calculation was done using a high-performance AGCM (AFES). Courtesy of K. Goto.

24.3 Aqua-planet experiment

The dynamical core experiment proposed by Held and Suarez requires such artificial processes as Newtonian cooling and Rayleigh damping, which are designed to produce a climatology similar to the observed one. Although the damping time of Newtonian cooling and that of Rayleigh damping are appropriately tuned, some unrealistic features still remain (i.e., too stable stratification near the surface). The distribution of diabatic heating also greatly depends on the assumed profile of Newtonian cooling. To obtain more realistic stratification or the distribution of diabatic heating without artificial assumptions, one needs to introduce the hydrological process. For this kind of test case that has a degree of complexity and includes the physical processes of the hydrological process, an aqua-planet experiment is proposed (Neale and Hoskins, 2001a). As already explained an *aqua-planet* is a hypothesized earth-like planet covered completely by ocean. Although the aqua-planet experiment is still highly idealized, it can be used to argue the differences between atmospheric general circulation models without introducing uncertainties derived from the surface land process. Since there is no contrast between land and ocean, the complex features of monsoon circulation that are driven by the land-sea contrast can also be excluded. In addition, one can ascertain how the choice of hydrological process, such as cumulus parameterization, affects the model results.

The most important parameter of the aqua-planet model is the distribution of sea surface temperature (SST). While planetary radiation is internally calculated through the radiation process, the appropriate profile of solar radiation must be externally specified. One of the simplest forms of SST may be given by

$$T_s(\lambda, \varphi) = T_E - \Delta T_s \sin^2 \varphi, \tag{24.3.1}$$

with $T_E = 300$ K and $\Delta T_s = 40$ K, for instance. ΔT_s is the equator-pole surface temperature difference. This profile comes from (24.2.3) of the Held and Suarez experiment and also corresponds to (16.2.1) in the study of Hadley circulation.

Neale and Hoskins (2001a) proposed the following SST distributions: control experiment is given by

$$T_s(\lambda, \varphi) = T_0 + \Delta T_s \left(1 - \sin^2 \frac{3\varphi}{2}\right), \tag{24.3.2}$$

with $\Delta T_s = 27$ K and $T_0 = 0°$C at latitudes $|\varphi| < \frac{\pi}{3}$. At other latitudes, a constant SST $T_s = T_0$ is used. Modified profiles of SST are also proposed: peaked and flat SST profiles at the equator are respectively given by

$$T_s(\lambda, \varphi) = T_0 + \Delta T_s \left(1 - \frac{3|\varphi|}{\pi}\right), \tag{24.3.3}$$

and

$$T_s(\lambda, \varphi) = T_0 + \Delta T_s \left(1 - \sin^4 \frac{3\varphi}{4}\right), \tag{24.3.4}$$

at latitudes $|\varphi| < \frac{\pi}{3}$. The asymmetric profile with maximum SST at 5 N is given by

$$T_s(\lambda, \varphi) = \begin{cases} T_0 + \Delta T_s \left\{1 - \sin^2 \left[\dfrac{90}{55}\left(\varphi - \dfrac{\pi}{36}\right)\right]\right., & \text{for } \dfrac{\pi}{36} < \varphi < \dfrac{\pi}{3}, \\ T_0 + \Delta T_s \left\{1 - \sin^2 \left[\dfrac{90}{65}\left(\varphi - \dfrac{\pi}{36}\right)\right]\right., & \text{for } -\dfrac{\pi}{3} < \varphi < \dfrac{\pi}{36}. \end{cases} \tag{24.3.5}$$

In this case, again, the temperature is uniform $T_s = T_0$ at high latitudes with $|\varphi| > \frac{\pi}{3}$. As the longitudinal dependence of SST, the deviation of the surface temperature is given in the rectangular domain $|\lambda - \lambda_0| < \lambda_d$ and $|\varphi| < \varphi_d$ by

$$T_s'(\lambda, \varphi) = \Delta T \cos^2 \left[\frac{\pi}{2}\left(\frac{\lambda - \lambda_0}{\lambda_d}\right)\right] \cos^2 \left(\frac{\pi}{2}\frac{\varphi}{\varphi_d}\right), \tag{24.3.6}$$

or in the zonal belt $|\varphi| < \varphi_d$ by

$$T_s'(\lambda, \varphi) \quad = \quad \Delta T \cos(\lambda - \lambda_0) \cos^2\left(\frac{\pi}{2}\frac{\varphi}{\varphi_d}\right), \tag{24.3.7}$$

where ΔT is a magnitude of temperature deviation, λ_0 is the longitudinal position of the SST maximum, and λ_d and φ_d are decay widths of the SST in the longitudinal and latitudinal directions, respectively.

Fig. 24.6 shows the latitudinal SST profiles of (24.3.1)–(24.3.5). Fig. 24.7 shows the horizontal distribution of SST given by the latitudinal profile (24.3.2) with the longitudinal deviation (24.3.6) or (24.3.7) with $\Delta T = 3$ K. These SST distributions are intended for an idealized experiment on Walker circulation driven by the warm and cold SST contrast along the equator.

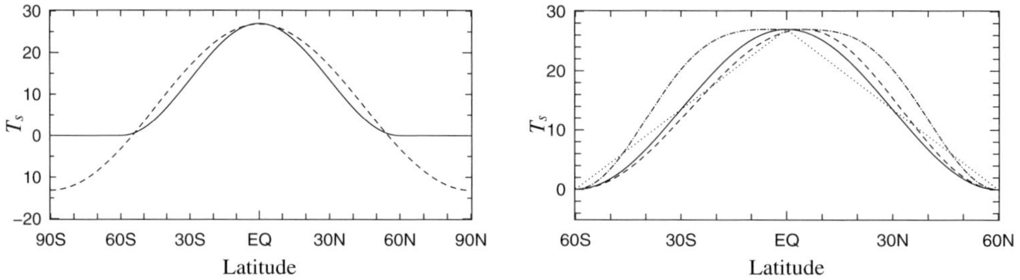

FIGURE 24.6: The latitudinal profiles of SST for the aqua-planet experiment. (Left) dashed: (24.3.1) and solid: (24.3.2). (Right) solid: (24.3.2), dotted: (24.3.3), dashed-dotted: (24.3.4), and dashed: (24.3.5).

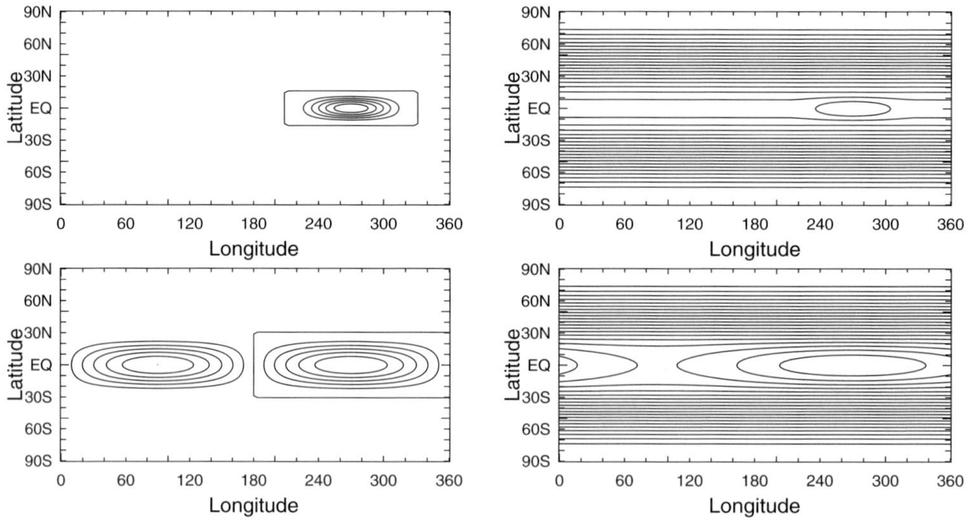

FIGURE 24.7: The horizontal SST distributions of the aqua-planet experiment. Top: T_s' given by (24.3.6) and $T_s + T_s'$ given by (24.3.2) + (24.3.6). $\lambda_d = 60°$, $\varphi_d = 15°$. Bottom: T_s' given by (24.3.7) and $T_s + T_s'$ given by (24.3.2) + (24.3.7). $\varphi_d = 30°$. The contour intervals of T_s' (left figures) are 0.5 K, and those of $T_s + T_s'$ (right figures) are 2 K.

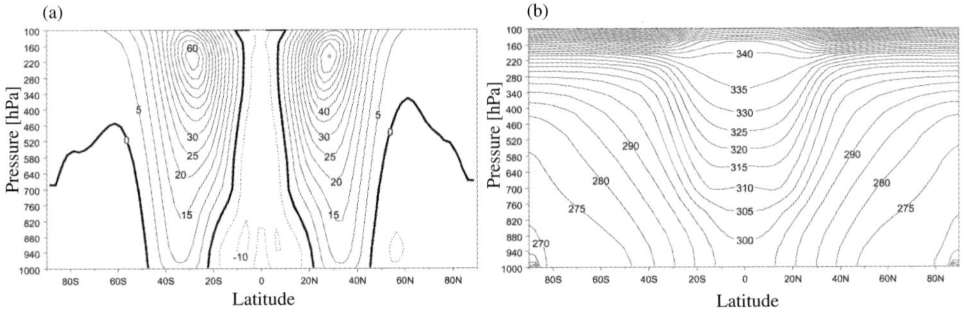

FIGURE 24.8: (a) Zonal-mean zonal wind and (b) potential temperature of the climatology of the aqua-planet experiment. After Neale and Hoskins (2001b) by permission of the *Quarterly Journal of the Royal Meteorological Society.*

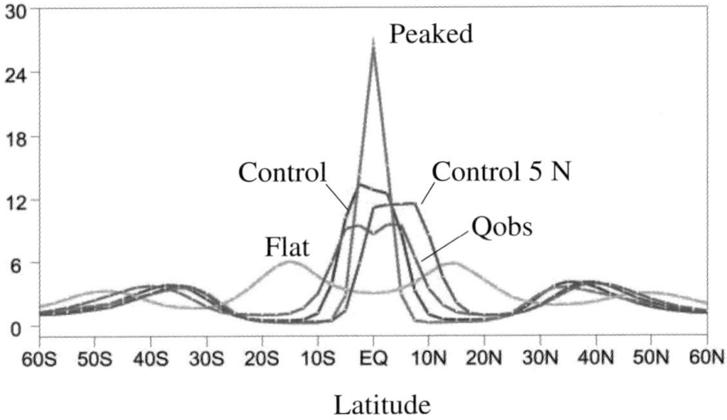

FIGURE 24.9: The latitudinal profiles of zonal-mean precipitation [mm day^{-1}]. After Neale and Hoskins (2001b) by permission of the *Quarterly Journal of the Royal Meteorological Society.*

Fig. 24.8 shows an example of the results from the aqua-planet experiment in the case when the SST distribution is given by (24.3.2) (Neale and Hoskins, 2001b). The meridional distributions of the zonal-mean zonal wind and potential temperature of climatology are shown. The latitudinal profiles of precipitation for the experiments with (24.3.2)–(24.3.5) are compared in Fig. 24.9. A strong precipitation maximum exists in the case of the peaked experiment. This precipitation maximum corresponds to the intertropical convergence zone (ITCZ). In contrast, there are two precipitation maxima near latitude 5° in both hemispheres. These are often called the *double ITCZ.* The observed profile denoted by Q_{obs} is the intermediate case, and is close to the control experiment.

24.4 Atmospheric Model Intercomparison Project (AMIP)

More realistic standard experiments of atmospheric general circulation models are proposed in the Atmospheric Model Intercomparison Project (AMIP). In order to compare the performance of different general circulation models, a set of realistic conditions is prepared. These experiments are used to investigate the extent to which general circulation models reproduce the observed climatology and the extent of differences between general circulation models. The experiments are carried out by giving observed distributions to global SST and ozone. It is reasonable to expect that the distribution of heating could be calculated internally using general circulation models and that realistic general circulation could be reproduced if such appropriate physical processes as radiation, moist convection, and the surface process were implemented. After integration of an initial spin up, the model atmosphere should behave like a similar seasonal sequence of the general circulation. The complete information of AMIP is given at `http://www-pcmdi.llnl.gov/amip` and will not be further explained here.

References and suggested reading

Many standard experiments are proposed for a different hierarchy of model complexity. The most basic are the seven test suits for global shallow water models proposed by Williamson et al. (1992). For three-dimensional general circulation models, linear or weakly nonlinear wave propagation problems can be thought of as fundamental test cases. A more useful fully nonlinear experiment with a deterministic phase is the life cycle experiments of extratropical cyclones originally proposed by Simmons and Hoskins (1978) and Thorncroft et al. (1993), and refined as a test case by Polvani et al. (2003). For a climatology of the dynamical core, or a statistically equilibrium state of the dry atmosphere for long time integration, two representative experiments proposed by Held and Suarez (1994) and Boer and Denis (1997) are well known. While the aqua-planet experiment was introduced as a standard test by Neale and Hoskins (2001a), its original idea can be traced back to Hayashi and Sumi (1986). The purpose of and further stories about AMIP are described by Gates (1992) and Phillips (1996).

Boer, G. J. and Denis, B., 1997: Numerical convergence of the dynamics of a GCM. *Climate Dynamics*, **13**, 359–374.

Gates, W. L., 1992: AMIP: The atmospheric model intercomparison project. *Bull. Am. Meteorol. Soc.*, **73**, 1962–1970.

Hayashi, Y.-Y. and Sumi, A., 1986: The 30–40 day oscillations simulated in an aqua-planet model. *J. Meteorol. Soc. Japan*, **64**, 451–467.

Held, I. M. and Suarez, M. J., 1994: A proposal for the intercomparison of the dynamical cores of atmospheric general circulation models. *Bull. Am. Meteorol. Soc.*, **75**, 1825–1830.

Neale, R. B. and Hoskins, B. J., 2001a: A standard test for AGCMs including their physical parameterizations. I: The proposal. *Atmospheric Science Letter*, **1**, doi:10.1006/asle.2000.0019.

Neale, R. B. and Hoskins, B. J., 2001b: A standard test for AGCMs including their physical parameterizations. II: Results for the Met Office Model. *Atmospheric Science Letter*, **1**, doi:10.1006/asle.2000.0020.

Phillips, T. J., 1996: Documentation of the AMIP models on the World Wide Web. *Bull. Am. Meteorol. Soc.*, **77**, 1191–1196.

Polvani, L. M., Scott, R. K., and Thomas, S. J., 2003: Numerically converged solutions of the global primitive equations for testing the dynamical core of atmospheric GCMs models. Submitted to *Mon. Wea. Rev.*

Simmons, A. J. and Hoskins, B. J., 1978: The life cycles of some nonlinear baroclinic waves. *J. Atmos. Sci.*, **35**, 414–432.

Thorncroft, C. D., Hoskins, B. J., and McIntyre, M. E., 1993: Two paradigms of baroclinic-wave life-cycle behavior. *Q. J. Roy. Meteorol. Soc.*, **119**, 17–55.

Tomita, H., Satoh, M., and Goto, K., 2003: Development of a nonhydrostatic general circulation model using an icosahedral grid. In: K. Matsuno, A. Ecer, J. Periaux, N. Satofuka, and P. Fox (eds.), *Parallel Computational Fluid Dynamics*. Elsevier Science, pp. 115–123.

Williamson, D. L., Drake, J. B., Hack, J. J., Jakob, R., and Swarztrauber, P. N., 1992: A standard test set for numerical approximations to the shallow water equations in spherical geometry. *J. Comp. Phys.*, **102**, 211–224.

Notes and suggested readings

A1

Transformation of coordinates

In this appendix, we summarize the transformation formulas of curvilinear coordinates. We particularly derive the formulas of such orthogonal curvilinear coordinates as cylindrical coordinates and spherical coordinates. Transformation of the equations of motion and other basic equations will be described. Although theories of the geometry of manifolds underlies the basics of these subjects, we do not propose to the discuss them here. We simply describe the technical methods for transformations in three-dimensional Euclidean space.

A1.1 General transformation formulas

A position vector $\boldsymbol{x}$ in three-dimensional space is expressed by three components of coordinates ξ^1, ξ^2, and ξ^3:

$$\boldsymbol{x} = (\xi^1, \xi^2, \xi^3).$$

Tangential vectors in the direction of increasing ξ^μ ($\mu = 1, 2, 3$) are defined as

$$\boldsymbol{x}_\mu \equiv \frac{\partial \boldsymbol{x}}{\partial \xi^\mu} \equiv \partial_\mu \boldsymbol{x}.$$

These vectors form a basis at the point $\boldsymbol{x}$. The metric tensor is defined as

$$g_{\mu\nu} \equiv \boldsymbol{x}_\mu \cdot \boldsymbol{x}_\nu. \tag{A1.1.1}$$

Let $(g^{\mu\nu})$ denote the inverse matrix of $(g_{\mu\nu})$. If we define a vector as

$$\boldsymbol{x}^\mu = g^{\mu\nu} \boldsymbol{x}_\nu,$$

we have

$$\boldsymbol{x}^\mu \cdot \boldsymbol{x}_\lambda = g^{\mu\nu} g_{\nu\lambda} = \delta^\mu_\lambda,$$

where (δ^μ_λ) is the identity matrix. Thus, vector $\boldsymbol{x}^\mu$ is orthogonal to vector $\boldsymbol{x}_\mu$. Any vector $\boldsymbol{v}$ can be decomposed with vectors $\boldsymbol{x}_\mu$ or $\boldsymbol{x}^\mu$ such that

$$\boldsymbol{v} = v^\mu \boldsymbol{x}_\mu = v_\mu \boldsymbol{x}^\mu.$$

v^μ arc called *contravariant components* and v_μ are called *covariant components.*

A line element, an area element, and a volume element are written as

$$\begin{aligned} d\boldsymbol{x} &= \boldsymbol{x}_\mu d\xi^\mu, \\ ds^2 &= d\boldsymbol{x}\cdot d\boldsymbol{x} = g_{\mu\nu}d\xi^\mu d\xi^\nu, \\ dV &= \sqrt{det(g_{\mu\nu})}\, d\xi^1 d\xi^2 d\xi^3 = \sqrt{g} d\xi^1 d\xi^2 d\xi^3, \end{aligned}$$

where $g = det(g_{\mu\nu})$. We write the exterior product of the basis vectors $\boldsymbol{x}_\mu$ and $\boldsymbol{x}_\nu$ as

$$\boldsymbol{x}_\mu \times \boldsymbol{x}_\nu = E_{\mu\nu}{}^\lambda \boldsymbol{x}_\lambda.$$

Then, the scalar triple product of the basis vectors are written as

$$(\boldsymbol{x}_\mu \times \boldsymbol{x}_\nu)\cdot \boldsymbol{x}_\lambda = E_{\mu\nu}{}^\kappa g_{\kappa\lambda} = E_{\mu\nu\lambda}.$$

If the basis $\boldsymbol{x}_\mu$ ($\mu = 1,\ 2,\ 3$) forms right-handedness in this order, the sign of the above product is the same as that of the antisymmetric tensor $\varepsilon_{\mu\nu\lambda}$. When μ, ν and λ are all different, the absolute value of the scalar triple product is equal to the volume of the parallelepiped formed by the three basis vectors:

$$E_{\mu\nu\lambda} = \sqrt{g}\,\varepsilon_{\mu\nu\lambda},$$

and

$$E^{\mu\nu\lambda} = \frac{\varepsilon_{\mu\nu\lambda}}{\sqrt{g}}.$$

The change in tangential vector $\partial_\mu \boldsymbol{x}_\nu$ is expressed by a linear combination of the basis vectors $\boldsymbol{x}_\nu$ as

$$\partial_\mu \boldsymbol{x}_\nu = \Gamma^\lambda_{\mu\nu}\boldsymbol{x}_\lambda,$$

where $\Gamma^\lambda_{\mu\nu}$ are the *connection coefficients* or *Christoffel's symbols.* In general, the symmetry $\Gamma^\lambda_{\mu\nu} = \Gamma^\lambda_{\nu\mu}$ is satisfied. Since $\boldsymbol{x}_\nu \cdot \boldsymbol{x}^\mu = \delta^\mu_\nu$, we have

$$\partial_\mu \boldsymbol{x}^\lambda \cdot \boldsymbol{x}_\nu = -\boldsymbol{x}^\lambda \cdot \partial_\mu \boldsymbol{x}_\nu = -\Gamma^\lambda_{\mu\nu},$$

thus

$$\partial_\mu \boldsymbol{x}^\nu = -\Gamma^\nu_{\mu\lambda}\boldsymbol{x}^\lambda.$$

$\Gamma^\lambda_{\mu\nu}$ is expressed in terms of $g_{\mu\nu}$ as follows. First, taking derivatives from each side of (A1.1.1), we have

$$\partial_\mu g_{\nu\lambda} = \partial_\mu \boldsymbol{x}_\nu \cdot \boldsymbol{x}_\lambda + \boldsymbol{x}_\nu \cdot \partial_\mu \boldsymbol{x}_\lambda = g_{\kappa\lambda}\Gamma^\kappa_{\mu\nu} + g_{\nu\kappa}\Gamma^\kappa_{\mu\lambda}. \tag{A1.1.2}$$

By replacing the indices, we have

$$\partial_\lambda g_{\mu\nu} = g_{\kappa\nu}\Gamma^\kappa_{\lambda\mu} + g_{\mu\kappa}\Gamma^\kappa_{\lambda\nu}, \tag{A1.1.3}$$
$$\partial_\nu g_{\lambda\mu} = g_{\kappa\mu}\Gamma^\kappa_{\nu\lambda} + g_{\lambda\kappa}\Gamma^\kappa_{\nu\mu}. \tag{A1.1.4}$$

Subtracting (A1.1.4) from the sum of (A1.1.2) and (A1.1.3), and dividing by two, we obtain

$$g_{\kappa\nu}\Gamma^{\kappa}_{\lambda\mu} = \frac{1}{2}(\partial_\mu g_{\nu\lambda} + \partial_\lambda g_{\mu\nu} - \partial_\nu g_{\lambda\mu}),$$

where we have used $g_{\mu\nu} = g_{\nu\mu}$ and $\Gamma^{\lambda}_{\mu\nu} = \Gamma^{\lambda}_{\nu\mu}$. Since $g_{\kappa\nu}g^{\lambda\nu} = \delta^{\lambda}_{\kappa}$, we finally obtain

$$\Gamma^{\lambda}_{\mu\nu} = \frac{1}{2}g^{\lambda\kappa}(\partial_\mu g_{\nu\kappa} + \partial_\nu g_{\kappa\mu} - \partial_\kappa g_{\mu\nu}).$$

In particular, we note that contracting for indices μ and λ gives

$$\Gamma^{\mu}_{\mu\nu} = \frac{1}{\sqrt{g}}\partial_\nu \sqrt{g}.$$

The gradient operator ∇ of a scalar f is defined as

$$df = d\boldsymbol{x} \cdot \nabla f. \tag{A1.1.5}$$

We expand the gradient by the basis as

$$\nabla f = (\nabla_\mu f)\boldsymbol{x}^\mu,$$

where $\nabla_\mu f$ are the components of the gradient vector of f. Comparing (A1.1.5) with

$$df = \partial_\mu f d\xi^\mu,$$

we have

$$\boldsymbol{x}_\mu \cdot \nabla f = \partial_\mu f.$$

Thus, we have the expression

$$\nabla_\mu f = \partial_\mu f. \tag{A1.1.6}$$

In the same manner, the gradient of a vector $\boldsymbol{v}$ is expressed as

$$d\boldsymbol{v} = d\boldsymbol{x} \cdot \nabla \boldsymbol{v},$$

where $\nabla \boldsymbol{v}$ is a tensor of degree two. Since

$$\begin{aligned} \boldsymbol{x}_\mu \cdot \nabla \boldsymbol{v} &= \partial_\mu \boldsymbol{v} = \partial_\mu(v^\nu \boldsymbol{x}_\nu) = (\partial_\mu v^\nu)\boldsymbol{x}_\nu + v^\nu(\partial_\mu \boldsymbol{x}_\nu) \\ &= (\partial_\mu v^\nu)\boldsymbol{x}_\nu + v^\nu \Gamma^{\lambda}_{\mu\nu}\boldsymbol{x}_\lambda, \end{aligned}$$

then we have

$$\begin{aligned} \nabla_\mu v^\nu &\equiv \boldsymbol{x}^\nu \cdot (\boldsymbol{x}_\mu \cdot \nabla \boldsymbol{v}) = \partial_\mu v^\nu + \Gamma^{\nu}_{\mu\lambda} v^\lambda, \\ \nabla_\mu v_\nu &\equiv \boldsymbol{x}_\nu \cdot (\boldsymbol{x}_\mu \cdot \nabla \boldsymbol{v}) = \partial_\mu v_\nu - \Gamma^{\lambda}_{\mu\nu} v_\lambda. \end{aligned} \tag{A1.1.7}$$

From the same manipulation of the tensor of degree two, the gradient of $T^{\nu\lambda}$ and $T_{\nu\lambda}$ is given as

$$\begin{aligned} \nabla_\mu T^{\nu\lambda} &= \partial_\mu T^{\nu\lambda} + \Gamma^{\nu}_{\mu\kappa} T^{\kappa\lambda} + \Gamma^{\nu}_{\mu\kappa} T^{\lambda\kappa}, \\ \nabla_\mu T_{\nu\lambda} &= \partial_\mu T_{\nu\lambda} - \Gamma^{\kappa}_{\mu\nu} T_{\kappa\lambda} - \Gamma^{\kappa}_{\mu\nu} T_{\lambda\kappa}. \end{aligned} \tag{A1.1.8}$$

Using these expressions, we have the following relations:

$$\nabla_\mu v_\nu - \nabla_\nu v_\mu = \partial_\mu v_\nu - \partial_\nu v_\mu, \tag{A1.1.9}$$

$$\nabla_\mu v_\nu + \nabla_\nu v_\mu = \partial_\mu v_\nu + \partial_\nu v_\mu + 2\Gamma^\lambda_{\mu\nu} v_\lambda, \tag{A1.1.10}$$

$$\begin{aligned} \nabla_\mu v^\mu &= \partial_\mu v^\mu + \Gamma^\mu_{\mu\nu} v^\nu = \partial_\mu v^\mu + \frac{1}{\sqrt{g}}(\partial_\nu \sqrt{g}) v^\nu \\ &= \frac{1}{\sqrt{g}} \partial_\mu (\sqrt{g} v^\mu), \end{aligned} \tag{A1.1.11}$$

$$\begin{aligned} \nabla_\nu T^{\nu\mu} &= \partial_\nu T^{\nu\mu} + \Gamma^\nu_{\nu\kappa} T^{\kappa\mu} + \Gamma^\mu_{\nu\kappa} T^{\nu\kappa} \\ &= \frac{1}{\sqrt{g}} \partial_\nu (\sqrt{g} T^{\nu\mu}) + \Gamma^\mu_{\nu\kappa} T^{\nu\kappa}, \end{aligned} \tag{A1.1.12}$$

$$\begin{aligned} \nabla^2 f &= \nabla_\mu \nabla^\mu f \\ &= \frac{1}{\sqrt{g}} \partial_\mu (\sqrt{g} \nabla^\mu f) = \frac{1}{\sqrt{g}} \partial_\mu (\sqrt{g}\, g^{\mu\nu} \partial_\mu f), \end{aligned} \tag{A1.1.13}$$

$$\nabla^2 v^\mu = \nabla_\nu \nabla^\nu v^\mu = \frac{1}{\sqrt{g}} \partial_\nu (\sqrt{g}\, \nabla^\nu v^\mu) + \Gamma^\nu_{\mu\kappa} \nabla^\nu v^\kappa \tag{A1.1.14}$$

$$= \nabla^\mu \nabla_\nu v^\nu - E^{\mu\nu\lambda} \nabla^\nu (E_{\lambda\kappa\pi} \nabla^\kappa v^\pi), \tag{A1.1.15}$$

$$(\nabla \times \boldsymbol{v})^\mu = E^{\mu\nu\lambda} \nabla_\nu v_\lambda. \tag{A1.1.16}$$

A1.2 Transformation of orthogonal coordinates

Hereafter, we concentrate on the transformation of orthogonal coordinates. The metric tensors of orthogonal coordinates have only diagonal components:

$$(g_{\mu\nu}) = \begin{pmatrix} h_1^2 & 0 & 0 \\ 0 & h_2^2 & 0 \\ 0 & 0 & h_3^2 \end{pmatrix}, \qquad (g^{\mu\nu}) = \begin{pmatrix} h_1^{-2} & 0 & 0 \\ 0 & h_2^{-2} & 0 \\ 0 & 0 & h_3^{-2} \end{pmatrix},$$

where[†]

$$h_\alpha = \sqrt{g_{\alpha\alpha}} = |\boldsymbol{x}_\alpha|, \qquad \text{for} \quad \alpha = 1, 2, 3.$$

Unit vectors of the tangential directions of coordinates are given by

$$\boldsymbol{e}_\alpha = \frac{\boldsymbol{x}_\alpha}{|\boldsymbol{x}_\alpha|} = \frac{\boldsymbol{x}_\alpha}{h_\alpha}.$$

The line elements and the volume elements are

$$\begin{aligned} d\boldsymbol{x} &= \boldsymbol{x}_\mu d\xi^\mu = h_\mu \boldsymbol{e}_\mu d\xi^\mu, \\ dV &= \sqrt{g}\, d\xi^1 d\xi^2 d\xi^3 = h_1 h_2 h_3 d\xi^1 d\xi^2 d\xi^3, \end{aligned}$$

[†]Hereafter, summation convention is not used for the repeated indices α, β, and γ, whereas summation is implied for μ, ν, and λ. Whenever h_α appears, no summation is assumed for any indices.

where $\sqrt{g} = h_1 h_2 h_3$. We assume that the three unit vectors $\boldsymbol{e}_\mu$ ($\mu = 1, 2, 3$) possess right-handedness in this order. We therefore have

$$(\boldsymbol{e}_\mu \times \boldsymbol{e}_\nu) \cdot \boldsymbol{e}_\lambda = \varepsilon_{\mu\nu\lambda}.$$

Christoffel's symbols are written as

$$\Gamma^\alpha_{\alpha\alpha} = \frac{\partial_\alpha h_\alpha}{h_\alpha}, \quad \Gamma^\alpha_{\alpha\beta} = \frac{\partial_\beta h_\alpha}{h_\alpha}, \quad \Gamma^\alpha_{\beta\beta} = -\frac{h_\beta}{h_\alpha^2}\partial_\alpha h_\beta, \tag{A1.2.1}$$

and the other components of $\Gamma^\alpha_{\beta\gamma}$ are zero. Then, the sum of the three symbols is

$$\Gamma^\mu_{\mu\alpha} = \sum_{i=1}^{3} \Gamma^i_{i\alpha} = \frac{1}{\sqrt{g}}\partial_\alpha \sqrt{g}.$$

The derivatives of the unit vector are calculated as

$$\partial_\mu \boldsymbol{e}_\nu = \partial_\mu \frac{\boldsymbol{x}_\nu}{h_\nu} = -\frac{\partial_\mu h_\nu}{h_\nu^2}\boldsymbol{x}_\nu + \frac{1}{h_\nu}\partial_\mu \boldsymbol{x}_\nu = -\frac{\partial_\mu h_\nu}{h_\nu}\boldsymbol{e}_\nu + \frac{h_\lambda}{h_\nu}\Gamma^\lambda_{\mu\nu}\boldsymbol{e}_\lambda.$$

In particular, using (A1.2.1), we have

$$\partial_\alpha \boldsymbol{e}_\alpha = -\frac{\partial_\beta h_\alpha}{h_\beta}\boldsymbol{e}_\beta - \frac{\partial_\gamma h_\alpha}{h_\gamma}\boldsymbol{e}_\gamma, \quad \partial_\beta \boldsymbol{e}_\alpha = \frac{\partial_\alpha h_\beta}{h_\alpha}\boldsymbol{e}_\beta. \tag{A1.2.2}$$

Now, we expand a vector $\boldsymbol{v}$ by the basis $\boldsymbol{x}_\mu$ or $\boldsymbol{e}_\mu$:

$$\boldsymbol{v} = v^\mu \boldsymbol{x}_\mu = v_\mu \boldsymbol{x}^\mu = \overline{v}_\mu \boldsymbol{e}_\mu. \tag{A1.2.3}$$

$\overline{v}_\mu$ is the length of $\boldsymbol{v}$ in the direction of ξ^μ, and is called a *physical component* or *ordinary component*. Physical components are related to contravariant and covariant components as

$$\overline{v}_\mu = v^\mu h^\mu = \frac{v_\mu}{h^\mu}.$$

The derivative of $\boldsymbol{v}$ is expressed as

$$\begin{aligned}\partial_\mu \boldsymbol{v} &= \partial_\mu(\overline{v}_\nu \boldsymbol{e}_\nu) = (\partial_\mu \overline{v}_\nu)\boldsymbol{e}_\nu + \overline{v}_\nu(\partial_\mu \boldsymbol{e}_\nu) \\ &= (\partial_\mu \overline{v}_\nu)\boldsymbol{e}_\nu - \overline{v}_\nu \frac{\partial_\mu h_\nu}{h_\nu}\boldsymbol{e}_\nu + \overline{v}_\nu \frac{h_\lambda}{h_\nu}\Gamma^\lambda_{\mu\nu}\boldsymbol{e}_\lambda,\end{aligned} \tag{A1.2.4}$$

in which $(\partial_\mu \boldsymbol{e}_\nu)$ is expanded by the basis $\boldsymbol{e}_\mu$ using (A1.2.2).

By using rule (A1.2.3), the gradient, divergence, rotation, and Laplacian operators, which are given as (A1.1.6), (A1.1.11), (A1.1.16), and (A1.1.13), are expressed by orthogonal components:

$$\begin{aligned}(\overline{\nabla f})_\mu &= \frac{1}{h_\mu}\nabla_\mu f = \frac{1}{h_\mu}\partial_\mu f, \\ \nabla \cdot \boldsymbol{v} &= \frac{1}{\sqrt{g}}\partial_\mu\left(\frac{\sqrt{g}}{h_\mu}\overline{v}_\mu\right), \\ (\overline{\nabla \times \boldsymbol{v}})_\mu &= h_\mu \frac{\varepsilon_{\mu\nu\lambda}}{\sqrt{g}}\nabla_\nu v_\lambda = \frac{\varepsilon_{\mu\nu\lambda}}{h_\nu h_\lambda}\partial_\nu(h_\lambda \overline{v}_\lambda), \\ \nabla^2 f &= \frac{1}{\sqrt{g}}\partial_\mu\left(\frac{\sqrt{g}}{h_\mu^2}\partial_\mu f\right).\end{aligned}$$

or their explicit expressions are

$$\overline{\nabla f} = \left(\frac{1}{h_1}\partial_1 f, \frac{1}{h_2}\partial_2 f, \frac{1}{h_3}\partial_3 f\right), \tag{A1.2.5}$$

$$\nabla \cdot \boldsymbol{v} = \frac{1}{h_1 h_2 h_3}\left[\partial_1(h_2 h_3 \overline{v}_1) + \partial_2(h_3 h_1 \overline{v}_2) + \partial_3(h_1 h_2 \overline{v}_3)\right], \tag{A1.2.6}$$

$$\overline{\nabla \times \boldsymbol{v}} = \left(\frac{1}{h_2 h_3}\left[\partial_2(h_3\overline{v}_3) - \partial_3(h_2\overline{v}_2)\right], \frac{1}{h_3 h_1}\left[\partial_3(h_1\overline{v}_1) - \partial_1(h_3\overline{v}_3)\right],\right.$$
$$\left.\frac{1}{h_1 h_2}\left[\partial_1(h_2\overline{v}_2) - \partial_2(h_1\overline{v}_1)\right]\right), \tag{A1.2.7}$$

$$\nabla^2 f = \frac{1}{h_1 h_2 h_3}\left[\partial_1\left(\frac{h_2 h_3}{h_1}\partial_1 f\right) + \partial_2\left(\frac{h_3 h_1}{h_2}\partial_2 f\right) + \partial_3\left(\frac{h_1 h_2}{h_3}\partial_3 f\right)\right]. \tag{A1.2.8}$$

From (A1.1.7), the gradient of a vector, or the deformation tensor, is written as

$$(\overline{\nabla \boldsymbol{v}})_{\mu\nu} = \frac{1}{h_\mu h_\nu}\left[\partial_\mu(h_\nu \overline{v}_\nu) - \Gamma^\lambda_{\mu\nu}(h_\lambda \overline{v}_\lambda)\right],$$

or, using (A1.2.1),

$$(\overline{\nabla \boldsymbol{v}})_{\alpha\beta} = \frac{1}{h_\alpha h_\beta}\left[\partial_\alpha(h_\beta \overline{v}_\beta) - (\partial_\beta h_\alpha \cdot \overline{v}_\alpha + \partial_\alpha h_\beta \cdot \overline{v}_\beta)\right] \tag{A1.2.9}$$

$$= \frac{1}{h_\alpha h_\beta}\left[h_\beta^2 \partial_\alpha \frac{\overline{v}_\beta}{h_\beta} - (\partial_\beta h_\alpha \cdot \overline{v}_\alpha - \partial_\alpha h_\beta \cdot \overline{v}_\beta)\right], \tag{A1.2.10}$$

$$(\overline{\nabla \boldsymbol{v}})_{\alpha\alpha} = \frac{1}{h_\alpha}\left[\partial_\alpha \overline{v}_\alpha + \frac{1}{h_\beta}\partial_\beta h_\alpha \cdot \overline{v}_\beta + \frac{1}{h_\gamma}\partial_\gamma h_\alpha \cdot \overline{v}_\gamma\right]. \tag{A1.2.11}$$

α, β, and γ take different values of indices. From (A1.2.9) and (A1.2.10), we have

$$(\overline{\nabla \boldsymbol{v}})_{\alpha\beta} - (\overline{\nabla \boldsymbol{v}})_{\beta\alpha} = \frac{1}{h_\alpha h_\beta}\left[\partial_\alpha(h_\beta \overline{v}_\beta) - \partial_\beta(h_\alpha \overline{v}_\alpha)\right],$$

$$(\overline{\nabla \boldsymbol{v}})_{\alpha\beta} + (\overline{\nabla \boldsymbol{v}})_{\beta\alpha} = \frac{h_\beta}{h_\alpha}\partial_\alpha \frac{\overline{v}_\beta}{h_\beta} + \frac{h_\alpha}{h_\beta}\partial_\beta \frac{\overline{v}_\alpha}{h_\alpha}.$$

From (A1.1.12), the divergence of a tensor of degree two, $\nabla_\nu T^{\nu\mu}$, is given by

$$(\overline{\nabla \cdot T})_\mu = \frac{h_\mu}{\sqrt{g}}\partial_\nu\left(\frac{\sqrt{g}}{h_\nu h_\mu}\overline{T}_{\nu\mu}\right) + h_\mu \Gamma^\mu_{\nu\kappa}\frac{\overline{T}_{\nu\kappa}}{h_\nu h_\kappa}, \tag{A1.2.12}$$

or, using (A1.2.1),

$$(\overline{\nabla \cdot T})_\alpha = \frac{1}{h_\alpha h_\beta h_\gamma}\left[\partial_\alpha(h_\beta h_\gamma \overline{T}_{\alpha\alpha}) + \partial_\beta(h_\gamma h_\alpha \overline{T}_{\beta\alpha}) + \partial_\gamma(h_\alpha h_\beta \overline{T}_{\gamma\alpha})\right]$$
$$- \frac{\partial_\alpha h_\beta}{h_\alpha h_\beta}\overline{T}_{\beta\beta} - \frac{\partial_\alpha h_\gamma}{h_\alpha h_\gamma}\overline{T}_{\gamma\gamma} + \frac{\partial_\beta h_\alpha}{h_\alpha h_\beta}\overline{T}_{\alpha\beta} + \frac{\partial_\gamma h_\alpha}{h_\alpha h_\gamma}\overline{T}_{\alpha\gamma}.$$

In particular, if T is antisymmetric $\overline{T}_{\alpha\beta} = -\overline{T}_{\beta\alpha}$, we have

$$(\nabla \cdot T)_\alpha = \frac{1}{h_\beta h_\gamma}\left[\partial_\beta(h_\gamma \overline{T}_{\beta\alpha}) + \partial_\gamma(h_\beta \overline{T}_{\gamma\alpha})\right].$$

The Laplacian of a vector is, from (A1.1.14),

$$(\nabla^2 \boldsymbol{v})_\mu = \frac{h_\mu}{\sqrt{g}}\partial_\nu\left[\frac{\sqrt{g}}{h_\nu h_\mu}(\nabla \boldsymbol{v})_{\nu\mu}\right] + h_\mu \Gamma^\nu_{\mu\kappa}\frac{(\nabla \boldsymbol{v})_{\nu\kappa}}{h_\nu h_\mu},$$

or, using (A1.1.15) and (A1.2.7),

$$\begin{aligned}(\nabla^2 \boldsymbol{v})_\alpha &= [\nabla(\nabla \cdot \boldsymbol{v})]_\alpha - [\nabla \times (\nabla \times \boldsymbol{v})]_\alpha \\ &= \frac{1}{h_\alpha}\partial_\alpha\left[\frac{1}{\sqrt{g}}\partial_\nu\left(\frac{\sqrt{g}}{h_\nu}\overline{v}_\nu\right)\right] \\ &\quad + \frac{1}{h_\beta h_\gamma}\left\{\partial_\beta\left(\frac{h_\gamma}{h_\alpha h_\beta}[\partial_\alpha(h_\beta \overline{v}_\beta) - \partial_\beta(h_\alpha \overline{v}_\alpha)]\right)\right. \\ &\quad \left. -\partial_\gamma\left(\frac{h_\beta}{h_\alpha h_\gamma}[\partial_\gamma(h_\alpha \overline{v}_\alpha) - \partial_\alpha(h_\gamma \overline{v}_\gamma)]\right)\right\}.\end{aligned}$$

Finally, let us summarize the transformation laws of the time derivative. For a scalar, we have

$$\frac{df}{dt} = \frac{\partial f}{\partial t} + v^\mu \nabla_\mu f = \frac{\partial f}{\partial t} + \frac{\overline{v}^\mu}{h_\mu}\partial_\mu f.$$

A time derivative of a vector $\boldsymbol{s}$ is

$$\begin{aligned}\left(\frac{d\boldsymbol{s}}{dt}\right)_\mu &= h_\mu \frac{ds^\mu}{dt} = h_\mu\left(\frac{\partial s^\mu}{\partial t} + v^\nu \nabla_\nu s^\mu\right) \\ &= h_\mu\left(\frac{\partial s^\mu}{\partial t} + v^\nu \partial_\nu s^\mu + v^\nu \Gamma^\mu_{\nu\lambda} s^\lambda\right) \\ &= h_\mu\left(\frac{\partial}{\partial t}\frac{\overline{s}_\mu}{h_\mu} + \frac{\overline{v}_\nu}{h_\nu}\partial_\nu \frac{\overline{s}_\mu}{h_\mu} + \frac{\overline{v}_\nu}{h_\nu}\Gamma^\mu_{\nu\lambda}\frac{\overline{s}_\lambda}{h_\lambda}\right) \\ &= \frac{d}{dt}\overline{s}_\mu - \frac{\partial_\nu h_\mu}{h_\mu h_\nu}\overline{v}_\nu \overline{s}_\mu + \frac{h_\mu}{h_\nu h_\lambda}\Gamma^\mu_{\nu\lambda}\overline{v}_\nu \overline{s}_\lambda, \end{aligned} \tag{A1.2.13}$$

where

$$\frac{d}{dt}\overline{s}_\mu = \frac{\partial}{\partial t}\overline{s}_\mu + \frac{\overline{v}_\nu}{h_\nu}\partial_\nu \overline{s}_\mu.$$

Substituting (A1.2.1) into (A1.2.13), we can express

$$\left(\frac{d\boldsymbol{s}}{dt}\right)_\alpha = \frac{d}{dt}\overline{s}_\alpha - \frac{\partial_\alpha h_\beta}{h_\alpha h_\beta}\overline{v}_\beta \overline{s}_\beta - \frac{\partial_\alpha h_\gamma}{h_\alpha h_\gamma}\overline{v}_\gamma \overline{s}_\gamma + \frac{\partial_\beta h_\alpha}{h_\alpha h_\beta}\overline{v}_\alpha \overline{s}_\beta + \frac{\partial_\gamma h_\alpha}{h_\alpha h_\gamma}\overline{v}_\alpha \overline{s}_\gamma.$$

A1.3 Basic equations in orthogonal coordinates

The governing equations of fluid motion (i.e., the continuity equation (1.2.8), the equations of motion (1.2.18), and the entropy equation (1.2.53)), are written in the flux form of orthogonal coordinates as

$$
\begin{aligned}
\frac{\partial}{\partial t}\rho + \frac{1}{\sqrt{g}}\partial_\mu\left(\frac{\sqrt{g}}{h_\mu}\rho\overline{v}_\mu\right) &= 0,\\
\frac{\partial}{\partial t}(\rho\overline{v}_\mu) + \frac{h_\mu}{\sqrt{g}}\partial_\nu\left[\frac{\sqrt{g}}{h_\mu h_\nu}(\rho\overline{v}_\mu\overline{v}_\nu - \overline{\sigma}_{\mu\nu})\right] &+ \Gamma^\mu_{\nu\lambda}\frac{h_\mu}{h_\nu h_\lambda}(\rho\overline{v}_\mu\overline{v}_\nu - \overline{\sigma}_{\mu\nu})\\
= -\frac{\rho}{h_\mu}\partial_\mu\Phi,\\
\frac{\partial}{\partial t}(\rho s) + \frac{1}{\sqrt{g}}\partial_\mu\left(\frac{\sqrt{g}}{h_\mu}\rho s\overline{v}_\mu\right) &= \frac{1}{T}\left[\varepsilon - \frac{1}{\sqrt{g}}\partial_\mu\left(\frac{\sqrt{g}}{h_\mu}\overline{q}_\mu\right)\right].
\end{aligned}
\tag{A1.3.1}
$$

We have used (A1.2.12) in the derivation of the equation of motion. In the advective form, the equation set is written as

$$
\begin{aligned}
\frac{d\rho}{dt} + \frac{\rho}{\sqrt{g}}\partial_\mu\left(\frac{\sqrt{g}}{h_\mu}\overline{v}_\mu\right) &= 0,\\
\frac{d\overline{v}_\mu}{dt} - \frac{\partial_\nu h_\mu}{h_\mu h_\nu}\overline{v}_\nu\overline{v}_\mu + \frac{h_\mu}{h_\nu h_\lambda}\Gamma^\mu_{\nu\lambda}\overline{v}_\nu\overline{v}_\lambda &= -\frac{1}{\rho h_\mu}\partial_\mu p - \frac{1}{h_\mu}\partial_\mu\Phi + \overline{f}_\mu,\\
\frac{ds}{dt} &= \frac{1}{\rho T}\left[\varepsilon - \frac{1}{\sqrt{g}}\partial_\mu\left(\frac{\sqrt{g}}{h_\mu}\overline{q}_\mu\right)\right].
\end{aligned}
\tag{A1.3.2}
$$

We have used (A1.2.13) for derivation of the equation of motion. The stress tensor is written as

$$
\begin{aligned}
\overline{\sigma}_{\mu\nu} &= -p\delta_{\mu\nu} + \overline{\sigma'}_{\mu\nu},\\
\overline{\sigma'}_{\mu\nu} &= \eta(\overline{\nabla_\mu v_\nu} + \overline{\nabla_\nu v_\mu}) + \delta_{\mu\nu}\left(\zeta - \frac{2}{3}\right)\nabla\cdot\boldsymbol{v},
\end{aligned}
\tag{A1.3.3}
$$

and the friction force $\boldsymbol{f}$ is

$$
\overline{f}_\mu = \frac{1}{\rho}\overline{\nabla_\nu\sigma'_{\mu\nu}} = \frac{1}{\rho}\left[\frac{h_\mu}{\sqrt{g}}\partial_\nu\left(\frac{\sqrt{g}}{h_\nu h_\mu}\overline{\sigma'}_{\mu\nu}\right) + h_\mu\Gamma^\mu_{\nu\kappa}\frac{\overline{\sigma'}_{\nu\kappa}}{h_\nu h_\kappa}\right].
\tag{A1.3.4}
$$

In particular, if η and ζ are constant, (1.2.21) becomes

$$
\overline{f}_\mu = \frac{\eta}{\rho}(\overline{\nabla^2\boldsymbol{v}})_\mu + \frac{1}{\rho}\left(\zeta + \frac{1}{3}\eta\right)\frac{1}{h_\mu}\partial_\mu(\nabla\cdot\boldsymbol{v}).
\tag{A1.3.5}
$$

The dissipation rate ε given by (1.2.37) is written as

$$
\varepsilon = \overline{\sigma'}_{\mu\nu}\overline{\nabla_\nu\boldsymbol{v}_\mu} = \frac{1}{2}\eta\left(\overline{\nabla_\mu\boldsymbol{v}_\nu} + \overline{\nabla_\nu\boldsymbol{v}_\mu} - \frac{2}{3}\nabla\boldsymbol{v}\delta_{\mu\nu}\right)^2 + \zeta(\nabla\cdot\boldsymbol{v})^2.
$$

The flux form and advective form of the vorticity equation are, respectively,

$$\frac{\partial \overline{\omega}_\mu}{\partial t} + \frac{h_\mu}{\sqrt{g}}\partial_\nu \left\{ \frac{\sqrt{g}}{h_\nu h_\mu} \left[\overline{\omega}_\mu \overline{v}_\nu - \overline{\omega}_\nu \overline{v}_\mu - \varepsilon_{\mu\nu\lambda} g_\lambda \right] \right\} = 0, \tag{A1.3.6}$$

$$\frac{d\overline{\omega}_\mu}{dt} - \frac{\partial_\nu h_\mu}{h_\mu h_\nu}\overline{v}_\nu \overline{\omega}_\mu + \frac{h_\mu}{h_\nu h_\lambda}\Gamma^\mu_{\nu\lambda}\overline{v}_\nu \overline{\omega}_\lambda$$
$$= -\overline{\omega}_\mu \nabla \cdot \boldsymbol{v} + \overline{\omega}_\nu \overline{\nabla_\nu v_\mu} + \varepsilon_{\mu\nu\lambda}\overline{\nabla_\nu g_\lambda}, \tag{A1.3.7}$$

where we have defined the force by

$$g_\lambda = -\frac{1}{\rho h_\lambda}\partial_\lambda p + f_\lambda .$$

Vorticity is expressed in the form (A1.2.7); that is,

$$\overline{\omega}_\alpha = \frac{1}{h_\beta h_\gamma}\left[\partial_\beta (h_\gamma \overline{v}_\gamma) - \partial_\gamma (h_\beta \overline{v}_\beta) \right].$$

A1.4 Cylindrical coordinates

In cylindrical coordinates, we use†

$$\xi^1 = \xi^r = r, \qquad \xi^2 = \xi^\varphi = \varphi, \qquad \xi^3 = \xi^z = z,$$

where r is the radius, φ is the azimuth, and z is the normal axis. The line element and metrics are

$$ds^2 = dr^2 + r^2 d\varphi^2 + dz^2,$$
$$h_r = 1, \qquad h_\varphi = r, \qquad h_z = 1,$$
$$\sqrt{g} = h_r h_\varphi h_z = r.$$

Christoffel's symbols are written as

$$\Gamma^\varphi_{\varphi r} = \Gamma^\varphi_{r\varphi} = \frac{1}{r}, \qquad \Gamma^r_{\varphi\varphi} = -r.$$

The other components of Christoffel's symbols are zero.

Expressions of the gradient, divergence, rotation, and Laplacian operators are, from (A1.2.5)–(A1.2.8),

$$\nabla f = \left(\partial_r f,\ \frac{1}{r}\partial_\varphi f,\ \partial_z f \right), \tag{A1.4.1}$$

$$\nabla \cdot \boldsymbol{v} = \frac{1}{r}\partial_r (r v_r) + \frac{1}{r}\partial_\varphi v_\varphi + \partial_z v_z . \tag{A1.4.2}$$

$$\nabla \times \boldsymbol{v} = \left(\frac{1}{r}\partial_\varphi v_z - \partial_z v_\varphi,\ \partial_z v_r - \partial_r v_z,\ \frac{1}{r}\partial_r (r v_\varphi) - \frac{1}{r}\partial_\varphi v_r \right), \tag{A1.4.3}$$

$$\nabla^2 f = \frac{1}{r}\partial_r (r \partial_r f) + \frac{1}{r^2}\partial_{\varphi\varphi} f + \partial_{zz} f. \tag{A1.4.4}$$

†In the following description, the overbar $(\bar{\ })$ for orthogonal components is omitted.

The gradient and the Laplacian of a vector $\boldsymbol{v}$ are given by

$$\begin{aligned}\nabla \boldsymbol{v} &= \begin{pmatrix} \partial_r v_r & \partial_r v_\varphi & \partial_r v_z \\ \frac{1}{r}\partial_\varphi v_r - \frac{1}{r}v_\varphi & \frac{1}{r}\partial_\varphi v_\varphi + \frac{1}{r}v_r & \frac{1}{r}\partial_\varphi v_z \\ \partial_z v_r & \partial_z v_\varphi & \partial_z v_z \end{pmatrix}, \\ \nabla^2 \boldsymbol{v} &= \left(\nabla^2 v_r - \frac{1}{r^2}v_r - \frac{2}{r^2}\partial_\varphi v_\varphi,\ \nabla^2 v_\varphi - \frac{1}{r^2}v_\varphi + \frac{2}{r^2}\partial_\varphi v_r,\ \nabla^2 v_z\right).\end{aligned}$$

The time derivative of a vector $\boldsymbol{s}$ is

$$\frac{d\boldsymbol{s}}{dt} = \left(\frac{ds_r}{dt} - \frac{v_\varphi s_\varphi}{r},\ \frac{ds_\varphi}{dt} + \frac{v_\varphi s_r}{r},\ \frac{ds_z}{dt}\right), \tag{A1.4.5}$$

where

$$\frac{d}{dt} = \frac{\partial}{\partial t} + v_r\partial_r + \frac{v_\varphi}{r}\partial_\varphi + v_z\partial_z.$$

The continuity equation and the equations of motion are, in flux form,

$$\begin{aligned}\frac{\partial\rho}{\partial t} + \frac{1}{r}\partial_r(r\rho v_r) + \frac{1}{r}\partial_\varphi(\rho v_\varphi) + \partial_z(\rho v_z) &= 0, \\ \frac{\partial(\rho v_r)}{\partial t} + \frac{1}{r}\partial_r[r(\rho v_r^2 - \sigma_{rr})] + \frac{1}{r}\partial_\varphi(\rho v_r v_\varphi - \sigma_{\varphi r}) + \partial_z(\rho v_r v_z - \sigma_{zr}) & \\ -\frac{1}{r}(\rho v_\varphi^2 - \sigma_{\varphi\varphi}) &= -\rho\partial_r\Phi, \\ \frac{\partial(\rho v_\varphi)}{\partial t} + \frac{1}{r^2}\partial_r[r^2(\rho v_\varphi v_r - \sigma_{r\varphi})] + \frac{1}{r}\partial_\varphi(\rho v_\varphi^2 - \sigma_{\varphi\varphi}) + \partial_z(\rho v_\varphi v_z - \sigma_{z\varphi}) & \\ = -\frac{\rho}{r}\partial_\varphi\Phi, & \\ \frac{\partial(\rho v_z)}{\partial t} + \frac{1}{r}\partial_r[r(\rho v_z v_r - \sigma_{rz})] + \frac{1}{r}\partial_\varphi(\rho v_z v_\varphi - \sigma_{\varphi z}) + \partial_z(\rho v_z^2 - \sigma_{zz}) & \\ = -\rho\partial_z\Phi. &\end{aligned} \tag{A1.4.6}$$

and, in advective form,

$$\begin{aligned}\frac{d\rho}{dt} + \rho\left[\frac{1}{r}\partial_r(rv_r) + \frac{1}{r}\partial_\varphi v_\varphi + \partial_z v_z\right] &= 0, \\ \frac{dv_r}{dt} &= \frac{1}{r}v_\varphi^2 - \frac{1}{\rho}\partial_r p - \partial_r\Phi + f_r, \\ \frac{dv_\varphi}{dt} &= -\frac{1}{r}v_r v_\varphi - \frac{1}{\rho r}\partial_\varphi p - \frac{1}{r}\partial_\varphi\Phi + f_\varphi, \\ \frac{dv_z}{dt} &= -\frac{1}{\rho}\partial_z p - \partial_z\Phi + f_z,\end{aligned} \tag{A1.4.7}$$

where the stress tensor is given by $\sigma_{\mu\nu} = -p\delta_{\mu\nu} + \sigma'_{\mu\nu}$ with

$$\begin{aligned}\sigma'_{rr} &= 2\eta\partial_r v_r + \left(\zeta - \frac{2}{3}\eta\right)\nabla\cdot\boldsymbol{v}, \\ \sigma'_{\varphi\varphi} &= 2\eta\left(\frac{1}{r}\partial_\varphi v_\varphi + \frac{v_r}{r}\right) + \left(\zeta - \frac{2}{3}\eta\right)\nabla\cdot\boldsymbol{v},\end{aligned}$$

$$
\begin{aligned}
&& \sigma'_{zz} &= 2\eta\partial_z v_z + \left(\zeta - \frac{2}{3}\eta\right)\nabla\cdot\boldsymbol{v}, \\
\sigma'_{r\varphi} &= & \sigma'_{\varphi r} &= \eta\left(r\partial_r\frac{v_\varphi}{r} + \frac{1}{r}\partial_\varphi v_r\right), \\
\sigma'_{\varphi z} &= & \sigma'_{\varphi z} &= \eta\left(\frac{1}{r}\partial_\varphi v_z + \partial_z v_\varphi\right), \\
\sigma'_{zr} &= & \sigma'_{rz} &= \eta\left(\partial_z v_r + \partial_z v_r\right).
\end{aligned} \tag{A1.4.8}
$$

The friction force is

$$
\begin{aligned}
f_r &= \frac{1}{\rho}\left[\frac{1}{r}\partial_r(r\sigma'_{rr}) + \frac{1}{r}\partial_\varphi\sigma'_{\varphi r} + \partial_z\sigma'_{zr} - \frac{\sigma'_{\varphi\varphi}}{r}\right] \\
&= \frac{1}{\rho}\left[\eta\left(\nabla^2 v_r - \frac{v_r}{r^2} - \frac{2}{r^2}\partial_\varphi v_\varphi\right) + \left(\zeta + \frac{1}{3}\eta\right)\partial_r\nabla\cdot\boldsymbol{v}\right], \\
f_\varphi &= \frac{1}{\rho}\left[\frac{1}{r}\partial_r(r\sigma'_{r\varphi}) + \frac{1}{r}\partial_\varphi\sigma'_{\varphi\varphi} + \partial_z\sigma'_{z\varphi} + \frac{\sigma'_{\varphi r}}{r}\right] \\
&= \frac{1}{\rho}\left[\eta\left(\nabla^2 v_\varphi - \frac{v_\varphi}{r^2} - \frac{2}{r^2}\partial_\varphi v_r\right) + \left(\zeta + \frac{1}{3}\eta\right)\frac{1}{r}\partial_\varphi\nabla\cdot\boldsymbol{v}\right], \\
f_z &= \frac{1}{\rho}\left[\frac{1}{r}\partial_r(r\sigma'_{rz}) + \frac{1}{r}\partial_\varphi\sigma'_{\varphi z} + \partial_z\sigma'_{zz}\right] \\
&= \frac{1}{\rho}\left[\eta\nabla^2 v_z + \left(\zeta + \frac{1}{3}\eta\right)\partial_z\nabla\cdot\boldsymbol{v}\right],
\end{aligned} \tag{A1.4.9}
$$

where η and ζ are assumed to be constant in each of the second lines.

Three components of vorticity $\boldsymbol{\omega} = (\omega_r, \omega_\varphi, \omega_z)$ are given by (A1.4.3). The flux-form vorticity equations are

$$
\begin{aligned}
\frac{\partial\omega_r}{\partial t} + \frac{1}{r}\partial_\varphi(v_\varphi\omega_r - v_r\omega_\varphi - g_z) - \partial_z(v_r\omega_z - v_z\omega_r - g_\varphi) &= 0, \\
\frac{\partial\omega_\varphi}{\partial t} + \partial_z(v_z\omega_\varphi - v_\varphi\omega_z - g_r) - \partial_r(v_\varphi\omega_r - v_r\omega_\varphi - g_z) &= 0, \\
\frac{\partial\omega_z}{\partial t} + \frac{1}{r}\partial_r\left[r(v_r\omega_z - v_z\omega_r - g_\varphi)\right] - \frac{1}{r}\partial_\varphi(v_z\omega_\varphi - v_\varphi\omega_z - g_r) &= 0,
\end{aligned} \tag{A1.4.10}
$$

where $\boldsymbol{g} = -\nabla p/\rho + \boldsymbol{f}$. In advective form, the vorticity equations are written as

$$
\begin{aligned}
\frac{d\omega_r}{dt} &= -\,\omega_r\left(\frac{1}{r}\partial_\varphi v_\varphi + \frac{v_r}{r} + \partial_z v_z\right) + \frac{\omega_\varphi}{r}\partial_\varphi v_r \\
&\quad + \omega_z\partial_\varphi v_r + \frac{1}{r}\partial_\varphi g_z - \partial_z g_\varphi, \\
\frac{d\omega_\varphi}{dt} &= -\,\omega_\varphi(\partial_r v_r + \partial_z v_z) + r\omega_r\partial_r\frac{v_\varphi}{r} \\
&\quad + \omega_z\partial_z v_\varphi + \partial_z g_r - \partial_r g_z,
\end{aligned}
$$

$$\begin{aligned}\frac{d\omega_z}{dt} &= -\omega_z\left(\partial_r v_r + \frac{1}{r}\partial_\varphi v_\varphi + \frac{v_r}{r}\right)\\ &\quad + \omega_r\partial_r v_z + \frac{\omega_\varphi}{r}\partial_\varphi v_z + \frac{1}{r}\partial_r(rg_\varphi) - \frac{1}{r}\partial_\varphi g_r.\end{aligned} \tag{A1.4.11}$$

Angular momentum about the axis z is given by $l = rv_\varphi$. The conservation of angular momentum is given from (A1.4.6) and (A1.4.7) as

$$\begin{aligned}&\frac{\partial(\rho l)}{\partial t} + \frac{1}{r}\partial_r[r(\rho l v_r - r\sigma_{r\varphi})] + \frac{1}{r}\partial_\varphi(\rho l v_\varphi - r\sigma_{\varphi\varphi}) + \partial_z(\rho l v_z - r\sigma_{z\varphi})\\ &\quad = -\rho\partial_\varphi\Phi,\\ &\frac{dl}{dt} = -\frac{1}{\rho}\partial_\varphi p - r\partial_\varphi\Phi + rf_\varphi.\end{aligned} \tag{A1.4.12}$$

A1.5 Spherical coordinates

Spherical coordinates are expressed by longitude λ, latitude φ, and radius r:

$$\xi^1 = \xi^\lambda = \lambda, \qquad \xi^2 = \xi^\varphi = \varphi, \qquad \xi^3 = \xi^r = r.$$

The use of longitude and latitude as coordinates is different from normal spherical coordinates where the co-latitude is used instead of φ and the direction of azimuthal coordinate is opposite to that of longitude. To differentiate between the two types of spherical coordinates, we refer to the spherical coordinates we use as *latitude-longitude coordinates*. The line element and the metrics are given by

$$\begin{aligned}ds^2 &= r^2\cos^2\varphi d\lambda^2 + r^2 d\varphi^2 + dr^2,\\ h_\lambda &= r\cos\varphi, \qquad h_\varphi = r, \qquad h_r = 1,\\ \sqrt{g} &= h_\lambda h_\varphi h_r = r^2\cos\varphi.\end{aligned}$$

Christoffel's symbols are

$$\begin{aligned}&\Gamma^r_{\varphi\varphi} = -r, \quad \Gamma^r_{\lambda\lambda} = -r\cos^2\varphi, \quad \Gamma^\varphi_{\varphi r} = \Gamma^\varphi_{r\varphi} = \frac{1}{r},\\ &\Gamma^\varphi_{\lambda\lambda} = \sin\varphi\cos\varphi, \quad \Gamma^\lambda_{\lambda r} = \Gamma^\lambda_{r\lambda} = \frac{1}{r}, \quad \Gamma^\lambda_{\lambda\varphi} = \Gamma^\lambda_{\varphi\lambda} = -\tan\varphi.\end{aligned}$$

The other components of Christoffel's symbols are zero.

Expressions of the gradient, divergence, rotation, and Laplacian operators are, from (A1.2.5)–(A1.2.8),

$$\nabla f = \left(\frac{1}{r\cos\varphi}\partial_\lambda f,\ \frac{1}{r}\partial_\varphi f,\ \partial_r f\right) \tag{A1.5.1}$$

$$\nabla\cdot\boldsymbol{v} = \frac{1}{r\cos\varphi}\partial_\lambda v_\lambda + \frac{1}{r\cos\varphi}\partial_\varphi(\cos\varphi v_\varphi) + \frac{1}{r^2}\partial_r(r^2 v_r). \tag{A1.5.2}$$

$$\begin{aligned}\nabla\times\boldsymbol{v} = &\left(\frac{1}{r}\partial_\varphi v_r - \frac{1}{r}\partial_r(rv_\varphi),\ \frac{1}{r}\partial_r(rv_\lambda) - \frac{1}{r\cos\varphi}\partial_\lambda v_r,\right.\\ &\left.\frac{1}{r\cos\varphi}\partial_\lambda v_\varphi - \frac{1}{r\cos\varphi}\partial_\varphi(\cos\varphi v_\lambda)\right),\end{aligned} \tag{A1.5.3}$$

$$\nabla^2 f = \frac{1}{r^2\cos^2\varphi}\partial_{\lambda\lambda} f + \frac{1}{r^2\cos\varphi}\partial_\varphi(\cos\varphi\partial_\varphi f) + \frac{1}{r^2}\partial_r(r^2\partial_r f). \tag{A1.5.4}$$

The gradient and the Laplacian of a vector $\boldsymbol{v}$ are given respectively by

$$\nabla\boldsymbol{v} = \begin{pmatrix} \frac{1}{r\cos\varphi}\partial_\lambda v_\lambda - \frac{\sin\varphi}{r\cos\varphi}v_\varphi + \frac{1}{r}v_r & \frac{1}{r\cos\varphi}\partial_\lambda v_\varphi + \frac{\sin\varphi}{r\cos\varphi}v_\lambda & \frac{1}{r\cos\varphi}\partial_\lambda v_r - \frac{1}{r}v_\lambda \\ \frac{1}{r}\partial_\varphi v_\lambda & \frac{1}{r}\partial_\varphi v_\varphi + \frac{1}{r}v_r & \frac{1}{r}\partial_\varphi v_r - \frac{1}{r}v_\varphi \\ \partial_r v_\lambda & \partial_r v_\varphi & \partial_r v_r \end{pmatrix},$$

$$\begin{aligned}\nabla^2\boldsymbol{v} = \Big(&\nabla^2 v_\lambda - \frac{2\sin\varphi}{r^2\cos^2\varphi}\partial_\lambda v_\varphi + \frac{2}{r^2\cos\varphi}\partial_\lambda v_r - \frac{1}{r^2\cos^2\varphi}v_\lambda, \\ &\nabla^2 v_\varphi - \frac{1}{r^2\cos^2\varphi}v_\varphi + \frac{2}{r^2}\partial_\varphi v_r + \frac{2\sin\varphi}{r^2\cos^2\varphi}\partial_\lambda v_\lambda, \\ &\nabla^2 v_r - \frac{2}{r^2}v_r - \frac{2}{r^2\cos\varphi}\partial_\lambda v_\lambda - \frac{2}{r^2\cos\varphi}\partial_\varphi(\cos\varphi v_\varphi)\Big).\end{aligned}$$

The time derivative of a vector $\boldsymbol{s}$ is

$$\begin{aligned}\frac{d\boldsymbol{s}}{dt} = \Big(&\frac{ds_\lambda}{dt} - \frac{\sin\varphi}{\cos\varphi}\frac{v_\lambda s_\varphi}{r} + \frac{v_\lambda s_r}{r}, \frac{ds_\varphi}{dt} + \frac{\sin\varphi}{\cos\varphi}\frac{v_\lambda s_\lambda}{r} + \frac{v_\varphi s_r}{r}, \\ &\frac{ds_r}{dt} - \frac{v_\lambda s_\lambda}{r} - \frac{v_\varphi s_\varphi}{r}\Big),\end{aligned} \tag{A1.5.5}$$

where

$$\frac{d}{dt} = \frac{\partial}{\partial t} + \frac{v_\lambda}{r\cos\varphi}\partial_\lambda + \frac{v_\varphi}{r}\partial_\varphi + v_r\partial_r.$$

The continuity equation and the equations of motion are given by, in flux form,

$$\begin{aligned}\frac{\partial\rho}{\partial t} + \frac{1}{r\cos\varphi}\partial_\lambda(\rho v_\lambda) + \frac{1}{r\cos\varphi}\partial_\varphi(\cos\varphi\rho v_\varphi) + \frac{1}{r^2}\partial_r(r^2\rho v_r) &= 0, \\ \frac{\partial(\rho v_\lambda)}{\partial t} + \frac{1}{r\cos\varphi}\partial_\lambda(\rho v_\lambda^2 - \sigma_{\lambda\lambda}) + \frac{1}{r\cos\varphi}\partial_\varphi[\cos\varphi(\rho v_\lambda v_\varphi - \sigma_{\varphi\lambda})] & \\ + \frac{1}{r^2}\partial_r[r^2(\rho v_\lambda v_r - \sigma_{r\lambda})] + \frac{1}{r}(\rho v_\lambda v_r - \sigma_{\lambda r}) & \\ - \frac{\sin\varphi}{r\cos\varphi}(\rho v_\lambda v_\varphi - \sigma_{\lambda\varphi}) &= -\frac{\rho}{r\cos\varphi}\partial_\lambda\Phi, \\ \frac{\partial(\rho v_\varphi)}{\partial t} + \frac{1}{r\cos\varphi}\partial_\lambda(\rho v_\varphi v_\lambda - \sigma_{\lambda\varphi}) + \frac{1}{r\cos\varphi}\partial_\varphi[\cos\varphi(\rho v_\varphi^2 - \sigma_{\varphi\varphi})] & \\ + \frac{1}{r^2}\partial_r[r^2(\rho v_\varphi v_r - \sigma_{r\varphi})] + \frac{\sin\varphi}{r\cos\varphi}(\rho v_\lambda^2 - \sigma_{\lambda\lambda}) & \\ + \frac{1}{r}(\rho v_\varphi v_r - \sigma_{\varphi r}) &= -\frac{\rho}{r}\partial_\varphi\Phi,\end{aligned}$$

$$\begin{aligned}\frac{\partial(\rho v_r)}{\partial t} &+ \frac{1}{r\cos\varphi}\partial_\lambda(\rho v_r v_\lambda - \sigma_{\lambda r}) + \frac{1}{r\cos\varphi}\partial_\varphi[\cos\varphi(\rho v_r v_\varphi - \sigma_{\varphi r})] \\ &+ \frac{1}{r^2}\partial_r[r^2(\rho v_r^2 - \sigma_{rr})] - \frac{1}{r}(\rho v_\varphi^2 - \sigma_{\varphi\varphi}) \\ &- \frac{1}{r}(\rho v_\lambda^2 - \sigma_{\lambda\lambda}) \quad = \quad -\rho\partial_r\Phi, \end{aligned} \tag{A1.5.6}$$

or, in advective form,

$$\begin{aligned}\frac{d\rho}{dt} + \rho\left[\frac{1}{r\cos\varphi}\partial_\lambda v_\lambda + \frac{1}{r\cos\varphi}\partial_\varphi(\cos\varphi\rho v_\varphi) + \frac{1}{r^2}\partial_r(r^2\rho v_r)\right] &= 0, \\ \frac{dv_\lambda}{dt} &= \frac{\sin\varphi}{r\cos\varphi}v_\lambda v_\varphi - \frac{v_\lambda v_r}{r} - \frac{1}{\rho r\cos\varphi}\partial_\lambda p - \frac{1}{r\cos\varphi}\partial_\lambda\Phi + f_\lambda, \\ \frac{dv_\varphi}{dt} &= \frac{\sin\varphi}{r\cos\varphi}v_\lambda^2 - \frac{v_\varphi v_r}{r} - \frac{1}{\rho r}\partial_\varphi p - \frac{1}{r}\partial_\varphi\Phi + f_\varphi, \\ \frac{dv_r}{dt} &= \frac{v_\varphi^2}{r} + \frac{v_\lambda^2}{r} - \frac{1}{\rho}\partial_r p - \partial_r\Phi + f_r. \end{aligned} \tag{A1.5.7}$$

The stress tensor reads

$$\begin{aligned}\sigma'_{\lambda\lambda} &= 2\eta\left(\frac{1}{r\cos\varphi}\partial_\lambda v_\lambda - \frac{\sin\varphi}{r\cos\varphi}v_\varphi + \frac{1}{r}v_r\right) + \left(\zeta - \frac{2}{3}\eta\right)\nabla\cdot\boldsymbol{v}, \\ \sigma'_{\varphi\varphi} &= 2\eta\left(\frac{1}{r}\partial_\varphi v_\varphi + \frac{v_r}{r}\right) + \left(\zeta - \frac{2}{3}\eta\right)\nabla\cdot\boldsymbol{v}, \\ \sigma'_{rr} &= 2\eta\partial_z v_z + \left(\zeta - \frac{2}{3}\eta\right)\nabla\cdot\boldsymbol{v}, \\ \sigma'_{\varphi r} = \sigma'_{r\varphi} &= \eta\left(\frac{1}{r}\partial_\varphi v_r + r\partial_r\frac{v_\varphi}{r}\right), \\ \sigma'_{r\lambda} = \sigma'_{\lambda r} &= \eta\left(r\partial_r\frac{v_\lambda}{r} + \frac{1}{r\cos\varphi}\partial_\lambda v_r\right), \\ \sigma'_{\lambda\varphi} = \sigma'_{\varphi\lambda} &= \eta\left(\frac{1}{r\cos\varphi}\partial_\lambda v_\varphi + \frac{\cos\varphi}{r}\partial_\varphi\frac{1}{\cos\varphi}v_\lambda\right), \end{aligned} \tag{A1.5.8}$$

and the friction force is

$$\begin{aligned}f_\lambda &= \frac{1}{\rho}\left[\frac{1}{r\cos\varphi}\partial_\lambda\sigma'_{\lambda\lambda} + \frac{1}{r\cos\varphi}\partial_\varphi(\cos\varphi\sigma'_{\lambda\varphi}) + \frac{1}{r^2}\partial_r(r^2\sigma'_{\lambda r})\right. \\ &\quad\left. + \frac{1}{r}\sigma'_{\lambda r} - \frac{\sin\varphi}{r\cos\varphi}\sigma'_{\lambda\varphi}\right] \\ &= \frac{1}{\rho}\left[\eta\left(\nabla^2 v_\lambda - \frac{2\sin\varphi}{r^2\cos\varphi}\partial_\lambda v_\varphi + \frac{2}{r^2\cos\varphi}\partial_\lambda v_r - \frac{1}{r^2\cos^2\varphi}v_\lambda\right)\right. \\ &\quad\left. + \left(\zeta + \frac{1}{3}\eta\right)\frac{1}{r\cos\varphi}\partial_\lambda\nabla\cdot\boldsymbol{v}\right], \end{aligned}$$

$$
\begin{aligned}
f_\varphi &= \frac{1}{\rho}\left[\frac{1}{r\cos\varphi}\partial_\lambda\sigma'_{\lambda\varphi} + \frac{1}{r\cos\varphi}\partial_\varphi(\cos\varphi\sigma'_{\varphi\varphi}) + \frac{1}{r^2}\partial_r(r^2\sigma'_{r\varphi})\right.\\
&\quad \left. + \frac{\sin\varphi}{r\cos\varphi}\sigma'_{\lambda\lambda} + \frac{1}{r}\sigma'_{\varphi r}\right]\\
&= \frac{1}{\rho}\left[\eta\left(\nabla^2 v_\varphi - \frac{1}{r^2\cos^2\varphi}v_\varphi + \frac{2}{r^2}\partial_\varphi v_r - \frac{2\sin\varphi}{r^2\cos\varphi}\partial_\lambda v_\lambda\right)\right.\\
&\quad \left. + \left(\zeta + \frac{1}{3}\eta\right)\frac{1}{r}\partial_\varphi\nabla\cdot\boldsymbol{v}\right],\\
f_r &= \frac{1}{\rho}\left[\frac{1}{r\cos\varphi}\partial_\lambda\sigma'_{\lambda r} + \frac{1}{r\cos\varphi}\partial_\varphi(\cos\varphi\sigma'_{\varphi r}) + \frac{1}{r^2}\partial_r(r^2\sigma'_{rr})\right.\\
&\quad \left. - \frac{1}{r}\sigma'_{\varphi\varphi} - \frac{1}{r}\sigma'_{\lambda\lambda}\right]\\
&= \frac{1}{\rho}\left[\eta\left(\nabla^2 v_r - \frac{2}{r^2}v_r - \frac{2}{r^2\cos\varphi}\partial_\lambda v_\lambda - \frac{2}{r^2\cos\varphi}\partial_\varphi(\cos\varphi v_\varphi)\right)\right.\\
&\quad \left. + \left(\zeta + \frac{1}{3}\eta\right)\partial_r\nabla\cdot\boldsymbol{v}\right],
\end{aligned}
\tag{A1.5.9}
$$

where η and ζ are assumed to be constant in each of the second lines.

Vorticity $\boldsymbol{\omega} = (\omega_\lambda, \omega_\varphi, \omega_r)$ is given by (A1.5.3). The vorticity equations are, in flux form,

$$
\begin{aligned}
&\frac{\partial\omega_\lambda}{\partial t} + \frac{1}{r}\partial_\lambda(v_\lambda\omega_\varphi - v_\varphi\omega_\lambda - g_r)\\
&\qquad -\frac{1}{r}\partial_r[r(v_\lambda\omega_r - v_r\omega_\lambda - g_\varphi)] \;=\; 0,\\
&\frac{\partial\omega_\varphi}{\partial t} + \frac{1}{r\cos\varphi}\partial_\lambda(v_\varphi\omega_\lambda - v_\lambda\omega_\varphi - g_r)\\
&\qquad -\frac{1}{r}\partial_r[r(v_r\omega_\varphi - v_\varphi\omega_r - g_\lambda) \;=\; 0,\\
&\frac{\partial\omega_r}{\partial t} + \frac{1}{r\cos\varphi}\partial_\lambda(v_\lambda\omega_r - v_r\omega_\lambda - g_\varphi)\\
&\qquad -\frac{1}{r\cos\varphi}\partial_\lambda[\cos\varphi(v_r\omega_\lambda - v_\lambda\omega_r - g_\lambda)] \;=\; 0,
\end{aligned}
\tag{A1.5.10}
$$

and, in advective form,

$$
\begin{aligned}
\frac{d\omega_\lambda}{dt} &= -\omega_\lambda\left(\frac{1}{r}\partial_\varphi v_\varphi + \frac{v_r}{r} + \partial_r v_r\right)\\
&\quad + \frac{\omega_\varphi}{r}\left(\partial_\varphi v_\lambda + \frac{\sin\varphi}{\cos\varphi}v_\lambda\right) + \omega_r\left(\partial_r v_\lambda - \frac{v_\lambda}{r}\right) + \frac{1}{r}\partial_\varphi g_r - \frac{1}{r}\partial_r(r g_\varphi),\\
\frac{d\omega_\varphi}{dt} &= -\omega_\varphi\left(\frac{1}{r\cos\varphi}\partial_\lambda v_\lambda - \frac{\sin\varphi}{r\cos\varphi}v_\varphi + \frac{v_r}{r} + \partial_r v_r\right)\\
&\quad + \frac{\omega_\lambda}{r\cos\varphi}\partial_\lambda v_\varphi + \omega_r\left(\partial_r v_\varphi - \frac{v_\varphi}{r}\right) + \frac{1}{r}\partial_r(r g_\lambda) - \frac{1}{r\cos\varphi}\partial_\lambda g_r,
\end{aligned}
$$

$$\frac{d\omega_r}{dt} = -\omega_r\left(\frac{1}{r\cos\varphi}\partial_\lambda v_\lambda - \frac{\sin\varphi}{r\cos\varphi}v_\varphi + \frac{1}{r}\partial_\varphi v_\varphi + \frac{2}{r}v_r\right)$$
$$+ \frac{\omega_\lambda}{r\cos\varphi}\partial_\lambda v_r + \frac{\omega_\varphi}{r}\partial_\varphi v_r + \frac{1}{r\cos\varphi}\partial_\lambda g_\varphi - \frac{1}{r\cos\varphi}\partial_\varphi(\cos\varphi g_\lambda). \tag{A1.5.11}$$

The component of angular momentum about the pole-to-pole axis (or the rotation axis) is denoted by $l = r\cos\varphi v_\varphi$. Using (A1.5.6) and (A1.5.7), the conservation of angular momentum of this component is given by

$$\frac{\partial(\rho l)}{\partial t} + \frac{1}{r\cos\varphi}\partial_\lambda(\rho l v_\lambda - r\sigma_{\lambda\lambda}\cos\varphi)$$
$$+ \frac{1}{r\cos\varphi}\partial_\varphi[\cos\varphi(\rho l v_\varphi - r\sigma_{\lambda\varphi}\cos\varphi)]$$
$$+ \frac{1}{r^2}\partial_r[r^2(\rho l v_r - r\sigma_{\lambda r}\cos\varphi)] = -\rho\partial_\lambda\Phi,$$
$$\frac{dl}{dt} = -\frac{1}{\rho}\partial_\lambda p - \partial_\lambda\Phi + r\cos\varphi f_\lambda. \tag{A1.5.12}$$

We note that the absolute angular momentum in a rotating frame is defined by

$$l = r\cos\varphi(v_\varphi + \Omega r\cos\varphi). \tag{A1.5.13}$$

References and suggested reading

The mathematical expressions of tensors summarized in this chapter are based on the geometry of manifolds. Flanders (1963) and Schutz (1980) are introductory textbooks to the mathematical methods of manifolds. Weinberg (1972), which is a textbook on the theory of the universe, concisely describes tensor analysis in a slightly general form and is very useful. Many textbooks on dynamic meteorology have an appendix similar to this to summarize the expressions of orthogonal coordinates, but some errors are frequently found, particularly for the derivatives of a tensor of degree two. I suggest that readers should derive these expressions for themselves and do not take them for granted.

Flanders, H., 1963: *Differential Forms with Applications to the Physical Sciences.* Dover, New York, 205 pp.

Schutz, B. F., 1980: *Geometrical Methods of Mathematical Physics.* Cambridge University Press, Cambridge, UK, 264 pp.

Weinberg, S., 1972: *Gravitation and Cosmology: Principles and Applications of the General Theory of Relativity.* John Wiley & Sons, New York, 657 pp.

A2

Physical constants

The following values of physical constants are based on Committee on Data for Science and Technology (CODATA, 1998).

Name	Symbol and value
Atomic mass constant	$m_u = 1.660\ 538\ 73 \times 10^{-27}$ kg
Avogadro constant	$N_A = 6.022\ 141\ 99 \times 10^{23}$ mol^{-1}
Boltzmann constant	$k_B = 1.380\ 6503 \times 10^{-23}$ J K^{-1}
Molar volume of ideal gas[†]	$v^I = 22.413\ 996 \times 10^{-3}$ m^3 mol^{-1}
Newtonian constant of gravitation	$G = 6.673 \times 10^{-11}$ m^3 kg^{-1} s^{-2}
Planck constant	$h = 6.626\ 068\ 76 \times 10^{-34}$ J s
Planck constant over 2π	$\hbar = 1.054\ 571\ 596 \times 10^{-34}$ J s
Speed of light in vacuum	$c = 299\ 792\ 458$ m s^{-1}
Standard acceleration of gravity	$g_n = 9.806\ 65$ m s^{-2}
Standard atmosphere	$p_0 = 101{,}325$ Pa
Stefan-Boltzmann constant	$\sigma_B = 5.670\ 400 \times 10^{-8}$ W m^{-2} K^{-4}
Wien displacement law constant	$2.897\ 7686 \times 10^{-3}$ m K
Mean radius of the earth[‡]	$R = 6{,}371.01$ km
Mass of the earth[‡]	$M = 5.9736 \times 10^{24}$ kg
Mean sidereal day[‡]	$\tau = 86{,}164.090\ 54$ s
Rotation rate of the earth	$\Omega = 2\pi/(24 \times 60 \times 60) = 7.27221 \times 10^{-5}$ rad s^{-1}

† Values at the standard atmospheric state (273.15 K, 1013.25 hPa).
‡ Yoder (1995).

The following values of thermodynamic constants are representative. The temperature dependency of the thermodynamic constants of the moist air is summarized in List (1951). Iribarne and Godson (1981) and Emanuel (1994) are also useful references.

Name	Symbol and value
Molecular weight of dry air	$m_d = 28.966$ kg
Molecular weight of water vapor	$m_w = 18.0160$ kg
Gas constant of ideal gas	$R^* = 8.314\ 36$ J mol^{-1} K^{-1}
Gas constant of dry air	$R_d = 287.04$ J kg^{-1} K^{-1}
Gas constant of water vapor	$R_v = 461.50$ J kg^{-1} K^{-1}
Specific heat at constant volume of dry air	$C_{vd} = 5R_d/2 = 717.6$ J kg^{-1} K^{-1}
Specific heat at constant pressure of dry air	$C_{pd} = 7R_d/2 = 1{,}004.6$ J kg^{-1} K^{-1}
Specific heat at constant volume of water vapor	$C_{vv} = 3R_v = 1{,}390$ J kg^{-1} K^{-1}
Specific heat at constant pressure of water vapor	$C_{pv} = 4R_v = 1{,}846$ J kg^{-1} K^{-1}
Specific heat of liquid water†	$C_l = 4{,}218$ J kg^{-1} K^{-1}
Specific heat of ice†	$C_i = 2{,}106$ J kg^{-1} K^{-1}
Latent heat of vaporization†	$L_v = 2.501 \times 10^6$ J kg^{-1}
Latent heat of sublimation†	$L_s = 2.834 \times 10^6$ J kg^{-1}
Latent heat of fusion†	$L_f = 3.337 \times 10^5$ J kg^{-1}
Saturation vapor pressure†	$e_{s0} = 6.1078 \times 10^2$ Pa

† Values at 0°C.

References

Committee on Data for Science and Technology, 1998: online at http://physics.nist.gov/cuu/Constants/bibliography.html

Emanuel, K., 1994: *Atmospheric Convection.* Oxford University Press, New York, 580 pp.

Iribarne, J. V. and Godson, W. L., 1981: *Atmospheric Thermodynamics*, 2nd ed. D. Reidel, Dordrecht, The Netherlands, 259 pp.

List, R.J., 1951: *Smithsonian Meteorological Tables*, 6th rev. ed., Smithsonian Institution Press, Washington D. C., 527 pp.

Yoder, C. F., 1995: Astrometric and geodetic properties of earth and the solar system. In: T. J. Ahrens (ed.), *Global Earth Physics. A Handbook of Physical Constants*, AGU Reference Shelf 1, American Geophysical Union, Washington D. C., pp. 1–31.

A3

Meridional structure

The following figures are produced by using the reanalysis data of NCEP/NCAR (Kalnay et al., 1996). The monthly mean data are averaged in time from 1982 to 1994 and in the zonal direction on pressure surfaces.

References

Kalnay, E., and coauthors, 1996: The NCEP/NCAR 40-year reanalysis project. *Bull. Amer. Meteorol. Soc.*, **77**, 437–471.

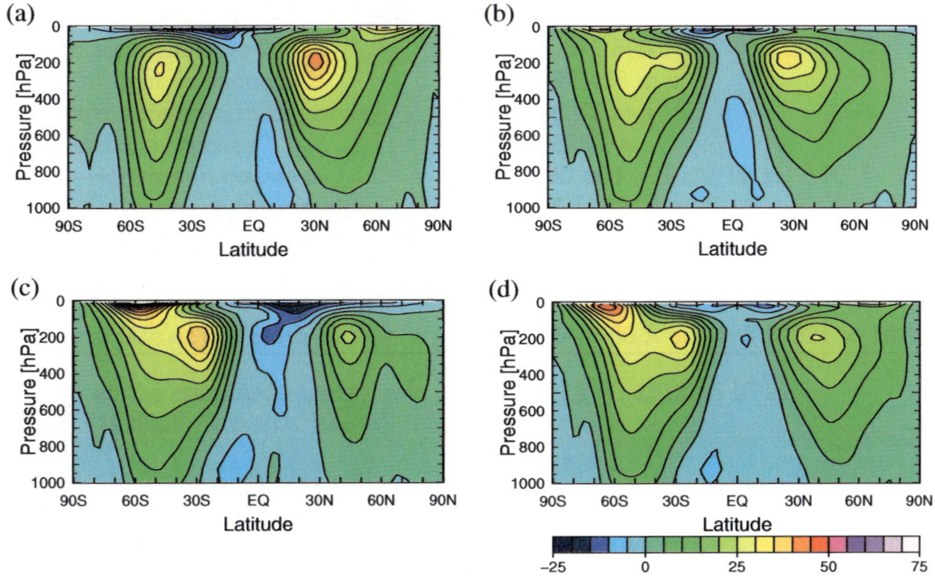

FIGURE A3.1: The meridional distribution of zonal winds. The contour interval is 5 m s^{-1}. (a) January, (b) April, (c) July, and (d) October.

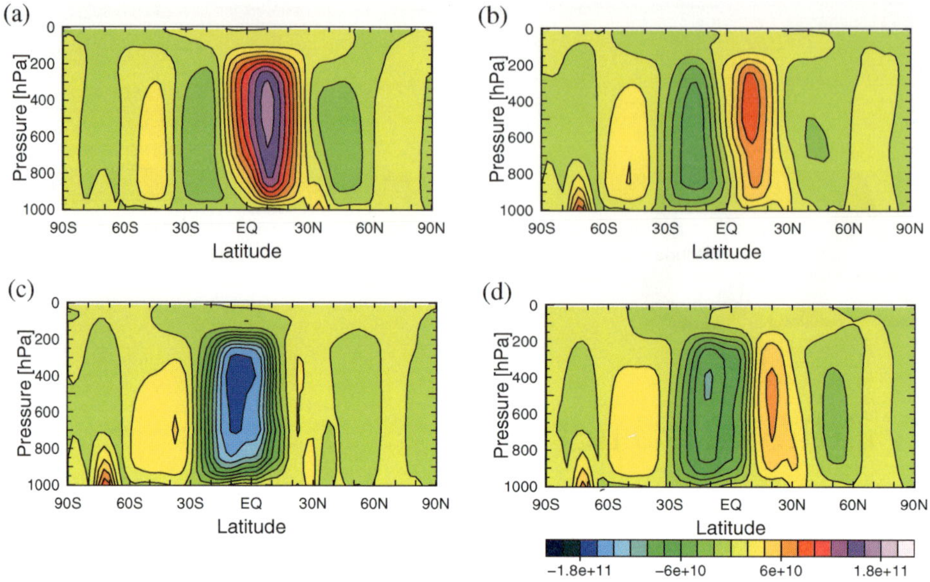

FIGURE A3.2: The meridional distribution of streamfunctions. The contour interval is 10^{10} kg s^{-1}. (a) January, (b) April, (c) July, and (d) October.

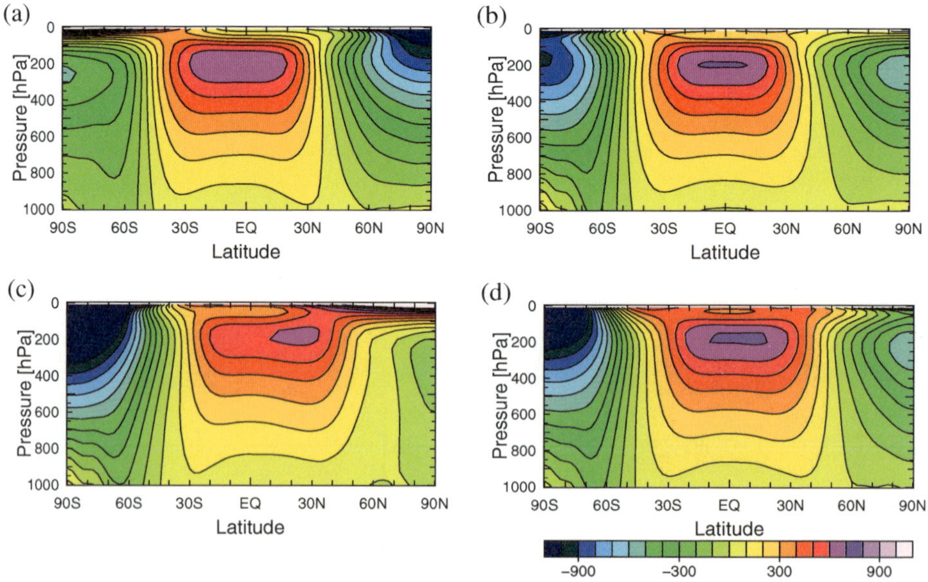

FIGURE A3.3: The meridional distribution of deviation of geopotential height from the horizontal average. The contour interval is 100 m. (a) January, (b) April, (c) July, and (d) October.

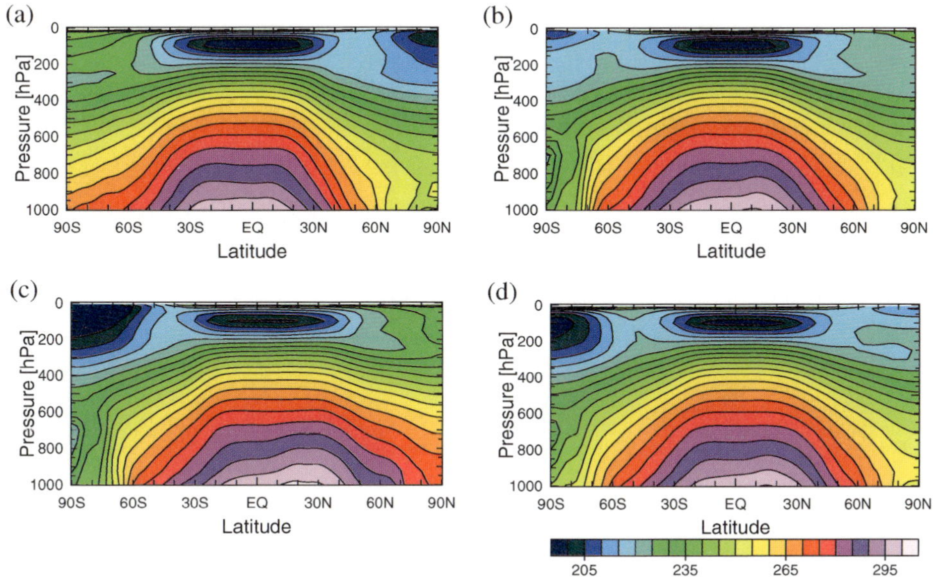

FIGURE A3.4: The meridional distribution of temperature. The contour interval is 5 K. (a) January, (b) April, (c) July, and (d) October.

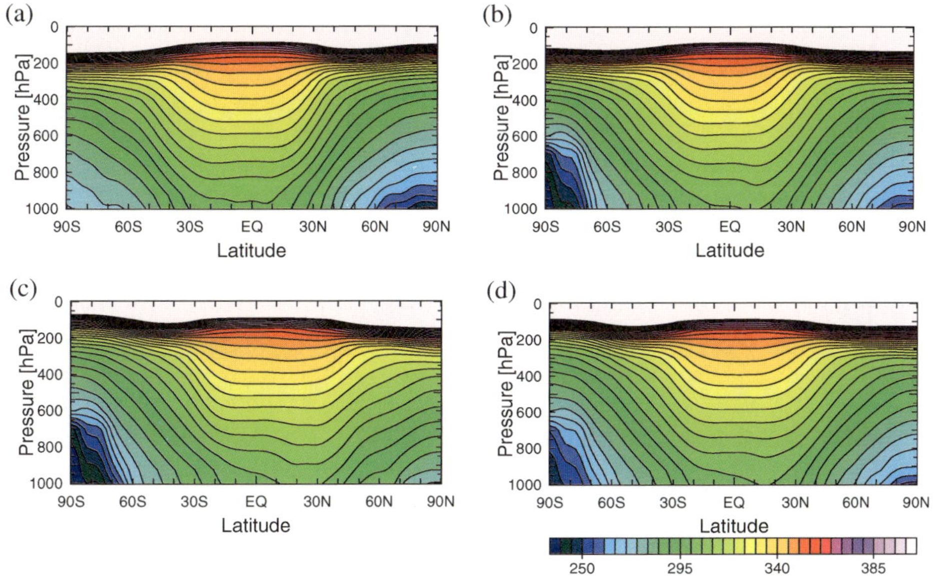

FIGURE A3.5: The meridional distribution of potential temperature. The contour interval is 5 K. (a) January, (b) April, (c) July, and (d) October.

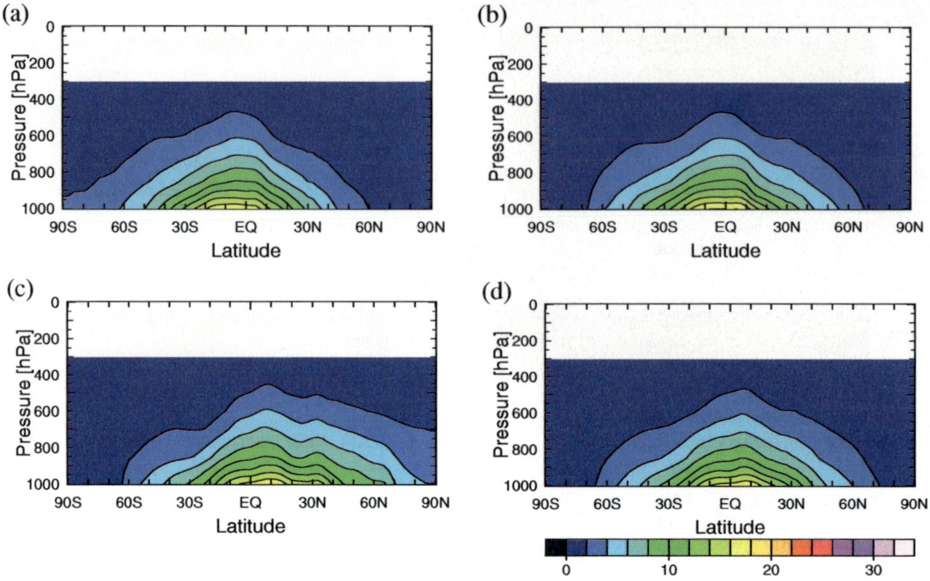

FIGURE A3.6: The meridional distribution of specific humidity The contour interval is 1 g kg^{-1}. (a) January, (b) April, (c) July, and (d) October.

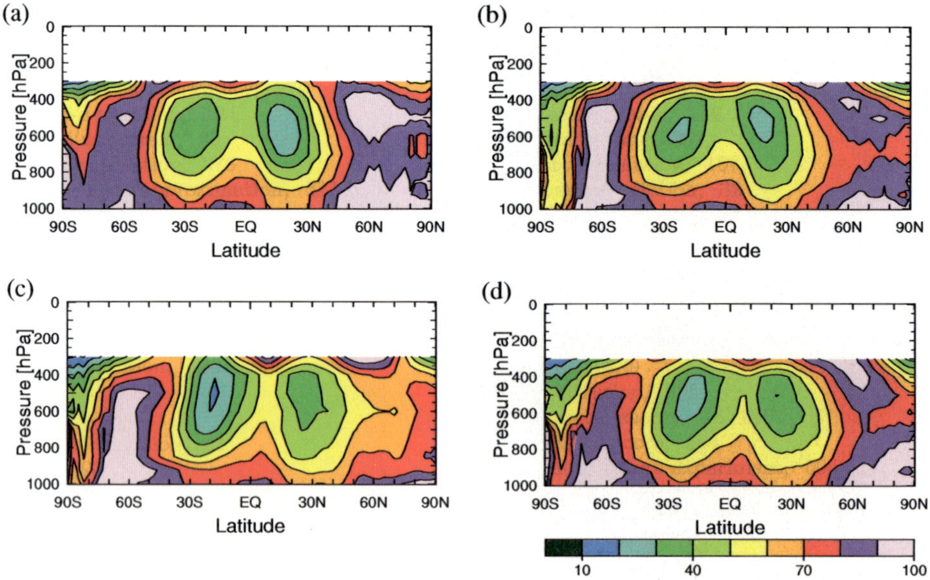

FIGURE A3.7: The meridional distribution of relative humidity calculated from time-averaged values of specific humidity and temperature. The contour interval is 5%. (a) January, (b) April, (c) July, and (d) October.

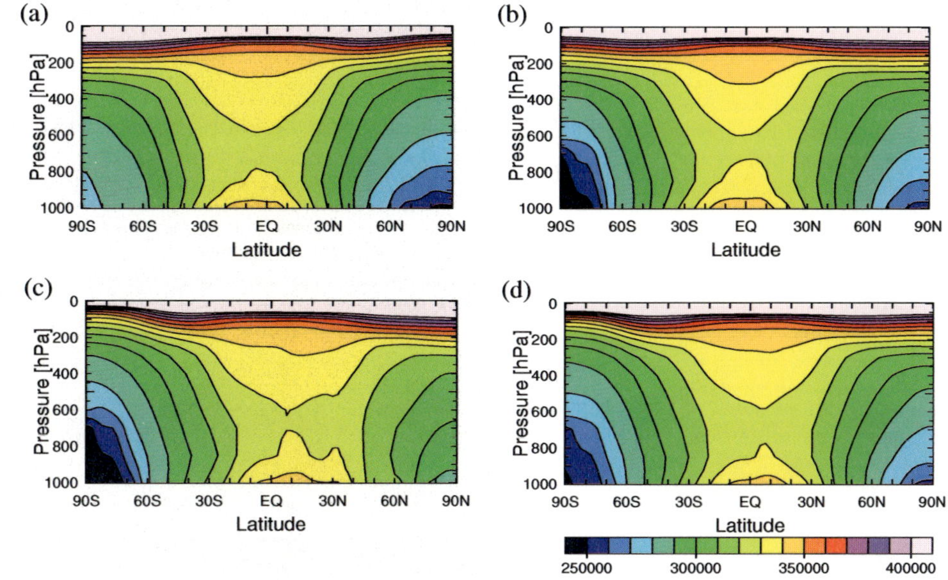

FIGURE A3.8: The meridional distribution of moist static energy calculated from time-averaged values of specific humidity and temperature. The contour interval is 10,000 J kg^{-1}. (a) January, (b) April, (c) July, and (d) October.

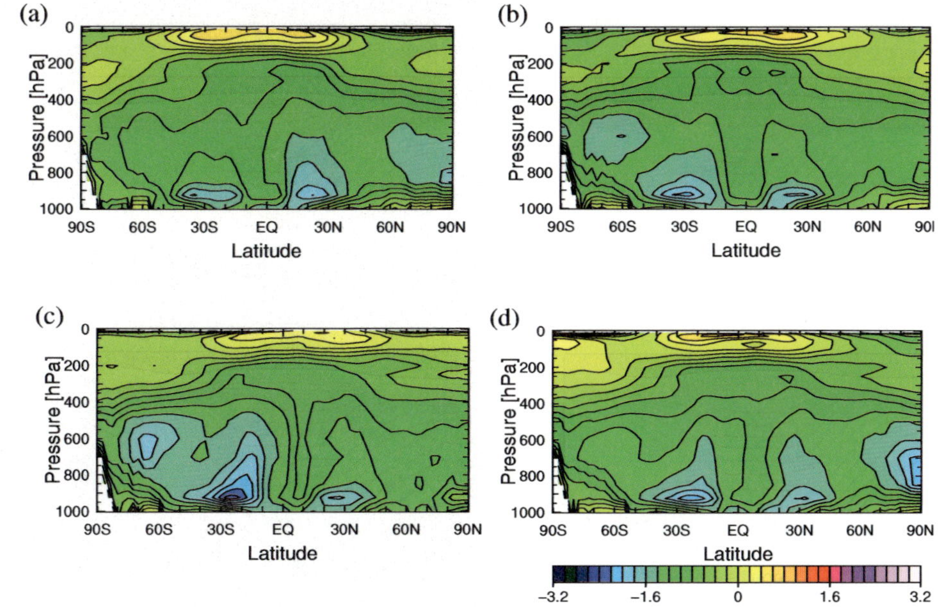

FIGURE A3.9: The meridional distribution of total radiative heating rate (sum of short-wave and long-wave radiative heating rates). The contour interval is 0.2 K day^{-1}. (a) January, (b) April, (c) July, and (d) October.

Index

Printing: Mercedes-Druck, Berlin
Binding: Stein + Lehmann, Berlin